Applied Principles of Chemometrics

Applied Principles of Chemometrics

Edited by **Jina Redlin**

NY RESEARCH
P R E S S

New York

Published by NY Research Press,
23 West, 55th Street, Suite 816,
New York, NY 10019, USA
www.nyresearchpress.com

Applied Principles of Chemometrics
Edited by Jina Redlin

International Standard Book Number: 978-1-63238-055-5 (Hardback)

Printed in the United States of America.

Contents

Preface   VII

Part 1   Methods   1

Chapter 1   **Model Population Analysis
for Statistical Model Comparison**   3
Hong-Dong Li, Yi-Zeng Liang and Qing-Song Xu

Chapter 2   **Exploratory Data Analysis with Latent Subspace Models**   21
José Camacho

Chapter 3   **Critical Aspects of Supervised Pattern Recognition
Methods for Interpreting Compositional Data**   49
A. Gustavo González

Chapter 4   **Analysis of Chemical Processes,
Determination of the Reaction Mechanism
and Fitting of Equilibrium and Rate Constants**   69
Marcel Maeder and Peter King

Chapter 5   **Experimental Optimization and Response Surfaces**   91
Veli-Matti Tapani Taavitsainen

Part 2   Biochemistry   139

Chapter 6   **Chemometric Study
on Molecules with Anticancer Properties**   141
João Elias Vidueira Ferreira, Antonio Florêncio de Figueiredo,
Jardel Pinto Barbosa and José Ciríaco Pinheiro

Chapter 7   **Metabolic Biomarker Identification with Few Samples**   157
Pietro Franceschi, Urska Vrhovsek, Fulvio Mattivi
and Ron Wehrens

Chapter 8   **Kinetic Analyses of Enzyme Reaction Curves**
              with New Integrated Rate Equations and Applications   173
              Xiaolan Yang, Gaobo Long, Hua Zhao and Fei Liao

Chapter 9   **Electronic Nose Integrated with Chemometrics for Rapid**
              Identification of Foodborne Pathogen   201
              Yong Xin Yu and Yong Zhao

Part 3   **Technology**   215

Chapter 10   **Chemometrics in Food Technology**   217
              Riccardo Guidetti, Roberto Beghi and Valentina Giovenzana

Chapter 11   **Metabolomics and Chemometrics**
              as Tools for Chemo(bio)diversity Analysis
              - Maize Landraces and Propolis   253
              Marcelo Maraschin, Shirley Kuhnen, Priscilla M.M. Lemos,
              Simone Kobe de Oliveira, Diego A. da Silva, Maíra M. Tomazzoli,
              Ana Carolina V. Souza, Rúbia Mara Pinto, Virgílio G. Uarrota,
              Ivanir Cella, Antônio G. Ferreira, Amélia R.S. Zeggio,
              Maria B.R. Veleirinho, Ivone Delgadillo and Flavia A. Vieira

Chapter 12   **PARAFAC Analysis**
              for Temperature-Dependent NMR
              Spectra of Poly(Lactic Acid) Nanocomposite   271
              Hideyuki Shinzawa, Masakazu Nishida, Toshiyuki Tanaka,
              Kenzi Suzuki and Wataru Kanematsu

Chapter 13   **Using Principal Component Scores and Artificial**
              Neural Networks in Predicting Water Quality Index   287
              Rashid Atta Khan, Sharifuddin M. Zain, Hafizan Juahir,
              Mohd Kamil Yusoff and Tg Hanidza T.I.

Chapter 14   **Application of Chemometrics to**
              the Interpretation of Analytical Separations Data   305
              James J. Harynuk, A. Paulina de la Mata and Nikolai A. Sinkov

              Permissions

              List of Contributors

Preface

Over the recent decade, advancements and applications have progressed exponentially. This has led to the increased interest in this field and projects are being conducted to enhance knowledge. The main objective of this book is to present some of the critical challenges and provide insights into possible solutions. This book will answer the varied questions that arise in the field and also provide an increased scope for furthering studies.

A comprehensive account on chemometrics has been provided in this book. It elaborates the diverse uses of chemometric methods in various spheres like chemistry, biochemistry and chemical technology. Selected techniques of chemometry have been described in a lucid and comprehensive manner. This book is dedicated to bridging the distance between textbooks and science journals on chemometrics and chemoinformatics.

I hope that this book, with its visionary approach, will be a valuable addition and will promote interest among readers. Each of the authors has provided their extraordinary competence in their specific fields by providing different perspectives as they come from diverse nations and regions. I thank them for their contributions.

Editor

Part 1

Methods

Model Population Analysis for Statistical Model Comparison

Hong-Dong Li[1], Yi-Zeng Liang[1] and Qing-Song Xu[2]

[1]College of Chemistry and Chemical Engineering, Central South University, Changsha,
[2]School of Mathematic Sciences, Central South University, Changsha,
P. R. China

1. Introduction

Model comparison plays a central role in statistical learning and chemometrics. Performances of models need to be assessed using a given criterion based on which models can be compared. To our knowledge, there exist a variety of criteria that can be applied for model assessment, such as Akaike's information criterion (AIC) [1], Bayesian information criterion (BIC) [2], deviance information criterion (DIC),Mallow's Cp statistic, cross validation [3-6] and so on. There is a large body of literature that is devoted to these criteria. With the aid of a chosen criterion, different models can be compared. For example, a model with a smaller AIC or BIC is preferred if AIC or BIC are chosen for model assessment.

In chemometrics, model comparison is usually conducted by validating different models on an independent test set or by using cross validation [4, 5, 7], resulting in a single value, *i.e.* root mean squared error of prediction (RMSEP) or root mean squared error of cross validation (RMSECV). This single metrics is heavily dependent on the selection of the independent test set (RMSEP) or the partition of the training data (RMSECV). Therefore, we have reasons to say that this kind of comparison is lack of statistical assessment and also at the risk of drawing wrong conclusions. We recently proposed model population analysis (MPA) as a general framework for designing chemometrics/bioinformatics methods [8]. MPA has been shown to be promising in outlier detection and variable selection. Here we hypothesize that reliably statistical model comparison could be achieved via the use of model population analysis.

2. Model population analysis

2.1 The framework of model population analysis

Model population analysis has been recently proposed for developing chemometrics methods in our previous work [8]. As is shown in **Figure 1**, MPA works in three steps which are summarized as (1) randomly generating N sub-datasets using Monte Carlo sampling (2) building one sub-model on each sub-dataset and (3) statistically analyzing some interesting output of all the N sub-models.

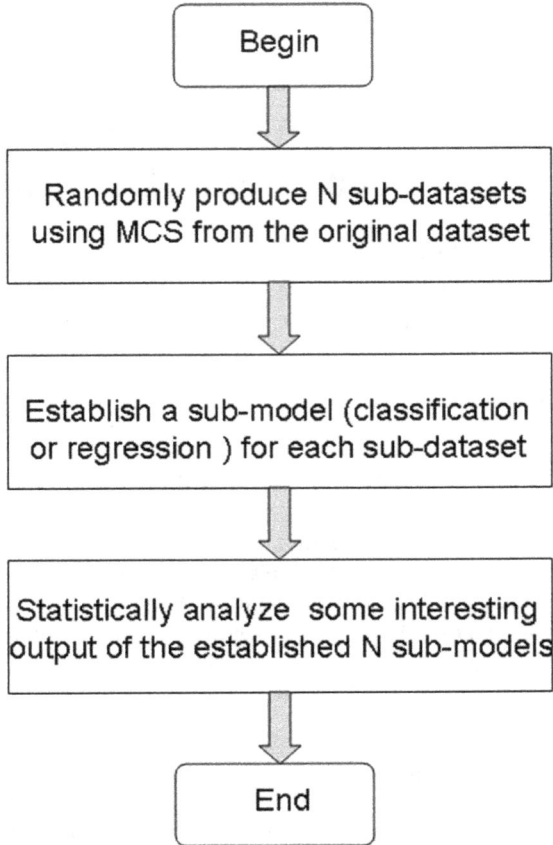

Fig. 1. The schematic of MPA. MCS is the abbreviation of Monte Carlo Sampling.

2.1.1 Monte Carlo sampling for generating a sub-dataset

Sampling plays a key role in statistics which allows us to generate replicate sub-datasets from which an interested unknown parameter could be estimated. For a given dataset $(\mathbf{X}, \mathbf{y})$, it is assumed that the design matrix $\mathbf{X}$ contains m samples in rows and p variables in columns, the response vector $\mathbf{y}$ is of size $m \times 1$. The number of Monte Carlo samplings is set to N. In this setting, N sub-datasets can be drawn from N Monte Carlo samplings with or without replacement [9, 10], which are denoted as $(\mathbf{X}_{sub}, \mathbf{y}_{sub})_i$, $i = 1, 2, 3, \ldots N$.

2.1.2 Establishing a sub-model using each sub-dataset

For each sub-dataset $(\mathbf{X}_{sub}, \mathbf{y}_{sub})_i$, a sub-model can be constructed using a selected method, $e.g.$ partial least squares (PLS) [11] or support vector machines (SVM) [12]. Denote the sub-model established as $f_i(\mathbf{X})$. Then, all these sub-models can be put into a collection:

$$C = (f_1(X), f_2(X), f_3(X), \ldots, f_N(X)) \tag{1}$$

All these N sub-models are mutually different but have the same goal that is to predict the response value **y**.

2.1.3 Statistically analyzing an interesting output of all the sub-models

The core of model population analysis is statistical analysis of an interesting output, *e.g.* prediction errors or regression coefficients, of all these sub-models. Indeed, it is difficult to give a clear answer on what output should be analyzed and how the analysis should be done. Different designs for the analysis will lead to different algorithms. As proof-of-principle, it was shown in our previous work that the analysis of the distribution of prediction errors is effective in outlier detection [13].

2.2 Insights provided by model population analysis

As described above, Monte Carlo sampling serves as the basics of model population analysis that help generate distributions of interesting parameters one would like to analyze. Looking on the surface, it seems to be very natural and easy to generate distributions using Monte Carlo sampling. However, here we show by examples that the distribution provided by model population analysis can indeed provide very useful information that gives insights into the data under investigation.

2.2.1 Are there any outliers?

Two datasets are fist simulated. The first contains only normal samples, whereas there are 3 outliers in the second dataset, which are shown in Plot A and B of **Figure 2**, respectively. For each dataset, a percentage (70%) of samples are randomly selected to build a linear regression model of which the slope and intercept is recorded. Repeating this procedure 1000 times, we obtain 1000 values for both the slope and intercept. For both datasets, the intercept is plotted against the slope as displayed in Plot C and D, respectively. It can be observed that the joint distribution of the intercept and slope for the normal dataset appears to be multivariate normally distributed. In contrast, this distribution for the dataset with outliers looks quite different, far from a normal distribution. Specifically, the distributions of slopes for both datasets are shown in Plot E and F. These results show that the existence of outliers can greatly influence a regression model, which is reflected by the odd distributions of both slopes and intercepts. In return, a distribution of a model parameter that is far from a normal one would, most likely, indicate some abnormality in the data.

2.2.2 Are there any interfering variables?

In this study, we first simulate a design matrix **X** of size 50 × 10, the response variable **Y** is simulated by multiplying X with a 10-dimensional regression vector. Gussian noises with standard deviation equal to 1 are then added to **Y**. That is to say, all the variables in **X** are "true variables" that collectively predict **Y**. This dataset (**X**, **Y**) is denoted SIMUTRUE. Then another design matrix **F** is simulated of size 50 × 10. Denote the combination of **X** and **F** as **Z**=[**X F**]. This dataset (**Z**, **Y**) is called SIMUINTF, which contains variables that are not predictive of **Y**. For both datasets, we randomly choose 70% samples to first build a regression model which is then used to make predictions on the remaining 30% samples, resulting in a RMSEP value. Repeating this procedure 1000 times, we, for both datasets,

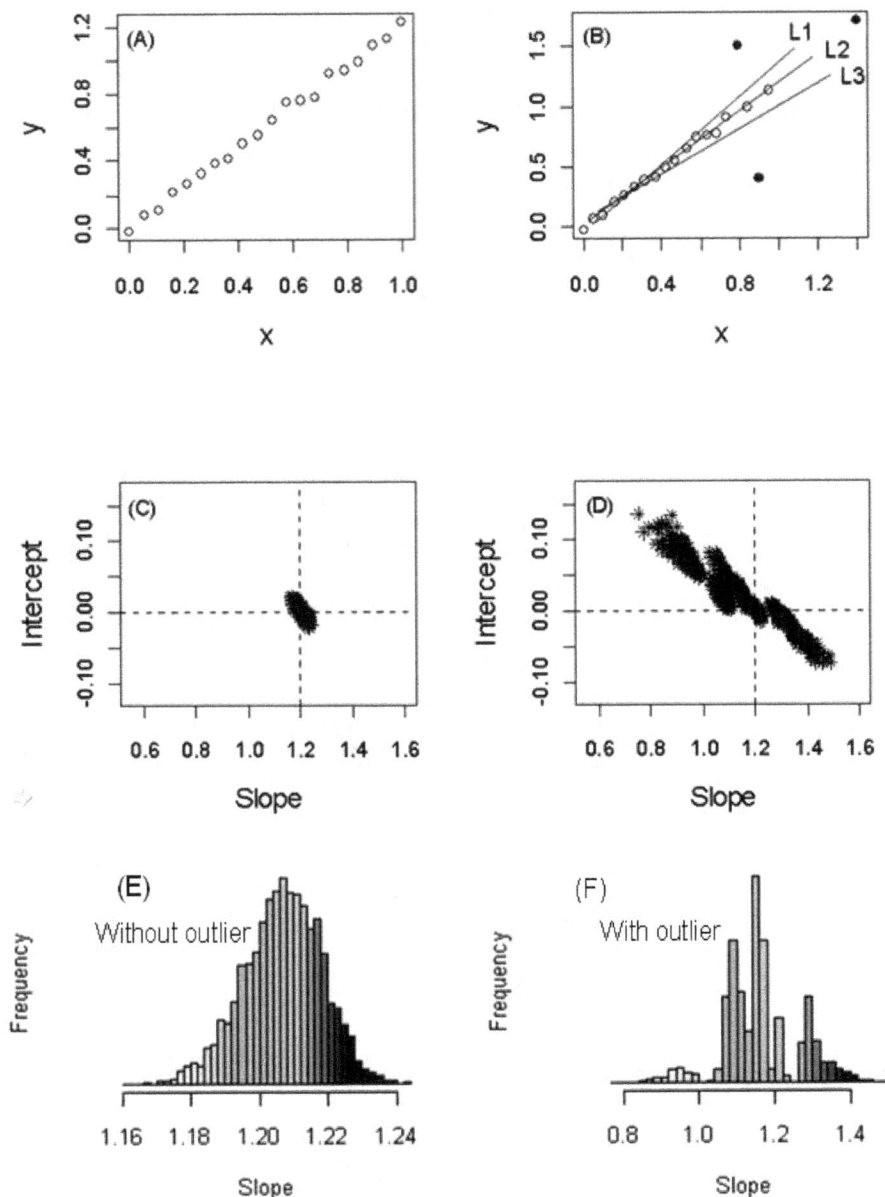

Fig. 2. A simulation study illustrating the use of model population analysis to detect whether a dataset contains outliers. Plot A and Plot B shows the data simulated without and with outliers, respectively. 1000 linear regression models computed using 1000 sub-datasets randomly selected and the slope and intercept are presented in Plot C and D. Specifically, the distribution of slope for these two simulated datasets are displayed in Plot E and Plot F.

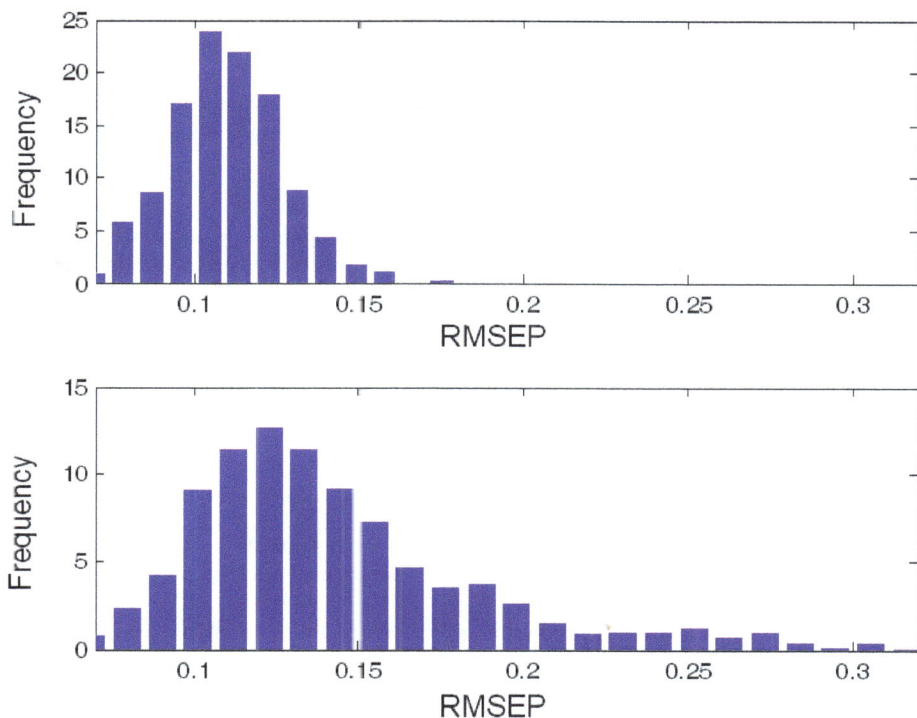

Fig. 3. The distribution of RMSEPs using the variable set that contains only "true variables" (upper panel) and the variable set that includes not only "true variables" but also "interfering variables" (lower panel).

obtain 1000 RMSEP values, of which the distributions are given in **Figure 3**. Clearly, the distribution of RMSEP of the SIMUINTF is right shifted, indicating the existence of variables that are not predictive of Y can degrade the performance of a regression model. We call this kind of variables "interfering variables". Can you tell whether a dataset contains interfering variables for a real world dataset? Curious readers may ask a question like this. Indeed, we can. We can do replicate experiments to estimate the experimental error that could serve as a reference by which it is possible to judge whether interfering variables exist. For example, if a model containing a large number of variables (with true variables included) shows a large prediction error compared to the experimental error, we may predict that interfering variables exist. In this situation, variable selection is encouraged and can greatly improve the performance of a model. Actually, when interfering variables exist, variable selection is a must. Other methods that use latent variables like PCR or PLS cannot work well because latent variables have contributions coming from interfering variables.

2.3 Applications of model population analysis

Using the idea of model population analysis, we have developed algorithms that address the fundamental issues in chemical modeling: outlier detection and variable selection. For outlier detection, we developed the MC method [13]. For variable selection, we developed subwindow permutation analysis (SPA) [14], noise-incorporated subwindow permutation analysis (NISPA) [15] and margin influence analysis (MIA) [16]. Here, we first give a brief description of these algorithms, aiming at providing examples that could help interested readers to understand how to design an algorithm by borrowing the framework of model population analysis.

As can be seen from **Figure 1**, These MPA-based methods share the first two steps that are (1) generating N sub-datasets and (2) building N sub-models. The third step "statistical analysis of an interesting output of all these N sub-models" is the core of model population analysis that underlines different methods. The key points of these methods as well as another method Monte Carlo uninformative variable elimination (MCUVE) that also implements the idea of MPA are summarized in Table 1. In a word, the distribution from model population analysis contains abundant information that provides insight into the data analyzed and by making full use of these information, effective algorithms can be developed for solving a given problem.

Methods*	What to statistically analyze
MC method	Distribution of prediction errors of each sample
SPA	Distribution of prediction errors before and after each variable is permuted
NISPA	Distribution of prediction errors before and after each variable is permuted with one noise variable as reference
MIA	Distribution of margins of support vector machines sub-models
MCUVE	Distribution of regression coefficients of PLS regression sub-models

*: The MC method, SPA, NISPA, MIA and MCUVE are described in references [13], [14], [15] [16] and [27].

Table 1. Key points of MPA-based methods.

2.4 Model population analysis and bayesian analysis

There exist similarities as well as differences between model population analysis and Bayesian analysis. One important similarity is that both methods consider the parameter of interest not as a single number but a distribution. In model population analysis, we generate distributions by causing variations in samples and/or variables using Monte Carlo sampling [17]. In contrast, in Bayesian analysis the parameter to infer is first assumed to be from a prior distribution and then observed data are used to update this prior distribution to the posterior distribution from which parameter inference can be conducted and predictions can be made [18-20]. The output of Bayesian analysis is a posterior distribution of some interesting parameter. This posterior distribution provides a natural link between Bayesian analysis and model population analysis. Taking Bayesian linear regression (BLR) [20] as an example, the output can be a large number of regression coefficient vectors that are sampled from its posterior distribution. These regression coefficient vectors actually represent a population of sub-models that can be used directly for model population analysis. Our future work will be constructing useful algorithms by borrowing merits of both Bayesian analysis and model population analysis.

2.5 Model population analysis and ensemble learning

Ensemble learning methods, such as bagging[21], boosting [22] and random forests [23], have emerged as very promising strategies for building a predictive model and these methods have found applications in a wide variety of fields. Recently, a new ensemble technique, called feature-subspace aggregating (Feating) [24], was proposed that was shown to have nice performances. The key point of these ensemble methods is aggregating a large number of models built using sub-datasets randomly generated using for example bootstrapping. Then ensemble models make predictions by doing a majority voting for classification or averaging for regression. In our opinion, the basic idea of ensemble learning methods is the same as that in model population analysis. In this sense, ensemble learning methods can also be formulated into the framework of model population analysis.

3. Model population analysis for statistical model comparison

Based on model population analysis, here we propose to perform model comparison by deriving an empirical distribution of the difference of RMSEP or RMSECV between two models (variable sets), followed by testing the null hypothesis that the difference of RMSEP or RMSECV between two models is zero. Without loss of generality, we describe the proposed method by taking the distribution of difference of RMSEP as an example. We assume that the data X consists of m samples in row and p variables in column and the target value Y is an m-dimensional column vector. Two variable sets, say V_1 and V_2, selected from the p variables, then can be compared using the MPA-based method described below.

First, a percentage, say 80%, from the m samples with variables in V_1 and V_2 is randomly selected to build two regression models using a preselected modeling method such as PLS [11] or support vector machines (SVMs) [12], respectively. Then an RMSEP value can be computed for each model by using the remaining 20% samples as the test set. Denote the two RMSEP values as $RMSEP_1$ and $RMSEP_2$, of which the difference can be calculated as

$$D = RMSEP_1 - RMSEP_2 \qquad (2)$$

By repeating this procedure N, say 1000, times, N D values are obtained and collected into a vector D. Now, the model comparison can be formulated into a hypothesis test problem as:

Null hypothesis: the mean of D is zero.

Alternative hypothesis: the mean of D is not zero.

By employing a statistical test method, *e.g.* *t*-test or Mann-Whitney U test [25], a P value can be computed for strictly assessing whether the mean of D is significantly different from zero ($P<0.05$) or not ($P>0.05$). If $P<0.05$, the sign of the mean of D is then used to compare which model (variable set) is of better predictive performance. If $P>0.05$, we say two models have the same predictive ability.

4. Results and discussions

4.1 Comparison of predictive performances of variables subsets

The corn NIR data measured on *mp5* instrument is used to illustrate the use of the proposed method (http://software.eigenvector.com/Data/index.html). This data contain NIR spectra measured at 700 wavelengths on 80 corn samples. The original NIR spectra are shown in **Figure 4**. The chemical property modeled here is the content of protein. As was demonstrated in a large body of literature [26-30], variable selection can improve the predictive performance of a model. Here we would like to investigate whether the gain in predictive accuracy using variable subsets identified by variable selection methods is significant.

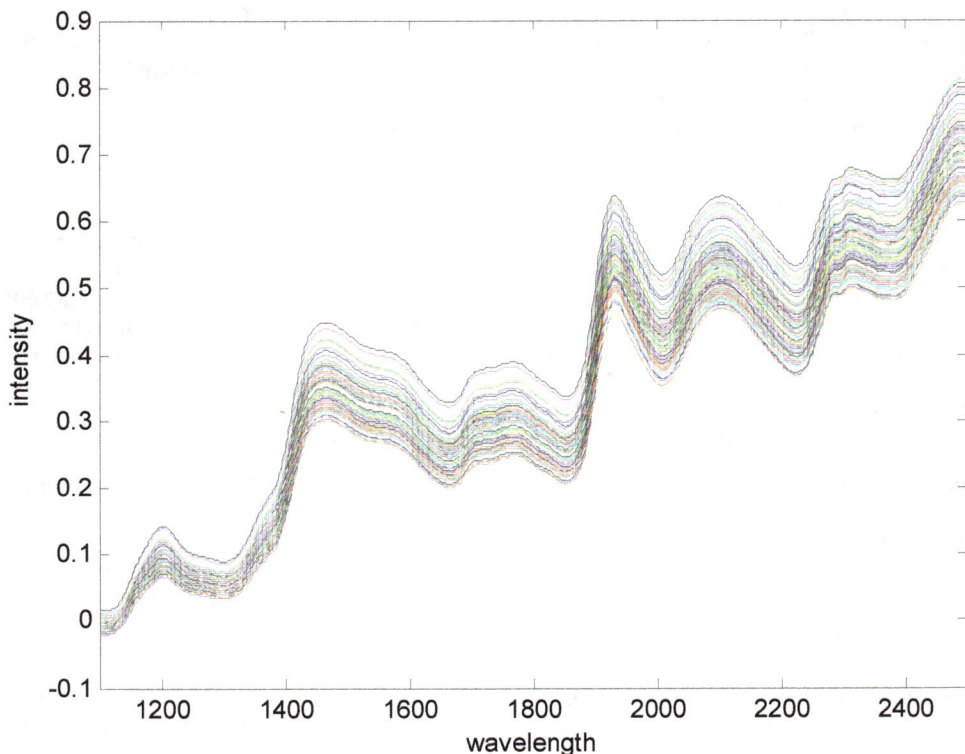

Fig. 4. Original near infrared spectra of corn on the mp5 instrument.

Uninformative variable elimination (UVE) is a widely used method for variable selection in chemometrics [26]. Its extended version, Monte Carlo UVE (MCUVE), was recently proposed [27, 31]. Mimicking the principle of "survival of the fittest" in Darwin's evolution theory, we developed a variable selection method in our previous work, called competitive adaptive reweighted sampling (CARS) [8, 28, 32, 33], which was shown to have the potential to identify an optimal subset of variables that show high predictive performances. The source codes of CARS are freely available at [34, 35].

In this study, MCUVE and CARS is chosen to first identify two variable sets, named V_1 and V_2, respectively. The set of the original 700 variables are denoted as V_0. Before data analysis, each wavelength of the original NIR spectra is standardized to have zero mean and unit variance. Regarding the pretreatment of spectral data, using original spectra, mean-centered

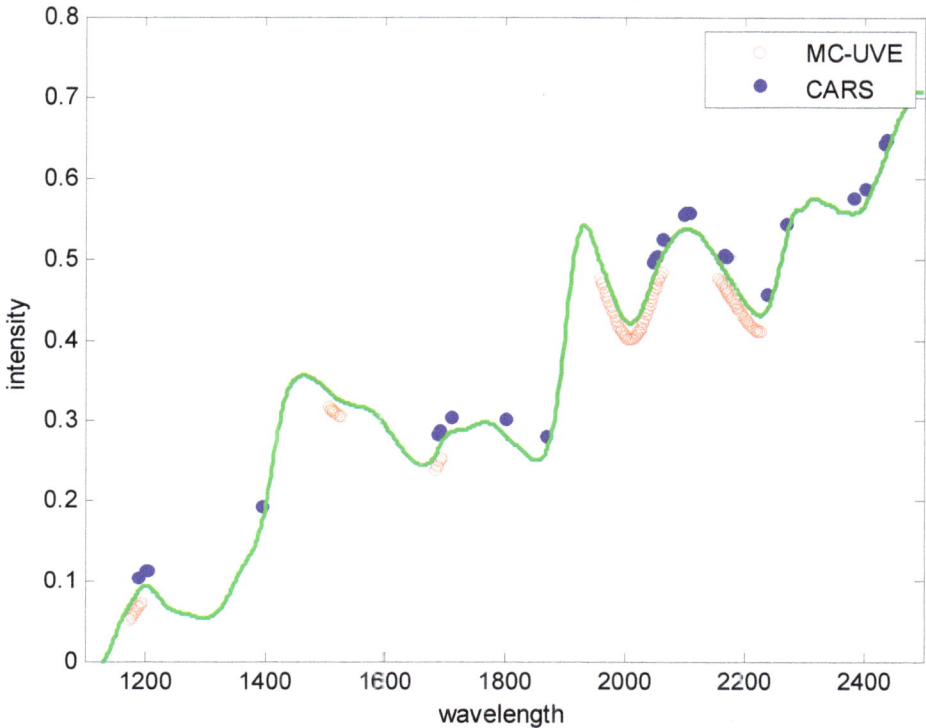

Fig. 5. Comparison of selected wavelengths using MC-UVE (red circles) and CARS (blue dots). The green line denotes the mean of the 80 corn NIR spectra.

spectra or standardized spectra indeed would lead to different results. But the difference is usually not big according to our experience. The reason why we choose standardization is to remove the influence of each wavelength's variance on PLS modeling because the decomposition of spectrum data X using PLS depends on the magnitude of covariance between wavelengths and the target variable Y. The number of PLS components are optimized using 5-fold cross validation. For MCUVE, the number of Monte Carlo simulations is set to 1000 and at each simulation 80% samples are selected randomly to build a calibration model. We use the reliability index (RI) to rank each wavelength and the number of wavelengths (with a maximum 200 wavelengths allowed) is identified using 5-fold cross validation. Using MCUVE. 115 wavelengths in 5 bands (1176-1196nm, 1508-1528nm, 1686-1696nm, 1960-2062nm and 2158-2226nm) are finally selected and shown in **Figure 5** as red circles. For CARS, the number of iterations is set to 50. Using CARS, altogether 28 variables (1188, 1202, 1204, 1396, 1690, 1692, 1710, 1800, 1870, 2048, 2050, 2052, 2064, 2098, 2102, 2104, 2106, 2108, 2166, 2168, 2238, 2270, 2382, 2402, 2434, 2436, 2468 and 2472 nm) are singled out and these variables are also shown in **Figure 5** as blue dots. Intuitively, MCUVE selects 5 wavelength bands while the variables selected by CARS are more diverse and scattered at different regions. In addition, the Pearson correlations variables selected by both methods are shown in **Figure 6**.

(A) MCUVE   (B) CARS

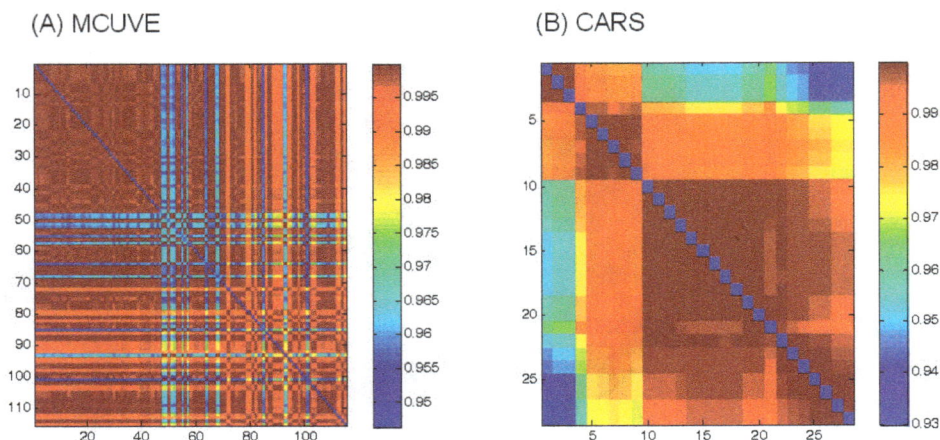

Fig. 6. The Pearson pair-wise correlations of variables selected using MCUVE (115 variables, left) and CARS (28 variables, right).

We choose PLS for building regression models. For the MPA-based method for model comparison, the number of Monte Carlo simulations is set to 1000 and at each simulation 60% samples are randomly selected as training samples and the rest 40% work as test samples. The number of PLS components is chosen based on 5-fold cross validation. In this setting, we first calculated 1000 values of $RMSEP_0$, $RMSEP_1$ and $RMSEP_2$ using V_0, V_1 and V_2, respectively. The distributions of $RMSEP_0$, $RMSEP_1$ and $RMSEP_2$ are shown in **Figure 7**. The mean and standard deviations of these three distributions are 0.169±0.025 (full spectra), 0.147±0.018 (MCUVE) and 0.108±0.015 (CARS). On the whole, both variable selection methods improve the predictive performance in terms of lower prediction errors and smaller standard deviations. Looking closely, the model selected by CARS has smaller standard deviation than that of MCUVE. The reason may be that CARS selected individual wavelengths and these wavelengths display lower correlations (see **Figure 6**) than those wavelength bands selected by MCUVE. The lower correlation results in better model stability which is reflected by smaller standard deviations of prediction errors. Therefore from the perspective of prediction ability, we recommend to adopt methods that select individual wavelengths rather than continuous wavelength bands.

Firstly, we compare the performance of the model selected by MCUVE to the full spectral model. The distribution of D values (MCUVE – Full spectra) is shown in Plot A of **Figure 8**. The mean of D is -0.023 and is shown to be not zero ($P < 0.000001$) using a two-side t test, indicating that MCUVE significantly improves the predictive performance. Of particular note, it can be observed that a percentage (83.1%) of D values are negative and the remaining (16.9%) is positive, which indicates model comparison based on a single split of the data into a training set and a corresponding test set may have the potential risk of drawing a wrong conclusion. In this case, the probability of saying that MCUVE does not improve predictive performances is about 0.169. However, this problem can be solved by the proposed MPA-based method because the model performance is tested on a large number of sub-datasets, rendering the current method potentially useful for reliably

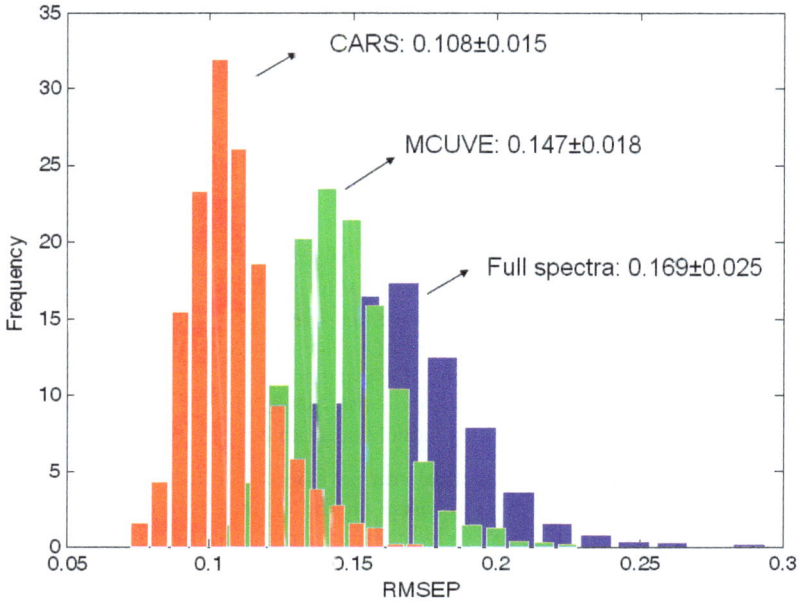

Fig. 7. Distributions of root mean squared errors of prediction (RMSEP) from 1000 test sets (32 samples) randomly selected from the 80 corn samples using full spectra and variables selected by MCUVE and CARS, respectively.

statistical model comparison. With our method, we have evidence showing that the improvement resulting from MCUVE is significant.

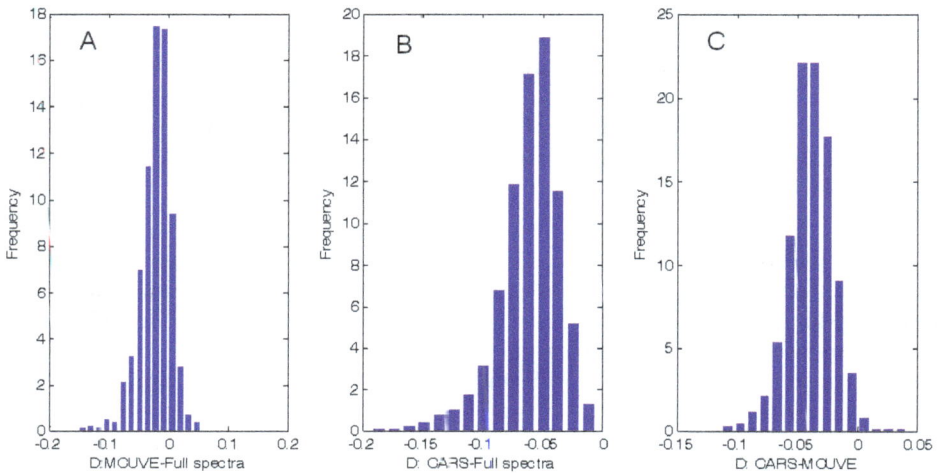

Fig. 8. The distributions of D values. The P values of t test for these three distributions are 5.36×10^{-120}, 0 and 0, respectively.

Further, the performance of the model selected by CARS is compared to the full spectral model. The distribution of D values (CARS – Full spectrum) is shown in Plot B of **Figure 8**. The mean of D is -0.061 which is much smaller that that from MCUVE (-0.023). Using a two-side t test, this mean is shown to be significantly different from zero ($P = 0$), indicating that the improvement over the full spectral model is significant. Interestingly, it is found that all the D values are negative, which implies the model selected by CARS is highly predictive and there is little evidence to recommend the use of a full spectral model, at least for this dataset.

Finally, we compare the models selected by MCUVE and CARS, respectively. The distribution of D values (CARS – MCUVE) is shown in Plot C of **Figure 8**. The mean of D values is -0.039. Using a two-side t test, this mean is shown to be significantly different from zero ($P = 0$), indicating that the improvement of CARS over MCUVE is significant. We find that 98.9% of D values are negative and only 1.1% are positive, which suggests that there is a small probability to draw a wrong conclusion that MCUVE performs better than CARS. However, with the help of MPA, this risky conclusion can be avoided, indeed.

Summing up, we have conducted statistical comparison of the full spectral model and the models selected by MCUVE and CARS based on the distribution of D values calculated using RMSEP. Our results show that model comparison based on a single split of the data into a training set and a corresponding test set may result in a wrong conclusion and the proposed MPA approach can avoid drawing such a wrong conclusion thus providing a solution to this problem.

4.2 Comparison of PCR, PLS and an ECR model

In chemometrics, PCR and PLS seem to be the most widely used method for building a calibration model. Recently, we developed a method, called elastic component regression (ECR), which utilizes a tuning parameter $\alpha \in [0,1]$ to supervise the decomposition of X-matrix [36], which falls into the category of continuum regression [37-40]. It is demonstrated theoretically that the elastic component resulting from ECR coincides with principal components of PCA when $\alpha = 0$ and also coincides with PLS components when $\alpha = 1$. In this context, PCR and PLS occupy the two ends of ECR and $\alpha \in (0,1)$ will lead to an infinite number of transitional models which collectively uncover the model path from PCR to PLS. The source codes implementing ECR in MATLAB are freely available at [41]. In this section, we would like to compare the predictive performance of PCR, PLS and an ECR model with $\alpha = 0.5$.

We still use the corn protein data described in Section 4.1. Here we do not consider all the variables but only the 28 wavelengths selected by CARS. For the proposed method, the number of Monte Carlo simulations is set to 1000. At each simulation 60% samples selected randomly are used as training samples and the remaining serve as test samples. The number of latent variables (LVs) for PCR, PLS and ECR ($\alpha = 0.5$) is chosen using 5-fold cross validation.

Figure 9 shows the three distributions of RMSEP computed using PCR, PLS and ECR ($\alpha = 0.5$). The mean and standard deviations of these distributions are 0.1069±0.0141, 0.1028±0.0111 and 0.0764±0.0108, respectively. Obviously, PLS achieves the lowest prediction errors as well as the smallest standard deviations. In contrast, PCR performs the worst. As a transitional model that is between PCR and PLS, ECR with α = 0.5 achieves the medium level performance.

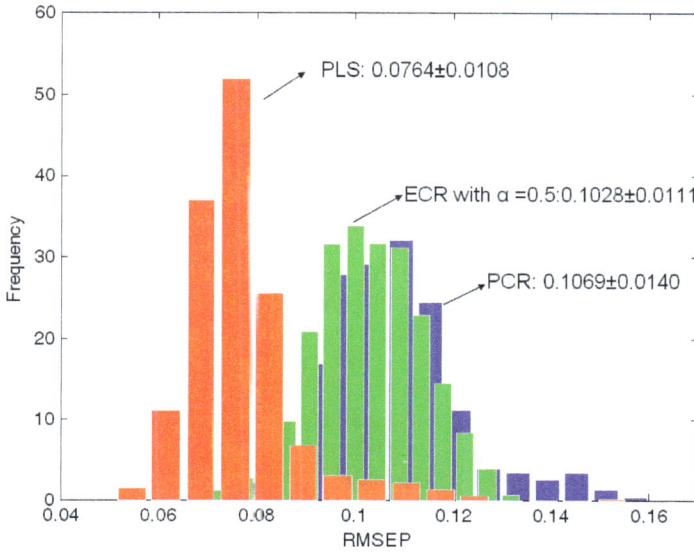

Fig. 9. The distributions of RMSEP from PCR, PLS and an ECR model with α =0.5

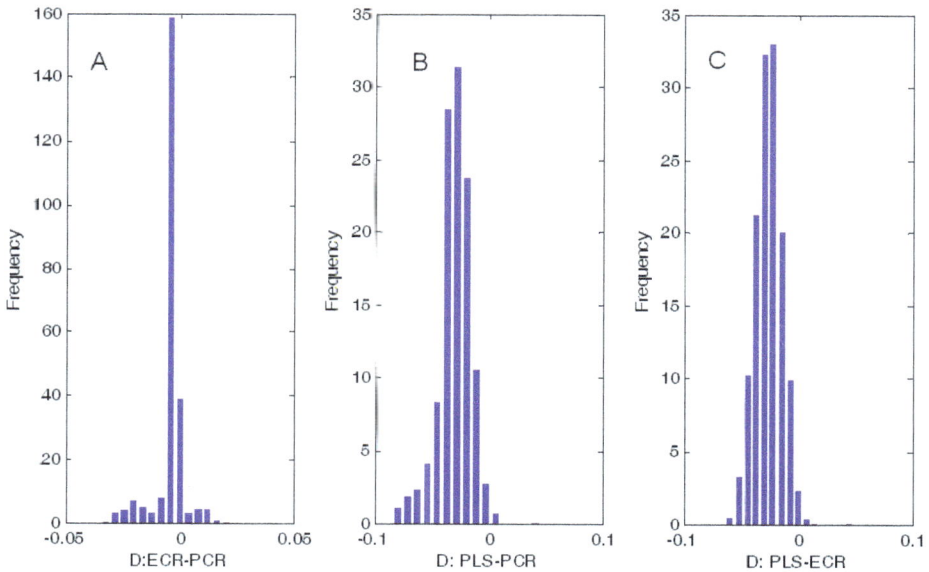

Fig. 10. The distributions of D values. The P values of t test for these three distributions are 0, 0 and 0, respectively.

The distributions of D values are displayed in **Figure 10**. The means of these three distributions are -0.0041 (Plot A), -0.0305 (Plot B) and -0.0264 (Plot C), respectively. Using a two-side t test, it is shown that all these three distributions of D values have a mean value that is significant not zero with P values equal to 0 , 0 and 0 for Plot A, Plot B and Plot C. To conclude, this section provides illustrative examples for the comparison of different modeling methods. Our example demonstrates that PLS (an ECR model associated with α = 1) performs better than PCR (an ECR model associated with α = 0) and a specific transitional ECR model associated with α = 0.5 has the moderate performance.

4.3 Comparison of PLS-LDA models before and after variable selection

Partial least squares-linear discriminant analysis (PLS-LDA) is frequently used in chemometrics and metabolomics/metabonomics for building predictive classification models and/or biomarker discovery [32, 42-45]. With the development of modern high-throughput analytical instruments, the data generated often contains a large number of variables (wavelengths, m/z ratios etc). Most of these variables are not relevant to the problem under investigation. Moreover, a model constructed using this kind of data that contain irrelevant variables would not be likely to have good predictive performance. Variable selection provides a solution to this problem that can help select a small number of informative variables that could be more predictive than an all-variable model.

In the present work, two methods are chosen to conduct variable selection. The first is t-test, which is a simple univariate method that determines whether two samples from normal distributions could have the same mean when standard deviations are unknown but assumed to be equal. The second is subwindow permutation analysis (SPA) which was a model population analysis-based approach proposed in our previous work [14]. The main characteristic of SPA is that it can output a conditional P value by implicitly taking into account synergistic effects among multiple variables. With this conditional P value, important variables or conditionally important variables can be identified. The source codes in Matlab and R are freely available at [46].We apply these two methods on a type 2 diabetes mellitus dataset that contains 90 samples (45 healthy and 45 cases) each of which is characterized by 21 metabolites measured using a GC/MS instrument. Details of this dataset can be found in reference [32].

Using t-test, 13 out of the 21 variables are identified to be significant (P < 0.01). For SPA, we use the same setting as described in our previous work [14]. Three variables are selected with the aid of SPA. Let V_0, V_1 and V_2 denote the sets containing all the 21 variables, the 13 variables selected by t-test and the 3 variables selected by SPA, respectively. To run the proposed method, we set the number of Monte Carlo simulations to 1000. At each simulation 70% samples are randomly selected to build a PLS-LDA model with the number of latent variables optimized by 10-fold cross validation. The remaining 30% samples working as test sets on which the misclassification error is computed.

Figure 11 shows the distributions of misclassification errors computed using these three variable sets. The mean and standard deviations of these distributions are 0.065±0.048 (all variables), 0.042±0.037 (t-test) and 0.034±0.034 (SPA), respectively. It can be found that the models using selected variables have lower prediction errors as well as higher stability in terms of smaller standard deviations, indicating that variable selection can improve the

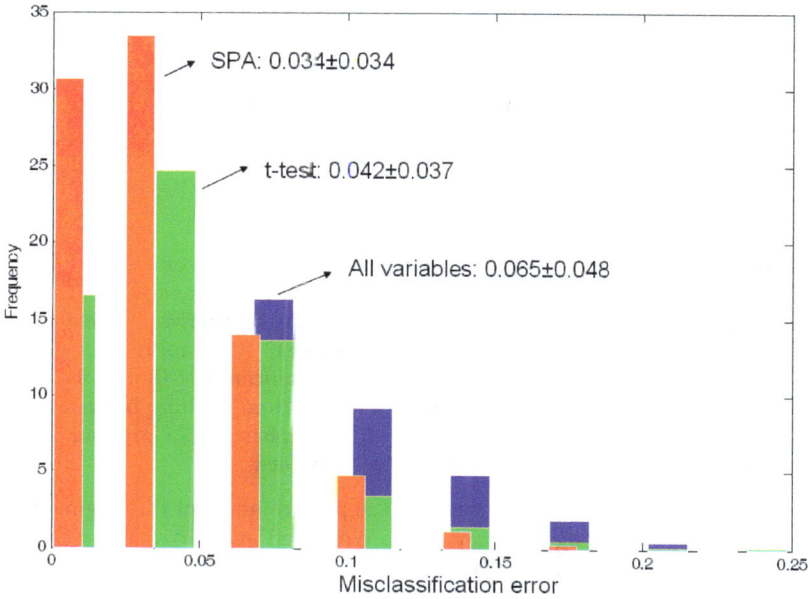

Fig. 11. The distributions of misclassification error on 1000 test sets using all variables and variables selected by t test and SPA. respectively.

performance of a classification model. The reason why SPA performs better than t-test is that synergistic effects among multiple variables are implicitly taken into account in SPA while t-test only considers univariate associations.

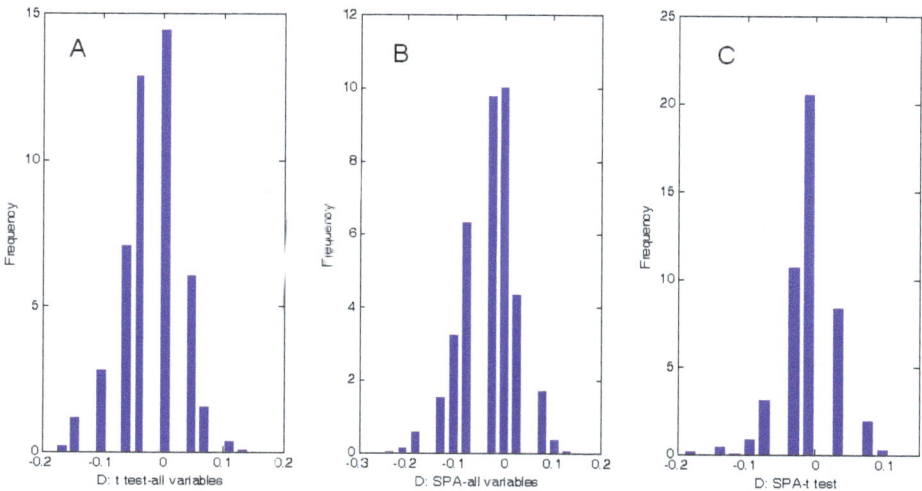

Fig. 12. The distributions of D values. The P values of t test for these three distributions are 1.66×10^{-46}, 1.02×10^{-57} and 1.27×10^{-8}.

We conducted pair-wise comparison of performances of the three variable sets described above. The distribution of D values (t-test – all variables) is shown in Plot A of **Figure 12**. The mean of D is -0.023 and is demonstrated to be significantly not zero ($P=0$) using a two-side t test, suggesting the improvement of variable selection. In spite of this improvement, we should also notice that a percentage (17.3%) of D values is positive, which again imply that model comparison based on a single split of the data into a training set and a corresponding test set is risky. However, with the aid of this MPA-based approach, it is likely to reliably compare different models in a statistical manner.

The distribution of D values (SPA – all variables) is shown in Plot B of **Figure 12**. The mean of D is -0.031 and is shown to be not zero ($P = 0$). Also, 171 D values are positive, again indicating the necessity of the use of a population of models for model comparison. In analogy, Plot C in **Figure 12** displays the distributions of D values (SPA-t-test). After applying a two-side t-test, we found that the improvement of SPA over t-test is significant (P= 0). For this distribution, 22.7% D values is positive, indicating that based on a random splitting of the data t-test will have a 22.7% chance to perform better than SPA. However, based on a large scale comparison, the overall performance of SPA is statistically better than t-test.

To conclude, in this section we have compared the performances of the original variable set and variable sets selected using t-test and SPA. We found evidences to support the use of the proposed model population analysis approach for statistical model comparison of different classification models.

5. Conclusions

A model population analysis approach for statistical model comparison is developed in this work. From our case studies, we have found strong evidences that support the use of model population analysis for the comparison of different variable sets or different modeling methods in both regression and classification. P values resulting from the proposed method in combination with the sign of the mean of D values clearly shows whether two models have the same performance or which model is significantly better.

6. Acknowledgements

This work was financially supported by the National Nature Foundation Committee of P.R. China (Grants No. 20875104 and No. 21075138) and the Graduate degree thesis Innovation Foundation of Central South University (CX2010B057).

7. References

[1] H. Akaike, A new look at the statistical model identification, IEEE Transactions on Automatic Control, 19 (1974) 716.
[2] G.E. Schwarz, Estimating the dimension of a model, Annals of Statistics, 6 (1978) 461.
[3] S. Wold, Cross-validatory estimation of the number of components in factor and principal component analysis, Technometrics, 20 (1978) 397.
[4] Q.-S. Xu, Y.-Z. Liang, Monte Carlo cross validation, Chemometr. Intell. Lab., 56 (2001) 1.
[5] P. Filzmoser, B. Liebmann, K. Varmuza, Repeated double cross validation, J Chemometr, 23 (2009) 160.

[6] J. Shao, Linear Model Selection by Cross-Validation, J Am. Stat. Assoc., 88 (1993) 486.

[7] M. Stone, Cross-validatory choice and assessment of statistical predictions, J. R. Stat. Soc. B, 36 (1974) 111.

[8] H.-D. Li, Y.-Z. Liang, Q.-S. Xu. D.-S. Cao, Model population analysis for variable selection, J. Chemometr., 24 (2009) 418.

[9] B. Efron, G. Gong, A Leisurely Look at the Bootstrap, the Jackknife, and Cross-Validation, Am. Stat., 37 (1983) 36.

[10] B. Efron, R. Tibshirani, An introduction to the bootstrap, Chapman&Hall, (1993).

[11] S. Wold, M. Sjöström, L. Eriksson, PLS-regression: a basic tool of chemometrics, Chemometr. Intell. Lab., 58 (2001) 109.

[12] H.-D. Li, Y.-Z. Liang, Q.-S. Xu, Support vector machines and its applications in chemistry, Chemometr. Intell. Lab., 95 (2009) 188

[13] D.S. Cao, Y.Z. Liang, Q.S. Xu, H.D. Li, X. Chen, A New Strategy of Outlier Detection for QSAR/QSPR, J. Comput. Chem., 31 (2010) 592.

[14] H.-D. Li, M.-M. Zeng, B.-B. Tan, Y.-Z. Liang, Q.-S. Xu, D.-S. Cao, Recipe for revealing informative metabolites based on model population analysis, Metabolomics, 6 (2010) 353.

[15] Q. Wang, H.-D. Li, Q.-S. Xu, Y.-Z. Liang, Noise incorporated subwindow permutation analysis for informative gene selection using support vector machines, Analyst, 136 (2011) 1456.

[16] H.-D. Li, Y.-Z. Liang, Q.-S. Xu, D.-S. Cao, B.-B. Tan, B.-C. Deng, C.-C. Lin, Recipe for Uncovering Predictive Genes using Support Vector Machines based on Model Population Analysis, IEEE/ACM T Comput Bi, 8 (2011) 1633.

[17] A.I. Bandos, H.E. Rockette, D. Gur, A permutation test sensitive to differences in areas for comparing ROC curves from a paired design, Statistics in Medicine, 24 (2005) 2873.

[18] Y. Ai-Jun, S. Xin-Yuan, Bayesian variable selection for disease classification using gene expression data, Bioinformatics, 26 (2009) 215.

[19] A. Vehtari, J. Lampinen, Bayesian model assessment and comparison using cross-validation predictive densities, Neural Computation, 14 (2002) 2439.

[20] T. Chen, E. Martin, Bayesian linear regression and variable selection for spectroscopic calibration, Anal. Chim. Acta, 631 (2009) 13.

[21] L. Breiman, Bagging Predictors, Mach. Learn., 24 (1996) 123.

[22] Y. Freund, R. Schapire, Experiments with a new boosting algorithm, Machine Learning: Proceedings of the Thirteenth International Conference, (1996) 148.

[23] L. Breiman, Random Forests, Mach. Learn., 45 (2001) 5.

[24] K. Ting, J. Wells, S. Tan, S. Teng, G. Webb, Feature-subspace aggregating: ensembles for stable andÂ unstable learners, Mach. Learn., 82 (2010) 375.

[25] H.B. Mann, D.R. Whitney, On a test of whether one of two random variables is stochastically larger than the other, Ann. Math. Statist., 18 (1947) 50.

[26] V. Centner, D.-L. Massart, O.E. de Noord, S. de Jong, B.M. Vandeginste, C. Sterna, Elimination of Uninformative Variables for Multivariate Calibration, Anal. Chem., 68 (1996) 3851.

[27] W. Cai, Y. Li, X. Shao, A variable selection method based on uninformative variable elimination for multivariate calibration of near-infrared spectra, Chemometr. Intell. Lab., 90 (2008) 188.

[28] H.-D. Li, Y.-Z. Liang, Q.-S. Xu, D.-S. Cao, Key wavelengths screening using competitive adaptive reweighted sampling method for multivariate calibration, Anal. Chim. Acta, 648 (2009) 77.

[29] J.-H. Jiang, R.J. Berry, H.W. Siesler, Y. Ozaki, Wavelength Interval Selection in Multicomponent Spectral Analysis by Moving Window Partial Least-Squares Regression with Applications to Mid-Infrared and Near-Infrared Spectroscopic Data, Anal. Chem., 74 (2002) 3555.

[30] C. Reynes, S. de Souza, R. Sabatier, G. Figueres, B. Vidal, Selection of discriminant wavelength intervals in NIR spectrometry with genetic algorithms, J. Chemometr., 20 (2006) 136.

[31] Q.-J. Han, H.-L. Wu, C.-B. Cai, L. Xu, R.-Q. Yu, An ensemble of Monte Carlo uninformative variable elimination for wavelength selection, Anal. Chim. Acta, 612 (2008) 121.

[32] B.-B. Tan, Y.-Z. Liang, L.-Z. Yi, H.-D. Li, Z.-G. Zhou, X.-Y. Ji, J.-H. Deng, Identification of free fatty acids profiling of type 2 diabetes mellitus and exploring possible biomarkers by GC–MS coupled with chemometrics, Metabolomics, 6 (2009) 219.

[33] W. Fan, H.-D. Li, Y. Shan, H.-Y. Lv, H.-X. Zhang, Y.-Z. Liang, Classification of vinegar samples based on near infrared spectroscopy combined with wavelength selection, Analytical Methods, 3 (2011) 1872.

[34] Source codes of CARS-PLS for variable selection: http://code.google.com/p/carspls/

[35] Source codes of CARS-PLSLDA for variable selection: http://code.google.com/p/cars2009/

[36] H.-D. Li, Y.-Z. Liang, Q.-S. Xu, Uncover the path from PCR to PLS via elastic component regression, Chemometr. Intell. Lab., 104 (2010) 341.

[37] M. Stone, R.J. Brooks, Continuum Regression: Cross-Validated Sequentially Constructed Prediction Embracing Ordinary Least Squares, Partial Least Squares and Principal Components Regression, J. R. Statist. Soc. B, 52 (1990) 237.

[38] A. Bjorkstrom, R. Sundberg, A generalized view on continuum regression, Scand. J. Statist., 26 (1999) 17.

[39] B.M. Wise, N.L. Ricker, Identification of finite impulse response models with continuum regression, J. Chemometr., 7 (1993) 1.

[40] J.H. Kalivas, Cyclic subspace regression with analysis of the hat matrix, Chemometr. Intell. Lab., 45 (1999) 215.

[41] Source codes of Elastic Component Regression: http://code.google.com/p/ecr/

[42] M. Barker, W. Rayens, Partial least squares for discrimination, J. Chemometr., 17 (2003) 166.

[43] J.A. Westerhuis, H.C.J. Hoefsloot, S. Smit, D.J. Vis, A.K. Smilde, E.J.J. van Velzen, J.P.M. van Duijnhoven, F.A. van Dorsten, Assessment of PLSDA cross validation, Metabolomics, 4 (2008) 81.

[44] L.-Z. Yi, J. He, Y.-Z. Liang, D.-L. Yuan, F.-T. Chau, Plasma fatty acid metabolic profiling and biomarkers of type 2 diabetes mellitus based on GC/MS and PLS-LDA, FEBS Letters, 580 (2006) 6837.

[45] J. Trygg, E. Holmes, T.r. Lundstedt, Chemometrics in Metabonomics, Journal of Proteome Research, 6 (2006) 469.

[46] Source codes of Subwindow Permutation Analysis: http://code.google.com/p/spa2010/

2

Exploratory Data Analysis with Latent Subspace Models

José Camacho

Department of Signal Theory, Telematics and Communication,
University of Granada, Granada
Spain

1. Introduction

Exploratory Data Analysis (EDA) has been employed for decades in many research fields, including social sciences, psychology, education, medicine, chemometrics and related fields (1) (2). EDA is both a data analysis philosophy and a set of tools (3). Nevertheless, while the philosophy has essentially remained the same, the tools are in constant evolution. The application of EDA to current problems is challenging due to the large scale of the data sets involved. For instance, genomics data sets can have up to a million of variables (5). There is a clear interest in developing EDA methods to manage these scales of data while taking advantage of *the basic importance of simply looking at data* (3).

In data sets with a large number of variables, collinear data and missing values, projection models based on latent structures, such as Principal Component Analysis (PCA) (6) (7) (1) and Partial Least Squares (PLS) (8) (9) (10), are valuable tools within EDA. Projection models and the set of tools used in combination simplify the analysis of complex data sets, pointing out to special observations (outliers), clusters of similar observations, groups of related variables, and crossed relationships between specific observations and variables. All this information is of paramount importance to improve data knowledge.

EDA based on projection models has been successfully applied in the area of chemometrics and industrial process analysis. In this chapter, several standard tools for EDA with projection models, namely score plots, loading plots and biplots, are revised and their limitations are elucidated. Two recently proposed tools are introduced to overcome these limitations. The first of them, named Missing-data methods for Exploratory Data Analysis or MEDA for short (11), is used to investigate the relationships between variables in projection subspaces. The second one is an extension of MEDA, named observation-based MEDA or oMEDA (33), to discover the relationships between observations and variables. The EDA approach based on PCA/PLS with scores and loading plots, MEDA and oMEDA is illustrated with several real examples from the chemometrics field.

This chapter is organized as follows. Section 2 briefly discusses the importance of subspace models and score plots to explore the data distribution. Section 3 is devoted to the investigation of the relationship among variables in a data set. Section 4 studies the relationship between observations and variables in latent subspaces. Section 5 presents a EDA case study of Quantitative Structure–Activity Relationship (QSAR) modelling and section 6 proposes some concluding remarks. Examples and Figures were computed using the MATLAB programming environment, with the PLS-Toolbox (32) and home-made software. A MATLAB toolbox with the tools employed in this chapter is available at http://wdb.ugr.es/ josecamacho/.

2. Patterns in the data distribution

The distribution of the observations in a data set contains relevant information for data understanding. For instance, in an industrial process, one outlier may represent an abnormal situation which affects the process variables to a large extent. Studying this observation with more detail, one may be able to identify if it is the result of a process upset or, very commonly, a sensor failure. Also, clusters of observations may represent different operation points. Outliers, clusters and trends in the data may be indicative of the degree of control in the process and of assignable sources of variation. The identification of these sources of variation may lead to the reduction of the variance in the process with the consequent reduction of costs.

The distribution of the observations can be visualized using scatter plots. For obvious reasons, scatter plots are limited to three dimensions at most, and typically to two dimensions. Therefore, the direct observation of the data distribution in data sets with several tens, hundreds or even thousands of variables is not possible. One can always construct scatter plots for selective pairs or thirds of variables, but this is an overwhelming and often misleading approach. Projection models overcome this problem. PCA and PLS can be used straightforwardly to visualize the distribution of the data in the latent subspace, considering only a few latent variables (LVs) which contain most of the variability of interest. Scatter plots of the scores corresponding to the LVs, the so-called score plots, are used for this purpose.

Score plots are well known and accepted in the chemometric field. Although simple to understand, score plots are paramount for EDA. The following example may be illustrative of this. In Figure 1, three simulated data sets of the same size (100×100) are compared. Data simulation was performed using the technique named Approximation of a DIstribution for a given COVariance matrix (15), or ADICOV for short. Using this technique, the same covariance structure was simulated for the three data sets but with different distributions: the first data set presents a multi-normal distribution in the latent subspace, the second one presents a severe outlier and the third one presents a pair of clusters. If the scatter plot of the observations in the plane spanned by the first two variables is depicted (first row of Figure 1), the data sets seem to be almost identical. Therefore, unless an extensive exploration is performed, the three data sets may be though to come from a similar data generation procedure. However, if a PCA model for each data set is fitted and the score plots corresponding to the first 2 PCs are shown (second row of Figure 1), differences among the three data sets are made apparent: in the second data set there is one outlier (right side of Figure 1(e)) and in the third data set there are two clusters of observations. As already discussed, the capability to find these details is paramount for data understanding, since outliers and clusters are very informative of the underlaying phenomena. Most of the times these details are also apparent in the original variables, but finding them may be a tedious work. Score plots after PCA modelling are perfectly suited to discover large deviations among the observations, avoiding the overwhelming task of visualizing each possible pair of original variables. Also, score plots in regression models such as PLS are paramount for model interpretation prior to prediction.

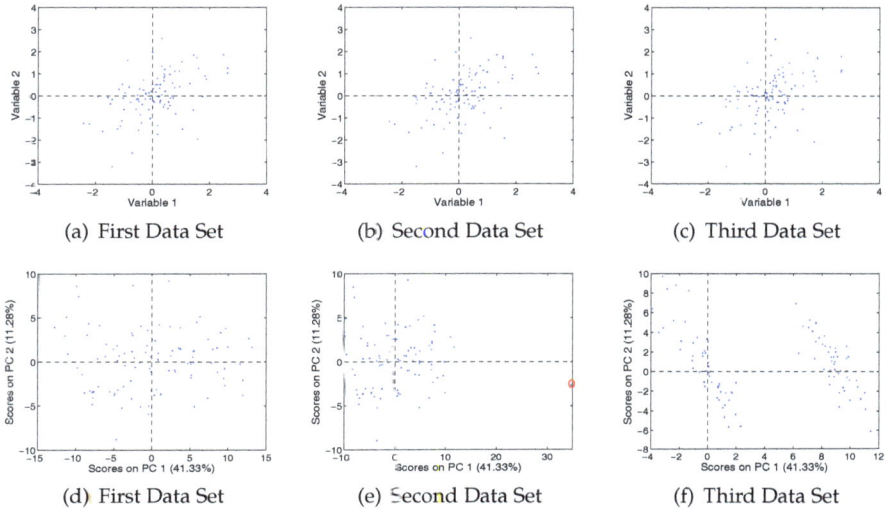

(a) First Data Set   (b) Second Data Set   (c) Third Data Set

(d) First Data Set   (e) Second Data Set   (f) Third Data Set

Fig. 1. Experiment with three simulated data sets of dimension 100×100. Data simulation was performed using the ADICOV technique (15). In the first row of figures, the scatter plots corresponding to the first two variables in the data sets are shown. In the second row of figures, the scatter plots (score plots) corresponding to the first two PCs in the data sets are shown.

3. Relationships among variables

PCA has been often employed to explore the relationships among variables in a data set (19; 20). Nevertheless, it is generally accepted that Factor Analysis (FA) is better suited than PCA to study these relationships (1; 7). This is because FA algorithms are designed to distinguish between shared and unique variability. The shared variability, the so-called communalities in the FA community, reflect the common factors–common variability–among observable variables. The unique variability is only present in one observable variable. The common factors make up the relationship structure in the data. PCA makes no distinction between shared and unique variability and therefore it is not suited to find the structure in the data.

When either PCA or FA are used for data understanding, a two step procedure is typically followed (1; 7). Firstly, the model is calibrated from the available data. Secondly, the model is rotated to obtain a so-called simple structure. The second step is aimed at obtaining loading vectors with as much loadings close to 0 as possible. That way, the loading vectors are easier to interpret. It is generally accepted that oblique transformations are preferred to the more simple orthogonal transformations (19; 20), although in many situations the results are similar (7).

The limitation of PCA to detect common factors and the application of rotation methods will be illustrated using the pipelines artificial examples (14). Data for each pipeline are simulated according to the following equalities:

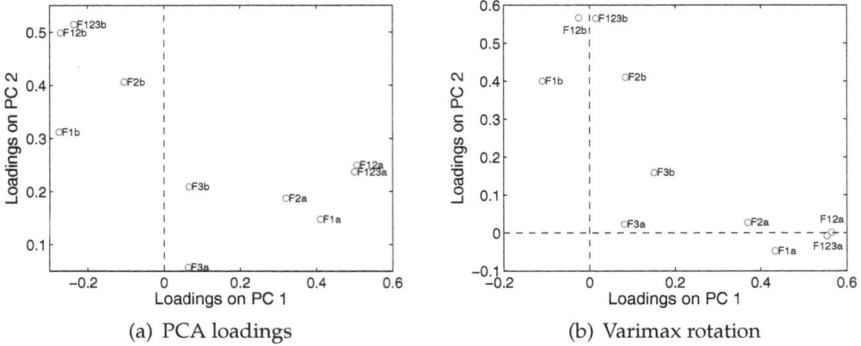

(a) PCA loadings   (b) Varimax rotation

Fig. 2. Loading plots of the PCA model fitted from the data in the two pipelines example: (a) original loadings and (b) loadings after varimax rotation.

$$F12 = F1 + F2$$

$$F123 = F1 + F2 + F3$$

where F1, F2 and F3 represent liquid flows which are generated independently at random following a normal distribution of 0 mean and standard deviation 1. A 30% of measurement noise is generated for each of the five variables in a pipeline at random, following a normal distribution of 0 mean:

$$\tilde{x}_i = (x_i + \sqrt{0.3} \cdot n)/(\sqrt{1.3})$$

where $\tilde{x}_i$ is the contaminated variable, x_i the noise-free variable and n the noise generated. This simulation structure generates blocks of five variables with three common factors: the common factor F1, present in the observed variables F1, F12 and F123; the common factor F2, present in the observed variables F2, F12 and F123; and the common factor F3, present in F3 and F123. Data sets of any size can be obtained by combining the variables from different pipelines. In this present example, a data set with 100 observations from two pipelines for which data are independently generated is considered. Thus, the size of the data set is 100 × 10 and the variability is built from 6 common factors.

Figure 2 shows the loading plots of the PCA model of the data before and after rotation. Loading plots are interpreted so that close variables, provided they are far enough from the origin of coordinates, are considered to be correlated. This interpretation is not always correct. In Figure 2(a), the first component separates the variables corresponding to the two pipelines. The second component captures variance of most variables, specially of those in the second pipeline. The two PCs capture variability corresponding to most common factors in the data at the same time, which complicates the interpretation. As already discussed, PCA is focused on variance, without making the distinction between unique and shared variance. The result is that the common factors are not aligned with the PCs. Thus, one single component reflects several common factors and the same common factor may be reflected in several components. As a consequence, variables with high and similar loadings in the same subset of components do not necessarily need to be correlated, since they may present very different loadings in others components. Because of this, inspecting only a pair of components may lead to incorrect conclusions. A good interpretation would require inspecting and interrelating all pairs of components with relevant information, something which may be challenging in many situations. This problem affects the interpretation and it is the reason why FA is generally preferred to PCA.

Figure 2(b) shows the resulting loadings after applying one of the most used rotation methods: the varimax rotation. Now, the variables corresponding to each pipeline are grouped towards one of the loading vectors. This highlights the fact that there are two main and orthogonal sources of variability, each one representing the variability in a pipeline. Also, in the first component variables collected from pipeline 2 present low loadings whereas in the second component variables collected from pipeline 1 present low loadings. This is the result of applying the notion of simple structure, with most of the loadings rotated towards 0. The interpretation is simplified as a consequence of improving the alignment of components with common factors. This is especially useful in data sets with many variables.

Although FA and rotation methods may improve the interpretation, they still present severe limitations. The derivation of the structure in the data from a loading plot is not straightforward. On the other hand, the rotated model depends greatly on the normalization of the data and the number of PCs (1; 21). To avoid this, several alternative approaches to rotation have been suggested. The point in common of these approaches is that they find a trade-off between variance explained and model simplicity (1). Nevertheless, imposing a simple structure has also drawbacks. Reference (11) shows that, when simplicity is pursued, there is a potential risk of simplifying even the true relationships in the data set, missing part of the data structure. Thus, the indirect improvement of data interpretation by imposing a simple structure may also report misleading results in certain situations.

3.1 MEDA

MEDA is designed to find the true relationships in the data. Therefore, it is an alternative to rotation methods or in general to the simple structure approach. A main advantage of MEDA is that, unlike rotation or FA methods, it is applied over any projection subspace without actually modifying it. The benefit is twofold. Firstly, MEDA is straightforwardly applied in any subspace of interest: PCA (maximizing variance), PLS (maximizing correlations) and any other. On the contrary, FA methods are typically based on complicated algorithms, several of which have not been extended to regression. Secondly, MEDA is also useful for model interpretation, since common factors and components are easily interrelated. This is quite useful, for instance, in the selection of the number of components.

MEDA is based on the capability of missing values estimation of projection models (22–27). The MEDA approach is depicted in Figure 3. Firstly, a projection model is fitted from the calibration $N \times M$ matrix $\mathbf{X}$ (and optionally $\mathbf{Y}$). Then, for each variable m, matrix $\mathbf{X}_m$ is constructed, which is a $N \times M$ matrix full with zeros except in the m-th column where it contains the m-th column of matrix $\mathbf{X}$. Using $\mathbf{X}_m$ and the model, the scores are estimated with a missing data method. The known data regression (KDR) method (22; 25) is suggested at this point. From the scores, the original data is reconstructed and the estimation error computed. The variability of the estimation error is compared to that of the original data according to the following index of goodness of prediction:

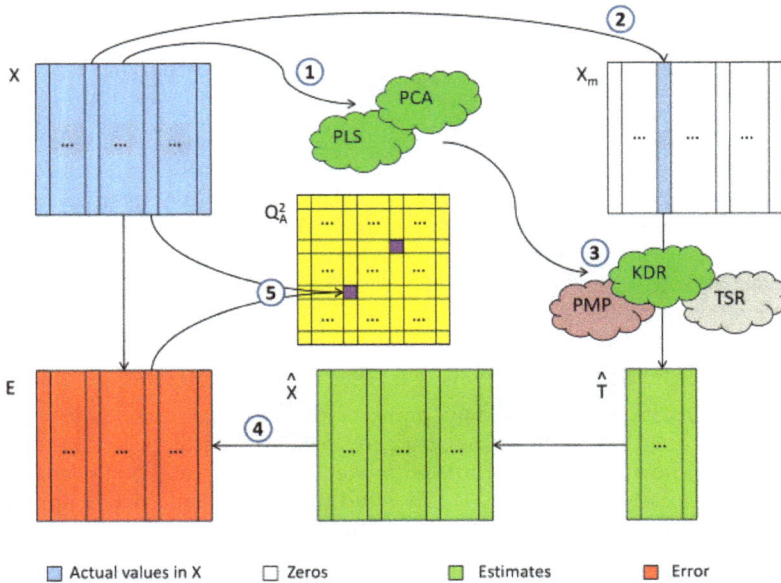

Fig. 3. MEDA technique: (1) model calibration, (2) introduction of missing data, (3) missing data imputation, (4) error computation, (5) computation of matrix $\mathbf{Q}_A^2$.

$$q_{A,(m,l)}^2 = 1 - \frac{\|\hat{\mathbf{e}}_{A,(l)}\|^2}{\|\mathbf{x}_{(l)}\|^2}, \quad \forall l \neq m. \tag{1}$$

where $\hat{\mathbf{e}}_{A,(l)}$ corresponds to the estimation error for the l-th variable and $\mathbf{x}_{(l)}$ is its actual value. The closer the value of the index is to 1, the more related variables m and l are. After all the indices corresponding to each pair of variables are computed, matrix $\mathbf{Q}_A^2$ is formed so that $q_{A,(m,l)}^2$ is located at row m and column l. For interpretation, when the number of variables is large, matrix $\mathbf{Q}_A^2$ can be shown as a color map. Also, a threshold can be applied to $\mathbf{Q}_A^2$ so that elements over this threshold are set to 1 and elements below the threshold are set to 0.

The procedure depicted in Figure 3 is the original and more general MEDA algorithm. Nevertheless, provided KDR is the missing data estimation technique, matrix Q_A^2 can be computed from cross-product matrices following a more direct procedure. The value corresponding to the element in the i-th row and j-th column of matrix $\mathbf{Q}_A^2$ in MEDA is equal to:

$$q_{A,(m,l)}^2 = \frac{2 \cdot S_{ml} \cdot S_{ml^A} - (S_{ml^A})^2}{S_{mm} \cdot S_{ll}}. \tag{2}$$

where S_{lm} stands for the cross-product of variables $\mathbf{x}_l$ and $\mathbf{x}_m$, i.e. $S_{lm} = \mathbf{x}_l^T \cdot \mathbf{x}_m$, and S_{lm^A} stands for the cross-product of variables $\mathbf{x}_l$ and $\mathbf{x}_m^A$, being $\mathbf{x}_m^A$ the projection of $\mathbf{x}_m$ in the model sub-space in coordinates of the original space. Thus, S_{lm} is the element in the l-th row

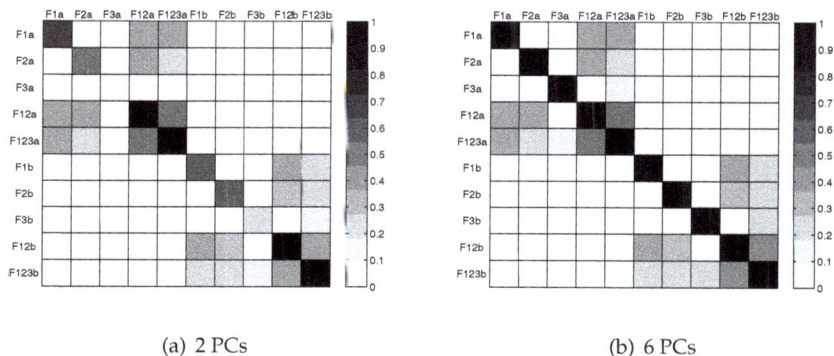

(a) 2 PCs   (b) 6 PCs

Fig. 4. MEDA matrix from the PCA model fitted from the data in the two pipelines example.

and m-th column of the cross-product matrix $\mathbf{XX} = \mathbf{X}^T \cdot \mathbf{X}$ and S_{lmA} corresponds to the element in the l-th row and m-th column of matrix $\mathbf{XX} \cdot \mathbf{P}_A \cdot \mathbf{P}_A^t$ in PCA and of matrix $\mathbf{XX} \cdot \mathbf{R}_A \cdot \mathbf{P}_A^t$ in PLS, with:

$$\mathbf{R}_A = \mathbf{W}_A \cdot (\mathbf{P}_A^T \cdot \mathbf{W}_A)^{-1}. \tag{3}$$

The relationship of the MEDA algorithm and cross-product matrices was firstly pointed out by Arteaga (28) and it can also be derived from the original MEDA paper (11). Equation (2) represents a direct and fast procedure to compute MEDA, similar in nature to the algorithms for model fitting from cross-product matrices, namely the eigendecomposition (ED) for PCA and the kernel algorithms (29) (30) (31) for PLS.

In Figure 4(a), the MEDA matrix corresponding to the 2 PCs PCA model of the example in the previous section, the two independent pipelines, is shown. The structure in the data is elucidated from this matrix. The separation between the two pipelines is shown in the fact that upper-right and lower-left quadrants are close to zero. The relationship among variables corresponding to factors F1 and F2 are also apparent in both pipelines. Since the variability corresponding to factors F3 is barely captured by the first 2 PCs, these are not reflected in the matrix. Nevertheless, if 6 PCs are selected. (Figure 4(b)) the complete structure in the data is clearly found.

MEDA improves the interpretation of both the data set and the model fitted without actually pursuing a simple structure. The result is that MEDA has better properties than rotation methods: it is more accurate and its performance is not deteriorated when the number of PCs is overestimated. Also, the output of MEDA does not depend on the normalization of the loadings, like rotated models do, an it is not limited to subspaces with two or three components at most. A comparison of MEDA with rotation methods is out of the scope of this chapter. Please refer to (11) for it and also for a more algorithmic description of MEDA.

5.2 Loading plots and MEDA

The limitations of loading plots and the application of MEDA were introduced with the pipelines artificial data set. This is further illustrated in this section with two examples provided with the PLS-toolbox (32): the Wine data set, which is used in the documentation of the cited software to show the capability of PCA for improving data understanding, and the

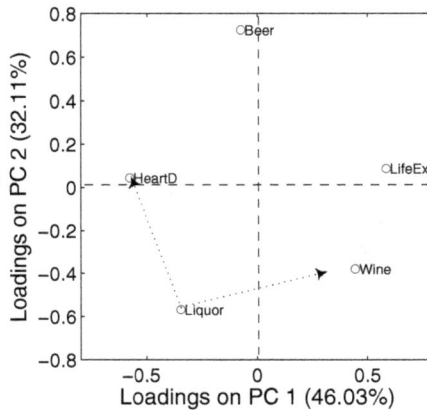

Fig. 5. Loading plot of the first 2 PCs from the Wine Data set provided with the PLS-toolbox (32).

PLSdata data set, which is used to introduce regression models, including PLS. The reading of the analysis and discussion of both data sets in (32) is recommended.

As suggested in (32), two PCs are selected for the PCA model of the Wine data set. The corresponding loading plot is shown in Figure 5. According to the reference, this plot shows that variables HeartD and LifeEx are negatively correlated, being this correlation captured in the first component. Also, "wine is somewhat positively correlated with life expectancy, likewise, liquor is somewhat positively correlated with heart disease". Finally, bear, wine and liquor form a triangle in the figure, which "suggests that countries tend to trade one of these vices for others, but the sum of the three tends to remain constant". Notice that although these conclusions are interesting, some of them are not so evidently shown by the plot. For instance, Liquor is almost as close to HeartD than to Wine. Is Liquor correlated to Wine as it is to HeartD?

MEDA can be used to improve the interpretation of loading plots. In Figure 6(a), the MEDA matrix for the first PC is shown. It confirms the negative correlation between HeartD and LifeEx, and the–lower–positive correlation between HeartD and Liqour and LifeEx and Wine. Notice that these three relationships are three different common factors. Nevertheless, they all manifest in the same component, making the interpretation with loading plots more complex. The MEDA matrix for the second PC in Figure 6(b) shows the relationship between the three types of drinks. The fact that the second PC captures this relationship was not clear in the loading plot. Furthermore, the MEDA matrix shows that Wine and Liquor are not correlated, answering to the question in the previous paragraph. Finally, this absence of correlation refutes that countries tend to trade wine for liquor or viceversa, although this effect may be true for bear.

In the PLSdata data set, the aim is to obtain a regression model that relates 20 temperatures measured in a Slurry-Fed Ceramic Melter (SFCM) with the level of molten glass. The x-block contains 20 variables which correspond to temperatures collected in two vertical thermowells. Variables 1 to 10 are taken from the bottom to the top in thermowell 1, and variables 11 to 20 from the bottom to the top in thermowell 2. The data set includes 300 training observations

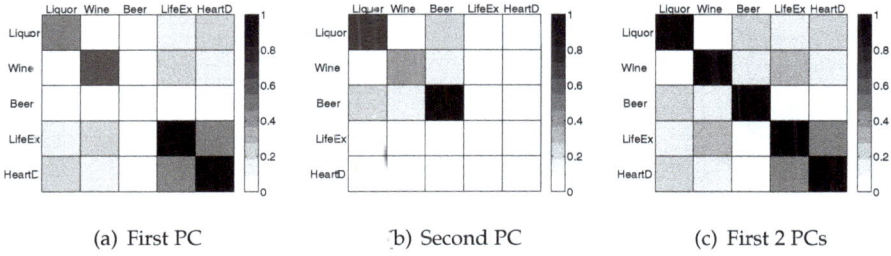

(a) First PC (b) Second PC (c) First 2 PCs

Fig. 6. MEDA matrices of the first PCs from the Wine data set provided with the PLS-toolbox (32).

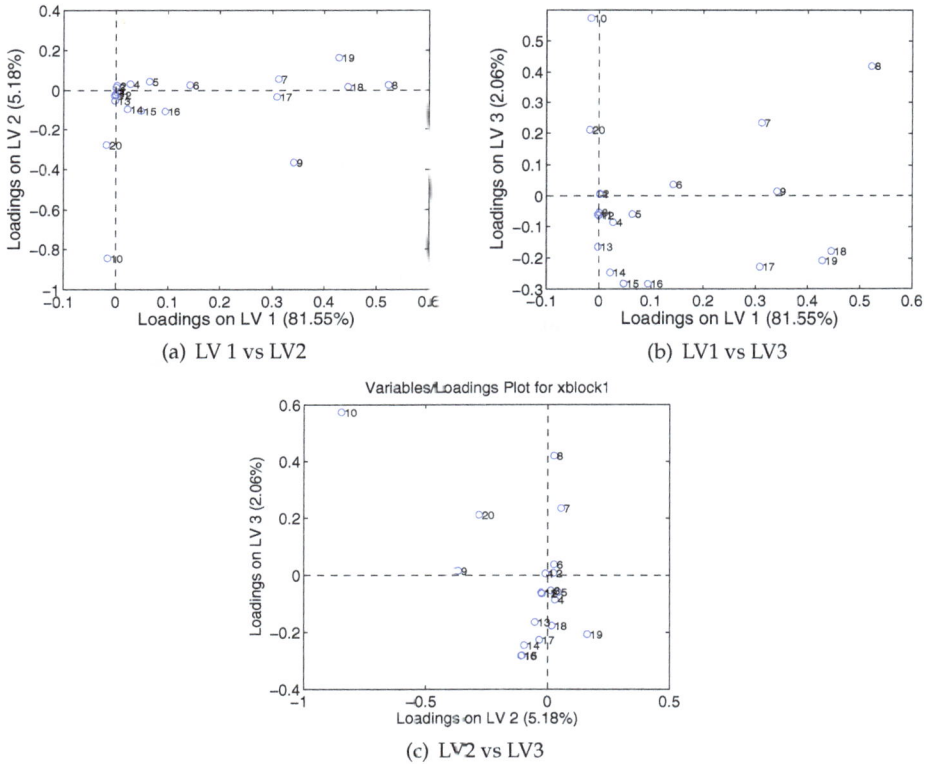

(a) LV 1 vs LV2 (b) LV1 vs LV3

(c) LV2 vs LV3

Fig. 7. Loading plots from the PLS model in the Slurry-Fed Ceramic Melter data set.

and 200 test observations. This same data set was used to illustrate MEDA with PCA in (11) with the temperatures and the level of molten glass together in the same block of data [1].

---

[1] There are some results of the analysis in (11) which are not coherent with those reported here, since the data sets used do not contain the same observations.

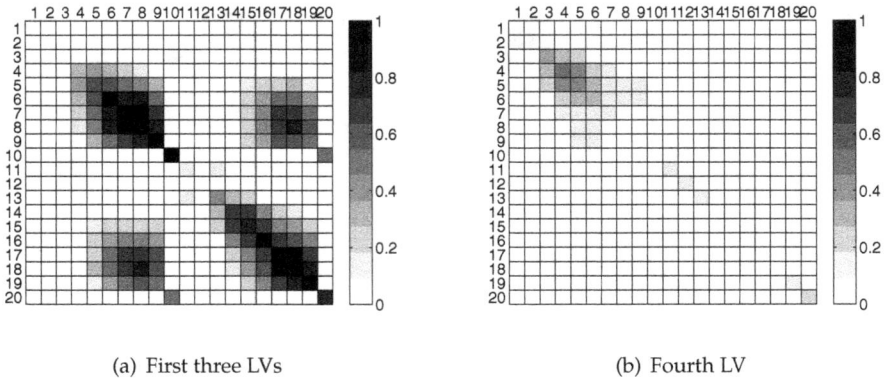

(a) First three LVs   (b) Fourth LV

Fig. 8. MEDA matrices from the PLS model in the Slurry-Fed Ceramic Melter data set.

Following recommendations in (32), a 3 LVs PLS model from a mean-centered x-block was fitted. The loading plots corresponding to the three possible pairs of LVs are shown in Figure 7. The first LV captures the predictive variance in the temperatures, with higher loadings for higher indices of the temperatures in each thermowell. This holds exception made on temperatures 10 and 20, which present their predictive variance in the second and third LVs, being the variance in 10 between three and four times higher than that in 20. The third LVs also seems to discriminate between both thermowells. For instance, in Figure 7(c) most temperatures of the first thermowell are in the upper middle of the Figure, whereas most temperatures of the second thermowell are in the lower middle. Nevertheless, it should be noted that the third LV only captures 2% of the variability in the x-block and 1% of the variability in the y-block.

The information gained with the loading plots can be complemented with that in MEDA, which brings a much clearer picture of the structure in the data. The MEDA matrix for the PLS model with 3 LVs is shown in Figure 8(a). There is a clear auto-correlation effect in the temperatures, so that closer sensors are more correlated. This holds exception made on temperatures 10 and 20. Also, the corresponding temperatures in both thermowells are correlated, including 10 and 20. Finally, the temperatures at the bottom do not contain almost any predictive information of the level of molten glass. In (32), the predictive error by cross-validation is used to identify the number of LVs. Four LVs attain the minimum predictive error, but 3 LVs are selected since the fourth LV does not contribute much to the reduction of this error. In Figure 8(b), the contribution of this fourth LV is shown. It is capturing the predictive variability from the third to the fifth temperature sensors in the first thermowell, which are correlated. The corresponding variability in the second thermowell is already captured by the first 3 LVs. This is an example of the capability of MEDA for model interpretation, which can be very useful in the determination of the number of LVs. In this case and depending on the application of the model, the fourth LV may be added to the model in order to compensate the variance captured in both thermowells, even if the improvement in prediction performance is not high.

The information provided by MEDA can also be useful for variable selection. In this example, temperatures 1-2 and 11-12 do not contribute to a relevant degree to the regression model. As shown in Table 1, if those variables are not used in the model, its prediction performance

Variables	Complete model	[3 : 10 13 : 20]	[3 : 10]	[13 : 20]
LVs	3	3	3	3
X-block variance	88.79	89.84	96.36	96.93
Y-block variance	87.89	87.78	84.61	83.76
RMSEC	0.1035	0.1039	0.1166	0.1198
RMSECV	0.1098	0.1098	0.1253	0.1271
RMSEP	0.1396	0.1394	0.1522	0.1631

Table 1. Comparison of three PLS models in the Slurry-Fed Ceramic Melter data set. The variance in both blocks of data and the Root Mean Square Error of Calibration (RMSEC), Cross-validation (RMSECV) and Prediction (RMSEP) are compared.

remains the same. Also, considering the correlation among thermowells, one may be tempted to use only one of the thermowells for prediction, reducing the associated costs of maintaining two thermowells. If this is done, only 8 predictor variables are used and the prediction performance is reduced, but not to a large extent. Correlated variables in a prediction model help to better discriminate between true structure and noise. For instance, in this example, when only the sensors of one thermowell are used, the PLS model captures more x-block variance and less y-block variance. This is showing that more specific–noisy–variance in the x-block is being captured. Using both thermowells reduces this effect. Another example of variable selection with MEDA will be presented in Section 5.

3.3 correlation matrices and MEDA

There is a close similarity between MEDA and correlation matrices. To this regard, equation (2) simplifies the interpretation of the MEDA procedure. The MEDA index combines the original variance with the model subspace variance in S_{ml} and S_{mlA}. Also, the denominator of the index in eq. (2) is the original variance. Thus, those pairs of variables where a high amount of the total variance of one of them can be recovered from the other are highlighted. This is convenient for data interpretation, since only factors of high variance are highlighted. On the other hand, it is easy to see that when the number of LVs, A, equals the rank of $\mathbf{X}$, then Q_A^2 is equal to the element-wise squared correlation matrix of $\mathbf{X}$, C^2 (11). This can be observed in the following element-wise equality:

$$q^2_{Rank(X),(m,l)} = \frac{S^2_{ml}}{S_{mm} \cdot S_{ll}} = c^2_{(m,l)}.$$ (4)

This equivalence shows that matrix Q_A^2 has a similar structure than the–element-wise squared–correlation matrix. To elaborate this similarity, a correlation matrix can be easily extended to the notion of latent subspace. The correlation matrix in the latent subspace, C_A, can de defined as the correlation matrix of the reconstruction of $\mathbf{X}$ with the first A LVs. Thus, $C_A = \mathbf{P}_A \cdot \mathbf{P}_A^t \cdot \mathbf{C} \cdot \mathbf{P}_A \cdot \mathbf{P}_A^t$ in PCA and $C_A = \mathbf{P}_A \cdot \mathbf{R}_A^t \cdot \mathbf{C} \cdot \mathbf{R}_A \cdot \mathbf{P}_A^t$ in PLS. If the elements of C_A are then squared, the element-wise squared correlation in the latent subspace, noted as C_A^2, is obtained. Strictly speaking, each element of C_A^2 is defined as:

$$c^2_{A,(m,l)} = \frac{S^2_{mAlA}}{S_{mAmA} \cdot S_{lAlA}}.$$ (5)

However, for the same reason explained before, if C_A^2 is aimed at data interpretation, the denominator should be original variance:

$$c_{A,(m,l)}^2 = \frac{S_{mA|A}^2}{S_{mm} \cdot S_{ll}}. \tag{6}$$

If this equation is compared to equation (2), we can see that the main difference between MEDA and the–projected and element-wise squared–correlation matrix is the combination of original and projected variance in the numerator of the former. This combination is paramount for interpretation. Figure 9 illustrates this. The example of the pipelines is used again, but in this case ten pipelines and only 20 observations are considered, yielding a dimension of 20×50 in the data. Two data sets are simulated. In the first one, the pipelines are correlated. As a consequence, the data present three common factors represented by the three biggest eigenvalues in Figure 9(a). In the second one, each pipeline is independently generated, yielding a more distributed variance in the eigenvalues (Figure 9(b)). For matrices Q_A^2 and C_A^2 to infer the structure in the data, they should have large values in the elements which represent real structural information (common factors) and low values in the rest of the elements. Since in both data sets it is known a-priori which elements in the matrices represent actual common factors and which not, the mean values for the two groups of elements in matrices Q_A^2 and C_A^2 can be computed. The ratio of these means, computed by dividing the mean of the elements with common factors by the mean of the elements without common factors, is a measure of the discrimination capability between structure and noise of each matrix. The higher this index is, the better the discrimination capability is. This ratio is shown in Figures 9(c) and 9(d) for different numbers of PCs. Q_A^2 outperforms C_A^2 until all relevant eigenvalues are incorporated to the model. Also, Q_A^2 presents maximum discrimination capability for a reduced number of components. Notice that both alternative definitions of C_A^2 in equations (5) and (6) give exactly the same result, though equation (6) is preferred for visual interpretation.

4. Connection between observations and variables

The most relevant issue for data understanding is probably the connection between observations and variables. It is almost useless to detect certain details in the data distribution such as outliers or clusters, if the set of variables related to these details are not identified. Traditionally, biplots (12) have been used for this purpose. In biplots, the scatter plot of loadings and scores are combined in a single plot. Apart from relevant considerations regarding the comparability of the axes in the plot, which is also important for any scatter plots, and of the scales in scores and loadings (18), biplots may be misleading just because of the loading plot included. In this point, a variant of MEDA, named observation-based MEDA or oMEDA, can be used to unveil the connection between observations and variables without the limitations of biplots.

4.1 oMEDA

oMEDA is a variant of MEDA to connect observations and variables. Basically, oMEDA is a MEDA algorithm applied over a combination of the original data and a dummy variable designed to cover the observations of interest. Take the following example: a number of subsets of observations $\{C_1, ..., C_N\}$ form different clusters in the scores plot which are located

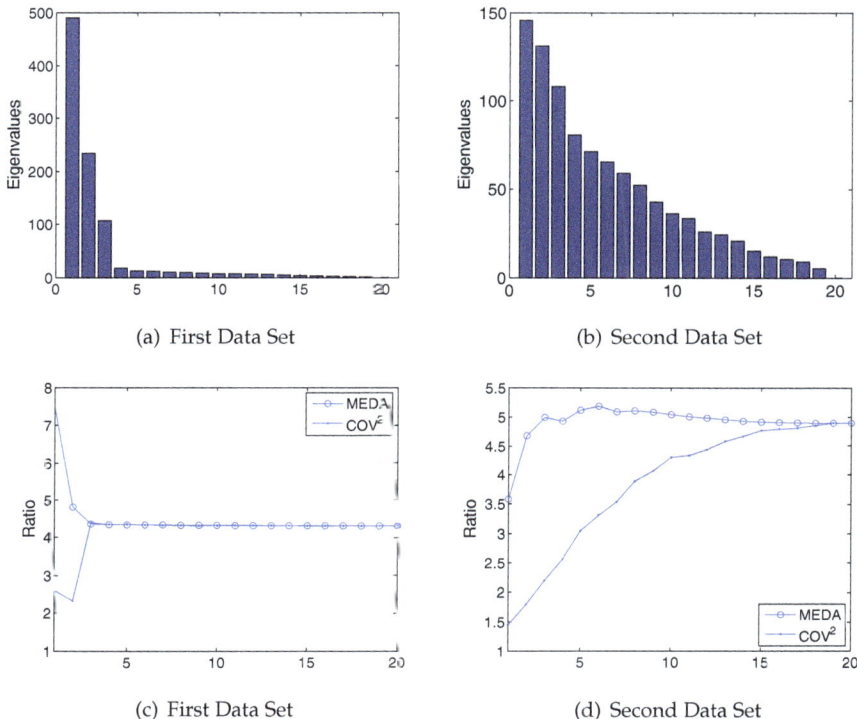

(a) First Data Set   (b) Second Data Set

(c) First Data Set   (d) Second Data Set

Fig. 9. Comparison of MEDA and the projected and element-wise squared covariance matrix in the identification of the data structure from the PCA model fitted from the data in the ten pipelines example: (a) and (b) show the eigenvalues when the pipelines are correlated and independent, respectively, and (c) and (d) show the ratio between the mean of the elements with common factors and the mean of the elements without common factors in the matrices.

far from the bulk of the data, L. One may be interested in identifying, for instance, the variables related to the deviation of C_1 from L without considering the rest of clusters. For that, a dummy variable d is created so that observations in C_1 are set to 1, observations in L are set to -1, while the remaining observations are left to 0. Also, values other than 1 and -1 can be included in the dummy variable if desired. $oMEDA$ is then performed using this dummy variable.

The $oMEDA$ technique is illustrated in Figure 10. Firstly, the dummy variable is designed and combined with the data set. Then, a MEDA run is performed by predicting the original variables from the dummy variable. The result is a single vector, $\mathbf{d}_A^2$, of dimension $M \times 1$, being M the number of original variables. In practice, the $oMEDA$ index is slightly different to that used in MEDA. Being d the dummy variable, designed to compare a set of observations with value 1 (or in general positive values) with another set with value -1 (or in general negative values), then the $oMEDA$ index follows:

$$d_{A,(l)}^2 = \|\mathbf{x}_{(l)}^d\|^2 - \|\hat{\mathbf{e}}_{A,(l)}^d\|^2, \quad \forall l. \tag{7}$$

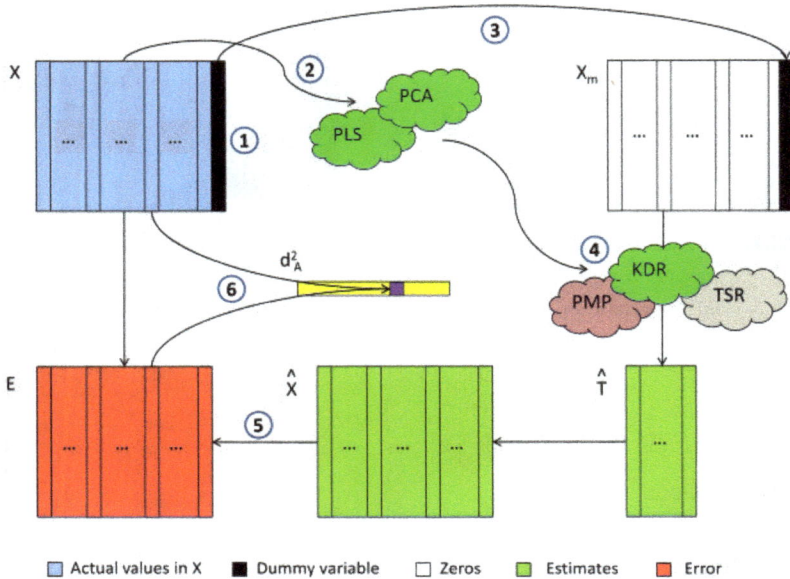

Fig. 10. *o*MEDA technique: (1) introduction of the dummy variable, (2) model calibration, (3) introduction of missing data, (4) missing data imputation, (5) error computation, (6) computation of vector $\mathbf{d}_A^2$.

where $\mathbf{x}_{(l)}^d$ represents the values of the l-th variable in the original observations different to 0 in $\mathbf{d}$ and $\hat{\mathbf{e}}_{A,(l)}^d$ is the corresponding estimation error. The main difference between the computation of index $d_{A,(l)}^2$ in *o*MEDA and that of MEDA is the absence of the denominator in the former. This modification is convenient to avoid high values in $d_{A,(l)}^2$ when the amount of variance of a variable in the reduced set of observations of interest is very low. Once $\mathbf{d}_A^2$ is computed for a given dummy variable, sign information can be added from the mean vectors of the two groups of observations considered (33).

In practice, in order to avoid any modification in the PCA or PLS subspace due to the introduction of the dummy variable, the *o*MEDA algorithm is slightly more complex than the procedure shown in Figure 10. For a description of this algorithm refer to (33). However, like in MEDA, the *o*MEDA vector can be computed in a more direct way by assuming KDR (26) is used as the missing data estimation procedure. If this holds, the *o*MEDA vector follows:

$$d_{A,(l)}^2 = 2 \cdot \mathbf{x}_{(l)}^t \cdot \mathbf{D} \cdot \mathbf{x}_{A,(l)} - \mathbf{x}_{A,(l)}^t \cdot \mathbf{D} \cdot \mathbf{x}_{A,(l)}, \tag{8}$$

where $\mathbf{x}_{(l)}$ represents the l-th variable in the–complete–set of original observations and $\mathbf{x}_{A,(l)}$ its projection in the latent subspace in coordinates of the original space and:

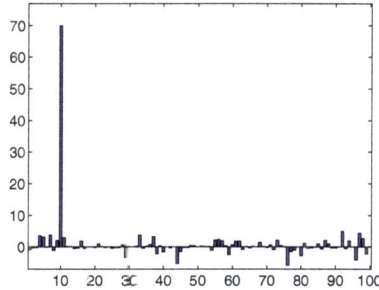

Fig. 11. *o*MEDA vector of two clusters of data from the 10 PCs PCA model of a simulated data set of dimension 100×100. This model captures 30% of the variability. Data present two clusters in variable 10.

$$\mathbf{D} = \frac{\mathbf{d} \cdot (\mathbf{d})^T}{\|\mathbf{d}\|^2}. \tag{9}$$

Finally, the equation can also be reexpressed as follows:

$$d^2_{A,(l)} = \frac{1}{N} \cdot (2 \cdot \Sigma^d_{(l)} - \Sigma^d_{A,(l)}) \cdot |\Sigma^d_{A,(l)}| \tag{10}$$

with $\Sigma^d_{(l)}$ and $\Sigma^d_{A,(l)}$ being the weighted sum of elements in $\mathbf{x}(l)$ and $\mathbf{x}_{A,(l)}$ according to the weights in $\mathbf{d}$, respectively. Equation (10) has two advantages. Firstly, it presents the *o*MEDA vector as a weighted sum of values, which is easier to understand. Secondly, it has the sign computation built in, due to the absolute value in the last element. Notice also that *o*MEDA inherits the combination of total and projected variance present in MEDA.

In Figure 11 an example of *o*MEDA is shown. For this, a 100×100 data set with two clusters of data was simulated. The distribution of the observations was designed so that both clusters had significantly different values only in variable 10 and then data was auto-scaled. The *o*MEDA vector clearly highlights variable 10 as the main difference between both clusters.

4.2 Biplots vs oMEDA

Let us return to the discussion regarding the relationship between the common factors and the components. As already commented, several common factors can be captured by the same component in a projection model. As a result, a group of variables may be located close in a loading plot without the need to be correlated. This is also true for the observations. Thus, two observations closely located in a score plot may be quite similar or quite different depending on their scores in the remaining LVs. However, score plots are typically employed to observe a general distribution of the observations. This exploration is more aimed at finding differences among observations rather than similarities. Because of this, the problem described for loading plots is not so relevant for the typical use of score plots. However, this is a problem when interpreting biplots. In biplots, deviations in the observations are related to deviations in the variables. Like loading plots, biplots may be useful to perform a fast view on the data, but any conclusion should be confirmed with another technique. *o*MEDA is perfectly suited for this.

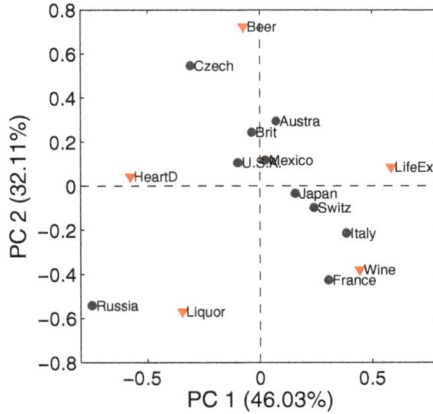

Fig. 12. Biplot of the first 2 PCs from the Wine Data set provided with the PLS-toolbox (32).

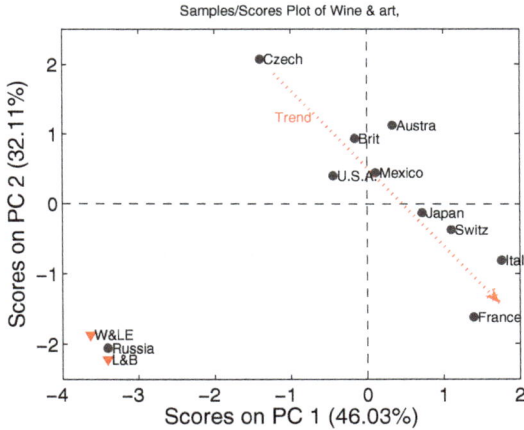

Fig. 13. Score plot of the first 2 PCs from the Wine Data set provided with the PLS-toolbox (32) and two artificial observations.

In Figure 12, the biplot of the Wine data set is presented. The distribution of the scores show that all countries but Russia follow a trend from the Czech Republic to France, while Russia is located far from this trend. The biplot may be useful to make hypothesis on the variables related to the special nature of Russia or to the trend in the rest of countries. Nevertheless, this hypothesis making is not straightforward. To illustrate this, in Figure 13 the scores are shown together with two artificial observations. The artificial observations were designed to lay close to Russia in the score plot of the first two PCs. In both cases, three of the five variables were left to their average value in the Wine data set and only two variables are set to approach Russia. Thus, observation W&LE only uses variables Wine and LifeEx to yield a point close to Russia in the score plot while the other variables are set to the average. Observation L&B only uses Liquor and Beer. With this example, it can be concluded that very little can be said about Russia only by looking at the biplot.

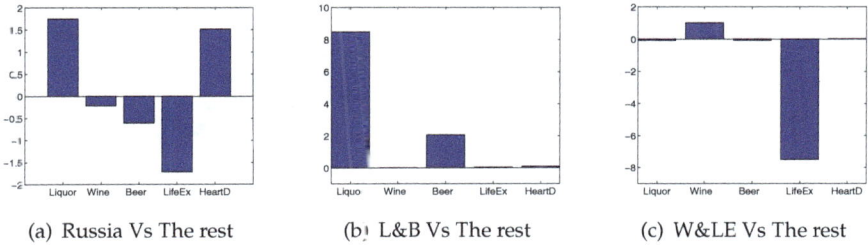

(a) Russia Vs The rest   (b) L&B Vs The rest   (c) W&LE Vs The rest

Fig. 14. oMEDA vectors of the first 2 PCs from the Wine Data set provided with the PLS-toolbox (32). Russia (a), L&B (b) and W&LE (c) compared to the rest of countries.

In Figure 14(a), the oMEDA vector to discover the differences between Russia and the rest of countries is shown. For this, a dummy variable is built where all observations except Russia are set to -1 and Russia is set to 1. oMEDA shows that Russia has in general less life expectancy and more heart disease and liquor consumption than the rest of countries. The same experiment is repeated for artificial observations L&B and W&LE in Figures 14(b) and 14(c). oMEDA clearly distinguishes among the three observations, while in the biplot they seem to be very similar.

To analyze the trend shown by all countries except Russia in Figure 12, the simplest approach is to compare the most separated observations, in this case France and the Czech Republic. The oMEDA vector is shown in Figure 15(a). In this case, the dummy variable is built so that France has value 1, the Czech Republic has value -1 and the rest of the countries have 0 value. Thus, positive values in the oMEDA vector identify variables with higher value in France than in the Czech Republic and negative values the opposite. oMEDA shows that the French consume more wine and less beer than Czech people. Also, according to the data, the former seem to be more healthy.

Comparing the two most separated observations may be misleading in certain situations. Another choice is to use the capability of oMEDA to unveil the variables related to any direction in a score plot. For instance, let us analyze the trend of the countries incorporating the information in all the countries. For this, different weights are considered in the dummy variable. We can think of these weights as-approximate–projections of the observations in the direction of interest. Following this approach, the weights listed in Table 2 are assigned, which approximate the projection of the countries in the imaginary line depicted by the arrow in Figure 13. Since Russia is not in the trend, it is left to 0. Using these weights, the resulting oMEDA vector is shown in Figure 15(b). In this case, the analysis of the complete set of observations in the trend resembles the conclusions in the analysis of the two most separated observations.

Country	Weight	Country	Weight
France	3	Mexico	-1
Italy	2	U.S.A.	-1
Switz	1	Austra	-1
Japan	1	Brit	-1
Russia	0	Czech	-3

Table 2. Weights used in the dummy variable for the oMEDA vector in Figure 15(b).

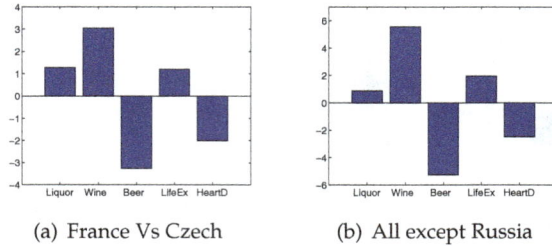

(a) France Vs Czech   (b) All except Russia

Fig. 15. oMEDA vectors of the first 2 PCs from the Wine Data set provided with the PLS-toolbox (32). In (a), France and Czech Republic are compared. In (b), the trend shown in the score plot by all countries except Russia is analyzed.

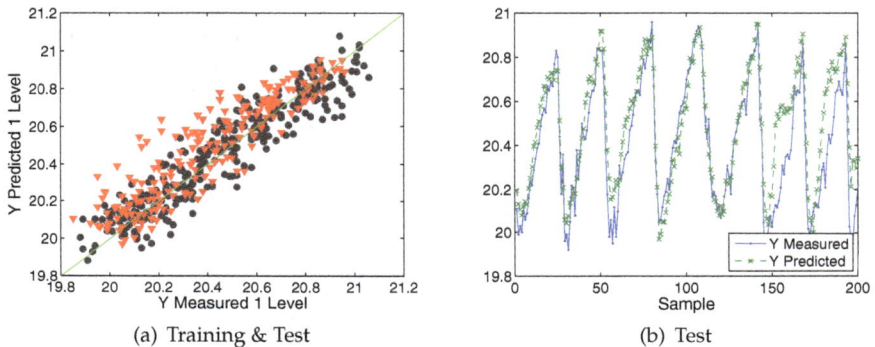

(a) Training & Test   (b) Test

Fig. 16. Measured vs predicted values of molten glass level in the PLS model with 3 LVs fitted from the Slurry-Fed Ceramic Melter (SFCM) data set in (32).

Let us return to the PLSdata data set. A PLS model relating temperatures with the level of molten glass was previously fitted. As already discussed, the data set includes 300 training observations and 200 test observations. The measured and predicted values of both sets of observations are compared in Figure 16(a). The predicted values in the test observations (inverted triangles) tend to be higher than true values. This is also observed in Figure 16(b). The cause for this seems to be that the process has slightly moved from the operation point where training data was collected. oMEDA can be used to identify this change of operation point by simply comparing training and test observations in the model subspace. Thus, training (value 1) and test observations (value -1) are compared in the subspace spanned by the first 3 LVs of the PLS model fitted only from training data. The resulting oMEDA vector is shown in Figure 17. According to the result, considering the test observations have value -1 in the dummy variable, it can be concluded that the process has moved to a situation in which top temperatures are higher than during model calibration.

5. Case study: Selwood data set

In this section, an exploratory data analysis of the Selwood data set (34) is carried out. The data set was downloaded from http://michem.disat.unimib.it/chm/download/datasets.htm. It

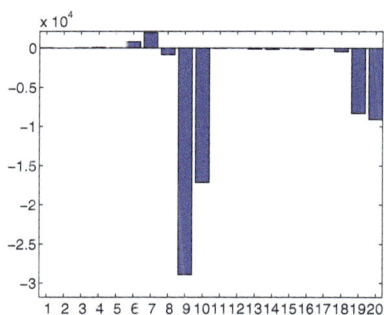

Fig. 17. oMEDA vector comparing training and test observations in the PLS model with 3 LVs fitted from the Slurry-Fed Ceramic Melter (SFCM) data set in (32).

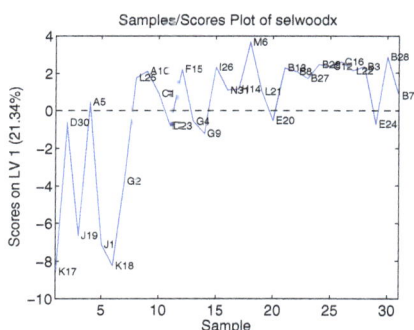

Fig. 18. Scores corresponding to the first LV in the PLS model with the complete set of descriptors of the Selwood dataset.

consists of 31 antifilarial antimycin A_1 analogues for which 53 physicochemical descriptors were calculated for Quantitative Structure-Activity Relationship (QSAR) modelling. The set of descriptors is listed in Table 3. These descriptors are used for predicting in vitro antifilarial activity (-LOGEC50). This data set has been employed for testing variables selection methods, for instance in (35; 36), in order to find a reduced number of descriptors with best prediction performance. Generally speaking, these variable selection methods are based on complex optimization algorithms which make use of heuristics to reduce the search space.

Indices	Descriptors
1:10	ATCH1 ATCH2 ATCH3 ATCH4 ATCH5 ATCH6 ATCH7 ATCH8 ATCH9 ATCH10
11:20	DIPV_X DIPV_Y DIPV_Z DIPMOM ESDL1 ESDL2 ESDL3 ESDL4 ESDL5 ESDL6
21:30	ESDL7 ESDL8 ESDL9 ESDL10 NSDL1 NSDL2 NSDL3 NSDL4 NSDL5 NSDL6
31:40	NSDL7 NSDL8 NSDL9 NSDL10 VDWVOL SURF_A MOFI_X MOFI_Y MOFI_Z PEAX_X
41:50	PEAX_Y PEAX_Z MOL_WT S8_1DX S8_1DY S8_1DZ S8_1CX S8_1CY S8_1CZ LOGP
51:53	M_PNT SUM_F SUM_R

Table 3. Physicochemical descriptors of the Selwood dataset.

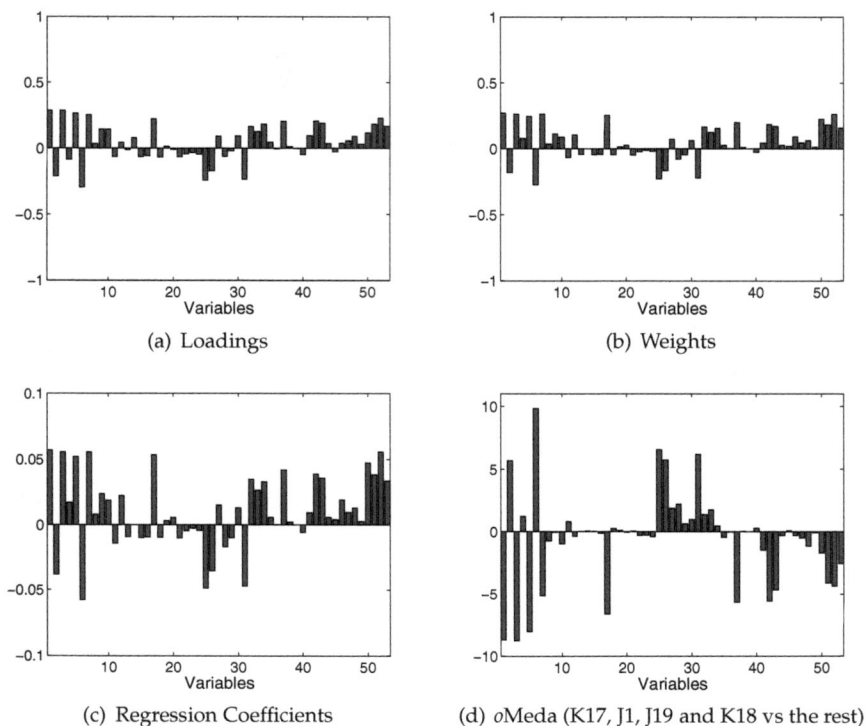

(a) Loadings

(b) Weights

(c) Regression Coefficients

(d) oMeda (K17, J1, J19 and K18 vs the rest)

Fig. 19. Several vectors corresponding to the first LV in the PLS model with the complete set of descriptors of the Selwood dataset.

First of all, a PLS model is calibrated between the complete set of descriptors and -LOGEC5$0$. Leave-one-out cross-validation suggests one LV. The score plot corresponding to the 0 analogues in that LV are shown in Figure 18. Four of the compounds, namely K17, J1, J19 and K18, present an abnormally low score. This deviation is highly contributing to the variance in the first LV and the reason for it should be investigated. Two of these compounds, K17 and K18, were catalogued as outliers by (34), where the authors stated that "Chemically, these compounds are distinct from the bulk of the training set in that they have an n-alkyl side chain as opposed to a side chain of the phenoxy ether type". Since the four compounds present an abnormally low score in the first LV, typically the analyst may interpret the coefficients of that LV to try to explain this abnormality. In Figure 19, the loadings, weights and regression coefficients of the PLS model are presented together with the oMEDA vector. The latter identifies those variables related to the deviation of the four compounds from the rest. The oMEDA vector is similar, but with opposite sign, to the other vectors in several descriptors, but quite different in others. Therefore, the loadings, weights or coefficient vectors should not be used in this case for the investigation of the deviation, or otherwise one may arrive to incorrect conclusions. On the other hand, it may be worth to check whether the oMEDA vector is representative of the deviation in the four compounds. Performing oMEDA individually on each of the compounds confirm this fact (see Figure 20)

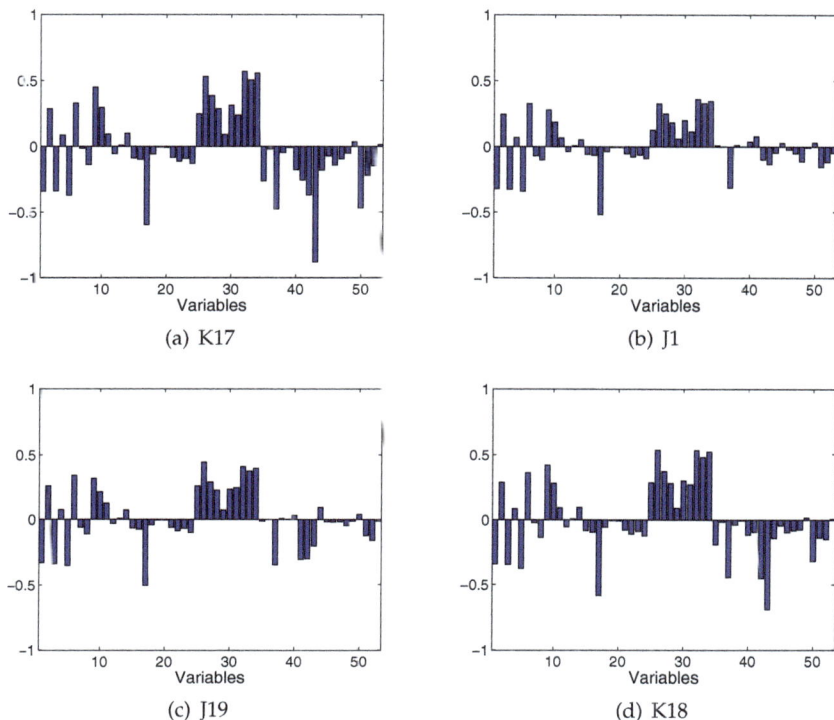

(a) K17

(b) J1

(c) J19

(d) K18

Fig. 20. oMEDA vectors corresponding to the first LV in the PLS model with the complete set of descriptors of the Selwood dataset. To compute each of the vectors, one of the four compounds K17, J1, J19 and K18 are set to 1 in the dummy variable and the other three are set to 0, while the rest of compounds in the data set are set to -1.

The subsequent step is to search for relevant descriptors (variable selection) For this, MEDA will be employed. In this point, there are two choices. The four compounds with low score in the first LV may be treated as outliers and separated from the rest of the data (34) or the complete data set may be modelled with a single QSAR model (35; 36). It should be noted that differences among observations in one model may not be found in a different model, so that the same observation may be an outlier or a normal observation depending on the model. Furthermore, as discussed in (35), the more general the QSAR model is, so that it models a wider set of compounds, the better. Therefore, the complete set of compounds will be considered in the remaining of the example. On the other hand, regarding the analysis tools used, there are different possibilities. MEDA may be applied over the PLS model relating the descriptors in the x-block and -LOGEC50 in the y-block. Alternatively, both blocks may be joined together in a single block of data and MEDA with PCA be applied. The second choice will be generally preferred to avoid over-fitting, but typically both approaches may lead to the same conclusions, like it happens in the present example.

The application of MEDA requires to select the number of PCs in the PCA model. Considering that the aim is to understand how the variability in -LOGEC50 is related to the descriptors in

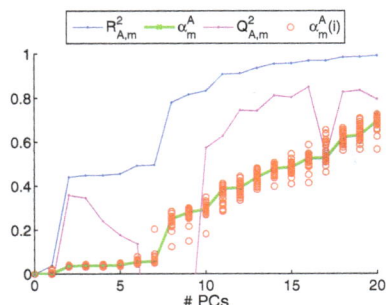

Fig. 21. Structural and Variance Information (SVI) plot of in vitro antifilarial activity (-LOGEC50). The data set considered combines -LOGEC50 with the complete set of descriptors of the Selwood dataset.

the data set, the Structural and Variance Information (SVI) plots are the adequate analysis tool (14). The SVI plots combine variance information with structural information to elucidate how a PCA model captures the variance of a single variable. The SVI plot of a variable v reveals how the following indices evolve with the addition of PCs in the PCA model:

- The R^2 statistic, which measures the variance of v.
- The Q^2 statistic, which measures the performance of the missing data imputation of v, o otherwise stated its prediction performance.
- The α statistic, which measures the portion of variance of v which is identified as unique variance, i.e. variance not shared with other variables.
- The stability of α, as an indicator of the stability of the model calibration.

Figure 21 shows the SVI plot of -LOGEC50 in the PCA model with the complete set of descriptors. The plot shows that the model remains quite stable until 5-6 PCs are included. This is seen in the closeness of the circles which represents the different instances of α computed on a leave-one-out cross-validation run. The main portion of variability in -LOGEC50 is captured in the second and eighth PCs. Nevertheless, is not until the tenth PC that the missing data imputation (Q^2) yields a high value. For more PCs, the captured variability is only slightly augmented. Since MEDA makes use of the missing data imputation of a PCA model, Q^2 is a relevant index. At the same time, from equation (2) is clear that MEDA is also influenced by captured variance. Thus, 10 PCs are selected. In any case, it should be noted that MEDA is quite robust to the overestimation in the number of PCs (11) and very similar MEDA matrices are obtained for 3 or more PCs in this example.

The MEDA matrix corresponding to the PCA model with 10 PCs from the data set which combines -LOGEC50 with the complete set of descriptors of the Selwood dataset is presented in Figure 22. For variable selection, the most relevant part of this matrix is the last column (or row), which corresponds to -LOGEC50. This vector is shown in Figure 23(a). Those descriptors with high value in this vector are the ones from which -LOGEC50 can be better predicted. Nevertheless, the selection of, say, the first n variables with higher value is not an adequate strategy because the relationship among the descriptors should also be considered. Let us select the descriptor with better prediction performance, in this case ATCH6, though

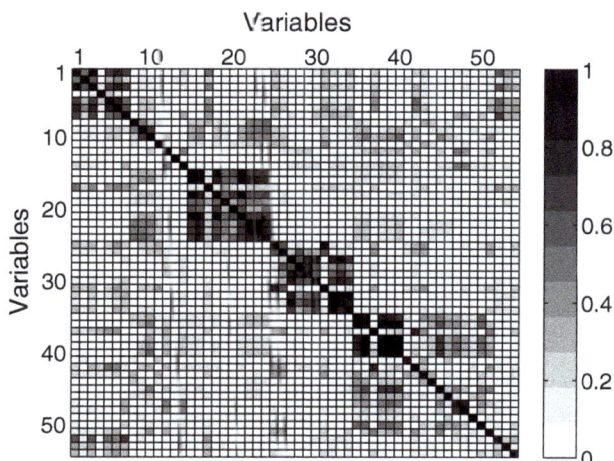

Fig. 22. MEDA matrix of the PCA model with 10 PCs from the data set which combines the in vitro antifilarial activity (-LOGEC50) with the complete set of descriptors of the Selwood dataset.

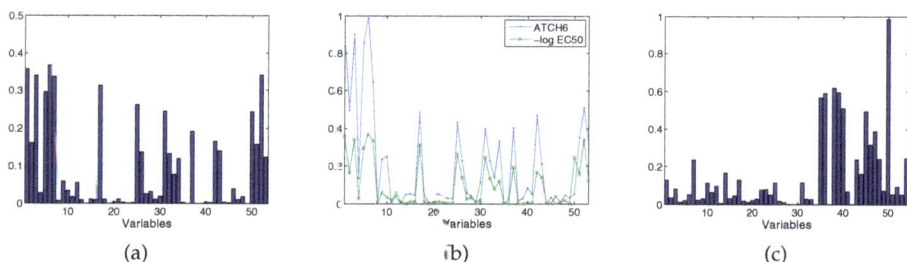

(a)   (b)   (c)

Fig. 23. MEDA vector corresponding to the in vitro antifilarial activity (-LOGEC50) (a) comparison between this vector and that corresponding to ATCH6 (b) and MEDA vector corresponding to LOGP (c) in the PCA model with 10 PCs from the data set which combines -LOGEC50 with the complete set of descriptors of the Selwood dataset.

ATCH1, ATCH3, ATCH7 or SUM_F have a very similar prediction performance. The least-squares regression model with ATCH6 as regressor attains a Q^2 equal to 0.30, more than the Q^2 attained by any number of LVs in the PLS model with the complete set of descriptors. If for instance ATCH6 and ATCH1 are used as regressors, $Q^2 = 0.26$ is obtained for least squares regression and $Q^2 = 0.31$ for PLS with 1 LV, which means an almost negligible improvement with the addition of ATCH1. The facts that the improvement is low and that the 1 LV PLS model outperforms the least squares model are caused by the correlation between ATCH6 and ATCH1, correlation clearly pointed out in the MEDA matrix (see the element at the sixth column and first row or the first column and sixth row) Clearly, both ATCH6 and ATCH1 are related to the same common factor in -LOGEC50. However, the variability in -LOGEC50 is the result of several sources of variability, which may be common factors with other descriptors.

Therefore, in order to introduce a new common factor in the model other than that in ATCH6, we need to find a descriptor related to -LOGEC50 but not to ATCH6. Also, the model may be improved by introducing a descriptor related to ATCH6 but not to -LOGEC50. For this, Figure 23(b) compares the columns in the MEDA matrix corresponding to ATCH6 and -LOGEC50. The comparison should not be performed in terms of direct differences between values. For instance, ATCH1 and ATCH6 are much more correlated than ATCH1 and -LOGEC50. It is the difference in shape which is informative. Thus, we find that -LOGEC50 present a high correlation with LOGP (variable 50) which is not found in ATCH6. Thus, LOGP presents a common factor with -LOGEC50 which is not present in ATCH6. Using LOGP and ATCH6 as regressors, the least squares model presents $Q^2 = 0.37$.

If an additional descriptor is to be added to the model, again it should present a different common factor with any of the variables in the model. The MEDA vector corresponding to LOGP is shown in Figure 23(c). This descriptor is related to a number of variables which are not related to -LOGEC50. This relationship represents a common factor in LOGP but not in -LOGEC50. The inclusion of a descriptor containing this common factor, for instance MOFI_Y (variable 38) may improve prediction because it may help to distinguish the portion of variability in LOGP which is useful to predict -LOGEC50 from the portion which is not. Using LOGP, ATCH6 and MOFI_Y as regressors yields $Q^2 = 0.56$, illustrating that the addition of a descriptor which is not related to the predicted variable may be useful for prediction.

In Figure 24, the two common factors described before, the one present in ATCH6 and -LOGEC50 and the one present in LOGP and MOFI_Y, are approximately highlighted in the MEDA matrix. If variables ATCH6 and MOFI_Y are replaced by others with the same common factors, the prediction performance of the model remains similar. However, LOGP is utmost for the model since is the only descriptor which relates the second common factor and -LOGEC50. These results are coherent with findings in the literature. Both (35) and (36) highlight the relevance of LOGP, and justify it with the results in several more publications. Furthermore, the top 10 models found in (35), presented in Table 4, follow the same pattern of the solution found here. The models with three descriptors contain LOGP with one descriptor from the first and second common factors. The models with two descriptors contain LOGP and a variable with the second common factor.

Descriptors	Q^2
SUM_F (52) LOGP (50) MOFI_Y (38)	0.647
ESDL3 (17) LOGP (50) SURF_A (36)	0.645
SUM_F (52) LOGP (50) MOFI_Z (39)	0.644
LOGP (50) MOFI_Z (39)	0.534
ESDL3 (17) LOGP (50) MOFI_Y (38)	0.605
ESDL3 (17) LOGP (50) MOFI_Z (39)	0.601
LOGP (50) MOFI_Y (38)	0.524
LOGP (50) PEAX_X (40)	0.518
LOGP (50) SURF_A (36)	0.501
SUM_F (52) LOGP (50) PEAX_X (40)	0.599

Table 4. Top 10 models obtained after variable selection of the Selwood data set in (35)

Finally, in Figure 25 the plot of measured vs predicted values of -LOGEC50 in the model resulting from the exploration is shown. No outliers are identified, though the four

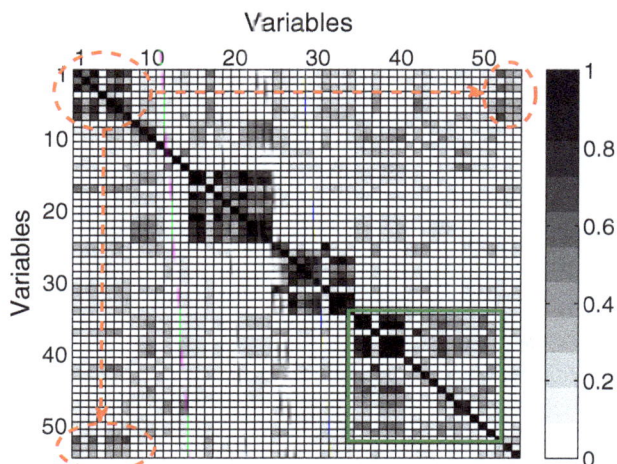

Fig. 24. MEDA matrix of the PCA model with 10 PCs from the data set which combines the in vitro antifilarial activity (-LOGEC50) with the complete set of descriptors of the Selwood dataset. Two common factors are highlighted. The first one is mainly found in descriptors 1 to 3, 5 to 7, 17, 52 and 53. The second one is mainly found in descriptors 35, 36, 38 to 40, 45, 47 and 50. Though the second common factor is not present in -LOGEC50, it is in LOGP.

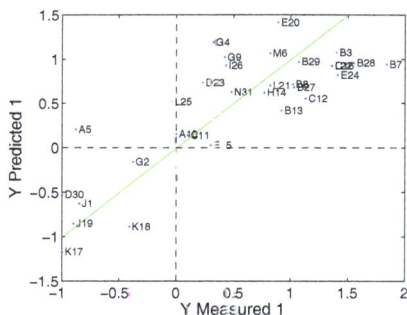

Fig. 25. Plot of measured vs predicted values of -LOGEC50 in the model with regressors LOGP, ATCH6 and MOFI_Y of the Selwood dataset.

compounds previously highlighted are found at the bottom left corner. This result support the non-convenience of isolating these compounds.

Notice that MEDA is not a variable selection technique per se and therefore other methods may be more powerful for this purpose. Nevertheless, being an exploratory method, the benefit of using MEDA for variable selection is that the general solution can be identified and understood, like in the present example. On the contrary, most variable selection approaches are based on highly complex algorithms which can only report a set of possible alternative solutions (e.g. Table 4).

6. Conclusion

In this chapter, new tools for exploratory data analysis are presented and combined with already well known techniques in the chemometrics field, such as projection models, score and loading plots. The shortcomings and potential pitfalls in the application of common tools are elucidated and illustrated with examples. Then, the new techniques are introduced to overcome these problems.

The Missing-data methods for Exploratory Data Analysis technique, MEDA for short, studies the relationships among variables. As it is discussed in the chapter, while chemometric models such as PCA and PLS are quite useful for data understanding, they have a main problem which complicates its interpretation: a single component captures several sources of variability or common factors and at the same time a single common factor is captured in several components. MEDA, like rotation methods or Factor Analysis (FA), is a tool for the identification of the common factors in subspace models, in order to elucidate the structure in the data. The output of MEDA is similar to a correlation matrix but with better properties associated. MEDA is the perfect complement of loading plots. It gives a different picture of the relationships among variables which is especially useful to find groups of related variables. Using a Quantitative Structure-Activity Relationship (QSAR) example, it was shown that the understanding of the relationships among variables in the data may lead to perform variable selection with similar performance of highly sophisticated algorithms, with the extra benefi that the global solution is not only found but also understood.

The second technique introduced in this chapter is a variant of MEDA, named observation-based MEDA or oMEDA. oMEDA was designed to identify the variables which differ between two groups of observations in a latent subspace, but it can be used for the more general problem of identifying the variables related to a given direction in the score plot. Thus, when a number of observations are located in a specific direction in the score plot, oMEDA gives the variables related to that distribution. oMEDA is the perfect complement of score plots and much more reliable than biplots. It can also be seen as an extension of contribution plots to groups of observations. It may be especially useful to check whether the distribution of a new set of observations agree with a calibration model.

Though MEDA and oMEDA are grounded on missing-data imputation methods and their original algorithms are complex to a certain extent, both tools can be computed with very simple equations. A MATLAB toolbox with the tools employed in this chapter, including MEDA, oMEDA, ADICOV and SVI plots, is available at http://wdb.ugr.es/ josecamacho .

7. Acknowledgement

Research in this work is partially supported by the Spanish Ministry of Science and Technology through grant CEI BioTIC GENIL (CEB09-0010).

8. References

[1] Jolliffe I.T.. *Principal component analysis*. EEUU: Springer Verlag Inc. 2002.
[2] Han J., Kamber M.. *Data Mining: Concepts and Techniques*. agora.cs.illinois.edu: Morgan Kaufmann Publishers, Elsevier 2006.
[3] Keren Gideon, Lewis Charles. *A Handbook for data analysis in the behavioral sciences: statistical issues*. Hillsdale, NJ: Lawrence Erlbaum Associates 1993.

[4] Tukey John W. *Exploratory data analysis.* Addison-Wesley Series in Behavioral ScienceReading, MA: Addison-Wesley 1977.

[5] Teo Yik Y.. Exploratory data analysis in large-scale genetic studies *Biostatistics.* 2010;11:70-81.

[6] Pearson K.. On Lines and Planes of Closest Fit to Systems of Points in Space *Philosophical Magazine.* 1901;2:559–572.

[7] Jackson J.E.. *A User's Guide to Principal Components.* England: Wiley-Interscience 2003.

[8] Wold H., Lyttkens E.. Nonlinear iterative partial least squares (NIPALS) estimation procedures in *Bull. Intern. Statist. Inst. Proc., 37th session, London*:1–15 1969.

[9] Geladi P., Kowalski B.R.. Partial Least-Squares Regression: a tutorial *Analytica Chimica Acta.* 1986;185:1–17.

[10] Wold S., om M. Sj Eriksson L.. PLS-regression: a basic tool of chemometrics *Chemometrics and Intelligent Laboratory Systems.* 2001;58:109–130.

[11] Camacho J.. Missing-data theory in the context of exploratory data analysis *Chemometrics and Intelligent Laboratory Systems.* 2010;103:8–18.

[12] Gabriel K.R.. The biplot graphic display of matrices with application to principal component analysis *Biometrika.* 1971;58:453–467.

[13] Westerhuis J.A., Gurden S.P., Smilde A.K.. Generalized contribution plots in multivariate statistical process monitoring *Chemometrics and Intelligent Laboratory Systems.* 2000;51:95–114.

[14] Camacho J., Picó J., Ferrer A.. Data understanding with PCA: Structural and Variance Information plots *Chemometrics and Intelligent Laboratory Systems.* 2010;100:48–56.

[15] Camacho J., Padilla P., Díaz-Verdejo J., Smith K., Lovett D.. Least-squares approximation of a space distribution for a given covariance and latent sub-space *Chemometrics and Intelligent Laboratory Systems.* 2011;105 171–180.

[16] Kosanovich K.A., Dahl K.S., Piovoso M.J.. Improved Process Understanding Using Multiway Principal Component Analysis *Engineering Chemical Research.* 1996;35:138–146.

[17] Ferrer A.. Multivariate Statistical Process Control based on Principal Component Analysis (MSPC-PCA): Some Reflections and a Case Study in an Autobody Assembly Process *Quality Engineering.* 2007;19:311–325.

[18] Kjeldahl K., Bro R.. Some common misunderstandings in chemometrics *Journal of Chemometrics.* 2010;24:558–564.

[19] L. Fabrigar, D. Wegener, R. MacCallum. E. Strahan, Evaluating the use of exploratory factor analysis in psychological research, Psycological Methods 4 (3) (1999) 272–299.

[20] A. Costello, J. Osborne, Best practices in exploratory factor analysis: Four recommendations for getting the most from your analysis, Practical Assessment, Research & Evaluation 10 (7) (2005) 1–9.

[21] I. Jolliffe, Rotation of principal components: choice of normalization constraints, Journal of Applied Statistics 22 (1) (1995) 29–35.

[22] P. Nelson, P. Taylor, J. MacGregor, Missing data methods in pca and pls: score calculations with incomplete observations, Chemometrics and Intelligent Laboratory Systems 35 (1996) 45–65.

[23] D. Andrews, P. Wentzell, Applications of maximum likelihood principal component analysis: incomplete data sets and calibration transfer, Analytica Chimica Acta 350 (1997) 341Û–352.

[24] B. Walczak, D. Massart, Dealing with missing data: Part i, Chemometrics and Intelligent Laboratoy Systems 58 (2001) 15–27.

[25] F. Arteaga, A. Ferrer, Dealing with missing data in mspc: several methods, different interpretations, some examples, Journal of Chemometrics 16 (2002) 408–418.

[26] F. Arteaga, A. Ferrer, Framework for regression-based missing data imputation methods in on-line mspc, Journal of Chemometrics 19 (2005) 439–447.

[27] M. Reis, P. Saraiva, Heteroscedastic latent variable modelling with applications to multivariate statistical process control, Chemometrics and Intelligent Laboratoy Systems 80 (2006) 57–66.

[28] F. Arteaga, Unpublished results.

[29] F. Lindgren, P. Geladi, S. Wold, The kernel algorithm for pls, Journal of Chemometrics 7 (1993) 45–59.

[30] S. de Jong, C. ter Braak, Comments on the pls kernel algorithm, Journal of Chemometrics 8 (1994) 169–174.

[31] B. Dayal, J. MacGregor, Improved pls algorithms, Journal of Chemometrics 11 (1997) 73–85.

[32] B. Wise, N. Gallagher, R. Bro, J. Shaver, W. Windig, R. Koch, PLSToolbox 3.5 for use with Matlab, Eigenvector Research Inc., 2005.

[33] Camacho J. Observation-based missing data methods for exploratory data analysis to unveil the connection between observations and variables in latent subspace models *Journal of Chemometrics* 25 (2011) 592 - 600.

[34] D.L. Selwood, D.J. Livingstone, J.C.W. Comley, A.B. O'Dowd, A.T. Hudson, P. Jackson, K.S. Jandu, V.S. Rose, J.N. Stables Structure-Activity Relationships of Antifiral Antimycin Analogues: A Multivariate Pattern Recognition Study, Journal of Medical Chemistry 33 (1990) 136–142.

[35] S.J. Cho, M.A. Hermsmeier, Genetic Algorithm Guided Selection: Variable Selection and Subset Selection, J. Chem. Inf. Comput. Sci. 42 (2002) 927–936.

[36] S.S. Liu, H.L. Liu, C.S. Yin, L.S. Wang, VSMP: A Novel Variable Selection and Modelling Method Based on the Prediction, J. Chem. Inf. Comput. Sci. 43 (2003) 964–969.

3

Critical Aspects of Supervised Pattern Recognition Methods for Interpreting Compositional Data

A. Gustavo González

Department of Analytical Chemistry, University of Seville, Seville
Spain

1. Introduction

A lot of multivariate data sets of interest to scientists are called compositional or "closed" data sets, and consists essentially of relative proportions. A recent search on the web by entering "chemical compositional data', led to more than 2,730,000 results within different fields and disciplines, but specially, agricultural and food sciences (August 2011 using Google searcher). The driving causes for the composition of foods lie on four factors (González, 2007): Genetic factor (genetic control and manipulation of original specimens), Environmental factor (soil, climate and symbiotic and parasite organisms), Agricultural factor (cultures, crop, irrigation, fertilizers and harvest practices) and Processing factor (post-harvest manipulation, preservation, additives, conversion to another food preparation and finished product). But the influences of these factors are hidden behind the analytical measurements and only can be inferred and uncover by using suitable chemometric procedures.

Chemometrics is a term originally coined by Svante Wold and could be defined as "The art of extracting chemically relevant information from data produced in chemical experiments" (Wold, 1995). Besides, chemometrics can be also defined as the application of mathematical, statistical, graphical or symbolic methods to maximize the chemical information which can be extracted from data (Rock, 1985). Within the jargon of chemometrics some other terms are very common; among them, multivariate analysis and pattern recognition are often used. Chemometricians use the term Multivariate Analysis in reference to the different approaches (mathematical, statistical, graphical...) when considering samples featured by multiple descriptors simultaneously. Pattern recognition is a branch of the Artificial Intelligence that seeks to identify similarities and regularities present in the data in order to attain natural classification and grouping. When applied to chemical compositional data, pattern recognition methods can be seen as multivariate analysis applied to chemical measurements to find classification rules for discrimination issues. Depending on our knowledge about the category or class membership of the data set, two approaches can be applied: Supervised or unsupervised learning (pattern recognition).

Supervised learning methods develop rules for the classification of unknown samples on the basis of a group of samples with known categories (known set). Unsupervised learning methods instead do not assume any known set and the goal is to find clusters of objects which may be assigned to classes. There is hardly any quantitative analytical method that does not make use of chemometrics. Even if one confines the scope to supervised learning pattern recognition, these chemometric techniques are increasingly being applied to of compositional data for classification and authentication purposes. The discrimination of the geographical origin, the assignation to Denominations of Origin, the classification of varieties and cultivars are typical issues in agriculture and food science.

There are analytical techniques such as the based on sensors arrays (electronic nose and electronic tongue) that cannot be imaginable without the help of chemometrics. Thus, one could celebrate the triumph of chemometrics...so what is wrong? (Pretsch & Wilkin, 2006). The answer could be supported by a very well-known quotation attributed to the british politician D'Israeli (Defernez & Kemsley, 1997) but modified by us as follows: "There are three kinds of lies: Lies, damned lies and chemometrics". Here we have changed the original word "statistics" into "chemometrics" in order to point out the suspicion towards some chemometric techniques even within the scientific community. There is no doubt that unwarranted reliance on poorly understood chemometric methods is responsible for such as suspicion.

By the way, chemometric techniques are applied using statistical/chemometric software packages that work as black boxes for final users. Sometimes the blindly use of "intelligen" problem solvers" or similar wizards with a lot of hidden options as default may lead to misleading results. Accordingly, it should be advisable to use software packages with full control on parameters and options, and obviously, this software should be used by a chemometric *connaisseur*.

There are special statistical methods intended for closed data such as compositional ones (Aitchison, 2003; Egozcue et al., 2003; Varmuza & Filzmoser, 2009), but a detailed description of these methods may be outside the scope of this chapter.

2. About the data set

Supervised learning techniques are used either for developing classification rules which accurately predict the classification of unknown patterns or samples (Kryger, 1981) or or finding calibration relationships between one set of measurements which are easy or cheap to acquire, and other measurements which are expensive or labour intensive, in order to predict these later (Naes et al., 2004). The simplest calibration problem consists of predicting a single response (y-variable) from a known predictor (x-variable) and can be solved by using ordinary linear regression (OLR). When fitting a single response from several predictive variables, multiple linear regression (MLR) may be used; but for the sake of avoiding multicollinearity drawbacks, some other procedures such as principal component regression (PCR) or partial least squares regression (PLS) are a good choice. If face to several response variables, multivariate partial least squares (PLS2) techniques have to be used. These procedures are common in multivariate calibration (analyte concentrations from NIR data) or linear free energy relationships (pKa values from molecular descriptors). However, in some instances like quantitative structure-activity relationships (QSAR) non linear strategies are needed, such as quadratic partial least squares regression (QPLS), or regression procedures based on artificial neural networks or support vector machines. All these calibration procedures have to be suitably validated by using validation procedures based on the knowledge of class memberships of the objects, in a similar way as discussed below in this chapter that is devoted to supervised learning for classification.

Let us assume that a known set of samples is available, where the category or class membership of every sample is *a priori* known. Then a suitable planning of the data-acquisition process is needed. At this point, chemical experience, *savoir faire* and intuition are invaluable in order to decide which measurements should be made on the samples and which variables of these measurements are most likely to contain class information.

In the case of compositional data, a lot of analytical techniques can be chosen. Analytical procedures based on these techniques are then selected to be applied to the samples. Selected analytical methods have to be fully validated and with an estimation of their uncertainty (González & Herrador, 2007) and carried out in Quality Assurance conditions (equipment within specifications, qualified staff, and documentation written as Standard Operational Procedures...). Measurements should be carried out at least duplicate and according to a given experimental design to ensure randomization and avoid systematic trends.

Difficulties arise when the concentration of an element is below the detection limit (DL). It is often standard practice to report these data simply as '<DL' values. Such 'censoring' of data, however, can complicate all subsequent statistical analyses. The best method to use generally depends on the amount of data below the detection limit, the size of the data set, and the probability distribution of the measurements. When the number of '< DL' observations is small, replacing them with a constant is generally satisfactory (Clarke, 1998). The values that are commonly used to replace the '< DL' values are 0, DL, or DL/2. Distributional methods such as the marginal maximum likelihood estimation (Chung, 1993) or more robust techniques (Helsel, 1990) are often required when a large number of '< DL' observations are present.

After all measurements are done we can built the corresponding data table or data matrix. A sample, object or pattern is described by a set of "p" variables, features or descriptors. So, all descriptors of one pattern form a 'pattern vector' and accordingly, a given pattern "i" can be seen as a vector $\vec{x}_i$ whose components are $x_{i1}, x_{i2}, ... x_{ij}, ... x_{ip}$ in the vectorial space defined by the features. In matricial form, pattern vectors are row vectors. If we have n patterns, we can build a data matrix $X_{n \times p}$ by assembling the different row pattern vectors. A change in perspective is also possible: A given feature "j" can be seen as a column vector $\vec{x}_j$ with components $x_{1j}, x_{2j}, ... x_{ij}, ... x_{nj}$ in the vectorial space defined by the patterns. We can also construct the data matrix by assembling the different feature column vectors. Accordingly, the data matrix can be considered as describing the patterns in terms of features or *vice versa*. This lead to two main classes of chemometric techniques called Q- and R-modes respectively. R-mode techniques are concerned with the relationships amongst the features of the experiment and examine the interplay between the columns of the data matrix. A starting point for R-mode procedures is the covariance matrix of mean centered variables $C = X^T X$ whose elements are given by

$$c_{jk} = \frac{1}{n-1} \sum_{l=1}^{n} \left(x_{lj} - \overline{x}_j \right) \left(x_{lk} - \overline{x}_k \right)$$ where $\overline{x}_j$ and $\overline{x}_k$ are the mean of the observations on the j*th*

and kth feature. If working with autoscaled data, the sample correlation matrix and the covariance matrix are identical. The element r_{jk} of correlation matrix R represents the cosine between each pair of column vectors and is given by $r_{jk} = \dfrac{c_{jk}}{\sqrt{c_{jj}c_{kk}}}$. The diagonal elements of R are always unity. The alternative viewpoint considers the relationships between patterns, the Q-mode technique. This way normally starts with a matrix of distances between the objects in the n-dimensional pattern space to study the clustering of samples. Typical metric measurements are Euclidean, Minkowski, Manhattan, Hamming, Tanimoto and Mahalanobis distances (Varmuza, 1980).

3. Inspect data matrix

Once the data matrix has been built, it should be fully examined in order to ensure the suitable application of Supervised Learning methodology. A typical undesirable issue is the existence of missing data. Holes in the data matrix must be avoided; however some measurements may not have been recorded or are impossible to obtain experimentally due to insufficient sample amounts or due to high costs. Besides, data can be missing due to various malfunctions of the instruments, or responses can be outside the instrument range.

As stated above, most chemometric techniques of data analysis do not allow for data gaps and thereof different methods have been applied for handling missing values in data matrix. Aside from the extreme situations of casewise deletion or mean substitution, the use of iterative algorithms (IA) is a promising tool. Each iteration consists of two steps. The first step performs estimation of model parameters just as if there were no missing data. The second step finds the conditional expectation of the missing elements given the observed data and current estimated parameters. The detailed procedure depends on the particular application. The typical IA used in Principal Component Analysis can be summarized as (Walczak & Massart, 2001):

1. Fill in missing elements with their initial estimates (expected values, calculated as the mean of the corresponding row's and column's means)
2. Perform singular value decomposition of the complete data set
3. Reconstruct X with the predefined number of factors
4. Replace the missing elements with the predicted values and go to step 2 until convergence.

Replacement of missing values or censored data with any value is always risky since this can substantially change the correlation in the data. It is possible to deal with both missing values and outliers simultaneously (Stanimirova et al., 2007). An excellent revision dealing with zeros and missing values in compositional data sets using non-parametric imputation has been performed by Martin-Fernández et al. (2003).

On the other hand, a number of chemometric procedures are based on the assumption of normality of features. Accordingly, features should be assessed for normality. The well-known Kolmogorov-Smirnov, Shapiro-Wilks and Lilliefors tests are often used (González,

2007), although the data about the skewness and kurtosis of the distribution are also of interest in order to consider parametric or non parametric descriptive statistics. Some simple presentations such as the Box-and-whisker plots help in the visual identification of outliers and other unusual characteristics of the data set. The box-and-whisker plot assorted with a numerical scale is a graphical representation of the five-number summary of the data set (Miller & Miller, 2005) where it is described by its extremes, its lower and upper quartiles and the median and gives at a glance the spread and the symmetry of the data set. Box-and-whisker plots may reveal suspicious patterns that should be tested for outliers. Abnormal data can road chemometric techniques leading to misleading conclusions, especially when outliers are present in the training set. Univariate outlier tests such as Dean and Dixon (1951) or Grubbs (1969) assays are not suitable. Instead, multivariate criteria for outlier detection are more advisable. The techniques based on the Mahalanobis distance (Gemperline & Boyer, 1995), and the hat matrix leverage (Hoaglin & Welsch, 1978) have been often used for decades. Hat matrix $H = X\left(X^T X\right)^{-1} X^T$ has diagonal values h_{ii} called leverage values. Patterns having leverage values higher than $2p/n$ are commonly considered outliers. However these methods are unreliable for multivariate outlier detection. Numerous methods for outlier detection have been based on the singular value decomposition or Principal Component Analysis (PCA) (Jollife, 2002). Soft Independent Modelling of Class Analogy (SIMCA) has been also applied to outlier detection (Mertens et al. , 1994). Once outliers have been deleted, researchers usually remove them from the data set, but outliers could be corrected before applying the definite mathematical procedures by using robust algorithms (Daszykowski et al., 2007). Robust methods give better results, specially some improved algorithms such as resampling by the half-means (RHM) and smallest half-volume (SHV) (Egan & Morgan , 1998).

Within the field of food authentication, the wrong conclusions are, however, mostly due to data sets that do not keep all aspects of the food characterisation (partial or skewed data set) or do merge data measured with different techniques (Aparicio & Aparicio-Ruiz, 2002). A classical example is the assessment of the geographical origin of Italian virgin olive oils by Artificial Neural Networks (ANN). The differences between oils were mainly due to the use of different chromatographic columns (packed columns against capillary columns) when quantifying the free fatty acid (FFA) profile of the oils from Southern and Northern Italy respectively (Zupan et al. , 1994). The neural network thus mostly learned to recognise the chromatographic columns.

4. Data pre-treatment

Data transformation (scaling) can be applied either for statistical dictates to optimise the analysis or based on chemical reasons. Raw compositional data are expressed in concentration units that can differ by orders of magnitude (e.g., percentage, ppm or ppb), the features with the largest absolute values are likely to dominate and influence the rule development and the classification process. Thus, for statistical needs, transformation of raw data can be applied to uniformize feature values. Autoscaling and column range scaling are the most common transformation (Sharaf et al., 1986). In the autoscaling or Z-transformation, raw data are transformed according to

$$x'_{ij} = \frac{x_{ij} - \overline{x}_j}{s_j} \text{ with } s_j = \sqrt{\frac{\sum_{i=1}^{n}(x_{ij} - \overline{x}_j)^2}{n-1}} \tag{1}$$

and can be seen as a parametric scaling by column standardisation leading to $\overline{x}'_j = 0$ and $s'_j = 1$.

Column range scaling or minimax scaling involves the following transformation

$$x'_{ij} = \frac{x_{ij} - \min_j(x_{ij})}{\max_j(x_{ij}) - \min_j(x_{ij})} \tag{2}$$

Now, the transformed data verify $0 \le x'_{ij} \le 1$. Range scaling can be considered as an interpolation of data within the interval (0,1). It is a non-parametric scaling but sensitive to outliers. When the data contain outliers and the preprocessing is necessary, one should consider robust way of data preprocessing in the context of robust Soft Independent Modelling of Class Analogy (SIMCA) method (Daszykowski et al., 2007).

The second kind of transformations are done for chemical reasons and comprise the called "constant-row sum" and "normalization variable" (Johnson & Ehrlich, 2002). Dealing with compositional data, concentrations can vary widely due to dilution away from a source. In the case of contaminated sediment investigations, for example, concentrations may decrease exponentially away from the effluent pipe. However, if the relative proportions of individual analytes remain relatively constant, then we would infer a single source scenario coupled with dilution far away from the source. Thus, a transformation is needed to normalize concentration/dilution effects. Commonly this is done using a transformation to a fractional ratio or percent, where each concentration value is divided by the total concentration of the sample: Row profile or constant row-sum transformation because the sum of analyte concentrations in each sample (across rows) sums unity or 100%:

$$x'_{ij} = \frac{x_{ij}}{\sum_{j=1}^{p} x_{ij}} \tag{3}$$

leading to $\sum_{j=1}^{p} x'_{ij} = 1$ (or 100%). An alternative is to normalize the data with respect to a single species or compound set as reference congener, the normalization variable. This transformation involves setting the value of the normalization feature to unity, and the values of all other features to some proportion of 1.0, such that their ratios with respect to the normalization feature remain the same in the original metric.

When n > p, the rank of data matrix is p (if all features are independent). Thus, autoscaling and minimax transformations do not change data dimensionality because these treatments do not induce any bound between features. Row profiles instead build a relationship between features scores (the constant-row sum) that decreases the data dimensionality by 1 and the rank of data matrix is then p-1. Accordingly, the patterns fall on a hypersurface in the feature space and it is advisable to remove one feature to avoid problems involving matrix inversion when the rank of the matrix is less than p.

As a final remark it should be realized that when using some Supervised Learning techniques like SIMCA, the scaling of the data set is carried out only over the samples belonging to the same class (separate scaling). This is due because the own fundamentals of the methodology and has a beneficial effect on the classification (Derde et al., 1982).

5. Feature selection and extraction

Irrelevant features are very expensive ones, because they contribute to the chemical information with noise only; but even more expensive may be simply wrong features. Accordingly, they should be eliminated in order to circumvent disturb in classification. In almost all chemical applications of pattern recognition the number of original raw features is too large and a reduction of the dimensionality is necessary. Features which are essential for classification purposes are often called intrinsic features. Thus, a common practise to avoid redundant information consists of computing the correlation matrix of features R. Pair of most correlated features can be either combined or one of them is deleted. Researchers should be aware that the number of independent descriptors or features, p, must be much smaller than that of patterns, n. Otherwise, we can build a classification rule that even separates randomly selected classes of the training set (Varmuza, 1980). This assumes that the set of features is linearly independent (actually, it is the basis of the vectorial space) and the number of features is the dimensionality of the vectorial space. Accordingly, the true dimensionality should be evaluated for instance from an eigenanalysis (PCA) of the data matrix and extract the proper number of factors which correspond to the true dimensionality (d) of the space. Most efficient criteria for extracting the proper number of underlying factors are based on the Malinowski indicator function (Malinowski, 2002) and the Wold's Cross-Validation procedure (Wold, 1978). For most classification methods, a ratio $n/d > 3$ is advised and > 10, desirable. However, PCA-based methods like SIMCA or Partial Least Squares Discriminant Analysis (PLS-DA) can be applied without problem when $p \gg n$. However, even in these instances there are suitable methods for selecting a subset of features and to build a final model based on it.

Therefore, when it is advisable the feature selection, weighing methods determine the importance of the scaled features for a certain classification problem. Consider a pattern vector $\vec{x}_i \equiv (x_{i1}, x_{i2}, \dots x_{ip})$. Assuming that the data matrix X can be partitioned into a number Q of classes, let $\vec{x}_i^{(C)}$ a pattern vector belonging to class C. The averaged patterns $\vec{m}$ and $\vec{m}^{(C)}$ represent the general mean vector and the C-class mean, according to:

$$\vec{m} = \frac{\sum_{i=1}^{n} \vec{x}_i}{n} \quad \text{and} \quad \vec{m}^{(C)} = \frac{\sum_{i=1}^{n(C)} \vec{x}_i^{(C)}}{n} \tag{4}$$

When they are applied to a selected j feature we have

$$m_j = \frac{\sum_{i=1}^{n} x_{ij}}{n} \quad \text{and} \quad m_j^{(C)} = \frac{\sum_{i=1}^{n(C)} x_{ij}^{(C)}}{n(C)} \tag{5}$$

Where n(C) is the number of patterns of class C.

Accordingly, we can explore the inter-class scatter matrix as well as the intra-class scatter matrix. The total scatter matrix can be obtained as

$$T = \sum_{i=1}^{n} (\vec{x}_i - \vec{m})(\vec{x}_i - \vec{m})^T \tag{6}$$

and its elements as

$$T_{jk} = \sum_{i=1}^{n} (x_{ij} - m_j)(x_{ik} - m_k) \tag{7}$$

The within classes scatter matrix, together with its element is given by

$$W = \sum_{C=1}^{Q} \sum_{i=1}^{n(C)} (\vec{x}_i^{(C)} - \vec{m}^{(C)})(\vec{x}_i^{(C)} - \vec{m}^{(C)})^T$$

$$W_{jk} = \sum_{C=1}^{Q} \sum_{i=1}^{n(C)} (x_{ij}^{(C)} - m_j^{(C)})(x_{ik}^{(C)} - m_k^{(C)}) \tag{8}$$

And the between classes matrix and element,

$$B = \sum_{C=1}^{Q} n(C)(\vec{m}^{(C)} - \vec{m})(\vec{m}^{(C)} - \vec{m})^T$$

$$B_{jk} = \sum_{C=1}^{Q} n(C)(m_j^{(C)} - m_j)(m_k^{(C)} - m_k) \tag{9}$$

For a case involving two classes 1 and 2 and one feature j we have the following:

$$T_{jj} = \sum_{i=1}^{p} (x_{ij} - m_j)^2$$

$$W_{jj} = \sum_{i=1}^{n(1)} (x_{ij}^{(1)} - m_j^{(1)})^2 + \sum_{i=1}^{n(2)} (x_{ij}^{(2)} - m_j^{(2)})^2 \tag{10}$$

$$B_{jj} = n(1)(m_j^{(1)} - m_j)^2 + n(2)(m_j^{(2)} - m_j)^2$$

Weighting features in Supervised Learning techniques can be then extracted from its relative importance in discriminating classes pairwise. The largest weight corresponds to the most important feature. The most common weighting factors are:

- Variance weights (VW) (Kowalski & Bender, 1972): $VW_j = \dfrac{B_{jj}}{W_{jj}}$

- Fisher weights (FW) (Duda et al., 2000): $FW_j = \dfrac{(m_j^{(1)} - m_j^{(2)})^2}{W_{jj}}$

- Coomans weights (g) (Coomans et al., 1978):

$$g_j = \frac{\left| m_j^{(1)} - m_j^{(2)} \right|}{s_j^{(1)} + s_j^{(2)}} \quad \text{with} \quad s_j^{(C)} = \sqrt{\frac{\sum_{i=1}^{n(C)} (x_{ij}^{(C)} - m_j^{(C)})^2}{n(C)}}$$

A multi-group criterion is the called Wilks' λ or McCabe U statistics (McCabe, 1975). This is a general statistic used as a measure for testing the difference among group centroids. All classes are assumed to be homogeneous variance-covariance matrices and the statistic is defined as

$$\lambda = \frac{\det W}{\det T} = \frac{SSW}{SST} = \frac{SSW}{SSB + SSW} \tag{11}$$

Where SSW, SSB and SST refer to the sum of squares corresponding to the scatter matrices W, B and T, respectively, as defined above. Remembering that the ratio $\eta = \sqrt{\dfrac{BSS}{WSS}}$ is the coefficient of canonical correlation, $\eta = \sqrt{1 - \lambda}$, and hence when $\eta \to 1$ for intrinsic features, $\lambda \to 0$ and more significant are the centroid difference. Before calculation of the statistic, data should be autoscaled. This later criterion as well as the largest values of Rao's distance or Mahalanobis distance is generally used in Stepwise Discriminant Analysis (Coomans et al., 1979). Certain Supervised Learning techniques enable feature selection according to its own philosophy. Thus, for instance, SIMCA test the intrinsic features according the values of two indices called discriminating power and modelling power (Kvalheim & Karstang, 1992). Using ANNs for variable selection is attractive since one can globally adapt the variables selector together with the classifier by using the called "pruning" facilities. Pruning is a heuristic method to feature selection by building networks that do not use those variables as inputs. Thus, various combinations of input features can be added and removed, building new networks for each (Maier et al., 1998).

Genetic algorithms are also very useful for feature selection in fast methods such as PLS (Leardi & Lupiañez, 1998).

6. Development of the decision rule

In order to focus the commonly used Supervised Learning techniques of pattern recognition we have selected the following methods: K-Nearest Neighbours (KNN) (Silverman & Jones, 1989), Linear Discriminant Analysis (LDA) (Coomans et al., 1979), Canonical Variate Analysis (CVA) (Cole & Phelps, 1979), Soft Independent Modelling of Class Analogy (SIMCA) (Wold, 1976), Unequal dispersed classes (UNEQ) (Derde & Massart, 1986), PLS-DA (Stahle & Wold, 1987), Procrustes Discriminant Analysis (PDA) (González-Arjona et al., 2001), and methods based on ANN such as Multi-Layer Perceptrons (MLP) (Zupan & Gasteiger, 1993; Bishop, 2000), Supervised Kohonen Networks (Melssen et al., 2006),

Kohonen Class-Modelling (KCM) (Marini et al., 2005), and Probabilistic Neural Networks (PNN) (Streit & Luginbuhl, 1994). Recently, new special classification techniques arose. A procedure called Classification And Influence Matrix Analysis (CAIMAN) has been introduced by Todeschini *et al* (2007). The method is based on the leverage matrix and models each class by means of the class dispersion matrix and calculates the leverage of each sample with respect to each class model space. Since about two decades another new classification (and regression) revolutionary technique based on statistical learning theory and kernel latent variables has been proposed: Support Vector Machines (SVM) (Vapnik, 1998; Abe, 2005; Burges, 1998). The purpose of SVM is separate the classes in a vectorial space independently on the probabilistic distribution of pattern vectors in the data set (Berrueta et al., 2007). This separation is performed with the particular hyperplane which maximizes a quantity called margin. The margin is the distance from a hyperplane separating the classes to the nearest point in the data set (Pardo & Sberveglieri, 2005). The training pattern vectors closest to the separation boundary are called *support vectors*. When dealing with a non linear boundary, the kernel method is applied. The key idea of kernel method is a transformation of the original vectorial space (input space) to a high dimensional Hilbert space (feature space), in which the classes can be separated linearly. The main advantages of SVM against its most direct concurrent method, ANN, are the easy avoiding of overfitting by using a penalty parameter and the finding of a deterministic global minimum against the non deterministic local minimum attained with ANN.

Some of the mentioned methods are equivalent. Let us consider some couples: CVA and LDA and PLS-DA and PDA. CVA attempts to find linear combinations of variables from each set that exhibit maximum correlation. These may be referred to as canonical variates, and data can be displayed as scatterplot of one against the other. The problem of maximizing the correlation can be formulated as an eigenanalysis problem with the largest eigenvalue providing the maximized correlation and the eigenvectors giving the canonical variates. Loadings of original features in the canonical variates and cumulative proportions of eigenvalues are interpreted, partly by analogy with PCA. Note that if one set of features are dummy variables giving group indicators, and then CVA is mathematically identical to LDA (González-Arjona et al., 2006). PLS-DA finds latent variables in the feature space which have a maximum covariance with the y variable. PDA may be considered equivalent to PLS-DA. The only difference is that in PDA, eigenvectors are obtained from the covariance matrix $Z^T Z$ instead of $X^T X$, with $Z = Y^T X$ where Y is the membership target matrix constructed with ones and zeros: For a three classes problem, sample labels are 001, 010 and 100. Accordingly, we can consider CVA equivalent to LDA and PLS-DA equivalent to PDA.

Researchers should be aware of apply the proper methods according to the nature and goals of the chemical problem. As Daszykowski and Walczak pointed out in his excellent survey (Daszykowski & Walczak, 2006), in many applications, unsupervised methods such as PCA are used for classification purposes instead of the supervised approach. If the data set is well structured, then PCA-scores plot can reveal grouping of patterns with different origin, although the lack of these groups in the PCA space does not necessarily mean that there is no statistical difference between these samples. PCA by definition maximizes data variance, but the main variance cannot be necessarily associated with the studied effect (for instance, sample origin). Evidently, PCA can be used for exploration, compression and visualization of data trends, but it cannot be used as Supervised Learning classification method.

On the other hands, according to the nature of the chemical problem, some supervised techniques perform better than others, because its own fundamentals and scope. In order to consider the different possibilities, four paradigms can be envisaged:

1. *Parametric/non-parametric techniques*: This first distinction can be made between techniques that take account of the information on the population distribution. Non parametric techniques such as KNN, ANN, CAIMAN and SVM make no assumption on the population distribution while parametric methods (LDA, SIMCA, UNEQ, PLS-DA) are based on the information of the distribution functions. LDA and UNEQ are based on the assumption that the population distributions are multivariate normally distributed. SIMCA is a parametric method that constructs a PCA model for each class separately and it assumes that the residuals are normally distributed. PLS-DA is also a parametric technique because the prediction of class memberships is performed by means of model that can be formulated as a regression equation of Y matrix (class membership codes) against X matrix (González-Arjona et al., 1999).

2. *Discriminating (hard)/Class-Modelling (soft) techniques*: Pure classification, discriminating or hard classification techniques are said to apply for the first level of Pattern Recognition, where objects are classified into either of a number of defined classes (Albano et al., 1978). These methods operate dividing the hyperspace in as many regions as the number of classes so that, if a sample falls in the region of space corresponding to a particular category, it is classified as belonging to that category. These kinds of methods include LDA, KNN, PLS-DA, MLP, PNN and SVM. On the other hands, Class-Modelling techniques build frontiers between each class and the rest of the universe. The decision rule for a given class is a class box that envelopes the position of the class in the pattern space. So, three kinds of classification are possible: (i) an object is assigned to a category if it is situated inside the boundaries of only a class box, (ii) an object can be inside the boundaries (overlapping region) of more than one class box, or (iii) an object is considered to be an outlier for that class if it falls outside the class box. These are the features to be covered by methods designed for the so called second level of Pattern Recognition: The first level plus the possibility of outliers and multicategory objects. Thus, typical class modelling techniques are SIMCA and UNEQ as well as some modified kind of ANN as KCM. CAIMAN method is developed in different options: D-CAIMAN is a discriminating classification method and M-CAIMAN is a class modelling one

3. *Deterministic/Probabilistic techniques*: A deterministic method classifies an object in one and only one of the training classes and the degree of reliability of this decision is not measured. Probabilistic methods provide an estimate of the reliability of the classification decision. KNN, MLP, SVM and CAIMAN are deterministic. Other techniques, including some kind of ANN are probabilistic (e.g., PNN where a Bayesian decision is implemented).

4. *Linear/Non-Linear separation boundaries*: Here our attention is focused on the mathematical form of the decision boundary. Typical non-linear classification techniques are based on ANN and SVM, specially devoted to apply for classification problems of non-linear nature. It is remarkable that CAIMAN method seems not to suffer of nonlinear class separability problems.

7. Validation of the decision rule

A very important issue is the improper model validation. This pitfall even appears in very simple cases, such as the fitting of a series of data points by using a polynomial function. If we use a parsimonic fitting where the number of points is higher than the number of polynomial coefficients, the fitting train the generalities of the data set. Overparametrized fitting where the number of points becomes equal to the number of polynimial coefficients, trains idiosyncrasies and leads to overtraining or overfitting. Thus, a complex fitting function may fit the noise, not just the signal. Overfitting is a Damocles' sword that gravitates over any attempt to model the classification rule. We are interested to an intermediate behaviour: A model which is powerful enough to represent the underlying structure of the data (generalities), but not so powerful that it faithfully models the noise (idiosyncrasies) associated to data. This balance is known as the bias-variance tradeoff . The bias-variance tradeoff is most likely to become a problem when we have relatively few data points. In the opposite case, there is no danger of overfitting, as the noise associated with any single data point plays an immaterial role in the overall fit.

If we transfer the problem of fitting a polynomial to data into the use of another functions, such as the discriminant functions of canonical variates issued from LDA, the number of discriminant functions will be p (the number of features) or Q-1 (Q is the number of classes), whichever is smaller. As a rule of thumb (Defernez & Kemsley, 1997), the onset of overfitting should be strongly suspected when the dimensionality $d > \dfrac{n-Q}{3}$. One of the simplest and most widely used means of preventing overfitting is to split the known data set into two sets: the training set and the validation, evaluation, prediction or test set.

Commonly, the known set is generally randomly divided into the training and validation sets, containing about P% and 100-P% samples of every class. Typical values are 75-25% or even 50-50% for training and validation sets. The classification performance is computed in average. Thus, the randomly generation of training and validation sets is repeated a number of times, 10 times for instance. Once the classification rule is developed, some workers consider as validation parameters the recalling efficiency (rate of training samples correctly classified by the rule) and, specially, the prediction ability (rate of evaluation samples correctly classified by the rule).

An alternative to the generation of training and validation sets are the cross-validation and the bootstrapping method (Efron and Gong, 1983). In the called k-fold cross validation, the know set is split into k subsets of approximately equal size. Then the training is performed k times, each time leaving out one of k the subsets, but using only the omitted subset to predict its class membership. From all predictions, the percentage of hits gives an averaged predictive ability. A very common and simple case of cross-validation is the leave-one-out method: At any given time, only a pattern is considered and tested and the remaining patterns form the training set. Training and prediction is repeated until each pattern was treated as test once. This later procedure is easily confused with jacknifing because both techniques involve omitting each pattern in turn, but cross-validation is used just for validation purposes and jacknife is applied in order to estimate the bias of a statistic.

In bootstrapping, we repeatedly analyze subsamples, instead of subsets of the known set. Each subsample is a random sample with replacement from the full sample (known set). Bootstrapping seems to perform better than cross-validation in many instances (Efron, 1983).

However, the performance rate obtained for validating the decision rule could be misleading because they do not consider the number of false positive and false negative for each class. These two concepts provide a deep knowledge of the classes' space. Accordingly, it seems to be more advisable the use of terms sensitivity (SENS) and specificity (SPEC) (González-Arjona et al., 2006) for validating the decision rule. The SENS of a class corresponds to the rate of evaluation objects belonging to the class that are correctly classified, and the SPEC of a class corresponds to the rate of evaluation objects not belonging to the class that are correctly considered as belonging to the other classes. This may be explained in terms of the first and second kind of risks associated with prediction. The first kind of errors (a) corresponds to the probability of erroneously reject a member of the class as a non-member (rate of false negative, FN). The second kind of errors (β) corresponds to the probability of erroneously classify a non-member of the class as a member (rate of false positive, FP). Accordingly, for a given class A, and setting n_A as the number of members of class A, $\bar{n}_A$ as the number of non-members of class A, $\langle n_A \rangle$ as the number of members of class A correctly classified as "belonging to class A" and $\langle \bar{n}_A \rangle$ as the number of non-members of class A classified as "not belonging to class A", we have (Yang et al., 2005):

$$TP = \langle n_A \rangle \qquad FP = \bar{n}_A - \langle \bar{n}_A \rangle$$
$$TN = \langle \bar{n}_A \rangle \qquad FN = n_A - \langle n_A \rangle \tag{12}$$

TP and TN being the number of True Positive and True Negative members of the considered class. Accordingly,

$$SENS = \frac{\langle n_A \rangle}{n_A} = 1 - \alpha = 1 - \frac{FN}{n_A} = \frac{TP}{TP + FN}$$
$$SPEC = \frac{\langle \bar{n}_A \rangle}{\bar{n}_A} = 1 - \beta = 1 - \frac{FP}{\bar{n}_A} = \frac{TN}{TN + FP} \tag{13}$$

It is clear that values close to unity for both parameters indicates a successfully validation performance.

With these parameters it can be built the called *confusion matrix* for class A:

$$^C M_A = \begin{bmatrix} TN & FP \\ FN & TP \end{bmatrix} \tag{14}$$

As it has been outlined, the common validation procedure consists of dividing the known set into two subsets, namely training and validation set. However, the validation procedure has to be considered with more caution in case of some kinds of ANN such as MLP because they suffer a special overfitting damage. The MLP consists of formal neurons and connection (weights) between them. As it is well known, neurons in MLP are commonly arranged in three layers: an input layer, one hidden layer (sometimes plus a bias neuron) and an output layer. The number of hidden nodes in a MLP indicates the complexity of the relationship in a way very similar to the fitting of a polynomial to a data set. Too many connections have the risk of a network specialization in training noise and poor prediction ability. Accordingly, a first action should be minimizing the number of neurons of the hidden layer. Some authors (Andrea & Kalayeh, 1991) have proposed the parameter ρ which plays a major role in determining the best architecture:

$$\rho = \frac{\text{Number of data points in the training set}}{\text{Sum of the number of connections in the network}} \tag{15}$$

In order to avoid overfitting it is recommended that $1 < \rho < 2.2$.

Besides, the overfitting problem can be minimized by monitoring the performance of the network during training by using an extra verification set different from training set. This verification set is needed in order to stop the training process before the ANN learns idiosyncrasies present in the training data that leads to overfitting (González, 2007).

8. Concluding remarks

The selection of the supervised learning technique depends on the nature of the particular problem. If we have a data set composed only by a given number of classes and the rule is going to be used on test samples that we know they may belong to one of the former established classes only, then we can select a discriminating technique such as LDA, PLS-DA, SVM or some kind of discriminating ANN (MLP or PNN). Otherwise, class modelling techniques such as SIMCA, UNEQ or KCM are useful. Class modelling tools offer at least two main advantages: To identify samples which do not fall in any of the examined categories (and therefore can be either simply outlying observations or members of a new class not considered in the known set) and to take into account samples that can simultaneously belong to more than one class (multiclass patterns).

If the idiosyncrasy of the problem suggests that the boundaries could be of non-linear nature, then the use of SVM or ANN is the best choice.

In cases where the number of features in higher than the number of samples (p > n), a previous or simultaneous step dealing with feature selection is needed when non-PCA based techniques are used (KNN, LDA, ANN, UNEQ). PCA-based methods such as SIMCA and PLS-DA can be applied without need of feature selection. This characteristic is very interesting beyond of compositional analysis, when samples are characterized by a spectrum, like in spectrometric methods (FT-IR, FT-Raman, NMR...). A different behaviour of these two methods against the number of FP and FN has been noticed (Dahlberg et al., 1997). SIMCA is focused on class specificities, and hence it detects strangers with high accuracy (only when the model set does not contain outliers. Otherwise, robust SIMCA model can be used), but sometimes fails to recognize its own members if the class is not homogeneous enough or the training set is not large enough. PLS-DA, on the contrary, deals with an implicitly closed universe (since the Y variables have a constant sum) so that it ignores the possibility of strangers. However, this has the advantage to make the method more robust to class inhomogeneities, since what matters most in class differences.

In compositional data, as pointed out Berrueta et al. (2007), the main problem is class overlap, but with a suitable feature selection and adequate sample size, good classification performances can be achieved. In general, non-linear methods such as ANN or SVM are rarely needed and most classification problems can be solved using linear techniques (LDA, CVA, PLS_DA).

Sometimes, several different types of techniques can be applied to the same data set. Classification methods are numerous and then the main problem is to select the most suitable one, especially dealing with quantitative criteria like prediction ability or misclassification percentage. In order to carry out the comparison adequately, the McNemar's test is a good choice (Roggo et al., 2003). Two classification procedures A and B are trained and the same validation set is used. Null hypothesis is that both techniques lead to the same misclassification rate. McNemar's test is based on a χ^2 test with one degree of freedom if the number of samples is higher than 20. The way to obtain the McNemar's statistic is as follows:

$$\text{McNemar's value} = \frac{\left(\left|n_{01} - n_{10}\right| - 1\right)^2}{n_{01} + n_{10}} \tag{16}$$

with

n_{00}: number of samples misclassified by both methods A and B

n_{01}: number of samples misclassified by method A but not by B

n_{10}: number of samples misclassified by method B but not by A

n_{11}: number of samples misclassified by neither method A nor B

$n_{val} = n_{00} + n_{01} + n_{10} + n_{11}$ = number of patterns in the validation set

The critical value for a 5% significance level is 3.84. In order to get insight about this procedure, the paper of Roggo et al (2006) is very promising.

Finally, a last consideration about problems with the data set representativeness. As it has been claimed in a published report a LDA was applied to differentiate 12 classes of oils on the basis of the chromatographic data, where some classes contained two or three members only (and besides, the model was not validated). There is no need of being an expertise chemometrician to be aware of two or three samples are insufficient to draw any relevant conclusion about the class to which they belong. There are more sources of possible data variance than the number of samples used to estimate class variability (Daszykowski & Walczak, 2006). The requirements of a sufficient number of samples for every class could be envisaged according to a class modelling technique to extract the class dimensionality and consider, for instance, a number of members within three to ten times this dimensionality.

Aside from this representativity context, it should be point out that when the aim is to classify food products or to build a classification rule to check the authentic origin of samples, they have to be collected very carefully according to a well established sampling plan. Often not enough care is taken about it, and thus is it hardly possible to obtain accurate classification models.

9. References

Abe, S. (2005). Support vector machines for pattern classification. Springer, ISBN:1852339299, London, UK

Aitchison, J. (2003). The statistical analysis of compositional data. The Blackburn Press, ISBN:1930665784, London, UK

Albano, C.; Dunn III, W.; Edlund, U.; Johansson, E.; Norden, B.; Sjöström, M. & Wold, S. (1978). Four levels of Pattern Recognition. Analytica Chimica Acta. Vol. 103, pp. 429-443. ISSN:0003-2670

Andrea, T.A.; Kalayeh, H. (1991). Applications of neural networks in quantitative structure-activity relationships of dihydrofolate reductase inhibitors. Journal of Medicinal Chemistry. Vol. 34, pp. 2824-2836. ISSN: 0022-2623

Aparicio, R. & Aparicio-Ruíz, R. (2002). Chemometrics as an aid in authentication, In: Oils and Fats Authentication, M. Jee (Ed.), 156-180, Blackwell Publishing and CRC Press, ISBN:1841273309, Oxford, UK and FL, USA

Bishop, C.M. (2000). Neural Networks for Pattern Recognition, Oxford University Press, ISBN:0198538642, NY, USA

Berrueta, L.A.; Alonso-Salces, R.M. & Héberger, K. (2007). Supervised pattern recognition in food analysis. Journal of Chromatography A. Vol. 1158, pp. 196-214. ISSN:0021-9673

Burges, C.J.C. (1998). A tutorial on support vector machines for pattern recognition. Data Mining and Knowledge Discovery. Vol. 2, pp. 121-167. ISSN:1384-5810

Chung, C.F. (1993). Estimation of covariance matrix from geochemical data with observations below detection limits. Mathematical Geology. Vol. 25, pp. 851-865. ISSN:1573-8868

Clarke, J.U. (1998). Evaluation of censored data methods to allow statistical comparisons among very small samples with below detection limits observations. Environmental Science & Technology. Vol. 32, pp. 177-183. ISSN:1520-5851

Cole, R.A. & Phelps, K. (1979). Use of canonical variate analysis in the differentiation of swede cultivars by gas-liquid chromatography of volatile hydrolysis products. Journal of the Science of Food and Agriculture. Vol. 30, pp. 669-676. ISSN:1097-0010

Coomans, D.; Broeckaert, I.; Fonckheer, M; Massart, D.L. & Blocks, P. (1978). The application of linear discriminant analysis in the diagnosis of thyroid diseases. Analytica Chimica Acta. Vol. 103, pp. 409-415. ISSN:0003-2670

Coomans, D.; Massart, D.L. & Kaufman, L. (1979) Optimization by statistical linear discriminant analysis in analytical chemistry. Analytica Chimica Acta. Vol. 112, pp. 97-122. ISSN:0003-2670

Dahlberg, D.B.; Lee, S.M.; Wenger, S.J. & Vargo, J.A. (1997). Classification of vegetable oils by FT-IR. Applied Spectroscopy. Vol. 51, pp. 1118-1124. ISSN:0003-7028

Daszykowski, M. & Walczak, B. (2006). Use and abuse of chemometrics in chromatography. Trends in Analytical Chemistry. Vol. 25, pp. 1081-1096. ISSN:0165-9936

Daszykowski, M.; Kaczmarek, K.; Stanimirova, I.; Vander Heyden, Y. & Walczak, B. (2007). Robust SIMCA-bounding influence of outliers. Chemometrics and Intelligent Laboratory Systems. Vol. 87, pp. 95-103. ISSN:0169-7439

Dean, R.B. & Dixon, W.J. (1951). Simplified statistics for small number of observations. Analytical Chemistry. Vol. 23, pp. 636-638. ISSN:0003-2700

Defernez, M. & Kemsley, E.K. (1997). The use and misuse of chemometrics for treating classification problems. Trends in Analytical Chemistry. Vol. 16, pp. 216-221. ISSN:0165-9936

Derde, M.P.; Coomans, D. & Massart, D.L. (1982). Effect of scaling on class modelling with the SIMCA method. Analytica Chimica Acta. Vol. 141, pp. 187-192. ISSN:0003-2670

Derde, M.P. & Massart, D.L. (1986). UNEQ: A class modelling supervised pattern recognition technique. Microchimica Acta. Vol. 2, pp. 139-152. ISSN:0026-3672

Duda, R.O.; Hart, P.E. & Stork, D.G. (2000). Pattern classification. 2nd edition. Wiley, ISBN:0471056693, NY, USA

Efron, B. (1983). Estimating the error rate of a prediction rule: Improvement on cross validation. Journal of the American Statistical Association. Vol. 78, pp. 316-331. ISSN:0162-1459

Efron, B. & Gong, G. (1983). A leisurely look at the bootstrap, the jacknife and cross validation. The American Staticscian. Vol. 37, pp. 36-48. ISSN:0003-1305

Egan, W.J. & Morgan, S.L. (1998). Outlier detection in multivariate analytical chemical data. Analytical Chemistry. Vol. 70, pp. 2372-2379. ISSN:0003-2700

Egozcue, J.J.; Pawlowsky-Glahn, V.; Mateu-Figueros, G.; Barcelo-Vidal, C. (2003). Isometric logratio transformation for compositional data analysis. Mathematical Geology. Vol. 35, pp. 279-300. ISSN:1573-8868

Gemperline, P.J. & Boyer, N.R. (1995). Classification of near-infrared spectra using wavelength distances: Comparisons to the Mahalanobis distance and Residual Variance methods. Analytical Chemistry. Vol.67, pp. 160-166. ISSN:0003-2700

González, A.G. (2007). Use and misuse of supervised pattern recognition methods for interpreting compositional data. Journal of Chromatograpy A. Vol. 1158, pp. 215-225. ISSN:0021-9673

González, A.G. & Herrador, M.A. (2007). A practical guide to analytical method validation, including measurement uncertainty and accuracy profiles. Trends in Analytical Chemistry. Vol. 26, pp. 227-237. ISSN:0165-9936

González-Arjona, D.; López-Pérez, G. & González, A.G. (1999). Performing Procrustes discriminant analysis with HOLMES. Talanta. Vol. 49, pp. 189-197. ISSN:0039-9140

González-Arjona, D.; López-Pérez, G. & González, A.G. (2001). Holmes, a program for performing Procrustes Transformations. Chemometrics and Intelligent Laboratory Systems. Vol. 57, pp. 133-137. ISSN:0169-7439

González-Arjona, D.; López-Pérez, G. & González, A.G. (2006). Supervised pattern recognition procedures for discrimination of whiskeys from Gas chromatography/Mass spectrometry congener analysis. Journal of Agricultural and Food Chemistry. Vol. 54, pp. 1982-1989. ISSN:0021-8561

Grubbs. F. (1969). Procedures for detecting outlying observations in samples. Technometrics. Vol. 11, pp. 1-21. ISSN:0040-1706

Helsel, D.R. (1990).Less than obvious: Statistical treatment of data below the detection limit. Environmental Science & Technology. Vol. 24, pp. 1766-1774. ISSN: 1520-5851

Hoaglin, D.C. & Welsch, R.E. (1978). The hat matrix in regression and ANOVA. The American Statiscian. Vol. 32, pp. 17-22. ISSN:0003-1305

Holger, R.M.; Dandy, G.C. & Burch, M.D. (1998). Use of artificial neural networks for modelling cyanobacteria Anabaena spp. In the river Murray, South Australia. Ecological Modelling. Vol. 105, pp. 257-272. ISSN:0304-3800

Jollife, I.T. (2002). Principal Component Analysis. 2nd edition, Springer, ISBN:0387954422, NY, USA

Johnson, G.W. & Ehrlich, R. (2002). State of the Art report on multivariate chemometric methods in Environmental Forensics. Environmental Forensics. Vol. 3, pp. 59-79. ISSN:1527-5930

Kryger, L. (1981). Interpretation of analytical chemical information by pattern recognition methods-a survey. Talanta. Vol. 28, pp. 871-887. ISSN:0039-9140

Kowalski, B.R. & Bender, C.F. (1972). Pattern recognition. A powerful approach to interpreting chemical data. Journal of the American Chemical Society. Vol. 94, pp. 5632-5639. ISSN:0002-7863

Kvalheim, O.M. & Karstang, T.V. (1992). SIMCA-Classification by means of disjoint cross validated principal component models, In: Multivariate Pattern Recognition in Chemometrics, illustrated by case studies, R.G. Brereton (Ed.), 209-245, Elsevier, ISBN:0444897844, Amsterdam, Netherland

Leardi, R. & Lupiañez, A. (1998). Genetic algorithms applied to feature selection in PLS regression: how and when to use them. Chemometrics and Intelligent Laboratory Systems. Vol. 41, pp. 195-207. ISSN:0169-7439

Malinowski, E.R. (2002). Factor Analysis in Chemistry. Wiley, ISBN:0471134791, NY, USA

Marini, F.; Zupan, J. & Magrí, A.L. (2005). Class modelling using Kohonen artificial neural networks. Analytica Chimica Acta. Vol.544, pp. 306-314. ISSN:0003-2670

Martín-Fernández, J.A.; Barceló-Vidal, C. & Pawlowsky-Glahn, V. (2003). Dealing with zeros and missing values in compositional data sets using nonparametric imputation. Mathematical Geology. Vol. 35, pp. 253-278. ISSN:1573-8868

McCabe, G.P. (1975). Computations for variable selection in discriminant analysis. Technometrics. Vol. 17, pp. 103-109. ISSN:0040-1706

Melssen, W.; Wehrens, R. & Buydens, L. (2006). Supervised Kohonen networks for classification problems. Chemometrics and Intelligent Laboratory Systems. Vol. 83, pp. 99-113. ISSN:0169-7439

Mertens, B.; Thompson, M. & Fearn, T. (1994). Principal component outlier detection and SIMCA: a synthesis. Analyst. Vol. 119, pp. 2777-2784. ISSN:0003-2654

Miller, J.N. & Miller, J.C. (2005). Statistics and Chemometrics for Analytical Chemistry. 4th edition. Prentice-Hall, Pearson. ISBN:0131291920. Harlow, UK

Naes, T.; Isaksson, T.; Fearn, T.; Davies, T. (2004). A user-friendly guide to multivariate calibration and classification. NIR Publications, ISBN;0952866625, Chichester, UK

Pardo, M. & Sberveglieri, G. (2005). Classification of electronic nose data with support vector machines. Sensors and Actuators. Vol. 107, pp. 730-737. ISSN:0925-4005

Pretsch, E. & Wilkins, C.L. (2006). Use and abuse of Chemometrics. Trends in Analytical Chemistry. Vol. 25, p. 1045. ISSN:0165-9936

Rock, B.A. (1985). An introduction to Chemometrics, 130th Meeting of the ACS Rubber Division. October 1985. Available from http://home.neo.rr.com/catbar/chemo/int_chem.html

Roggo, Y.; Duponchel, L. & Huvenne, J.P. (2003). Comparison of supervised pattern recognition methods with McNemar's statistical test: Application to qualitative analysis of sugar beet by near-infrared spectroscopy. Analytica Chimica Acta. Vol. 477, pp. 187-200. ISSN:0003-2670

Sharaf, M.A.; Illman, D.A. & Kowalski, B.R. (1986). Chemometrics. Wiley, ISBN:0471831069, NY, USA

Silverman, B.W. & Jones, M.C. (1989). E. Fix and J.L. Hodges (1951): An important contribution to non parametric discriminant analysis and density estimation. International Statistical Review. Vol. 57, pp. 233-247. ISSN:0306-7734

So, S.S. & Richards, W.G. (1992). Application of Neural Networks: Quantitative structure activity relationships of the derivatives of 2,4-diamino-5-(substituted-benzyl) pyrimidines as DHFR inhibitors. Journal of Medicinal Chemistry. Vol. 35, pp. 3201-3207. ISSN:0022-2623

Stahle, L. & Wold, S. (1987). Partial least squares analysis with cross validation for the two class problem: A monte-Carlo study. Journal of Chemometrics. Vol. 1, pp. 185-196. ISSN:1099-128X

Stanimirova, I.; Daszykowski, M. & Walczak, B. (2007). Dealing with missing values and outliers in principal component analysis. Talanta. Vol. 72, pp. 172-178. ISSN:0039-9140

Streit, R.L. & Luginbuhl, T.E. (1994). Maximun likelihood training of probabilistic neural networks. IEEE Transactions on Neural Networks. Vol. 5, pp. 764-783. ISSN:1045-9227

Todeschini, R.; Ballabio, D.; Consonni, V.; Mauri, A. & Pavan, M. (2007). CAIMAN (Classification And Influence Matrix Analysis): A new approach to the classification based on leverage-scale functions. Chemometrics and Intelligent Laboratory Systems. Vol. 87, pp. 3-17. ISSN:0169-7439

Vapnik, V.N. (1998). Statistical learning theory. Wiley, ISBN:0471030031, NY, USA

Varmuza, K. (1980). Pattern recognition in chemistry. Springer, ISBN:0387102736, Berlin, Germany

Varmuza, K. & Filzmoser, P. (2009). Introduction to multivariate statistical analysis in chemometrics, CRC Press, Taylor & Francis Group, ISBN:14005975, Boca Ratón, FL, USA

Walczak, B. & Massart, D.L. (2001). Dealing with missing data: Part I and Part II. Chemometrics and Intelligent Laboratory System. Vol. 58, pp. 15-27 and pp. 29-42. ISSN:0169-7439

Wold, S. (1976). Pattern recognition by means of disjoint principal component models. Pattern Recognition. Vol. 8, pp. 127-139. ISSN:0031-3203

Wold, S. (1978). Cross validatory estimation of the number of components in factor and principal components models. Technometrics. Vol. 20, pp. 397-405. ISSN:0040-1706

Wold, S. (1995). Chemometrics, what do we mean with it, and what do we want from it? Chemometrics and Intelligent Laboratory Systems. Vol. 30, pp. 109-115. ISSN:0169-7439

Yang, Z.; Lu, W.;Harrison, R.G.; Eftestol, T.; Steen,P.A. (2005). A probabilistic neural network as the predictive classifier of out-of-hospital defibrillation outcomes. Resuscitation. Vol. 64, pp. 31-36. ISSN:0300-9572

Zupan, J. & Gasteiger, J. (1993). Neural Networks for chemists. VCH, ISBN:1560817917, Weinheim, Germany

Zupan, J.; Novic, M.; Li, X. & Gasteiger, J. (1994). Classification of multicomponent analytical data of olive oils using different neural networks. Analytica Chimica Acta. Vol. 292, pp. 219-234. ISSN:0003-2670

Analysis of Chemical Processes, Determination of the Reaction Mechanism and Fitting of Equilibrium and Rate Constants

Marcel Maeder and Peter King

Department of Chemistry, University of Newcastle, Australia
Jplus Consulting Ltd, Perth,
Australia

1. Introduction

This chapter is intended to demonstrate some recent approaches to the quantitative determination of chemical processes based on the quantitative analysis of experimental spectrophotometric measurements. In this chapter we will discuss kinetic processes, equilibrium processes and also processes that include a combination of kinetic and equilibrium steps.

We also emphasise the advantage of 'global' multivariate (multiwavelength) data analysis which has the advantage of allowing the robust determination of more complex mechanisms than single wavelength analysis and also has the benefit of yielding the spectra of all the participating species.

Rather than dwell on the mathematical derivation of the complex numerical algorithms and a repetition of the fundamentals of non-linear regression methods and least squares fitting which are available from a wide variety of sources (Martell and Motekaitis 1988; Polster and Lachmann 1989; Gans 1992; Press, Vetterling et al. 1995; Maeder and Neuhold 2007), we aim to show the experimentalist how to obtain the results they are interested, using purpose designed global analysis software and a variety of worked examples. We will be using ReactLab, a suite of versatile and powerful reaction modelling and analysis tools developed by the authors. Other academic and commercial applications exist for multivariate and related types of analysis and the reader is encouraged to explore these for comparative purposes. All offer different features and benefits but will not be discussed here.

2. Spectrophotometry, the ideal technique for process analysis

Any spectroscopic technique is ideal for the analysis of chemical processes as there is no interference in the underlying chemistry by the measurement technique. This is in sharp contrast to say chromatographic analysis or other separation methods which are totally unsuitable for the analysis of dynamic equilibrium systems. Such methods are also of very limited use for kinetic studies which often are too fast on the chromatographic time scale of typically tens of minutes to hours (except where reactions are first quenched and the intermediates stabilised). In contrast most forms of spectroscopy provide a completely non-invasive snapshot of a sample's composition at a single instant.

Amongst the different spectroscopies routinely available to the chemist, light absorption spectrophotometry in the UV-Visible (UV/Vis) is most common for several reasons: instruments are relatively inexpensive and accurate, they provide stable referenced signals as they are usually split or double beam instruments, there is a simple relationship between concentration and the measured absorbance signal (Beer-Lambert's law) and many compounds absorb somewhere in the accessible wavelength region. As a consequence there is a considerable amount of software available for the analysis of spectrophotometric data. This is the case both for kinetic and equilibrium investigations. NMR spectroscopy is a powerful method for structural investigations but it is less commonly used for quantitative analytical purposes. A theoretically very powerful alternative to UV/Vis absorption spectroscopy is FT-IR spectroscopy. The richness of IR spectra is very attractive as there is much more information contained in an IR spectrum compared with a relatively structureless UV/Vis spectrum. The main disadvantage is the lack of long term stability as FT-IR instruments are single beam instruments. Other difficulties include solvent absorption and the lack of non-absorbing cell materials, particularly for aqueous solutions. However, attenuated total reflection or ATR is a promising novel measurement technique in the IR. Near-IR spectroscopy is very similar to UV/Vis spectroscopy and is covered by the present discussions.

Of course fluorescence detection is a very sensitive and important tool particularly in kinetics studies and can yield important mechanistic information where intermediates do not possess chromophores and are therefore colourless or are studied at very low concentrations. In the main fluorescence studies are carried out at a single emission wavelength or adopting the total fluorescence approach (using cut-off filters), so there is no wavelength discrimination in the data. Whilst this type of measurement can be analysed by the methods described below and is essentially equivalent to analysing single wavelength absorption data. We will in the following discussion concentrate on the general case of processing multiwavelength measurements

3. The experiment, structure of the data

For kinetic investigations the absorption of the reacting solution is measured as a function of reaction time; for equilibrium investigations the absorption is recorded as a function of the reagent addition or another independent variable such as pH. Absorption readings can of course be taken at a single wavelength but with modern instrumentations it is routine and advantageous to record complete spectra vs. time or reagent addition. This is particularly prevalent with the use of photodiode array (PDA) based spectrophotometers and online detectors.

In the case of kinetics, depending on the rate of a chemical reaction the mixing of the reagents that undergo the reaction has to be done fast using a stopped-flow instrument or it can be done manually for slower reactions in the cuvette of a standard UV-Vis spectrometer with suitably triggered spectral data acquisition. A series of spectra are collected at time intervals following the mixing event to cover the reaction time of interest. The measured spectra change as the reaction proceeds from reagents to products (Wilkins 1991; Espenson 1995).

For equilibrium investigations the spectra of a series of pre-mixed and equilibrated solutions have to be recorded (Martell and Motekaitis 1988; Polster and Lachmann 1989). This is most commonly done as a titration where small amounts of a reagent are added stepwise to the solution under investigation. Titrations can be done in the cuvette, requiring internal stirring after each addition, prior to the absorption measurement, or the solutions can be mixed externally with transfer of the equilibrated solutions into the cuvette performed manually or using a flow cell and automatic pumping. In an alternative configuration optical probes can be coupled to the optical path in some spectrometers and placed into the solution contained in an external titration vessel (Norman and Maeder 2006). Often the pHs of the equilibrated titration solutions are recorded together with the absorption spectra where protonation equilibria are a feature of the mechanism.

For both kinetic and equilibrium investigations the measurement data can be arranged in a data matrix D which contains row-wise the recorded spectra as a function of time or reagent addition. The number of columns of D is the number of wavelengths, $nlam$, over which the spectra are taken. For single wavelength data the matrix reduces to a single column (vector). The number of rows, $nspectra$, corresponds to the number of spectra recorded during the process (one at each time interval for kinetics or reagent addition for an equilibrium titration). The dimensions of D thus are $nspectra \times nlam$. For spectra taken on a mechanical scanning instrument, the number of wavelengths can be 1 to typically some 10 or 20 but for diode array instruments it can easily be in excess of 1000 depending on the solid state detector pixel resolution (typically these provide a resolution progression of 256, 512 and 1024 pixels). The number of spectra taken is typically much larger on a stopped-flow instrument equipped with a fast diode array detector with a typical minimum spectrum acquisition time of the order of a millisecond. Frequently a logarithmic time base is an option which enables both fast and slower events to be resolved in a single kinetic experiment. A graphical representation of a date matrix D is given in Figure 1.

wavelength

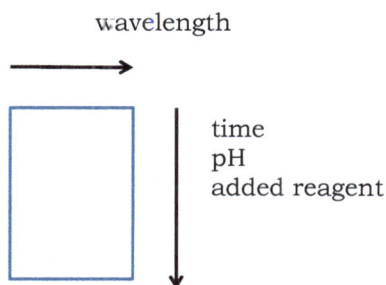

time
pH
added reagent

Fig. 1. Graphical representation of a data matrix D, the spectra are arranged as the rows.

For both kinetic and equilibrium investigations, we obtain a series of spectra each of which represent the solution at one particular moment during the process. The spectra are taken as a function of time or reagent addition.

4. Information to be gained from the measurements

The purpose of collecting this type of data is to determine the chemical reaction mechanism that describes the underlying process in terms of identifiable steps together with the associated key parameters; the rates and/or equilibrium constants which define the interconversions and stabilities of the various species. This may initially be a purely academic exercise to characterise a novel chemical reaction for publication purposes but ultimately defines the behaviour of the participating species for any future research into this or related chemistry as well as being the foundation for commercially important applications e.g. drug binding interactions in pharmaceutical development or reaction optimisation in industrial processes.

The objective is therefore to find the chemical model which best fits the data and validate and refine this model with subsequent experiments under other conditions. The clear benefit of multi-wavelength measurements is that the model must satisfy (fit) the data at all measurement wavelengths simultaneously and this significantly helps the accurate determination of multiple parameters and also allows determination of the individual spectra of the participating species.

5. Beer-Lambert's law

Before we can start the possible ways of extracting the useful parameters from the measured data set, the rate constants in the case of kinetics, the equilibrium constants in the case of equilibria, we need to further investigate the structure of the data matrix D. According to Beer-Lambert's law for multicomponent systems, the total absorption at any particular wavelength is the sum over all individual contributions of all absorbing species at this wavelength. It is best to write this as an equation:

$$D(i,j) = \sum_{k=1}^{ncomp} C(i,k) \times A(k,j) \qquad (1)$$

where:
$D(i,j)$: absorption of the i-th solution at wavelength j
$C(i,k)$; concentration of the k-th component in the i-th solution
$A(k,j)$: molar absorptivity of the k-th component at the j-th wavelength
$ncomp$: number of components in the system under investigation.

Thus equation (1) represents a system of $i \times j$ equations with many unknowns, i.e. all elements of C ($nspectra \times ncomp$) and all elements of A ($ncomp \times nlam$).

It is extremely useful to realise that the structure of Beer-Lambert's law allows the writing of Equation (1) in a very elegant matrix notion, Equation (2) and Figure 2

$$D = C \times A \qquad (2)$$

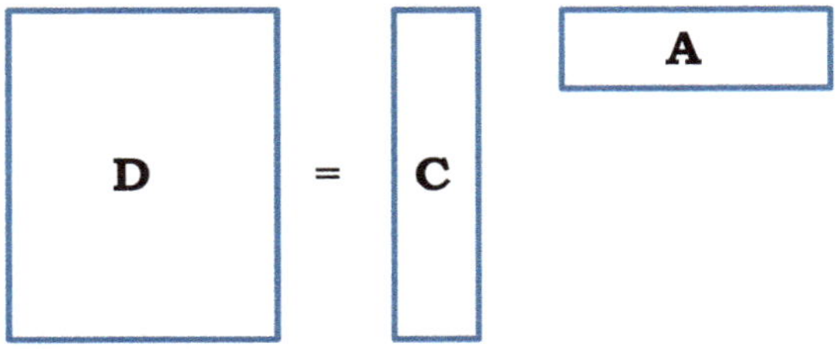

Fig. 2. Beer-Lambert's law, Equation (1) in matrix notation.

The matrix D(*nspectra×nlam*) is the product of a matrix of concentrations C(*nspectra×ncomp*) and a matrix A(*ncomp×nlam*). C contains as columns the concentration profiles of the reacting components and the matrix A contains, as rows, their molar absorption spectra.

Equations (1) and (2) and Figure 2 represent the ideal case of perfect absorption readings without any experimental noise. This of course is not realistic and both equations have to be augmented by an error term, $R(i,j)$ which is the difference between the ideal value and its measured counterpart, equation (3) and Figure 3.

$$D(i,j) = \sum_{k=1}^{ncomp} C(i,k) \times A(k,j) + R(i,j)$$

(3)

$$D = C \times A + R$$

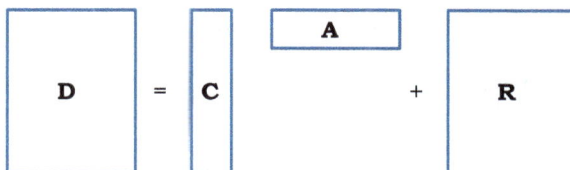

Fig. 3. Beer-Lambert's law including the residuals.

The goal of the fitting is to determine that set of matrices C and A for which the sum over all the squares of the residuals, *ssq*, is minimal,

$$ssq = \sum_{i=1}^{nspectra} \sum_{j=1}^{nlam} R(i,j)$$

(4)

At first sight this looks like a very daunting task. However, as we will see, it is manageable.

Ideally the final square sum achieved should be numerically equal to the sum of the squares of the Gaussian noise in the measurement – usually instrumental in origin. At this point the fit cannot be further improved, though this is not a guarantee that the model is the correct one.

6. The chemical model

The first and central step of any data analysis is the computation of the matrix C of concentration profiles based on the proposed chemical model and the associated parameters such as, but not exclusively, the rate and or equilibrium constants. Initially these key parameters may be only rough estimates of the true values.

So far the explanations are valid for kinetic and equilibrium studies. The difference between these two investigation lies in the different computations required for the calculation of the concentration profiles in the matrix C.

7. Kinetics

The chemical model for a kinetic investigation is a set of reaction equations which describe the process under investigation. Consider as an example the basic enzymatic reaction scheme

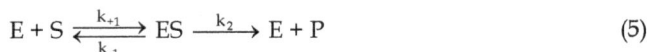

$$E + S \underset{k_{-1}}{\overset{k_{+1}}{\rightleftharpoons}} ES \xrightarrow{k_2} E + P \tag{5}$$

An enzyme E reacts rapidly and reversibly with the substrate S to form an enzyme substrate complex ES. This is followed by the 1st order chemical conversion of the substrate and release of product. The free enzyme is then available to undergo another catalytic cycle.

Before proceeding to a ReactLab based mechanistic analysis it is informative to briefly outline the classical approach to the quantitative analysis of this and similar basic enzyme mechanisms. The reader is referred to the many kinetics textbooks available for a more detailed description of these methods. The scheme in equation (5) was proposed by Michaelis and Menten in 1913 to aid in the interpretation of kinetic behaviour of enzyme-substrate reactions (Menten and Michaelis 1913). This model of the catalytic process was the basis for an analysis of measured initial rates (v) as a function of initial substrate concentration in order to determine the constants K_M (The Michaelis constant) and V_{max} that characterise the reaction. At low [S], v increases linearly, but as [S] increases the rise in v slows and ultimately reaches a limiting value V_{max}.

Analysis was based on the derived Michaelis Menten formula:

$$v = \frac{[E_0] \, [S] \, k_{cat}}{K_M + [S]} \tag{6}$$

Where $V_{max} = k_{cat}[E]_0$, and K_M is equal to the substrate concentration at which $v = \frac{1}{2} V_{max}$. The key to this derivation is that the enzyme substrate complex ES is in dynamic equilibrium with free E and S and the catalytic step proceeds with a first order rate constant k_{cat}. This 'turnover number' k_{cat} is represented by k_2 in the scheme in equation (5).

It can be shown that under conditions where $k_2 << k_{-1}$ then K_M is in fact equal to the equilibrium dissociation constant K_1,

$$K_1 = \frac{[ES]}{[E] \, [S]} = \frac{k_{+1}}{k_{-1}} \tag{7}$$

Importantly, however, the parameter K_M is not always equivalent to this fundamental equilibrium constant K_1 when this constraint ($k_2 << k_{-1}$) doesn't apply.

Furthermore though the Michealis Menten scheme can be extended to cover more complex mechanisms with additional intermediates, the K_M and k_{cat} parameters now become even more complex combinations of individual rate and equilibrium constants. The k_{cat} and K_M parameters determined by these classical approaches are therefore not the fundamental constants defining the mechanism and significant effort is required to determine the true underlying equilibrium and microscopic rate constants.

In contrast direct analysis using ReactLab to fit the core mechanism to suitable data delivers the true rate and equilibrium constants in a wholly generic way that can be applied without assumptions and also to more complex models.

This involves the modelling of the entire mechanism to deliver the matrix **C** comprising the concentration profiles of all the participating species. The reaction scheme in equation (5) defines a set of ordinary differential equations, ODE's, which need to be solved or integrated (Maeder and Neuhold 2007). Reaction schemes that only consist of first order reactions can be integrated analytically in which case the concentration can be calculated directly at any point using the resulting explicit function. Most other schemes, containing one or more second order reactions, require numerical integration. Numerical integration is usually done with variable step-size Runge-Kutta algorithms, unless the system is 'stiff' (comprising both very fast and slow reactions) for which special stiff solvers, such as Gear and Bulirsch-Stoer algorithms are available (Press, Vetterling et al. 1995).

Integration, explicit or numerical, requires the knowledge of the initial conditions, in the case of kinetics the initial concentrations of all interacting species. For the above example, equation

$$E + S \underset{k_{-1}}{\overset{k_{+1}}{\rightleftharpoons}} ES \xrightarrow{k_2} E + P \tag{5},$$

and using the rate constants (k_{+1}=10^3 M^{-1} sec^{-1}, k_{-1}=10^2 sec^{-1}, k_2=10^2 sec^{-1}) and initial concentrations, ([S]$_0$=1 M, [E]$_0$=10^{-4} M), the resulting concentration profiles generated by numerical integration and used to populate the columns of matrix **C** are shown in Figure 4.

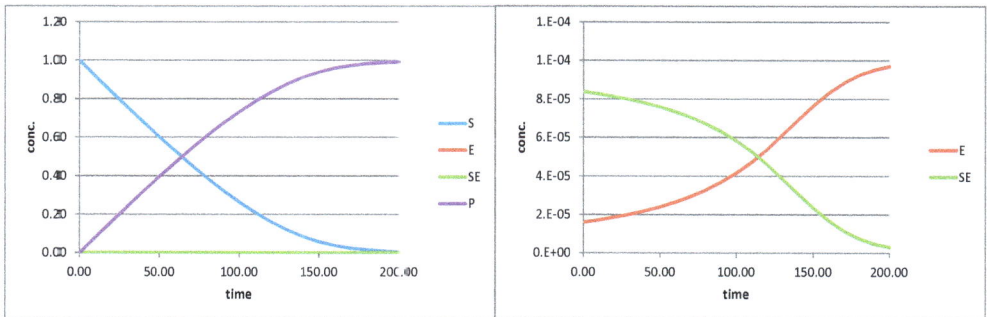

Fig. 4. Concentration profiles for the enzymatic reaction of equation (5); an expanded concentration axis is used in the right panel.

The transformation of the substrate into the product follows approximately zero-th order kinetics for most of the reaction whilst the substrate is in excess and all the enzyme sites are populated. Later in the reaction the substrate is exhausted and free enzyme released. The expanded plot in the right hand panel displays more clearly the small concentrations for the enzyme and the enzyme-substrate complex.

8. Equilibria

The chemical model for an equilibrium process is similar to the model of a kinetic process, only now there are exclusively equilibrium interactions, e.g.

$$A + H \xrightleftharpoons{K_{AH}} AH$$
$$AH + H \xrightleftharpoons{K_{AH_2}} AH_2 \tag{8}$$
$$B + H \xrightleftharpoons{K_{BH}} BH$$

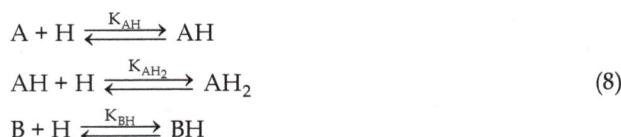

The chemistry in this example comprises the protonation equilibria of the di-protic acid AH_2 and the mono-protic acid BH. The key difference now is that the steps are defined in terms of instantaneous stability or equilibrium constants, and the fast processes of the attainment of the equilibria are not observed.

Equilibrium investigations require a titration, which consists of the preparation of a series of solutions with different but known total component concentrations. In equilibrium studies we distinguish between components and species. Components are the building blocks; in the example (6) they are A, B and H; species are all the different molecules that are formed from the components during the titration, the example they are A, AH, AH_2, B, BH, H and OH. Note, the components are also species.

Instead of utilising numerical integration to compute the concentration profiles of the individual species as we did with kinetic time courses we instead use an iterative Newton-Raphson algorithm to determine the speciation based on the total component concentrations for each sample and the estimated equilibrium constants (Maeder and Neuhold 2007).

For a titration of 10ml of a solution with total component concentrations $[A]_{tot}= 0.1M$, $[B]_{tot}=0.06$ M and $[H]_{tot}=0.4M$ with 5ml of 1.0M $NaOH$ the concentration profiles of Figure 5 result. The protonation constants are $\log(K_{AH})=9$, $\log(K_{AH2})=3$ and $\log(K_{BH})=4$.

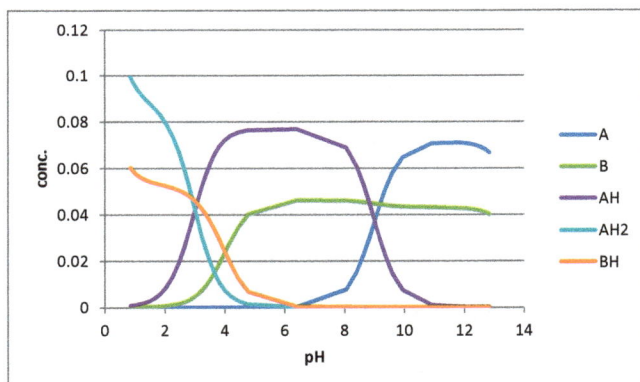

Fig. 5. Concentration profiles for the titration of an acidified solution of AH_2 and BH with $NaOH$.

8.1 Kinetics with coupled protonation equilibria

A significant recent development is the incorporation of instantaneous equilibria to kinetic analyses. Careful combination of numerical integration computations alongside the Newton-Raphson speciation calculations have made this possible (Maeder, Neuhold et al. 2002). This development has made the modelling of significantly more complex and realistic mechanisms possible. An example is the complex formation between ammonia and Ni^{2+} in aqueous solution as represented in equation (9).

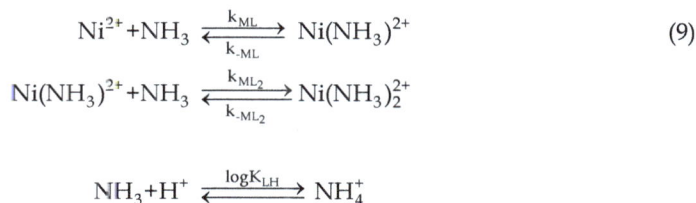

$$Ni^{2+}+NH_3 \xrightleftharpoons[k_{-ML}]{k_{ML}} Ni(NH_3)^{2+} \qquad (9)$$

$$Ni(NH_3)^{2+}+NH_3 \xrightleftharpoons[k_{-ML_2}]{k_{ML_2}} Ni(NH_3)_2^{2+}$$

$$NH_3+H^+ \xrightleftharpoons{logK_{LH}} NH_4^+$$

Ni^{2+} is interacting with NH_3 to form the 1:1 and subsequently the 1:2 complexes. Importantly the ammonia is also involved in a protonation equilibrium. As a result the pH changes during the reaction and the rates of the complex formation reactions appear to change. The classical approach to this situation is to add buffers that approximately maintain constant pH and thus also the protonation equilibrium. Since buffers often interfere with the process under investigation the possibility of avoiding them is advantageous. This has only been made possible by this more sophisticated method of mechanistic analysis.

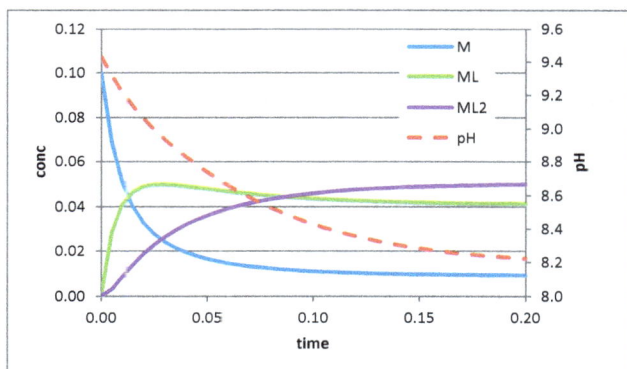

Fig. 6. The concentration profiles for the complex species in the reaction of Ni^{2+} and NH_3; also displayed is the pH of the reacting solution.

The concentration profiles for the complex species are shown in Figure 6. The patterns for a 'normal' consecutive reaction are distorted as the reaction slows down with the drop in pH from 9.5 to 8.2. The initial concentrations for this reaction are $[Ni^{2+}]_0$=0.1 M, $[NH_3]_0$=0.25 M and $[H^+]_0$=0.1 M.

9. Parameters

Any parameter that is used to calculate the matrix C of concentrations is potentially a parameter that can be fitted. Obvious parameters are the rate constants k and the equilibrium constants K; other less obvious parameters are the initial concentrations in kinetics or the total concentrations in equilibrium titrations. Concentration determinations are of course very common in equilibrium studies (quantitative titrations); for several reasons concentrations are not often fitted in kinetic studies. For first order reactions the concentrations are not defined at all unless there is additional spectroscopic information, i.e. molar absorptivities. For second order reactions they are in principle defined but only very poorly and thus kinetic determination is not a robust analytical technique in this case.

The parameters defining C are non-linear parameters and cannot be fitted explicitly, they need to be computed iteratively. Estimates are provided, a matrix C constructed and this is compared to the measurement according to the steps that follow below. Once this is complete it is possible to calculate shifts in these parameter estimates in a way that will improve the fit (i.e. reduce the square sum) when a new C is computed. This iterative improvement of the non-linear parameters is the basis of the non-linear regression algorithm at the heart of most fitting programs.

10. Calculation of the absorption spectra

The relationship between the matrix C and the measurement is based on equation (3). The matrix A contains the molar absorptivity for each species at each measured wavelength. All these molar absorptivities are unknown and thus also parameters to be determined. When spectra are collected the number of these parameters can be very large, but fortunately they are linear parameters and can be dealt with differently to the non-linear parameters discussed above.

Once the concentration profiles have been calculated, the matrix A of absorption spectra is computed. This is a linear least-squares calculation with an explicit solution

$$A = C^+ D \tag{10}$$

C^+ is the pseudo-inverse of the concentration matrix C, it can be calculated as $(C^t C)^{-1} C^t$, or better using a numerically superior algorithm (Press, Vetterling et al. 1995).

11. Non-linear regression: fitting of the non-linear parameters

Fitting of the parameters requires the software to systematically vary all non-linear parameters, the rate and equilibrium constants as well as others such as initial concentrations, with the aim of minimising the sum of squares over all residuals, as defined in equation (4).

There are several algorithms for that task, the simplex algorithm which is relatively easy to program and features robust convergence with a high price of slow computation times particularly for the fitting of many parameters. Significantly faster is the Newton-Gauss algorithm; additionally it delivers error estimates for the parameters and with implementation of the Marquardt algorithm it is also very robust (Gans 1992; Maeder and Neuhold 2007).

As mentioned earlier non-linear regression is an iterative process and, provided the initial parameter estimates are not too poor and the model is not under-determined by the data, will converge to a unique minimum yielding the best fit parameters. With more complex models it is often necessary to fix certain parameters (either rate constants, equilibrium constants or complete spectra) particularly if they are known through independent investigations and most fitting applications will allow this type of constraint to be applied.

12. ReactLab analysis tools

ReactLab™ (Jplus Consulting Ltd) is a suite of software which is designed to carry out the fitting of reaction models to either kinetic or equilibrium multiwavelength (or single wavelength) data sets. All the core calculations described above are handled internally and the user simply provides the experimental measurements and a reaction scheme that is to be fitted to the data. A range of relevant supporting options are available as well as comprehensive graphical tools for visualising the data and the result of the analysis. To facilitate this all data, models and results are provided in pre-formatted Excel workbooks to allow post processing of results or customised plots to be added after the main analysis is complete.

13. Representing the chemical model

As has been discussed, at the root of the analysis is the generation of the species concentration matrix **C**.

Fitting a proposed reaction mechanism, or part of it, to the data therefore requires the determination of **C** from a reaction scheme preferably as would be written by a chemist. The ReactLab representation of the Ni^{2+}/NH_3 complexation mechanism of Equation (9) is shown in Figure 7. Forward arrows, >, are used to represent rate constants and the equal sign, =, represents an instantaneous equilibrium, e.g. a protonation equilibrium.

Reactants	Reaction Type	Products	Label	Parameters k / log K	±	Fit ☑
M+L	>	ML	k_ML	5.845E+02		☑
ML	>	M+L	k_-ML	1.114E+00		☐
ML+L	>	ML2	k_ML2	2.965E+02		☑
ML2	>	ML +L	k_-ML2	2.051E+00		☐
L+H	=	LH	logK_LH	9.250E+00		☐

Fig. 7. The ReactLab definition of the mechanism in Equation (7) with rate and equilibrium constants used to compute the concentration profiles of Figure 6.

To get from this scheme to our intermediate matrix **C** involves a number of key computational steps requiring firstly the dissection of the mechanism into its fundamental mathematical building blocks. Many analysis tools require the user to take this initial step manually, and therefore understand some fairly sophisticated underlying mathematical principles. Whilst this is no bad thing it does complicate the overall process and provide a significant barrier to trying and becoming familiar with this type of direct data fitting using numerical integration based algorithms. A significant advance has been the development of model editors and translators which carry out this process transparently. These can be found in a variety of data fitting applications including ReactLab.

14. Examples

In the following we will guide the reader through the steps required for the successful analysis of a number of example data sets. This section consists of two examples each from kinetics and equilibrium studies.

Example 1: Consecutive reaction scheme $A \xrightarrow{k_1} B \xrightarrow{k_2} C$

The data set comprises a collection spectra measured as a function of time. These are arranged as rows of the 'Data' worksheet in Figure 8, the first spectrum in the cells D6:X6, the second in D7:X7, and so on. For each spectrum the measurement time is required and these times are collected in column C. The vector of wavelengths at which the spectra were acquired is stored in the row 5 above the spectra. The two inserted figures display the data as a function of time and of wavelength.

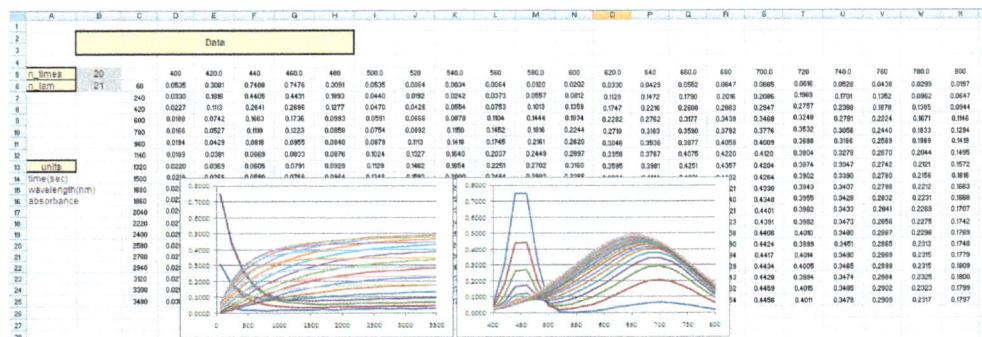

Fig. 8. The data arranged in an excel spreadsheet; the figure on the left displays the kinetic traces at all wavelengths, the figure on the right displays the measured spectra.

Prior to the fitting, the chemical reaction model on which the analysis will be based needs to be defined. As mentioned above ReactLab and other modern programs incorporate a model translator that allows the definition in a natural chemistry language and which subsequently translates automatically into internal coefficient information that allows the automatic construction of the mathematical expressions required by the numerical and speciation algorithms. Note for each reaction an initial guess for the rate constant has to be supplied. The ReactLab model is for this reaction is shown in Figure 9.

Reactants	Reaction Type	Products	Label	Parameters k / log K
A	>	B	k1	1.000E-02
B	>	C	k2	1.000E-04

Fig. 9. The definition of the chemical model for the consecutive reaction scheme $A \xrightarrow{k_1} B \xrightarrow{k_2} C$

The 'compiler' recognises that there are 3 reacting species, A, B, C, and 2 rate constants. For the initial concentrations the appropriate values have to be supplied by the user. In the example $[A]_{init}=0.001$ M $[B]_{init}$ and $[C]_{init}$ are zero. Further the spectral status of each species needs to be defined, in the example all 3 species are 'colored' i.e. they do absorb in the wavelength range of the data, see Figure 10. The alternative 'non-absorbing' indicates that the species does not absorb in the measured range. Advanced packages including ReactLab also allow the implementation of 'known' spectra which need to be introduced elsewhere in the workbook.

n_species	3
n_par	2
n_aux_par	0

Species	A	B	C
init []	1.00E-03	0.00E+00	0.00E+00
Spectrum	colored	colored	colored

Fig. 10. For the consecutive reaction scheme there are 3 reaction species for which initial concentrations need to be given.

The program is now in a position to first calculate the concentration of all species as a function of time and subsequently their absorption spectra. The results for the present initial guesses for the rate constants are displayed in Figure 11.

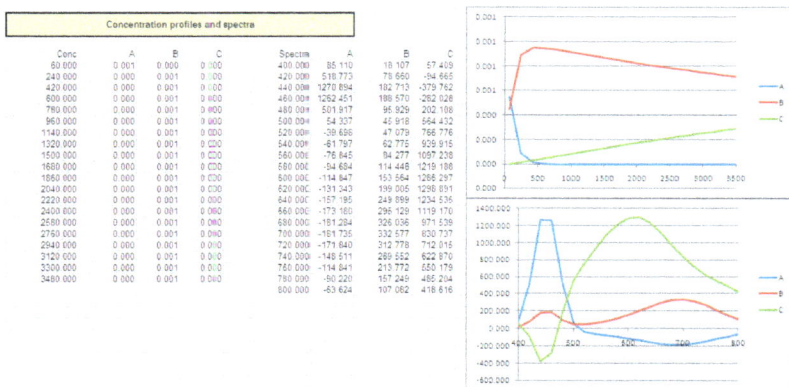

Fig. 11. The concentration profiles and absorption spectra, as calculated with initial guesses for the rate constants shown in Figure 9.

The concentration profiles indicate a fast first reaction $A>B$ and a much slower subsequent reaction $B>C$. However, the calculated, partially negative absorption spectra clearly demonstrate that there is 'something wrong', that the initial guesses for the rate constants are obviously not correct. In this example the deviations are not too severe indicating the model itself is plausible.

Clicking the Fit button initiates the iterative improvement of the parameters and after a few iterations the 'perfect' results are evident. This of course is supportive of the validity of the model itself. If the scheme is wrong and cannot account for the detail in the data, a good fit will be unobtainable. The ReactLab GUI at the end of the fit is given in Figure 12.

On the other hand an over complex model has to be carefully avoided as any data can usually be fitted with enough parameters (including artefacts!). Occam's razor should be assiduously applied accepting the simplest model that fits the data as the most likely.

Of course it is also a risk that a model is underdetermined by the data. Put simply the information in the measurement is not sufficient to deliver a unique solution, and the program will not converge properly and usually oscillate delivering one of an infinite number of solutions (usually combinations of rates). Whilst this does not imply the model is

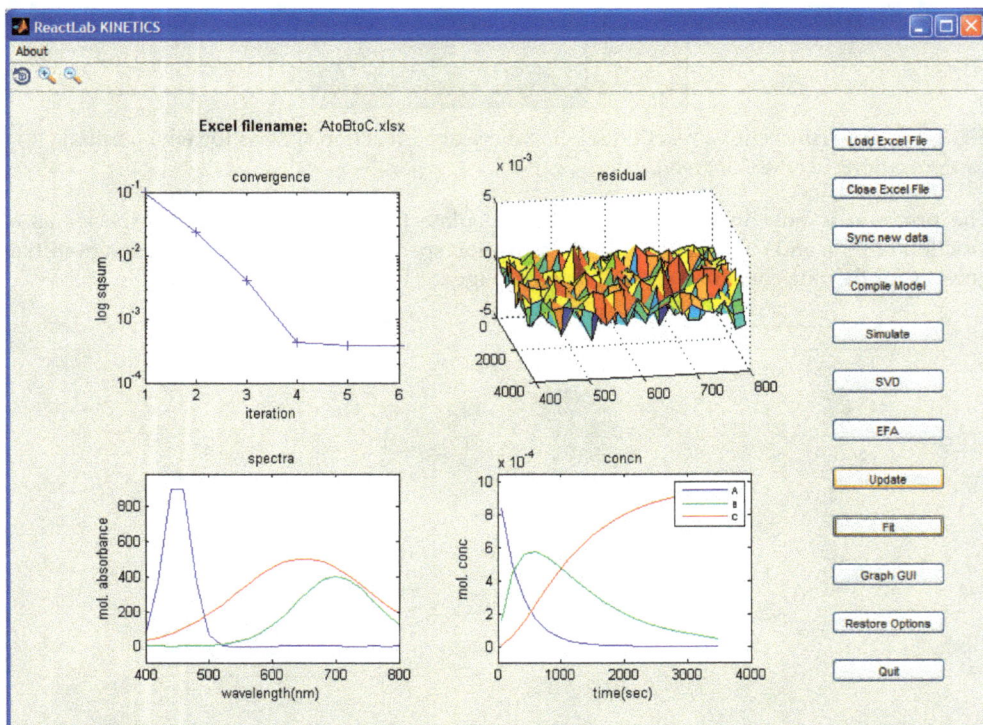

Fig. 12. The ReactLab GUI displaying the progress of *ssq*, the residuals, the absorption spectra and the concentration profiles after the fitting.

incorrect, further work will be required to determine and fix key parameters or spectra in order to resolve the problem.

Example 2: Kinetic analysis of the reaction of Ellmans reagent (*DTNB*) and thioglycerol *RS*.

This example illustrates the direct fitting of a simplified model followed by the correct and more complex model to a data set collected using a PDA on a stopped flow (data courtesy of TgK Scientific Ltd, UK).

Ellmans reagent, 5,5'-Dithio-bis(2-nitrobenzoic acid) or *DTNB* is a commercially available reagent for quantifying thiols both in pure and biological samples and measuring the number of thiol groups on proteins. The reaction yields a colored thiolate, *RS-TNB*, ion which absorbs at 412nm and can be used to quantify the original thiol concentration. In this particular case the reaction with thioglycerol, *RS-*, leads to a 2 step disulphide exchange reaction and is particularly suited for establishing the dead-time of stopped flow instruments (Paul, Kirschner et al. 1979). The reaction is represented in Figure 13. The model in the ReactLab definition is given in Figure 16.

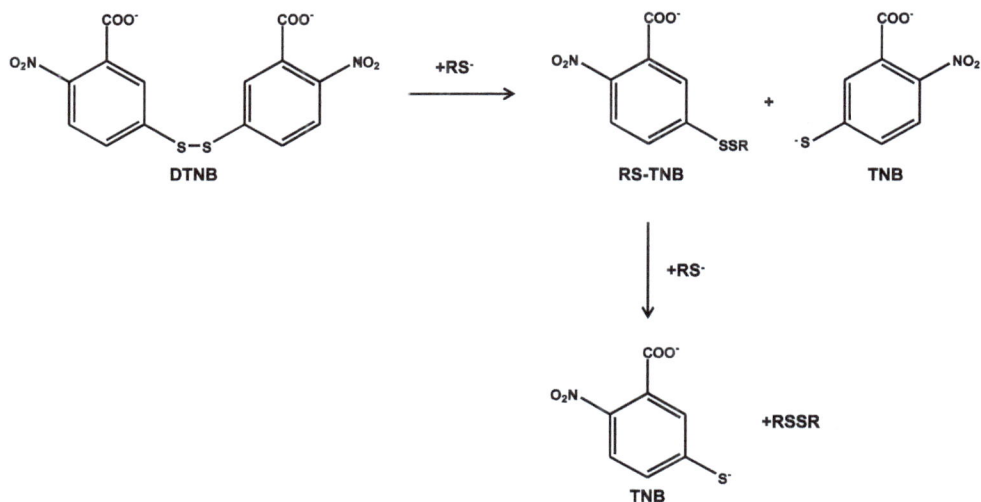

Fig. 13. The 2-step reaction of *DTNB* with a thiolate, *RS-*.

A 3-D representation of the spectra measured at different time intervals for a total of 1.5 sec. on a stopped-flow instrument is shown in Figure 14.

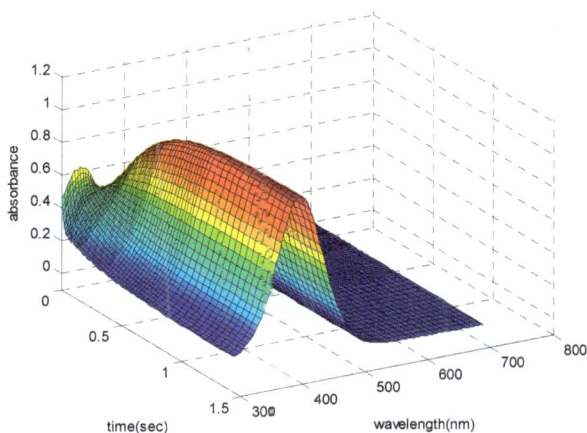

Fig. 14. Spectral changes resulting from the reactions of Ellmans reagent (*DTNB*) and thioglycerol (*RS-*).

In the experiment a large excess of thioglycerol was used and thus the two second order reactions can be approximated with a two-step pseudo first order sequential mechanism. Thus, we first attempt to fit this biphasic reaction with a simple consecutive reaction scheme with three colored species $A \xrightarrow{k_1} B \xrightarrow{k_2} C$ (Figure 15). The fitted rates are 75sec^{-1} and 3.5sec^{-1}.

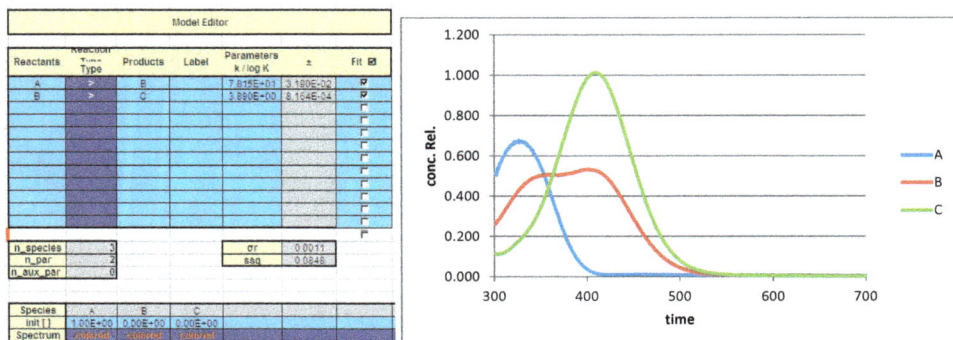

Fig. 15. A simple consecutive reaction model with best fit rates and spectra.

The problems with this fit are two-fold. First we know the reactions are second order and that the intermediate spectrum for species B is implausible with two peaks.

The complete model as defined in ReactLab, together with the fit at 412 nm, is displayed in Figure 16. Whilst the square sum is not significantly improved the spectra are now correct according to literature sources and the corresponding rates for the 2 steps are $1.57 \times 10^4 \pm 6$ $M^{-1}sec^{-1}$ and 780 ± 2 $M^{-1}sec^{-1}$.

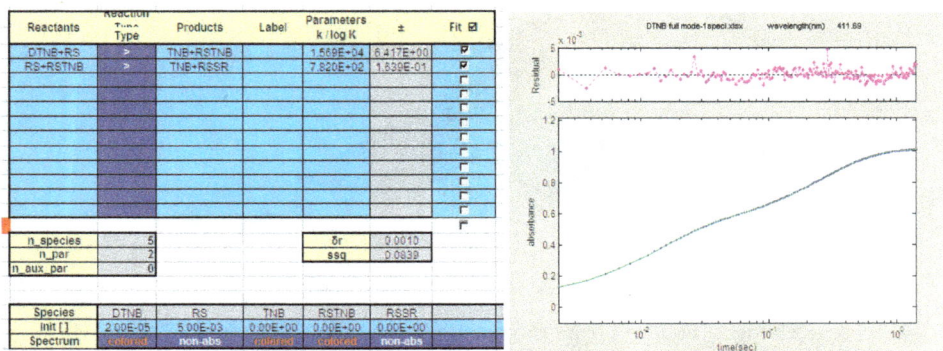

Fig. 16. Fitting the complete mechanism to the data yields the correct rates. The right panel displays the quality of the fit at 412 nm on a logarithmic time axis.

The concentration profiles and spectra resulting from a fit to the correct model are now as in Figure 17.

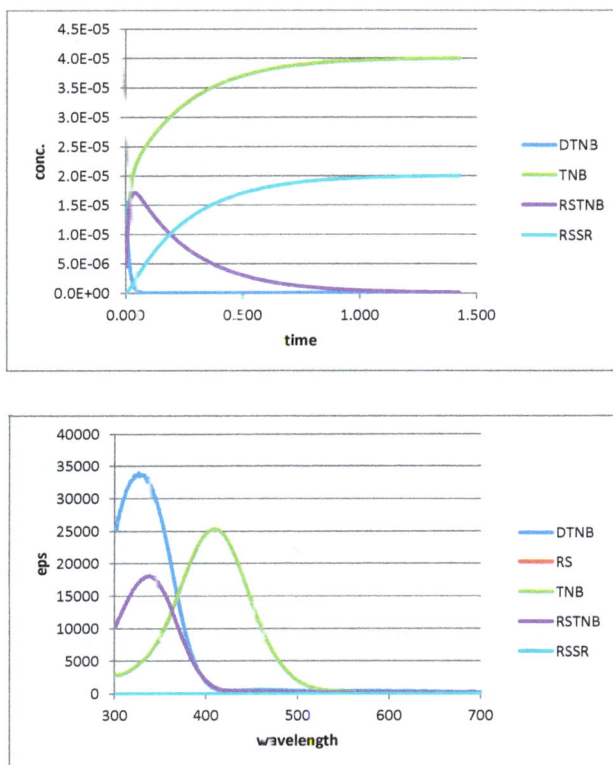

Fig. 17. Concentration profiles and molar absorption spectra for the analysis based on the complete reaction scheme.

This example does serve to demonstrate that good fits can be obtained with an incorrect or simplistic model and that some insight and care is required to establish the correct mechanism and obtain genuine parameters. What is certainly true is that the second model could only be fitted because of the numerical integration of the more complex second order mechanisms. This was a trivial change of model configuration in ReactLab and could not have been achieved using classical analysis approaches. Secondly the importance of dealing with whole spectra is highlighted in that the spectra resulting for the fit provide important insight into the underlying chemistry and must make sense in this respect. Single wavelength kinetic analysis has no such indirect reinforcement.

By way of a final comment on this example; we noted that the data was collected under pseudo first order conditions i.e. one reagent in excess. This ubiquitous approach was essential to enable the determination of second order rate constants using a first order fit by classical analysis using explicit functions (usually sums of exponentials). In the pseudo first order simplification a 2nd order rate constant is calculated from the observed pseudo first order rate constant.

Numerical integration methods eliminate the need for this constraint and therefore any requirement to work under pseudo first order conditions (or indeed the comparable constraint of keeping the reactant concentrations equal).

Example 3: Equilibrium investigation, concentration determination of a diprotic and a strong acid

Titrations can be used for the determination of equilibrium constants and insofar the analysis is very similar to a kinetic investigation. Titrations are also an important analytical tool for the determination of concentrations, in real terms this is probably the more common application.

Let us consider a titration of a solution of the diprotic acid AH_2 in the presence of an excess of a strong acid, e.g. HCl. The equilibrium model only includes the diprotic acid as the strong acid is always completely dissociated, see the ReactLab model in Fig. 18.

Reactants	Reaction Type	Products	Label	Parameters k / log K
A+H	=	AH		8.000E+00
AH+H	=	AH2		3.000E+00

Fig. 18. The model for a diprotic acid that undergoes two protonation equilibria.

The components are A and H, the species are A, AH, AH_2, H, OH. For the components the total concentrations have to be known in each solution during the titration. They are collected in the columns E and F of a spreadsheet. The measured spectra are collected from the cell N7 on, see Figure 19

	A	B	C	D	E	F	G	H	M	N	O	P	Q
1									Expand				
2				Data and Component concentrations									
3													
4													
5					Total component ☐					lam			
6	n_spectra	151	Vadd(ml)	Vtot(ml)	A	H				400.0	410.0	420.0	430.
7	n_lam	21	0.000	10.000	0.100	0.250				0.000	0.000	0.001	-0.00
8			0.100	10.100	0.099	0.244				0.000	0.001	0.000	0.00
9			0.200	10.200	0.098	0.239				0.001	0.001	-0.001	0.00
10			0.300	10.300	0.097	0.234				-0.001	-0.001	0.001	0.00
11			0.400	10.400	0.096	0.229				-0.001	0.000	0.000	-0.00
12			0.500	10.500	0.095	0.224				0.000	-0.001	0.000	0.00

Fig. 19. Data entry for a titration, the crucial total concentrations are stored in the columns E and F, the spectra to the right (incomplete in this Figure).

In the example 10ml of a solution containing A and H are titrated with 0.1 ml aliquots of base. The concentrations $[A]_{tot}$ and $[H]_{tot}$ are computed from the volumes and concentrations of the original solution in the 'beaker' and in the 'burette'. These concentrations are stored in the main sheet in the rows 37 and 38 as seen in Figure 20.

The definition of the total concentrations of A and H in the 'beaker' are defined in the cells C37, D37, the same component concentrations in the burette solution in the row below. Often these concentrations are known and then there is nothing to be added. In the example the component concentrations in the 'beaker' are to be fitted. This is achieved by defining

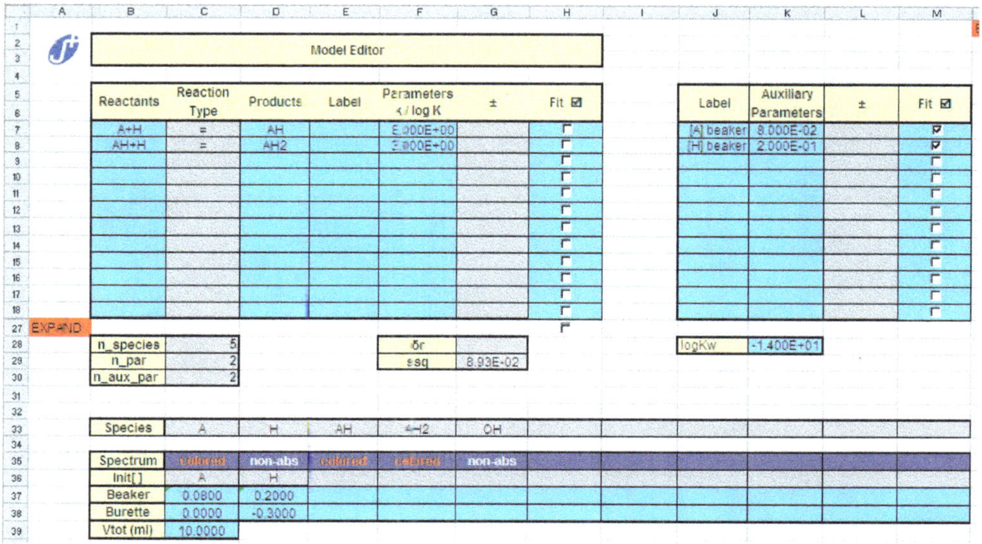

	Reactants	Reaction Type	Products	Label	Parameters K / log K	±	Fit ☑		Label	Auxiliary Parameters	±	Fit ☑
	A+H	=	AH		E.900E+00		☐		[A] beaker	8.000E-02		☑
	AH+H	=	AH2		3.900E+00		☐		[H] beaker	2.000E-01		☑
							☐					☐
							☐					☐
							☐					☐
							☐					☐
							☐					☐
							☐					☐
							☐					☐
							☐					☐
							☐					☐
							☐					☐

EXPAND

n_species	5			δr			logKw	-1.400E+01
n_par	2			ssq	8.93E-02			
n_aux_par	2							

Species	A	H	AH	AH2	OH				
Spectrum	colored	non-abs	colored	colored	non-abs				
Init[ ]	A	H							
Beaker	0.0800	0.2000							
Burette	0.0000	-0.3000							
Vtot (ml)	10.0000								

Fig. 20. Model entry and information on concentrations and spectral status.

them as auxiliary parameters in the cells K7, K8; the contents of cells C37, D37, are now references to the auxiliary parameters which are fitted.

Fitting results in values for the concentrations and their error estimates, Figure 21.

Label	Auxiliary Parameters	±	Fit ☑
[A] beaker	1.000E-01	2.126E-04	☑
[H] beaker	2.499E-01	2.152E-04	☑
			☐

Fig. 21. The result of the fitting of the concentrations, complete with error analysis.

Example 4: Equilibrium Interaction between Cu^{2+} and PHE (1,9-Bis(2-hydroxyphenyl)-2,5,8-triazanonane)

In this example we demonstrate the analysis of 'pH-metered' titrations, a mode of titration that is common in for the investigation of equilibria in aqueous solution. In this kind of titration the independent variable is the pH, rather than the added volume of reagent as has been the case in all previous examples. As before the measured spectra are the dependent variables. An important rationale for 'pH-metered' titrations is the fact that it is often difficult to completely exclude CO_2 during the whole titration. Unknown amounts of CO_2 result in the addition of unknown amounts of acid via formation of carbonate species. In 'pH-metered' titrations the effect of this impurity is minimal as long as none of the carbonate species interfere with the process under investigation; the effect in the 'default mode' can be much more pronounced. The price to pay for that advantage is more complex data acquisition as the pH has to be measured and recorded together with the spectra after each addition of reagent.

The example is the titration of *PHE*, 1,9-Bis(2-hydroxyphenyl)-2,5,8-triazanonane, with Cu^{2+} in aqueous solution.(Gampp, Haspra et al. 1984) The structure of the ligand is shown below. It forms several complexes: *ML*, where the ligand is penta-coordinated presumable via all three secondary amine groups as well as the deprotonated phenolates; and two partially protonated species *MLH* and *MLH2*, in these complexes one or both of the phenolates are protonated and most likely not or only very weakly coordinated.

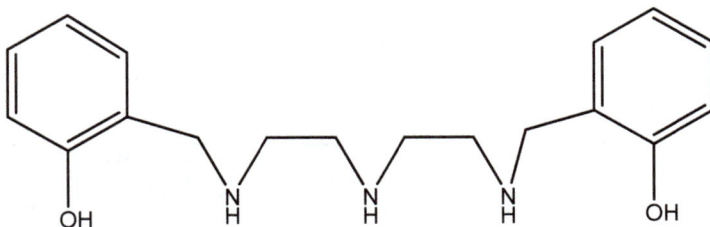

In this titration a solution of 7.23×10^{-4} M Cu^{2+} and 1.60×10^{-3} M *PHE* with an excess *HCl* were titrated with a total of approx. 750μL *NaOH* solution. After each addition of the base the pH and the spectrum were measured. The total concentrations of metal and ligand are entered for each sample in the '**Data**' worksheet. Note that the columns for the total concentration of the protons is left empty: the measured pH in column M is defining the free proton concentration which in turn is used to compute all species concentrations in conjunction with the total concentrations of in this case the metal ion and the ligand provided, see Figure 22.

Fig. 22. Only the total concentrations of the metal and ligand are required, the column for the protons is left empty. Column M contains the pH values and the entry 'pH' in cell M6 to indicate a 'pH-metered' titration.

The measurement is displayed in Figure 23, each curve is the measured absorption at one particular wavelength.

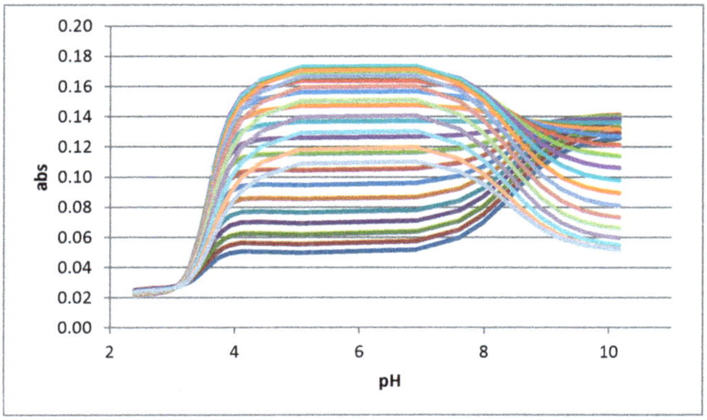

Fig. 23. The measurement, here displayed as a series of titration curves at the different wavelengths.

The ligand *PHE* has five protonation constants which have to be determined independently. The successive logK values are 10.06, 10.41, 9.09, 7.94 and 4.18. Note that in the model the protonations are defined as overall stabilities, see Figure 24.

The results of the analysis are summarised in Figure 24 and Figure 25. Again the protonation equilibria for the complex species are defined as formation constants, the logK values for the protonation equilibria $ML+H \rightleftharpoons MLH$ and $MLH+H \rightleftharpoons MLH_2$ are 8.42 and 3.92.

Reactants	Reaction Type	Products	Label	Parameters logK/logβ	±	Fit ☑
L+H	=	LH		11.060		☐
L+2H	=	LH2		21.470		☐
L+3H	=	LH3		30.560		☐
L+4H	=	LH4		38.500		☐
L+5H	=	LH5		42.680		☐
Cu+L	=	CuL		22.563	0.011	☑
Cu+L+H	=	CuLH		30.979	0.011	☑
Cu+L+2H	=	CuLH2		34.895	0.002	☑
						☐
						☐
						☐
						☐

n_species	12		or	4.63E-04
n_par	8		ssq	1.76E-04
n_aux_par	0			

Fig. 24. The fitted equilibrium constants for the formation of the *ML*, *MLH* and *MLH2* complexes.

The concentration profiles are represented in two different modes, the left part has the measured pH as the x-axis and only the metal species are shown, the right part shows all species concentrations as a function of the added volume of base. This figure reveals that a substantial excess of acid has been added to the initial solution and the first 0.2 mL of base are used to neutralise this excess

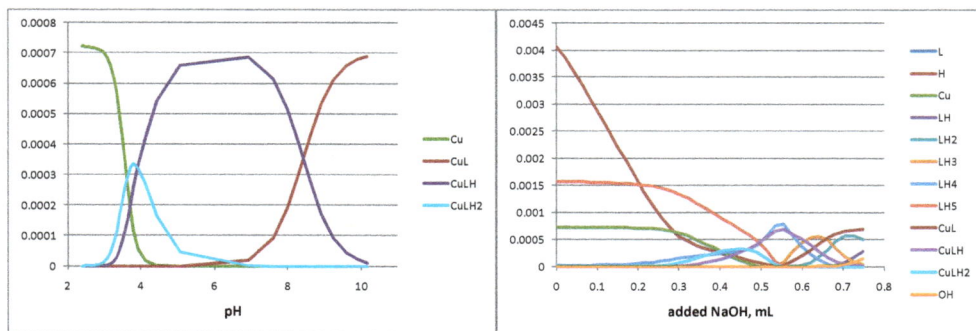

Fig. 25. The concentration profiles of the complexes as a function of pH and all species as a function of the volume of added base.

15. Conclusion

Data fitting is a well-established method that has been extensively used for the analysis of chemical processes since the beginnings of instrumental methods. Generally, increasing sophistication of instrumentation has inspired parallel developments in numerical methods for appropriate analysis of ever better and more plentiful data. This chapter concentrates on spectroscopic methods for the investigation of chemical processes; it details the structure of the data and the principles of model based data fitting. It is rounded of by a collection of typical and illustrative examples using a modern data analysis software package. The aim is to demonstrate the power and effortlessness of modern data analysis of typical data sets.

16. References

Espenson, J. H. (1995). *Chemical Kinetics and Reaction Mechanisms.* New York, McGraw-Hill.

Gampp, H., D. Haspra, et al. (1984). "Copper(II) Complexes with Linear Pentadentate Chelators." *Inorganic Chemistry* 23: 3724-4730.

Gans, P. (1992). Data Fitting in the Chemical Sciences, Wiley.

Maeder, M. and Y.-M. Neuhold (2007). *Practical Data Analysis in Chemistry.* Amsterdam, Elsevier.

Maeder, M., Y. M. Neuhold, et al. (2002). "Analysis of reactions in aqueous solution at non-constant pH: No more buffers?" *Phys. Chem. Chem. Phys.* 5: 2836-2841.

Martell, A. E. and R. J. Motekaitis (1988). *The Determination and Use of Stability Constants* New York, VCH.

Menten, L. and M. I. Michaelis (1913). "Die Kinetik der Invertinwirkung." *Biochem Z* 49: 333-369.

Norman, S. and M. Maeder (2006). "Model-Based Analysis for Kinetic and Equilibrium Investigations." *Critical Reviews in Analytical Chemistry* 36: 199-209.

Paul, C., K. Kirschner, et al. (1979). "Calibration of Stopped-Flow Spectrophotometers Using a Two-Step Disulfide Exchange Reaction." *Analytical Biochemistry* 101: 442-448.

Polster, J. and H. Lachmann (1989). *Spectrometric Titrations: Analysis of Chemical Equilibria.* Weinheim, VCH.

Press, W. H., W. T. Vetterling, et al. (1995). *Numerical Recipes in C.* Cambridge, Cambridge University Press.

Wilkins, R. G. (1991). *Kinetics and Mechanism of Reactions of Transition Metal Complexes.* Weinheim, VCH.

5

Experimental Optimization and Response Surfaces

Veli-Matti Tapani Taavitsainen
Helsinki Metropolia University of Applied Sciences
Finland

1. Introduction

Statistical design of experiments (DOE) is commonly seen as an essential part of chemometrics. However, it is often overlooked in chemometric practice. The general objective of DOE is to guarantee that the dependencies between experimental conditions and the outcome of the experiments (the responses) can be estimated reliably at minimal cost, i.e. with the minimal number of experiments. DOE can be divided into several subtopics, such as finding the most important variables from a large set of variables (screening designs), finding the effect of a mixture composition on the response variables (mixture designs), finding sources of error (variance component analysis) in a measurements system, finding optimal conditions in continuous processes (evolutionary operation, EVOP) or batch processes (response surface methodology, RSM), or designing experiments for optimal parameter estimation in mathematical models (optimal design).

Several good textbooks exist. Of the general DOE textbooks, i.e. the ones that are focused on any special field, perhaps (Box et. al., 2005), (Box & Draper, 2007) and (Montgomery, 1991) are the most widely used ones. Some of the DOE textbooks, e.g. (Bayne & Rubin, 1986), (Carlson & Carlson, 2005) and (Bruns et. al., 2006) focus on chemometric problems. Good textbooks covering other fields of applications include e.g. (Himmelblau, 1970) for chemical engineering, (Berthouex & Brown, 2002) and (Hanrahan, 2009) for environmental engineering, or (Haaland, 1989) for biotechnology. Many textbooks about linear models or quality technology also have good treatments of DOE, e.g. (Neter et. al., 1996), (Vardeman, 1994) and (Kolarik, 1995).

More extensive lists of DOE literature are given in many textbooks, see e.g. (Box & Draper, 2007) , or in the documentation of commercial DOE software packages, see e.g. (JMP, release 6)

This chapter focuses on common strategies of empirical optimization, i.e. optimization based on designed experiments and their results. The reader should be familiar with basic statistical concepts. However, for the reader's convenience, the key concepts needed in DOE will be reviewed. Mathematical prerequisites include basic knowledge of linear algebra, functions of several variables and elementary calculus. However, neither theory, nor the methodology is presented in a rigorous mathematical style; rather the style is relying on examples, common sense, and on pinpointing the key ideas.

The aim of this chapter is that the material could be used to guide chemists, chemical engineers and chemometricians in real applications requiring experimentation. Naturally, the examples presented have chemical/chemometric origin, but as with most statistical techniques, the field of possible applications is truly vast. The focus is on problems with quantitative variables and, correspondingly, on regression techniques. Qualitative (categorical) variables and analysis of variance (ANOVA) are merely mentioned.

Typical chemometric applications of RSM are such as optimization of chemical syntheses, optimization of chemical reactors or other unit operations of chemical processes, or optimization of chromatographic columns.

2. Optimization strategies

This section introduces the two most common empirical optimization strategies, the simplex method and the Box-Wilson strategy. The emphasis is on the latter, as it has a wider scope of applications. This section presents the basic idea; the techniques needed at different steps in following the given strategy are given in the subsequent sections.

2.1 The Nelder-Mead simplex strategy

The Nelder-Mead simplex algorithm was published already on 1965, and it has become a 'classic' (Nelder & Mead, 1965). Several variants and applications of it have been published since then. It is often also called the flexible polyhedron method. It should be noted that it has nothing to do with the so-called Dantzig's simplex method used in linear programming. It can be used both in mathematical and empirical optimization.

The algorithm is based on so-called simplices N-polytopes with $N+1$ vertices, where N is the number of (design) variables. For example, a simplex in two dimensions is a triangle, and a simplex in three dimensions is a tetrahedron. The idea behind the method is simple: a simplex provided with the corresponding response values (or function values in mathematical optimization) gives a minimal set of points to fit perfectly an N-dimensional hyperplane in a $(N+1)$-dimensional space. For example for two variables and the responses, the space is a plane in 3-dimensional space. Such a hyperplane is the simplest linear approximation of the underlying nonlinear function, often called a response surface, or rather a response hypersurface. The idea is to reflect the vertex corresponding to the worst response value along the hyperplane with respect to the opposing edge. The algorithm has special rules for cases in which the response at a reflected point doesn't give improvement, or if an additional expanded reflection gives improvement. These special rules make the simplex sometimes shrink, and sometimes expand. Therefore, it is also called the flexible simplex algorithm.

The idea is easiest understood graphically in a case with 2 variables: Fig. 1 depicts an ideal response surface the yield of a batch reactor with respect to the batch length in minutes and the reactor temperature in °C. The model is ideal in the sense that the response values are free from experimental error. We can see that first the simplex expands because the surface around the starting simplex is quite planar. Once the chain of simplexes attains the ridge going approximately from right, some of the simplexes are contracted, i.e. they shrink considerably. You can easily see, how a reflection would worsen the response (this is depicted as an arrow in the upper left panel). Once the chain finds the direction of the ridge, the simplexes expand again and approach the optimum effectively. The Nelder-Mead simplex algorithm is not very effective in final positioning of the optimal point, because that would require many contractions.

The Nelder-Mead Simplex algorithm

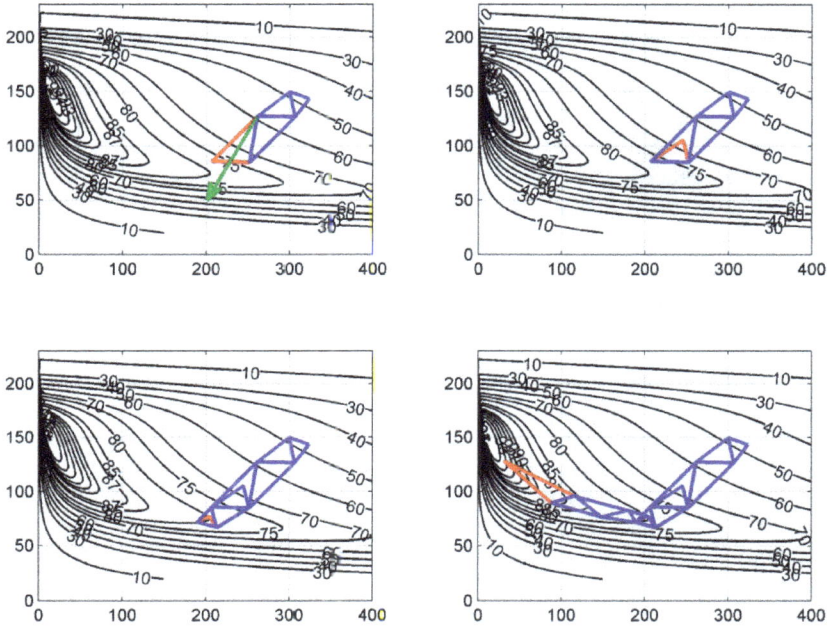

Fig. 1. Sequences of Nelder-Mead simplex experiments with respect to time and temperature based on an errorless reactor model. In all panels, the x-axis corresponds to reaction time in minutes and y-axis corresponds to reactor temperature in °C. Two edges of the last simplex are in red in all panels. Upper left panel: the first 4 simplexes and the reflection of the last simplex. Upper right panel: the first 4 simplexes and the first contraction of the last simplex. Lower right panel: the first 7 simplexes and the second contraction of the last simplex. Lower right panel: the first 12 simplexes and the expanded reflection of the last simplex.

The Nelder-Mead algorithm has been used successfully e.g. in optimizing chromatographic columns. However, its applicability is restricted by the fact that it doesn't work well if the results contain substantial experimental error. Therefore, in most cases another type of a strategy is a better choice, presented in the next section.

2.2 The Box-Wilson strategy (the gradient method)

In this section we try to give an overall picture of the Box-Wilson strategy, and the different types of designs used within the strategy will be explained in subsequent sections; the focus is on the strategy itself.

The basic idea behind the Box-Wilson strategy is to follow the path of the steepest ascent towards the optimal point. In determining the direction of the steepest ascent, mathematically speaking, the gradient vector, the method uses local polynomial modelling. It is a sequential method, where the sequence of main steps is: 1) make a design around the current best point, 2) make a polynomial model, 3) determine the gradient path, and 4) carry out experiments along the path as long as the results will improve. After step 4, return to step 1, and repeat the sequence of steps. Typically the steps 1-4 have to be repeated 2 to 3 times.

Normally the first design is a 2^N factorial design (see section 3.1) with an additional centre point, possibly replicated one or more times. The idea is that, at the beginning of the optimization, the surface within the design area is approximately linear, i.e. a hyperplane. A 2^N factorial design allows also modelling of interaction effects. Interactions are common in problems of chemical or biological origin. The additional centre point can be used to check for curvature. If the curvature is found to be statistically significant, the design should be upgraded into a second order design (see section 5), allowing building of a quadratic model. The replicate experiments are used to estimate the mean experimental error, and for testing model adequacy, i.e. the lack-of-fit in the model.

After the first round of steps 1-4 (see also Fig. 4), it is clear that a linear or linear plus interactions model cannot fit the results anymore, as the results first get better and then worse. Therefore, at this point, an appropriate design is a second order design, typically a central composite or a Box-Behnken design (explained in section 5), both allowing building of a quadratic polynomial model. The analysis of the quadratic model lets us estimate whether the optimum is located near the design area or further away. In the latter case, new experiments are again conducted along the gradient path, but in the first case, the new experiments will be located around the optimum predicted by the model.

The idea is best grasped by a graphical illustration given in Figs. 2 and 3 using the same reactor model as in section 2.1. Fig. 2 shows the theoretical errorless response surface, the region of the initial design (the black rectangle), and the theoretical gradient path. The contours inside the rectangle show that the response behaves approximately linearly inside the region of the first design (the contours are approximately parallel straight lines).

It is important to understand that the gradient path must be calculated using small enough steps. This is best seen graphically: Fig. 3 shows what happens using too large a step size: too large a step size creates the typical zigzag pattern. Obviously, this is inefficient, and such a path also misses the optimum.

Next we shall try to illustrate how the gradient method works in practice. In order to make the situation more realistic, we have added Gaussian noise ($\mu = 0, \sigma^2 = 1$) to the yield, given by the simulation model of the reactor, i.e. instead of carrying out real experiments, the results are obtained from the model. In addition, random experimental error is added to the modelled responses. The sequence of designs following the box-Wilson strategy and the corresponding gradient path experiments are depicted in Fig. 4. Notice that the gradient path based on the model of the first design is slightly curved due to the interaction between time and temperature.

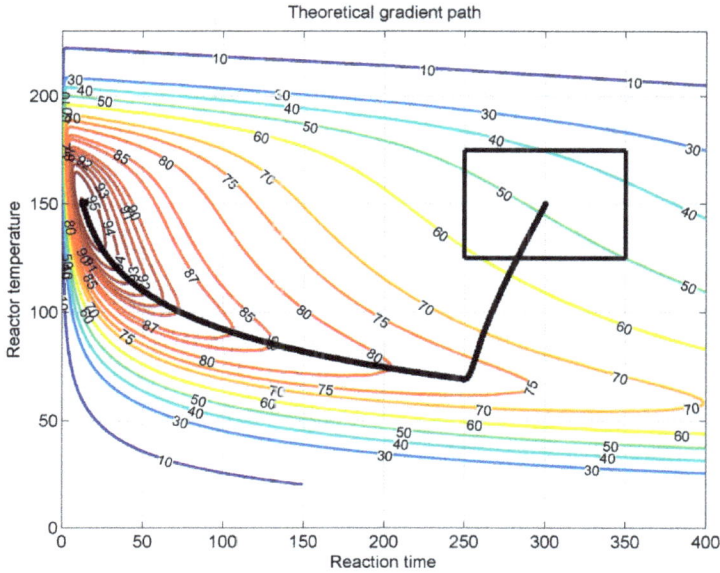

Fig. 2. The gradient path (black solid line with dots at points of calculation) of a reactor model yield surface. The solid line cannot be distinguished due to the small step size between the points of calculation.

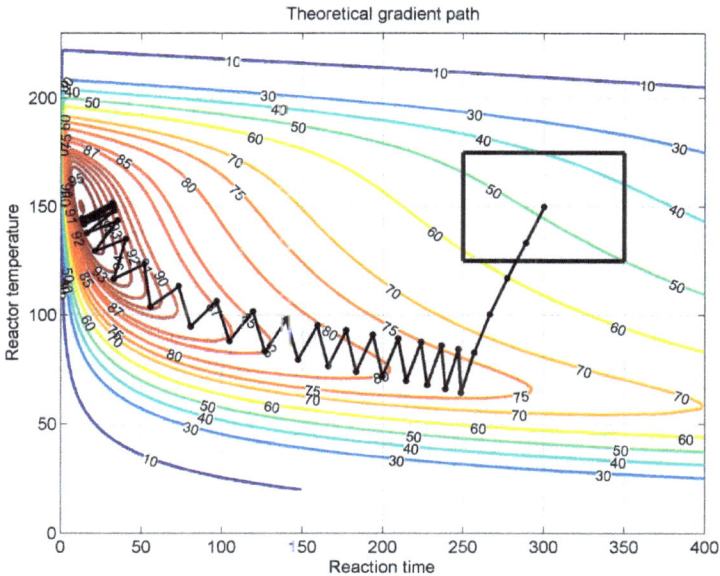

Fig. 3. The gradient path (black solid line with dots at the points of calculation) of a reactor model yield surface. The path is calculated using too large a step size causing the zigzag pattern.

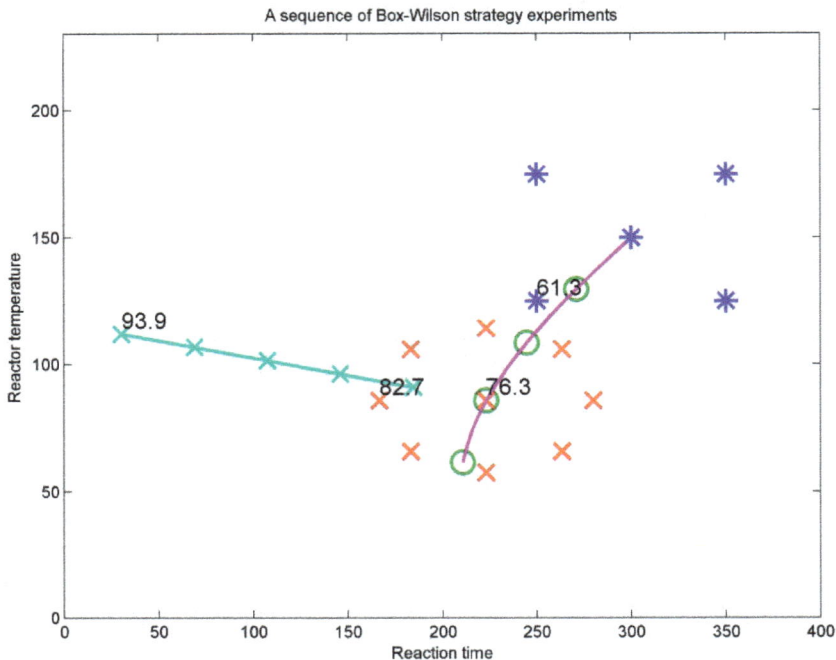

Fig. 4. The first design (blue asterisks) and the gradient path (magenta solid line) based on an empirical model estimated from the results of the experiments. Four points from the gradient path are chosen for the experiments (green circles). A second order design (red x's) and the gradient path based on its modelled results (turquoise solid line with x's at the experimental points). The best results of the four sets of experiments, showing excellent improvement, are 61.3, 76.3, 82.7 and 93.9.

Typically, the next step would be making of a new second order design around the best point. However, one should keep in mind that the sensitivity to changes in the design variables decreases. As a consequence, any systematic changes may be hidden under experimental errors. Therefore, the accurate location of the optimum is difficult to find, perhaps requiring repetitions of the whole design.

Simulation models like the one used in this example are very useful in practising Box-Wilson strategy. It can be obtained upon request from the author in the form of a Matlab, R or Excel VBA. For maximal learning, the user is advised to start the procedure at different locations.

3. Factorial designs

Factorial designs make the basis of all most common designs. The idea of factorial designs is simple: a factorial design is made up of all possible combinations of all chosen values, often called levels, of all design variables. Factorial designs can be used both for qualitative and quantitative variables. If variables $x_1, x_2, ..., x_N$ have $m_1, m_2, ..., m_N$ different levels, the number of experiments is $m_1 \cdot m_2 \cdot .... \cdot m_N$. As a simple example, let us consider a case where the variables and their levels are: x_1 the type of a catalyst (A, B and C), x_2 the catalyst concentration (1 ppm and 2 ppm), and x_3 the reaction temperature (60 °C, 70 °C and 80 °C). The corresponding factorial design is given in Table 1.

x_1	x_2	x_3
A	1	60
B	1	60
C	1	60
A	2	60
B	2	60
C	2	60
A	1	70
B	1	70
C	1	70
A	2	70
B	2	70
C	2	70
A	1	80
B	1	80
C	1	80
A	2	80
B	2	80
C	2	80

Table 1. A simple factorial design of three variables.

It is good to understand why factorial designs are good designs. The main reasons are that they are *orthogonal* and *balanced*. Orthogonality means that the factor (variable) effects can be estimated independently. For example, in the previous example the effect of the catalyst can be estimated independently of the catalyst concentration effect. In a balanced design, each variable combination appears equally many times. In order to understand why orthogonality is important, let us study an example of a design that is not orthogonal. This design, given in Table 2 below, has two design variables, x_1 and x_2, and one response variable, y.

x_1	x_2	y
1.18	0.96	0.91
1.90	2.12	1.98
3.27	2.98	2.99
4.04	3.88	3.97
4.84	5.10	5.03
5.88	6.01	5.96
7.14	7.14	7.07
8.05	8.08	7.92
9.04	8.96	9.09
9.98	10.19	10.02
2.30	2.96	3.76
2.94	4.10	4.85
4.29	5.01	5.95
5.18	5.80	6.88
5.84	7.14	7.98
6.85	7.90	9.16
8.33	8.98	10.26
8.96	10.05	10.98
10.00	11.09	12.21
10.82	12.08	12.94
10.13	8.88	8.20
10.91	9.94	9.15
12.10	10.83	10.30
12.57	11.56	11.16
13.53	13.04	12.13
14.85	14.00	12.72
16.19	15.16	13.99
17.01	16.22	14.72
18.31	16.86	16.28
19.21	18.47	17.35

Table 2. A non-orthogonal design.

Now, if we plot the response against the design variables we get the following plot:

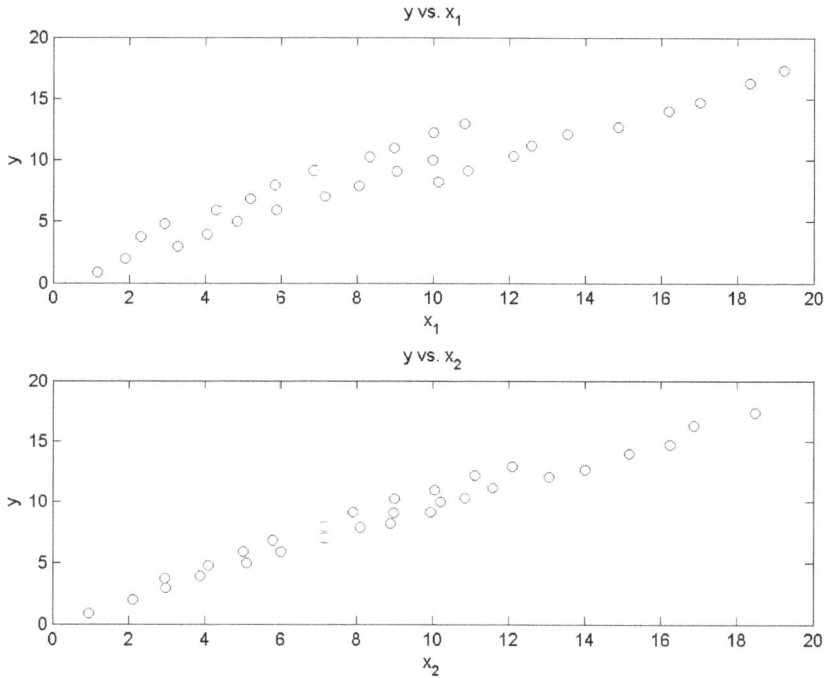

Fig. 5. Response (y) against the design variables (x_1 and x_2) in a non-orthogonal design; upper panel: y against, lower panel: y against.

Now, Fig. 5 clearly gives the illusion that the response depends approximately linearly both on x_1 and x_2 with positive slopes. However, the true slope between y and x_1 is negative. To see this, let us plot the design variable against each other and show the response values as text, as shown in Fig. 6.

Now, careful inspection of Fig. 6 reveals that actually yield decreases when x_1 increases. The reason for the wrong illusion that Fig. 5 gives is that x_1 and x_2 are strongly correlated with each other, i.e. the design variables are collinear. Although fitting a linear regression model using both design variables would give correct signs for the regression coefficients, collinearity will increase the confidence intervals of the regression coefficients. Problems of this kind can be avoided by using factorial designs.

After this example, it is obvious that orthogonality, or near orthogonality, is a desired property of a good experimental design. Other desired properties are

- The design contains as few experiments as possible for reliable results.
- The design gives reliable estimates for the empirical model fitted to the data.

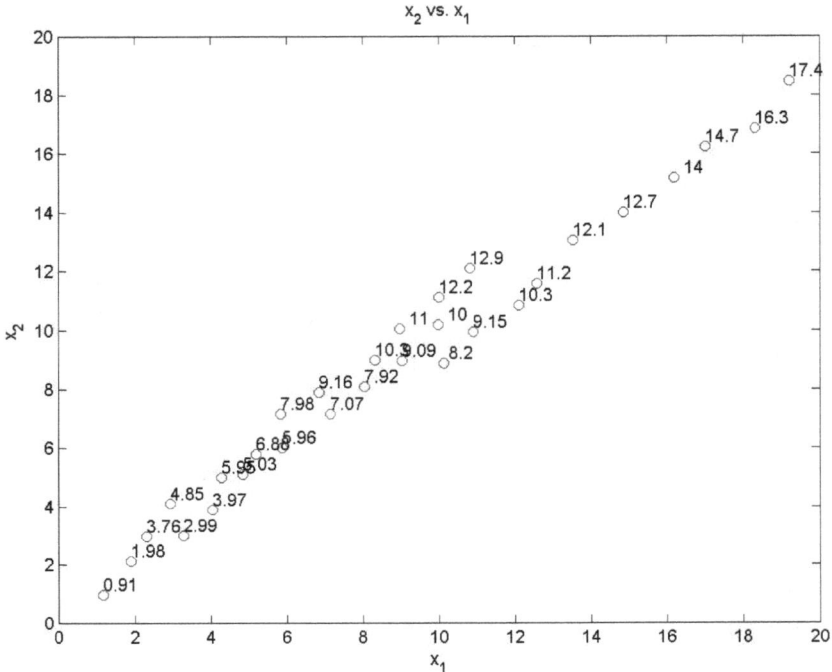

Fig. 6. Response (y) against the design variables (x_1 and x_2) in a non-orthogonal design; the response values (y values) are shown on upper right position with respect to each point.

- The design allows checking the reliability of the model fitted to the data, typically by statistical tests about the model parameters and the model adequacy (lack-of-fit), and by cross validation.

In general, factorial designs have these properties, except for the minimal number of the experiments. This topic will be treated in the subsequent sections.

3.1 Two level factorial designs

Factorial designs with only two values (levels) for all design variables are the most frequently used designs. This is mainly due to the following facts: 1) the number of experiments is less than with more levels, and 2) the results can be analysed using regression analysis for both qualitative and quantitative variables. The natural limitation is that only linear (main) effects and interactions effects can be detected. A drawback of full two-level factorial designs with a high number of variables is that the number of experiments is also high. Fortunately, this problem can be solved by using so-called fractional factorial designs, explained in section 3.3.

Two-level factorial designs, usually called 2^N designs, are typically tabulated using dimensionless coded variables having only values -1 or +1. For example, for a variable that represents a catalyst type, type A might correspond to -1 and type B might correspond to +1, or coarse raw material might be -1 and fine raw material might be +1. For quantitative variables, coding can be performed by the formula

$$X_i = \frac{x_i - \overline{x}_i}{\frac{1}{2}\Delta x_i} \tag{1}$$

where X_i stands for the coded value of the i'th variable, x_1 stands for the original value of the i'th variable, $\overline{x}_i$ stands for the centre point value of the original i'th variable, and Δx_i stands for the difference of the original two values of the i'th variable. The half value of the difference is called the step size. All statistical analyses of the results are usually carried out using the coded variables. Quite often, we need to convert also coded dimensionless values into the original physical values. For quantitative variables, we can simply use the inverse of Eq. 1, i.e.

$$x_i = \overline{x}_i + \frac{1}{2}\Delta x_i \cdot X_i \tag{2}$$

Tables of two-level factorial designs can be found in most textbooks of DOE. A good source is also NIST SEMATECH e-Handbook of Statistical Methods (NIST SEMATCH). Another way to create such tables is to use DOE-software e.g. (JMP, MODDE, MiniTab,…). It is also very easy to create tables of two-level factorial designs in any spreadsheet program. For example in Excel, you can simply enter the somewhat hideous formula

=2*MOD(FLOOR((ROW($B3)-ROW($B$3))/2^(COLUMN(C$2)-COLUMN(C2));1);2)-1

into the cell C3, and then first copy the formula to the right as many time as there are variables in the design (N) and finally copy the whole first row down 2^N times. Of course, you can enter the formula anywhere in the spreadsheet, e.g. if you enter it into the cell D7 the references in the ROW functions must be changed into $C7 and C7, and the references in the column function must be changed into D$6 and D6, respectively.

If all variables are quantitative it is advisable to add a centre point into the design, i.e. an experiment where all variables are set to their mean values. Consequently, in coded units, all variables have value 0. The centre point experiment can be used to detect nonlinearities within the design area. If the mean experimental error is not known, usually the most effective way to find it out is to repeat the centre point experiment. All experiments, including the possible centre point replicates, should be carried out in random order. The importance of randomization is well explained in e.g. (Box, Hunter & Hunter).

3.1.1 Empirical models related to two-level factorial designs

2^N designs can be used only for linear models with optional interaction terms up order N. By experience, it is known that interaction of order higher than two are seldom significant. Therefore, it is common to consider those terms as random noise, giving extra degrees of freedom for error estimation. However, one should be careful about such interpretations, and models should always be carefully validated. It should also be noted that the residual errors always contain both experimental error and modelling error. For this reason, independent replicate experiments are of utmost importance, and only having a reliable estimate of the experimental error gives a possibility to check for lack-of-fit, i.e. the model adequacy.

The general form of a model for a response variable y with linear terms and interaction terms up to order N is

$$y = \sum_{i=1}^{N} b_i X_i + \sum_{i<j}^{N} b_{ij} X_i X_j + \sum_{i<j<k}^{N} b_{ijk} X_i X_j X_k + \cdots \tag{3}$$

The number of terms in the second sum is $\binom{N}{2}$, and in the third sum it is $\binom{N}{3}$, and so on.

The most common model types used are models with linear terms only, or models with linear terms and pairwise interaction terms.

If all terms of model (3) are used, and there are no replicate experiments in the corresponding 2^N design, there are as many unknown parameters in the model as there are experiments in the design, leaving no degrees of freedom for the residual error, i.e. all residuals are zero. In such cases, the design is called saturated with respect to the model, or just saturated, if it is obvious what the model is. In these cases traditional statistical tests cannot be employed. Instead, the significant terms can often be detected by inspecting the estimated model parameter values using normal probability plots.

Later we need to differentiate between the terms "design matrix" and "model matrix". A design matrix is a $N_{exp} \times N$ matrix whose columns are the values of design variables where N_{exp} is the number of experiments. A model matrix is a $N_{exp} \times p$ matrix that is the design matrix appended with columns corresponding to the model terms. For example, a model matrix for a linear plus interaction model for two variables has a column of ones (corresponding to the intercept), columns for values of X_1 and X_2 and a column for values of the product $X_1 X_2$.

It is good to understand the nature of the pairwise interaction terms. Let us consider a model for two variables, i.e. $y = b_0 + b_1 X_1 + b_2 X_2 + b_{12} X_1 X_2$, and let us rearrange the terms as $y = b_0 + b_1 X_1 + (b_2 + b_{12} X_1) X_2$. This form reveals that the interaction actually means that the slope of X_2 depends linearly on X_1. Taking X_1 as the common factor instead of X_2 shows that the slope of X_1 depends linearly on X_2. In other words, a pairwise interaction between two variables means that the other variable affects the effect of the other one. If two variables don't interact, their effects are said to be additive. Fig. 7 depicts additive and interacting variables.

In problems of chemical or biological nature, it is more a rule than an exception that interactions between variables exist. Therefore, main effect models serve only as rough approximations, and are used typically in cases with a very high number of variables. It is also quite often useful to try to model some transformation of the response variable,

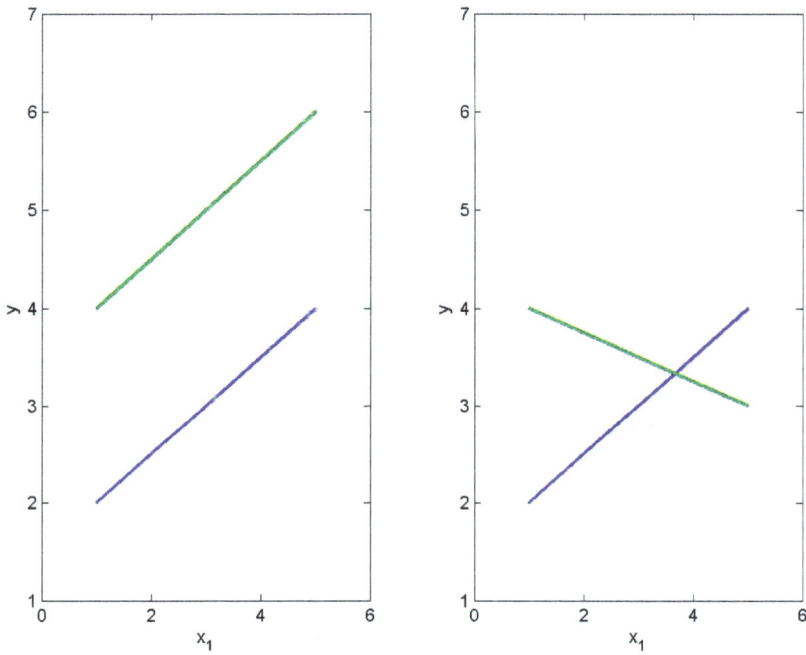

Fig. 7. Linear dependency of y on x_1 and on x_2. Left panel: an additive case without interaction, right panel: a non-additive case. Blue line: x_2 has a constant value, green line: x_2 has another constant value.

typically a logarithm, or a Box-Cox transformation. Usually the aim is to find a transformation that makes the residuals as normal as possible.

3.1.2 Analysing the results of two level factorial designs

Two level factorial designs can be analysed either by analysis of variance (ANOVA) or by regression analysis. Using regression analysis is more straightforward, and we shall concentrate on it in the sequel. However, one should bear in mind that the interpretation of the estimated model parameter is different between quantitative and qualitative variables. Actually, due to the orthogonality of 2^N designs, regression analysis could be carried out quite easily even by hand using the well-known Yates algorithm, see e.g. (Box & Draper, 2007).

Ordinary least squares regression (OLS) is the most common way to analyse results of orthogonal designs, but sometimes more robust techniques, e.g. minimizing the median of absolute values of the residuals, can be employed. Using latent variable techniques, e.g. PLS, doesn't give any extra benefit with orthogonal designs. However, in some other kind of designs, typically optimal designs or mixture designs, latent variable techniques can be useful.

The mathematics and statistical theory behind regression analysis can be found in any basic textbook about regression analysis or statistical linear models see e.g. (Weisberg, 1985) or (Neter et. al., 1996). In this chapter we shall concentrate on applying and interpreting the results of OLS in DOE problems. However, it is always good to bear in mind the statistical assumptions behind the classical regression tests, i.e. approximately normally distributed and independent errors. The latter is more important, and dependencies between random errors can make the results of the tests completely useless. This fact is well illustrated in (Box et. al., 2005). Even moderate deviations from normality are usually not too harmful, unless they are caused by gross errors, i.e. by outliers. For this reason, normal probability plots, or other tools for detecting outliers, should always be included in validating the model.

Since OLS is such a standard technique, a plethora of software alternatives exists for carrying out the regression analyses of the results of a given design. One can use general mathematics software like Matlab, Octave, Mathematica, Maple etc., or general purpose statistical software like S-plus, R, Statistica, MiniTab, SPSS, etc, or even spreadsheet programs like Excel or Open Office Calc. However, it is advisable to use software that contains those model validation tools that are commonly used with designed experiments. Practically all general mathematical or statistical software packages contain such tools.

Quite often there are more than one response variables. In such cases, it is typical to estimate models for each response separately. If a multivariate response is a 'curve', e.g. a spectrum or a distribution, it may be simpler to use latent variable methods, typically PLS or PCR.

This example is taken from Box & Draper (Box & Draper, 2007).

3.2 Model validation

Model validation is an essential part of analysing the results of a design. It should be noted that most of the techniques presented in this section can be used with all kinds of designs, not only with 2^N designs.

In the worst case, the validation yields the conclusion that the design variables have no effect on the response(s), significantly different from random variation. In such a case, one has to consider the following alternatives: 1) to increase the step sizes in the design variables, 2) to replicate the experiments one or more times, or 3) to make a new design with new design variables. In the opposite case, i.e. the model and at least some of the design variables are found to be statistically significant, the continuation depends on the scope of the design, and on the results of the (regression) analysis. The techniques used for optimization tasks are presented in subsequent sections.

3.2.1 Classical statistical tests

Classical statistical tests can be applied mainly to validate regression models that are linear with respect to the model parameters. The most common empirical models used in DOE are linear models (main effect models), linear plus interactions models, and quadratic models. They all are linear with respect to the parameters. The most useful of these (in DOE context) are 1) t-tests for testing the significance of the individual terms of the model, 2) the lack-of-fit test for testing the model adequacy, and 3) outlier tests based on so-called externally studentized residuals, see e.g. (Neter et. al., 1996).

The t-test for testing the significance of the individual terms of the model is based on the test statistic that is calculated by dividing a regression coefficient by its standard deviation. This statistic can be shown to follow the t-distribution with $n-p-1$ degrees of freedom where n is the number experiments, and p is the number of model parameters. If the model doesn't contain an intercept, the number of degrees of freedom is $n-p$. Typically, a term in the model is considered significant if the p-value of the test statistic is below 0.05.

The standard errors of the coefficients are usually based on the residual error. If the design contains a reasonable number of replicates this estimate can also be based on the standard error of the replicates. The residual based standard error of the i'th regression coefficient s_{b_i} can be easily transformed into replicate error based ones by the formula $\sqrt{MS_E / MS_R} \cdot s_{b_i}$. In this case the degrees of freedom are $n_r - 1$ (the symbols are explained in the next paragraph).

The lack-of fit test can be applied only if the design contains replicate experiments which permit estimation of the so-called pure error, i.e. an error term that is free from modelling errors. Assuming that the replicate experiments are included in regression, the calculations are carried out according to the following equations. First calculate the pure error sum of squares SS_E

$$SS_E = \sum_{i=1}^{n_r} (y_i - \overline{y})^2 , \qquad (4)$$

where n_r is the number of replicates, y_i's are outcomes of the replicate experiments, and $\overline{y}$ is the mean value of the replicate experiments. The number of degrees of freedom of SS_E is $n_r - 1$. Then calculate the residual sum of squares SS_R:

$$SS_R = \sum_{i=1}^{n} (y_i - \hat{y})^2 , \qquad (5)$$

where n is the number of experiments and $\hat{y}$'s are the fitted values, i.e. the values calculated using the estimated model. The number of degrees of freedom of SS_R is $n-p-1$, or $n-p$ if the model doesn't contain an intercept. Then calculate the lack-of-fit sum of squares SS_{LOF}:

$$SS_{LOF} = SS_R - SS_E \qquad (6)$$

The number of degrees of freedom of SS_{LOF} is $n-n_r-p-1$, or $n-n_r-p$ if the model doesn't contain an intercept. Then, calculate the lack-of-fit mean squares MS_{LOF} and the pure error mean squares MS_E by dividing the corresponding sums of squares by their degrees of freedom. Finally, calculate the lack-of-fit test statistic $F_{LOF} = MS_{LOF} / MS_E$. It can be shown that F_{LOF} follows an F-distribution with $n-n_r-p-1$ (or $n-n_r-p$) and n_r-1 degrees of freedom. If F_{LOF} is significantly greater than 1, it is said that the model suffers from lack-of-fit, and if it is significantly less than 1, it is said that the model suffers from over-fit.

An externally studentized residual is a deleted residual, i.e. residual calculated using leave-one-out cross-validation, divided the standard error of deleted residuals. It can be shown that the externally studentized residuals follow a t-distribution with $n - p - 2$ degrees of freedom, or $n - p - 1$ degrees of freedom if the model doesn't contain an intercept. If the p-value of an externally studentized residual is small enough, the result of the corresponding experiment is called an outlier. Typically, outliers should be removed, or the corresponding experiments should be repeated. If the result of a repeated experiment still gives an outlying value, it is likely that model suffers from lack-of-fit. Otherwise, the conclusion is that something went wrong in the original experiment.

3.2.2 Cross-validation

Cross-validation is familiar to all chemometricians. However, in using cross-validation for validating results of designed experiments some important issues should be considered. First, cross-validation requires extra degrees of freedom, and consequently all candidate models cannot be cross-validated. For example, in a 2^2 design, a model containing linear terms and the pairwise interaction cannot be cross-validated. Secondly, often the designs become severely unbalanced, when observations are left out. For example, in a 2^2 design with a centre point, the model containing linear terms and the pairwise interaction can be cross-validated, but when the corner point (+1, +1) is left out, the design is very weak for estimating the interaction term; in such cases the results of cross-validation can be too pessimistic. On the other hand, replicated experiments may give too optimistic results in cross-validation, as the design variable combinations corresponding to replicate experiments are never left out. This problem can be easily avoided by using the response averages instead of individual responses of the replicated experiments.

Usually only statistically significant terms are kept in the final model. However, it is also common to include mildly non-significant terms in the model, if keeping such terms improves cross-validated results.

3.2.3 Normal probability plots

Normal probability plots, also called normal qq-plots, can be used to study either the regression coefficients or the residuals (or deleted residuals). The former is typically used in saturated models where ordinary t-tests cannot be applied. Normal probability plots are constructed by first sorting the values from the smallest to largest. Then the proportions $p_i = (i - 0.5) / n$ are calculated, where n is the number of the values, and i is the ordinal number of a sorted value, i.e. 1 for the smallest value and n for the largest value (subtracting 0.5 is called the continuity correction). Then the normal score, i.e. inverse of p_i using the standard normal distribution, is calculated. Finally, the values are plotted against the normal scores. If the distribution of the values is normal, the points lie approximately on a straight line. The interpretation in the former case is that the leftmost or the rightmost values that do not follow a linear pattern represent significant terms. In the latter case, the same kind of values represent outlying residuals.

3.2.4 Variable selection

If the design is orthogonal, or nearly orthogonal, removing or adding terms into the model doesn't affect the significance of the other terms. This is also the case if the estimates of the standard error of the coefficients are based on the standard error of the estimates (cf. 3.2.1). Therefore, variable selection based on significance is very simple; just take the variables significant enough, without worrying about e.g. the order of taking terms into a model. Because models based on designed experiments are often used for extrapolatory prediction, one should, whenever possible, test the models using cross-validation. However, one should bear in mind the limitations of cross-validation when it is applied to designed experiments (cf. 3.2.2). In addition, it is wise also to test models with almost significant variables using cross-validation, since sometimes such models have better predictive power.

If the design is not orthogonal, traditional variable (feature) selection techniques can be used, e.g. forward, backward, stepwise, or all checking possible models (total search). Naturally, the selection can be based on different criteria, e.g. Mallows C_p, PRESS, R^2, Q^2, Akaike's information etc., see e.g. (Weisberg, 1985). If models are used for extrapolatory prediction, a good choice for a criterion is to minimize PRESS or maximize Q^2. In many cases of DOE modelling, the number of possible model terms, typically linear, pair-wise interaction, and quadratic terms, is moderate. For example, a full quadratic model for 4 variables has 14 terms, plus the intercept. Thus the number of all possible sub-models is 2^{14} which is 16384. In such cases, with the speed of modern computers, it is easy to test all sub-models with respect to the chosen criterion. However, if the number of variables is greater, going through all possible regression models becomes impossible in practice. In such cases, one can use genetic algorithms, see e.g. (Koljonen & al., 2008).

Another approach is to use latent variable techniques, e.g. PLS or PCR, in which the selection of the dimension replaces the selection of variables. Although variable selection seems more natural, and is more commonly used in typical applications of DOE than latent variable methods, neither of the approaches have been proved generally better. Therefore, it is good to try out different approaches, combined with proper model validation techniques.

A third alternative is to use shrinkage methods, i.e. different forms of ridge regression. Recently, new algorithms based on L_1 norm have been developed, including such as LASSO (Tibshirani, 1996), LARS (Efron & al., 2004), or elastic nets (Zou & al., 2005). Elastic nets use combinations of L_1 and L_2 norm penalties. Penalizing the least squares solution by the L_1 norm of the regression coefficient tends to make the non-significant terms zero which effectively means selecting variables.

In a typical application of DOE, the responses are multivariate in a way that they represent individual features which, in turn, typically depend on different variable combinations of the design variables. In such cases, it is better to build separate models for each response, i.e. the significant variables have to be selected separately for each response. However, if the response is a spectrum, or an object of similar nature, variable selection should usually be carried out for all responses simultaneously, using e.g. PLS regression or some other multivariate regression technique. In such cases, there's an extra problem of combining the individual criteria of the goodness of fit into a single criterion. In many cases, a weighted average of e.g. the RMSEP values, i.e. the standard residual errors in cross-validation, of the individual responses is a good choice, e.g. using signal to noise ratios as weights.

3.2.5 An example of a 2^N design

As a simple example of a 2^N design we take a 2^2 design published in the Brazilian Journal of Chemical Engineering (Silva et. al., 2011). In this study the ethanol production by Pichia stipitis was evaluated in a stirred tank bioreactor using semi defined medium containing xylose (90.0 g/l) as the main carbon source. Experimental assays were performed according to a 2^2 full factorial design to evaluate the influence of aeration (0.25 to 0.75 vvm) and agitation (150 to 250 rpm) conditions on ethanol production. The design contains also a centre point (0.50 vvm and 200 rpm), and in a replication of the (+1, +1) experiment. It should be noted that this design is not fully orthogonal due the exceptional selection of the replication experiment (the design would have been orthogonal, if the centre point had been replicated).

The results of the design are given in Table 3 below (X_1 and X_2 refer to aeration and agitation in coded levels, respectively).

Assay	Aeration	Agitation	X_1	X_2	Production (g/l)
1	0.25	150	-1	-1	23.0
2	0.75	150	1	-1	17.7
3	0.25	250	-1	1	26.7
4	0.75	250	1	1	16.2
5	0.75	250	1	1	16.1
6	0.50	200	0	0	19.4

Table 3. A 2^2 design.

Fig. 8 shows the effect of aeration at the two levels of agitation. From the figure, it is clear that aeration has much greater influence on productivity (Production) than agitation. It also shows an interaction between the variables. Considering the very small difference in the response between the two replicate experiments, it is plausible to consider both aeration and the interaction between aeration and agitation significant effects.

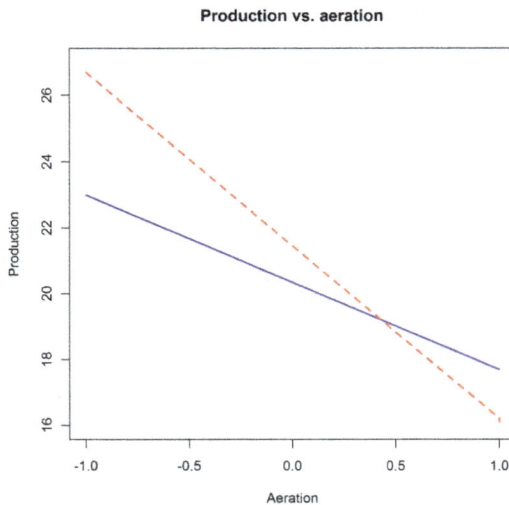

Production vs. aeration

Fig. 8. Production vs. aeration. Blue solid line: Agitation = 150; red dashed line: Agitation = 250.

If we carry out classical statistical tests used in regression analysis, it should be remembered that the design has only one replicated experiment, and consequently very few degrees of freedom for the residual error. Testing lack-of-fit, or nonlinearity, is also unreliable because we can estimate the (pure) experimental error with only one degree of freedom. However, it is always possible to use common sense and investigations about the effects on relative basis. For example, the difference between the centre point result and the mean value of the other (corner point) results is only 0.45 which is relatively small compared to differences between the corner points. Therefore, it is highly unlikely that the behaviour would be nonlinear within the experimental region. Consequently, it is likely that the model doesn't suffer from lack-of-fit, and a linear plus interaction model should suffice.

Now, let us look at the results of the regression analyses of a linear plus interaction model (model 1), the same model without the agitation main effect (model 2), and the model with aeration only (model 3). The regression analyses are carried out using basic R and some additional DOE functions written by the author (these DOE functions, including the R-scripts of all examples of this chapter, are available from the author upon request).

The R listing of the summary of the regression models 1, 2 and 3 are given in Tables 4-6 below. Note that values of the regression coefficients of the same effects vary a little between the models. This is due to the fact that design is not fully orthogonal. In an orthogonal design, the estimates of the same regression coefficients will not change when terms are dropped out.

	Estimate	Std. Error	t value	p value
(Intercept)	20.6205	0.4042	51.015	0.000384
X1	-3.9244	0.4454	-8.811	0.012638
X2	0.5756	0.4454	1.292	0.325404
I(X1 * X2)	-1.2744	0.4454	-2.861	0.325404

Residual standard error: 0.9541 on 2 degrees of freedom
Multiple R-squared: 0.9796,   Adjusted R-squared: 0.9489
F-statistic: 31.95 on 3 and 2 DF, p-value: 0.03051

Table 4. Regression summary of model 1 (I(X1 * X2) denotes interaction between X_1 and X_2).

	Estimate	Std. Error	t value	p value
(Intercept)	20.6882	0.4433	46.668	2.17e-05
X1	-3.8397	0.4873	-7.879	0.00426
I(X1 * X2)	-1.1897	0.4873	-2.441	0.09238

Residual standard error: 1.055 on 3 degrees of freedom
Multiple R-squared: 0.9625,   Adjusted R-squared: 0.9375
F-statistic: 38.48 on 2 and 3 DF, p-value: 0.007266

Table 5. Regression summary of model 2 (I(X1 * X2) denotes interaction between X_1 and X_2).

	Estimate	Std. Error	t value	p value
(Intercept)	20.5241	0.6558	31.3	6.21e-06
X1	-4.0448	0.7184	-5.63	0.0049

Residual standard error: 1.579 on 4 degrees of freedom
Multiple R-squared: 0.8879,   Adjusted R-squared: 0.8599
F-statistic: 38.48 on 1 and 4 DF, p-value: 0.004896

Table 6. Regression summary of model 3.

The residual standard error is approximately 1 g/l in models 1 and 2. This seems quite high compared to the variation in replicate experiments (16.2 and 16.1 g/l) corresponding to the pure experimental pure error standard deviation of ca. 0.071 g/l. The calculations of a lack-of-fit test for model 2 are the following: The residual sum of squares (SS_{RES}) is $3 \cdot 1.055^2 = 3.339$. The pure error sum of squares (SS_E) is $1 \cdot 0.071^2 = 0.005$. The lack-of-fit sum of squares (SS_{LOF}) is $3.339 - 0.005 = 3.334$. The corresponding mean squares are $SS_{LOF}/(df_{RES} - df_E) = SS_{LOF}/(3-1) = 3.334/2 = 1.667$ and the lack-of-fit F-statistic is $MS_{LOF}/MS_E = 1.667/0.005 = 333.4$ having 2 and 1 degrees of freedom. The corresponding p-value is 0.039 which is significant at the 0.05 level of significance. Thus, a formal lack-of-fit test exhibits significant lack-of-fit, but one must keep in mind that estimating standard deviation from only two observations is very unreliable. The lack-of-fit p-values for models 1 and 3 are 0.033 and 0.028, respectively, i.e. the lack-of-fit is least significant in model 2.

The effect of aeration (X_1) is significant in all models, and according to model 1 it is obvious that agitation doesn't have a significant effect on productivity. The interaction term is not significant in any of the models; however, it is not uncommon to include terms whose p-values are between 0.05 and 0.10 in models used for designing new experiments. The results of the new experiments would then either support or contradict the existence of an interaction.

Carrying out the leave-one-out (loo) cross-validation, gives the following Q^2 values (Table 7).

Model	R^2	Q^2
1	98.0	-22.0
2	96.2	84.5
3	88.8	68.1

Table 7. Comparison of R^2 and Q^2 values between model 1-3.

Fig. 9 shows the fitted and CV-predicted production values and the corresponding residual normal probability plots of models 1-3. By cross-validation, the model 2, i.e. $y = b_0 + b_1 X_1 + b_{12} X_1 X_2$, is the best one. Finally, Fig. 10 shows the contour plot of the best model, model 2.

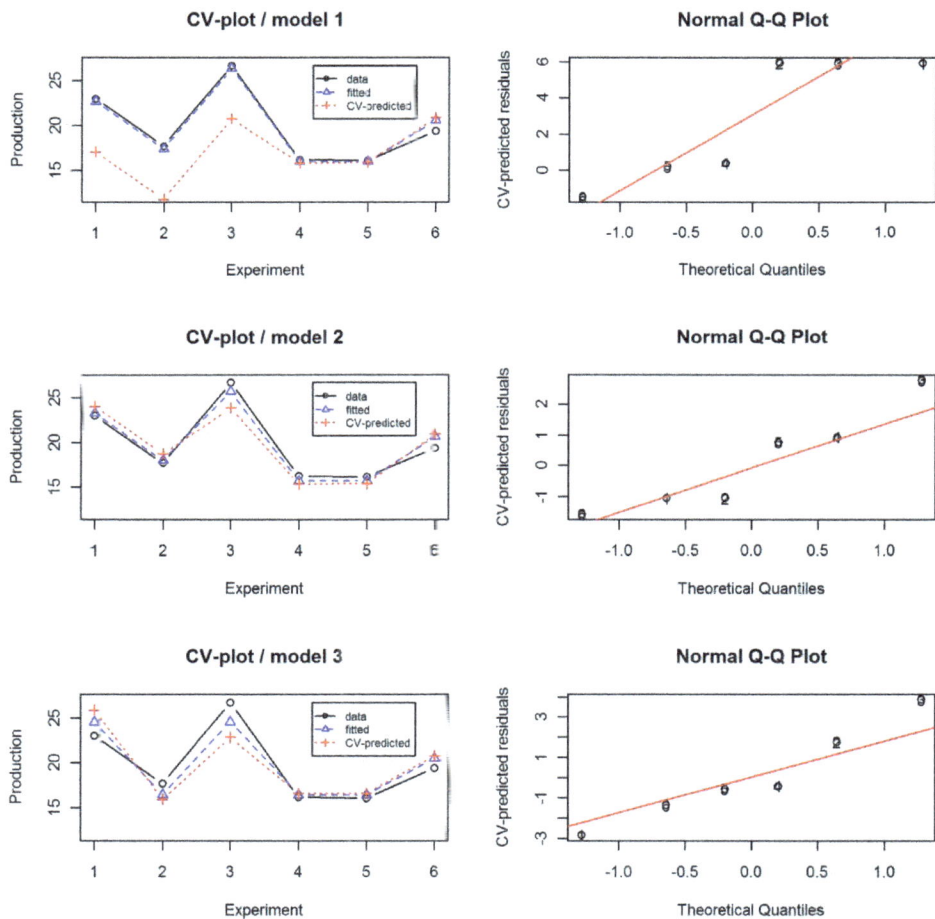

Fig. 9. Cross-validation of models 1-3. Left panel: Production vs. the number of experiment; black circles: data; blue triangles: fitted values; red pluses: cross-validated leave-one-out prediction. Right panel: Normal probability plots of the cross-validated leave-one-out residuals.

Production vs. Aeration and Agitation

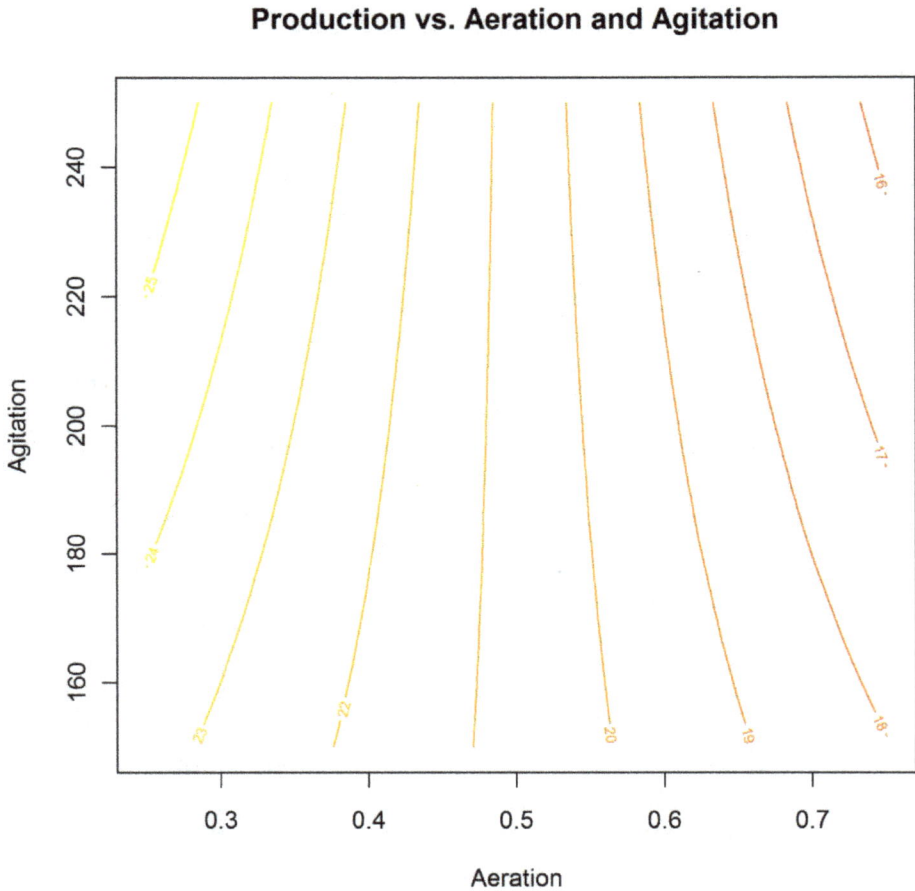

Fig. 10. Production vs. Aeration and Agitation.

3.3 Fractional 2^N designs (2^{N-p} designs)

The number of experiments in 2^N designs grows rapidly with the number of variables N. This problem can be avoided by choosing only part of the experiments of the full design. Naturally, using only a fraction of the full design, information is lost. The idea behind fractional 2^N designs is to select the experiments in a way that the information lost is related only to higher order interactions which seldom represent significant effects.

3.3.1 Generating 2^{N-p} designs

The selection of experiments in 2^{N-p} designs can be accomplished by using so-called generators (see e.g. Box & al., 2005). A generator is an equation between algebraic elements that represent variable effects, typically denoted by bold face numbers or upper case letters. For example **1** denotes the effect of variable 1. If there are more than 9 variables in the design, brackets are used to avoid confusion, i.e. we would use **(12)** instead of **12** to represent the effect of the variable 12. The bold face letter **I** represents the average response, i.e. the intercept of the model when coded variables are used. The generator elements (effects) follow the following algebraic rules of 'products' between the effects.

The effects are commutative, e.g. **12** = **21**
The effects are associative, e.g. **(12)3** = **1(23)**
I is a neutral element, e.g. **I2** = **2**
Even powers produce the neutral element, e.g. **22** = **I** or **2222** = **I** (naturally, for example **222** = **2**)

Now, a generator of a design is an equation between a product of effects and **I**, for example **123** = **I**. The interpretation of a product, also called a word, is that of a corresponding interaction between the effects. Thus, for example, **123** = **I** means that the third order interaction between variables 1-3 is confounded with the mean response in a design generated using this generator. Confounding (sometimes called aliasing) means that the confounding effects cannot be estimated unequivocally using this design. For example, in a design generated by **123** = **I** the model cannot contain both an intercept and a third order interaction. If the model is deliberately chosen to have both an intercept and the third order interaction term, there is no way to tell whether the estimate of the intercept really represents the intercept or the third order interaction.

Furthermore, any equation derived from the original generator, using the given algebraic rules, gives a confounding pattern. For example multiplying both sides of **123** = **I** by **1** gives **1123** = **1I**. Using the given rules this simplifies into **I23** = **1I** and then into **23** = **1**. Thus, in a design with this generator the pairwise interaction between variable 2 and 3 is confounded with variable 1. Multiplying the original generator by **2** and **3** it is easy to see that all pairwise interactions are confounded with main effects (**2** with **13** and **3** with **12**) in this design. Consequently, the only reasonable model whose parameters can be estimated unequivocally, is the main effect model $y = b_0 + b_1 X_1 + b_2 X_2 + b_3 X_3$. Technically possible alternative models, but hardly useful in practice, would be e.g. $y = b_0 + b_1 X_1 + b_2 X_2 + b_{12} X_1 X_2$ or $y = b_{123} X_1 X_2 X_3 + b_1 X_1 + b_2 X_2 + b_3 X_3$.

A design can be generated using more than one generator. Each generator halves the number of experiments. For example, a design with two generators has only ¼ of the original number of experiments in the corresponding full 2^N design. If p is the number of generators, the corresponding fractional 2^N design is denoted by 2^{N-p}.

In practice, 2^{N-p} designs are constructed by first making a full 2^N design table and then adding columns that contain the interaction terms corresponding to the generator words. Then only those experiments (rows) are selected where all interaction terms are +1. Alternatively one can choose the experiments where all interaction terms are -1. As an example, let us construct a 2^{3-1} design with the generator **123** = **I**. The full design table with an additional column containing the three-way interaction term is given in Table 8.

x^1	x^2	x^3	$x_1 x_2 x_3$
-1	-1	-1	-1
-1	-1	+1	+1
-1	+1	-1	+1
-1	+1	+1	-1
+1	-1	-1	+1
+1	-1	+1	-1
+1	+1	-1	-1
+1	+1	+1	+1

Table 8. A table for constructing a 2^{3-1} design.

Now, the desired design table is obtained by deleting the rows 1, 4, 6 and 7. An alternative design is obtained by deleting the rows 2, 3, 5 and 8.

3.3.2 Confounding (aliasing) and resolution

An important concept related to 2^{N-p} designs is the resolution of a design, denoted by roman numerals. Technically, resolution is the minimum word length of all possible generators derived from the original set of generators. For design with a single generator, finding out the resolution is easy. For example, the resolution of the 2^{3-1} design with the generator **123** = **I** is III because the length of the word **123** is 3 (note that e.g. (**12**) would be counted as a single letter in a generator word). If there are more generators than one, the situation is more complicated. For example, if the generators in a 2^{5-2} design were **1234** = **I** and **1235** = **I**, then the equation **1234** = **1235** would be true which after multiplying both sides **1235** gives **45** = **I**. Thus the resolution of this design would be II. Naturally, this would be a really bad design with confounding main effects.

The interpretation of the resolution of a design is (designs of resolution below III are normally not used)

- If the resolution is III, only a main effect model can be used
- If the resolution is IV, a main effect model with half of all the pairwise interaction effects can be used
- If the resolution is V or higher, a main effect model with all pairwise interaction effects can be used

If the resolution is higher than V also at least some of the higher order interaction can be estimated. There are many sources of tables listing 2^{N-p} designs and their confounding patterns, e.g. Table 3.17 in (NIST SEMATCH). Usually these tables give so-called minimum aberration designs, i.e. designs that minimize the number of short words in all possible generators of a design with given N and p.

3.3.3 Example

This example is taken from (Box & Draper, 2007) (Example 5.2 p. 189), but the analysis is not completely identical to the one given in the book.

The task was to improve the yield (y) (in percentage) of a laboratory scale drug synthesis. The five design variables were the reaction time (t), the reactor temperature (T), the amount of reagent B (B), the amount of reagent C (C), and the amount of reagent D (D). The chosen design levels in a two level fractional factorial design are given in Table 9 below.

Coded	Original	Lower (-1)	Upper (+1)	Formula
X_1	t	6 h	10 h	$X_1 = \dfrac{t-8}{2}$
X_2	T	85°C	90°C	$X_2 = \dfrac{T-87.5}{2.5}$
X_3	B	30 ml	60 ml	$X_3 = \dfrac{B-45}{15}$
X_4	C	90 ml	115 ml	$X_4 = \dfrac{C-102.5}{12.5}$
X_5	D	40 g	50 g	$X_5 = \dfrac{D-45}{5}$

Table 9. The variable levels of example 3.3.3.

The design was chosen to be a fractional resolution V design (2^{5-1}) with the generator I = 12345. The design table in coded units, including the yields and the run order of the experiments is given in Table 10 (y stands for the yield).

order	X_1	X_2	X_3	X_4	X_5	y
16	-1	-1	-1	-1	1	51.8
2	1	-1	-1	-1	-1	56.3
10	-1	1	-1	-1	-1	56.8
1	1	1	-1	-1	1	48.3
14	-1	-1	1	-1	-1	62.3
8	1	-1	1	-1	1	49.8
9	-1	1	1	-1	1	49.0
7	1	1	1	-1	-1	46.0
4	-1	-1	-1	1	-1	72.6
15	1	-1	-1	1	1	49.5
13	-1	1	-1	1	1	56.8
3	1	1	-1	1	-1	63.1
12	-1	-1	1	1	1	64.6
6	1	-1	1	1	-1	67.8
5	-1	1	1	1	-1	70.3
11	1	1	1	1	1	49.8

Table 10. The design of example 3.3.3 in coded units.

Since the resolution of this design is V, we can estimate a model containing linear and pairwise interaction effects. However the design is saturated with respect to this model, and thus the model cannot be validated by statistical tests, or by cross-validation. The regression summary is given Table 11.

	Estimate	Std. Error	t value	p value
(Intercept)	57.1750	NA	NA	NA
t	-3.3500	NA	NA	NA
T	-2.1625	NA	NA	NA
B	0.2750	NA	NA	NA
C	4.6375	NA	NA	NA
D	-4.7250	NA	NA	NA
I(t * T)	0.1375	NA	NA	NA
I(t * B)	-0.7500	NA	NA	NA
I(T * B)	-1.5125	NA	NA	NA
I(t * C)	-0.9125	NA	NA	NA
I(T * C)	0.3500	NA	NA	NA
I(B * C)	1.0375	NA	NA	NA
I(t * D)	0.2500	NA	NA	NA
I(T * D)	0.6875	NA	NA	NA
I(B * D)	0.5750	NA	NA	NA
I(C * D)	-1.9125	NA	NA	NA

Residual standard error: NaN on 0 degrees of freedom
Multiple R-squared: 1,   Adjusted R-squared: NaN
F-statistic: NaN on 15 and 0 DF, p-value: NA

Table 11. Regression summary of the linear plus pairwise interactions model. NA stands for "not available".

Because the design is saturated with respect to the linear plus pairwise interactions model there are no degrees of freedom for any regression statistics. Therefore, for selecting the significant terms we have to use either a normal probability plot of the estimated values of the regression coefficient or variable selection techniques. We chose to use forward selection based on the Q^2 value. This technique gave the maximum Q^2 value in a model with 4 linear terms and 7 pairwise interaction terms. However, after 6 terms the increase in the Q^2 value is minimal, and in order to avoid over-fitting we chose to use the model with 6 terms. The chosen terms were the main effects of t, T, C and D, and the interaction effects between C and D and between T and B. This model has a Q^2 value 83.8 % and the regression summary for this model is given in Table 12.

All terms in the model are now statistically significant at 5 % significance level, and the predictive power of the model is fairly good according the Q^2 value . Section 4.3 shows how this model has been used in search for improvement.

3.4 Plackett-Burman (screening) designs

If the number of variables is high, and the aim is to select the most important variables for further experimentation, usually only the main effects are of interest. In such cases the most cost effective choice is to use designs that have as many experiments as there are parameters in

	Estimate	Std. Error	t value	p value
(Intercept)	57.1750	0.6284	90.980	1.19e-14
t	-3.3500	0.6284	-5.331	0.000474
T	-2.1625	0.6284	-3.441	0.007378
C	4.6375	0.6284	7.379	4.19e-05
D	-4.7250	0.6284	-7.519	3.62e-05
I(T * B)	-1.5125	0.6284	-2.407	0.039457
I(C * D)	1.9125	0.6284	-3.043	0.013944
Residual standard error: NaN on 0 degrees of freedom				
Multiple R-squared: 1, Adjusted R-squared: NaN				
F-statistic: NaN on 15 and 0 DF, p-value: NA				

Table 12. Regression summary of the 6 terms model.

the corresponding main effect model, i.e. $N+1$ experiments. It can be proved that such designs that are also orthogonal exist in multiples of 4, i.e. for 3, 7, 11, ... variables having 4, 8, 12, ... experiments respectively. The ones in which the number of experiments is a power of 2 are actually 2^{N-p} designs. Thus for example a Plackett-Burman design for 3 variables that has $8 = 2^3$ experiments is a 2^{3-1} design. General construction of Plackett-Burman designs is beyond the scope of this chapter. The interested reader can refer to e.g. section 5.3.3.5 in (NIST SEMATECH). Plackett-Burman designs are also called 2-level Taguchi designs or Hadamard matrices.

3.5 Blocking

Sometimes uncontrolled factors, such as work shifts, raw material batches, differences in pieces of equipment, etc., may affect the results. In such cases the effects of such variables should be taken into account in the design. If the design variables are qualitative, such classical designs as randomized blocks design, Latin square design, or Graeco-Latin square design can be used, see e.g. (Montgomery, 1991). If the design variables are quantitative, a common technique is to have extra columns (variables) for the uncontrolled variables. For 2^N and CC-designs, tables of different blocking schemes exist, see e.g. section 5.3.3.3.3. in (NIST SEMATECH).

3.6 Sizing designs

An important issue in DOE is the total number of experiments, i.e. the size of a design. Sizing can be based on predictive power, or on the power of detecting differences of predefined size Δ. The latter is more commonly used, and many commercial DOE software packages have tools for determining the required number of estimates in such a way that the statistical power, i.e. $1 - \beta$ (β is the probability of type II error), has a desired value at a given level of significance α. For pairwise comparisons, exact methods based on the non-central t-distribution exist. For example, in R the function called power.t.test can be used to find the number of experiments needed in pairwise comparisons. For multiple comparisons, one can use the so-called Wheeler's formula (Wheeler, 1974) for an estimate of the required number of experiments n: $n = (4r\sigma / \Delta)^2$ where r is the number of levels of a factor, σ is the experimental standard deviation, and Δ is size of the difference. The formula assumes that the level of significance α is 0.05, and the power $1-\beta$ is 0.90. Wheeler gives also formulas for several other common design/model combinations (Wheeler, 1974).

4. Improving results by steepest ascent

If the goal of the experimentation has been to optimize something, the next step after analysing the results of a 2^N or a fractional 2^N design is to try to make improvement using knowledge provided by the analysis. The most common technique is the method of steepest ascent, also called the gradient (path) method.

4.1 Calculating the gradient path

It is well known from calculus that the direction of the steepest ascent on a response surface is given by the gradient vector, i.e. the vector of partial derivatives with respect to the design variables at a given point. The basic idea has been presented in section 3.2, and now we shall present the technical details.

In principle, the procedure is simple. First we choose a starting point, say $\mathbf{X}_0$, which typically is the centre point of the design. Then we calculate the gradient vector, say ∇ at this point. Note that it is important to use coded variables in gradient calculations. Next, the gradient vector has to be scaled small enough in order to avoid zigzagging (see 2.2). This can be done by multiplying the corresponding unit vector, $\nabla^0 = \nabla / \|\nabla\|$, by a scaling factor, say c. Now, the gradient path points are obtained by calculating $\mathbf{X}_i = \mathbf{X}_{(i-1)} + c\nabla^0, i = 1, 2, \ldots, n$ where n is the number of points. Once the points have been calculated, the experimental points are chosen from the path so that the distance between the points matches the desired step size, typically 1 in coded units. Naturally, the coded values have to be decoded into physical values having the original units before experimentation.

4.2 Alternative improvement techniques

Another principle in searching optimal new experiments is to use direct optimization techniques using the current model. In this approach, first the search region has to be defined. There are basically two different alternatives: 1) a hypercube whose centre is at the design centre with a given length for the sides of the hypercube, or 2) a hypersphere whose centre is at the design centre with a given radius. In the first alternative, typically the length of the side is first set to a value slightly over 2, say 3, giving mild extrapolation outside the experimental region. In the latter, typically the length of the radius is first set to a value slightly over 1, say 1.5, giving mild extrapolation outside the experimental region.

If the model is a linear plus pair-wise interactions model, the solution can easily be shown to be one of the vertices of the hypercube in the hypercube approach. If the model is a quadratic one, and the optimum (according to the model) is not inside the hypercube, the solution is a point on one of the edges of the hypercube and a point on the hypersphere in the hypersphere approach. In both approaches, the solution is found most easily using some iterative constrained optimization tool, e.g. Excel's Solver Tool. In the latter (hypersphere) approach, it is easy to show, using the Lagrange multiplier technique of constrained optimization, that the optimal point $\mathbf{X}_{opt}$ on the hypersphere of radius r is obtained by $\mathbf{X}_{opt} = (\mathbf{B} + 2\lambda \mathbf{I})^{-1} \mathbf{b}$, where λ is solved from the equation $(\mathbf{B} + 2\lambda \mathbf{I})^{-1} \mathbf{b}^2 = r^2$. The notation is explained in section 5.2. Unfortunately, λ must be solved numerically unless the model is linear. The benefit of using (numerical) iterative optimization in both approaches, or using the gradient path technique, is that they all work for all kind of models, not only for quadratic ones.

4.3 Example

Let us continue with the example of section 3.3.3 and see how the model can be used to design new experiments along the gradient path. The model of the previous example can be written (in coded variables)

$$y = b_0 + b_1 X_1 + b_2 X_2 + b_4 X_4 + b_5 X_5 + b_{23} X_2 X_3 + b_{45} X_4 X_5 \qquad (7)$$

The coefficients (b's) refer to the values given in Table 12. The gradient, i.e. the direction of steepest ascent, is the vector of partial derivatives of the model with respect to the variables. Differentiating the expression given in Eq. 7 gives in matrix notation

$$\nabla = \begin{bmatrix} b_1 & b_2 - b_{23} X_3 & b_{23} X_2 & b_4 + b_{45} X_5 & b_5 + b_{45} X_4 \end{bmatrix}^T \qquad (8)$$

Because this is a directional vector, it can be scaled to have any length. If we want it to have unit length, it must be divided by its norm, i.e. we use $\nabla / \|\nabla\|$. Now, let us start the calculation of the gradient path from the centre of the design, where all coded values are zeros. Substituting numerical values into Eq. 8 gives

$$\nabla \approx \begin{bmatrix} -3.35 & -2.16 & 0.00 & 4.64 & -4.72 \end{bmatrix}^T \qquad (9)$$

The norm of this vector is ca. 7.73. Dividing Eq. 9 by its norm gives

$$\frac{\nabla}{\|\nabla\|} \approx \begin{bmatrix} -0.43 & -0.28 & 0.00 & 0.60 & -0.61 \end{bmatrix}^T \qquad (10)$$

These are almost the same values as in the example 6.3.2 in (Box & Draper, 2007) though we have used a different model with significant interaction terms included. The reason for this is that the starting point is the centre point where the interaction terms vanish because the coefficients are multiplied by zeros.

The vector of Eq. 10 tells us that we should decrease the time by 0.43 coded units, the temperature by 0.28 coded units, and the amount of reagent D by 0.61 coded units and increase the amount of reagent C by 0.60 coded units. Of course, there isn't much sense to carry out this experiment because it is inside the experimental region. Therefore we shall continue from this point onwards in the direction of the gradient. Now, because of the interactions, we have to recalculate the normed gradient at the new point where $X_1 = -0.43$, $X_2 = -0.28$, $X_3 = 0.00$, $X_4 = 0.60$, and $X_5 = -0.61$.

When this is added to the previous values, we get $X_1 = -0.80$, $X_2 = -0.52$, $X_3 = 0.05$, $X_4 = 1.23$, and $X_5 = -1.25$. These values differ slightly more from the values in the original source; however the difference has hardly any significance. The difference becomes more substantial if we continue the procedure because the interactions start to bend the gradient path. Box and Draper calculated the new points at distances 2, 4 , 6 and 8 in coded units from the centre point. If we do the same we get the points given in Table 13.

Distance	X_1	X_2	X_3	X_4	X_5
2	-0.80	-0.52	0.05	1.23	-1.25
4	-1.38	-0.91	0.21	2.55	-2.57
6	-1.82	-1.25	0.40	3.90	-3.93
8	-2.18	-1.55	0.62	5.26	-5.23

Table 13. Four new experiments (in coded units) along the gradient path.

For comparison, the values in the book together with the reported yields of these experiments are given in Table 14.

Distance	X_1	X_2	X_3	X_4	X_5	yield
2	-0.86	-0.56	0.08	1.20	-1.22	72.6
4	-1.72	-1.12	0.16	2.40	-2.44	85.1
6	-2.58	-1.68	0.24	3.60	-3.66	82.4
8	-3.44	-2.24	0.32	4.80	-4.88	80.8

Table 14. Four new experiments (in coded units) along the gradient path given in (Box & Draper, 2007).

The differences in the design variables in the last two rows start to be significant, but unfortunately we cannot check whether they had been any better than the ones used in the actual experiments. The actual experiments really gave substantial improvement; see Example 6.3.2 in (Box & Draper, 2007).

Before going to quadratic designs and models, let us recall what was said about calculation step in section 3.2 and let us calculate the gradient using a step size 0.1 instead of 1.0, but tabulating only those points where the sum of the steps is two, i.e. the arc length along the path between two sequential points is approximately 2. These points are given in Table 15.

Distance	X_1	X_2	X_3	X_4	X_5
2	-0.74	-0.48	0.08	1.26	-1.27
4	-1.28	-0.87	0.24	2.58	-2.61
6	-1.70	-1.20	0.43	3.94	-3.96
8	-2.04	-1.50	0.64	5.30	-5.34

Table 15. Four new experiments (in coded units) along the gradient path using a small step size in the gradient path calculation.

If you compare these values with our first table, the differences are not big. The reason is that the model has not quadratic terms and the zigzag effect, explained in section 3.2, would take place only with really large step sizes. In any case, the best way to do these calculations is to use appropriate software, and then it doesn't matter if you calculate more accurately using a small step size.

Of course, before experimentation, one has to convert the coded units back to physical units. This could be easily done by solving for the variables in physical units from the equations given in the column Formula in Table 9. However, the easiest way is to use appropriate software. Table 16 gives the values in Table 15 in physical units.

Distance	t	T	B	C	D
2	6.5	86.3	46.1	118.2	38.6
4	5.4	85.3	48.6	134.8	32.0
6	4.6	84.5	51.5	151.7	25.2
8	3.9	83.7	54.6	168.8	18.3

Table 16. Experiments of Table 15 in physical units.

5. Second and higher order designs and response surface modelling

When the response doesn't depend linearly on the design variables, two-level designs are not adequate. Nonlinear behaviour can typically be detected by comparing the centre point results with the actual 2^N design point results. Alternatively, a nonlinear region is found by steepest ascent experiments. Sometimes nonlinearity can be assumed by prior knowledge about the system under study. There are several alternative designs for empirical nonlinear modelling. Before going to the different design alternatives let us review the most common nonlinear empirical model types. The emphasis is on so-called quadratic models, commonly used in the Box-Wilson strategy of empirical optimization. We shall first introduce typical models used with these designs, and after that, introduce the most common designs used for creating such models.

5.1 Typical nonlinear empirical models

The most common nonlinear empirical model is a second order polynomial of the design variables, often called a quadratic response surface model, or simply, a quadratic model. It is a linear plus pairwise interactions model added with quadratic terms, i.e. design variables raised to power 2. For example, a quadratic model for two variables is $y = b_0 + b_1 X_1 + b_2 X_2 + b_{12} X_1 X_2 + b_{11} X_1^2 + b_{22} X_2^2$. In general, we use the notation that b_i is the coefficient of X_i, b_{ii} is the coefficient of X_i^2, and $b_{ij}, i < j$ is the coefficient of $X_i X_j$. Fig. 11 depicts typical quadratic surfaces of two variables X_1 and X_2. Now, let $\mathbf{B}$ be a matrix whose diagonal elements B_{ii} are defined by $B_{ii} = 2b_{ii}$, and the other elements $B_{ij}, i \neq j$ are defined by $B_{ij} = b_{ij}, i < j$ and $B_{ji} = b_{ij}, i < j$. By definition, the matrix $\mathbf{B}$ is symmetric. Also, let $\mathbf{b}$ be the vector of main effect coefficients, i.e. $\mathbf{b} = [b_1, b_2, \cdots, n_N]^T$. Using this notation, a quadratic model can expressed in matrix notation as $y = b_0 + \mathbf{x}^T \mathbf{b} + \frac{1}{2}\mathbf{x}^T \mathbf{B} \mathbf{x}$ where $\mathbf{x}$ is the vector $[X_1, X_2, \cdots, X_N]^T$.

If a quadratic model has more than 2 variables, any 2 variables can be chosen as free variables corresponding to the x and y axes of the plot, and the other variables are kept at constant levels. Varying the values of the other variables in a systematic way, a good overview of the dependencies can be obtained till up to 4 or 5 variables. With more variables, one must rely on computational techniques.

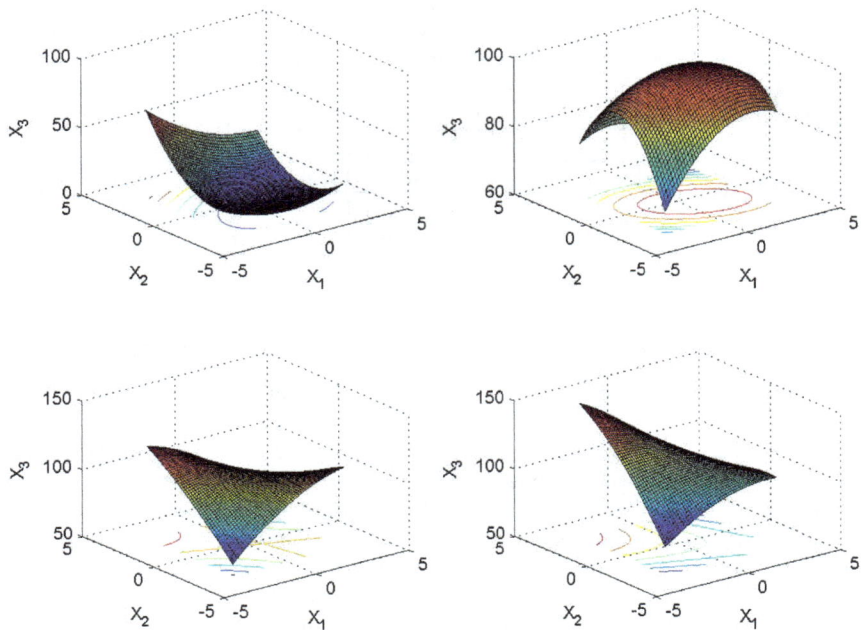

Fig. 11. Typical quadratic surfaces of 2 variables (X_1 and X_2): a surface having a minimum (upper left), a surface having a maximum (upper right), a surface having a saddle point (lower left), and a surface having a ridge (lower right).

Other model alternatives are higher order polynomials, rational functions of several variables, nonlinear PLS, neural networks, nonlinear SVM etc. With higher order polynomials, or with linearized rational functions, it advisable to use ridge regression, PLS, or some other constrained regression technique, see e.g. (Taavitsainen, 2010). These alternatives are useful typically in cases where the response is bounded in the experimental region; see e.g. (Taavitsainen et. al., 2010).

5.2 Estimation and validation of nonlinear empirical models

Basically the analyses and techniques presented in sections 3.1.2 and 3.2 are applicable to nonlinear models as well. Actually, polynomial models are linear in parameters, and thus the theory of linear regression applies. Normally, nonlinear regression refers to regression analysis of models that are nonlinear in parameters. This topic is not treated in this chapter, and the interested reader may see e.g. (Bard, 1973)

It should be noted that some of the designs presented in section 5.3 are not orthogonal, and therefore PLS or ridge regression are more appropriate methods than OLS for parameter estimation, especially in so-called mixture designs.

For quadratic models, a special form of analysis called canonical analysis is commonly used for gaining better understanding of the model. However, this topic is beyond the scope of this chapter, and the reader is advised to see e.g. (Box & Draper, 2007). Part of the canonical analysis is to calculate the so called stationary point of the model. A stationary point is a point where the gradient with respect to the design variables vanishes. Solving for the stationary point is straightforward. The stationary point is the solution of the linear system of equations $\mathbf{Bx} = -\mathbf{b}$, obtained by differentiation from the model in matrix form given in section 5.1. A stationary point can represent a minimum point, a maximum point, or a saddle point depending on the model coefficients.

5.3 Common higher order designs

Next we shall introduce the most common designs used for response surface modelling (RSM).

5.3.1 Factorial M^N designs

Full factorial designs with M levels can be used for estimating polynomials of order at most $M-1$. Naturally, these designs are feasible only with very few variables, say maximum 3, and typically for only few levels, say at most 4. For example, a 4^4 design would contain 256 which would be seldom feasible. However, the recent development in parallel microreactor systems having e.g. 64 simultaneously operating reactors at different conditions can make such designs reasonable.

5.3.2 Fractional factorial M^N designs, and mixed level factorial design.

Sometimes it is known that one or more variables act nonlinearly and the others linearly. For such cases a mixed level factorial design is a good choice. A simple way to construct e.g. a 3 or a 4 level mixed level factorial design is to combine a pair of variables in a 2^N design into a single new variable (Z) having 3 or 4 levels using the coding given in Table 17 (x_1, x_2, x_3 and x_4 represent the levels of the variable constructed from a pair of variables (X_i, X_j) in the original 2^N design).

X_i	X_j	Z (3 levels)	Z (4 levels)
-1	-1	x_1	x_1
-1	+1	x_2	x_2
+1	-1	x_2	x_3
+1	+1	x_3	x_4

Table 17. Construction of a 3, or 4 level variable from two variables of a 2^N design.

There are also fractional factorial designs which are commonly used in Taguchi methodology. The most common such designs are the so-called Taguchi L9 and L27 orthogonal arrays, see e.g. (NIST SEMATECH).

5.3.3 Box-Behnken designs

The structure and construction of Box-Behnken designs (Box & Behnken, 1960) is simple.

First, a 2^{N-1} design is constructed, say $\mathbf{X_0}$, then a $N \cdot 2^{N-1}$ by N matrix of zeros $\mathbf{X}$ is created. After this, $\mathbf{X}$ is divided into N blocks of 2^{N-1} rows and all columns, and in each block the columns, omitting the i'th column, is replaced by $\mathbf{X_0}$. Finally one or more rows of N zeros are appended to $\mathbf{X}$. This is easy to program e.g. in Matlab or R starting from a 2^{N-1} design. The following R commands will do the work for any number of variables (mton is a function that generates M^N designs, and nrep is the number of replicates at the centre point):

```
X0 <- as.matrix(mton(2,N-1))
M   <- 2^(N-1)
X   <- matrix(0,N*M,N)
for(i in 1:N) X[((i-1)*M+1):(M*i),(1:N)[-i]] <- X0
X   <- rbind(X,rep(nrep,N))
```

As an example, Table 18 shows a Box-Behnken design of 3 variables and 3 centre point replicate experiments.

X_1	X_2	X_3
0	-1	-1
0	-1	+1
0	+1	-1
0	+1	+1
-1	0	-1
-1	0	+1
+1	0	-1
+1	0	+1
-1	-1	0
-1	+1	0
+1	-1	0
+1	+1	0
0	0	0
0	0	0
0	0	0

Table 18. A Box-Behnken design with 3 variables.

5.3.4 Central composite designs

The so-called central composite (CC) designs are perhaps the most common ones used in RSM, perhaps due their simple structure (for other possible reasons, see section 5.3). As the name suggests they are composed of other designs, namely, of a factorial or fractional 2^N

part, of so-called axial points, and of centre points. Sometimes the latter two parts together are called a star design. As an example, Table 19 shows a CC design for two variables.

$\bar{X}_1$	X_2	
-1	-1	Factorial 2^2 part
-1	+1	
+1	-1	
+1	+1	
$-\alpha$	0	Axial points
0	$-\alpha$	
$+\alpha$	0	
0	$+\alpha$	
0	0	Centre points
0	0	

Table 19. A CC design with 2 variables.

The value α depends on the kind of properties we want the design to have. Typical desired properties are orthogonality, rotatability, and symmetry. A rotatable design is such that the prediction variance of a point in the design space does depend only on its distance from. design centre, not on its direction. Let us denote the number of the centre points by N_{cp}. Then, for an orthogonal design α is given by the following Eq. 11.

$$\alpha = \left[\left(\sqrt{2^N - 2N + N_{cp}} - \sqrt{2^N} \right)^2 \cdot \frac{2^N}{4} \right]^{\frac{1}{4}} \tag{11}$$

The derivation of this rather formidable looking equation, and of the two following ones, are given e.g. in (Box & Draper, 2007). It should be noted that the model matrix obtained with this choice for α is not strictly orthogonal, because the intercept column (the column of ones) vector is not orthogonal to the column vector of the quadratic terms. However, all the other columns of the model matrix are orthogonal to each other. This can also be expressed by saying that the quadratic effects are partially confounded with the intercept.

For a rotatable design, the appropriate value for α is given by the equation Eq. 12.

$$\alpha = 2^{\frac{N}{4}} \tag{12}$$

For maximal symmetry, i.e. all points except for the centre point, lie on a hypersphere of radius α, the appropriate value for α is given by the equation Eq. 13

$$\alpha = \sqrt{N} \tag{13}$$

A common fourth choice is to set α to 1. Such CC design is called a face centred CC design (CCF). For α's greater than 1 the designs are called circumscribed, CCC. Some other alternatives, e.g. compromising between orthogonality and rotatability, exist too. Sometimes a CCC design is scaled so that α is scale to 1, and the coordinates of the factorial points are scaled to $1/\alpha$. Such designs are called inscribed (CCI), though they actually are CCC designs with a different coding of variables.

Figs. 12 and 13 depict a CCC and a CCF design of 3 variables.

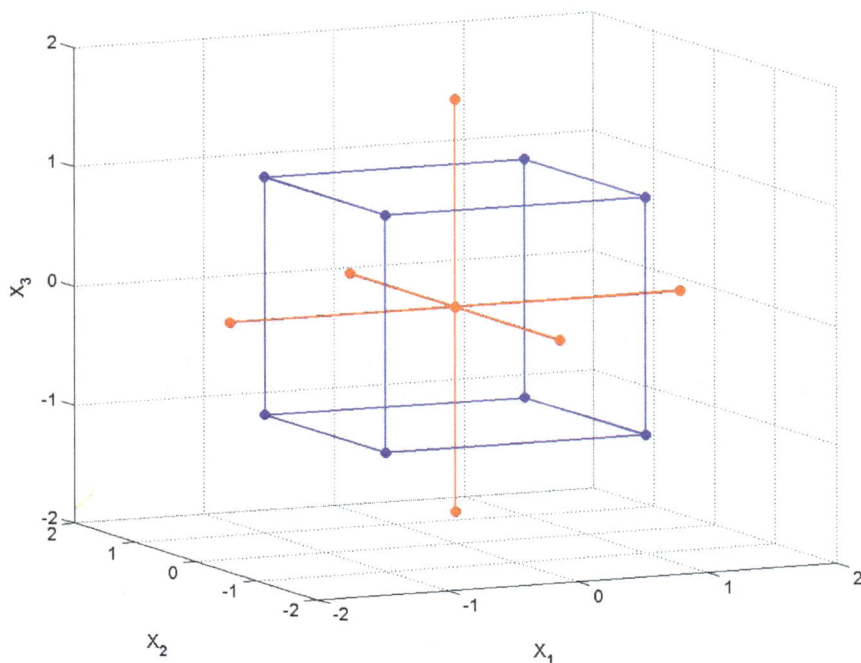

Fig. 12. A CCC design in coded units for 3 variables (X_1, X_2 and X_3) with $\alpha = 2^{\frac{3}{4}}$.

5.3.5 Doehlert designs

Doehlert designs are constructed from so-called regular simplexes. For example, a regular triangle and a regular tetrahedron represent regular simplexes in 2D and 3D, respectively. A Doehlert design for two variables consists of the vertexes of 6 adjacent regular triangles. Thus it comprises the vertexes of a regular hexagon plus the centre point. Doehlert designs fill the experimental space in a regular way in the sense that distances between the experimental points are constant. Doehlert designs have $1+N+N^2$ experimental points, which is less than in CC designs. Thus they are typically used in cases where the experiments are either very expensive or time consuming. The interested reader may refer to e.g. (Doehlert, 1970). Construction of Doehlert designs for more than 2 variables is rather tedious, and use of appropriate software, or tables of Doehlert designs, are recommended, see e.g. (Bruns & al, 2006)

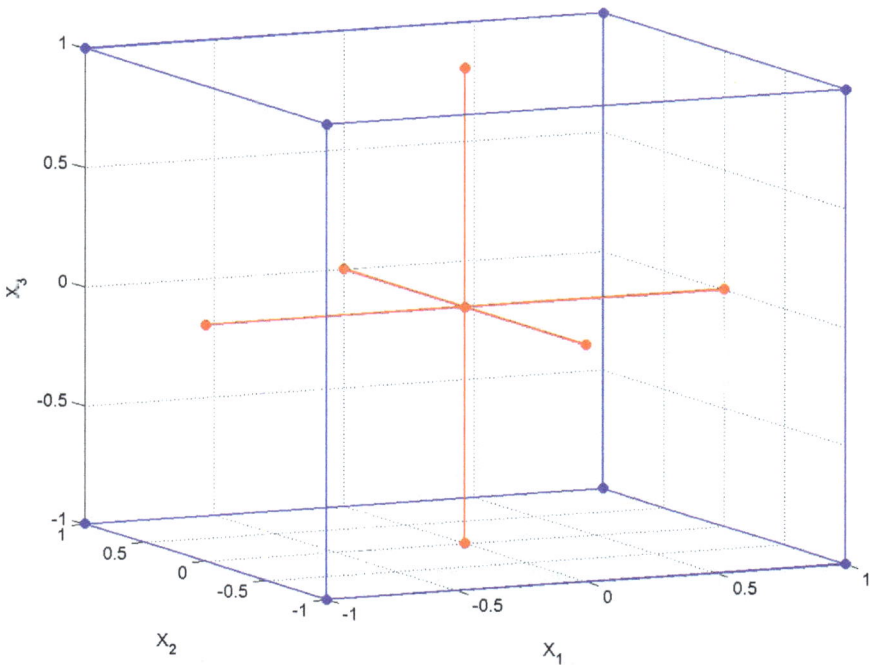

Fig. 13. A CCF design in coded units for 3 variables (X_1, X_2 and X_3) with $\alpha = 1$.

5.3.6 Mixture designs

If the design variables are proportions of constituents in a mixture, then in each experiment the values of the design variables sum up to 1 (100 %). In such cases, ordinary designs cannot be applied, since the row sums of ordinary designs vary irrespective of the coding used. If there are no other constraints than the closure constraint, the most common designs are the so-called simplex lattice, and simplex centroid designs. If some constraints are imposed as well, a good choice is to use optimal designs (see the next section), though other alternatives exist as well; see e.g. (Cornell, 1981), or (Montgomery, 1991).

The closure constraint has to be taken into account also in modelling results of mixture experiments. The closure means that the columns of the model matrix are linearly dependent making the matrix singular. One way to overcome this problem is to make the model using only $N-1$ variables, because we need to know only the values of $N-1$ variables, and the value of the N'th variable is one minus the sum of the others. However, this may make the interpretation of the model coefficients quite difficult. Another alternative is to use the so-called Scheffe polynomials, i.e. polynomials without the intercept and the quadratic terms. It can be shown that Scheffe polynomials of N variables represent the same model as an ordinary polynomial of $N-1$ variables, naturally with different values for the polynomial coefficients. For example the quadratic polynomial of two variables $y = b_1 X_1 + b_2 X_2 + b_{12} X_1 X_2$ can be simplified into $y = b_2 + (b_1 - b_2 + b_{12}) X_1 - b_{12} X_1^2$ if $X_2 = 1 - X_1$. This shows that it is a quadratic function of X_1 only; for more details see e.g. (Cornell, 1981).

The model matrices of mixture designs are not orthogonal, and they are usually quite ill-conditioned. For this reason, it is commonly recommended to use PLS or ridge regression for estimating the model parameters.

5.3.7 Optimal designs

The idea behind so-called optimal designs is to select the experimental points so that they satisfy some optimality criterion about the model to be used. It is important to notice that the optimality of such designs is always dependent on the model. For this reason, optimal designs are often used in designing experiments for mechanistic modelling problems. In empirical modelling we don't know the model representing the 'true' behaviour, and even a good empirical model is just an approximation of the true behaviour. Of course, if it has been decided to use e.g. a quadratic approximation, using a design that is optimal for a quadratic model is perfectly logical. However, the design still should have extra experiments that allow assessing the lack-of-fit.

Typically optimal designs are planned for quadratic models. Probably the most common optimality criterion is the D-optimality criterion. A D-optimal design is a design that minimizes the determinant of the information matrix, i.e. $\left| X^T X \right|^{-1}$ where X is the model matrix. There are several other optimality criteria, typically related to minimizing the variance of predictions, or to minimizing the variances of the model parameter estimates. In many cases, a design that is optimal according to one criterion is also optimal or nearly optimal according to several other criteria as well.

A nice feature in optimal designs is that it is easy take into account constraints in the design space, e.g. a mixture constraint, or a constraint in which one variable always has to have a greater value than some other variable. Constraints can sometimes be handled by some 'tricks', e.g. instead of using x_1 and x_2 when $x_1 < x_2$, one could use in design x_1 and x_3 and set $x_2 = x_1 + x_3$, i.e. to use a variable that tells how much greater to the value of x_1 the value of x_2 is. In general, using optimal designs is the most straightforward approach for constrained problems.

In practice, constructing optimal designs requires suitable software. Optimal design routines are available in most commercial statistical software packages containing tools for DOE. There is also an R package for creating optimal designs, called AlgDesign (http://cran.r-project.org/web/packages/AlgDesign/index.html). See also (Fedorov, 1972) or (Atkinson et. al., 2007).

5.4 Choosing an appropriate second order design

As we have seen, there are many types of designs for nonlinear empirical (usually quadratic) models. How does a practitioner know which one to choose? A good strategy is to try first a simple design that has extra degrees of freedom for validation and for checking model adequacy. Of course, if the problem at hand is a mixture problem, one has to rely on mixture designs or optimal designs. If the experiments are very expensive, one may have to use saturated, or almost saturated designs, e.g. optimal designs or Doehlert designs. In other cases CC or Box-Behnken designs are better choices. For 3 variables, a Box-Behnken design contains fewer experiments than a CC design for 3 variables, but for more variables it is the other way round. For example, a 4 variable Box-Behnken design (without replicates) contains 33 experiments, as the corresponding CC design contains 25 experiments. Thus, except for mixture problems or constrained problems, a CC design is usually the best choice. In general, CCC designs should be preferred to CCF designs, but otherwise choosing the value for α is usually not a big issue from the practical point of view; the differences in performance are minor. CCF designs should be used only in cases where there is a real benefit of having fewer variable levels than the 5 variable levels of CCC designs (CCF designs use only 3 variable levels).

5.5 Example: Analysis of a Doehlert design for two variables

This example comes from (Dos Santos et. al., 2008). The aim was to optimize the recovery percentage of several elements with respect to the temperature and the volume of concentrated nitric acid from which we take only the recovery percentage of manganese (for details, see (Dos Santos et. al., 2008). The design is a Doehlert design with 3 replicates, and it is given in physical units in Table 20.

Temperature	Volume	Recovery %
135	5	89.0
165	5	90.2
120	3	90.4
150	3	94.3
150	3	91.6
150	3	91.2
180	3	91.0
135	1	82.6
165	1	88.0

Table 20. A Doehlert design with 2 variables.

Next, the variables are coded so that the maximum values are set to +1 and the minimum values are set to -1. Thus the coding formulas will be $X_1 = \dfrac{T - 150}{30}$, and $X_2 = \dfrac{V - 3}{2}$. The design in coded units is given in Table 21.

X_1	X_2	Recovery %
-0.5	+1	89.0
0.5	+1	90.2
-1	0	90.4
0	0	94.3
0	0	91.6
0	0	91.2
+1	0	91.0
-0.5	-1	82.6
+0.5	-1	88.0

Table 21. A Doehlert design with 2 variables in coded units.

Fig. 14 shows the design together with the recoveries visualizing the hexagonal structure of the design.

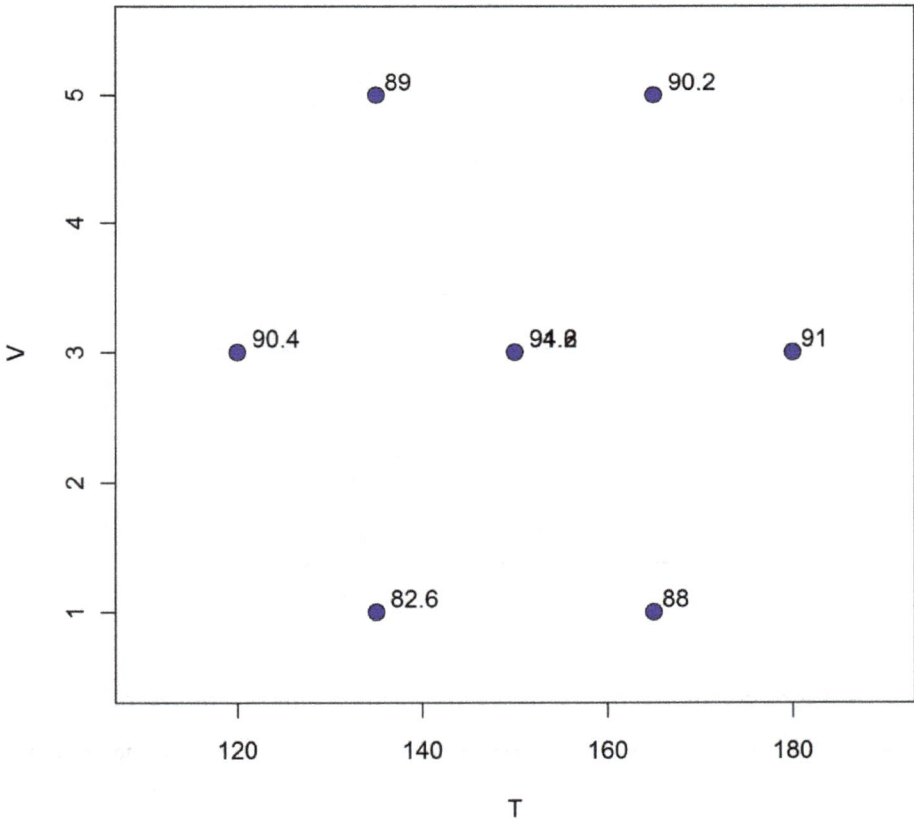

Fig. 14. The design of example 5.4.1 together with the measured recovery percentages.

Next a quadratic model is fitted to the data. The parameter estimates and the related statistics are given in Table 22.

	Estimate	Std. Error	t value	p value
(Intercept)	92.37	1.14	81.06	4.14e-06
X1	1.30	1.14	1.14	0.337
X2	2.15	0.99	2.18	0.118
I(X1^2)	-1.67	1.80	-0.93	0.423
I(X1 * X2)	-2.10	1.97	-1.06	0.365
I(X2^2)	-4.50	1.35	-3.33	0.045

Residual standard error: 1.974 or 3 degrees of freedom
Multiple R-squared: 0.8594,   Adjusted R-squared: 0.6251
F-statistic: 3.668 on 5 and 3 DF, p-value: 0.1569

Table 22. Regression summary of the quadratic model.

According to Table 22 only the intercept and the quadratic effect of x_2 are significant. The p-value of the lack-of-fit test based on the 3 replicates is ca. 0.28. Thus the lack-of-fit is not significant. The apparent reason for the low significance is the rather poor repeatability of the experiments. The standard deviation of the recoveries of the replicate experiments is ca. 1.68 which is relatively high compared to the overall variation in the recoveries.

Next, let us see the results of cross-validation. Before cross-validation, the 3 replicates are replaced by the average of them. Fig. 15 shows the cross-validation results.

According to the cross-validation the predictions of the model are not very good. Due to the poor repeatability, i.e. large experimental error, it is hard to tell whether the reason for unreliable prediction is the large experimental error or something else, e.g. more complicated nonlinearity than quadratic one. According to the model, the optimum lies inside the experimental region and it corresponds to the stationary point. The optimal point in coded units is $X_1 = 0.25$ and $X_2 = 0.17$ which corresponds to $T = 158$ and $V = 3.35$ in physical units. This should be compared to Fig. 16 which shows the corresponding response surface.

6. Multi-response optimization

A common problem is to optimize the values of several responses simultaneously. This occurs quite frequently, because many products have to meet several different goodness criteria. The problem in such applications is that the individual optima can be contradictory, i.e. the optimal values for one of the responses may be far from the optimal values for some other response. Several different techniques, such as finding the so-called Pareto optimal result, exist. By far the simplest approach to this problem is to use so-called desirability functions, presented in the next section. The idea was first presented by (Derringer & Suich, 1980) in an application of product development in rubber industry.

Optimization using the desirability function technique can be divided into the following steps:

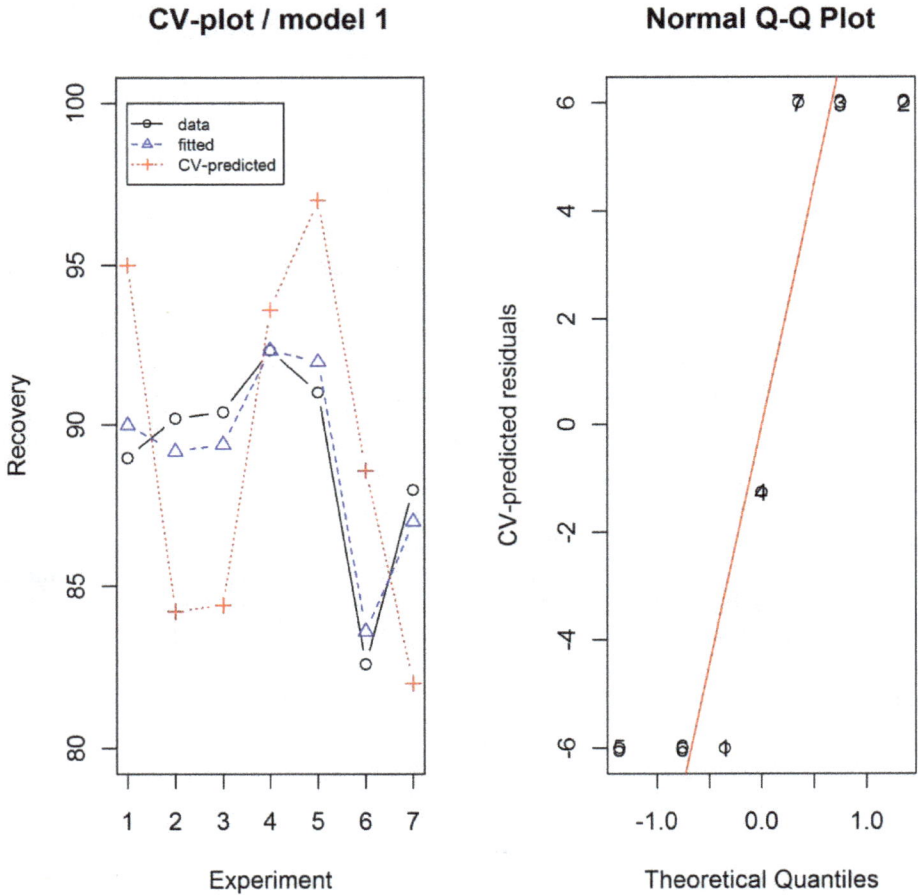

Fig. 15. Cross-validation of the quadratic model. Left panel: Recovery % vs. the number of experiment; black circles: data; blue triangles: fitted values; red pluses: cross-validated leave-one-out prediction. Right panel: Normal probability plots of the cross-validated leave-one-out residuals.

Recovery %

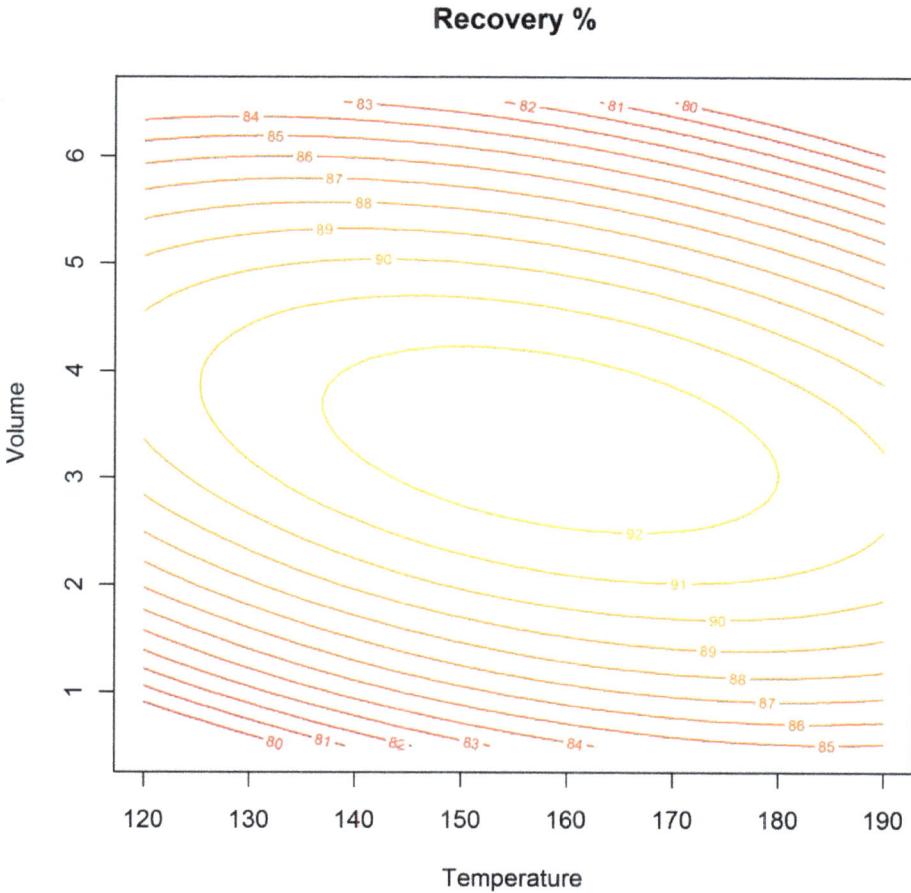

Fig. 16. Recovery % vs. Temperature and Volume.

First make a regression model, based on the designed experiments, individually for each response. Validate the models and proceed to step 2 after all models are satisfactory.

Make a desirability function $d_i = D_i(y_i)$ for each response separately (i goes from 1 to the number responses, say q). Remember that the responses have been modelled as functions of the design variables, i.e. $y_i = f_i(X_1, X_2, \cdots, X_N)$.

Building the desirabilities should be done together with a person who knows what the customers want from the product, and it is typically team work. How to build such functions in practice is explained later. Note that combining the two functions, desirabilities can be expressed as functions of design variables only.

Use an optimizer to maximize the combined desirability D which is the geometric mean of the individual desirabilities, i.e. $D = \left(D_1 \cdot D_2 \cdot \cdots \cdot D_q\right)^{\frac{1}{q}}$, with respect to the design variables.

Check by experimentation that the found optimum really gives a good product.

There are many ways to produce suitable desirability functions, one of which is explained in (Derringer & Suich, 1980). Any function that gives the 1 value for a perfect response and the value 0 for an unacceptable product and continuously values between 0 and 1 for responses whose goodness is in-between unacceptable and perfect can be used. One of the simplest alternatives is to use the following functions: $d_i = \left(1 + e^{-\frac{y_i - a}{b}}\right)^{-1}$ for one-sided desirabilities,

and $d_i = e^{-\left|\frac{y_i - a}{b}\right|^c}$ for two-sided desirabilities. The parameters a, b and c are user-defined parameters chosen with the help of an expert on the product quality.

The idea is best illustrated by an example. Let us consider an example where the product would be the better the higher its elasticity is. Let us also assume that elasticity from 0.60 upwards would mean a practically perfect product and elasticity below 0.30 would mean a totally unacceptable product. Then the one-sided desirability function looks like (with $a = 0.46$ and $b = 0.028$) the one given in Fig. 17.

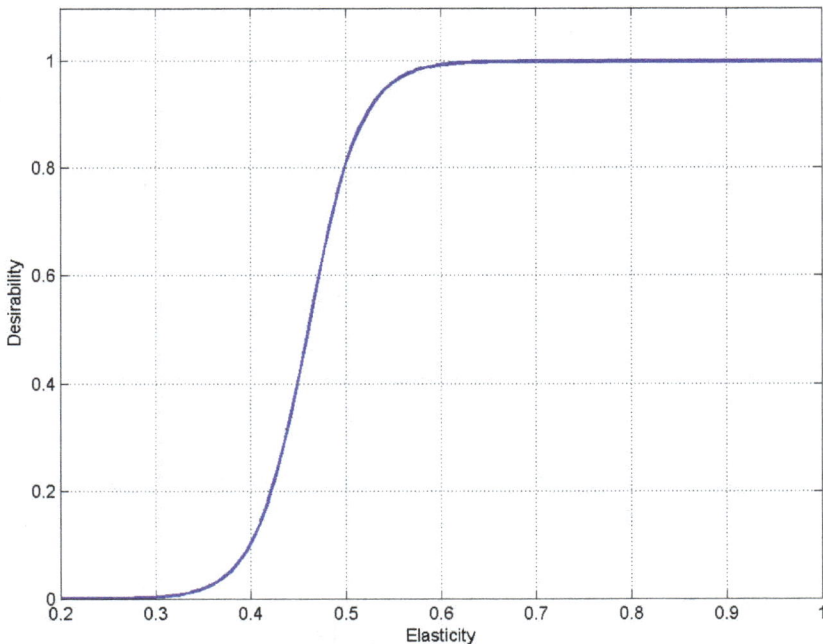

Fig. 17. A one-sided desirability function for elasticity that should be 0.60 or more and that would be totally unacceptable below 0.30.

If for some reason, the elasticity should not be higher than 0.60, and the elasticity over 0.90 or elasticity below 0.30 meant an unacceptable product, we would need a two-sided desirability function, e.g. like the one given in Fig. 18 (with $a = 0.60$, $b = 0.028$ and $c = 2.5$).

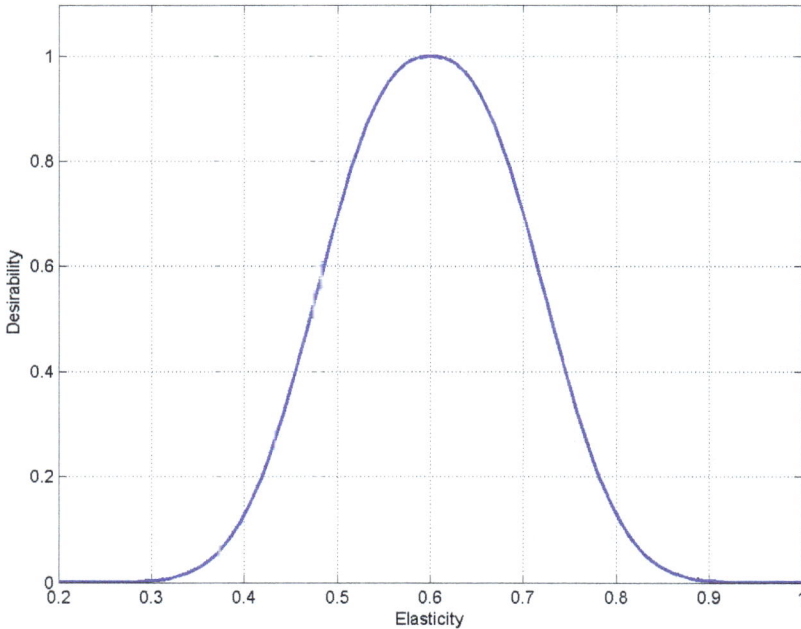

Fig. 18. A two-sided desirability function for elasticity that should be 0.60 and that would be totally unacceptable below 0.30 or above 0.90.

For practical examples see e.g. (Taavitsainen et. al., 2010) or (Laine et. al., 2011).

7. Conclusion

Design of experiments is as much of an art as of science. Becoming an expert in the field requires both theoretical studies and experience in practical applications. Although many problems can be solved in principle by hand calculations, in practice use of suitable software is needed. If the person involved is not familiar with command line style programs whose use is essentially that of programming, he or she is recommended to use some commercial software that typically also guide the user in the design and in the analysis of the results. The use of simulation models, where artificial experimental error is added into the results of the simulation, is highly recommended.

8. Acknowledgment

This work has been supported by the Helsinki Metropolia University of Applied Sciences.

9. References

Atkinson, A., Donev, A. & Tobias, R. (2007). *Optimum experimental designs, with SAS*. Oxford University Press, ISBN 978-0-19-929660-6, New York, USA

Bard, Y. (1973) *Nonlinear Parameter Estimation*, Academic Press, ISBN 978-0120782505, New York, USA

Bayne, K. & Rubin, I. (1986). *Practical Experimental Designs and Optimization Methods for Chemists*, VCH, ISBN 0-89573-136-3, Deerfield Beach, Florida, USA

Berthouex, P. & Brown, L. (2002). *Statistics for Environmental Engineers*, Lewis Publishers (CRC), ISBN 1-56670-592-4, Boca Raton, Florida, USA

Box, G. & Behnken, D. (1960). A simplex method for function minimization. *Technometrics*, Vol.2, No.4, pp. 455–475

Box, G. & Draper. (2007). *Response Surfaces, Mixtures, and Ridge Analyses*, Wiley (2nd ed.), ISBN 978-0-470-05357-7, Hoboken, New Jersey, USA

Box, G., Hunter, Hunter. (2005). *Statistics for Experimenters: Design, Innovation, and Discovery* (2nd ed.),Wiley, ISBN 978-0-471-71813-0, New York, USA

Bruns, R., Scarmiano, I. & de Barros Neto, B. (2006). *Statistical Design - Chemometrics* , Elsevier, ISBN 978-0-444-52181-1, Amsterdam, The Netherlands

Carlson, R. & Carlson, R. (2005). *Design and optimization in organic synthesis*, Elsevier

Cornell, J. (1981). *Experiments with Mixtures: Designs, Models, and the Analysis of Mixture Data* (3rd ed.), Wiley, ISBN 0-471-07916-2, New York, USA

Derringer, G. & Suich, R. (1980), Simultaneous Optimization of Several Response Variables. *Journal of Quality Technology*, Vol.12, pp. 214-219

Doehlert, D. (1970) Uniform shell designs. *Applied Statistics*, Vol.19, pp.231-239.

Dos Santos, W., Gramacho, D., Teixeira, A., Costa,A. & Korn, M. (2008), Use of Doehlert Design for Optimizing the Digestion of Beans for Multi-Element Determination by

Inductively Coupled Plasma Optical Emission Spectrometry, *J. Braz. Chem. Soc.*, Vol. 19, No. 1, pp.1-10.

Efron, B., Hastie, T., Johnstone. I. & Tibshirani, R. (2004), Least angle regression, *Ann. Statist.* Vol. 32, No. 2, pp. 407-499.

Fedorov, V. (1972). *Theory of Optimal Experiments.* Academic Press, ISBN 978-0824778811, New York, USA

Haaland, P. (1989). *Experimental Design in Biotechnology*, Marcel Dekker, ISBN 978-0824778811, New York, USA

Hanrahan, G. (2009). *Environmental Chemometrics*, CRC Press, ISBN 978-1420067965, Florida, USA

Himmelblau, D. (1970). *Process Analysis by Statistical Methods*, Wiley, ISBN 978-0471399858 , New York, USA

JMP, release 6, Design of Experiments. http://www.jmp.com/support/downloads/pdf/jmp_design_of_experiments.pdf, 9.9.2011

Kolarik, W. (1995). *Creating Quality*, McGraw-Hill, ISBN 0-07-113935-4

Koljonen, J., Nordling, T. & Alander, J. (2008), A review of genetic algorithms in near infrared spectroscopy and chemometrics: past and future, *Journal Of Near Infrared Spectroscopy*, Vol. 16, No. 3, pp. 189-197

Laine, P., Toppinen, E., Kivelä, R. Taavitsainen, V-M., Knuutila, O., Sontag-Strohm, T.,Jouppila, K. & Loponen, J.(2011), Emulsion preparation with modified oat bran: Optimization of the emulsification process for microencapsulation purposes, *Journal of Food Engineering*, Vol.104, pp.538-547

Montgomery, D. (1991). *Design and Analysis of Experiments* (3rd ed.), Wiley, ISBN 0-471-52994-X, Singapore

Nelder, J. & Mead, R. (1965) A simplex method for function minimization. *Computer Journal* Vol.7, No.4, pp. 308-313, ISSN 0010-4620

Neter, J., Kutner, M., Nachtsheim, C. & Wasserman, W. (1996), *Applied Linear Statistical Models* (4th ed.), WCB/McGraw-Hill, ISBN 0-256-11736-5, Boston, Massachusetts, USA

NIST/SEMATECH e-Handbook of Statistical Methods, http://www.itl.nist.gov/div898/handbook/, 9.8.2011

Taavitsainen, V-M. Ridge and PLS based rational function regression (2010), *Journal of Chemometrics.* Vol.24, No. 11-12, pp.665-673

Taavitsainen, V-M., Lehtovaara, A. & Lähteenmäki, M. (2010), Response surfaces, desirabilities and rational functions in optimizing sugar production, *Journal of Chemometrics.* Vol. 24, No. 7-8, pp. 505-513

Tibshirani, R. (1996), Regression shrinkage and selection via the lasso, *J. Royal. Statist. Soc. B.* Vol. 58, pp. 267-288

Vardeman, S. (1994). *Statistics for Engineering Problem Solving*, PWS , ISBN 978-0780311183, Boston, Massachusetts, USA

Weisberg, S. (1985). *Applied Linear Regression*, Wiley, ISBN 0-471-87957-6, New York, USA

Wheeler, R. (1974) Portable power , *Technometrics*, Vol. 16, No. 2, pp. 193-201

Zou, H. & Hastie, T. (2005), Regularization and Variable Selection via the Elastic Net, *Journal of the Royal Statistical Society, Series B*, Vol. 76, pp. 301-320.

Part 2

Biochemistry

6

Chemometric Study on Molecules with Anticancer Properties

João Elias Vidueira Ferreira[1], Antonio Florêncio de Figueiredo[2],
Jardel Pinto Barbosa[3] and José Ciríaco Pinheiro[3]
[1]*Universidade do Estado do Pará*
[2]*Instituto Federal de Educação, Ciência e Tecnologia do Pará*
[3]*Laboratório de Química Teórica e Computacional, Universidade Federal do Pará*
Brasil

1. Introduction

Cancer is a class of diseases characterized by uncontrolled growth of abnormal cells of an organism. All over the world millions of people die every year owing to one of the different types of cancer. Unfortunately cancer chemotherapy finds a serious limitation since treatment with drugs is followed by drug resistance in the tumorous cells and side effects (Efferth, 2005). So researches have been directed to make chemotherapy treatment more efficient.

In the late years literature has reported the research on natural products as a good strategy to discover new chemotherapy agents. One of the plants that have shown anticancer properties is *Artemisia annua L.* (*qinghao*). It has the active ingredient artemisinin, which is used as antimalarial. Artemisinin and derivatives have excellent efficacy against multidrug-resistant strains of *P. falciparum* and they are very well tolerated (Price et al., 1998). Recently the sensibility to artemisinin has been evaluated in some tumorous cells. Studies suggest that artemisinin is more toxic to cancerous cells than to normal cells, so giving a new perspective in cancer therapy (Lai et al, 2009).

However ... This book is on chemometrics and what has chemometrics to do with cancer chemotherapy? Well... understanding how these two different areas can be related to one another is the purpose of this chapter. You just must keep on reading this chapter and you will see the many ways chemometrics can be employed to investigate the "behavior" molecules exhibit considering anticancer activity and to make predictions about drugs that were not tested yet. The potential application of chemometrics to analytical data arising from problems in biology and medicine is enormous and, in fact, the applications of chemometrics have diversified substantially over the last few years (Brereton, 2007; 2009). At the end of the chapter you will note that, as in many areas of research, chemometrics plays an important role in medicinal chemistry, fortunately.

Firstly it is necessary to remember that producing a drug is something that takes time and money, so the process must be rationalized! However, in the past, drugs were discovered by synthesizing a lot of molecules, rather without rigorous criteria, and testing experimentally all of them to evaluate their capacity of cure of the disease or at least to control it. But in process of time this methodology became more and more inadequate, for the more new compounds are studied the less a new compound may be discovered to be potent against a disease. It has long been desired to design active structures on the basis of logic and calculations, not relying on chance or trial-and-error (Fujita, 1995).

Nowadays, in science, there is a basic assumpion that molecular properties and structural characteristics are closely connected to biological functions of the compounds. It is often assumed that compounds with similar properties and structures also display similar biological responses. Chemical structure encodes a large amount of information explaining why a certain molecule is active, toxic or insoluble (Rajarshi, 2008). Thus to understand the mechanism of action of a drug it is necessary to interpret the role played by its molecular and structural properties.

In the last decades, much scientific research has focused on how to capture and convert the information encoded in a molecular structure into one or more numbers used to establish quantitative relationships between structures and properties, biological activities or other experimental properties (Puzyn et al., 2010). Quantitative structure-activity relationship (QSAR) studies have been of great value in medicinal chemistry. Statistical tools can be used for the prediction of the biological activities of new compounds based only on the knowledge of their chemical structures, i.e., not depending on experimental data, which are unknown. Such a strategy gives very useful information for the understanding of the mechanisms of the action of drugs and proposals for syntheses, in this way rationalizing drug discovery. QSAR is alive and well (Doweyko, 2008), that is, QSAR has been used with success and so it is still of relevance today.

Moreover advances in computation brought software that made possible to get many different types of information (descriptors) about the molecules. Consequently data gathered through experiments and computers can produce a huge matrix whose elements are information related to molecules. But it seems that analyzing all them will require infinite patience!

What to do?

Chemometics has the solution!

That is true because chemometrics is the art of extracting chemically relevant information from data produced in chemical experiments (Wold, 1995). Most people only think of statistics when faced with a lot of quantitative information to process (Bruns et al., 2006). In this text we show a common and efficient methodology used in medicinal chemistry to rationalize the process of producing a new drug by employing chemometric methods. It is presented a molecular modeling and a chemometric study of 25 artemisinins, which involves artemisinin and derivatives (training set, Fig. 1) with different degrees of cytotoxicities against human hepatocellular carcinoma HepG2 (Liu et al, 2005), since among the malignant tumors in the liver, the hepatocellular carcinoma is very commom. Literature has showed the application of the methodology here described to investigate biological properties (antimalarial and anticancer) of artemisinin and derivatives (Barbosa et al., 2011); (Cardoso et al., 2008); (Pinheiro et al., 2003).

2. Methodology

Any chemometric study requires data. In this study data are obtained from molecular descriptors calculated through computation. The start point is the molecular modeling step, which consists on the construction of the structures and the complete optimization of their geometries through a quantum chemistry approach implemented in computer. This is necessary to represent molecules as real as possible and thus to compute their molecular descriptors. The B3LYP/6-31G** method (Levine, 1991) as implemented in the Gaussian 98 program was employed (Frisch et al., 1998), considering this strategy is suitable for optimizing well all structures since a good description of the geometrical parameters of artemisinin is achieved.

The 25 compounds investigated include artemisinin, amides, esters, alcohols, ketones, and five-membered ring derivatives. All compounds have been associated to their in vitro bioactivity against a human hepatocellular carcinoma cell line, HepG2, and were labeled previously into two classes according with their activities: (-) less active (those with $IC_{50} \geqslant 97$ μM) and (+) more active (those with $IC_{50} < 97$ μM) derivatives. The criteria for choosing this value of IC_{50} are rather subjective. Nevertheless it is convenient to say that 97 μM is the IC_{50} for artemisinin and the higher IC_{50} the less active is the compound.

After molecular modeling, 1700 descriptors (independent variables) were computed for each molecule in the training set. They represent different source of chemical information (features) regarding the molecules and include geometric, electronic, quantum-chemical, physical-chemical, topological descriptors and others. They are assumed to be important to understand molecular characteristics such as bioactivity against cancer. In fact one of the purposes of a research like this is to find which descriptors of the molecules are better related to the disease under study, in this example cancer. The software used to compute these descriptors were e-Dragon (Virtual Computational Laboratory , 2010), a product from the Virtual Computational Laboratory and Gaussian 98 (Frisch et al., 1998).

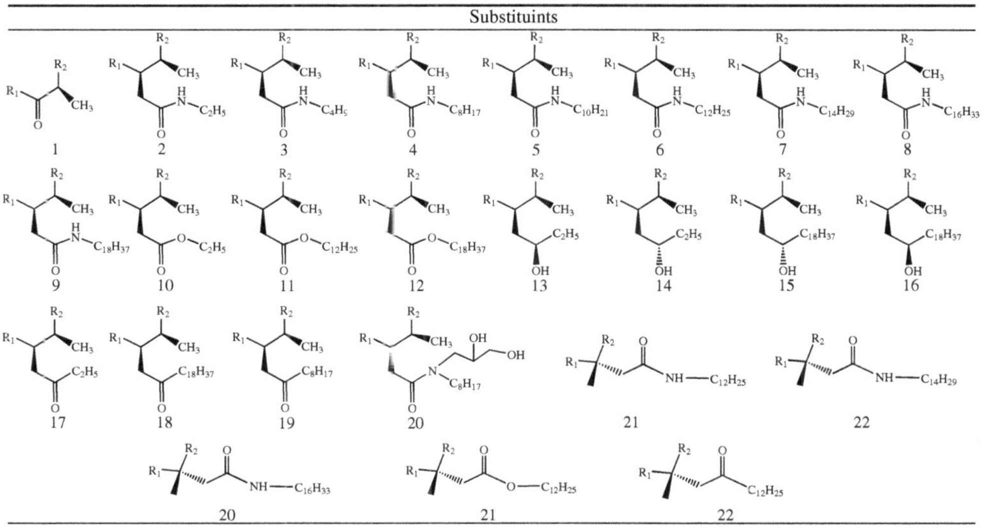

Fig. 1. Artemisinin and derivatives (training set) with different degrees of cytotoxicities against human hepatocellular carcinoma HepG2

However, a crucial point to be considered in any data analysis is preprocessing. The original data matrix usually does not have optimal value distribution for the analysis (for example it has different units and variances in variables), which requires some pretreatment prior to multivariate analysis. In general, the autoscale preprocessing, which results in scaled variables with zero mean and unit variance, is used (Ferreira, 2002). Then, all variables were auto-scaled as a preprocessing so that they could be standardized and this way could have the same importance regarding the scale.

Then the next step consists on application of multivariate statistical methods to find key features involving molecules, descriptors and anticancer activity. The methods include principal component analysis (PCA), hiererchical cluster analysis (HCA), K-nearest neighbor method (KNN), soft independent modeling of class analogy method (SIMCA) and stepwise discriminant analysis (SDA). The analyses were performed on a data matrix with dimension 25 lines (molecules) x 1700 columns (descriptors), not shown for convenience. For a further study of the methodology applied there are standard books available such as (Varmuza & Filzmoser, 2009) and (Manly, 2004).

2.1 PCA

Suppose that in your study, like in the example exhibited in this chapter, you have a large set of data, certainly it will not be a simple task to analyze so many variables and extract useful information from them. It would be a "revolution" in your research if you could confidently interpret all data in a simpler way. Fortunately, with the aid of PCA technique, this "revolution" can happen. Through PCA you can reduce the total number of variables to a smaller set while maintaining as much of the original information as is possible. No matter your area of research this is a great advantage.

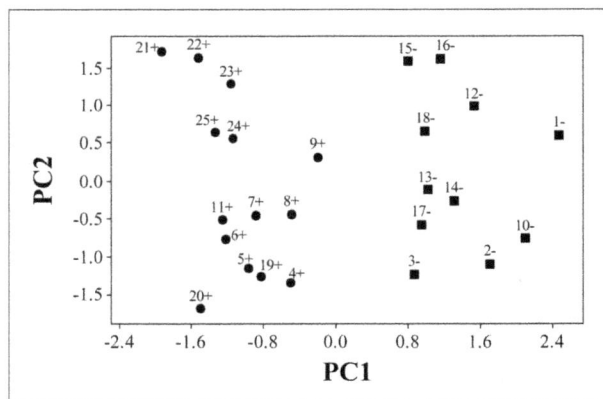

Fig. 2. Plot of PC1-PC2 scores for artemisinin and derivates (training set) with activity against human hepatocellular carcinoma HepG2. More active compounds displayed on the left side (plus sign) while less active ones on the right side (minus sign)

Now considering our data matrix, PCA was employed looking for a small group of descriptors so that they alone were responsible for classifying all 25 samples into two distinct classes: more active and less active. Besides it is desirable to choose uncorrelated descriptors that could be easier to interpret and analyze, trying to associate them to cytotoxicities against human hepatocellular carcinoma HepG2.

Furthermore, given the large quantity of multivariate data available, it was necessary to reduce the number of variables. Thus, if two any descriptors had a high Pearson correlation coefficient (r > 0.8), one of the two was randomly excluded from the matrix, since theoretically they describe the same property to be modeled (biological response). Therefore it is sufficient to use only one of them as an independent variable in a predictive model (Ferreira, 2002). Moreover those descriptors that showed the same values for most of the samples were eliminated too.

Compound	IC5	Mor29m	O1	MlogP	Activity
1	4.862	-0.305	-0.246	2.845	97
2	5.253	-0.308	-0.200	2.630	>100
3	5.389	-0.372	-0.202	3.080	>100
4	5.628	-0.445	-0.194	4.845	9.5
5	5.684	-0.474	-0.205	5.250	2.8
6	5.624	-0.525	-0.214	5.644	1.2
7	5.501	-0.514	-0.211	6.027	0.46
8	5.364	-0.518	-0.191	6.400	0.79
9	5.225	-0.501	-0.210	6.765	4.2
10	5.217	-0.236	-0.205	3.036	>100
11	5.597	-0.526	-0.218	6.050	0.72
12	5.197	-0.179	-0.225	7.171	>100
13	5.253	-0.364	-0.246	3.141	>100
14	5.253	-0.322	-0.237	3.141	>100
15	5.159	-0.294	-0.259	7.095	>100
16	5.159	-0.232	-0.258	7.095	>100
17	5.180	-0.443	-0.219	2.996	>100
18	5.168	-0.307	-0.209	7.131	>100
19	5.624	-0.485	-0.186	5.644	1.8
20	5.856	-0.518	-0.218	3.941	3.5
21	5.543	-0.562	-0.344	5.449	1.3
22	5.419	-0.560	-0.320	5.837	0.77
23	5.280	-0.591	-0.281	6.215	0.74
24	5.516	-0.498	-0.269	5.855	3.7
25	5.488	-0.545	-0.273	5.815	0.47
Mean	5.378	-0.425	-0.234	5.164	
Stardard Deviation	0.225	0.121	0.040	1.570	

Table 1. Values of the four descriptors selected through PCA for compounds from the training set

After this step, PCA was performed in order to continue reducing the dimensionality of the data, find descriptors that could be useful in characterizing the behavior of the compounds acting against cancer and look for natural clustering in the data and outlier samples. While processing PCA, several attempts to obtain a good classification of the compounds are made. At each attempt, one or more variables are removed, PCA is run and the score and loading plots are analyzed.

The score plot gives information about the compounds (similarities and differences). The loading plot gives information about the variables (how they are connected to each other and which are the best to describe the variance in the original data). Depending on the results displayed by the plots, variables remain removed or included in the data matrix. If a removal of a variable contributes to separate compounds showed by the score plot into two classes (more and less active), then in the next attempt PCA is run without this variable. But if no improvement is achieved, then the variable removed is inserted in the data matrix, another variable is selected to be removed and PCA is run again. The loadings plot gives good clues on which variables must be excluded. Variables that are very close to one another indicate they are correlated and, as stated before, only one of them needs to remain.

This methodology comprises part of the art of variable selection: patience and intuition are the fundamental tools here. It is not necessary to mention that the more you know about the system you are investigating (samples and variables and how they are connected), the more you can have success in the process of finding variables that really are important to your investigation. Variable selection does not occur like magic, at least, not always!

The descriptors selected in PCA were *IC5*, *Mor29m*, *O1* and *MlogP*, which represent four distinct types of interactions related to the molecules, especially between the molecules and the biological receptor. These descriptors are classified as steric (*IC5*), 3D-morse (*Mor29m*), electronic (*O1*) and molecular (*MlogP*). The main properties of a drug that appear to influence its activity are its lipophilicity, the electronic effects within the molecule and the size and shape of the molecule (steric effects) (Gareth, 2003).

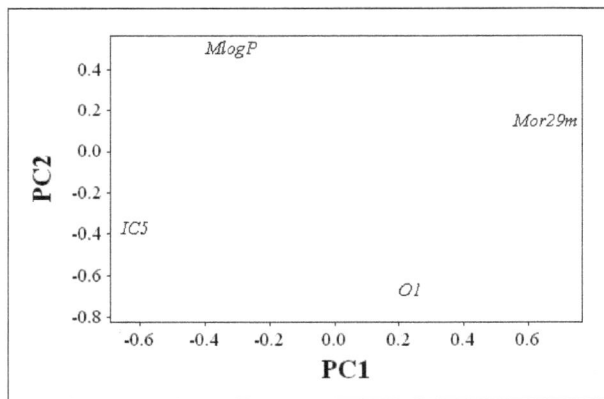

Fig. 3. Plot of the PC1-PC2 loadings for the four descriptors selected through PCA

The PCA results show the score plot (Fig. 2) relative to the first and second principal components. In PC1, there is a distinct separation of the compounds into two classes. More active compounds are on the left side, while less active are on the right side. They were chosen among all data set (1700 descriptors) and they are assumed to be very important to investigate anticancer mechanism involving artemisinins. Table 1 displays the values computed for these four descriptors. This step was crucial since a matrix with 1700 columns was reduced to only 4 columns. No doubt it is more appropriate to deal with a smaller matrix. The first three principal components, PC1, PC2 and PC3 explained 43.6%, 28.7% and 20.9% of the total variance, respectively. The Pearson correlation coefficient between the variables is in general low (less than 0.25, in absolute values); exception occurs between *Mor29m* and *IC5*, which is -0.65).

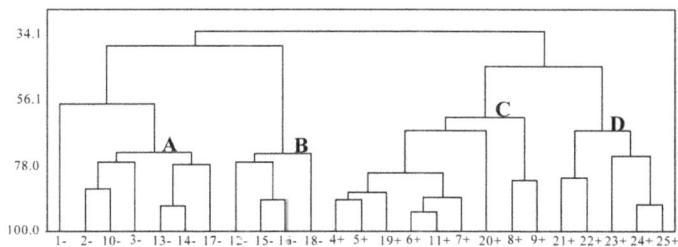

Fig. 4. HCA dendogram for artemisinin and derivatives (training set) with biological activity against human hepatocellular carcinoma HepG2. Plus sign for more active compounds while minus sign for less active ones

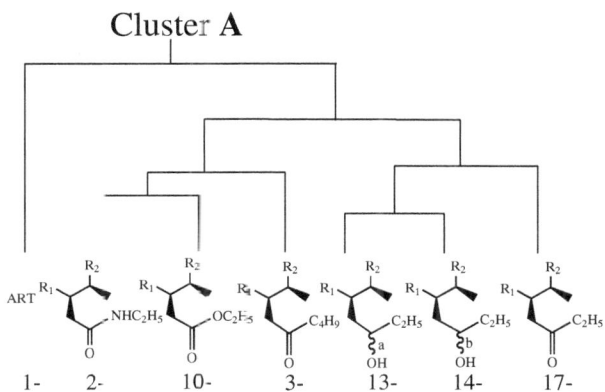

Fig. 5. Cluster **A**

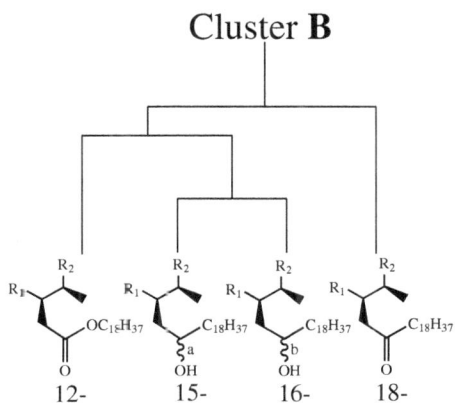

Fig. 6. Cluster **B**

The loading plot relative to the first and second principal components can be seen in Fig. 3. PC1 and PC2 are expressed in Equations 1 and 2, respectively, as a function of the four selected descriptors. They represent quantitative variables that provide the overall predictive ability

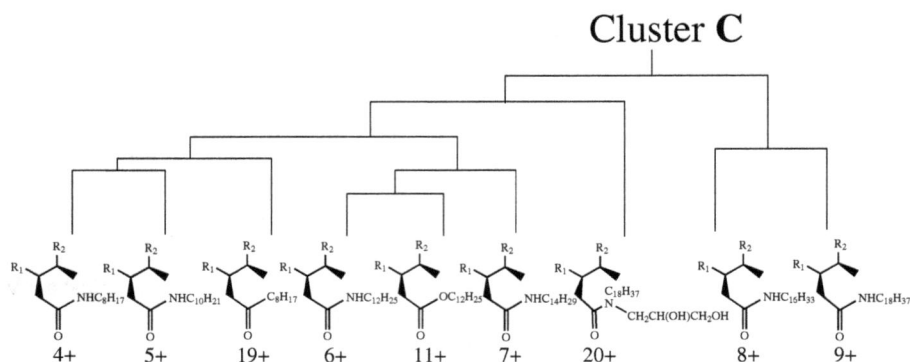

Fig. 7. Cluster **C**

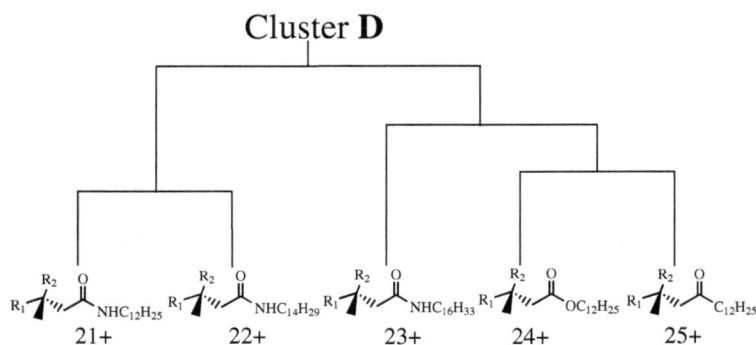

Fig. 8. Cluster **D**

of the different sets of molecular descriptors selected. In Equation 1 the loadings of *IC5* and *MlogP* are negative whereas they are positive for *Mor29m* and *O1*. Among all of them *IC5* and *Mor29m* are the most important to PC1 due to the magnitude of their coefficients (-0.613 and 0.687, respectively) in comparison to *O1* and *MlogP* (0.234 and -0.313, respectively). For a compound to be more active against cancer, it must generally be connected to negative values for PC1, that is, it must present high values for *IC5* and *MlogP*, but more negative values for *Mor29m* and *O1*.

$$PC1 = -0.613IC5 + 0.687Mor29m + 0.234O1 - 0.313MlogP \tag{1}$$

$$PC2 = -0.445IC5 + 0.081Mor29m - 0.743O1 + 0.493MlogP \tag{2}$$

2.2 HCA

Considering the necessity of grouping molecules of similar kind into respective categories (more and less active ones), HCA is suitable for this purpose since it is possible to visualize the disposition of molecules with respect to their similarities and so make suppositions of how they may act against the disease. When performing HCA many approaches are available. Each one differs basically by the way samples are grouped.

Compound	K1	K2	K3	K4	K5	K6
1	−	−	−	−	−	−
2	−	−	−	−	−	−
3	−	−	−	−	−	−
4	+	+	+	+	+	+
5	+	+	+	+	+	+
6	+	+	+	+	+	+
7	+	+	+	+	+	+
8	+	+	+	+	+	+
9	+	+	+	+	+	+
10	−	−	−	−	−	−
11	+	+	+	+	+	+
12	−	−	−	−	−	−
13	−	−	−	−	−	−
14	−	−	−	−	−	−
15	−	−	−	−	−	−
16	−	−	−	−	−	−
17	−	−	−	−	−	−
18	−	−	−	−	−	−
19	+	+	+	+	+	+
20	+	+	+	+	+	+
21	+	+	+	+	+	+
22	+	+	+	+	+	+
23	+	+	+	+	+	+
24	+	+	+	+	+	+
25	+	+	+	+	+	+

Table 2. Classification of compounds from the training set according to KNN method

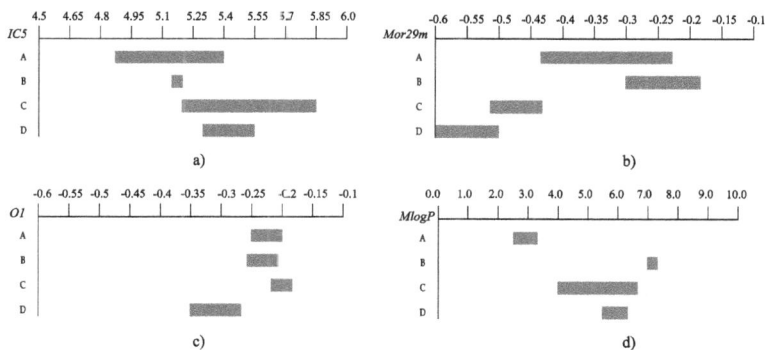

Fig. 9. Variations in descriptors: a) Variations in *IC5* for each cluster; b) Variations in *Mor29m* for each cluster; c) Variations in *O1* for each cluster; d) Variations in *MlogP* for each cluster

In this work, classification through HCA was based on the Euclidean distance and the average group method. This method established links between samples/cluster. The distance between two clusters was computed as the distance between the average values (the mean vector or centroids) of the two clusters. The descriptors employed in HCA were the same selected in

PCA, that is, *IC5*, *Mor29m*, *O1* and *MlogP*. The representation of clustering results is shown by the dendogram in Fig. 4, which depicts the similarity of samples. The branches on the bottom of the dendogram represent single samples. The length of the branches linking two clusters is related to their similarity. Long branches are related to low similarity while short branches mean high similarity. On the scale of similarity, a value of 100 is assigned to identical samples and a value of 0 to the most dissimilar samples. For a better interpretation of the dendogram, the clusters are also analyzed alone (Figs. 5, 6, 7 and 8), and variations in descriptors in each cluster are presented in Fig. 9. The scale above each figure is associated to the property considered and the letters indicate the cluster in the dendogram. It is easily recognized that descriptors in clusters in general have different pattern of variations, a characteristic supported by the fact that clusters have different groups of molecules.

Group or class	Number of Compounds	Compounds wrongly classified					
		K1	K2	K3	K4	K5	K6
Less active	11	0	0	0	0	0	0
More active	14	0	0	0	0	0	0
%Correct information	25	100	100	100	100	100	100

Table 3. Classification matrix obtained by using KNN

The dendogram shows compounds classified into two different classes according to their activities with no sample incorrectly classified. Less active compounds are on the left side and are divided into clusters **A** (Fig. 5) and **B** (Fig. 6). In cluster **A** substituints have either C_2H_5 (**2, 10, 13, 14** and **17**) or C_4H_9 (**3**). Here the lowest values for *IC5* (Fig. 9a) and *MlogP* are found (Fig. 9d). In cluster **B** (**12, 15, 16** and **18**) all substituints have $C_{18}H_{37}$ and are present the highest values for *MlogP* (Fig. 9d). Considering more active samples, right side of the figure, in cluster **C** (Fig. 7) compounds have amide group (exception is **11**, ester, and **19**, ketone) and attached to this group there is an alkyl chain of 8 to 18 carbon atoms. Here the descriptor *IC5* displays the highest values (Fig. 9a). In Cluster **D** (Fig. 8) substituints have an alkyl chain of 12 to 16 carbon atoms and the six-membered ring molecules with oxygen O_{11} are replaced by five-membered ring molecules. Compounds display the lowest values for *Mor29m* (Fig. 9b) and *O1* (Fig. 9c).

Besides these two methods of classification (PCA and HCA), others (KNN, SIMCA and SDA) were applied to data. They are important to construct reliable models useful to classify new compounds (test set) regarding their ability to face cancer. This is certainly the ultimate purpose of many researches in planning a new drug.

2.3 KNN

This method categorizes an unknown object based on its proximity to samples already placed in categories. After built the model, compounds from the test set are classified and their classes predicted taking into account the multivariate distance of the compound with respect to K samples in the training set. The model built for KNN in this example employs leave one out method, has 6 (six) as a maximum k value and autoscaled data. Table 2 shows classification for each sample at each k value. Column number corresponds to k setting so that the first column of this matrix holds the class for each training set sample when only one neighbor (the nearest) is polled whereas the last column holds the class for the samples when the kmax nearest neighbors are polled. Tables 2 and 3 summarizes the results for KNN analysis. All 6-nearest neighbors classified samples correctly.

2.4 SIMCA

The SIMCA method develops principal component models for each training set category. The main goal is the reliable classification of new samples. When a prediction is made in SIMCA, new samples insufficiently close to the PC space of a class are considered non-members. Table 4 shows classification for compounds from the training set. Here sample **9** was classified incorrectly since its activity is 4.2 (more active) but it is classified by SIMCA as less active.

	Compound												
	1	2	3	4	5	6	7	8	9	10	11	12	13
Class	-	-	-	+	+	+	+	+	-	-	+	-	-

	Compound										
	14	15	16	17	18	19	20	21	22	23	24	25
Class	-	-	-	-	-	+	+	+	+	+	+	+

Table 4. Classification of compounds from the training set according to SIMCA method

Probably the reason for this misclassification lies in the fact that compound **9** may not be "well grouped" into one of the two classes. In fact when you analyze Fig. 2 you note that **9** is the compound classified as more active that is closer to compounds classified as less active.

Group or Class	Number of Compounds	True group	
		More active	Less active
Less active	11	0	11
More active	14	14	0
Total	25		
%Correct information		100	100

Table 5. Classification matrix obtained by using SDA

2.5 SDA

SDA is also a multivariate method that attempts to maximize the probability of correct allocation. The main objectives of SDA are to separate objects from distinct populations and to allocate new objects into populations previously defined.

The discrimination functions for less active and more active classes are, respectively, Equations 3 and 4, given below:

$$Y_{LESS} = -5.728 - 2.825 MlogP - 0.682O1 - 3.243IC5 + 7.745Mor29m \tag{3}$$
$$Y_{MORE} = -3.536 + 2.220 MlogP + 0.536O1 + 2.548IC5 - 6.086Mor29m \tag{4}$$

The way the method is used is based on the following steps:

(a) Initially, for each molecule, the values for descriptors ($IC5$, $Mor29m$, $O1$ and $MlogP$) are computed;

(b) The values from (a) are inserted in the two discrimination functions (Equation 3 and Equation 4). However, since these equations were obtained from autoscaled values from Table 1 (training set), it is necessary that values from Table 7 (test set) are autoscaled before inserted into the equations;

(c) The two values computed from (b) are compared. In case the value calculated from Equation 3 is higher than that from Equation 4, then the molecule is classified as less active. Otherwise, the molecule is classified as more active.

Group or Class	Number of Compounds	True group	
		More active	Less active
Less active	11	0	11
More active	14	14	0
Total	25		
%Correct information		100	100

Table 6. Classification matrix obtained by using SDA with Cross Validation

Through SDA all compounds of the training set were classified as presented in Table 5. The classification error rate was 0% resulting in a satisfactory separation between more and less active compounds.

The reliability of the model is determined by carrying out a cross-validation test, which uses the leave-one-out technique. In this procedure, one compound is omitted of the data set and the classification functions are built based on the remaining compounds. Afterwards, the omitted compound is classified according to the classification functions generated. In the next step, the omitted compound is included and a new compound is removed, and the procedure goes on until the last compound is removed. The obtained results with the cross-validation methodology are summarized in Table 6. Since the total of correct information was 100%, the model can be believed as being a good model.

Compound	IC5	Mor29m	O1	MlogP
26	5.371	-0.437	-0.238	2.461
27	5.526	-0.544	-0.249	2.496
28	5.402	-0.516	-0.241	1.649
29	5.336	-0.481	-0.239	2.461
30	5.572	-0.553	-0.238	2.305
31	5.464	-0.411	-0.226	3.117
32	5.584	-0.323	-0.244	3.328
33	5.282	-0.496	-0.226	3.225
34	5.483	-0.570	-0.345	2.090
35	5.583	-0.667	-0.262	2.922

Table 7. Values of the four descriptors for the compounds from the test set

2.6 Classification of unknown compounds

The models built from compounds from the training set through PCA, HCA, KNN, SIMCA and SDA now can be used to classify others compounds (test set, Fig. 10) whose anticancer activities are unknown. So ten compounds were proposed here to verify if they must be classified as less active or more active against a human hepatocellular carcinoma cell line, HepG2. In fact, they were not selected from any literature, so it is supposed that they have not been tested against this carcinoma. These compounds were selected so that they have substituitions at the same positions as those for the training set (R1 and R2) and the same type of atoms. It is important to keep the main characteristics of the compounds that generated the models. This way good predictions can be achieved. The classification of the test set was based on the four descriptors used in the models: *IC5*, *Mor29m*, *O1* and *MlogP*, according to Table 7.

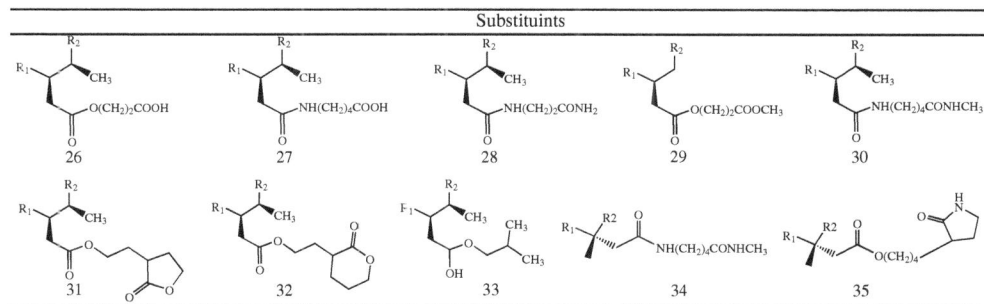

Fig. 10. Compounds from the test set which must be classified as either less active or more active

Compound	PCA	HCA	KNN	SIMCA	SDA
26	-	-	-	-	-
27	+	+	+	+	+
28	-	-	-	-	-
29	-	-	-	-	-
30	+	+	+	+	+
31	-	-	-	-	-
32	-	-	-	-	-
33	-	-	-	-	-
34	+	+	+	+	+
35	+	+	+	+	+

Table 8. Predicted classification for unknown compounds from the test set through different methods. Minus sign (-) for a compound classified as less active while plus sign (+) for a compound classified as more active

The result presented in Table 8 reveal that all samples (test set) receive the same classification by the four methods. Compounds **26, 28, 29, 31, 32** and **33** were classified as less active while compounds **27, 30, 34** and **35** were classified as more active. If you look for an explanation for such a pattern you will note that **26** and **27** present carboxylic acid group at the end of the chain, but only **27** is classified as more active. So it is possible that the change of an ester group by an amide group causes increase in activity. However when two amide groups are considered as occurs in **28** and **30** more carbon atoms in substituent means more active. Now comparing **26, 29, 31** and **32**, all of them have ester group associated with another different group and they all are classified as less active. The presence of the second group seams not to modify activity too much. The same effect is found in **34** and **35**, both more active.

3. Conclusion

All multivariate statistical methods (PCA, HCA, KNN, SIMCA and SDA) classified the 25 compounds from the training set into two distinct classes: more active and less active according to their degree of anticancer HepG2 activity. This classification was based on *IC5*,

Mor29m, O1 and *MlogP* descriptors. They represent four distinct classes of interactions related to the molecules, especially between the molecules and the biological receptor: steric (*IC5*), 3D-morse (*Mor29m*), electronic (*O1*) and molecular (*MlogP*).

A test set with ten molecules with unknown anticancer activity has its molecules classified, according to their biological response, into more active or less active compound. The results reveal in which classes they are grouped. In general molecules classified as more active must be seen as more efficient in cancer treatment than those classified as less active. Then the developed studies with PCA, HCA, KNN, SIMCA and SDA can provide valuable insight into the experimental process of syntheses and biological evaluation of the new artemisinin derivatives with activity against cancer HepG2. Without chemometrics no model and, consequently, no classification could be possible unless you are a prophet!

The interfacioal location of chemometrics, falling between measurements on the one side and statistical and computational theory and methods on the other, poses a challenge to the new practioner (Brow et al., 2009). The future of chemometrics lies in the development of innovative solutions to interesting problems. Some of the most exciting opportunities for innovation and new developments in the field of chemometrics lie at the interface between chemical and biological sciences. These opportunities are made possible by the exciting new scientific advances and discoveries of the past decade (Gemperline, 2006).

Finally, after reading this chapter you certainly must have noticed that chemometrics is a useful tool in medicinal chemistry, mainly when the great diversity of data is taken into account, because a lot of conclusions can be achieved. A study like this one here presented, where different methods are employed, is one of the examples of how chemometrics is important in drug design. Thus applications of statistics in chemical data analysis looking for the discovery of more efficacious drugs against diseases must continue and will certainly help researches.

4. References

Barbosa, J.; Ferreira, J.; Figueiredo, A.; Almeida, R.; Silva, O.; Carvalho, J.; Cristino, M.; Ciriaco-Pinheiro, J.; Vieira, J. & Serra, R. (2011). Molecular Modeling and Chemometric Study of Anticancer Derivatives of Artemisini. *Journal of the Serbian Chemical Society*, Vol. 76, No. 9, (September 2011), pp. 1263-1282, ISSN 0352-5139

Brereton, R. (2009). *Chemometrics for Pattern Recognition*, John Wiley & Sons, Ltd, ISBN 978-0-470-74646-2, West Sussex,England

Brereton, R. (2007). *Applied Chemometrics for Scientists*, John Wiley & Sons, Ltd, ISBN 978-0-470-01686-2, West Sussex, England

Brown, S.; Tauler, R. & Walczak, B. (Ed(s)) (2009). *Compreensive Chemometrics: Chemical and Biochemical Data Analysis*, Vol. 1, Elsevier, ISBN 978-0-444-52702-8, Amsterdam, The Netherlands

Bruns, R.; Scarminio, I. & Barrros Neto, B. (2006) *Statistical Design - Chemometrics*, Elsevier, ISBN 978-0-444-52181-1, Amsterdam, The Netherlands

Cardoso, F.; Figueiredo, A.; Lobato, M.; Miranda, R.; Almeida, R. & Pinheiro, J. (2008). A Study on Antimalarial Artemisinin Derivatives Using MEP Maps and Multivariate QSAR. *Journal of Molecular Modeling*, Vol. 14, No. 1, (January 2008), pp. 39-48, ISSN 0948-5023

Doweyko, A. (2008). QSAR: Dead or Alive? *Journal of Computer-Aided Molecular Design*, Vol. 22, No. 2, (February 2008), pp. 81-89, ISSN 1573-4951

Efferth, T. (2005). Mechanistic Perspectives for 1,2,4-trioxanes in Anti-cancer Therapy. *Drug Resistance. Updat*, Vol. 8, No. 1-2, (February 2005), pp. 85-97, ISSN 1368-7646

Ferreira, M. (2002). Multivariate QSAR. *Journal of the Brazilian Chemical Society*, Vol.13, No. 6, (November/December 2002), pp. 742-753, ISSN 1678-4790

Fujita, T. (1995). *QSAR and Drug Design: New Developments and Applications*, Elsevier, ISBN 0-444-88615-X, Amsterdan, The Netherlands

Gareth, T. (2003). *Fundamental of Medicinal Chemistry*, John Wiley & Sons, Ltd, ISBN 0-470-84307-1, West Sussex, England

Frisch, M. J.; Trucks, G. W.; Schlegel, H. B.; Scuseria, G. E.; Robb, M. A.; Cheeseman, J. R.; Zakrzewski, V. G.; Montgomery, Jr J. A.; Stratmann, R. E.; Burant, J. C.; Dapprich, S.; Millam, J. M.; Daniels, A. D.; Kudin, K.N.; Strain, M. C.; Farkas, O.; Tomasi, J.; Barone, V.; Cossi, M.; Cammi, R.; Mennucci, B.; Pomelli, C.; Adamo, C.; Clifford, S.; Ochterski, J.; Petersson, G. A.; Ayala, P. Y.; Cui, Q.; Morokuma, K.; Salvador, P.; Dannenberg, J. J.; Malick, D. K.; Rabuck, A. D.; Raghavachari, K. J.; Foresman, B.; Cioslowski, J.; Ortiz, J. V.; Baboul, A. G.; Stefanov, B. B.; Liu, G.; Liashenko, A.; Piskorz, P.; Komaromi, I.; Gomperts, R.; Martin, R. L.; Fox, D. J.; Keith, T.; Al-Laham, M. A.; Peng, C. Y.; Nanayakkara, A.; Challacombe, M.; Gill, P. M. W.; Johnson, B.; Chen, W.; Wong, M. W.; Andres, J. L.; Gonzalez, C.; Head-Gordon, M.; Replogle, E. S. & Pople, J. A. (1998) *Gaussian, Inc.*, Gaussian 98 Revision A.7, Pittsburgh PA

Gemperline, P. (2006). *Practical Guide to Chemometrics* (2nd), CRC Press, ISBN 1-57444-783-1, Florida, USA

Varmuza, K. & Filzmoser, P. (2009). *Introduction to Multivariate Statistical Analysis in Chemometrics*, CRC Press, ISBN 9781420059472, Florida, USA

Lai, H.; Nakasi, I.; Lacoste, E.; Singh, N. & Sasaki (2009). T. Artemisinin-Transferrin Conjugate Retards Growth of Breast Tumors in the Rat. *Anticancer Research*, Vol. 29, No. 10, (October 2009), pp. 3807-3810, ISSN 1791-7530

Levine, I. (1991). *Quantum Chemistry* (4th), Prentice Hall, ISBN 0-205-12770-3, New Jersey, USA

Liu,Y.; Wong, V.; Ko, B.; Wong, M. & Che, C. (2005). Synthesis and Cytotoxicity Studies of Artemisinin Derivatives Containing Lipophilic Alkyl Carbon Chains. *Organic Letters*, Vol. 7, No. 8, (March 2005), pp. 1561-1564. ISSN 1523-7052

Pinheiro, J.; Kiralj, R.; & Ferreira, M. (2003). Artemisinin Derivatives with Antimalarial Activity against Plasmodium falciparum Designed with the Aid of Quantum Chemical and Partial Least Squares Methods. *QSAR & Combinatorial Science*, Vol. 22, No. 8, (November 2003), pp. 830-842, ISSN 1611-0218

Manly, B. (2004). *Multivariate Statistical Methods: A Primer* (3), Chapman and Hall/CRC, ISBN 9781584884149, London, England

Price, R.; van Vugt, M.; Nosten, F.; Luxemburger, C.; Brockman, A.; Phaipun, L.; Chongsuphajaisiddhi, T. & White, N. (1998). Artesunate versus Artemether for the Treatment of Recrudescent Multidrug-resistant Falciparum Malaria. *The American Journal of Tropical Medicine and Hygiene*, Vol. 59, No. 6, (December 1998), pp. 883-888, ISSN 0002-9637

Puzyn, T.; Leszczynski, J. & Cronin, M. (Ed(s)). (2010). *Recent Advances in QSAR Studies: Methods and Applications*, Springer, ISBN 978-1-4020-9783-6, New York, USA

Rajarshi, G. (2008). On the interpretation and interpretability of quantitative structure-activity relationship models. *Journal of Computer-Aided Molecular Design*, Vol. 22, No. 12, (December 2008), pp. 857-871, ISSN 1573-4951

Wold, S. (1995). *Chemometrics, what do we mean with it, and what do we want from it? Chemometrics and Intelligent Laboratory Systems,* Vol. 30, No. 1, (November 1995), pp. 109-115, ISSN 0169-7439

Virtual Computational Laboratory, VCCLAB In: e-Dragon,   13.05.2010,   Available   from http://www.vcclab.org

7

Metabolic Biomarker Identification with Few Samples

Pietro Franceschi, Urska Vrhovsek, Fulvio Mattivi and Ron Wehrens
IASMA Research and Innovation Centre
Via E. Mach, 1 38010 S. Michele all'Adige (TN)
Italy

1. Introduction

Biomarker selection represents a key step in bioinformatic data processing pipelines; examples range from DNA microarrays (Tusher et al., 2001; Yousef et al., 2009) to proteomics (Araki et al., 2010; Oh et al., 2011) to metabolomics (Chadeau-Hyam et al., 2010). Meaningful biological interpretation is greatly aided by identification of a "short-list" of features – biomarkers – characterizing the main differences between several states in a biological system. In a two-class setting the biomarkers are those variables (metabolites, proteins, genes ...) that allow discrimination between the classes. A class or group tag can be used to distinguish many situations: it can be used to discriminate between treated and non-treated samples, to mark different varieties of the same organism, etcetera. In the following, we will – for clarity – restrict the discussion to metabolomics, and the variables will constitute concentration levels of metabolites, but similar arguments hold *mutatis mutandis* for other -omics sciences, such as proteomics and transcriptomics, where the variables correspond to protein levels or expression levels, respectively.

There are several reasons why the selection of biomarker short-lists can be beneficial:

- Predictive purposes: using only a small number of biomarkers in predictive class modeling in general leads to better, i.e., more robust and more accurate predictions.

- Interpretative purposes: it makes sense to first concentrate on those metabolites that show clear differences in levels in the different classes, since our knowledge of metabolic networks in many cases is only scratching the surface.

- Discovery purposes: the complete characterization of unknown compounds identified in untargeted experiments is time- and resource-consuming. The primary focus should thus be placed on a carefully selected group of "unknowns" to be characterized at structural and functional level.

Two fundamentally different statistical approaches to biomarker selection are possible. With the first, experimental data can be used to construct multivariate statistical models of increasing complexity and predictive power – well-known examples are Partial Least Square Discriminant Analysis (PLS-DA) (Barker & Rayens, 2003; Kemsley, 1996; Szymanska et al., 2011) or Principal Component Linear Discriminant Analysis (PC-LDA) (Smit et al., 2007; Werf et al., 2006). Inspection of the model coefficients then should point to those variables that are important for class discrimination. As an alternative, univariate statistical tests can be applied to individual variables, treating each one independent of the others and indicating which of them show significant differences between groups (see, e.g., Guo et al. (2007); Reiner et al. (2003); Zuber & Strimmer (2009)). Multivariate techniques are potentially more powerful in pin-pointing weak differences because they take into account correlation among the variables, but the models can be too much adapted to the experimental data, leading to poor generalization capacity. Univariate approaches, in contrast, both could miss important "weak" details and could overestimate the importance of certain variables, because correlation between variables is not taken into account.

As for many sciences with the "omics" suffix, in metabolomics the number of experimental variables usually greatly exceeds the number of objects, especially with the development of new mass-spectrometry-based technologies. In MS-based metabolomics, high resolution mass spectrometers are often coupled with high performance chromatographic techniques, like Ultra Performance Liquid Chromatography (UPLC). In these experiments, the variables, i.e., the metabolites, are represented by mass/retention-time combinations, and it is typical to have numbers of features varying from several hundreds to several thousands, depending on the experimental and analytical conditions. This increase in experimental possibilities, however, does not correspond to a proportional increase in the number of available samples, which can be limited by the availability of biological samples, by laboratory practice, in particular when complex protocols are required, and also by ethical issues, when, for example, experiments on animals have to be planned.

All these constraints produce *small sample sets*, presenting serious challenges for the statistical analysis, mainly because there is simply not enough information to model the natural biological variability. The situation is critical for multivariate approaches where the parameters of the statistical model need to be optimized (e.g., the number of components in a PLS-DA model). For this purpose, the classical approach is to use sub-sampling in combination with estimates of predictive power, like crossvalidation (Stone, 1974). In extreme conditions, i.e., really small sample sizes, this sub-sampling can give rise to inconsistent sub-models and tuning in the classical way becomes virtually impossible. In Hanczar et al. (2010), as an example, conclusions are focussed on ROC-based statistics (see below), but they are equally relevant for classical error estimates like the root-mean-square error of prediction, RMSEP) multivariate techniques can be still applied to the full data set, but it is not possible to assess the reliability of the biomarker selection pipeline, even if it is still reasonable to think that the biomarkers are strongly contributing to the statistical model. In these situations, univariate methods seem the best solution, also considering the presence of several strategies able to determine cut-off values in t-test based techniques (e.g., thresholding of p values subjected to some form of multiple testing correction (Benjamini & Hochberg, 1995; Noble, 2009; Reiner et al., 2003)). Regardless of the statistical strategy, for the "biomarkers" extracted in these conditions there is no obvious validation possible in the statistical sense; however, the results of the experiments are extremely important in the hypothesis generation phase to plan more informative investigations.

Interestingly, there is no literature on the effect of sample size on biomarker identification in the "omics" sciences, and the objective of this contribution is to fill this gap. We focus on a two-class problem, and in particular on small data sets. In our approach, real class differences have been introduced by spiking apple extracts with selected compounds, analyzing them using UPLC-TOF mass spectrometry, and comparing the feature lists to those of unspiked apple extracts. Using these data we are able to run a comparison between two multivariate methods (PLS-DA and PC-LDA) and the univariate t-test, leading to at least a rough estimate of how consistent biomarker discovery can be when small sample sizes are considered. In particular, we compare the effect of sample size reduction on multivariate and univariate models on the basis of Receiver Operating Characteristics (ROC) (Brown & Davis, 2005).

2. Material and methods

2.1 Biomarker Identification

There are many strategies for identifying differentially expressed variables in two-class situations – a recent overview can be found in Saeys et al. (2007). A general approach is to construct a model with good predictive properties, and to see which variables are important in such a model. Given the low sample-to-variable ratio, however, one can not expect to be able to fit very complicated models, and in many cases a linear model is the best one can do (Hastie et al., 2001). The oldest, and most well-known technique is Linear Discriminant Analysis (LDA, McLachlan (2004)). One formulation of this technique, dating back to R.A. Fisher, is to find a linear combination of variables a that maximizes the ratio of the between-groups sums of squares, B, and the within-groups sums of squares W:

$$a^T B a / a^T W a \qquad (1)$$

That is, a is the direction that maximizes the separation between the classes, both by having compact classes (a small within-groups variance) and by having the class centers far apart (a large between-groups variance). Large values in a indicate which variables are important in the discrimination. Another formulation is to calculate the Mahalanobis distance of a new sample x to the class centers μ_i:

$$d(x, i) = (x - \mu_i)^T \Sigma^{-1} (x - \mu_i) \qquad (2)$$

The new sample is then assigned to the class of the closest center. This approach is equivalent to Fisher's criterion for two classes (but not for more than two classes). In this equation, Σ is the (estimated) pooled covariance matrix of the classes. If the Mahalanobis distance to each class center is calculated using the individual class covariance matrices, the result is Quadratic Discriminant Analysis (QDA), which as the name suggests, no longer leads to linear class boundaries. A final formulation is to use regression using indicator variables for the class. In a two-class situation one can use, e.g., the values of -1 and 1 for the two classes; positive predictions will be assigned to class one, and negative predictions to class -1. In many other cases, 0 and 1 are used, and the class threshold is put at 0.5. When there are more than two classes, one can use a separate column in the dependent variable for every class – if a sample belongs to that class the column should contain 1, else 0. Again, the size of the regression coefficients indicates which of the variables contribute most to the discrimination.

For most applications in the "omics" fields, even the most simple multivariate techniques such as Linear Discriminant Analysis (LDA) cannot be applied directly. From Equation 2 it is clear that an inverse of the the covariance matrix Σ needs to be calculated, which is impossible in cases where the number of variables exceeds the number of samples. In practice, the number of samples is nowhere near the number of variables. For QDA, the situation is even worse: to allow a stable matrix inversion, every single class should have at least as many samples as variables (and preferably quite a bit more). A common approach is to compress the information in the data into a low number of latent variables (LVs), either using PCA (leading to PC-LDA, e.g. Smit et al. (2007); Werf et al. (2006)) or PLS (which gives PLS-DA; see Barker & Rayens (2003); Kemsley (1996)), and to perform the discriminant analysis on the resulting score matrices. These are not only of low dimension, but also orthogonal so that the matrix inversion, the calculation of Σ^{-1}, can be performed very fast and reliably. Both for PC-LDA and PLS-DA, the problem is more often usually cast in a regression context, where again the response variable Y can take values of either 0 or 1. The model thus becomes:

$$Y = XB + \mathcal{E} \approx TP^T B + \mathcal{E} \tag{3}$$

where $\mathcal{E}$ is the matrix of residuals. Matrix X is decomposed into a score matrix T and a loading matrix P, both consisting of a very low number of latent variables, typically less than ten or twenty. The coefficients for the scores, $A = P^T B$, can therefore be easily be calculated in the normal way of least-squares regression:

$$A = (T^T T)^{-1} T^T Y \tag{4}$$

which by premultiplication with P lead to estimates for the overall regression coefficients B:

$$B = PA \tag{5}$$

These equations are the same for both PLS-DA and PC-LDA. The difference lies in the decomposition of X. In PC-LDA, T and P correspond to the scores and loadings, respectively, from PCA. That is, the class of the samples is completely ignored, and the only criterion is to capture as much variance as possible from X. In PLS-DA, on the other hand, the scores and loadings are taken from a PLS model and the decomposition of X *does* take into account class information: the first PLS components by definition explain more, often much more, variance of Y than the first PCA components.

Both methods, PC-LDA as well as PLS-DA, are usually very sensitive to the choice of the number of LVs. Taking too few LVs will lead to bad predictions since important information is missed. Taking too many, the model will be too flexible and will show a phenomenon known as *overtraining*: it is more or less learning all the examples in the training set by heart but is not able to generalize and to make good predictions for new, unseen samples. As discussed, the assessment of the optimal number of LVs is neigh impossible with small sample sets. In the case under consideration, the extent of this effect is investigated by constructing several models with increasing numbers of LVs. Using real and simulated data sets (see below), models with 1–4, 6, and 8 LVs, respectively, are compared.

A simplification of statistical modeling can be obtained by ignoring all possible correlations between variables and assuming a diagonal covariance matrix, which leads to diagonal discriminant analysis (DDA). It can be shown that using the latter for feature selection corresponds to examining regular t-statistics (Zuber & Strimmer, 2009), and this is the approach we will take in this paper. For each variable, the difference between the class means $\bar{x}_{1i}$ and $\bar{x}_{2i}$ is transformed into a z-score by dividing by the appropriate standard deviation estimate s_i:

$$z_i = |\bar{x}_{1i} - \bar{x}_{2i}| / s_i \tag{6}$$

Using the appropriate number of degrees of freedom, these z-scores can be transformed into p values, which have the usual interpretation of the probability under the null hypothesis of encountering an observation with a value that is at least as extreme. In biomarker identification, p values can be used to sort the variables in order of importance and it is also possible to decide a cut-off value to identify variables which show "significant" differences from the null hypothesis.

Generally speaking, the absolute size of coefficients is taken as a measure for the likelihood of being a true marker: the variable with the largest coefficient, in a PLS-DA model for example, is the first biomarker candidate, the second largest the second candidate, and so on. Note that this approach assumes that all variables have been standardized, i.e., scaled to mean zero and unit variance. This is often done in metabolomics to prevent dominance of highly abundant metabolites. Statistics from a t-test can be treated in the same way.

2.2 Quality assessment

To evaluate the performance of biomarker selections one typically relies on quantities like the fraction of true positives, i.e., that fraction of the real biomarkers that is actually identified by the selection method, and the false positives – those variables that have been selected but do not correspond to real differences. Similarly, true and false negatives can be defined. These statistics can be summarized graphically in an ROC plot (Brown & Davis, 2005), where the fraction of true positives (y-axis) is plotted against the fraction of false positives (x-axis). These two characteristics are also known as the sensitivity and the (complement of) specificity. An ideal biomarker identification method would lead to a position in the top left corner: all true biomarkers would be found (the fraction of true positives would be one, or close to one) with no or only very few false positives. Gradually relaxing the selection criterion, allowing more and more variables to be considered as biomarkers, generally leads to an increase in the true positive fraction (upwards in the plot), but also to an increase in the false positive fraction (in the plot to the right). The best biomarker selection method is obviously the one that finds all biomarkers very quickly, leading to a very steep ROC curve at the beginning.

A quantitative measure of the efficiency of a method can be obtained by calculating the area under the ROC curve (AUC). A value of one (or close to one) indicates that the method does a very good job in identifying biomarkers – all true biomarkers are found almost immediately. A value of one half indicates a completely random selection (this corresponds to the diagonal in the ROC plot). Values significantly lower than one half should not occur. In many cases, the most important area in the ROC plot is the left side, which indicates the efficiency of the model in selecting the most important biomarkers. Consequently, it is common to calculate a partial area under the curve (pAUC), for instance up to twenty percent of false positives (pAUC.2). In a method with higher pAUC, the true biomarkers will be present in the first positions of the candidate biomarkers list, hence this is the quantity that will be considered in the current paper.

2.3 Apple data set

Twenty apples, variety Golden Delicious, were purchased at the local store. Extracts of every single fruit were prepared according to Vrhovsek et al. (Vrhovsek et al., 2004). The core of the fruit was removed with a corer and each apple was cut into equal slices. Three slices (cortex and skin) from the opposite side of each fruit were used for the preparation of aqueous acetone extracts. The samples were homogenized in a blender Osterizer model 847-86 at speed one in a mixture of acetone/water (70/30 w/w). Before the injection, acetone was evaporated by rotary evaporation, the samples were brought back to the original volume with ethanol and were filtered with a 0.22 μm filter (Millipore, Bedford, USA). UPLC-MS spectra were

HPLC	ACQUITY UPLC (Waters)
Column	BEH C18 1.7 μm, 2.1*50 mm
Column temperature	40°C
Injection volume	5μl
Eluent flux	0.8 mlmin^{-1}
Solvent A	0.1% formic acid in H_2O
Solvent B	0.1% formic acid in MeOH
Gradient	linear gradient
	from 0 to 100% of solvent B in 10 minutes
	100% of B for 2 minutes
	100% A within 0.1 minutes
	Equilibration for 2.9 minutes.

Mass Spectrometer	SYNAPT Q-TOF (Waters)
Mass range	50-3000 Da.
Capillary	3 kV
Sampling cone	25 V
Extraction cone	3 V
Source temperatures	150°C
Desolvation temperatures	500°C
Cone gas flow	50 Lh^{-1}
Desolvation gas flow	1000 Lh^{-1}

Table 1. Chromatographic and spectrometric conditions of the spiked-apple data set.

acquired on a ACQUITY - SYNAPT Q-TOF (Waters, Milford, USA) in positive and negative ion mode with the chromatographic conditions summarized in Table 1. No technical replicates were performed. Raw data were transformed to the open NetCDF format by the DataBridge built-in utility of the MassLynx software.

Class differences were introduced by spiking ten of the twenty extracts with a number of selected compounds, leaving the other ten as "untreated" controls. The majority of the spiked compounds are known to be commonly present in apples, while two of them (*trans*-resveratrol and cyanidin-3-galactoside) are not naturally present in the chosen matrix. The concentrations of the specific compounds in the pooled extract are presented in Table 2; markers were added in different concentrations to test the identification pipeline in conditions which mimic those found in a typical metabolomic experiment, where variation is usually present at different concentration levels. As an example of what the data look like, the first control sample, measured in positive mode, is shown in Figure 1. The horizontal axis shows the chromatographic dimension, and the vertical axis the mass-to-charge ratio. Circles indicate features that have been identified in this plane. In the remainder only the extracted triplets for the features, consisting of retention time, mass-to-charge ratio and intensity, will be used.

Feature extraction is performed with XCMS (Smith et al., 2006) and all statistical analyses are carried out in R (R Development Core Team, 2011). The CentWave peak-picking algorithm (Tautenhahn et al., 2008) is applied, using the following parameter settings: ppm = 20, peakwidth = c(3,15), snthresh = 2, prefilter = c(3,5). The average numbers of detected features per chromatogram are 1179 and 610 for positive and negative ion mode, respectively.

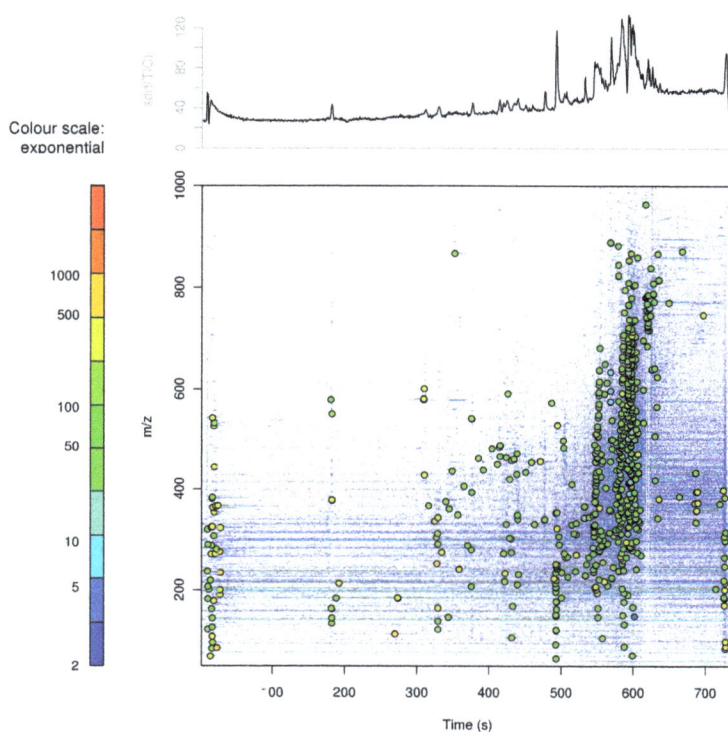

Fig. 1. Visualization of the data of the first control sample, measured in positive mode. The top of the figure shows the square root of the Total Ion Current (TIC); background color indicates the intensity of the signal in the plane formed by retention time and m/z axes. Circles indicate features found by the peak picking; the fill colour of these circles indicates the intensity of the features.

Compound	mgl^{-1} pool	Δ Conc. (mgl^{-1})
quercetin-3-galactoside (querc-3-gal)	5.69	1.48
quercetin	0.006	0.008
quercetin-3-glucoside (querc-3-glc)	1.05	0.3
quercetin-3-rhamnoside (querc-3rham)	3.64	3.55
phloridzin	2.92	2.3
cyanidin-3-galactoside (cy-3-gal)	n.d.	0.57
trans-resveratrol	n.d.	0.4

Table 2. Spiked compound summary. The difference in concentration is relative to the one measured in the pooled extract. Cyanidin-3-galactoside and *trans*-resveratrol are not normally found in Golden Delicious.

After grouping across samples, features are screened for isotopes, clusters and common adducts with in-house developed software.

Due to fragmentation occurring in the ionization source, it is common for a single neutral molecule to give rise to several ionic species. A single spiked compound can then generate several "biomarkers" in the MS peak table. Adducts, isotopes and common clusters are automatically screened, but fragments must be included in the biomarker list, as in real metabolomic experiments no a priori knowledge can be used to distinguish molecular from fragment ions. For the apple data set, the characteristic couples mass/retention time for all spiked metabolites were identified by manual inspection of the UPLC-MS profiles of standards. For negative ions the following numbers of features have been associated with the spike-in compounds: querc-3-gal/querc-3-glc (1 feature), phloridzin (2 features), *trans*-resveratrol (1), querc-3-rham (1). In the case of positive ion mode the numbers are cy-3-gal (1), *trans*-resveratrol (1), querc-3-rham (1), quercetin (1) and phloridzin (4). These feature are now taken to be the "true" biomarkers and they are used to construct ROC curves. The data set, as well as a more extended version including different concentrations of spiked-in compounds is publicly available in the R package BioMark (see `http://cran.r-project.org/web/packages/BioMark`, Wehrens & Franceschi (2011)) and has been used to evaluate a novel stability-based biomarker selection method (Wehrens et al., 2011).

In this application, the effects of decreasing sample size are investigated by subsampling the original set of twenty samples: sample sizes of 16, 12, 8 and 6 apples, respectively, are considered. In all cases, both classes (spiked and control) have equal sizes, which is the most easy case for detecting significant differences. Results are summarized by analysis of ROC curves – to prevent effects from accidentally easy or difficult subsets, the final ROC curves are obtained by averaging the results of 100 repeated re-samplings.

2.4 Simulated data sets

To assess the behaviour of biomarker selection for larger data sets, we resort to simulation. Simulated data sets have been constructed as multivariate normal distributions, using the means and covariance matrices of the experimental data: both classes (untreated and spiked) have been simulated separately. Simulations are performed for both positive and negative modes; in every simulation, one hundred data sets are created. The outcomes reported here are the averages of the results for the one hundred simulations. Data sets consisting of 10, 25, 50 and 200 biological samples per class have been synthesized.

3. Results and discussion

As a first step, the data are visualized using Principal Component Analysis (PCA). Since the intensities of the features can vary enormously, standardized data are used. The score plots for the positive and negative data sets are shown in Figure 2 for the positive ion mode, and in Figure 3 for the negative mode. In both cases, control and spiked data sets are not completely separated and the same is also true for the other PCs (not shown). This fact indicates that the "inherent" variability of the data set is not perturbed to a significant extent by spiking, as could be expected considering the small number of affected variables.

Even with this data structure, biomarker selection strategies can still perform efficiently. Figure 2 and Figure 3 also display the score plots of a PCA analysis performed considering only the top 10 variables selected by univariate *t*-testing. In these conditions, the separation

Full dataset   Top 10 biomarkers

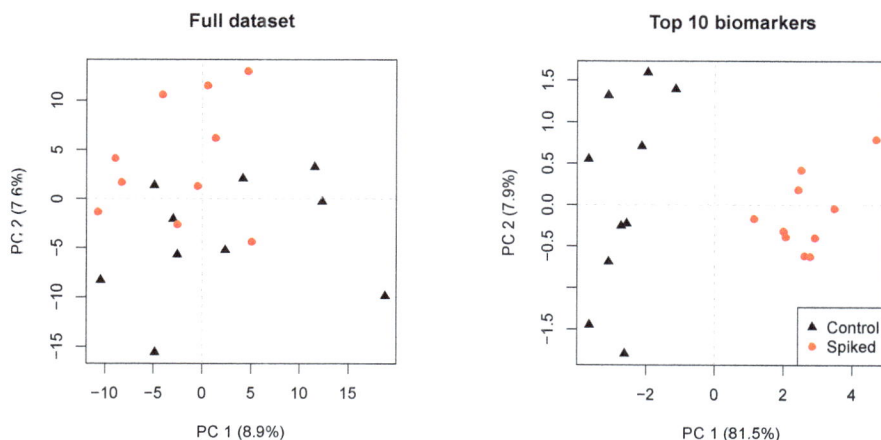

Fig. 2. PCA score plot (PC1 vs PC2) for the positive ion mode data set after standardization. In the left plot the principal components have been calculated on the full data set. In the right panel PCA analysis has been performed considering only the top 10 variables selected by a t-test.

between control and spiked samples is evident, thus indicating that this subset of the variables separates the two classes. Whether these ten variables contain the true biomarkers remains to be seen: especially in small data sets there may be chance correlations causing false positives, and seeing differences between the two groups in the score plots after t-testing in fact is trivial. The score plot is merely showing that the variables, selected on the bases of their discriminating power, are separating the two classes. As already discussed, small data sets will in general not capture all relevant biological variability, which implies that the predictive power of statistical models based on small data sets usually is very low. To illustrate this effect, the predictive power, i.e., the fraction of correct predictions for PC-LDA and PLS-DA models is presented in Figure 4. Four subsets of different sizes are considered as training sets, and the estimate of predictive power is based on predictions for the apples not in the training set. Again, the results are the average over 100 different subsamplings. Even if the control and spiked subsets are different, it can be seen that the predictive power of the multivariate methods is comparable to random selection, meaning that for every subset different variables will be important in the models and no consistency can be achieved. However, it is important to point out that this fact does not mean that some of the true biomarkers are not consistently selected upon subsetting, but rather that the more important variables are changing from a subset to another: even with models that are unpredictive it is possible to extract relatively good lists of putative biomarkers. Obviously, with very different characteristics for the two classes there *will* be predictive power, but for realistic data sets like the one used in this paper, where differences are small, it is unwise to focus solely on prediction.

To evaluate the efficiency of the different methods as far as biomarker selection is concerned, ROC curves for the t-test and two-component PLS-DA and PC-LDA models are presented in Figure 5, for 3, 4, 6 and 8 biological samples per class, respectively. The ROC curves indicate that all three variables selection methods perform significantly better than random selection.

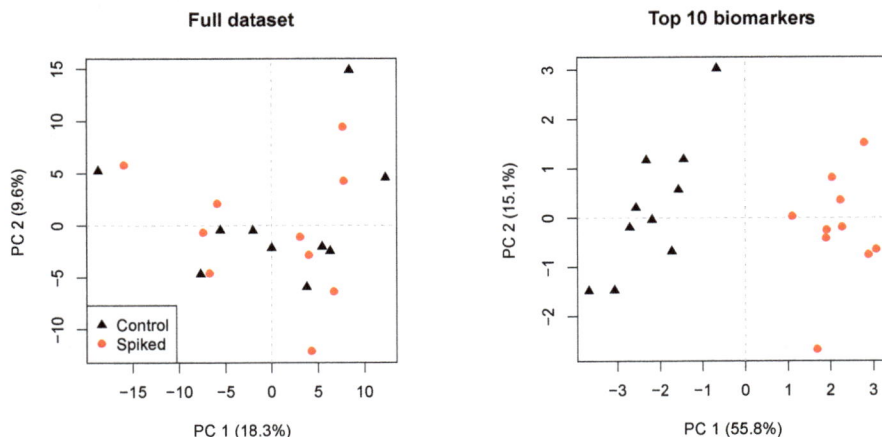

Fig. 3. PCA score plot (PC1 vs PC2) for the negative ion mode data set after standardization. In the left plot the Principal Components have been calculated on the full data set. In the right panel PCA analysis has been performed considering only the top 10 variables selected by a *t*-test.

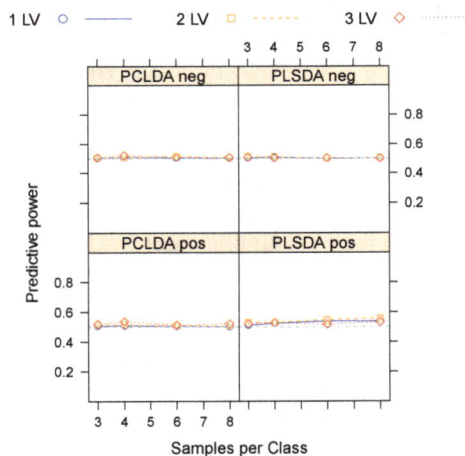

Fig. 4. Predictive power of multivariate PLS-DA and PC-LDA on a subset of the initial data set for positive and negative ion mode. Different lines are relative to models constructed with an increasing number of LVs. The horizontal dashed line indicates random selection.

Of the three, PC-LDA is always the least efficient, while PLS-DA and the *t*-test have a very similar performance. In absolute terms, the efficiency of the three methods increases with the number of biological samples. ROC curves for all possible conditions were constructed and the results are summarized in terms of early AUC (pAUC.2) in Figure 6, for positive and negative ion mode, respectively. From these figures it is possible to extract some clear trends:

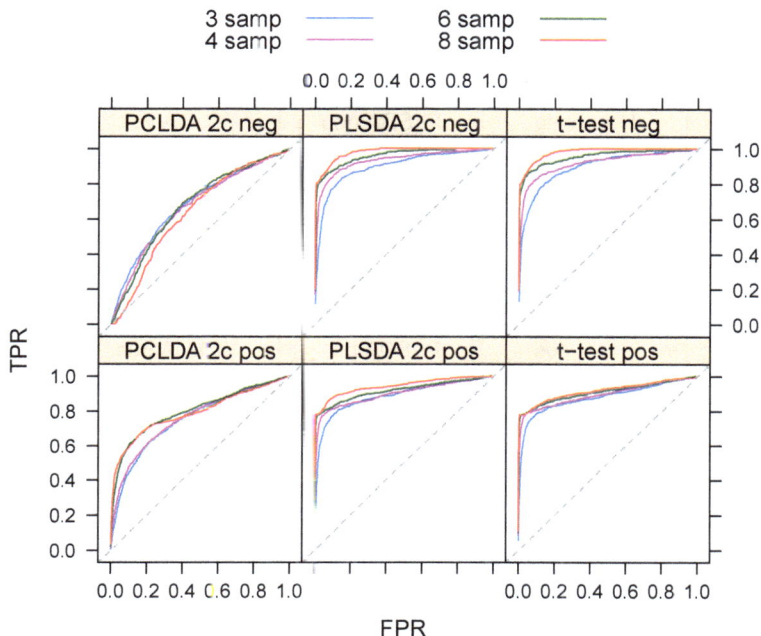

Fig. 5. ROC curves for the *t*-test and two component PLS-DA and PC-LDA as a function of the number of samples per class.

1. The performance of the methods improves by increasing the number samples per class.

2. The performance of PLS-DA is not particularly sensitive to the number of components.

3. PC-LDA does not show top class performance in any of the conditions considered.

4. The performance of PC-LDA is very much dependent on the number of components.

5. Multivariate approaches do not show a definitive advantage over univariate *t*-testing.

As expected, the performances of all the methods in terms of biomarker identification decrease with a reduction of the data set size. However, it is important to point out that even in the worst possible case (3 samples per class) early AUC for PLS-DA and the *t*-test are significantly greater than that obtained for completely random selection. This indicates that both methods can be used effectively in the biomarker selection phase, even with a low number of samples. In other words, features related to spiked compounds are consistently present in the top positions of the ordered list of experimental variables, which implies that also models constructed with very few samples can be relied upon to recognize these features.

The performance of PC-LDA is very much dependent on the number of components taken into account. This behavior can be explained by considering that in PC-LDA the variable reduction step is performed without any knowledge of class labels, only selecting the directions of greater variance. If these directions show little discriminating power, their supervised linear combination leads to poor modeling. However, performance improves with

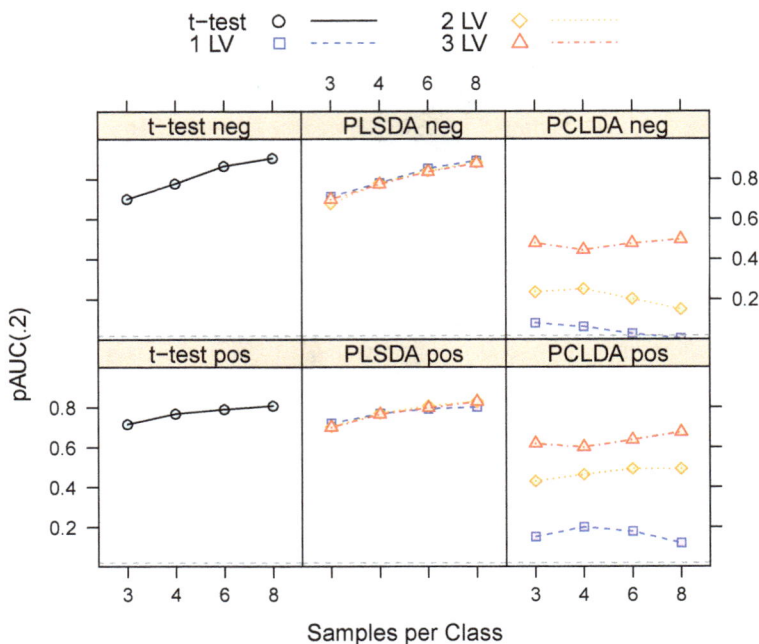

Fig. 6. pAUC.2 for PLS-DA, PC-LDA and *t*-test as a function of the number of samples per class and the number of LVs. The gray dashed line indicates the pAUC.2 of random selection.

the number of components, as an increase of the number of LVs leads to a better "coverage" of the data space. These limitations do not affect PLS-DA, as the variable reduction step is already performed in a supervised framework, where discriminating power is the main request. This means that the first PLS components are by definition more relevant than the first PCA components in biomarker identification. The other side of the coin is the danger of overfitting, very real in the application of PLS-DA (Westerhuis et al., 2008) – we will come back to this point later.

In this small-sample set, the *t*-test does as well as the best multivariate methods. This shows that modeling the correlation structure is not necessarily an advantage if the number of samples is low, or, alternatively, that the true correlation structure has not been captured well enough from the few samples that are available to allow meaningful inference. A definite advantage of the *t*-test is that it has no tunable parameters and can be applied without further optimization. It should be noted that we do not need to apply multiple-testing corrections in this context since we only use the order of the absolute size of the *t*-statistics to construct the ROC curves, and not a specific cut-off level α. In other applications, however, this aspect should be taken into account.

To extend the comparison between different models beyond the limits imposed by the apple experiment, ROC curves and early AUC were calculated for the simulated data using larger sample sizes (10, 25, 50, 200), both for positive and negative ion modes. The dependence of

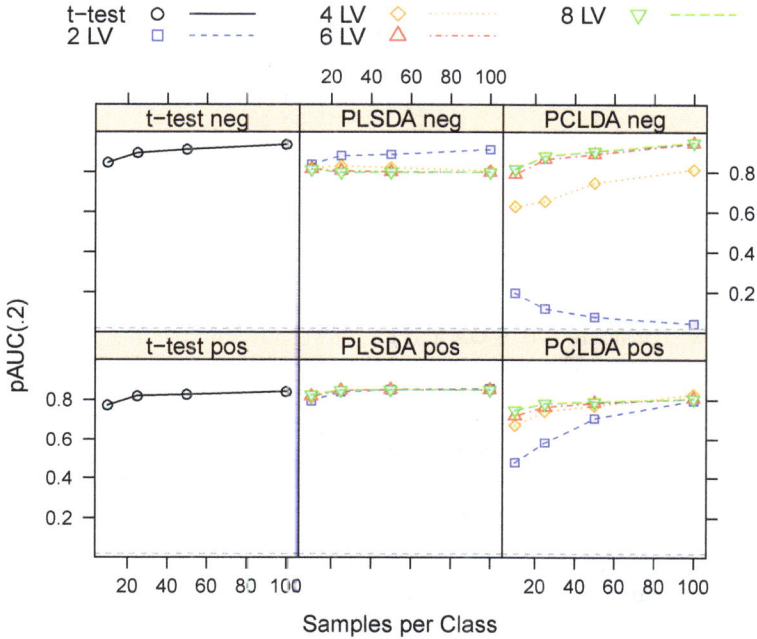

Fig. 7. pAUC.2 for PLS-DA, PC-LDA and *t*-test as a function of the number of samples per class and the number of LVs. Simulated data set. Gray dashed line indicates the pAUC.2 of random selection

pAUC.2 on the number of replicates and of components is presented in Figure 7, comparing the multivariate methods to the *t*-test and to "random" selection.

This analysis shows that PC-LDA only becomes effective if a large number of LVs is considered: the true biomarkers should have appreciable weight in the latent variables and it is by no means certain that this is the case for the first couple of LVs. Is it worth noting that for negative ion mode, the model with 2 LVs is comparable to random selection. In the case of PLS-DA, this dependence on the number of LVs is less evident and shows an opposite trend: the best performance is obtained with the smallest number of LVs. This is in agreement with the explanation given earlier: the relevant variables are captured in the very first PLS components, and the effect of overtraining leads to deterioration if more components are added. If anything, it is surprising that the overtraining effect is relatively small for these data.

The results on the simulated data sets are in agreement with the conclusions from the apple data. Differences between the methods decrease with increasing sample sizes, but even with the largest number of objects (200 in each group) the *t*-test still performs as well as PLS-DA. Multivariate testing is slightly more effective for the positive ion mode, while the *t*-test shows a slight advantage for the negative ion mode. This behaviour is probably due to the different characteristics of both ionization modes, leading to different levels of correlation between biomarkers. Indeed, in positive ion mode, the ionization shows a more pronounced fragmentation (phloridzine, for example, gives rise to four different biomarkers).

4. Conclusions

In this paper we have investigated the effects of sample set size on the performance of some popular strategies for biomarker identification (PLS-DA, PC-LDA and the t-test). The experiments are performed on a spiked metabolomic data set measured in apple extracts by UPLC-QTOF. The efficiency of the different statistical approaches is compared in terms of ROC curves, and in order to assess general trends, simulated data have been used to extend the data set. The experimental results clearly show that Linear Discriminant Analysis carried out on the Principal Components (PC-LDA) is the least efficient strategy for biomarker identification among the ones we considered. PLS-DA and the t-test show comparable performances in all the considered conditions. These results, and the observation that PLS-DA based selection is relatively consistent for different numbers of components, indicate that multivariate and univariate approaches are equally efficient for the apple data set. It is perhaps surprising that relatively good results in terms of biomarker selection are obtained, even for models that have very poor predictive performance. One should realise, however, that this is not a paradox at all: it merely is the result from the low sample-to-variable ratio, leading to chance correlations of intensities of metabolite signals with class. The true biomarkers are often present among the most significant variables in, e.g., a PLS-DA model, but many other false positives are, too, destroying the predictive power. One recently published approach actually utilizes this variability by focusing only on those variables that are *consistently* present in the most important variables upon disturbance of the data by jackkifing or bootstrapping (Wehrens et al., 2011).

The main point of this contribution, however, is the relation between data set size and reliability of biomarker identification. As expected, all the methods become less efficient as the number of biological replicates decreases, but even in these conditions the use of PLS-DA and the t-test offer effective biomarker identification strategies. This observation is fundamentally important in all studies where it is impossible to acquire more samples, and suggests that small sample sizes can still allow reliable selection of biomarkers.

5. References

Araki, Y., Yoshikawa, K., Okamoto, S., Sumitomo, M., Maruwaka, M. & Wakabayashi, T. (2010). Identification of novel biomarker candidates by proteomic analysis of cerebrospinal fluid from patients with moyamoya disease using SELDI-TOF-MS, *BMC Neurology* 10: 112.

Barker, M. & Rayens, W. (2003). Partial least squares for discrimination, *J. Chemom.* 17: 166–173.

Benjamini, Y. & Hochberg, Y. (1995). Controlling the false discovery rate: a practical and powerful approach to multiple testing, *J. Royal. Stat. Soc. B* 57: 289–300.

Brown, C. D. & Davis, H. T. (2005). Receiver operating characteristics curves and related decision measures: A tutorial, *Chemom. Intell. Lab. Syst.* 80: 24–38.

Chadeau-Hyam, M., Ebbels, T., Brown, I., Chan, Q., Stamler, J., Huang, C., Daviglus, M., Ueshima, H., Zhao, L., Holmes, E., Nicholson, J., Elliott, P. & Iorio, M. D. (2010). Metabolic profiling and the metabolome-wide association study: significance level for biomarker identification, *J. Proteome Res.* 9(9): 4620–4627.

Guo, Y., Hastie, T. & Tibshirani, R. (2007). Regularized discriminant analysis and its application in microarrays, *Biostatistics* 8: 86–100.

Hanczar, B., Hua, J., Sima, C., Weinstein, J., Bittner, M. & Dougherty, E. (2010). Small-sample precision of ROC-related estimates, *Bioinformatics* 28: 822–830.

Hastie, T., Tibshirani, R. & Friedman, J. (2001). *The Elements of Statistical Learning*, Springer Series in Statistics, Springer, New York.

Kemsley, E. K. (1996). Discriminant analysis of high-dimensional data: a comparison of principal components analysis and partial least squares data reduction methods, *Chemom. Intell. Lab. Syst.* 33: 47–61.

McLachlan, G. (2004). *Discriminant Analysis and Statistical Pattern Recognition*, Wiley-Interscience.

Noble, W. S. (2009). How does multiple testing correction work?, *Nat. Biotechnol.* 27: 1135–1137.

Oh, J., Craft, J., Townsend, R., Deasy, J., Bradley, J. & Naqa, I. E. (2011). A bioinformatics approach for biomarker identification in radiation-induced lung inflammation from limited proteomics data, *J. Proteome Res.* 10(3): 1406–1415.

R Development Core Team (2011). *R: A Language and Environment for Statistical Computing*, R Foundation for Statistical Computing, Vienna, Austria. ISBN 3-900051-07-0.
URL: *http://www.R-project.org*

Reiner, A., Yekutieli, D. & Benjamini, Y. (2003). Identifying differentially expressed genes using false discovery rate controlling procedures, *Bioinformatics* 19(3): 368–375.

Saeys, Y., Inza, I. & Larranaga, P. (2007). A review of feature selection techniques in bioinformatics, *Bioinformatics* 23: 2507–2517.

Smit, S., Breemen, M. J. v., Hoefsloot, H. C. J., Aerts, J. M. F. G., Koster, C. G. d. & Smilde, A. K. (2007). Assessing the statistical validity of proteomics based biomarkers, *Anal. Chim. Acta* 592: 210–217.

Smith, C. A., Want, E. J., Tong, G. C., Abagyan, R. & Siuzdak, G. (2006). XCMS: Processing Mass Spectrometry Data for Metabolite Profiling Using Nonlinear Peak Alignment, Matching, and Identification, *Anal. Chem.* 78: 779–787.

Stone, M. (1974). Cross-validatory choice and assessment of statistical predictions, *J. R. Statist. Soc. B* 36: 111–147. Including discussion.

Szymanska, E., Saccenti, E., Smilde, A. & Westerhuis, J. (2011). Double-check: validation of diagnostic statistics for PLS-DA models in metabolomics studies, *Metabolomics* .

Tautenhahn, R., Bottcher, C. & Neumann, S. (2008). Highly sensitive feature detection for high resolution LC/MS, *BMC Bioinformatics* 9: 504.

Tusher, V., Tibshirani, R. & Chu, G. (2001). Significance analysis of microarrays applied to the ionizing radiation response, *PNAS* 98: 5116–5121.

Vrhovsek, U., Rigo, A., Tonon, D. & Mattivi, F. (2004). Quantitation of polyphenols in different apple varieties, *J. Agr. Food. Chem.* 52(21): 6532–6538.

Wehrens, R. & Franceschi, P. (2011). *BioMark: finding biomarkers in two-class discrimination problems.* R package version 0.3.0.

Wehrens, R., Franceschi, P., Vrhovsek, U. & Mattivi, F. (2011). Stability-based biomarker selection, *Anal. Chim. Acta* 705: 15–23.

Werf, M. J. v. d., Pieterse, B., Luijk, N. v., Schuren, F., Vat, B. v. d. W.-v. d., Overkamp, K. & Jellema, R. H. (2006). Multivariate analysis of microarray data by principal component discriminant analysis: prioritizing relevant transcripts linked to the degradation of different carbohydrates in Pseudomonas putida S12, *Microbiology* 152: 257–272.

Westerhuis, J., Hoefsloot, H., Smit, S., D.J., V., Smilde, A. K., van Velzen, E., van Duijnhoven, J. & van Dorsten, F. A. (2008). Assessment of PLSDA cross validation, *Metabolomics* 4: 81–89.

Yousef, M., Ketany, M., Manevitz, L., Showe, L. & Showe, M. (2009). Classification and biomarker identification using gene network modules and support vector machines, *BMC Bioinformatics* 10: 337.

Zuber, V. & Strimmer, K. (2009). Gene ranking and biomarker discovery under correlation, *Bioinformatics* 25: 2700–2707.

8

Kinetic Analyses of Enzyme Reaction Curves with New Integrated Rate Equations and Applications

Xiaolan Yang, Gaobo Long, Hua Zhao and Fei Liao[*]
College of Laboratory Medicine, Chongqing Medical University,
Chongqing,
China

1. Introduction

A reaction system of Michaelis-Menten enzyme on single substrate can be characterized by the initial substrate concentration before enzyme action (S_0), the maximal reaction rate (V_m) and Michaelis-Menten constant (K_m), besides some other required parameters. The estimations of S_0, V_m and K_m can be used to measure enzyme substrates, enzyme activities, epitope or hapten (enzyme-immunoassay), irreversible inhibitors and so on. During enzyme reaction, the changes of substrate or product concentrations can be monitored; continuous monitor of such changes provides a reaction curve while discontinuous monitor of such changes provides signals just for the starting point and the terminating point of enzyme reaction. It is an end-point method when only signals for the starting point and the terminating point are analyzed. It is a kinetic method when a range of data from a reaction curve are analyzed, and can be classifieid into the initial rate method and kinetic analysis of reaction curve. The initial rate method only analyzes data for initial rate reaction whose instantaneous rates are constants; kinetic analysis of reaction curve analyzes data whose instantaneous rates show obvious deviations from the initial rate (Bergmeyer, 1983; Guilbault, 1976; Marangoni, 2003). To estimate those parameters of an enzyme reaction system, kinetic analysis of reaction curve is favoured because the analysis of one reaction curve can concomitantly provide V_m, S_0 and K_m. Hence, methods for kinetic analysis of reaction curve to estimate parameters of enzyme reaction systems are widely studied.

An enzyme reaction curve is a function of dependent variables, which are proportional to concentrations of a substrate or product, with respect to reaction time as the predictor variable. In general, there are two types of enzyme reaction curves. The first type involves the action of just one enzyme, and employs either a selective substrate to detect the activity of one enzyme of interest or a specific enzyme to act on a unique substrate of interest. The second type involves the actions of at least two enzymes, and requires at least one auxiliary enzyme as a tool to continuously monitor a reaction curve. The second type is an enzyme-coupled reaction system. For kinetic analysis of reaction curve, there are many reports on

---

[*] Corresponding Author

one enzyme reaction system, but are just a few reports on enzyme-coupled reaction system (Atkins & Nimmo, 1973; Liao, et al., 2005; Duggleby, 1983, 1985, 1994; Walsh, 2010).

In theory, enzyme reactions may tolerate reversibility, the activation/inhibition by substrates/products, and even thermo-inactivation of enzyme. From a mathematic view, it is still feasible to estimate parameters of an enzyme reaction system by kinetic analysis of reaction curve if the roles of all those factors mentioned above are included in a kinetic model (Baywenton, 1986; Duggleby, 1983, 1994; Moruno-Davila, et al., 2001; Varon, et al., 1998). However, enzyme kinetics is usually so complex due to the effects of those mentioned factors that there are always some technical challenges for kinetic analysis of reaction curve. Hence, most methods for kinetic analysis of reaction curve are reported for enzymes whose actions suffer alterations by those mentioned factors as few as possible.

In practice, kinetic analysis of reaction curve usually employs nonlinear-least-square-fitting (NLSF) of the differential or integrated rate equation(s) to either the reaction curve *per se* or data set(s) transformed from the reaction curve (Cornish-Bowden, 1995; Duggleby, 1983, 1994; Orsi & Tipton, 1979). The use of NLSF rather than matrix inversion is due to the existence of multiple minima of the sum of residual squares with respect to some nonlinear parameters (Liao, et al., 2003a, 2007a). When a differential rate equation is used, numerical differentiation of data from the reaction curve has to be employed to derive instantsneous reaction rates. In this case, there must be intervals as short as possible to monitor reaction curves (Burden & Faires, 2001; Dagys, 1990; Hasinoff, 1985; Koerber & Fink, 1987). However, the instantaneous reaction rates from reaction curves inherenetly exhibit narrow distribution ranges and large errors; the strategy by numerical differentiation of data in a reaction curve is unfavourable for estimating V_m and S_0 because of their low reliaiblity and unsatisfactory working ranges. On the other hand, when an integrated rate equation of an enzyme reaction is used for kinetic analysis of reaction curve, there is no prerequisites of short intervals to record reaction curves so that automated analyses in parallel can be realized for enhanced performance with a large number of samples. As a result, integrated rate equations of enzymes are widely studied for kinetic analysis of reaction curve to estimate parameters of enzyme reaction systems (Duggleby, 1994;Liao, et al, 2003a, 2005a; Orsi & Tipton, 1979).

Due possibly to the limitation on computation resources, integrated rate equations of enzymes in such methods are usually rearranged into special forms to facilitate NLSF after data transformation (Atkins & Nimmo, 1973; Orsi & Tipton, 1979). In appearance, the uses of different forms of the same integrated rate equation for NLSF to data sets transformed from the same reaction curve can give the same parameters. However, kinetic analysis of reaction curve with rearranged forms of an integrated rate equation always gives parameters with uncertainty too large to have practical roles (Newman, et al, 1974). Therefore, proper forms of an integrated rate equation should be selected carefully for estimating parameters by kinetic analysis of reaction curve.

In the past ten years, our group studied chemometrics for kinetic analysis of reaction curve to estimate parameters of enzyme reaction systems; the following results were found. (a) In terms of reliability and performance for estimating parameters, the use of the integrated rate equations with the predictor variable of reaction time is superior to the use of the integrated rate equations with predictor variables other than reaction time (Liao, et al., 2005a); (b) the integration of kinetic analysis of reaction curve with other methods to quantify initial rates and substrates has more absorbing advantages (Liu, et al., 2009; Yang, et al., 2010); such integration strategies can be applied to enzyme-coupled reaction systems and enzymes sufferring inhibition by substrates/products. Herein, we discuss chemometrics for both kinetic analysis of reaction curve and its integration with other methods, and demonstrate their applications to quantify enzyme initial rates and substrates with some typical enzymes.

2. Kinetic analysis of enzyme reaction curve: chemometrics and application

To estimate parameters by kinetic analysis of reaction curve, the desired parameters are included in a set of parameters for the best fitting. Regardless of the number of enzymes involved in a reaction curve, there are the following two approaches for kinetic analysis of reaction curve based on different ways to realize NLSF and their data transformation.

In the first approach, with a differential or integrated rate equation, a series of dependent variables are derived from data in a reaction curve with each set of preset parameters. Such dependent variables should follow a predetermined response to predictor variables that are either reaction time or data transformed from those in the reaction curve. The goodness of the predetermined response is the criterion for the best fitting. In this approach, NLSF is realized with a model for data transformed from a reaction curve (Burguillo, 1983; Cornish-Bowden, 1995; Liao, 2005; Liao, et al., 2003a, 2003b, 2005a, 2005b; Orsi & Tipton, 1979).

In the second approach, reaction curves are calculated with sets of preset parameters by iterative numerical integration from a preset staring point. Such calculated reaction curves are fit to a reaction curve of interest; the least sum of residual squares indicates the best fitting (Duggleby, 1983, 1994; Moruno-Davila, et al., 2001; Varon, et al., 1998; Yang, et al., 2010). In this approach, calculated reaction curves still utilize reaction time as the predictor variable and become discrete at the same intervals as the reaction curve of interest. Clearly, there is no transformation of data from a reaction curve in this approach.

With any enzyme, iterative numerical integration of the differential rate equation(s) from a starting point with sets of preset parameters can be universally applicable regardless of the complexity of the kinetics. Thus, the second approach exhibits better universality and there are few technical challenges to kinetic analysis of reaction curve *via* NLSF. In fact, however, the second approach is occasionally utilized while the first approach is widely practiced.

In the following subsections, the differential rate equation of simple Michaelis-Menten kinetics on single substrate is integrated; the prerequisites for kinetic analysis of reaction curve with integrated rate equations, kinetic analysis of enzyme-coupled reaction curve, the integrations of kinetic analysis of reaction curve with other methods, and the applications of such integration strategies to some typical enzymes are discussed.

2.1 Integrated rate equation for one enzyme on single substrate

Assigning instantaneous substrate concentration to S, instantaneous reaction time to t, steady-state kinetics of Michaelis-Menten enzyme on single substrate follows Equ.(1).

$$-dS/dt = (V_m \times S)/(K_m + S) \tag{1}$$

Assigning the substrate concentration at the first point for analysis to S_1, Equ.(1) is integrated into Equ.(2) when the enzyme is stable, the substrate and product do not alter the intrinsic activity of the enzyme and the reaction is irreversible (Atkins & Nimmo, 1973; Marangoni, 2003; Orsi & Tipton, 1979; Zou & Zhu, 1997). In Equ.(2), t_{lag} accounts for the lag time of steady-state reaction. After transformation of data in a reaction curve according to Equ.(3), there should be a linear response of the left part in Equ.(2) to reaction time, as in Equ.(4). The goodness of this linear response is judged based on regression analysis. However, to estimate parameters by kinetic analysis of reaction curve *via* NLSF, there are the following general prerequisites for Equ.(2) or any of its equivalency.

$$(S_1 - S)/K_m + \ln(S_1/S) = (V_m/K_m) \times (t - t_{lag}) \tag{2}$$

$$y = (S_1 - S)/K_m + \ln(S_1/S) \tag{3}$$

$$y = a + b \times t \tag{4}$$

The first prerequisite is that enzyme reaction should apparently follow kinetics on single substrate. For enzyme reactions with multiple substrates whose concentrations are all changing during reactions, kinetic analysis of reaction curve always give parameters of too low reliability to have practical roles no matter what methods are used for NLSF (data unpublished). From our experiences to estimate parameters by kinetic analyses of reaction curves, any substrate at levels below 10% of its K_m can be considered negligible; the use of one substrate at levels below 10% of those of other substrates can make enzyme reactions follow single substrate kinetics (Liao, et al, 2001; Liao, et al, 2003a, 2003b; Li et al., 2011; Zhao et al., 2006). For any enzyme on multiple substrates, therefore, there are two approaches to make it apparently follow kinetics on single substrate. The first is the use of one substrate of interest at levels below 10% of those of the other substrates; this approach has universal applicability to common enzymes such as hydrolases in aqueous buffers and oxidases in air-saturated buffers. The second is the utilization of special reaction systems to regenerate the substrate of the enzyme of interest by actions of some auxiliary enzymes, and indeed this approach usually yields enzyme-coupled reaction curves of complicated kinetics.

The second prerequisite is that enzyme reaction should be irreversible. In theory, the estimation of parameters by kinetic analysis of reaction curve is still feasible when reaction reversibility is considered, but the estimated parameters possess too low reliability to have practical roles (data unpublished). Generally, a preparation of a substance with contaminants less than 1% in mass content can be taken as a pure substance. Namely, a reagent leftover in a reaction accounting for less than 1% of that before reaction can be negligible. For convenience, therefore, an enzyme reaction is considered irreversible when the leftover level of a substrate of interest in equilibrium is much less than 1% of its initial one. To promote the consumption of the substrate of interest, the concentrations of other substrates should be preset at levels much over 10 times the initial level of the substrate of interest. In this case, the enzyme reaction is apparently irreversible and follows kinetics on single substrate. Or else, the use of scavenging reactions to remove products can drive the reaction forward. The concurrent uses of both approaches are usually better.

The third prerequisite is that there should be steady-state data for analysis (Atkins & Nimmo, 1973; Dixon & Webb, 1979; Liao, et al, 2005a; Marangoni, 2003; Orsi & Tipton, 1979).

For this prerequisite, the first and the last points of data in a reaction curve for analysis should be carefully selected. The first point should exclude data within the lag time of steady-state reaction. The last point should ensure data for analyses to have substrate concentrations high enough for steady-state reaction. Namely, substrate concentrations should be much higher than the concentration of the active site of the enzyme (Dixon & Webb, 1979). The use of special weighting functions for NLSF can mitigate the contributions of residual squares at low substrate levels that potentially obviate steady-state reaction.

The forth prerequisite is that the enzyme should be stable to validate Equ.(2), or else the inactivation kinetics of the enzyme should be included in the kinetic model. Enzyme stability should be checked before kinetic analysis of reaction curve. When the inactivation kinetics of an enzyme is included in a kinetic model for kinetic analysis of reaction curve, the integrated rate equation is usually quite complex or even inaccessible if the inactivation kinetics is too complex. For kinetic analysis of reaction curve of complicated kinetics, numerical integration to produce calculated reaction curves for NLSF to a reaction curve of interest, instead of NLSF with Equ.(4), can be used to estimate parameters (Duggleby, 1983, 1994; Moruno-Davila, et al., 2001; Varon, et al., 1998; Yang, et al., 2010).

The fifth prerequisite is that there should be negligible inhibition/activation of activity of an enzyme by products/substrates, or else such inhibition/activation on the activity of the enzyme by its substrate/product should be included in an integrated rate equation for kinetic analysis of reaction curve (Zhao, L.N., et al., 2006). For validating Equ.(2), any substrate that alters enzyme activity should be preset at levels low enough to cause negligible alterations; any product that alters enzyme activity can be scavenged by proper reactions. When such alterations are complex, numerical integration of differential rate equations for NLSF to a reaction curve of interest can be used (Duggleby, 1983, 1994; Moruno-Davila, et al., 2001; Varon, et al., 1998).

Obviously, the first three prerequisites are mandatory for the inherent reliability of parameters estimated by kinetic analysis of reaction curve; the later two prerequisites are required for the validity of Equ.(2) or its equivalency for kinetic analysis of reaction curve.

2.2 Realization of NLSF and limitation on parameter estimation

To estimate parameters by kinetic analysis of reaction curve based on NLSF, the main concerns are the satisfaction to the prerequisites for the quality of data under analysis, the procedure to realize NLSF, and the reliability of parameters estimated thereby.

For the estimation of parameters by kinetic analysis of reaction curve, there are two general prerequisites for the quality of data under analysis: (a) there should be a minimum number of the effective data whose changes in signals are over three times the random error; (b) there should be a minimum consumption percentage of the substrate in such effective data for analysis. In general, at least two parameters like V_m and S_0 should be estimated; the minimum number of the effective data should be no less than 7 (Atkins & Nimmo, 1973; Baywenton, 1986; Miller, J. C. & Miller, J. N., 1993). The minimum consumption percentage of the substrate can be about 40% if only V_m and S_0 are estimated while other parameters are fixed as constants. In general, the estimation of more parameters requires higher consumption percentages of the substrate in the effective data for analysis.

With a valid Equ.(2), data in a reaction curve can be transformed according to Equ.(3) to realize NLSF with Equ.(4). The use of Equ.(4) for NLSF needs no special treatment of the unknown t_{lag}. For any method to continuously monitor reaction curve, there may be an unknown but constant background in signals (Newman, et al., 1974; Liao, et al., 2003a, 2005a; Yang, et al., 2010). Thus, the background in the signal for S_1 in Equ.(2) is better to be treated as a nonlinear parameter to realize NLSF; this procedure gives the term of NLSF but causes the burden of computation; as a result, a rearranged form of Equ.(2) is suggested for kinetic analysis of reaction curve (Atkins & Nimmo, 1973; Liao, et al., 2005a).

In theory, Equ.(2) can be rearranged into Equ.(5) as a linear function of V_m and K_m. In Equ.(5), the instantaneous reaction time at the moment for S_1 is preset as zero so that there is no treatment of t_{lag}. When the signal for S_1 is not treated as a nonlinear parameter, kinetic analysis of reaction curve by fitting with Equ.(5) can be finished within 1 s with a pocket calculator. However, parameters estimated with Equ.(5) always have so large errors that Equ.(5) is scarcely practiced in biomedical analyses. Hence, the proper form of an integrated rate equation after validating should be selected carefully.

$$(S_1 - S)/(t - t_{lag}) = V_m - K_m \times (\ln(S_1/S)/(t - t_{lag}))$$ (5)

In principle, to reliably estimate parameters based on NLSF, the distribution ranges of both the dependent variables and the predictor variables in any kinetic model should be as wide as possible while their random errors should be as small as possible (Baywenton, 1986; del Rio, et al., 2001; Draper & Smith, 1998; Miller, J. C. & Miller, J. N., 1993). By serial studies with common enzymes, we found the use of Equ.(4) or similar forms of integrated rate equations with the predictor variables of reaction time for kinetic analysis of reaction curve could give reliable V_m and S_0, when K_m was fixed at a constant after optimization (Liao, 2005; Liao, et al, 2001, 2003a, 2003b, 2005a, 2005b, 2006, 2007b; Zhao, Y.S., et al., 2006, 2009). Reaction time as the predictor variable has the widest distribution and the smallest random errors, in comparison to the predictor variable in Equ.(5). The left part in Equ.(4) also possess a wider distribution range. Such differences in predictor variables and dependent variables should account for different reliability of parameters estimated with Equ.(2) and Equ.(5), and thus an integrated rate equation with the predictor variable of reaction time may be the proper form for kinetic analysis of reaction curve. However, when NLSF with Equ.(4) is realized with S_1 as a nonlinear parameter, there is nearly 10 s for computation with Celeron 300A CPU on a personal computer. Currently, computation resource is no longer a problem and thus Equ.(4) or its equivalent equations should always be adopted.

The selection of a weighting factor for kinetic analysis of reaction curve is also a concern. Based on error propagation and the principle for weighted NLSF with y defined in Equ.(3), squares of instantaneous rates can be the weighting factors (W_f) with Equ.(4) for NLSF to get the weighted sum of residual squares (Q), as described in Equ.(6), Equ.(7) and Equ.(8) (Baywenton, 1986; Draper & Smith, 1998; Gutierrez & Danielson, 2006; Miller, J. C. & Miller, J. N., 1993). The use of a weighting function like Equ.(7) can mitigate the effects of errors in substrate or product concentrations near the completion of reaction. The resistance of an estimated parameter (the variation within 3% in our studies) to reasonable changes in data ranges for analysis can be a criterion to judge the reliability of parameter estimated.

$$\partial y/\partial S = -(K_m + S)/(K_m \times S)$$ (6)

$$W_f = \partial S / \partial y = -K_m \times S / (K_m + S) \tag{7}$$

$$Q = \sum W_f^2 \times (y_{\text{predicted}} - y_{\text{calculated}})^2 \tag{8}$$

It is also concerned which parameter is suitable for estimation by kinetic analysis of reaction curve. In theory, all parameters of an enzyme reaction system can be simultaneously estimated by kinetic analysis of reaction curve. However, there is unknown covariance among some parameters to devalue their reliability; there is the limited accuracy of original data for analyses and the estimation of some parameters with narrow working ranges will have negligible practical roles. V_m is independent of all other parameters and so is S_0, and the assay of V_m and S_0 are already routinely practiced in biomedical analyses. Therefore, V_m and S_0 may be the parameters suitable for estimation by kinetic analysis of reaction curve. Additionally, K_m is used for screening enzyme mutants and enzyme inhibitors; but K_m estimated by kinetic analysis of reaction curve usually exhibits lower reliability and is preferred to be fixed for estimating V_m and S_0. If K_m is estimated as well, S_1 should be at least 1.5-fold K_m and there should be more than 85% consumption of the substrate in the data selected for analysis (Atkins & Nimmo, 1973; Liao, et al., 2005a; Newman, et al., 1974; Orsi & Tipton, 1979). To estimate K_m, the initial datum (S_1) and its corresponding ending datum from a reaction curve for analysis should be tried sequentially till the requirements for data range are met concurrently. In this case, the estimation of S_1 has no practical roles. In general, the resistance of V_m and S_0 to reasonable changes in ranges of data for analyses can be a criterion to select the optimized set of parameters that are fixed as constants.

In comparison to the low reliability to estimate K_m independently for screening enzyme inhibitors and enzyme mutants, the ratio of V_m to K_m as an index of enzyme activity can be estimated robustly by kinetic analysis of reaction curve. Reversible inhibitors of Michaelis-Menten enzyme include competitive, noncompetitive, uncompetitive and mixed ones (Bergmeyer, 1983; Dixon & Webb, 1979; Marangoni, 2003). The ratios of V_m to K_m will respond to concentrations of common inhibitors except uncompetitive ones that are very rare in nature. Thus, the ratio of V_m to K_m can be used for screening common inhibitors. More importantly, the ratio of V_m to K_m is an index of the intrinsic activity of an enzyme and the estimation of the ratios of V_m to K_m can also be a promising strategy to screen enzyme mutants of powerful catalytic capacity (Fresht, 1985; Liao, et al., 2001; Northrop, 1983).

For robust estimation of the ratio of V_m to K_m of an enzyme, S_0 can be preset at a value below 10% of K_m to simplify Equ.(2) into Equ.(9). Steady-state data from a reaction curve can be analyzed after data transformation according to the left part in Equ.(9). For validating Equ.(9), it is proposed that S_0 should be below 1% of K_m (Meyler-Almes & Auer, 2000). The use of extremely low S_0 requires special methods to monitor enzyme reaction curves and steady-state reaction can not always be achieved with enzymes of low intrinsic catalytic activities. On the other hand, the use of S_0 below 10% of K_m is reasonable to estimate the ratio of V_m to K_m (Liao, et al., 2001). To estimate the ratio of V_m to K_m, the use of Equ.(9) to analyze data is robust and resistant to variations of S_0 if Equ.(9) is valid; this property makes the estimation of the ratio of V_m to K_m for screening reversible inhibitors superior to the estimation of the half-inhibition concentrations (Cheng & Prusoff, 1973).

$$\ln(S_1/S) = a + (V_m/K_m) \times t \tag{9}$$

Kinetic analysis of reaction curve requires more considerations when activities of enzymes are altered by their substrates/products. In this case, more parameters can be included in kinetic models similar to Equ.(2) for kinetic analysis of reaction curve, but there must be complicated process to optimize reaction conditions and preset parameters. Based on the principle for kinetic analysis of reaction curve described above, we developed some new integration strategies to successfully quantify enzyme initial rates and substrate with absorbing performance even when the activities of enzymes of interest are altered significantly by substrates/products (Li, et al., 2011; Liao, 2007a; Zhao, L.N., et al., 2006).

2.3 Kinetic analysis of enzyme-coupled reaction curve

When neither substrate nor product is suitable for continuous monitor of reaction curve, a tool enzyme can be used to regenerate a substrate or consume a product of the enzyme of interest; the action of the tool enzyme should consume/generate a substrate/product as an indicator suitable for continuous monitor of reaction curve. Namely, the reaction of the tool enzyme is coupled to the reaction of an enzyme of interest for continuous monitor of reaction curve (Bergmeyer, 1983; Guilbault, 1976; Dixon & Webb, 1979). When such enzyme-coupled assays are used to measure initial rates of an enzyme, there are always unsatisfactory linear range because the activities of the tool enzyme is always limited and the concentration of the substrate of the tool enzyme is also limited (Bergmeyer, 1983; Dixon & Webb, 1979). It is expected that kinetic analysis of enzyme-coupled reaction curve may effectively enhance the upper limit of linear response. However, kinetics of enzyme-coupled reaction systems is described with a set of differential rate equations, which cause difficulty in accessing an integrated rate equation with the predictor variable of reaction time.

In this case, iterative numerical integration to obtain calculated reaction curves for NLSF to a reaction curve of interest can be used (Duggleby, 1983, 1994; Moruno-Davila, et al., 2001; Varon, et al., 1998; Yang, et al., 2010). Lactic dehydrogenase (LDH) is widely used as a tool enzyme for enzyme-coupled assay. The assay of activity of alanineaminotransferase (ALT) in sera has important biomedical roles and usually employs LDH-coupled assay. For LDH-coupled ALT assay, iterative numerical integration of the set of differential rate equations with each set of preset parameters from a preset starting point can produce a calculated reaction curve; such a calculated reaction curve can be made discrete at the same intervals as the reaction curve of interest and then be used for NLSF to the reaction curve of interest.

The process of iterative numerical integration for LDH-coupled ALT assay is given below (Yang, et al., 2010). In an LDH-coupled ALT reaction system, assigning instantaneous concentration of NADH to $C_{n,i}$, instantaneous concentration of pyruvate to $C_{p,i}$, instantaneous absorbance at 340 n for NADH to A_i, the molar absorptivity of NADH to ε, the initial rate of ALT under steady-state reaction to V_{1k}, the maximal activity of LDH to V_m, the integration step to Δt, there are Equ.(10), Equ.(11) and Equ.(12) to describe the iterative integration of the set of differential rate equations. Calculated reaction curves according to Equ. (12) using different sets of preset parameters become discrete and are fit to the reaction curve of interest, and background absorbance at 340 nm is treated as a parameter as well.

$$C_{n,i} = (A_i - A_b)/\varepsilon \tag{10}$$

$$C_{p,i+1} = C_{p,i} + V_{1k} \times \Delta t -$$
$$V_m \times \Delta t / (1 + K_a/C_{n,i} + K_b/C_{p,i} + K_{ab}/(C_{n,i} \times C_{p,i})) \tag{11}$$

$$A_{i+1} = A_i - \varepsilon \times V_m \times \Delta t / (1 + K_a/C_{n,i} + K_b/C_{p,i} + K_{ab}/C_{n,i} \times C_{p,i}) \tag{12}$$

By simulation with such a new approach for kinetic analysis of enzyme-coupled reaction curve recorded at 1-s intervals, the upper limit of linear response for measuring ALT initial rates is increased to about five times that by the classical initial rate method. This new approach is resistant to reasonable variations in data range for analysis. By experimentation using the sampling intervals of 10 s, the upper limit is about three times that by the classical initial rate method. Therefore, this new approach for kinetic analysis of enzyme-coupled reaction curve is advantageous, and can potentially be a universal approach for kinetic analysis of reaction curve of any system of much complicated kinetics.

The computation time for numerical integration is inversely proportional to the integration step, Δt; the use of shorter Δt is always better but Δt of 0.20 s at low cost on computation is sufficient for a desirable upper limit of linear response. This new approach with Celeron 300A CPU on a personal computer needs about 10 min for just 30 data in a LDH-coupled reaction curve, but it consumes just about 5 s with Lenovo Notebook S10e. The advancement of personal computers surely can promote the practice of this approach.

2.4 Integration of kinetic analysis of reaction curve with other methods

Any analytical method should have favourable analysis efficiency, wide linear range, low cost and strong robustness. Kinetic analysis of reaction curve for V_m and S_0 assay can have much better upper limit of linear response, but inevitably tolerates low analysis efficiency when wide linear range is required. Based on kinetic analysis of reaction curve, however, our group developed two integration strategies for enzyme initial rate and substrate assay, respectively, with both favourable analysis efficiency and ideal linear ranges.

2.4.1 New integration strategy for enzyme initial rate assay

The classical initial rate method to measure enzyme initial rates requires S_0 much higher than K_m to have desirable linear ranges (Bergmeyer, 1983; Dixon & Webb, 1979; Guilbault, 1976; Marangoni, 2003). Due to substrate inhibition, limited solubility and other causes, practical substrate levels are always relatively low and thus the linear ranges by the classical initial rate method are always unsatisfactory (Li, et al., 2011; Morishita, et al., 2000; Stromme & Theodorsen, 1976). As described above, kinetic analysis of reaction curve can measure enzyme V_m, and many approaches based on kinetic analysis of reaction curve are already proposed (Cheng, et al., 2008; Claro, 2000; Cornish-Bowden 1975, 1995; Dagys, et al., 1986, 1990; Duggleby, 1983, 1985, 1994; Hasinoff, 1985; Koerber, & Fink, 1987; Liao, et al., 2001; Lu & Fei, 2003; Marangoni, 2003; Walsh, et al. 2010). Such approaches all require substrate consumption percentage over 40% with K_m preset as a constant. As a result, there should be intolerably long reaction duration to monitor reaction curves for samples of low enzyme activities, or else the lower limits of linear response are unfavourable.

The integration of kinetic analysis of reaction curve using proper integrated rate equations with the classical initial rate method gives an integration strategy to measure enzyme initial rates with expanded linear ranges and practical analysis efficiency. This integration strategy is effective at substrate concentrations from one-eighth of K_m to three-fold of K_m (Li, et al., 2011; Liao, et al., 2009; Liu, et al., 2009; Yang, et al., 2011). The integration strategy for enzyme initial rate assay uses a special method to convert V_m into initial rates so that the indexes of enzyme activities by both methods become the same; it is applicable to enzymes suffering strong inhibition by substrates/products (Li, et al., 2011). Walsh et al. proposed an integration strategy to measure enzyme initial rate but they employed Equ.(9) that requires substrate levels below 10% of K_m (Walsh, et al. 2010). Our integration strategy is valid at any substrate level to satisfy Equ.(2) and hence can be a universal approach to common enzymes of different K_m. The principles and applications of the integration strategy to one enzyme reaction systems and enzyme-coupled reaction systems are discussed below.

As for one enzyme reaction systems, kinetic analysis of reaction curve can be realized with an integrated rate equation after data transformation; the integration strategy for enzyme initial rate assay requires enzyme kinetics on single substrate and an integrated rate equation with the predictor variable of reaction time (Liao, et al., 2003a, 2005a, Zhao, L.N., et al., 2006). Moreover, the integration strategy should solve the following challenges: (a) there should be an overlapped range of enzyme activities measurable by both methods with consistent results; (b) there should be consistent slopes of linear response for enzyme activities to enzyme quantities by both methods (Figure 1). After these two challenges are solved, the linear segment of response by the classical initial rate method is an extended line of the linear segment of response by kinetic analysis of reaction curve (Liu, et al., 2009).

To solve the first challenge, a practical S_0 and reasonable duration to monitor reaction curve for favourable analysis efficiency are required as optimized experimental conditions. By mathematic derivation and simulation analyses to solve the first challenge, it is demonstrated that a ratio of S_0 to K_m from 0.5 to 2.5, the duration of 5.0 min to monitor reaction curves at intervals no longer than 10 s can solve the first challenge for most enzymes, any ratio of S_0 to K_m smaller than 0.5 or larger than 2.5 requires longer duration to monitor reaction curves. The use of S_0 about one-eighth of K_m requires no shorter than 8.0 min to monitor reaction curves at 10-s intervals to solve the first challenge (Li, et al., 2011; Liu, et al., 2009). When S_0 is too much larger than three times K_m, reaction time to record reaction curves for analysis to solve the first challenge should be much longer than 5 min. Clearly, the first challenge can be solved with practical S_0 for favourable analysis efficiency.

To solve the second challenge, K_m and other parameters should be optimized and fixed as constants to estimate V_m by kinetic analysis of reaction curve, and a preset substrate concentration (PSC) should be optimized to covert V_m into initial rates according to the differential rate equation. In theory, a reliable V_m should be independent of ranges of data when they are reasonably restricted, and CVs for estimating parameters by enzymatic analysis are usually about 5%. Hence, the estimation of V_m with variations below 3% for the changes of substrate consumption percentages from 60% to 90% can be a criterion to select the optimized set of preset parameters. For converting V_m into initial rates, the optimized PSC is usually about 93% of S_0 and can be refined for different enzymes (Li, et al., 2011; Liao, et al., 2009; Liu, et al., 2009; Yang, et al., 2011). Optimized K_m and PSC to solve the second challenge are parameters for data processing while optimized S_0 and reaction duration to solve the first challenge are experimental conditions. The concomitant solution of the two challenges provides feasibility and potential reliability to the integration strategy.

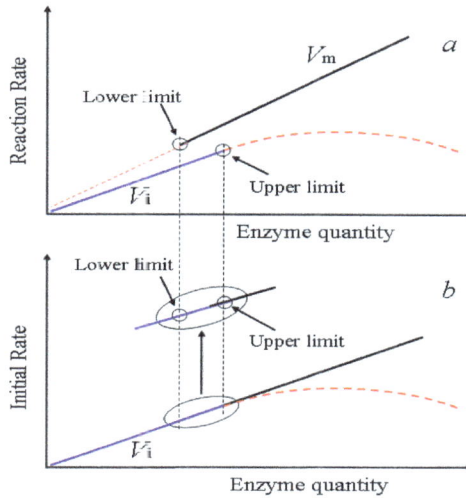

Fig. 1. The integration strategy for enzyme initial rate assay (Modified from Liu et al. (2009)).

After the integration strategy for enzyme initial rate assay is validated, a switch point should be determined for changing from the classical initial rate method to kinetic analysis of reaction curve. The estimation of V_m by kinetic analysis of reaction curve usually prefers substrate consumption percentages reasonably high. Therefore, the substrate consumption percentage that gives an enzyme activity from 90% to 100% of the upper limit of linear response by the classical initial rate method can be used as the switch point.

It should be noted that the lower limit of linear response is difficult to be defined for enzyme initial rate assay by an integration strategy. For most methods, their lower limits of linear response are usually defined as three times the standard errors of estimate (Miller, J. C. & Miller, J. N., 1993). Usually, enzyme initial rate assay utilizes just one method for data processing and the difference between the lower limit and the upper limit of linear response is seldom over 30-fold. By the integration strategy, the measurable ranges of enzyme quantities cover two magnitudes and the detection limit is reduced to that by the classical initial rate method. By manual operation, different dilution ratios of a stock solution of the enzyme have to be used and any dilution error will increase the standard error of estimate for regression analysis. The measurement of higher enzyme activities will inevitably have larger standard deviation. Thus, regression analysis of the response of all measurable enzyme initial rates by the integration strategy to quantities of the enzyme will give higher standard error of estimate and thus an unfavourable lower limit of linear response. By this new integration strategy, we arbitrarily use twice the lower limit of linear response by the classical initial rate method as the lower limit if the overall standard error of estimate is more than twice that by the classical initial rate method alone; or else, the lower limit of linear response is still three times the overall standard error of estimate.

Taken together, for measuring initial rates of enzyme acting on single substrate by the integration strategy based on NLSF and data transformation, there are the following basic steps different from those by the classical initial rate method. The first is to work out the integrated rate equation with the predictor variable of reaction time. The second is to optimize individually their parameters fixed as constants for kinetic analysis of reaction curve. The third is to optimize a ratio of S_0 to K_m and duration to monitor reaction curves; usually a ratio of S_0 to K_m from 0.5 to 2.5, the duration of 5.0 min and intervals of 10 s are effective. The forth is to refine PSC around 93% of S_0 to convert V_m into initial rates.

As for enzyme-coupled reaction system, initial rate itself is estimated by kinetic analysis of reaction curve based on numerical integration and NLSF of calculated reaction curves to a reaction curve of interest. Consequently, neither the conversion of indexes nor the optimization of parameters for such conversion is required and the integration strategy can be realized easily. By kinetic analysis of enzyme-coupled reaction curve, there still should be a minimum number of the effective data and a minimum substrate consumption percentage in the effective data for analysis; these prerequisites lead to unsatisfactory lower limits of linear response for favourable analysis efficiency (the use of reaction duration within 5.0 min). The classical initial rate method is effective to enzyme-coupled reaction systems when activities of the enzyme of interest are not too high. Therefore, this new approach for kinetic analysis of enzyme-coupled reaction curve can be integrated with the classical initial rate method to quantify enzyme initial rates potentially for wider linear ranges.

With enzyme-coupled reaction systems, only the first challenge should be solved to practice the integration strategy. Namely, reaction duration and sampling intervals to record reaction curve should be optimized so that there is an overlapped region of enzyme initial rates measurable by both methods with consistent results. The upper limit of the classical initial rate method should be high enough so that data after reaction of about 5.0 min for enzyme activity at such an upper limit are suitable for kinetic analysis of reaction curve. The integration strategy gives an approximated linear range from the lower limit of linear response by the classical initial rate method to the upper limit of linear response by kinetic analysis of LDH-coupled ALT reaction curve (Yang, et al., 2010).

2.4.2 New integration strategy for enzyme substrate assay

Analysis of a biochemical as the substrate of a typical tool enzyme, *i.e.*, enzymatic analysis of substrate in biological samples, is important in biomedicine (Bergmeyer, 1983; Dilena, 1986; Guilbault, 1976; Moss, 1980). In general, there are the kinetic method and the end-point method for enzyme substrate assay; the end-point method is called the equilibrium method, and it determines the difference between the initial signal for a reaction system before enzyme action and the last signal after the completion of enzyme reaction; such differences proportional to S_0 can serve as an index of substrate concentration (Dilena, et al., 1986; Guilbault, 1976; Moss, 1980; Zhao, et al., 2009). For better analysis efficiency and lower cost on tool enzymes, kinetic methods for enzyme substrate assay are preferred. Among available kinetic methods, the initial rate method based on the response of initial rates of an enzyme at a fixed quantity to substrate concentrations is conventional; however, it tolerates sensitivity to any factor affecting enzyme activities, requires tool enzymes of high K_m, and has narrow linear ranges. Kinetic analysis of reaction curve with a differential rate equation to estimate S_0 is proposed with favourable resistance to variations in enzyme activities and has upper limit of linear response over K_m, but it suffers from high sensitivity to background and has unfavourable lower limit of linear response (Dilena, et al., 1986; Hamilton & Pardue, 1982; Moss, 1980). Hence, new kinetic methods for enzyme substrate assay are still desired.

For enzymatic analysis of substrate, the equilibrium method can still be preferable as long as it has desirable analysis efficiency and favourable cost on tool enzyme. In theory, the last signal for the stable product or the background in the equilibrium method can be estimated by kinetic analysis of reaction curve with data far before the completion of reaction. This process can be a new kinetic method for enzyme substrate assay and is distinguished from the equilibrium method and other kinetic methods by its prediction of the last signal after the completion of enzyme reaction (Liao, 2005; Liao, et al., 2003, 2005a, 2006; Zhao, L.N., et al., 2006; Zhao, Y.S., et al., 2006, 2009). This new kinetic method should have resistance to factors affecting enzyme activities and upper limit of linear response higher than K_m besides all advantages of common kinetic methods.

An enzyme reaction curve can be monitored by absorbance of a stable product or the substrate itself (Figure 2). The initial absorbance before enzyme action (A_0) thus is the background (A_b) when absorbance for a stable product is quantified, or is the absorbance of the substrate plus background when absorbance of the substrate is quantified. The last absorbance after the completion of enzyme reaction, which is predicted by kinetic analysis of reaction curve, is the maximum absorbance of the stable product plus the background (A_m) or A_b itself. There is strong covariance between the initial signal and the last signal for the same reaction system; this assertion enhances precision of this kinetic method for substrate assay (Baywenton, 1986; Liao, et al, 2005b; Zhao, Y.S., et al., 2009).

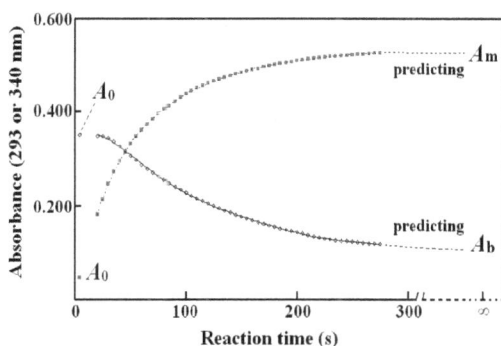

Fig. 2. Demonstration of reaction curves of uricase (293 nm) and glutathione-S-tranferase (340 nm), and the prediction of the last absorbance after infinite reaction time

However, this new kinetic method for substrate assay can by no means concomitantly have wider linear ranges and desirable analysis efficiency. Due to the prerequisites of the quality of data for kinetic analysis of reaction curve, the activity of a tool enzyme for enzymatic analysis of substrate should be reasonably high for higher upper limit of linear response, but it should be reasonably low for favourable lower limit of linear response. On the other hand, the duration to monitor reaction curves should be long enough to have higher upper limit at reasonable cost on a tool enzyme, but should be as short as possible for favourable analysis efficiency. Thus, this new kinetic method alone requires tough optimizations of conditions. Moreover, there are the inevitable random noises from any instrument to record an enzyme reaction curve; when there is much small difference between the initial signal before enzyme action and the last signal recorded after the preset reaction duration, this kinetic method for substrate assay always has unsatisfactory precision. Therefore, this new kinetic method itself is still much beyond satisfaction for substrate assay.

To concomitantly have wider linear ranges, desirable analysis efficiency and favourable precision for enzyme substrate assay, the integration of kinetic analysis of reaction curve with the equilibrium method can be used. The indexes of substrate quantities by the two methods have exactly the same physical meanings, and thus the integration strategy can be easily realized for enzyme substrate assay. By this integration strategy, there should still be an overlapped range of concentrations of the substrate measurable consistently by both methods, besides a switch threshold within such an overlapped region to change from the equilibrium method to kinetic analysis of reaction curve. Additionally, this overlapped region of substrate concentration measurable by both methods with consistent results should localize in a range of substrate concentration high enough for reasonable precision of substrate assay based on kinetic analysis of enzyme reaction curve. These requirements can be met as described below. (a) The upper limit of linear response by the equilibrium method should be optimized to be high enough, so that the difference between the initial signal before enzyme action and the last recorded signal for about 80% of this upper limit is 50 times higher than the random noise of an instrument to record enzyme reaction curves; such a difference can be used as the switch threshold. (b) The activity of a tool enzyme and the duration to monitor reaction curve as experimental conditions should be optimized; kinetic parameters except V_m for kinetic analysis of reaction curve are optimized as well. The resistance of the predicted last signal to reasonable variations in data ranges for analysis can be a criterion to judge the optimized set of preset parameters. For favourable analysis efficiency in clinical laboratories, reaction duration can be about 5.0 min. This reaction duration results in a minimum activity of the tool enzyme for the integration strategy so that the upper limit of linear response by the equilibrium method can be high enough to switch to kinetic analysis of reaction curve. This integration strategy after optimizations can simultaneously have wider linear ranges, higher analysis efficiency and lower cost, better precision and stronger resistance to factors affecting enzyme activities.

Similarly, with the integration strategy for enzyme substrate assay, we also use twice the lower limit of the equilibrium method as the lower limit by the integration strategy if the standard error of estimate is much larger; or else, three times the standard error of estimate by the integration strategy is taken as the lower limit of linear response.

In general, the following steps are required to realize this integration strategy for enzyme substrate assay: (a) to work out the integrated rate equation with the predictor variable of reaction time; (b) to optimize individually the (kinetic) parameters preset as constants for kinetic analysis of reaction curve; (c) to optimize the activity of the tool enzyme so that data for the upper limit of linear response by the equilibrium method within about 5.0-min reaction are suitable for kinetic analysis of reaction curve. As demonstrated later, this integration strategy is applicable to enzymes suffering from strong product inhibition.

2.5 Applications of new methods to some typical enzymes

We investigated kinetic analysis of reaction curve with arylesterase (Liao, et al., 2001, 2003a, 2007b), alcohol dehydrogenase (ADH) (Liao, et al., 2007a), gama-glutamyltransfease (Li, et al., 2011), uricase (Liao, 2005; Liao, et al., 2005a, 2005b, 2006; Liu, et al., 2009; Zhao, Y.S., et al., 2006, 2009), glutathione-S-tranferase (GST) (Liao, et al., 2003b; Zhao, L.N., et al., 2006), butylcholineasterase (Liao, et al., 2009; Yang, et al., 2011), LDH (Cheng, et al., 2008) and LDH-coupled ALT reaction systems (Yang, et al., 2010). Uricase of simple kinetics is a good example to study new methods for kinetic analysis of reaction curve; reactions of GGT and ADH suffer product inhibition and kinetic analyses of their reaction curves are complicated because they require unreported parameters. Hence, our new methods for kinetic analysis of reaction curve and the integration strategies for quantifying enzyme substrates and initial rates are demonstrated with uricase, GST and ADH as examples.

2.5.1 Uricase reaction

Uricase follows simple Michaelis-Menten kinetics on single substrate in air-saturated buffers, and suffers neither reversible reaction nor product inhibition (Liao, 2005; Liao, et al., 2005a, 2005b; Zhao, Y.S., et al., 2006). Uricase reaction curve can be monitored by absorbance at 293 nm. The potential interference from the intermediate 5-hydroxylisourate with uric acid absorbance at 293 nm can be alleviated by analyzing data of steady-state reaction in borate buffer at high reaction pH (Kahn & Tipton, 1998; Priest & Pitts, 1972). The integrated rate equation for uricase reaction with the predictor variable of reaction time is Equ.(4). Uricases from different sources have different K_m (Liao, et al., 2005a, 2006; Zhang, et al., 2010; Zhao, Y.S., et al., 2006). Using Equ.(4), K_m of *Candidate* utilis is estimated with reasonable reliability (Liao, et al., 2005a). Using Equ.(9) to estimate the ratio of V_m to K_m, uricase mutants of better catalytic capacity and their sensitivity to xanthine are routinely characterized (data unpublished). Thus, we used uricases of different K_m as models to test the two integration strategies for enzyme substrate assay and initial rate assay, respectively.

Uricase from *Bacillus* fastidiosus A.T.C.C. 29604 has high K_m to facilitate predicting A_b (Zhang, et al., 2010; Zhao, Y.S., et al , 2006, 2009). Reaction curves at low levels of uric acid with this uricase at 40 U/L are demonstrated in Fig. 3. Steady-state reaction is not reached within 30 s since reaction initiation; it is difficult to get more than 5 data with absorbance changes over 0.003 for kinetic analysis of reaction curve at uric acid levels below 3.0 µmol/L. At 40 U/L of this uricase, the absorbance after reaction for 5.0 min has negligible difference from that after reaction for 30 min for uric acid below 5.0 µmol/L. To quantify the difference between A_0 and A_b after reaction for 5.0 min, the equilibrium method has an upper limit of about 5.0 µmol/L, while kinetic analysis of reaction curve with K_m as a constant is feasible for S_0 of about 5.0 µmol/L. Thus, the change of absorbance over 0.050 between A_0 and the absorbance after reaction for 5.0 min can be the switch threshold to change from the equilibrium method to kinetic analysis of reaction curve.

This integration strategy for enzyme substrate assay gives the linear response from about 1.5 µmol/L up to 60 µmol/L uric acid at 40 U/L uricase (Fig.4, unpublished), and shows resistance to the action of xanthine at 30 µmol/L in reaction solutions (this level of xanthine always caused negative interference with all available kits commercialized for serum uric acid assay). Therefore, the integration strategy for uric acid assay is clearly superior to any other uricase method reported.

Uricases from *Candida* sp. with K_m of 6.6 µmol/L (Sigma U0880) and *Bacillus* fastidious uricase from A.T.C.C. 29604 with K_m of 0.22 mmol/L are used to test the integration strategy for initial rate assay. The use of uric acid at S_0 of 25 µmol/L to monitor reaction curves

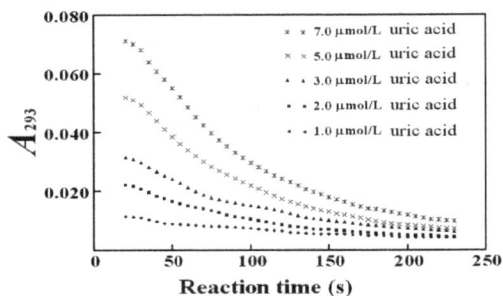

Fig. 3. Reaction curves (absorbance at 293 nm) at low levels of uric acid and 40 U/L uricase (recombinant uricase in *E. Coli* BL21 was as reported before (Zhang, et al., 2010)).

within 8.0 min or at S_0 of 75 μmol/L to monitor reaction curves within 5.0 min, the integration strategy to measure initial rates of both uricases is feasible; the use of PSC of 93% S_0 to convert V_m into initial rates gives the linear range of about two magnitudes (Liu, et al., 2009). Therefore, the integration strategy for enzyme initial rate assay is also advantageous.

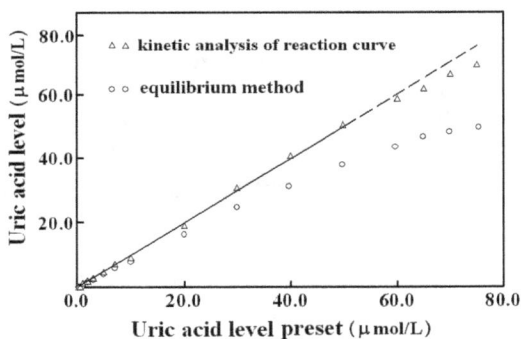

Fig. 4. Response of absorbance change at 293 nm to preset uric acid levels at 40 U/L uricase.

2.5.2 Glutathione-S-transferase reaction

Using purified alkaline GST isozyme from porcine liver as model on glutathione (GSH) and 2,4,-dinitrochlorobenzene (CDNB) as substrates, GST reaction curves are monitored by absorbance at 340 nm (Kunze, 1997; Pabst, et al, 1974; Zhao, L.N., et al., 2006). To promote reaction on single substrate, CDNB is fixed at 1.0 mmol/L while GSH concentrations are kept below 0.10 mmol/L (Zhao, L.N., et al., 2006). Because the concentration of product is calculated from absorbance at 340 nm, the background absorbance before GST reaction is adjusted to zero so that there is no need to treat A_b as a parameter. This treatment of background absorbance eliminate the estimation of A_b and thus confronts with no problem of covariance between A_b and A_m for NLSF. However, GST reaction is more complicated than uricase because it suffers strong product inhibition with an unreported inhibition constant (Kunze, 1997; Pabst, et al, 1974). Thus, the effectiveness of the two integration strategies is tested for measuring initial rate and GSH levels after the inhibition constant of the product is optimized for kinetic analysis of GST reaction curve.

The following symbols are assigned: C to instantaneous concentration of CDNB, B to instantaneous concentration of GSH. Q to instantaneous concentration of the product, K_{ma} to K_m of GST for CSNB, K_{mb} to K_m of GST for GSH, K_{ia} to the dissociation constant of GSH, K_{iq} to the dissociation constant of the product, A for instantaneous absorbance, A_m for the maximal absorbance of the product, ε to difference in absorptivity of product and CDNB, V_m for the maximal reaction rate of GST. The differential rate equation for GST reaction is Equ.(13). After the definition of M_1, M_2 and M_3, the integrated rate equation with the predictor variable of reaction time is Equ.(19) if GST reaction is irreversible and a process similar to that for Equ.(4) is employed (Zhao, L.N., et al., 2006).

$$\frac{1}{V} = \left(K_{mb}/V_m\right) \times \left[1 + K_{ib} \times K_{ma} \times Q/\left(K_{iq} \times K_{mb} \times C\right)\right]/B \\ + \left[1 + K_{ma} \times \left(1 + Q/K_{iq}\right)/C\right]/V_m \tag{13}$$

$$M_1 = K_{ma}/\left(\varepsilon \times K_{iq}\right) \tag{14}$$

$$M_2 = K_{ma} - K_{ib} \times K_{ma}/K_{iq} - A_m \times K_{ma}/\left(\varepsilon \times K_{iq}\right) \\ + C - A_0 \times K_{ma}/\left(\varepsilon \times K_{iq}\right) \tag{15}$$

$$M_3 = K_{ma} \times A_m + \varepsilon \times K_{mb} \times C + C \times A_m \\ - K_{ib} \times K_{mb} \times A_0/K_{iq} - K_{ma} \times A_m \times A_0/\left(K_{iq} \times \varepsilon\right) \tag{16}$$

$$\frac{M_1 \times A^2 - M_2 \times A - M_3}{A - A_m} \times dA = C \times \varepsilon \times V_m \times dt \tag{17}$$

$$Y = M_1 \times \left(A - A_m\right)^2/2 + \left(2 \times M_1 \times A_m + M_2\right) \times \left(A - A_m\right) \\ + \left(M_1 \times A_m^2 + M_2 \times A_m - M_3\right) \times Ln\left|A - A_m\right| \tag{18}$$

$$Y = C \times \varepsilon \times V_m \times \left(t - T_{lag}\right) = a + b \times t \tag{19}$$

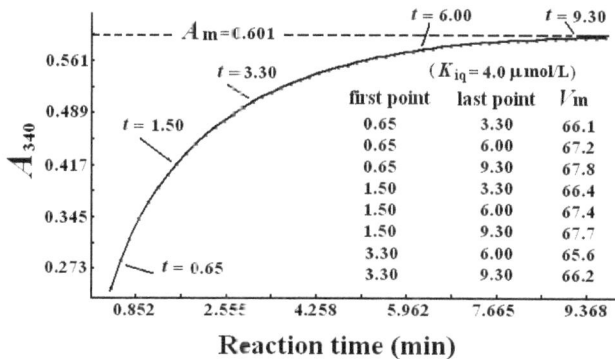

first point	last point	V_m
0.65	3.30	66.1
0.65	6.00	67.2
0.65	9.30	67.8
1.50	3.30	66.4
1.50	6.00	67.4
1.50	9.30	67.7
3.30	6.00	65.6
3.30	9.30	66.2

($K_{iq} = 4.0\ \mu mol/L$)

Reaction time (min)

Fig. 5. Estimated V_m to changes in data ranges for analyses with 60 μmol/L GSH.

As demonstrated in the definition of M_1, M_2 and M_3, kinetic parameters preset as constants for kinetic analysis of GST reaction curve should have strong covariance. Except K_{iq} as an unknown kinetic parameter for optimization, other kinetic parameters are those reported (Kunze, 1997; Pabst, et al, 1974). To optimize K_{iq}, two criteria are used. The first is the consistency of predicted A_m at a series of GSH concentrations using data of 6.0-min reaction with that by the equilibrium method after 40 min reaction (GST activity is optimized to complete the reaction within 40 min). The second is the resistance of V_m to reasonable changes in data ranges for analyses. After stepwise optimization, K_{iq} is fixed at 4.0 μmol/L; A_m predicted for GSH from 5.0 μmol/L to 50 μmol/L is consistent with that by the equilibrium method (Zhao, L.N., et al. 2006); the estimation of V_m is resistant to changes of data ranges (Fig. 5). Therefore, K_{iq} is optimized and fixed as a constant at 4.0 μmol/L.

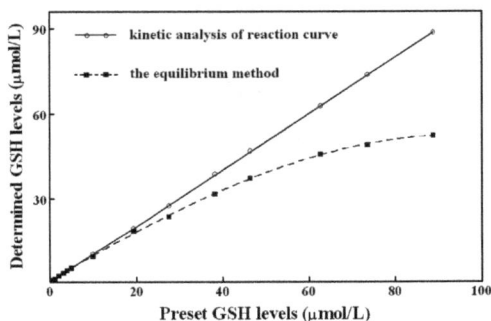

Fig. 6. Response of GSH concentration determined to preset GSH concentrations (the equilibrium method uses data with 6.0 min reaction).

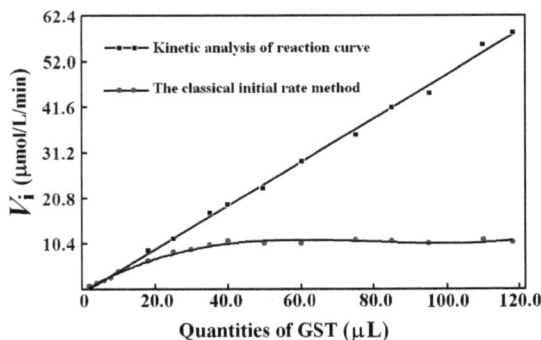

Fig. 7. Response of initial rates to quantities of purified porcine alkaline GST.

Kinetic analysis of GST reaction curve can predict A_m for GSH over 4.0 μmol/L, but there are no sufficient data for analyses at GSH below 3.0 μmol/L; after optimization of GST activity for complete conversion of GSH at 5.0 μmol/L within 6.0 min, reaction curve within 5.0 min for GSH at 5.0 μmol/L can be used for kinetic analysis of reaction curve to predict A_m. With the optimized GST activity for reaction within 5.0 min, the linear range for GSH assay is from 1.5 μmol/L to over 90.0 μmol/L by the integration strategy while it is from 4.0

μmol/L to over 90.0 μmol/L by kinetic analysis of reaction curve alone (Fig. 6, unpublished). By the equilibrium method alone for reaction within 5.0 min, the assay of 80.0 μmol/L GSH requires GST activity that is 50 folds higher due to the inhibition of GST by the accumulated product. Therefore, the integration strategy for GSH assay is obviously advantageous.

The integration strategy for measuring GST initial rates is tested. For convenience, S_0 of the final GSH is fixed at 50 μmol/L and the duration to monitor reaction curve is optimized. After the analyses of reaction curves recorded within 10 min, it is found that reaction for 6.0 min is sufficient to provide the required overlapped region of GST activities measurable by both methods. By using K_{ic} fixed at 4.0 μmol/L as a constant, the reaction duration of 6.0 min and PSC at 48 μmol/L to convert V_m to initial rates, the integration strategy gives a linear range from 2.0 U/L to 60 U/L; kinetic analysis of reaction curve alone gives the linear range from 5.0 U/L to 60 U/L while the classical initial rate method alone gives a linear range from 1.0 U/L to 5.0 U/L (Fig. 7, unpublished). Clearly, with enzyme suffering strong product inhibition, the integration strategy for enzyme initial rate assay is advantageous.

2.5.3 Alcohol dehydrogenase reaction

ADH is widely used for serum ethanol assay. ADH kinetics is sophisticated due to the reversibility of reaction and the inhibition by both acetaldehyde and NADH as products. To simplify ADH kinetics, some special approaches are employed to make ADH reaction apparently irreversible on single substrate (alcohol). Thus, reaction pH is optimized to 9.2 to scavenge hydrogen ion; semicabarzide at final 75 mmol/L is used to remove acetaldehyde as completely as possible; final nicotinamide adenine dinucleotide (NAD$^+$) is 3.0 mmol/L; final ADH is about 50 U/L (Liao, et al., 2007a). By assigning the maximal absorbance at 340 nm for reduced nicotinamide adenine dinucleotide (NADH) by the equilibrium method to A_{me} and that by kinetic analysis of reaction curve to A_{mk}, kinetic analysis of ADH reaction curve should predict A_{mk} consistent with A_{me}, but requires some special efforts.

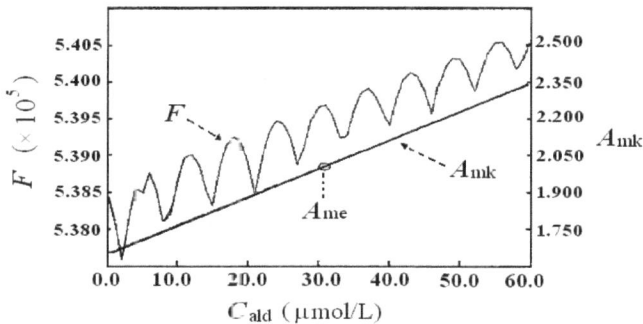

Fig. 8. Response of F values to preset C_{ald} for kinetic analysis of reaction curve for 0.31 mmol/L ethanol (reproduced with permission from Liao, et al, 2007a).

The use of semicabarzide reduces concentrations of acetaldehyde (C_{ald}) to unknown levels, and thus complicates the treatment of acetaldehyde inhibition on ADH. The integration rate equation with the predictor variable of reaction time can be worked out for ADH (Liao, et al., 2007a). All kinetic parameters and NAD^+ concentrations are preset as those used or reported (Ganzhorn, et al. 1987). However, there are multiple maxima of the goodness of fit with the continuous increase in steady-state C_{ald} for kinetic analysis of reaction curve (Fig. 8). Thus, C_{ald} can not be concomitantly estimated by kinetic analysis of reaction curve, and a special approach is used to approximate steady-state C_{ald} for predicting A_{mk}.

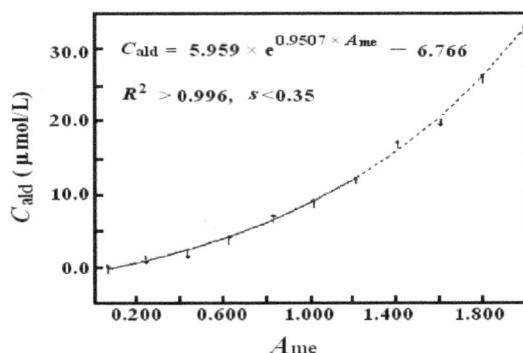

Fig. 9. Correlation function of the best steady-state C_{ald} with A_{me} (reproduced with permission from Liao, et al, 2007a).

Under the same reaction conditions, the equilibrium method can determine A_{me} for ethanol below 0.20 mmol/L after reaction for 50 min. For kinetic analyses of such reaction curves, the lag time for steady-state reaction is estimated to be over 40 s and is used to select data of steady-state reaction for analysis. Using the equilibrium method as the reference method, the best steady-state C_{ald} for data of 6.0-min reaction is obtained for consistency of A_{mk} with A_{me} at each tested ethanol level from 10 μmol/L to 0.17 mmol/L. After dilution and determination by the equilibrium method, A_{me} for each tested ethanol level from 0.17 mmol/L to 0.30 mmol/L is also available. Consequently, an exponential additive function is obtained to approximate the correlation of the best C_{ald} for predicting A_{mk} consistent with A_{me} (Fig. 9). This special correlation function for C_{ald} and A_{mk} is used as a restriction function to iteratively adjust C_{ald} for predicting A_{mk}; namely, iterative kinetic analysis of reaction curve with C_{ald} predicted from the restriction function using previous A_{mk} finally gives the desired A_{mk}. Such an artificial intelligence approach to the steady-state C_{ald} for kinetic analysis of reaction curve can hardly be found in publications.

To start kinetic analysis of an ADH reaction curve, the highest absorbance under analysis is taken as A_{mk} to predict the best C_{ald} for the current run of kinetic analysis of reaction curve. The estimated A_{mk} is then used to predict the second C_{ald} for the second run of kinetic analysis of reaction curve (Fig. 10). Such an iterative kinetic analysis of reaction curve can predict A_{mk} consistent with A_{me} for 0.31 mmol/L ethanol when reaction duration is just 6.0 min and the convergence criterion is set for absorbance change below 0.0015 in A_{mk}. Usually convergence is achieved with 7 runs of the iterative kinetic analysis of reaction. Moreover, it is resistant to the change of ADH activities by 50% and coefficients of variation (CV) are below 5% for final ethanol levels from 20 μmol/L to 310 μmol/L in reaction solutions.

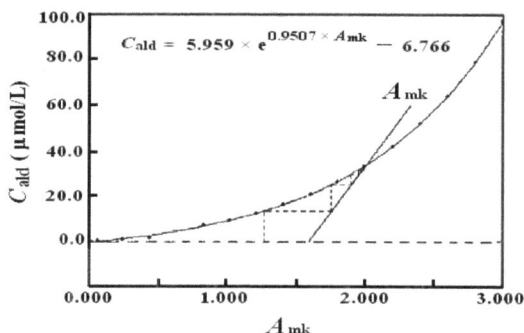

$$C_{ald} = 5.959 \times e^{0.9507 \times A_{mk}} - 6.766$$

Fig. 10. Iterative adjustment of C_{ald} to predict A_{mk} for 0.31 mmol/L ethanol at 50 U/L ADH (reproduced with permission from Liao, et al, 2007a).

Obviously, by this special approach for kinetic analysis of ADH reaction curve, the upper limit of linear response is excellent, but the lower limit of linear response is over 5.0 μmol ethanol. Under the stated reaction conditions, the equilibrium method after reaction for 8.0 min is effective to quantify ethanol up to final 6.0 μmol. Thus, the equilibrium method with reaction duration of 8.0 min can be integrated with iterative kinetic analysis of reaction curve for quantifying ethanol; this integration strategy gives the linear range from about final 2.0 μmol to about 0.30 mmol/L ethanol in reaction solutions; it has CVs below 8% for ethanol below 10 μmol/L, and CVs below 5% for ethanol over 20 μmol/L (Liao, et al., 2007a). These results clearly supported the advantage of the new integration strategy for substrate assay and the importance of chemometrics in kinetic enzymatic analysis of substrate.

2.6 Programming for kinetic analysis of enzyme reaction curve

Most software package like Origin, SAS, MATLAB can perform kinetic analysis of reaction curve, but they are usually ineffective to implicit functions for kinetic analysis of reaction curve. For convenience and the use of some complicated methods for kinetic analysis of reaction curve in widow-aided mode, self-programming is still favourable.

For simplicity in programming, we used Visual Basic 6.0 to write the source code and working windows (Liu, et al., 2011). The executable program has the main window to perform kinetic analysis of reaction curve (Fig. 11). Original data for each reaction curve is stored as a text file, and keywords are used to indicate specific information related to the reaction curve including sample numbering, the enzyme used, the quantification method, some necessary kinetic parameters, and usually initial signal before enzyme action. Such information is read into memory by the software for kinetic analysis of reaction curve.

On the main window to perform kinetic analysis of reaction curve, original data are listed and plotted for eyesight-checking of data for steady-state reaction. Text boxes are used to input some common parameters like K_m, and most parameters are read from the text file for the reaction curve. Subprogram for an enzyme reaction system is called for running; results are displayed on the main window and may be saved in text file for further analysis.

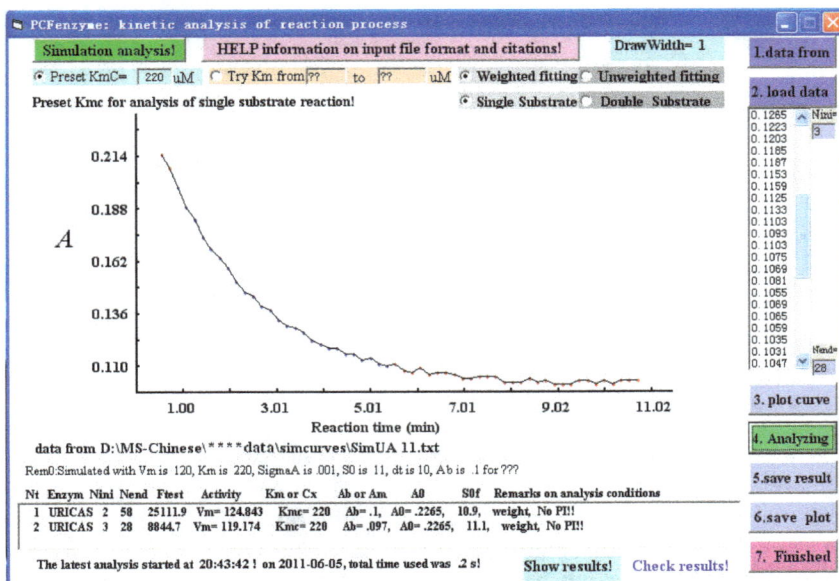

Fig. 11. Main window for the executable PCFenzyme

We called the software PCFenzyme. An old version of the executable PCFenzyme can be downloaded at http://dx.doi.org/10.1016/j.clinbiochem.2008.11.016. The latest version of the executable PCFenzyme with new methods included is available upon request by e-mail.

3. Conclusion

The following conclusions can be drawn. (a) Kinetic analysis of reaction curve can give the initial substrate concentration before enzyme action, the maximal reaction rate, Michaelis-Menten constant and other related parameters of an enzyme reaction system; for reliability, however, it is better to just estimate the initial substrate concentration before enzyme action and the maximal reaction rate with Michaelis-Menten constant and other parameters fixed as constants after optimization. (b) For an enzyme whose integrated rate equation with the predictor variable of reaction time is accessible, kinetic analysis of reaction curve can estimate parameters *via* nonlinear-least-square-fitting after transformation of data from the reaction curve under analysis. (c) For an enzyme reaction system whose kinetics is described by a set of differential rate equations or is difficult to be integrated with the predictor of reaction time, iterative numerical integration of the differential rate equation(s) with a series of preset parameters can produce serial calculated reaction curves; such calculated reaction curves can be fit to the reaction curve under analysis for estimating parameters based on nonlinear-least-square-fitting. This approach is applicable to enzyme-coupled reaction systems of sophisticated kinetics. (d) The integration of kinetic analysis of reaction curve with the equilibrium method can quantify enzyme substrates with expanded linear ranges, favourable analysis efficiency, low cost on tool enzyme, desirable resistance to factors affecting enzyme activities and enhanced precision; it can be applied to enzyme reaction suffering strong product inhibition. (e) The integration of kinetic analysis of reaction curve with the classical initial rate method can measure enzyme initial rates with wide linear ranges, favourable analysis efficiency and practical levels of substrates; it can be applicable to enzyme-coupled reaction curve or enzyme reaction suffering product inhibition.

Taken together, kinetic analysis of enzyme reaction curves under optimized conditions can screen common reversible inhibitors and enzyme mutants; the integration strategy for measuring enzyme activities can quantify serum enzymes and enzyme labels in enzyme-immunoassays to expand the quantifiable ranges, and can be applied to quantify irreversible inhibitors as environmental pollutants; the integration strategy to quantify enzyme substrate can be the second-generation approaches and potentially find wide applications in clinical laboratory medicine. Therefore, these new methodologies for enzymatic analyses based on chemometrics can potentially find their important applications in biomedical sciences.

4. Acknowledgment

This work is supported by the program for New Century Excellent Talent in University (NCET-09), high-technology-program "863" of China (2011AA02A108), National Natural Science Foundation of China (nos. 30200266, 30672009, 81071427), Chongqing Municipal Commission of Sciences and Technology (CQ CSTC2011BA5039), and Chongqing Education Commission (KJ100313).

5. References

Atkins, G.L. & Nimmo, I.A. (1973). The reliability of Michaelis–Menten constants and maximum velocities estimated by using the integrated Michaelis–Menten equation. *The Biochemical Journal*, vol. 135, no.4, (December 1973), pp. 779–784, ISSN 0264-6021

Baywenton, P. R. (1986). *Data process and error analysis* (Translated into Chinese by Weili Qiu, Genxin Xu, Enguang Zhao, and Shengzhong Chen), ISBN 13214.84, Knowledge Press, Beijing, China

Bergmeyer, H.U. (1983). Methods of Enzymatic Analysis, Vol. *I. Fundamentals (3rd Ed.)*, ISBN 978-3527260416, Wiley VCH, Weinheim, Germany

Burden, R.L. & Faires, J.D. (2001). *Numerical Analysis* (7th ed.) , ISBN 978-0534382162, Academic Internet Publishers, Ventura, Carliforlia, USA

Burguillo, J., Wright, A.J. & Bardsley, W. G. (1983). Use of the F test for determining the degree of enzyme-kinetic and ligand-binding data. *The Biochemical Journal*, vol. 211, no.1, (April 1983), pp. 23–34, ISSN0264-6021

Cheng, Y.C. & Prusoff, W.H. (1973). Relationship between the inhibition constant (KI) and the concentration of inhibitor which causes 50 per cent inhibition (I50) of an enzymatic reaction. Biochemical Pharmacology, vol.22, no. 23, (December 1973), pp. 3099-3108, ISSN 0006-2952.

Cheng, Z.L., Chen, H., Zhao, Y.S., Yang, X.L., Lu, W., Liao, H., Yu, M.A., & Liao, F. (2008). The measurement of the activity of rabbit muscle lactic dehydrogenase by integrating the classical initial rate method with an integrated method. *2nd International Conference on Bioinformatics and Biomedical Engineering, iCBBE 2008*, pp.1209-1212, ISBN 978-1-4244-1748-3, Shanghai, China, May 26-28, 2008

Claro, E. (2000). Understanding initial velocity after the derivatives of progress curves. *Biochemistry and Molecular Biology Education*, Vol.28, no.6, (November 2000), pp. 304-306, ISSN 1470-8175

Cornish-Bowden, A. (1975). The use of the direct linear plot for determining initial velocities. *The Biochemical Journal*, vol. 149, no.2, (August 1975), pp. 305-312, ISSN 0264-6021

Cornish-Bowden, A. (1995). *Analysis of enzyme kinetic data*, ISBN 978-0198548775, Oxford University Press, London, UK

Dagys, R., Tumas, S., Zvirblis, S. & Pauliukonis, A. (1990) Determination of first and second derivatives of progress curves in the case of unknown experimental error. *Computers and Biomedical Research*, Vol.23, no. 5, (October 1990), pp. 490-498, ISSN 0010-4809

Dagys, R., Pauliukonis, A., Kazlauskas, D., Mankevicius, M. & Simutis, R. (1986). Determination of initial velocities of enzymic reactions from progress curves. *The Biochemical Journal*, vol.237, no.3, (August 1986), pp. 821-825, ISSN 0264-6021

del Rio F.J., Riu, J. & Rius, F. X. (2001). Robust linear regression taking into account errors in the predictor and response variables. *Analyst*, vol. 126, no. , (July 2001), pp. 1113-1117, ISSN 0003-2654

Dilena, B.A., Peake, M.J., Pardue, H.L., Skoug, J.W. (1986). Direct ultraviolet method for enzymatic determination of uric acid, with equilibrium and kinetic data-processing options. *Clinical Chemistry*, vol. 32, no.3, (May 1986), pp. 486-491, ISSN 0009-9147

Dixon, M.C. & Webb, EC. (1979). *Enzymes* (3rd ed.), ISBN 0122183584, Academic Press, New York, USA

Draper, N.R. & Smith, H. (1998). *Applied regression analysis* (3rd ed.), ISBN 978-0471170822, Wiley-Interscience; New York, USA

Duggleby, R. G. (1983). Determination of the kinetic properties of enzymes catalysing coupled reaction sequences. *Biochimica et Biophysica Acta (BBA) - Protein Structure and Molecular Enzymology*, Vol.744, no. 3, (May 1983), pp. 249-259, ISSN 0167-4838

Duggleby, R.G. (1985). Estimation of the initial velocity of enzyme-catalysed reactions by non-linear regression analysis of progress curves. *The Biochemical Journal*, vol. 228, no.1, (May 1985), pp. 55-60, ISSN 0264-6021

Duggleby, R. G. (1994). Analysis of progress curves for enzyme-catalyzed reactions: application to unstable enzymes, coupled reactions, and transient state kinetics. *Biochimica et Biophysica Acta (BBA) – General subjects*, vol. 1205, no.2, (April 1994), pp. 268-274, ISSN 0304-4165

Fresht, A. (1985). *Enzyme structure and Mechanism* (2nd Ed.), ISBN 978-0716716143, Freeman WH, New York, USA

Ganzhorn, A.J., Green, D.W., Hershey, A.D., Gould, R.M., Plapp, B.V. (1987). Kinetic characterization of yeast alcohol dehydrogenases. Amino acid residue 294 and substrate specificity. The Journal of Biological chemistry,vol.262, no.8, (March 1987), pp. 3754-3761, ISSN 0021-9258

Guilbault, G. G. (1976). *Handbook of enzymatic methods of analysis*, ISBN 978-0824764258, Marcel Dekker, New York, USA

Gutierrez, O.A. & Danielson, U. H. (2006). Sensitivity analysis and error structure of progress curves. Analytical Biochemistry, vol.358, no.1, (August 2006), pp.1-10, ISSN 0003-2697

Hamilton, S. D. & Pardue, H. L. (1982). Kinetic method having a linear range for substrate concentration that exceed Michaelis–Menten constants. Clinical Chemistry, vol. 28, no.12, (December 1982), pp.2359–2365, ISSN 0009-9147

Hasinoff, B. B. (1985). A convenient analysis of Michaelis enzyme kinetic progress curves based on second derivatives. Biochimica et Biophysica Acta (BBA) - General Subjects, Vol. 838, no. 2, (February 1985), pp. 290-292, ISSN 0304-4165

Kahn, K. & Tipton, P.A. (1998). Spectroscopic characterization of intermediates in the urate oxidase reaction. Biochemistry, vol. 37, no. (August 1998), pp. 11651-11659, ISSN 0006-2960.

Koerber, S. C. & Fink, A. L. (1987). The analysis of enzyme progress curves by numerical differentiation, including competitive product inhibition and enzyme reactivation. Analytical Biochemistry, vol. 165, no.1, (December 2004), pp. 75-87, ISSN 0003-2697

Li, Z.R., Liu,Y., Yang, X.Y., Pu,J., Liu, B.Z., Yuan, Y.H., Xie, Y.L. & Liao, F. (2011). Kinetic analysis of gamma-glutamyltransferase reaction process for measuring activity via an integration strategy at low concentrations of gamma-glutamyl p-nitroaniline. Journal of Zhejiang University Sciecnce B, vol. 12, no.3, (March 2011), pp. 180-188, ISSN 1673-1581

Liao, F. (2005). The method for quantitative enzymatic analysis of uric acid in body fluids by predicting the background absorbance. China patent: ZL 03135649.4, 2005-08-31

Liao, F., Li, J.C., Kang, G.F., Zeng, Z.C., Zuo, Y.P. (2003a). Measurement of mouse liver glutathione-S- transferase activity by the integrated method. Journal of Medical Colleges of PLA, vol. 18, no.5, (October 2003), pp. 295-300, ISSN 1000-1948

Liao, F., Liu, W.L., Zhou, Q.X., Zeng, Z.C., Zuo, Y.P. (2001). Assay of serum arylesterase activity by fitting to the reaction curve with an integrated rate equation. Clinica Chimica Acta, vol. 314, no.1-2, (December 2001), pp.67-76, ISSN 0009-8981

Liao, F., Tian, K.C., Yang, X., Zhou, Q.X., Zeng, Z.C., Zuo, Y.P. (2003b). Kinetic substrate quantification by fitting to the integrated Michaelis-Menten equation. Analytical Bioanalytical Chemistry. vol. 375, no. 6, (Febrary 2003), pp. 756-762, ISSN1618-2642

Liao, F., Yang, D.Y., Tang, J.Q., Yang, X.L., Liu, B.Z., Zhao, Y.S., Zhao, L.N., Liao, H. & Yu, M.A. (2009). The measurement of serum cholinesterase activities by an integration strategy with expanded linear ranges and negligible substrate-activation. Clinical Biochemistry, vol.42, no 6, (December 2008), pp.926-928. ISSN 0009-9120

Liao, F., Zhao, L.N., Zhao, Y.S., Tao, J., Zuo, Y.P. (2007a). Integrated rate equation considering product inhibition and its application to kinetic assay of serum ethanol. Analytical Sciences, vol. 23, no.4, (April 2007), pp. 439-444 , ISSN 0910-6340

Liao, F., Zhao, Y.S., Zhao, L.N., Tao, J., Zhu, X.Y., Liu, L. (2006). The evaluation of a direct kinetic method for serum uric acid assay by predicting the background absorbance of uricase reaction solution with an integrated method. Journal of Zhejiang University Science B, vol. 7, no.6, pp. 497-502, ISSN 1673-1581

Liao, F., Zhao, Y.S., Zhao, L.N., Tao, J., Zhu, X.Y., Wang, Y.M., Zuo, Y.P. (2005b). Kinetic method for enzymatic analysis by predicting background with uricase reaction as model. *Journal of Medical Colleges of PLA*, vol.20, no.6, (Deember 2005), pp. 338-344, ISSN 1000-1948

Liao, F., Zhu, X.Y., Wang, Y.M., Zhao, Y.S, Zhu, L.P., Zuo, Y.P. (2007b). Correlation of serum arylesterase activity on phenylacetate estimated by the integrated method to common classical biochemical indexes of liver damage. *Journal of Zhejiang University Science B*, vol. 8, no.4, (April 2007), pp.237-241, ISSN 1673-1581

Liao, F., Zhu,X.Y., Wang, Y.M., Zuo, Y.P. (2005a). The comparison on the estimation of kinetic parameters by fitting enzyme reaction curve to the integrated rate equation of different predictor variables. *Journal of Biochemical Biophysical Methods*, vol. 62, no.1, (January 2005), pp. 13-24, ISSN 0165-022X

Liu, B.Z., Zhao, Y.S., Zhao, L.N., Xie, Y.L., Zhu,S., Li,Z.R., Liu,Y., Lu,W., Yang,X.L., Xie, G.M., Zhong, H.S., Yu, M.A., Liao,H. & Liao, F. (2009). An integration strategy to estimate the initial rates of enzyme reactions with much expanded linear ranges using uricases as models. *Analytica Chimica Acta*, vol.631, no.1, (October 2008), pp. 22-28. ISSN 0003-2670

Liu, M., Yang, X.L., Yuan, Y.H., Tao, J. & Liao, F. (2011). PCFenzyme for kinetic analyses of enzyme reaction processes. *Procedia Environmental Sciences*, vol. 8, (December 2011), pp.582-587, ISSN 1878-0296

Lu, W.P. & Fei, L. (2003). A logarithmic approximation to initial rates of enzyme reactions. *Analytical Biochemistry*, vol. 316, no. 1, (May 2003), pp.58-65, ISSN 0003-2697

Marangoni, A. G. (2003). *Enzyme kinetics:a modern approach*, ISBN 978-0471159858, Wiley-Interscience, New York, USA

Meyler-Almes, F.J. & Auer, M. (2000). Enzyme inhibition assay using fluorescence correlation spectroscopy: a new algorithm for the derivation of Kcat/KM.and Ki values at substrate concentration much lower than the Michaelis constant. *Biochemistry*, vol. 39, no.43 (October 2000), pp. 13261–13268, ISSN 0006-2960

Miller, J. C. & Miller, J. N. (1993). *Statistics for analytical chemistry* (3rd), ISBN 978-0130309907, Ellis Horwood, Chichester, New York, USA

Morishita, Y., Iinuma, Y., Nakashima, N., Majima, K., Mizuguchi, K. & Kawamura, Y. (2000). Total and pancreatic amylase measured with 2-chloro-4-nitrophenyl-4-O-ß-D-galactopyranosylmaltoside. *Clinical Chemistry*, vol. 46, no.7, (July 2000), pp. 928-933, ISSN 0009-9147

Moruno-Davila, M.A., Solo, C.G., Garcla-Moreno, M., Garcla-CAnovas, F. & Varon, R. (2001). Kinetic analysis of enzyme systems with suicide substrate in the presence of a reversible, uncompetitive inhibitor. *Biosystems*, vol. 61, no.1, (June 2001), pp.5-14, ISSN 0303-2647

Moss, D.W. (1980). Methodological principles in the enzymatic determination of substrates illustrated by the measurement of uric acid. *Clinica Chimica Acta*, Vol. 105, no. 3, (August 1980), pp. 351-360, ISSN 0009-8981

Newman, P.F.J., Atkins, G.L. & Nimmo, I. A. (1974). The effects of systematic error on the accuracy of Michaelis constant and maximum velocities estimated by using the integrated Michaelis–Menten equation. *The Biochemical Journal*, vol. 143, no. 3, (December 1974), pp. 779–781. ISSN 0264-6021

Northrop, D. B. (1983). Fitting enzyme-kinetic data to V/K. *Analytical Biochemistry*, vol. 132, No. 2, (July 1983), pp. 457–61, ISSN 0003-2697

Orsi, B.A. & Tipton, K. F. (1979). Kinetic analysis of progress curves. In: *Methods in Enzymology*, vol. 63, D. L. Purich, (Ed.), 159-183, Academic Press, ISBN 978-0-12-181963-7, New York, USA

Priest, D.G. & Pitts, O.M. (1972). Reaction intermediate effects on the spectrophotometric uricase assay. *Analytical Biochemistry*, vol.50, no.1, (November 1972), pp. 195-205, ISSN 0003-2697

Stromme,J.H. & Theodorsen, L. (1976). Gamma-glutamyltransferase: Substrate inhibition, kinetic mechanism, and assay conditions. *Clinical Chemistry*, vol. 22, no.4, (April 1976), pp. 417-421, ISSN 0009-9147

Varon, R., Garrido-del Solo, C., Garcia-Moreno, M., Garcoa-Canovas, F., Moya-Garcia, G., Vidal de Labra, J., Havsteen BH. (1998). Kinetics of enzyme systems with unstable suicide substrates. *Biosystems*, vol. 47, no.3, (August 1998), pp.177-192, ISSN 0303-2647

Walsh, R., Martin, E., Darvesh, S. (2010). A method to describe enzyme-catalyzed reactions by combining steady state and time course enzyme kinetic parameters. *Biochimica et Biophysica Acta-General Subjects*, vol.1800, no.1, (October 2009), pp1-5, ISSN 0304-4165.

Yang, D., Tang, J., Yang, X., Deng, P., Zhao, Y., Zhu, S., Xie, Y., Dai, X., Liao, H., Yu, M., Liao, J. & Liao, F. (2011). An integration strategy to measure enzyme activities for detecting irreversible inhibitors with dimethoate on butyrylcholinesterase as model. *International Journal of Environmental Analytical Chemistry*, vol.91, no.5, (March 2011), pp.431-439, ISSN 0306-7319

Yang, X. L., Liu, B.Z., Sang,Y., Yuan, Y.H., Pu, J., Liu,Y., Li, Z.R., Feng,J., Xie, Y.L, Tang, R. K., Yuan, H.D. & Liao, F. (2010). Kinetic analysis of lactate-dehydrogenase-coupled reaction process and measurement of alanine transaminase by an integration strategy. *Analytical Sciences*, vol.26, no. 11, (November 2010), pp. 1193-1198, ISSN0910-6340

Zhang, C., Yang, X.L., Feng,J., Yuan, Y.H., Li,X., Bu, Y.Q., Xie, Y.L., Yuan, H.D. & Liao, F. (2010). Effects of modification of amino groups with poly(ethylene glycol) on a recombinant uricase from Bacillus fastidiosus. *Bioscience Biotechnology Biochemistry*, 2010; vol.74, no.6, (June 2010), pp. 1298-1301, 0916-8451. ISSN 0916-8451.

Zhao, L.N., Tao, J., Zhao, Y.S., Liao, F. (2006). Quantification of reduced glutathione by analyzing glutathine-S-transferase reaction process taking into account of product inhibition. *Journal of Xi'an Jiaotong University (Medical Sciences)*, vol. 27, no.3, (June 2006), pp.300-303, ISSN 1671-8259

Zhao,Y.S., Yang, X.Y., Lu,W., Liao,H. & Liao, F. (2009). Uricase based method for determination of uric acid in serum. *Microchimica Acta*, vol. 164, no.1, (May 2008), pp.1-6, ISSN 0026-3672

Zhao,Y.S., Zhao,L.N., Yang,G.Q., Tao, J., Bu, Y.Q. & Liao, F. (2006). Characterization of an intracellular uricase from Bacillus fastidious ATCC 26904 and its application to serum uric acid assay by a patented kinetic method. *Biotechnology Applied Biochemistry*, vol. 45, no.2, (September 2006), pp. 75-80, ISSN 0885-4513

Zou, G.L. & Zhu, R.F. (1997). *Enzymology*, Wuhan University Press, ISBN 7-307-02271-0/Q, Wuhan, China

9

Electronic Nose Integrated with Chemometrics for Rapid Identification of Foodborne Pathogen

Yong Xin Yu and Yong Zhao*
*College of Food Science and Technology,
Shanghai Ocean University, Shanghai,
China*

1. Introduction

Diseases caused by foodborne pathogens have been a serious threat to public health and food safety for decades and remain one of the major concerns of our society. There are hundreds of diseases caused by different foodborne pathogenic microorganisms, including pathogenic viruses, bacteria, fungi, parasites, marine phytoplankton, and cyanobacteria, etc (Hui, 2001). Among these, bacteria such as *Salmonella* spp., *Shigella* spp., *Escherichia coli*, *Staphylococcus aureus*, *Campylobacter jejuni*, *Campylobacter coli*, *Bacillus cereus*, *Vibrio parahaemolyticus* and *Listeria monocytogenes* are the most common foodborne pathogens (McClure, 2002), which can spread easily and rapidly under requiring food, moisture and a favorable temperature (Bhunia, 2008).

Identification and detection pathogens in clinical, environmental or food samples usually involves time-consuming growth in selective media, subsequent isolation and laborious biochemical and molecular diagnostic procedures (Gates, 2011). Many of these techniques are also expensive or not sensitive enough for the early detection of bacterial activity (Adley, 2006). The development of alternative analytical techniques that are rapid and simple has become increasingly important to reduce sample preparation time investment and to conduct real time analyses.

It is well known that microorganisms can produce species-specific microbial volatile organic compounds (MVOCs), or odor compounds, which characterize as odor fingerprinting (Turner & Magan, 2004). Early in this research area, the question arose as to can we use odor fingerprinting like DNA fingerprinting to identify or detect microbe in pure culture or in food samples. To date it is still a very interesting scientific question. Many studies (Bjurman, 1999, Kim et al., 2007, Korpi et al., 1998, Pasanen et al., 1996, Wilkins et al., 2003), especially those using analytical tools such as gas chromatography (GC) or gas chromatography coupled with mass spectrometry (GC–MS) for headspace analysis, have shown that microorganisms produce many MVOCs, including alcohols, aliphatic acids and terpenes, some of which have characteristic odors (Schnürer et al., 1999).

---

* Corresponding Author, mail address: yzhao@shou.edu.cn

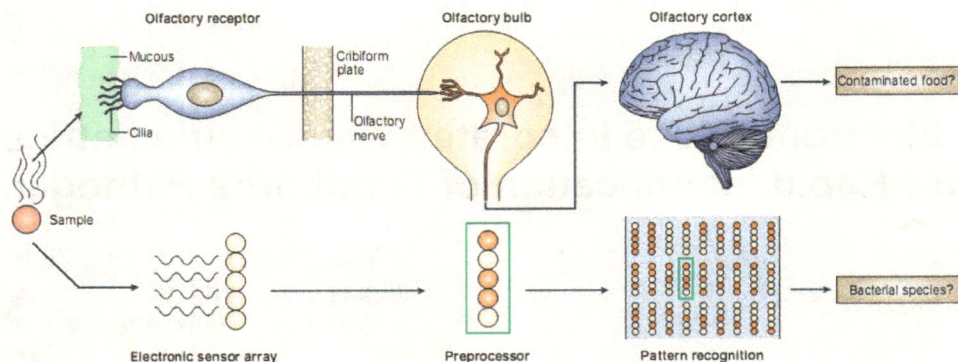

Fig. 1. Electronic nose devices mimic the human olfactory system.

The electronic devices simulate the different stages of the human olfactory system, resulting in volatile odor recognition, which can now be used to discriminate between different bacterial infections. (Turner & Magan, 2004)

During the past three decades there has been significant research interest in the development of electronic nose (E-nose) technology for food, agricultural and environmental applications (Buratti et al., 2004, Pasanen et al., 1996, Romain et al., 2000, Wilkins et al., 2003). The term E-nose describes a machine olfaction system, which successfully mimics human olfaction and intelligently integrates of multitudes of technologies like sensing technology, chemometrics, microelectronics and advanced soft computing (see Fig. 1). Basically, this device is used to detect and distinguish complex odor at low cost. Typically, an electronic nose consists of three parts: a sensor array which is exposed to the volatiles, conversion of the sensor signals to a readable format, and software analysis of the data to produce characteristic outputs related to the odor encountered. The output from the sensor array may be interpreted via a variety of chemometrics methods (Capone et al., 2001, Evans et al., 2000, Haugen & Kvaal, 1998) such as principal component analysis (PCA), discriminant function analysis (DFA), cluster analysis (CA), soft independent modelling by class analogy (SIMCA), partial least squares (PLS) and artificial neural networks (ANN) to discriminate between different samples. The data obtained from the sensor array are comparative and generally not quantitative or qualitative in any way. It has the potential to be a sensitive, fast, one-step method to characterize a wide array of different volatile chemicals. Since the first model of an intelligent electronic gas sensing model was described, a significant amount of gas sensing research has been focused on several industrial applications.

Recently, some novel microbiological applications of E-nose have been reported, such as the characterization of fungi (Keshri et al., 1998, Pasanen et al., 1996, Schnürer et al., 1999), bacteria (Dutta et al., 2005, Pavlou et al., 2002a) and the diagnosis of disease (Gardner et al., 2000, Pavlou et al., 2002b, Zhang et al., 2000). It is more and more clear that E-nose techniques coupled with different chemometrics analyses of the odor fingerprinting offer a wide range of applications for food microbiology, including identification of foodborne pathogen.

2. Detection strategies

Several E-nose devices have been developed, all of which comprise three basic building blocks: a volatile gas odor passes over a sensor array, the conductance of the sensors changes owing to the level of binding and results in a set of sensor signals, which are coupled to data-analysis software to produce an output (Turner & Magan, 2004).

The main strategy of foodborne pathogen identification based on E-nose, which is composed of three steps: headspace sampling, gas sensor detection and chemometrics analysis (see Fig. 2).

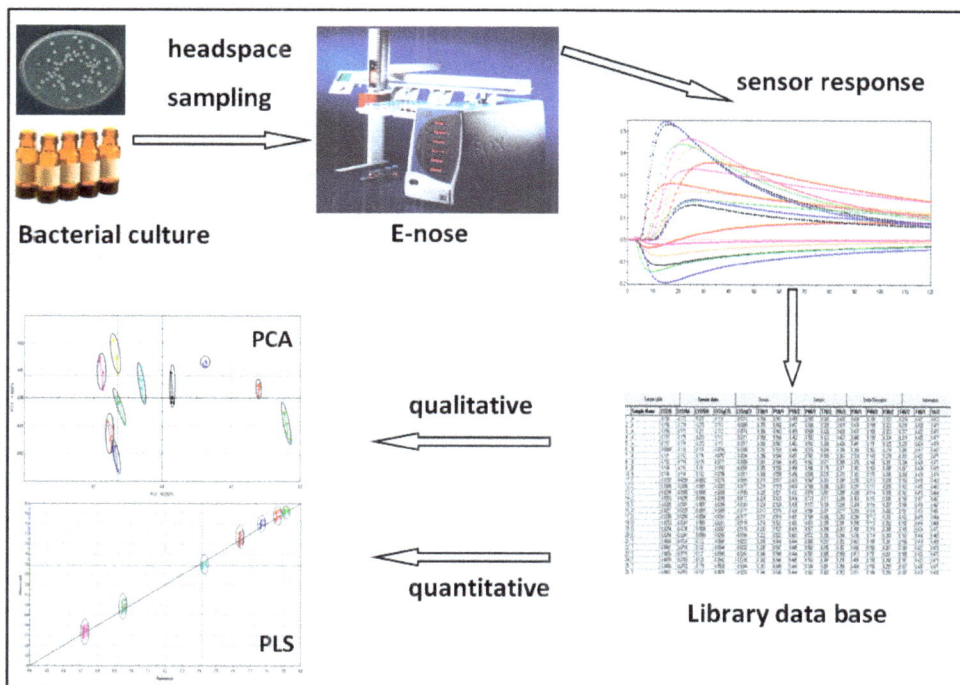

Fig. 2. Electronic nose and chemometrics for the identification of foodborne pathogen. The main strategy of foodborne pathogen identification based on E-nose.

2.1 Headspace sampling

Before analysis, the bacterial cultures should be transferred into standard 20 ml headspace vials and sealed with PTFE-lined Teflon caps to equilibrate the headspace. Sample handling is a critical step affecting the analysis by E-nose. The quality of the analysis can be improved by adopting an appropriate sampling technique. To introduce the volatile compounds present in the headspace (HS) of the sample into the E-nose's detection system, several headspace sampling techniques have been used in E-nose. Typically, the methods of headspace sampling (Ayoko, 2004) include static headspace (SHS) technique, purge and trap (P&T) technique, stir bar sorptive extraction (SBSE) technique, inside-needle dynamic extraction (INDEX) technique, membrane introduction mass spectrometry (MIMS) technique and solid phase micro extraction (SPME) technique.

Unlike the other techniques, SPME has a considerable concentration capacity and is very simple because it does not require especial equipment. The principle involves exposing a silica fibre covered with a thin layer of adsorbent in the HS of the sample in order to trap the volatile components onto the fibre. The adsorbed compounds are desorbed by heating and introduced into the detection system. A SPME sampler consists of a fused silica fiber that is coated by a suitable polymer (e.g. PDMS, PDMS/divinylbenzene, carboxen/PDMS) and housed inside a needle. The fiber is exposed to headspace volatile and after sampling is complete, it is retracted into the needle. Apart from the nature of the adsorbent deposited on the fiber, the main parameters to optimize are the equilibration time, the sample temperature and the duration of extraction. Compared with other sampling methods, SPME is simple to use and reasonably sensitive, so it is a user-friendly pre-concentration method.

In our studies, the headspace sampling method of E-nose was optimized for MVOCs analysis. The samples were placed in the HS100 auto-sampler in arbitrary order. The automatic injection unit heated the samples to 37°C with an incubation time of 600 seconds. The temperature of the injection syringe was 47°C. The delay time between two injections was 300 seconds. Then the adsorbed compounds are desorbed by heating and introduced into the detection system (Yu Y. X., 2010a, Yu Y. X., 2010b).

2.2 Gas sensor detection

The most complicated part of electronic olfaction process is odor capture and sensor technology to be deployed for such capturing. Once the volatile compounds of samples are introduced into the gas sensor detection system, the sensor array is exposed to the volatile compounds and then the odor fingerprint of samples is generated from sensor respond. By chemical interaction between the volatile compounds and the gas sensors, the state of the sensors is altered giving rise to electrical signals that are registered by the instrument of E-nose. In this way the signals from the individual sensor represent a pattern that is unique for the gas mixture measured and those data based on sensors is transformed to a matrix. The ideal sensors to be integrated in an electronic nose should fulfill the following criteria (Barsan & Weimar, 2001, James et al., 2005): high sensitivity toward the volatile chemical compounds, that is, the chemicals to be detected may be present in the concentration range of ppm or ppb, and the sensor should be sufficiently sensitive to small concentration level of gaseous species within a volatile mixture, similar to that of the human nose (down to 10^{-12} g/ml); low sensitivity toward humidity and temperature; medium selectivity, that is, they must respond to a range of different compounds present in the headspace of the sample; high stability; high reproducibility and reliability; high speed of response, short reaction and recovery time, that is, in order to be used for online measurements, the response time of the sensor should be in the range of seconds; reversibility, that is, the sensor should be able to recover after exposure to gas; robust and durable; easy calibration; easily processable data output; and small dimensions.

The E-nose used in our studies is a commercial equipment (FOX4000, Alpha M.O.S., Toulouse, France), with 18 metal oxide sensors (LY2/AA, LY2/G, LY2/gCT, LY2/gCTl, LY2/Gh, LY2/LG, P10/1, P10/2, P30/1, P30/2, P40/1, P40/2, PA2, T30/1, T40/2, T70/2,

T40/1, TA2), and this sensor array system is used for monitoring the volatile compounds produced by microorganism, and so on. The descriptors associated with the sensors are shown in Table 1. FOX4000 E-nose assay measurements showed signal with maximum intensities changing with the type of samples, which indicate that discrimination is obtained.

Sensors	Volatile description	Sensors	Volatile description
LY2/LG	Fluoride, chloride, oxynitride, sulphide	P30 /1	Hydrocarbons, ammonia, ethanol
LY2 /G	Ammonia, amines, Carbon oxygen compounds	T70 /2	Toluene, xylene, carbon monoxide
LY2 / AA	Alcohol, acetone, ammonia	T40 /1	Fluorine
LY2 /GH	Ammonia, amines compounds	P40 /1	Fluorine, chlorine
P40 /2	Chlorine, hydrogen sulfide, fluoride	LY2 /gCTL	hydrogen sulfide
P30 /2	Hydrogen sulphide, ketone	LY2 /gCT	Propane, butane
T30 /1	Polar compound, hydrogen chloride	T40 /2	chlorine
P10 /1	Nonpolar compound: hydrocarbon, Ammonia, chlorine	PA /2	Ethanol, ammonia, amine compounds
P10 /2	Nonpolar compound: Methane, ethane	TA /2	ethanol

Table 1. Sensor types and volatile descriptors of FOX4000 E-nose.

Each sensor element changes its electrical resistance (R_{max}) when exposed to volatile compounds. In order to produce consistent data for the classification, the sensor response is presented with a volatile chemical relative to the baseline electrical resistance in fresh air, which is the maximum change in the sensor electrical resistance divided by the initial electrical resistance, as follows:

$$\text{Relative electrical resistance change} = (R_{max} - R_0) / R_0$$

where R_0 is the initial baseline electrical resistance of the sensor and $R_{max} - R_0$ is the maximum change of the sensor electrical resistance. The baseline of the sensors was acquired in a synthetic air saturated steam at fixed temperature. The relative electrical resistance change value was used for data evaluation because it gives the most stable result, and is more robust against sensor baseline variation (Siripatrawan, 2008).

Data of the relative electrical resistance changes from the 18 sensors can combine with every sample to form a matrix (see Fig. 2: The library data base) and the data is without preprocessing prior to chemometrics analysis. The sensor response is stored in the computer through data acquisition card and these data sets are analyzed to extract information.

2.3 Chemometrics analysis

The matrix of signal is interpreted by multivariate chemometrics techniques like the PCA, PLS, ANN, and so on. Samples with similar odor fingerprinting generally give rise to similar sensor response patterns, while samples with different odor fingerprinting show differences in their patterns. The sensors of an E-nose can respond to both odorous and odorless volatile compounds.

These various chemometrics methods are used in those works, according to the aim of the studies. Generally speaking, the chemometrics methods can be divided into two types: unsupervised and supervised methods(Mariey et al., 2001). The objective of unsupervised methods is to extrapolate the odor fingerprinting data without a prior knowledge about the bacteria studied. Principal component analysis (PCA) and Hierarchical cluster analysis (HCA) are major examples of unsupervised methods. Supervised methods, on the other hand, require prior knowledge of the sample identity. With a set of well-characterized samples, a model can be trained so that it can predict the identity of unknown samples. Discriminant analysis (DA) and artificial neural network (ANN) analysis are major examples of supervised methods.

PCA is used to reduce the multidimensionality of the data set into its most dominant components or scores while maintaining the relevant variation between the data points. PCA identifies the natural clusters in the data set with the first principal component (PC) expressing the largest amount of variation, followed by the second PC which conveys the second most important factor of the remaining analysis, and so forth(Di et al., 2009, Huang et al., 2009, Ivosev et al., 2008). Score plots can be used to interpret the similarities and differences between bacteria. The closer the samples are within a score plot, the more similar they are with respect to the principal component score evaluated(Mariey et al., 2001). In our studies, each sample data of 18 sensors is then compared to the others in order to make homogeneous groups. A scatter plot can then be drawn to visualize the results, each sample being represented by a plot.

3. Application of E-nose and chemometrics for bacteria identification

With the success of the above applications of the E-nose have been published, the authors were interested in determining whether or not an E-nose would be able to identify bacteria. A series of experiments were designed to determine this. In this part, bacteria identification at different levels (genus, species, strains) was cited as an example to illustrate using this integrated technology to foodborne bacteria effective identification.

3.1 At genus level

In this study, three bacteria, *Listeria monocytogenes*, *Staphylococcus lentus* and *Bacillus cereus*, which from three different genus, were investigated for the odor fingerprint by E-nose. The result of PCA (Fig.3a) shows that, the fingerprints give a good difference between the blank culture and the bacterial culture, and the three bacteria can be classified from each other by the odor fingerprints. Using the cluster analysis to represent the sensor responses (Fig. 3b), it is also possible to obtain a clear separation between the blank control and culture inoculated with bacteria. And the CA result also reveals that successful discrimination between the bacteria at different genus is possible(Yu Y. X., 2010a).

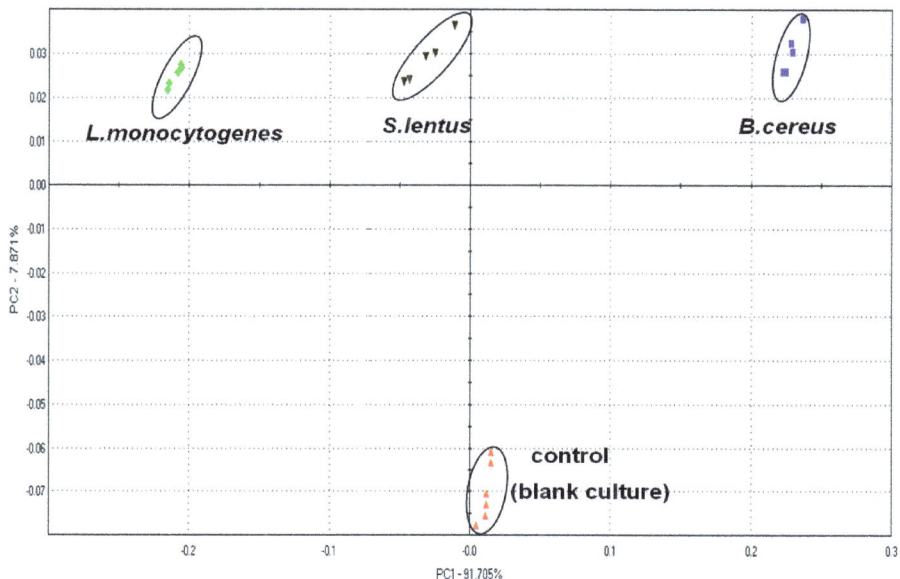

Fig. 3(a). Principal components analysis (PCA) for the discrimination of three bacteria from different genus on the basis of E-nose. The plot displays clear discrimination between the four groups, accounting for nearly 99% of the variance within the dataset.

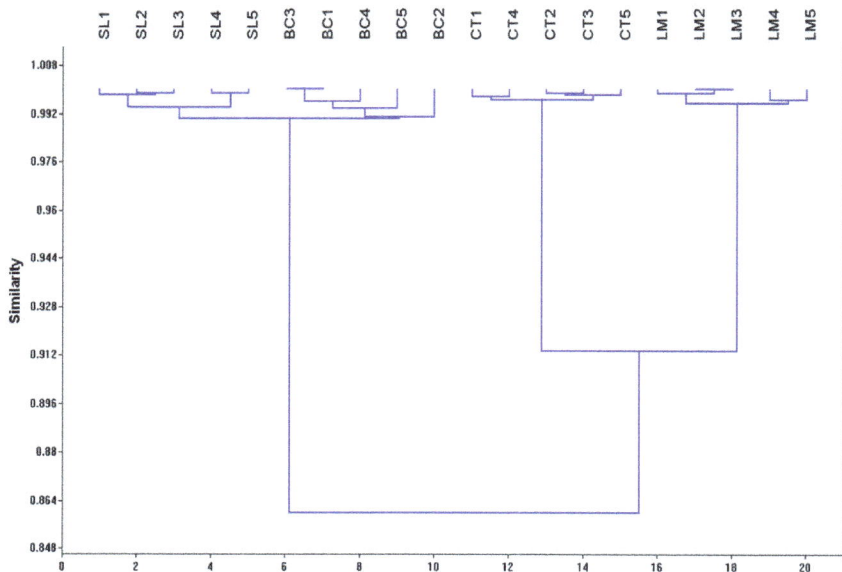

Fig. 3(b). Cluster analysis (CA) for the discrimination of three bacteria from different genus on the basis of E-nose. (*S. lentus*: SL1-SL5, *B. cereus*: BC1-BC5, *L. monocytogenes*: LM1-LM5, control blank culture: CT1-CT5).

3.2 At species level

In this study, using the same collection methodology, the E-nose was tested for its ability to distinguish among bacterial pathogens at species levels. Four species bacteria selected from *Pseudomonas* sp, named *Pseudomonas fragi, Pseudomonas fluorescens, Pseudomonas putida* and *Pseudomonas aeruginosa*, were investigated for the odor fingerprint by E-nose. It is clear that the E-nose was able to distinguish amongst all specimens tested. The PCA result in Fig.4(a) shows a representative experiment, where individual species of bacteria clustered in individual groups, separate from each other and the bacteria *Pseudomonas fragi* is given a great difference form the three other bacteria by the odor fingerprints. The result of cluster analysis in Fig. 4(b) also reveals that successful discrimination between the different bacteria at strains level is possible.

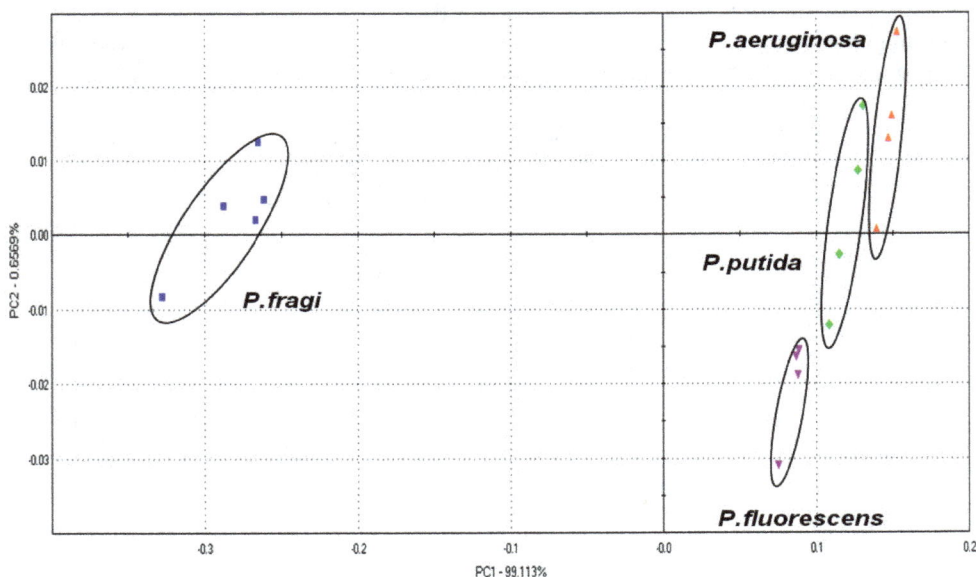

Fig. 4(a). Principal components analysis (PCA) for the discrimination of four different species of Pseudomonas sp on the basis of E-nose. The plot displays clear discrimination between the four groups, accounting for nearly 99% of the variance within the dataset.

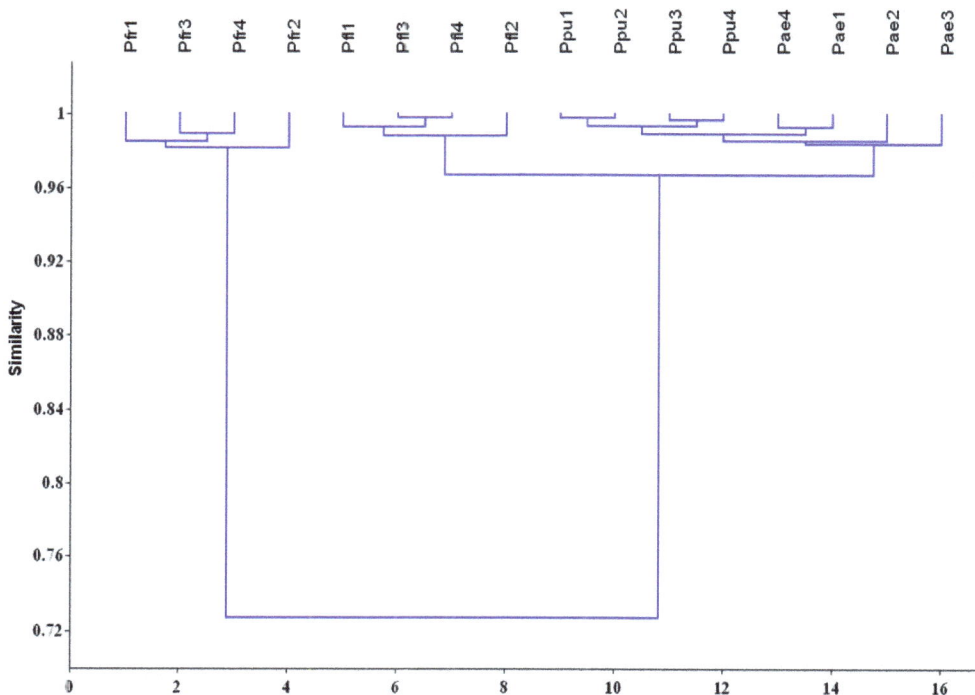

(*P. fragi*: Pfr1-Pfr4, *P. fluorescens*: Pfl1-Pfl4, *P. putida*: Ppu1-Ppu4, *P. aeruginosa*: Pae1-Pae4).

Fig. 4(b). Cluster analysis (CA) for the discrimination of four different species of Pseudomonas sp on the basis of E-nose.

3.3 At strains level

The next set of experiments involved testing the integrated method to see whether it could correctly differentiate bacteria samples as different strains. In this study, four strains of *Vibrio parahaemolyticus*, named *V. parahaemolyticus* F01, *V. parahaemolyticus* F13, *V. parahaemolyticus* F38 and *V. parahaemolyticus* F54, were compared with the odor fingerprint by E-nose. As shown in a representative data set in Fig. 5(a), the four strains of *V. parahaemolyticus* are separated from each other. However, the result from cluster analysis in Fig. 5(b) shows that some overlap appeared between *V. parahaemolyticus* F01 and *V. parahaemolyticus* F13, and it indicate that the odor fingerprints of these two strains may be too similar to identify by this method.

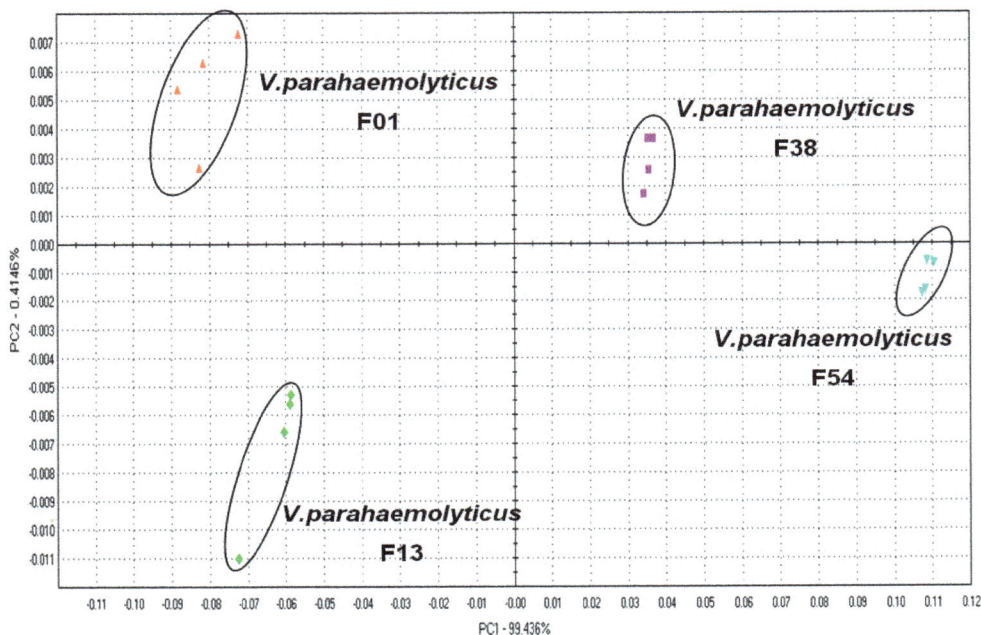

Fig. 5(a). Principal components analysis (PCA) for the discrimination of four different strains of *V. parahaemolyticus* on the basis of E-nose. The plot displays clear discrimination between the four groups, accounting for nearly 99% of the variance within the dataset.

4. Future perspectives

Electronic nose technology is relatively new and holds great promise as a detection tool in food safety area because it is portable, rapid and has potential applicability in foodborne pathogen identification or detection. On the basis of the work described above, we have demonstrated that the E-nose integrated with chemometrics can be used to identify pathogen bacteria at genus, species and strains levels.

As to know, bacteria respond to environmental triggers by switching to different physiological states. If such changes can be detected in the odor fingerprints, then E-nose analysis can produce information that can be very useful in determining virulence,

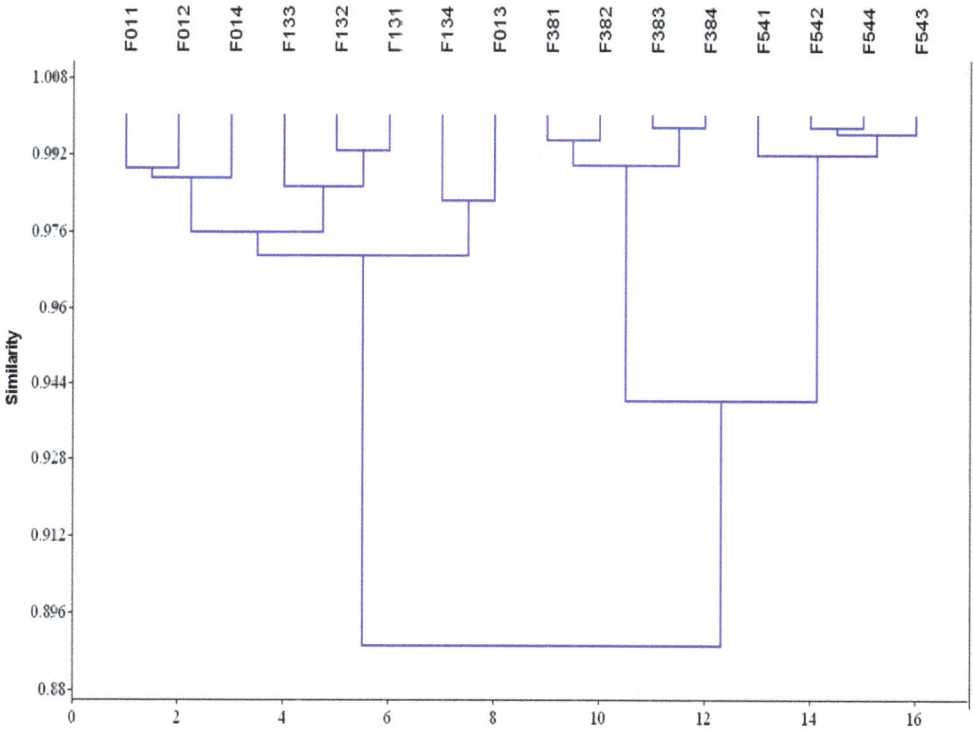

(*V.p* F01: F011-F014, *V.p* F13: F131-F134, *V.p* F38: F381-F384, *V.p* F54: F541-F544).

Fig. 5(b). Cluster analysis (CA) for the discrimination of four different strains of *V. parahaemolyticus* on the basis of E-nose.

conducting epidemiological studies, or determining the source of a food poisoning outbreak. Of course the ability to produce information on the physiological state of a microorganism offers many potential benefits. Nevertheless, a variety of different fingerprints, produced under a variety of growth conditions, must be developed for each pathogen, for inclusion in the reference database. To avoid this complication, we should culture the pathogens under controlled conditions. Otherwise, the identification algorithm must be capable of sorting through them all, to find a single, reliable, positive identification for the unknown.

Recently developed chemometrics algorithms are particularly suited to the rapid analysis and depiction of this data. Chemometrics is one approach that may offer novel insights into our understanding of the difference of microbiology. Adopting appropriate chemometrics methods will improve the quality of analysis.

Odor fingerprinting method based on E-nose is still in its infancy. Many recent technological advances, which are outside the scope of this chapter, can be used to transform the odor fingerprinting concept into user-friendly, automated systems for high-throughput analyses. The introduction of smaller, faster and smarter instrumentation of E-nose to the market could also depend much on the embedding of chemometrics. In addition, more and more classification techniques based on odor fingerprinting may be developed to classify the pathogens into exact levels such as genus, species and stains. Further investigation may contribute to make a distinction between the pathogen and non-pathogen bacterial.

In short, E-nose integrated with chemometrics is a reliable, rapid, and economic technique which could be explored as a routine diagnostic tool for microbial analysis.

5. Acknowledgments

The authors acknowledge the financial support of the project of Shanghai Youth Science and Technology Development (Project No: 07QA14047), the Leading Academic Discipline Project of Shanghai Municipal Education Commission (Project No: J50704), Shanghai Municipal Science, Technology Key Project of Agriculture Flourishing plan (Grant No: 2006, 10-5; 2009, 6-1), Public Science and Technology Research Funds Projects of Ocean (Project No: 201105007), Project of Science and Technology Commission of Shanghai Municipality (Project No: 11310501100), and Shanghai Ocean University youth teacher Fund (Project No: A-2501-10-011506).

6. References

Adley C, 2006. Food-borne pathogens: methods and protocols. *Humana Pr Inc*.

Ayoko GA, 2004. Volatile organic compounds in indoor environments. *Air Pollution*, 1-35.

Barsan N, Weimar U, 2001. Conduction model of metal oxide gas sensors. *Journal of Electroceramics* 7, 143-67.

Bhunia AK, 2008. Foodborne microbial pathogens: mechanisms and pathogenesis. *Springer Verlag*.

Bjurman J, 1999. Release of MVOCs from microorganisms. *Organic Indoor Air Pollutants*, 259-73.

Buratti S, Benedetti S, Scampicchio M, Pangerod E, 2004. Characterization and classification of Italian Barbera wines by using an electronic nose and an amperometric electronic tongue. *Anal Chim Acta* 525, 133-9.

Capone S, Epifani M, Quaranta F, Siciliano P, Taurino A, Vasanelli L, 2001. Monitoring of rancidity of milk by means of an electronic nose and a dynamic PCA analysis. *Sensors and Actuators B: Chemical* 78, 174-9.

Di CZ, Crainiceanu CM, Caffo BS, Punjabi NM, 2009. Multilevel functional principal component analysis. *Annals of Applied Statistics* 3, 458-88.

Dutta R, Morgan D, Baker N, Gardner JW, Hines EL, 2005. Identification of Staphylococcus aureus infections in hospital environment: electronic nose based approach. *Sensors and Actuators B: Chemical* 109, 355-62.

Evans P, Persaud KC, Mcneish AS, Sneath RW, Hobson N, Magan N, 2000. Evaluation of a radial basis function neural network for the determination of wheat quality from electronic nose data. *Sensors and Actuators B: Chemical* 69, 348-58.

Gardner JW, Shin HW, Hines EL, 2000. An electronic nose system to diagnose illness. *Sensors and Actuators B: Chemical* 70, 19-24.

Gates KW, 2011. Rapid Detection and Characterization of Foodborne Pathogens by Molecular Techniques. *Journal of Aquatic Food Product Technology* 20, 108-13.

Haugen JE, Kvaal K, 1998. Electronic nose and artificial neural network. *Meat Sci* 49, S273-S86.

Huang SY, Yeh YR, Eguchi S, 2009. Robust kernel principal component analysis. *Neural Comput* 21, 3179-213.

Hui YH, 2001. Foodborne Disease Handbook: Plant Toxicants. *CRC*.

Ivosev G, Burton L, Bonner R, 2008. Dimensionality reduction and visualization in principal component analysis. *Anal Chem* 80, 4933-44.

James D, Scott SM, Ali Z, C'hare WT, 2005. Chemical sensors for electronic nose systems. *Microchimica Acta* 149, 1-17.

Keshri G, Magan N, Voysey P, 1998. Use of an electronic nose for the early detection and differentiation between spoilage fungi. *Lett Appl Microbiol* 27, 261-4.

Kim JL, Elfman L, Mi Y, Wieslander G, Smedje G, Norbäck D, 2007. Indoor molds, bacteria, microbial volatile organic compounds and plasticizers in schools–associations with asthma and respiratory symptoms in pupils. *Indoor Air* 17, 153-63.

Korpi A, Pasanen AL, Pasanen P, 1998. Volatile compounds originating from mixed microbial cultures on building materials under various humidity conditions. *Appl Environ Microbiol* 64, 2914.

Mariey L, Signolle J, Amiel C, Travert J, 2001. Discrimination, classification, identification of microorganisms using FTIR spectroscopy and chemometrics. *Vibrational Spectroscopy* 26, 151-9.

Mcclure PJ, 2002. Foodborne pathogens: hazards, risk analysis, and control. *Woodhead Pub Ltd*.

Pasanen AL, Lappalainen S, Pasanen P, 1996. Volatile organic metabolites associated with some toxic fungi and their mycotoxins. *Analyst* 121, 1949-53.

Pavlou A, Turner A, Magan N, 2002a. Recognition of anaerobic bacterial isolates in vitro using electronic nose technology. *Lett Appl Microbiol* 35, 366-9.

Pavlou AK, Magan N, Mcnulty C, *et al.*, 2002b. Use of an electronic nose system for diagnoses of urinary tract infections. *Biosensors and Bioelectronics* 17, 893-9.

Romain AC, Nicolas J, Wiertz V, Maternova J, Andre P, 2000. Use of a simple tin oxide sensor array to identify five malodours collected in the field. *Sensors and Actuators B: Chemical* 62, 73-9.

Schnürer J, Olsson J, Börjesson T, 1999. Fungal volatiles as indicators of food and feeds spoilage. *Fungal Genetics and Biology* 27, 209-17.

Siripatrawan U, 2008. Rapid differentiation between E. coli and Salmonella Typhimurium using metal oxide sensors integrated with pattern recognition. *Sensors and Actuators B: Chemical* 133, 414-9.

Turner APF, Magan N, 2004. Electronic noses and disease diagnostics. *Nature Reviews Microbiology* 2, 161-6.

Wilkins K, Larsen K, Simkus M, 2003. Volatile metabolites from indoor molds grown on media containing wood constituents. *Environmental Science and Pollution Research* 10, 206-8.

Yu Y. X., Liu Y., Sun X. H., Pan Y. J., Zhao Y., 2010a. Recognition of Three Pathogens Using Electronic Nose Technology. *Chinese Journal of Sensors and Actuators* 23, 10-3.

Yu Y. X., Sun X. H., Pan Y. J., Zhao Y., 2010b. Research on Food-borne Pathogen Detection Based on Electronic Nose. *chemistry online (in Chinese)*, 154-9.

Zhang Q, Wang P, Li J, Gao X, 2000. Diagnosis of diabetes by image detection of breath using gas-sensitive laps. *Biosensors and Bioelectronics* 15, 249-56.

Part 3

Technology

10

Chemometrics in Food Technology

Riccardo Guidetti, Roberto Beghi and Valentina Giovenzana
Department of Agricultural Engineering,
Università degli Studi di Milano, Milano,
Italy

1. Introduction

The food sector is one of the most important voices in the economic field as it fulfills one of the main needs of man. The changes in the society in recent years have radically modified the food industry by combining the concept of globalization with the revaluation of local production. Besides the production needs to be global, in fact, there are always strong forces that tend to re-evaluate the expression of the deep local production like social history and centuries-old tradition.

The increase in productivity, in ever-expanding market, has prompted a reorganization of control systems to maximize product standardization, ensuring a high level of food security, promote greater compliance among all batches produced. The protection of large quantities of production, however, necessarily passes through systems to highlight possible fraud present throughout the production chain: from the raw materials (controlled by the producer) to the finished products (controlled by large sales organizations). The fraud also concern the protection of local productions: the products of guaranteed origin must be characterized in such a way to identify specific properties easily and detectable by objective means.

The laboratories employ analytical techniques that are often inadequate because they require many samples, a long time to get the response, staff with high analytical ability. In a context where the speed is an imperative, technology solutions must require fewer samples or, at least no one (non-destructive techniques); they have to provide quick answers, if not immediate, in order to allow the operator to decide quickly about further steps to control or release the product to market; they must be easy to use, to promote their use throughout the production chain where it is not always possible to have analytical laboratories. The technologies must therefore be adapted to this new approach to production: the sensors and the necessary related data modeling, which allows the "measure", are evolving to meet the needs of the agri-food sector. The trial involves, often, Research Institutions on the side of Companies, a sign of a great interest and a high level of expectations. The manufacturers of technologies, often, provide devices that require calibration phases not always easy to perform, but that are often the subject of actual researches. These are particularly complex when the modeling approach must be based on chemometrics.

This chapter is essentially divided into two parts: the first part analyzes the theoretical principles of the most important technologies, currently used in the food industry, that used a chemometric approach for the analysis of data (spectrophotometry Vis/NIR (Visible and Near InfraRed) and NIR (Near InfraRed), Image Analysis with particular regard to Hyperspectral Image Analysis and Electronic Nose); the second part will present some case studies of particular interest related to the same technologies (fruit and vegetables, wine, meat, fish, dairy, olive, coffee, baked goods, etc.) (Frank & Todeschini, 1994; Massart et al., 1997 and 1998; Basilevsk, 1994; Jackson, 1991).

2. Technologies used in the food sector combined with chemometrics

2.1 NIR and Vis/NIR spectroscopy

Among the non-destructive techniques has met a significant development in the last 20 years the optical analysis in the region of near infrared (NIR) and visible-near infrared (Vis/NIR), based on the use of information arising from the interaction between the structure of food and light.

2.1.1 Electromagnetic radiation

Spectroscopic analysis is a group of techniques allowing to get information on the structure of matter through its interaction with electromagnetic radiation.

Radiation is characterized by (Fessenden & Fessenden, 1993):

- a wavelength (λ), which is the distance between two adjacent maxima and is measured in nm;
- a frequency (v), representing the number of oscillations described by the wave per unit of time and is measured in hertz (cycles/s);
- a wave number (n), which represents the number of cycles per centimeter and is measured in cm^{-1}.

The entire electromagnetic spectrum is divided into several regions, each characterized by a range of wavelengths (Fig.1)

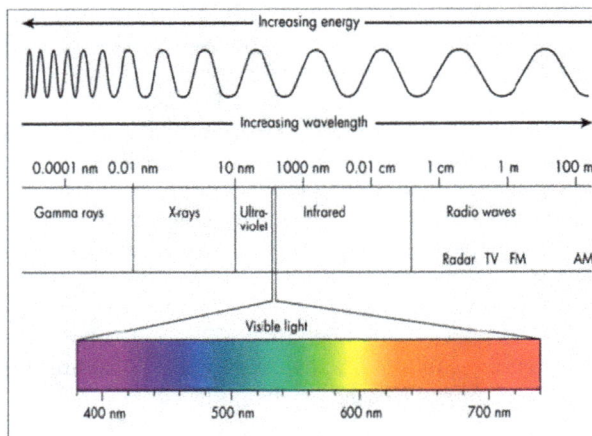

Fig. 1. The electromagnetic spectrum (Lunadei, 2008).

2.1.2 Transitions in the near infrared region (NIR)

The radiation from the infrared region is able to promote transitions at vibrational level. The infrared spectroscopy is used to acquire information about the nature of the functional groups present in a molecule. The infrared region is conventionally divided into three sub-regions: near (750-2500 nm), medium (2500-50000 nm) and far infrared (50-1000 µm).

Fundamental vibrational transitions, namely between the ground state and first excited state, take place in the mid-infrared, while in the region of near-infrared absorption bands are due to transitions between the ground state and the second or the third excited state. This type of transitions are called overtones and their absorption bands are generally very weak. The absorption bands associated with overtones can be identified and correlated to the corresponding absorption bands arising from the fundamental vibrational transitions because they fall at multiple wavelengths of these.

Following the process of absorption of photons by molecules the intensity of the radiation undergoes a decrease. The law that governs the absorption process is known as the Beer-Lambert Law:

$$A = \log (I_0/I) = \log (1/T) = \varepsilon \cdot l \cdot c \qquad (1)$$

where:

A = absorbance [log (incident light intensity/transmitted beam intensity)];
T = transmittance [beam intensity transmitted/incident light intensity];
I_0 = radiation intensity before interacting with the sample;
I = radiation intensity after interaction with the sample;
ε = molar extinction coefficient characteristic of each molecule ($l \bullet mol^{-1} \bullet cm^{-1}$);
l = optical path length crossed by radiation (cm);
c = sample concentration (mol/l).

The spectrum is a graph where in the abscissa is reported a magnitude related to the nature of radiation such as the wavelength (λ) or the wave number (n) and in the Y-axis a quantity related to the change in the intensity of radiation as absorbance (A) or transmittance (T).

2.1.3 Instruments

Since '70s producers developed analysis instruments specifically for NIR analysis trying to simplify them to fit also less skillful users, thanks to integrated statistical software and to partial automation of analysis.

Instruments built in this period can be divided in three groups: desk instruments, compact portable instruments and on-line compatible devices.

Devices evolved over the years also for the systems employed to select wavelength. First instruments used filter devices able to select only some wavelength (Fig. 2). These devices are efficient when specific wavelength are needed. Since the second half of '80s instruments capable to acquire simultaneously the sample spectrum in a specific interval of wavelength were introduced, recording the average spectrum of a single defined sample area (diode array systems and FT-NIR instruments) (Stark & Luchter, 2003). At the same time, chemometric data analysis growth helped to diffuse NIR analysis.

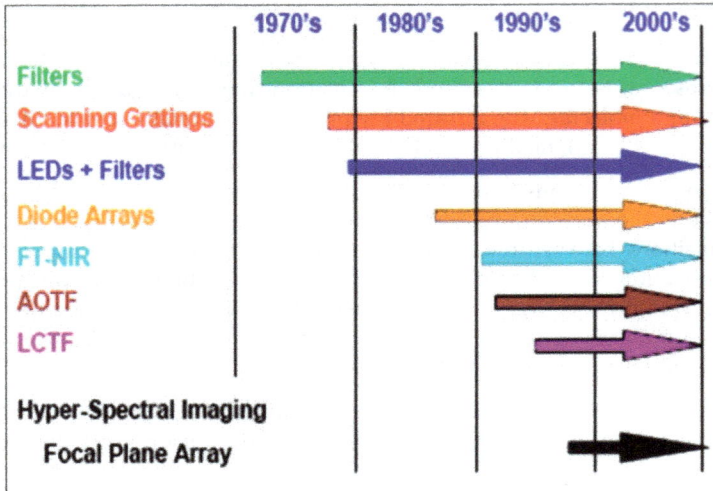

Fig. 2. Development of the different analysis technologies scheme (Stark & Luchter, 2003).

Particularly, food sector showed interest towards NIR and Vis/NIR instruments, both mobile and on-line. Devices based on diode array spectrophotometers and FT-NIR desk systems proved to be the best for this sector.

Both in the case of portable and stationary instruments, the fundamental components of these systems are common and are four:

- Light source;
- Light radiation transport system;
- Sample compartment and measurement zone;
- Spectrophotometer and Personal Computers.

Light source

Tungsten filament halogen lamps are chosen as the light source by most of the instruments. This is due to a good compromise between good performance and relatively low cost. This type of lamps are particularly suitable for use in low voltage. A little drawback may be represented by sensitivity to vibration of the filament.

Halogen bulbs are filled with halogen gas to extend their lives by using the return of evaporated tungsten to the filament. The life of the lamp depends on the design of the filament and the temperature of use, on average ranges from a minimum of 50 hours and a maximum of 10000 hours at rated voltage. The lamp should be chosen according to the use conditions and the spectral region of interest. An increase in the voltage of the lamp may cause a shift of the peaks of the emission spectrum towards the visible region but can also lead to a reduction of 30% of its useful life. On the contrary, use of lower voltages can increase the lamp life together, however, with an intensity reduction of light radiation, especially in the visible region. Emission spectrum of the tungsten filament changes as a function of temperature and emissivity of the tungsten filament. The spectrum shows high intensity in the VNIR region (NIR region close to the area of the visible).

Even if less common, alternative light sources are available. For example, LED light sources and ad laser sources could be used. LED sources (light emitting diodes) are certainly interesting sources thank to their efficiency and their small size. They meet, however, a limited distribution due to limited availability of LEDs emitting at wavelengths in the NIR region. Technology to produce LEDs to cover most of the NIR region already exists, but demand for this type of light sources is currently too low and the development of commercial product of this type is still in an early stage.

The use of laser sources guarantees very intense emission in a narrow band. But the reduced spectral range covered by each specific laser source can cause problems in some applications. In any case the complexity and high cost of these devices have limited very much their use so far, mostly restricted to the world of research.

Light radiation transport system

Light source must be very close to the sample to light it up with good intensity. This is not always possible, so systems able to convey light on the samples are needed. Thanks to optic fibers this problem was solved, allowing the development of different shapes devices.

The use of fiber optics allows to separate the area of placement of the instrument from the measuring proper area. There are indeed numerous circumstances on products sorting line in which environmental conditions do not fulfill direct installation of measure instruments. For example, high temperature, excessive vibrations or lack of space are restricting factors to the use of on-line NIR devices. In all these situations optic fibers are the solution to the problem of conveying light. They transmit light from lamp to sample and from sample to spectrophotometer. They allow to have an immediate measure on a localized sample area, thanks to their small dimensions, reaching areas difficult to access. Furthermore, they are made of a dielectric material that protects from electric and electromagnetic interferences. 'Optic fibers' means fibers optically transparent, purposely studied to transmit light thanks to total internal reflection phenomenon. Internal reflection is said to be total because it is highly efficient, in fact more than 99,999% radiation energy is transmitted in every reflection. This means that radiation can be reflected thousands of times during the way without suffer an appreciable attenuation of intensity (Osborne et al., 1993).

Optic fiber consists of an inner core, a covering zone and of an external protection cover. The core is usually made of pure silica, but can also be used plastics or special glasses. The cladding area consists of material with a lower refractive index, while the exterior is only to protect the fiber from mechanical, thermal and chemical stress.

In figure 3 are shown the inner core and the cladding of an optical fiber. Index of refraction of inner core have to be bigger than cladding one. Each ray of light that penetrates inside the fiber with an angle $\leq \theta_{max}$ (acceptance angle) is totally reflected with high efficiency within the fiber.

Sample compartment and measurement zone

Samples compartment and measurement zone are highly influenced by the technique of acquisition of spectra. Different techniques are employed, depending on type of samples, solid or liquid, small or large, to be measured in plan or in line, that influence the geometry of the measurement zone.

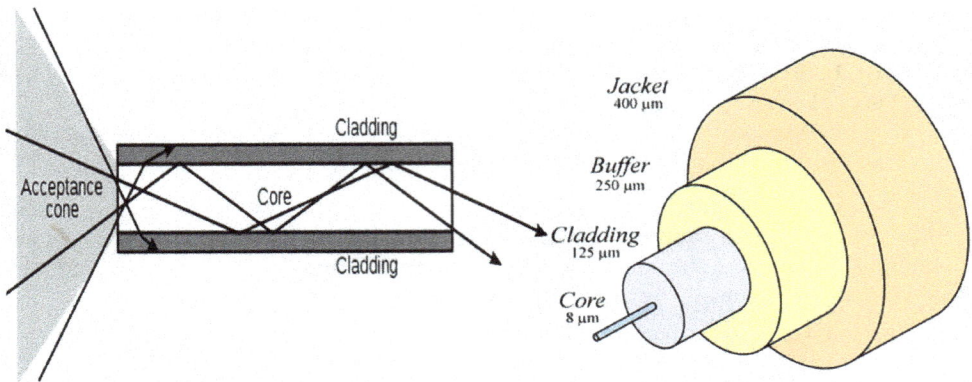

Fig. 3. Scheme of an optical fiber. The acceptance cone is determined by the critical angle for incoming light radiation. Buffer absorbs the radiation not reflected by the cladding. Jacket has a protective function.

The techniques to acquire spectra are four: transmittance, reflectance, transflectance and interactance. They are different mainly for the different positioning of the light source and of the measurement sensor around the sample (Fig. 4).

a. **Transmittance** - The transmittance measurements are based on the acquisition of spectral information by measuring the light that goes through the whole sample (Lu & Ariana, 2002). The use of analysis in transmittance can explore much of the internal structure of the product. This showed that is a technique particularly well suited to detect internal defects. To achieve significant results with this technique is required a high intensity light source and a high sensitivity measuring device. This because intensity of light able to cross the product is often very low. The transmittance measurements generally require a particular geometry of the measuring chamber, which can greatly influence the design of the instrument.

b. **Reflectance** - This technique measures the component of radiation reflected from the sample. The radiation is not reflected on the surface but penetrates into the sample a few millimeters, radiation is partly absorbed and partly reflected back again. Measuring this component of reflected radiation after interacting with the sample is possible to establish a relationship of proportionality between reflectance and analyte concentration in the sample. The reflectance measurement technique is well suited to the analysis of solid matrices because the levels of intensity of light radiation after the interaction with the sample are high.

This technique also allows to put in a limited space inside a tip the bundle of fibers that illuminate the sample and the fibers leading to the spectrophotometer the radiation after the interaction with the product. Therefore the use of this type of acquisition technique is particularly versatile and is suitable for compact, portable instruments, designed for use in field or on the process line. The major drawback using this technique is related to the possibility to investigate only the outer area of the sample without having the chance to go deep inside.

c. **Transflectance** – This technique is used in case it is preferable to have a single point of measurement, as in the case of acquisitions in reflectance. In this case, however, the incident light passes through the whole sample, is reflected by a special reflective surface, recross the sample and strikes the sensor located near the area of illumination. The incident light so makes a double passage through the sample. Obviously this type of technique can be used only in the case of samples very permeable to light radiation such as partially transparent fluid. It is therefore not applicable to solid samples.

d. **Interactance** - This technique is considered a hybrid between transmittance and reflectance, as it uses characteristics of both techniques previously seen. In this case the light source and sensor are located in areas near the sample but between them physically separated. So the radiation reaches the measurement sensor after interacting with part of internal structure of the sample. This technique is mainly used in the analysis of big solid samples, for example, a whole fruit. Interactance is thus a compromise between reflectance and transmittance and has good ability to detect internal defects of the product combined with a good intensity of light radiation. This analysis is widely used on static equipment where, through the use of special holders, is easily obtained the separation between the areas of incidence of light radiation and the area to which the sensor is placed. It is instead difficult to use this configuration on-line because is complicated to place a barrier between incident and returning light to the sensor directly on the process line.

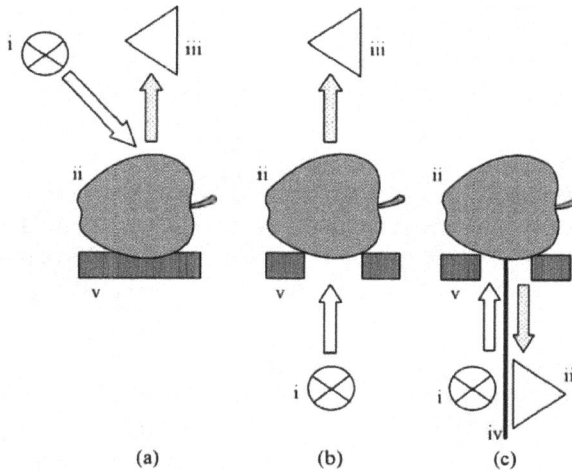

Fig. 4. Setup for the acquisition of (a) reflectance, (b) transmittance, and (c) interactance spectra, with (i) the light source, (ii) fruit, (iii) monochromator/detector, (iv) light barrier, and (v) support. In interactance mode, light due to specular reflection is physically prevented from entering the monochromator by means of a light barrier (Nicolai et al., 2007).

Spectrophotometer and Personal Computers

Spectrophotometer can be considered the heart of an instrument for NIR analysis. The employed technology for the wavelengths selection greatly influences the performance of the instrument. For example, the use of filters allows instruments to record the signal of a single wavelength at a time. Modern instruments (diode array instruments and interferometers) allow to record the spectrum of the entire wavelengths range.

Instruments equipped with a diode array spectrophotometer are those who have met the increased use for portable and online applications in food sector. This is due to their compact size, versatility and robustness, thanks to the lack of moving parts during operation and also thanks to a relatively low cost.

As seen before, fiber optic sensor collects the portion of the electromagnetic radiation after interaction with the internal structure of the sample and transfers it to the spectrophotometer. The optical fiber is connected to the optical bench of the instrument. The optical bench allows to decompose the electromagnetic radiation and recording the intensity at different wavelengths.

Optical bench of this type of instrument generally consists of five components:

a. Optical fiber connector: connects the fiber optic with the optical bench of the instrument.
b. First spherical mirror (collimating mirror), has the function of collimating the light and send it to the diffraction grating.
c. Diffraction grating: in this area of the instrument, the light is split into different wavelengths and sent to the second spherical mirror.
d. Second spherical mirror (focussing mirror), collects diffracted radiation from the grating and sends them to the CCD sensor.
e. Matrix CCD sensor (diode array): records the signal intensity at each wavelength.

High sensitivity of the CCD matrix sensor compensate the low intensity of light radiation input due to the reduced diameter of the optical fibers used. Sensors used are generally Si-diode array or InGaAs-diode array. The first ones are certainly the most common and cheap and allow the acquisition of the spectrum in the range between 400 and 1100 nm, so are used for Vis/NIR analysis. InGaAs sensors, more expensive, are used in applications requiring the acquisition of spectra at longer wavelengths, their use should range from 900 to 2300 nm.

Recorded signal by the CCD sensor is digitized and acquired by a PC using the software management tool of the instruments. Software records and allows to display graphically the spectrum of the analyzed sample. The management software also allows to interface with the spectrophotometer enabling to change some parameters during the acquisition of spectra.

2.2 Image analysis

In the food industry, since some time, there is a growing interest in image analysis techniques, since the appearance of a food contains a variety of information directly related to the quality of the product itself and this characteristics are difficult to measure through use of classical methods of analysis. In addition, image analysis techniques: provide information much more accurate than human vision, are objective and continuous over time and offer the great advantage of being non-destructive. These features, enable vision

Encyclopedia of Hepatic Surgery

Edited by **Amelia Foster**

FOSTER
ACADEMICS

New Jersey

Published by Foster Academics,
61 Van Reypen Street,
Jersey City, NJ 07306, USA
www.fosteracademics.com

Encyclopedia of Hepatic Surgery
Edited by Amelia Foster

International Standard Book Number: 978-1-63242-159-3 (Hardback)

Printed in the United States of America.

Contents

	Preface	VII
Chapter 1	**General Introduction: Advances in Hepatic Surgery** J.H.M.B. Stoot, R.J.S. Coelen, J.L.A. van Vugt and C.H.C. Dejong	1
Chapter 2	**Anesthetic Considerations for Patients with Liver Disease** Aparna Dalal and John D. Jr. Lang	41
Chapter 3	**Critical Care Issues After Major Hepatic Surgery** Ashok Thorat and Wei-Chen Lee	62
Chapter 4	**Essential Functional Hepatic and Biliary Anatomy for the Surgeon** Ronald S. Chamberlain	83
Chapter 5	**Experimental Models in Liver Surgery** M.B. Jiménez-Castro, M. Elias-Miró, A. Casillas-Ramírez and C. Peralta	103
Chapter 6	**Strategies to Decrease Morbidity After Hepatectomy for Hepatocellular Carcinoma** Hiroshi Sadamori, Takahito Yagi and Toshiyoshi Fujiwara	149
Chapter 7	**The Aim of Technology During Liver Resection — A Strategy to Minimize Blood Loss During Liver Surgery** Fabrizio Romano, Mattia Garancini, Fabio Uggeri, Luca Gianotti, Luca Nespoli, Angelo Nespoli and Franco Uggeri	165
Chapter 8	**The Role of Ultrasound in Hepatic Surgery** Mattia Garancini, Luca Gianotti, Fabrizio Romano, Vittorio Giardini, Franco Uggeri and Guido Torzilli	204

Chapter 9   **Two-Step Hanging Maneuver for an Isolated Resection of the
Dorsal Sector of the Liver**   220
Hideaki Uchiyama, Shinji Itoh and Kenji Takenaka

Chapter 10   **Segmental Oriented Liver Surgery**   233
O. Al-Jiffry Bilal and Khayat H. Samah

Chapter 11   **Right Anterior Sectionectomy for Hepatocellular
Carcinoma**   267
Hiromichi Ishii, Shimpei Ogino, Koki Ikemoto, Kenichi Takemoto,
Atsushi Toma, Kenji Nakamura and Tsuyoshi Itoh

Chapter 12   **Benign Hepatic Neoplasms**   275
Ronald S. Chamberlain and Kim Oelhafen

Chapter 13   **Liver Resection for Hepatocellular Carcinoma**   296
Mazen Hassanain, Faisal Alsaif, Abdulsalam Alsharaabi and Ahmad
Madkhali

Chapter 14   **Surgical Management of Primary Hepatocellular
Carcinoma**   321
Kun-Ming Chan and Ashok Thorat

Permissions

List of Contributors

Preface

I am honored to present to you this unique book which encompasses the most up-to-date data in the field. I was extremely pleased to get this opportunity of editing the work of experts from across the globe. I have also written papers in this field and researched the various aspects revolving around the progress of the discipline. I have tried to unify my knowledge along with that of stalwarts from every corner of the world, to produce a text which not only benefits the readers but also facilitates the growth of the field.

This book provides a comprehensive study on hepatic surgery. The liver was referred to as a hostile organ by Longmire, since it welcomes sepsis and malignant cells so warmly, bleeds so freely, and is usually the first organ to get injured in blunt abdominal trauma. To balance these negative factors, it has an excellent characteristic: its capability to regenerate after a great loss of substance. This book highlights a great range of topics including the surgical anatomy of the liver, history of liver surgery, portal hypertension, liver trauma, methods of liver resection, and malignant and benign liver tumors. This book would benefit as a good source of reference to students, researchers and surgeons.

Finally, I would like to thank all the contributing authors for their valuable time and contributions. This book would not have been possible without their efforts. I would also like to thank my friends and family for their constant support.

Editor

General Introduction: Advances in Hepatic Surgery

J.H.M.B. Stoot, R.J.S. Coelen, J.L.A. van Vugt and
C.H.C. Dejong

Additional information is available at the end of the chapter

1. Introduction

Hepatic resection is a commonly performed procedure for a variety of malignant and benign hepatic tumours [1, 2]. Historically, liver resection, irrespective of the indication, was associated with a high morbidity and mortality [2-4]. During the last decades however, perioperative outcome after hepatic resection has improved, due to increased knowledge of liver anatomy and function, improvement of operating techniques and advances in anaesthesia and postoperative care [1, 3, 4].

Hepatic resectional surgery is possible since the liver has the ability to regenerate. Although it is doubtful whether the ancient Greeks already appreciated this unique quality of the liver, it was first described in the myth of Prometheus (Προμηθεύς): he enraged the Gods for his disrespect (ὕβρις) after climbing the Mount Olympus and stealing the torch in order to give fire to the humans. He was punished by Zeus and chained to a rock in the Kaukasus Mountains. Every couple of days, an eagle came and ate part of his liver. As the liver regenerated every time, the eagle returned again and again to eat the liver and thereby torture poor Prometheus (figure 1). With this ancient knowledge it was considered possible to take parts of the liver, as this organ has enough capacity to work with a smaller part and is able to regenerate.

Apart from the eagle, no human dared to remove a part of the liver. In the ancient period of the Assyrian and Babylonian cultures of 2000 - 3000 BC the liver played an important role to predict the future by reading the surface of sacrificed animals [5]. This was also common in the Etruscan society, where the haruspices predicted the future from sheep livers. Hippocrates (460-377 BD), one of the founding fathers of ancient medicine, produced not only an oath with ethical rules, which is still used in modern times for all doctors. His careful observations also led to the recommendation to incise and drain abscesses of the liver with a knife [5]. Celsus documented the treatment of exposed liver in war wounds. Although he was not a physician,

Figure 1. Prometheus chained (243 x 210 cm), Peter Paul Rubens, ca. 1611-1618, Philadelphia, Philadelphia Museum of Art.

he described his observations in the first century AD from the Alexandrian school led by Herophilus of Chalcedon and Erisastratus of Chios [5]. In the same era, the Greek Galen became one of the emperor's physicians in Rome and wrote reports about the dissection of many species of animals, including primates. He described the central role of the liver in absorption and digestion and his work remained of great importance for the coming centuries [5]. In the centuries thereafter many reports were produced describing the treatment of war or trauma wounds.

Glisson performed extensive investigations of the vascular anatomy in 1654 (figure 2) [6]. It took more than two centuries before his work was rediscovered and further clarified by Rex (1888) in Germany and Cantlie (1897) in England [5, 7]. These contributions led to the division of the liver in a left and right lobe [5].

Figure 2. Francis Glisson (1599-1677).

2. History of hepatic surgery

It still took 17 centuries before Hildanus successfully performed the first partial liver resection for trauma [8]. The introduction of ether anaesthesia (1846) and the growing knowledge of antisepsis (1867) made successful elective abdominal operations possible (table 1) [5]. Langenbuch was the first to perform a successful elective liver resection in 1887 (figure 3) and Wendel did the first hemihepatectomy in 1911 [8]. The principles of liver haemostasis and regeneration were determined in the period 1880-1900 [8]. The knowledge of the principle of inflow and outflow of the liver and vascular control was one of the major advancements. Before that, wedge resections and mattress sutures were mostly used. This insight of inflow and outflow reduction was marked by the publication of James Hogart Pringle of Glasgow, Scotland (figure 4) [9]. He described the idea of digital control of the hilar ligament to reduce liver haemorrhage. In his famous report (1908) on liver haemorrhage after trauma, eight patients were included. Three died before the operation, one refused the operation and all four operated patients died; two died during the operation and two shortly thereafter [5, 9]. However, his idea of digital vascular control of the hilum was more success-ful in the laboratory setting, where he operated three rabbits with better results, which led to his publication. Nowadays, more than a century later, the 'Pringle manoeuvre' or 'Pringle's pinch' is still used worldwide in hepatic resectional surgery and taught to all young sur-geons to control haemorrhage of the liver.

1846	Introduction of Ether anaesthesia	Morton
1863	Bacterial fermentation of wine	Pasteur
1867	Antisepsis	Lister
1870	First successful excision of section of the liver	Bruns
1880	Discovery of Streptococci, staphylococci and pneumococci	Pasteur
1881	First successful gastrectomy	Billroth
1882	First successful cholecystectomy	Langenbuch
1883	First human colon anastomosis	Billroth and Senn
1884	Pancreas excised for cancer	Billroth
1886	Report on appendicitis	Fitz
	Introduction of sterilisation by steam	Von Bergmann
	First elective liver resection for adenoma	Lius
1887	First successful elective liver resection	Langenbuch
1887	Successful packing of stabwound of liver	Burckhardt
1888	First successful laparotomy for traumatic liver injury	Willet

Table 1. Advances in the beginning of surgery [5].

Figure 3. James Hogarth Pringle (1863-1941).

Figure 4. Carl Langenbuch (1846-1901).

Liver surgery became gradually more popular as a better understanding of anatomic segments was established after the work of Couinaud [10]. The classic morphological (outside) anatomy with two main lobes (left and right) was extended by the internal hepatic anatomy with several independent functional segments (figure 5). Each hepatic segment consists of liver parenchyma with an efferent hepatic vein branch and a portal triad; a hepatic artery branch, an afferent portal vein, and an efferent bile duct. The classic right lobe consists of four segments, the left lobe consists of three segments and the caudate lobe is segment 1.

With knowledge of the segmental anatomy of the liver, a safe transection plane could be chosen for resection without excessive blood loss and without necrosis of remnant liver. This specific anatomy of independent functional segments made it possible to resect parts of the liver without compromising the hepatic function of remnant segments. Moreover, as already described by the myth of Prometheus, the liver has regeneration capacity in contrast to other human organs. In other words after partial resections, the liver can recover its mass and function. The term 'function of the liver' is actually a collective term for a range of functions including amongst others ammonia detoxification, urea synthesis, bile synthesis and secretion, protein synthesis, gluconeogenesis and clearance or detoxification of drugs, bacterial toxins and bacteria [11]. As the liver is the main detoxifying organ in humans, adaptation of its function is crucial to survive. Regeneration however, takes time. After liver surgery with a reduction of the hepatic cell mass, a 'survival programme' may start for vital liver functions [12]. Some of these functions are increased rapidly in the remnant liver after resection [13]. In the light of major hepatic resections, it is conceivable that too little functional liver remnant may lead to liver failure, a lethal complication of liver surgery.

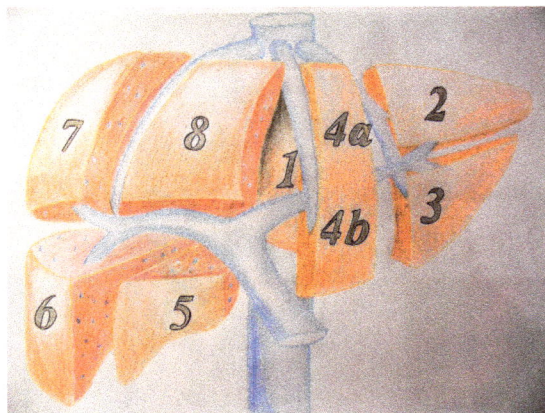

Figure 5. The anatomy of the liver with separate segments following Couinaud's classification. In this drawing only major venous vessels are displayed (portal vein, caval vein and hepatic veins).

3. Resectional hepatic surgery

Hepatobiliary surgery incorporates a wide range of indications for surgical treatment of the liver, varying from biopsy and resection to liver transplantation. The most important indications for surgical treatment are liver lesions: these comprise a wide range of both benign and malignant lesions, which can be either primary tumours (hepatocellular carcinoma) or secondary tumours (i.e. metastases). Also, some infectious diseases of the liver (such as echinococcosis) may be an indication for surgery. Irreversible liver dysfunction caused by acute or chronic liver diseases, may be an indication for transplantation of the liver. Other benign diseases of the liver such as symptomatic simple cysts and Polycystic Liver Disease (PCLD) may also warrant surgical treatment. Other reasons for surgery of the liver may be after severe injury or trauma of the liver. The latter indications are beyond the scope of this chapter. Since hepatic lesions form the main surgical indication for hepatic diseases, the focus will be on resectional liver surgery.

3.1. History of hepatic surgery for malignant lesions

The report of the first anatomical right hepatectomy for cancer by Lortat-Jacob in 1952 marked a new era in liver surgery [14]. In the beginning, however, blood loss and mortality were considerable. A multicentre analysis in 1977 of more than 600 hepatic resections for various indications showed an operative mortality of 13%, which rose to 20% for major resections [15]. Despite this, pioneers in liver surgery continued the quest for improving this challenging field of expertise and gradually mortality decreased to 5.6% [16]. The 5 year survival rates have increased from 20% in the beginning [16, 17] to as high as 67% in selected patients [18]. Earlier developments in liver surgery have been marked by major contributions of Starzl (USA), Bismuth (France) and Ton That Tung (Vietnam) [19-22]. With better knowledge of the segmental anatomy, it was shown that parenchyma-sparing segmental resections were equally effective as classic lobar resections, and in this way more functional remnant liver was preserved [3, 23, 24]. Also, anaesthetic care and liver transection techniques were modernized and improved over time [1, 3, 4, 25, 26].

Over the last decades, it was shown in several large series that perioperative results became more encouraging, with operative mortality rates less than 5% in high volume centres [3, 24, 25]. Due to these improvements in liver surgery which not only proved to prolong life but also to be a potentially curative treatment option for primary and metastatic cancers [27, 28], liver surgery became standard of care for selected patients with primary and secondary hepato-biliary malignancies. Moreover, with the increasing improvements in the safety of hepatic resections, this evolved to the most effective treatment for some benign diseases [29].

It is hard to pinpoint one discriminating factor that made the improvements in outcome possible [3]. Many factors contribute to the gradually improved outcome. Most important factors in this regard are probably the better knowledge of hepatic anatomy and thus ana-tomically based resections, better patient selection, general improvements in operative and anaesthetic care and the development of hepatobiliary surgery as a distinct area of specialisation [3].

3.2. Transection techniques in hepatic resection

Parenchymal transection is the most challenging part of liver resection. Due to the complicated vascular and biliary anatomy of the liver, haemorrhage is a great risk [30-35]. The firstly performed liver resections failed as a consequence of haemorrhage or patients died shortly after because of bleeding [31]. Before the 1980s, mortality after hepatic resection was 10 to 20% and haemorrhage was a common cause [30]. Moreover, blood transfusion in the perioperative period is associated with poorer outcome in the long term [33]. In contrast to patient- or tumour-related factors, surgical techniques can be changed in order to prevent blood loss and transfusion.

Parenchymal division was first described in 1958 when Lin and colleagues introduced the finger fracture technique (digitoclasy) in which liver tissue is crushed between the surgeon's fingers [30]. Vessels and bile ducts are exposed, identified and then divided. Soon this technique was improved by using surgical clamps (i.e. Kelly clamp) and called the crush-clamp technique [30, 31]. Division of the vessels and bile ducts can be achieved by suture ligation, bipolar electrocautery, vessel sealing devices or vascular clips. It is frequently combined with intermittent inflow occlusion by portal triad clamping (Pringle maneuver) [31].

Subsequently, many transection techniques have been developed in order to improve results. The Cavitron Ultrasonic Surgical Aspirator (CUSA, Tyco Healthcare, Mansfield, MA, USA) combines ultrasonic energy with aspiration and results in a more precise transection plane.

Vessels and bile ducts are exposed and can then be divided with a method according to the surgeon's preference [30, 31]. In a recent study, liver parenchyma transection using CUSA was associated with higher numbers of potentially dangerous air embolism although patients did not show clinical symptoms [36]. The Harmonic Scalpel (Ethicon Endo-Surgery, Cincinnati, OH, USA) is comparable to the CUSA, but it uses ultrasonic shears and vibration to cut through the parenchyma. It instantly coagulates blood vessels by protein denaturation and is mainly used in laparoscopic procedures, because of the difficulties using the other transection instruments in this setting. The hydro or water jet uses a high-pressure water jet to dissect liver parenchyma and expose vessels and bile ducts after which they can be divided. Like with the Harmonic Scalpel, less thermal damage is caused. In radiofrequency-assisted liver resection radiofrequent electrodes are inserted in the transection plane and radio frequent energy is applied for one to two minutes, followed by transection of the coagulated liver using a conventional scalpel. [30, 31].

In a review including seven randomized controlled trials with a total of 556 patients, the clamp-crush technique was quicker and associated with lower rates of blood loss and transfusion compared with CUSA, hydrojet and radiofrequency dissecting sealer. No significant differences in mortality, morbidity, liver dysfunction, ICU stay and length of hospital stay were found. The crush-clamp technique comes with low costs and does not need any extra advanced tools. However, not all techniques in the trials were combined with vascular occlusion. This may have led to a bias in favour of the clamp-crush technique [32, 34]. The CRUNSH trial will demonstrate whether vascular stapling is superior to the crush-clamp method in elective hepatic resection [37]. Palavecino and colleagues developed the so-called 'two-surgeon method', combining a saline-linked cautery and an ultrasonic dissector. Exposure of vessels and biliary ducts and haemostasis are performed simultaneously. Retrospectively, significantly lower transfusion rates were seen [33].

In conclusion, the clamp-crush technique seems to be superior especially as it is an easy method and comes with low costs. It might be regarded as the golden standard with which new devices or methods should be compared. However, high-quality randomized controlled trials are missing. Besides, the surgeon's experience plays an important role. Because of this, one could say that the method of choice is the clamp-crush technique and other techniques can be applied, or combined, dependent on the surgeon's experience and preference.

3.3. Malignant lesions

The liver has an important function as a detoxifying organ and due to the anatomical position in the abdomen; most gastro-intestinal organs drain their venous blood to the liver. This makes the liver a frequent location of metastases from a variety of intra-abdominal and sometimes even extra-peritoneal primary cancers. Also, primary cancers can arise in the liver. Of these the hepatocellular carcinoma is the most common malignancy. With a normal functioning liver, resection is the treatment of choice for most of these malignant lesions.

Metastases of colorectal origin are the most frequent malignant lesions in the liver. With nearly one million new cases diagnosed each year and around half a million deaths annually, colorectal cancer is one of the most common causes of cancer related death worldwide [38]. Over half of the patients with colorectal cancer will develop liver metastases [39]. Moreover, up to 25% of these patients present with liver metastases at the same time of the primary diagnosis [40]. Colorectal liver metastases may therefore be regarded as a major health problem [39].

The only chance of long-term survival in patients with liver metastases is provided by resection of these liver metastases, with 5-year survival rates around 30-40% [41]. Until recently, however, few patients with malignant liver lesions were considered for partial hepatic resection. Due to the restricted resection criteria, only 10-20% of the patients with malignant lesions were selected. Palliative chemotherapy was offered for the remaining proportion of the patients, resulting in a median survival of 6-12months [8, 42]. Due to the increased safety of liver surgery, liver resection is currently also used for other metastases such as neuroendocrine tumours [43], sarcoma's [44], melanoma [45-47], gastric cancer [48-50] and breast cancer [48, 51, 52].

The selection criteria for liver resections were initially fairly strict: unilobar distribution, less than four metastases, maximum tumour size of 5 cm and tumour free margin of 1 cm. These resection criteria have been evaluated over time and have gradually been abandoned, as these appeared to be not as important as previously assumed [53-55]. Even in elderly patients and poor prognostic groups, complete tumour resection results in a good long-term survival [56-58].

In the treatment of malignant liver disease, many improvements have been developed in recent years: new surgical strategies for safer resection (including two stage hepatectomy and portal vein embolisation), more effective chemotherapy, and additional techniques such as local ablation therapies to increase possible curative treatment [59-64]. The combination of these developments has led to an important progress and has resulted in more patients being considered suitable for liver resection to almost 30% [62]. Better survival of patients with primary or metastatic liver cancer has been reported in recent years and liver resection is currently the only potentially curative treatment option.

3.4. Benign hepatic lesions

In case of malignant hepatic disease, surgical resection is currently felt justified despite a morbidity and mortality, which may be as high as 42% and 6.5% respectively [1, 3, 65-67]. In case of benign hepatic disease, however, this decision remains more difficult. Due to the widespread use of imaging modalities such as ultrasonography, computed tomography (CT), and magnetic resonance imaging (MRI), benign hepatic masses are increasingly being identified. However, not all benign hepatic tumours require resection. Careful diagnosis with contrast enhanced CT or MRI needs to be performed first. Benign lesions can grossly be divided in solid and non-solid lesions (table 2).

Solid lesions	Symptoms	Treatment
Hepatocellular adenoma	Variable: from incidental finding to severe abdominal pain and shock in case of rupture	<5cm watchful waiting, stop oral contraceptives ≥5cm resection to prevent rupture and malignant degeneration
Focal Nodular Hyperplasia	Mostly incidental finding	Surgery rarely indicated
Angiomyolipoma	Mostly incidental finding	Surgery rarely indicated
Nodular regenerative hyperplasia	Mostly asymptomatic, should be considered in patients with clinical signs of portal hypertension without evidence of cirrhosis	No proven treatment
Non-solid lesions		
Simple hepatic cyst	Variable: from incidental finding to abdominal pain	Surgery indicated only in case of symptoms
Biliary cystadenoma	Variable: from incidental finding to abdominal pain	Surgery may be indicated (malignant degeneration)
Biliary hamartoma	None	Surgery not indicated
Cavernous haemangioma	Variable, depending on size	Surgery rarely indicated
Hydatid disease	Variable: from incidental finding to severe abdominal pain and shock	Surgery indicated to relieve symptoms and to prevent rupture

Table 2. Most important benign liver lesions, divided in solid and non-solid lesions.

3.5. History of hepatic surgery for benign lesions

The first case of surgical resection for a presumably benign liver tumour was described in 1886 by Antonio Lius in Italy [68]. Lius was the assistant of Theodore Escher who excised a pedunculated adenoma with the size of a child's head (15.5 cm in greatest diameter) from the left liver lobe of 67-year-old women. An uncontrollable bleeding was encountered during the operation and the patient died several hours following surgery. The German surgeon Von Langenbuch was the first to perform a successful resection of a benign solid pedicled liver mass weighing 370 gram of the left liver in a 30-year-old woman who complained of abdominal discomfort in the years following her first child's birth in 1887 [69]. Postoperatively, secondary haemorrhage occurred due to a bleeding hilar vessel. This was managed at re-exploration and the patient survived. The course of symptoms and events in the latter case suggests the tumour was most likely a hepatocellular adenoma.

It is nowadays well established that small benign lesions compatible with a diagnosis of haemangioma, focal nodular hyperplasia (FNH) or hepatocellular adenomas (HCAs) are no indication for liver resection [53]. Hepatocellular adenomas are considered the most important, albeit uncommon, benign tumours of the liver that mostly occur in women. They are known for their increased risk of haemorrhage and malignant transformation into hepatocellular carcinoma (HCC) if size exceeds 5 cm. Therefore, surgical resection of HCAs is recommended for larger lesions [53, 54]. Focal nodular hyperplasia and haemangiomas have not been regarded as potentially premalignant lesions.

The first case report of malignant transformation of a HCA was published in 1981 by Tesluk and Lawrie [70]. The patient was a 34–year-old female with a large HCA measuring 16 cm in diameter. She first presented with tumour haemorrhage after which her oral contraceptive use was discontinued and the tumour subsequently shrank to a stable 5 cm. Three years later a partial hepatectomy was performed when the tumour had reverted to its size at first presentation. Histological analysis revealed a well-differentiated HCC. The patient died of sepsis five weeks postoperatively.

Foster and Berman were the first to report an estimated risk of malignant transformation in 1994, as they found a frequency of 13% in their series of 13 patients [71]. More recently, a systematic review of the literature of the past 40 years containing more than 1600 HCAs worldwide identified 68 reports of malignant transformation resulting in an overall frequency of 4.2% among all adenoma cases [72]. Nowadays several other risk factors for malignant potential of HCAs apart from size have been identified [73-84]. These are listed in table 3.

Risk factors
Tumour size ≥5 cm
Presence of β-catenin activating mutation
Presence of liver cell dysplasia within HCA
Patients with glycogen storage disease
History of androgen or anabolic steroid intake
Male sex
Obesity/overweight

Table 3. Risk factors for malignant transformation of hepatocellular adenomas.

3.6. Surgical treatment of hepatocellular adenomas

The identification of several risk factors for malignant potential of HCAs in recent years, provides better indications for surgical treatment of these presumably benign tumours. Also, the Bordeaux adenoma tumour markers (table 4) have greatly contributed to the subtype classification of HCAs and have given clearer insights into the pathological mechanism of malignant evolvement [79]. More recently, MR imaging techniques have been shown to be of value in identifying premalignant HCAs [85, 86]. These advances in risk factor stratification, together with tumour subtyping prior to hepatic surgery, might aid in selecting HCAs at high risk of malignant evolvement for surgical resection. Unfortunately, routine performance of biopsy of an HCA has not been implemented yet owing to the risk of sampling error, bleeding, needle-track tumour seeding and the difficult interpretation of β-catenin staining. However, a change towards a more stringent selection process in the near future is inevitable and may imply a major reduction of the number of liver resections, and thus morbidity and even mortality, in a selected group of predominantly young patients.

HCA type	Frequency (%)	Malignant transformation	Markers
β-catenin activated	10-15	Yes	β-catenin+/GS+
HNF1α inactivated	30-50	Rarely	LFABP-
Inflammatory	35	No	SAA+/CRP+
Unclassified	5-10	No	None

CRP, C-reactive protein; GS, glutamine synthetase; HCA, hepatocellular adenoma; HNF1a, hepatocyte nuclear factor 1a; LFABP, liver-fatty acid binding protein; SAA, serum amyloid A; +, positive; -, negative. Table adapted with permission from Stoot et al. 2010 [72].

Table 4. Types of HCAs and their immunohistochemical markers.

Concerning the management of ruptured HCAs, emergency surgery is associated with high morbidity and mortality rates [73, 85]. Although this treatment is still suggested by some authors [86], the maximally invasive therapy of immediate liver resection has gradually been abandoned. Many liver surgeons prefer conservative management of ruptured HCAs consisting of immediate resuscitation with laparotomy and gauze packing [74]. Selective arterial embolisation for ruptured HCAs may be a valuable alternative although it has rarely been reported [55, 63, 70, 72, 87].

In conclusion, hepatic resection for benign tumours is mainly reserved for HCAs at risk for malignant evolvement or haemorrhage. Advances in pathological subtyping, radiological imaging and risk stratification have led to new insights and aid in justifying hepatic resection in a more selected population.

4. Advances in the surgical treatment of benign cystic lesions: hydatid disease

Surgical treatment may also be indicated for infectious diseases of the liver such as benign lesions caused by the parasitic infection called Echinococcosis. Human echinococcosis is a zoonosis caused by larval forms (metacestodes) of Echinococcus (E.) tapeworms found in the small intestine of carnivores. Two species are of clinical importance – *E. granulosus* and *E. multilocularis* – causing cystic echinococcosis (CE) and alveolar echinococcosis (AE) in humans, respectively [87]. Besides, in the beginning of the 20th century the so-called neotropical echinococcosis species *E. oligarthrus* and *E. vogeli* were discovered to cause polycystic echinococcosis (PE). *E. vogeli* causes disease similar to AE and *E. oligharthrus* has a more benign character [88]. Echinococcosis is endemic worldwide in large sheep-raising areas including Africa, the Mediterranean region of Europe, the Middle East, Asia, South America, Australia and New Zealand [89-96]. Human cystic echinococcosis is one of the most neglected parasitic diseases in the world. In many endemic regions most infected patients suffer considerably from this disease, usually because of the lack of treatment possibilities due to poor infrastructure and shortage of equipment and drugs [97, 98]. The incidence of hydatid disease in Western industrial nations is relatively low [93, 94, 99]. Migration and travelling has led to an increase of the prevalence of this disease in Northern parts of Europe and North America [96, 100]. The diagnosis of hepatic echinococcosis can be made with a combination of patients' symptoms, liver imaging findings, detection of Echinococcis-specific antibodies and microscopic or molecular examination of cyst fluid. The most frequent site for cystic lesions is the liver (60% of patients), followed by the lungs in about 20% of patients. The remaining lesions are found throughout the body [92, 95, 99, 101, 102].

The natural course of this infection can be extremely variable [101]. The hepatic cysts can spontaneously collapse, calcify or even disappear. These patients can remain symptom-free for years. It is not uncommon that the cysts are detected when abdominal imaging is performed for a different reason. On the other hand, the cysts can also steadily grow about 1-3 cm in diameter per year [96, 99]. They do not tend to grow infiltratively or destructively, but pressure or mass effects of the cysts can displace healthy tissue and organs. Thus, most patients present with symptoms from mechanical effects on other organs or structures, which can lead to pain in the upper right quadrant, hepatomegaly and jaundice, depending on the location and nature of the cysts [91, 96, 99, 101]. Infection of the cysts can result in sepsis and/or the formation of liver abscesses. A feared complication is rupture of hepatic hydatid cysts into the peritoneal cavity. This can result in serious anaphylaxis, sepsis and/or peritoneal dissemination. The content of the ruptured cyst can disseminate into the biliary tract leading to cholangitis or cholestasis, but also to the pleurae or lungs leading to pleural hydatidosis or bronchial fistula, respectively [91, 92, 102].

4.1. History of hepatic surgery for hydatid disease

Hydatid disease was already recognized by Hippocrates more than two millennia ago. This benign disease has been shown to act as a malignant disease as it has the tendency to disseminate to other organs and to cause a devastating disease sometimes even leading to death. The serious effects of this disease were known in the late 1880s, when Loretta performed the first left lateral liver resection for echinococcosis in Bologna [8]. Last years many developments have improved the course of hydatid disease: better medical therapy, improved surgical procedures and the development of minimally invasive techniques.

From a historical perspective, the main treatment option of hepatic hydatid disease was the open surgical approach with side packing and several radical or more conservative surgical techniques [96, 99]. This terminology in literature might be confusing. Conservative surgery means that tissue-sparing techniques are used; the hydatid cyst is evacuated and the pericyst is left in situ, while in radical procedures both the cyst and the pericyst are removed. The most common conservative techniques include simple tube drainage, marsupialization, capitonnage, deroofing, partial cystectomy or open or closed total cystectomy with or without omentoplasty. Conservative operations have good results regarding blood loss and length of hospital stay [103, 104]. In contrast, the cyst content and the entire pericystic membrane are removed in radical procedures; a total pericystectomy or liver resection (hemihepatectomy or lobectomy) is performed [90, 94, 101, 104].

In surgical interventions of hepatic hydatid cysts, complete removal of the parasite should be performed. Also, prevention of intraoperative spilling of cyst content and saving healthy hepatic issue is of utmost importance [91, 93, 96]. Spilling could not only lead to recurrence of hydatid disease, it could also lead to anaphylactic shock before the introduction of the antihelmintic drugs. Therefore, surgeons need to perform procedures with a focus on safe and complete exposure of the cyst, safe decompression of the cyst, safe evacuation of the cyst contents, sterilization of the cyst, treatment of biliary complications and management of the remaining cyst cavity. Especially in non-endemic areas where the number of operations is low, the technique needs to be safe and easily reproducible, with a low complication rate. In the former century, hydatid disease was operated with a high risk of morbidity and recurrence, possibly due to the spilling of cyst content during the operation. In the 1970s, Saidi developed a special cone, which was frozen to the cyst in order to reduce the risk of spilling cyst contents. This cone also simplified the disinfection of the cyst cavity [105]. Recently, this old treatment, also known as the 'frozen seal method', was evaluated in a non-endemic area and it was concluded to be an effective surgical treatment for hepatic hydatid disease [104]. In this retrospective study, 112 consecutive patients were treated surgically with the 'frozen seal' method for hydatid disease between 1981 and 2007. Recurrence rate was observed in 9 (8%) patients and morbidity occurred in twenty patients (17.9%). More importantly, no mortality was observed in this study of more than 25 years of surgically treated 'echinococcosis'. It was concluded that this surgical method used in the past century was still safe and effective in the new millennium. This technique is especially useful in non-endemic areas as it provides high efficacy and low morbidity rates.

Apart from the 'frozen-seal method', surgical treatment options may vary from conservative treatment (cystectomy) to radical treatment (complete open resection) to laparoscopic techniques. The debate on best surgical treatment is still ongoing: should this be conservative surgery or radical surgery in which the cyst is totally removed including the pericyst by total pericystectomy or partial hepatectomy or should it be the open or laparoscopic approach [101, 102].

4.2. Percutaneous treatments

With the introduction of antihelmintic drugs, new possibilities for treatment arose. By using this medication, the risk of anaphylaxis became smaller and percutaneous treatments were developed. One of these treatments for hydatid disease is PAIR: Percutaneous Aspiration, Injection and Re-aspiration. In a recent meta-analysis of operative versus non-operative treatment (PAIR) of hepatic echinococcosis [92], PAIR plus chemotherapy proved to be superior compared to surgery. The meta-analysis showed that PAIR was associated with improved efficacy, lower rates of morbidity, mortality, disease recurrence and shorter hospital stay [92].

In conclusion, the main treatment options for hepatic cystic echinococcosis are threefold: medical therapy, surgery and percutaneous drainage (Puncture Aspiration Injection and Reaspiration, also known as PAIR) or a combination of these therapies [91, 92, 100]. In the last revision of the WHO IWGE it was stated that surgery remains the cornerstone of treatment of hydatid disease, since it has the potential to remove the hydatid cyst and lead to complete cure. However, it is advised to evaluate surgical treatment carefully against other less invasive options such as percutaneous interventions. [88]

5. Improvements in pre-operative planning

An important way to improve the outcome in liver surgery is to prevent liver resection related complications. One of the main feared complications in liver surgery remains postresectional liver failure. This major complication may occur if the extent of tumour involvement requires major liver resection (3 or more segments), leaving a small postoperative remnant liver [3, 106, 107]. Due to impaired liver function this may even result in mortality. Obviously, limiting the liver resection, in order to leave enough liver remnant volume for proper function of the liver, can prevent this. However, major hepatectomies are performed increasingly often, mainly because indications for liver resection are continuously being extended. Former contraindications such as bilobar disease, number of metastases and even extrahepatic disease have been abandoned gradually and compromised liver function may be expected after aggressive induction chemotherapy. Consequently, postoperative remnant liver volume and function have become the main determinants of respectability [108-110]. In order to improve outcome in extended resections and thus to prevent postoperative liver failure after liver resection, a reliable volumetric assessment of the part of the liver to be resected as well as future residual liver volume should be a critical part of preoperative evaluation particularly. The safety of liver resection may increase if an estimate of minimal remnant liver volume is obtained via CT-volumetry [106, 111].

The utility of existing professional image-processing software is often limited by costs, lack of flexibility and specific hardware requirements such as coupling to a CT-scanner. In addition, the intended operation should be known to the investigator to predict the remnant liver volume accurately and requires the expertise of a liver surgeon. Therefore, CT-volumetry has hitherto been a multidisciplinary modality requiring the efforts of dedicated surgeons and radiologists and expensive software. Prospective CT-volumetric analysis of the liver on a Personal Computer performed by the operating surgeon in patients undergoing major liver would greatly enhance this preoperative assessment. ImageJ is a free, open-source Java-based image processing software programme developed by the National Institute of Health (NIH) and may be used for this purpose [112]. OsiriX® is Apple's version for image analysis and has been tested for CT volumetry of the liver [113]. It is also a freely available, user-friendly software system, which can be used for virtual liver resections and volumetric analysis [113].

As more major liver resections are performed, it is becoming more important to perform liver volumetry. Recently, these two open source image processing software packages were investigated to measure prospectively the remnant liver volume in order to reduce the risk of post-resectional liver failure. Volumes of total liver, tumour and future resection specimen of the included patients were measured preoperatively with ImageJ and OsiriX by two surgeons and a surgical trainee [114]. Results were compared with the actual weights of resected specimens and the measurements of the radiologist using professional CT scanner-linked Aquarius iNtuition® software. It was concluded that the prospective hepatic CT-volumetry with ImageJ or OsiriX® was reliable and can be accurately used on a Personal Computer by non-radiologists. ImageJ and OsiriX® yield results comparable to professional radiological software iNtuition®.

6. Minimally invasive surgery

To minimize the damage of treatment, laparoscopic surgery was introduced to avoid large incisions for many gastrointestinal operations in the previous century. After the first laparoscopic cholecystectomy in 1987 [115], the number of indications for this minimally invasive approach increased. The outcome has encouraged surgeons to develop a laparoscopic technique for many procedures including liver resections [116]. Although this type of surgery is technically more demanding and thereby time-consuming [117, 118], it proved to be beneficial for patients with less pain and better recovery compared to open liver surgery [119-121].

6.1. The history of laparoscopic surgery

The fundamentals of laparoscopic surgery were laid down in the early twentieth century when the German surgeon Kelling reported on the endoscopic visualization of the peritoneal cavity in an anesthetized dog using a Nitze cystoscope (1887) in 1902 [122]. Following the introduction of endoscopic inspection of the abdominal contents in an animal model, fellow countryman Jacobeus started experimenting with laparoscopy in human cadavers as well as living humans. In 1911 he reported on 80 laparoscopic examinations of the abdominal cavity [123, 124]. In the years thereafter the laparoscopic approach was enhanced with the introduction of illumination techniques, advancement in lens systems, the use of more than one single trocar and induction of pneumoperitoneum (Goetze and Veress). The era of therapeutic laparoscopy was then born, making it possible to minimize damage of treatment and avoid large incisions for many gastrointestinal operations. However, it was not until 1987 that the first laparoscopic cholecystectomy was performed [115].

At first, liver surgery was thought to be unsuitable for laparoscopic techniques since it might impose the risk of gas embolisms and major blood loss during transection of the liver. Also, sceptics pointed out the suspected risk of trocar site metastases in skin incisions. Gradually, as some expert centres progressively reported feasibility and safety, it became more popular.

This novel approach for liver resections was introduced during the 1990s. At first the procedure was only used for diagnostic laparoscopies and liver biopsies, later indications were extended to fenestration of liver cysts and anatomic liver resections. In 1992, Gagner et al. reported the first laparoscopic wedge resection of the liver. Only three years later, Cuesta et al. were the first to perform two cases of limited laparoscopic liver surgery of segment II and IV in the Netherlands [125]. The first laparoscopic left lateral bisegmentectomy of the liver was performed by the group of Azagra [126]. Since then, several studies have reported the feasibility and safety of laparoscopic resections for liver tumours in centres with extensive experience in both hepatobiliary surgery and laparoscopic surgery [116, 117, 127-130].

However, after its introduction, laparoscopic liver resection remained challenging because of the difficulties concerning safe mobilization and exposure of this fragile and heavy organ. Therefore, in the beginning only superficial and peripheral lesions in anterolateral segments were selected for the laparosopic approach. In recent times, centres with extensive experience in laparoscopy and hepatic surgery have also performed major hepatic resections laparoscopically with satisfactory outcomes. Importantly, no evidence of a compromised oncological clearance in laparoscopic liver resection has hitherto been found [120]

6.2. Advantages of the laparoscopic technique

The laparoscopic approach is said to have shifted the pain of the patient to the surgeon, as the latter had to obtain new operative skills and more demanding techniques. In fact laparoscopic surgery is a totally different concept of surgery. The conventional three-dimensional field is inherently two-dimensional, and the tactile feedback is impaired as compared to open surgery. Moreover, a full ambidexterity is required, as well as the skills to manipulate fragile structures with long instruments under minimal tactile feedback. Also, the surgeon becomes even more dependent on his team and instruments, as he will need experienced assistance for traction and camerawork and needs to trust the material even more compared to open surgery. For patients the most important presumed advantages of the laparoscopic procedure are reduced blood loss [119, 120], less postoperative pain [118, 127, 131], earlier functional recovery [127, 130], shorter postoperative hospital stay [118, 120, 121, 127, 130-132] and improved cosmetic aspects [127, 130]. Reoperations are reported to be easier due to reduced adhesions [127, 130-132]. Also, open-close procedures with large incisions can be avoided if peritoneal metastases are detected at laparoscopy.

However, up till now no randomised controlled trials comparing the open and laparoscopic liver resection technique have been reported. This may well be one of the reasons why many surgeons remained reluctant to incorporate this new laparoscopic approach. The currently available evidence is primarily based on case-series and identifies a technique that is reproducible with limited morbidity and mortality. In a consensus statement on laparoscopic liver resections, Buell J et al [133] concluded that resection of segments 2 and 3 by the laparoscopic approach should be the standard of care. In that same year a large international study reported comparable encouraging results concerning the superiority of laparoscopic liver resections in terms of complications from 109 patients: the complication rate was only 12% and there were no perioperative deaths [134]. Median hospital length of stay was 4 days. Negative margins were achieved in 94.4% of patients.

Overall survival rates and disease-free survival rates for the entire series were 50% and 43% at 5-year respectively. It was concluded that laparoscopic liver resection for colorectal metastases was safe, feasible and comparable to open liver resection for both minor and major liver resections in oncologic surgery. This is confirmed in a recent meta-analysis on short and long-term outcomes after laparoscopic and open resection. This study included a total of 26 studies, incorporating a population of 1678 patients [135]. Although laparoscopic liver resections resulted in longer operation time, most endpoints were superior for the laparoscopic approach compared with open resection, including reduced blood loss, portal clamp time, overall and liver specific complications, ileus and length of hospital stay. As for the long-term outcomes, no difference was found for oncologic outcomes between the laparoscopic and open surgical techniques. Therefore, it was concluded that the laparoscopic liver resection was a feasible alternative to open surgery in experienced hands [135].

7. Enhanced Recovery After Surgery (ERAS) or fast-track liver surgery

Another recent development in elective liver surgery is the introduction of Enhanced Recovery After Surgery (ERAS) programmes, also referred to as fast track perioperative care. These multimodal enhanced recovery programmes proved to be beneficial in open colonic and liver surgery [136, 137]. The multimodal recovery programme is evidence based and combines several interventions in perioperative care to reduce the stress response and organ dysfunction with a focus on enhancing recovery [137, 138]. In patients undergoing colorectal surgery, the ERAS® programme enabled earlier recovery and consequently shorter length of hospital stay [137-140]. Also, reduction of postoperative morbidity in patients undergoing intestinal resection was reported [141-144]. In other fields of elective surgery similar programmes have also shown a reduction in hospital stay of several days [145, 146].

One of the pioneers of the fast track colonic surgery is the Danish surgeon Henrik Kehlet. He treated 60 consecutive patients with colonic resection in a fast track surgery programme and reported a median postoperative hospital stay of 2 days. At that time, patients undergoing a colonic resection usually required 5 to 10 days postoperative hospital stay [147, 148]. Previously, he stressed the importance of a multimodal approach in order to improve rehabilitation after surgery (figure 6) [149]. This rehabilitation programme after surgery combined a number of interventions to reduce stress of the surgical intervention, risk of organ dysfunction and loss of functional capacity. Stress induced organ dysfunction, pain, nausea and vomiting, ileus, hypoxemia and sleep disturbances, immobilisation and semi-starvation had to be reduced.

Factors were identified that contribute to postoperative functional deterioration. These were actually traditional postoperative care principles such as use of drains, nasogastric tubes, fasting regimes and bed rest. Kehlet initiated a multimodal programme that abandoned the traditional care principles and introduced innovations such as: carbohydrate loading before surgery, regional anaesthetic techniques, maintenance of normal temperature during surgery, minimally invasive or laparoscopic surgical techniques, optimal treatment of postoperative pain and prophylaxis of nausea and vomiting [139, 150]. This programme improved postoperative recovery, physical performance and pulmonary function and reduced hospital length of stay [142].

In collaboration with Kehlet, the Enhanced Recovery After Surgery (ERAS) group was initiated to investigate the perioperative care in four other hospitals (Royal Infirmary, Edinburgh, UK, The Karolinska Institutet at Ersta Hospital, Stockholm, Sweden, the University Hospital of Nothern Norway, Tromso, Noway and Maastricht University Medical Centre) [151]. Thus,

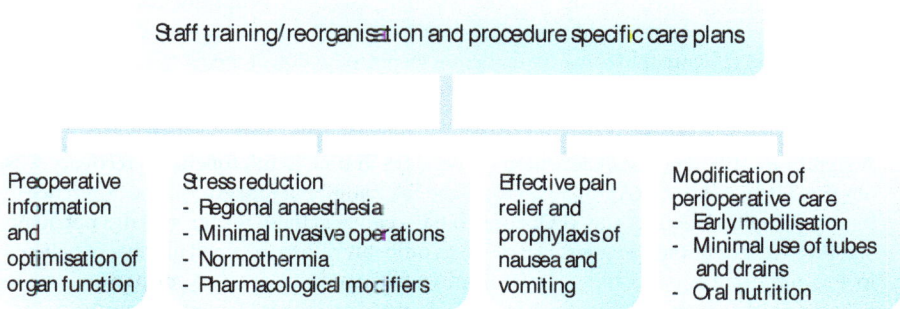

Figure 6. Multimodal interventions may lead to a reduction in postoperative morbidity and improved recovery. [149] Figure adapted with kind permission from Kehlet et al. 1997.

with Kehlet's programme as a starting point, a new evidence based programme was developed incorporating different aspects leading to faster recovery. Preoperative counselling, perioperative intravenous fluid restriction, optimal pain relief preferably without the use of opioid analgesia, early oral nutrition, enforced mobilisation, no nasogastric tubes and no drains are the key elements of this protocol (figure 7). Since the colonic programme showed improvements in recovery, the liver surgeons of the ERAS® group (Maastricht, Edinburgh and Tromso) set up an ERAS-programme for every patient undergoing open liver resection [136] (www.erassociety.org).

So far, the ERAS programmes have shown promising results with respect to improved recovery and outcome in open elective colorectal and liver surgery [136, 137]. One of the first studies on ERAS for liver surgery showed that the majority of patients treated within this multimodal enhanced recovery programme tolerated fluid within four hours of surgery and a normal diet one day after surgery. As an effect of the accelerated functional recovery, these patients were discharged two days earlier than the patients treated with traditional care, without significant differences in readmission, morbidity and mortality rates [136].

These results were confirmed in a recent systematic review including seven studies on fast-track programmes for hepatopancreatic resections, incorporating more than 550 patients treated in fast track setting [152]. This study showed that the primary hospital stay was reduced significantly after the introduction of a multimodal perioperative care programme for open liver surgery [152]. Moreover, there were no significant differences in rates of readmission, morbidity and mortality.

7.1. Synergy of ERAS and laparoscopic liver surgery

For solid tumours in the liver, the open approach for resection is gradually replaced by the laparoscopic technique in many expert centres worldwide. The results, mostly from cohort studies, suggest benefits with notably shorter postoperative stay [120]. Recently, the added value of a fast-track ERAS-programme in laparoscopic liver surgery specifically has been elucidated [153]. A group consisting of patients undergoing laparoscopic liver resections in an ERAS-setting was compared with historical data from consecutive laparoscopic liver resections performed either in that same centre before the introduction of the ERAS-programme or in other major liver centres in the Netherlands performing laparoscopic liver surgery in a traditional perioperative care programme.

• A significant difference with a median of two days in time to full functional recovery was observed between the ERAS-treated group and the traditional care group. The difference in median hospital length of stay (LOS) of two days between these two groups did not attain significance. The authors suggested that it was probably due to the small number of patients in this multicentre pilot-study. Apart from faster functional recovery in patients in the enhanced recovery group, this study also showed reduced blood loss in this group.

• As from a historical perspective, this multicentre fast-track laparoscopic liver resection study was the first study to explore the effect of ERAS and laparoscopic surgery. This small study suggests that a multimodal enhanced recovery programme for laparoscopic liver surgery is feasible, safe and may lead to accelerated functional recovery and reduction in length of hospital stay. With these findings it may be concluded that the additional effect of ERAS leads to an improvement of liver surgery and outcome.

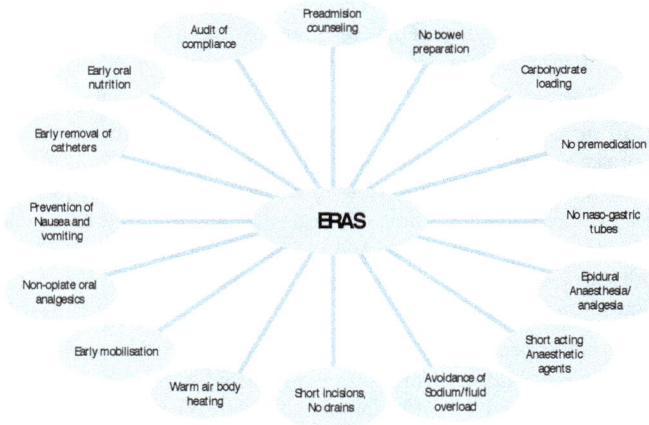

Figure 7. Important elements of the Enhanced Recovery After Surgery programme. [138] Figure adapted with kind permission from Fearon et al. 2005.

8. Recent developments in hepatic malignancies

As for the recent developments in the treatment of liver diseases, these can be mainly divided into surgical and non-surgical treatment modalities. Developments in surgical treatment can be divided in true surgical and perioperative care improvements. The focus is on the surgical treatments in this chapter, but some thoughts will also be spent on the non-surgical treatment modalities, an interesting and expanding field of expertise.

For malignant liver tumours, the majority of which are colorectal liver metastases, the main concern is the resectability if colorectal cancer is diagnosed. Colorectal cancer is one of the most common causes of cancer related death worldwide [38] and more than half of patients with colorectal cancer will develop liver metastases [39]. Unfortunately, only 20% of the patients can be treated with surgical resection of these liver metastases [154]. The remaining 80% of the patients present with lesions, which are not suitable for a safe resection. This can be caused by large diameters of the lesions, location of the lesion near vascular and biliary structures and extrahepatic disease. Also, the number of lesions can be the cause of non-resectability: resection can only be carried out safely if 25-30% of functioning liver remains after resection [155]. The non-surgical treatment by means of chemotherapy for the patients with unresected liver metastases has proven very successful in decreasing the size and number of liver lesions. It was shown that new chemotherapy regimens could change the previously unresectable liver metastases into resectable liver disease [156]. With neoadjuvant chemotherapy more patients with colorectal liver metastases can be offered a treatment with curative intent [156]. It was concluded that neoadjuvant chemotherapy enables liver resection in some patients with initially unresectable colorectal metastases. Long-term survival proved to be similar to that reported for a priori surgical candidates [56]. As for the future perspective of chemotherapy, neoadjuvant treatment will improve curability and long-term survival for selected patients.

Other non-surgical therapies for malignant liver disease are external irradiation (whole liver irradiation) [157, 158], stereotactic liver irradiation [159-162] and injectable small radioactive particles that irradiate the tumours within the liver (e.g. Yttrium-90(^{90}Y) radio-embolisation [163, 164], radioactive holmium microspheres [165, 166]). These modalities may have curative potential but future studies have to be awaited. Another attractive field of development are the thermal ablative therapies for unresectable liver metastases. These ablative thermal therapies can be used either percutaneously or in adjunct with surgery and have shown to decrease focal liver lesions [167-170]. Microwave ablation is a tumour destruction method to treat patients with unresectable liver lesions [169]. It can be used with a single insertion of the probe and it was shown to be a safe and effective method for treating unresectable hepatic tumours, with a low rate of local recurrence [170]. Overall survival is comparable to alternative ablation modalities [169].

8.1. Future perspectives

As for surgical treatments, different treatment strategies have been developed to increase the number of patients suitable for surgery as described earlier. Current research has focussed on improving resectability in terms of the quantity of resected liver tissue, but at the same time studies focussed on reducing perioperative distress in patients undergoing liver resections by multimodal perioperative treatment protocols and minimally invasive surgery. Since the introduction of laparoscopic liver surgery in 1992, more liver resections have been performed with this minimally invasive approach for primary and secondary malignant liver lesions [129,

134, 153]. For future perspectives, some gain might be expected from even less invasive modalities as the first reports on single incision laparoscopic resections have been presented [171-173]. Also, a two-stage laparoscopic approach for malignant liver disease and the robotic approach for liver resections have been published [174-176].

As discussed previously in this chapter, the recent developments in liver surgery include the introduction of laparoscopic surgery and enhanced recovery programmes, which focus on improvement of postoperative recovery and/or shorter hospital length of stay. A significantly accelerated recovery after open liver resection was previously reported if patients were managed within a multimodal ERAS protocol. Median hospital length of stay was reduced from 8 to 6 days (25%) [136]. Moreover, since there was a delay between recovery and discharge of the patients a further reduction of stay should be possible. Regarding the results of previous, non-randomised randomized studies and case series, it seems that laparoscopic left lateral liver sectionectomy is associated with shorter hospital length of stay, less postoperative pain, better quality of life and a faster recovery [177]. In most trials aiming at a reduction of hospital length of stay, surgery and/or perioperative management are not standardised. No randomised trials have hitherto been reported to study the added value of ERAS and/or laparoscopy for liver surgery. There is a need for a randomised controlled trial covering these aspects of improving the recovery and outcome of liver surgery.

9. Liver transplantation

Liver transplantation surgery is one of the main advances in hepatic surgery. Until recently, it was considered to be too complex, since artificial organ support, like haemodialysis in renal failure, was considered impossible. The term liver transplantation was first used in an article of Welch (NY, USA) in 1955 [178]. The first experimental liver transplantation surgery was performed on animals (dogs) in the 1950s and 1960s by Starzl (Denver, USA, figure 8) and Moore (Boston, USA). These transplantations failed as a result of the stagnation of blood in the mesenterial vessels and a lack of blood flow to the heart after clamping the inferior vena cava. Methods for a venovenous bypass to the superior vena cava were developed, whereupon transplantation seemed to be realizable. Despite the fact that immunosuppressive drugs became available at that time, most grafts were rejected though. As a result, only a few dogs survived [178-181].

9.1. The history of liver transplantation in humans

In 1963 the first three orthotropic liver transplantations in humans were performed by Starzl and colleagues. All livers came from non-heart beating donors (NHBDs). Although the first transplantation was performed in one session, the second and third took two sessions; the first session was designated for the preparation of the removal of the liver from the donor and in the second session the liver was removed and transplanted in the recipient after the donor died. In the donor patient extracorporeal perfusion was performed via the femoral vein and artery. The structures in the hepatoduodenal ligament were cut through and the liver was

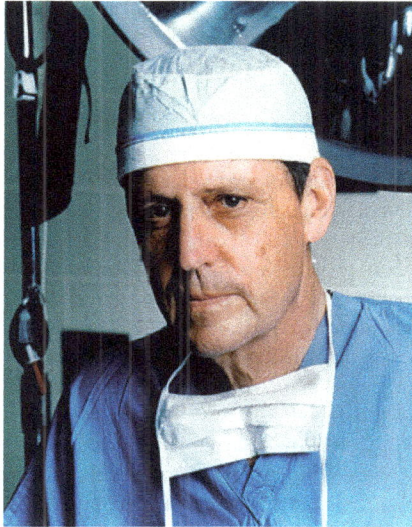

Figure 8. Thomas E. Starzl (1926).

taken out with the vena cava. In the recipient the liver was taken out likewise and a venovenous bypass was made to circumvent the hemodynamic effects of clamping the vena cava [182]. Immunosuppressive therapy, by ways of azathioprine and prednisone, was applied since these drugs were proven to be effective in renal transplantation [183]. The first patient was a three-year-old boy with biliary atresia who died during the operation due to haemorrhage, the second and third patient were adult males suffering from liver cancer who died 7 and 22 days postoperative, as a result of lung embolism [182]. Starzl then decided to take a break to have a period of reflection. Four years later, in 1967, he decided to try again and he then performed the first successful liver transplantation with a one-year-survival [184].

Infections were frequently occurring complications [185]. The most important complication of these early transplantations however, was severe blood loss. This was caused by manipulation of abdominal veins which had been under great pressures due to chronic liver diseases [179]. The first orthotropic liver transplantation in Europe was performed in Cambridge in 1968 by Calne [186]. In the same year consensus was achieved concerning the concept of cerebral death. From that moment on, heart-beating donation with donor organs originating from heart beating, brain dead donors was possible [184]. Nowadays the above described venovenous bypass has been abandoned in many centres in Europe. Since the beginning of the 1990's most centres use the so called 'piggyback' technique. The liver is exposed from the vena cava after which the vena cava is partially clamped longitudinally. After the liver has been flushed with albumin to remove ischemic waste products, a side-to-side cavocaval anastomosis is made. In doing so, the hemodynamic stability of the patient is guaranteed. Then, the portal liaison is made by an end-to-end anastomosis, the liver is perfused and the arterial anastomosis is made. Finally the biliary ducts are connected by way of end-to-end anastomosis and in case of sclerosis a Roux-en-Y-reconstruction [187, 188].

9.2. Immunsuppressive drugs

The discovery and appliance of immunosuppressive medication to prevent graft rejection has been an important development in transplantation surgery. Despite the fact that graft rejection has been a serious problem during the early years of liver transplantation, many transplanted patients survived more than 20 years as a result of this immunosuppressive therapy with an azathioprine-prednisone cocktail. Some time later, a third immunosuppressive drug, antilymphocyte-globulin (ALG), was added to the therapy [178, 189, 190]. Then Calne discovered the possibility to use cyclosporin A, a calcineurin inhibitor, as an immunosuppressive drug [191]. After cyclosporine A was first used in renal transplantations in 1980 [192], it was then applied in liver transplantation and the one-year-survival rate in liver transplantation turned out to have increased to 80% [193]. Currently Tacrolimus (FK 506), also a calcineurin inhibitor, is recommended [194-197]. A detailed overview of the development and the working mechanisms of immunosuppressive drugs is beyond the scope of this chapter.

9.3. Split liver transplantation

The concept of liver transplantation has been developed gradually, which made it a widely accepted treatment with an increasing number of indications and good survival rates. This caused a shortage of donor organs, especially among children, and long waiting lists. New techniques had to be developed to answer to this growing demand. In 1984 Bismuth developed the reduced-size adult liver transplantation; an adult left lobe was transplanted into a child. This is a unique method, only applicable in liver transplantation surgery because of its segmental anatomy with independently functioning parts [198]. Further development of segmental liver surgery resulted in the split liver transplantation (SLT); the donor liver is splitted, the left part (segment 2 and 3 with the common hepatic duct and common hepatic artery) is transplanted into a child and the right part (segment 1, 4-7 with the vena cava) into an adult. In the recipient of the left liver part, the vena cava is preserved and an anastomosis is made with the left hepatic vein. The other anastomoses are made in the usual way. In the recipient of the right liver part, an anastomosis is made between the right hepatic artery of the donor liver and the common hepatic artery of the recipient by means of a saphenous vein interposition graft. Two intrahepatic biliary ducts are connected with the jejunum through a Roux-en-Y loop, the other anastomosis are executed in the usual way [199]. There are two ways of splitting the liver, in situ and ex situ, both with its (dis)advantages. The main disadvantage of in situ splitting is a longer operation time and therefore the need for a haemodynamically stable patient. Splitting ex situ on the other hand, is done in blood vacuum. The time of cold ischemia is longer and it is harder to distinguish structures from each other. Hence, strict donor selection is essential and there is a trend to only select donors <50 years or who are heamodynamically stable. Bile spill is reported as the most common complication. Other complications are an insufficient hepatic artery, portal vein thrombosis, intra-abdominal haemorrhage and gastro-intestinal bleeding. Mortality rates of 11% have been reported [200, 201]. In Europe, in 2003, 89% of all liver transplantations consisted of full-size transplantations, 4% of SLT's and 5% of reduced-liver transplantations. In specialized centres, the survival rates of these techniques are comparable to the survival rates of regular transplantation [202].

9.4. Living-donor liver transplantation (LDLT)

In 1987 Raia (Brazil) developed the living-donor liver transplantation (LDLT) from an adult into a child. The operation itself was successful, but the recipient child died due to a transfusion reaction [203]. The first successful LDLT from mother to son with a left liver lobe was performed in Australia by Strong [204] after which this method was refined by many other pioneers. It is a very difficult operation technique in which precise knowledge of the anatomy is a prerequisite. Because of a great shortage of donor organs in Asia, most experience with the LDLT was gained there. Innovative surgery was the only possibility to tide over this shortage. These techniques seemed to be effective; waiting-list-related mortality among children was reduced to almost 0% [205, 206]. Since Fan (Honk Kong) introduced the adult-to-adult living liver transplantation with a hemi-liver (dependent on the size of donor and recipient either the right or left lobe is transplanted) in 1997, the availability of donor livers for adults increased [207].

The main advantage of LDLT is limitation of warm ischemia because operations can be planned simultaneously [208]. The results of LDLT are comparable to those of regular (orthotopic) liver transplantation. According to the Japanese Liver Transplantation Society the 5-year-survival rate in adults is 69%. In children this rate is significantly higher with 83% [205]. In the USA the reported survival rate in adults is 80% [209]. In Europe, a 5-year-survival of 75% (80% in children, 66% in adults) between 1991 and 2001 was reported [202, 205]. In Europe, in 2003, only 1.6% of all liver transplantations consisted of LDLT [202].

The main disadvantages of this technique are the potential complications in the healthy donor and the psychological impact [189, 210]. The number of postoperative complications in donors is reported to be 20%. Worldwide 10 (0.15%) donor deaths have been reported. The mortality rate in Europe, in 2010, was 0.2% (6/2906) [211]. The critical period for death and primary dysfunction is within 6 months from the operation. In a graft too small for the recipient, dysfunction will develop with hyperbilirubinemia, ascites and liver function failure resulting in coagulation disorders and renal failure. A graft which is too big for the recipient will result in necrosis because of shortage in blood supply. Besides good patient selection, proper calculation to determine the correct graft size has to be done to prevent these complications [189, 205].

9.5. Improving survival

In 1997 the Institute of Medicine (USA) declared NHBD-organs to be medically effective and ethically acceptable [178]. From that time on, the trend exists to use NHBD- and marginal organs (livers with steatosis) again to tide over the shortage of donor organs and shorten the waiting lists. Marginal livers are associated with primary non-function [212]. The main problem of NHBD's is the prolonged period of warm ischemia. A distinction between controlled NHBD's (Maastricht type I and II) and uncontrolled NHBD's (Maastricht type III and IV) is made. Controlled NHBD's provide organs with less chance on ischemic damage and a greater chance on good post-transplantation function. In this group of patients a controlled end of vital support takes place after which a circulation stop occurs. In most cases the patient is already in the operation theatre with a transplantation team on site. This way, the time of warm ischemia is minimalised. In uncontrolled NHBD's a non-foreseen circulation stop occurs, usually before arrival in the hospital, possibly followed by resuscitation. A variable period of warm ischemia occurs with a higher chance on complications [212, 213]. Cold ischemia causes damage of sinusoidal endothelial cells and warm ischemia of hepatocytes [214]. Besides, warm ischemia intensifies the effects of cold ischemia and predisposes for a higher incidence of ischemic biliary structures both on the short and the long term. In such cases, re-transplantation might be needed [215]. Since the University of Wisconsin Solution, introduced in 1988, has become the golden standard for cooling donor organs and the maximum period of cold ischemia has been limited to 12 hours, ischemic damage due to cold ischemia has been reduced drastically with increased graft survival [202]. However, as a consequence of warm ischemia graft survival is lower in NHBD's compared to heart-beating donors with a 3-year-survival of 63.3% versus 72.1%. The risk of primary non-function is also significantly higher among NHBD's: 11.8% versus 6.4% [189, 216]. For this reason NHBD's can be used to overcome organ shortage, on condition that strict criteria are maintained: strict donor (<60 years) and recipient (haemodynamically stable and not intubated) selection, minor warm (<30 minutes) and cold (<8 hours) ischemia, no extensive steatosis of the donor liver and the use of at most one inotropic drug (to prevent hypotension and thus hypoperfusion) [212].

With the gradual progression in surgical competences, management of postoperative complications and the development of immunosuppressive drugs to prevent graft rejection, liver transplantation has nowadays become a widely accepted treatment for an increasing number of indications and it has become the golden standard for patients with irreversible decompensated chronic liver failure (e.g. as a result of cirrhosis or hepatocellular cancer) and acute liver failure (e.g. as a result of hepatic viruses or intoxication with medication). In the early days cancer was the most common indication for liver transplantation. In Europe, however, with 50% the most important indication for liver transplantation was cirrhosis (of which 24% was caused by a virus (especially Hepatitis C) and 18% by alcohol abuse), followed by pathology of the biliary tract (13%), primary liver tumours (10%), of which hepatocellular cancer is the most common, and acute liver failure (9%), with fulminant viral hepatitis as the most important cause. The most important indications in children are biliary atresia (56%) and metabolic diseases (21%) [202]. Due to the development of different methods and techniques, organ shortage has been reduced and waiting lists have been shortened. Hence, one can conclude that liver transplantation is a recent and very important advancement, which has expanded in a short time. It is a perfect example of modern and innovative medical practice, in which the challenge remains to find solutions to new problems time after time.

Author details

J.H.M.B. Stoot[1,2,3*], R.J.S. Coelen[1,2], J.L.A. van Vugt[2] and C.H.C. Dejong[1,4]

*Address all correspondence to: jan@stoot.com

1 Department of Surgery, Maastricht University Medical Centre, Maastricht, The Netherlands

2 Department of Surgery, Orbis Medical Centre, Sittard, The Netherlands

3 Department of Surgery, Atrium Medical Centre, Heerlen, The Netherlands

4 NUTRIM School for Nutrition, Metabolism and Toxicology, Maastricht University Medical Centre, Maastricht, The Netherlands

References

[1] Poon, R. T, et al. Improving perioperative outcome expands the role of hepatectomy in management of benign and malignant hepatobiliary diseases: analysis of 1222 consecutive patients from a prospective database. Ann Surg, (2004). discussion 708-10., 698-708.

[2] Cescon, M, et al. Trends in perioperative outcome after hepatic resection: analysis of 1500 consecutive unselected cases over 20 years. Ann Surg, (2009). , 995-1002.

[3] Jarnagin, W. R, et al. Improvement in perioperative outcome after hepatic resection: analysis of 1,803 consecutive cases over the past decade. Ann Surg, (2002). discussion 406-7., 397-406.

[4] Tsao, J. I, et al. Trends in morbidity and mortality of hepatic resection for malignancy. A matched comparative analysis. Ann Surg, (1994). , 199-205.

[5] Foster, J. H. History of liver surgery. Arch Surg, (1991). , 381-387.

[6] Glisson, F. Anatomia Hepatis. London, England, 1654.

[7] Cantlie, J. On a new arrangement of the right and left lobes of the liver. J. Anat. Physiol., (1898). , 4-9.

[8] Hardy, K. J. Liver surgery: the past 2000 years. Aust N Z J Surg, (1990). , 811-817.

[9] Pringle, J. H. V. Notes on the Arrest of Hepatic Hemorrhage Due to Trauma. Ann Surg, (1908). , 541-549.

[10] Couinaud, C. Le Foie. Etudes anatomiques et chirurgicales. Paris: Masson, (1957).

[11] Guyton, A, & Hall, J. The liver as an organ, in Textbook of Medical Physiology(1996). Philadelphia: WB Saunders. , 883-888.

[12] Taub, R. Liver regeneration: from myth to mechanism. Nat Rev Mol Cell Biol, (2004). , 836-847.

[13] van de PollM.C., et al., Effect of major liver resection on hepatic ureagenesis in humans. Am J Physiol Gastrointest Liver Physiol, (2007). , G956-G962.

[14] Lortat-jacob, J. L, Robert, H. G, & Henry, C. Excision of the right lobe of the liver for a malignant secondary tumor]. Arch Mal Appar Dig Mal Nutr, (1952). , 662-667.

[15] Foster, J. H, & Berman, M. M. Solid liver tumors. Major Probl Clin Surg, (1977). , 1-342.

[16] Ekberg, H, et al. Determinants of survival in liver resection for colorectal secondaries. Br J Surg, (1986). , 727-731.

[17] Adson, M. A, et al. Resection of hepatic metastases from colorectal cancer. Arch Surg, (1984). , 647-651.

[18] Simmonds, P. C, et al. Surgical resection of hepatic metastases from colorectal cancer: a systematic review of published studies. Br J Cancer, (2006). , 982-999.

[19] Iwatsuki, S, Shaw, B. W, & Jr, T. E. Starzl, Experience with 150 liver resections. Ann Surg, (1983). , 247-253.

[20] Bismuth, H, Houssin, D, & Castaing, D. Major and minor segmentectomies "reglees" in liver surgery. World J Surg, (1982). , 10-24.

[21] Bismuth, H. Surgical anatomy and anatomical surgery of the liver. World J Surg, (1982). , 3-9.

[22] Tung, T. T. Les resections majeures et mineures du foie. Paris: Masson, (1979).

[23] Billingsley, K. G, et al. Segment-oriented hepatic resection in the management of malignant neoplasms of the liver. J Am Coll Surg, (1998). , 471-481.

[24] Fan, S. T, et al. Hepatectomy for hepatocellular carcinoma: toward zero hospital deaths. Ann Surg, (1999). , 322-330.

[25] Farid, H, & Connell, T. O. Hepatic resections: changing mortality and morbidity. Am Surg, (1994). , 748-752.

[26] Poon, R. T. Recent advances in techniques of liver resection. Surg Technol Int, (2004). , 71-77.

[27] Adam, R, et al. Patients with initially unresectable colorectal liver metastases: is there a possibility of cure? J Clin Oncol, (2009). , 1829-1835.

[28] Tomlinson, J. S, et al. Actual 10-year survival after resection of colorectal liver metastases defines cure. J Clin Oncol, (2007). , 4575-4580.

[29] Charny, C. K, et al. Management of 155 patients with benign liver tumours. Br J Surg, (2001). , 808-813.

[30] Poon, R. T. Current techniques of liver transection. HPB (Oxford), (2007). , 166-173.

[31] Aragon, R. J, & Solomon, N. L. Techniques of hepatic resection. J Gastrointest Oncol, (2012). , 28-40.

[32] Gurusamy, K. S, et al. Techniques for liver parenchymal transection in liver resection. Cochrane Database Syst Rev, (2009). , CD006880.

[33] Palavecino, M, et al. Two-surgeon technique of parenchymal transection contributes to reduced transfusion rate in patients undergoing major hepatectomy: analysis of 1,557 consecutive liver resections. Surgery, (2010). , 40-48.

[34] Pamecha, V, et al. Techniques for liver parenchymal transection: a meta-analysis of randomized controlled trials. HPB (Oxford), (2009). , 275-281.

[35] Rahbari, N. N, et al. Meta-analysis of the clamp-crushing technique for transection of the parenchyma in elective hepatic resection: back to where we started? Ann Surg Oncol, (2009). , 630-639.

[36] Koo, B. N, et al. Hepatic resection by the Cavitron Ultrasonic Surgical Aspirator increases the incidence and severity of venous air embolism. Anesth Analg, (2005). table of contents., 966-970.

[37] Rahbari, N. N, et al. Clamp-crushing versus stapler hepatectomy for transection of the parenchyma in elective hepatic resection (CRUNSH)--a randomized controlled trial (NCT01049607). BMC Surg, (2011). , 22.

[38] Boyle, P, & Leon, M. E. Epidemiology of colorectal cancer. Br Med Bull, (2002). , 1-25.

[39] Steele, G, & Jr, T. S. Ravikumar, Resection of hepatic metastases from colorectal cancer. Biologic perspective. Ann Surg, (1989). , 127-138.

[40] Manfredi, S, et al. Epidemiology and management of liver metastases from colorectal cancer. Ann Surg, (2006). , 254-259.

[41] Scheele, J, et al. Resection of colorectal liver metastases. World J Surg, (1995). , 59-71.

[42] Thirion, P, et al. Modulation of fluorouracil by leucovorin in patients with advanced colorectal cancer: an updated meta-analysis. J Clin Oncol, (2004). , 3766-3775.

[43] Mayo, S. C, et al. Surgical management of hepatic neuroendocrine tumor metastasis: results from an international multi-institutional analysis. Ann Surg Oncol, (2010). , 3129-3136.

[44] Rehders, A, et al. Hepatic metastasectomy for soft-tissue sarcomas: is it justified? World J Surg, (2009). , 111-117.

[45] Mondragon-sanchez, R, et al. Repeat hepatic resection for recurrent metastatic melanoma. Hepatogastroenterology, (1999). , 459-461.

[46] Pawlik, T. M, et al. Hepatic resection for metastatic melanoma: distinct patterns of recurrence and prognosis for ocular versus cutaneous disease. Ann Surg Oncol, (2006)., 712-720.

[47] Frenkel, S, et al. Long-term survival of uveal melanoma patients after surgery for liver metastases. Br J Ophthalmol, (2009)., 1042-1046.

[48] Karavias, D. D, et al. Liver resection for metastatic non-colorectal non-neuroendocrine hepatic neoplasms. Eur J Surg Oncol, (2002)., 135-139.

[49] Hirai, I, et al. Surgical management for metastatic liver tumors. Hepatogastroenterology, (2006)., 757-763.

[50] Makino, H, et al. Indication for hepatic resection in the treatment of liver metastasis from gastric cancer. Anticancer Res, (2010)., 2367-2376.

[51] Lermite, E, et al. Surgical resection of liver metastases from breast cancer. Surg Oncol, (2009)., e79-e84.

[52] Sakamoto, Y, et al. Hepatic resection for metastatic breast cancer: prognostic analysis of 34 patients. World J Surg, (2005)., 524-527.

[53] Figueras, J, et al. Effect of subcentimeter nonpositive resection margin on hepatic recurrence in patients undergoing hepatectomy for colorectal liver metastases. Evidences from 663 liver resections. Ann Oncol, (2007)., 1190-1195.

[54] Figueras, J, et al. Surgical resection of colorectal liver metastases in patients with expanded indications: a single-center experience with 501 patients. Dis Colon Rectum, (2007)., 478-488.

[55] Khatri, V. P, Petrelli, N. J, & Belghiti, J. Extending the frontiers of surgical therapy for hepatic colorectal metastases: is there a limit? J Clin Oncol, (2005)., 8490-8499.

[56] Adam, R, et al. Five-year survival following hepatic resection after neoadjuvant therapy for nonresectable colorectal. Ann Surg Oncol, (2001)., 347-353.

[57] Figueras, J, et al. Surgical treatment of liver metastases from colorectal carcinoma in elderly patients. When is it worthwhile? Clin Transl Oncol, (2007)., 392-400.

[58] Adam, R, et al. Liver resection of colorectal metastases in elderly patients. Br J Surg, (2010)., 366-376.

[59] De Haas, R. J, Wicherts, D. A, & Adam, R. Resection of colorectal liver metastases with extrahepatic disease. Dig Surg, (2008)., 461-466.

[60] Adam, R, et al. Is hepatic resection justified after chemotherapy in patients with colorectal liver metastases and lymph node involvement? J Clin Oncol, (2008)., 3672-3680.

[61] Wicherts, D. A, et al. Impact of portal vein embolization on long-term survival of patients with primarily unresectable colorectal liver metastases. Br J Surg, (2010)., 240-250.

[62] Choti, M. A, et al. Trends in long-term survival following liver resection for hepatic colorectal metastases. Ann Surg, (2002). , 759-766.

[63] De Haas, R. J, et al. R1 resection by necessity for colorectal liver metastases: is it still a contraindication to surgery? Ann Surg, (2008). , 626-637.

[64] Wicherts, D. A, et al. Long-term results of two-stage hepatectomy for irresectable colorectal cancer liver metastases. Ann Surg, (2008). , 994-1005.

[65] Virani, S, et al. Morbidity and mortality after liver resection: results of the patient safety in surgery study. J Am Coll Surg, (2007). , 1284-1292.

[66] Dixon, E, et al. Mortality following liver resection in US medicare patients: does the presence of a liver transplant program affect outcome? J Surg Oncol, (2007). , 194-200.

[67] Fong, Y, Blumgart, L. H, & Cohen, A. M. Surgical treatment of colorectal metastases to the liver. CA Cancer J Clin, (1995). , 50-62.

[68] Lius, A. Di un adenoma del fegato. Gazz delle cliniche, (1886).

[69] Langenbuch, C. Ein Fall von Resektion eines linksseitigen Schnurlappens der Leber. Berl Klin Woschenschr, (1888). , 37-38.

[70] Tesluk, H, & Lawrie, J. Hepatocellular adenoma. Its transformation to carcinoma in a user of oral contraceptives. Arch Pathol Lab Med, (1981). , 296-299.

[71] Foster, J. H, & Berman, M. M. The malignant transformation of liver cell adenomas. Arch Surg, (1994). , 712-717.

[72] Stoot, J. H, et al. Malignant transformation of hepatocellular adenomas into hepatocellular carcinomas: a systematic review including more than 1600 adenoma cases. HPB (Oxford), (2010). , 509-522.

[73] Bioulac-sage, P, et al. Hepatocellular adenoma subtypes: the impact of overweight and obesity. Liver Int, (2012).

[74] Dokmak, S, et al. A Single Center Surgical Experience of 122 Patients with Single and Multiple Hepatocellular Adenomas. Gastroenterology, (2009).

[75] Franco, L. M, et al. Hepatocellular carcinoma in glycogen storage disease type Ia: a case series. J Inherit Metab Dis, (2005). , 153-162.

[76] Gorayski, P, et al. Hepatocellular carcinoma associated with recreational anabolic steroid use. Br J Sports Med, (2008). discussion 75., 74-75.

[77] Labrune, P, et al. Hepatocellular adenomas in glycogen storage disease type I and III: a series of 43 patients and review of the literature. J Pediatr Gastroenterol Nutr, (1997). , 276-279.

[73] Velazquez, I, & Alter, B. P. Androgens and liver tumors: Fanconi's anemia and non-Fanconi's conditions. Am J Hematol, (2004). , 257-267.

[79] Zucman-rossi, J, et al. Genotype-phenotype correlation in hepatocellular adenoma: new classification and relationship with HCC. Hepatology, (2006). , 515-524.

[80] Anthony, P. P, Vogel, C. L, & Barker, L. F. Liver cell dysplasia: a premalignant condition. J Clin Pathol, (1973). , 217-223.

[81] Ho, J. C, Wu, P. C, & Mak, T. K. Liver cell dysplasia in association with hepatocellular carcinoma, cirrhosis and hepatitis B surface antigen in Hong Kong. Int J Cancer, (1981). , 571-574.

[82] Lee, R. G, Tsamandas, A. C, & Demetris, A. J. Large cell change (liver cell dysplasia) and hepatocellular carcinoma in cirrhosis: matched case-control study, pathological analysis, and pathogenetic hypothesis. Hepatology, (1997). , 1415-1422.

[83] Su, Q, et al. Human hepatic preneoplasia: phenotypes and proliferation kinetics of foci and nodules of altered hepatocytes and their relationship to liver cell dysplasia. Virchows Arch, (1997). , 391-406.

[84] Tao, L. C. Oral contraceptive-associated liver cell adenoma and hepatocellular carcinoma. Cytomorphology and mechanism of malignant transformation. Cancer, (1991). , 341-347.

[85] Van Aalten, S. M, et al. Hepatocellular adenomas: correlation of MR imaging findings with pathologic subtype classification. Radiology, (2011). , 172-181.

[86] Laumonier, H, et al. Hepatocellular adenomas: magnetic resonance imaging features as a function of molecular pathological classification. Hepatology, (2008). , 808-818.

[87] Brunetti, E, Kern, P, & Vuitton, D. A. Expert consensus for the diagnosis and treatment of cystic and alveolar echinococcosis in humans. Acta Trop, (2010). , 1-16.

[88] Tappe, D, Stich, A, & Frosch, M. Emergence of polycystic neotropical echinococcosis. Emerg Infect Dis, (2008). , 292-297.

[89] Ammann, R. W, & Eckert, J. Cestodes. Echinococcus. Gastroenterol Clin North Am, (1996). , 655-689.

[90] Dziri, C, Haouet, K, & Fingerhut, A. Treatment of hydatid cyst of the liver: where is the evidence? World J Surg, (2004). , 731-736.

[91] Gourgiotis, S, et al. Surgical techniques and treatment for hepatic hydatid cysts. Surg Today, (2007). , 389-395.

[92] Khuroo, M. S, et al. Percutaneous drainage compared with surgery for hepatic hydatid cysts. N Engl J Med, (1997). , 881-887.

[93] Smego, R. A, et al. Percutaneous aspiration-injection-reaspiration drainage plus albendazole or mebendazole for hepatic cystic echinococcosis: a meta-analysis. Clin Infect Dis, (2003). , 1073-1083.

[94] Smego, R. A, & Jr, P. Sebanego, Treatment options for hepatic cystic echinococcosis. Int J Infect Dis, (2005). , 69-76.

[95] Yagci, G, et al. Results of surgical, laparoscopic, and percutaneous treatment for hydatid disease of the liver: 10 years experience with 355 patients. World J Surg, (2005)., 1670-1679.

[96] Sayek, I, Tirnaksiz, M. B, & Dogar, R. Cystic hydatid disease: current trends in diagnosis and management. Surg Today, (2004). , 987-996.

[97] Seimenis, A. Overview of the epidemiological situation on echinococcosis in the Mediterranean region. Acta Trop, (2003). , 191-195.

[98] Menezes da SilvaA.M., Human echinococcosis: a neglected disease. Gastroenterol Res Pract, 2010. (2010). p. pii: 583297.

[99] Buttenschoen, K. and D. Carli Buttenschoen, Echinococcus granulosus infection: the challenge of surgical treatment. Langenbecks Arch Surg, (2003). , 218-230.

[100] Khuroo, M. S, et al. Percutaneous drainage versus albendazole therapy in hepatic hydatidosis: a prospective, randomized study. Gastroenterology, (1993). , 1452-1459.

[101] Dervenis, C, et al. Changing concepts in the management of liver hydatid disease. J Gastrointest Surg, (2005). , 869-877.

[102] Guidelines for treatment of cystic and alveolar echinococcosis in humansWHO Informal Working Group on Echinococcosis. Bull World Health Organ, (1996). , 231-242.

[103] Mueller, L, et al. A retrospective study comparing the different surgical procedures for the treatment of hydatid disease of the liver. Dig Surg, (2003). , 279-284.

[104] Stoot, J. H, et al. More than 25 years of surgical treatment of hydatid cysts in a nonendemic area using the "frozen seal" method. World J Surg, (2010). , 106-113.

[105] Saidi, F, & Nazarian, I. Surgical treatment of hydatid cysts by freezing of cyst wall and instillation of 0.5 per cent silver nitrate solution. N Engl J Med, (1971). , 1346-1350.

[106] Schindl, M. J, et al. The value of residual liver volume as a predictor of hepatic dysfunction and infection after major liver resection. Gut, (2005). , 289-296.

[107] Shoup, M, et al. Volumetric analysis predicts hepatic dysfunction in patients undergoing major liver resection. J Gastrointest Surg, (2003). , 325-330.

[108] Shah, S. A, et al. Surgical resection of hepatic and pulmonary metastases from colorectal carcinoma. J Am Coll Surg, (2006). , 468-475.

[109] Fusai, G, & Davidson, B. R. Management of colorectal liver metastases. Colorectal Dis, (2003). , 2-23.

[110] Scheele, J, et al. Resection of colorectal liver metastases. What prognostic factors determine patient selection?]. Chirurg, (2001). , 547-560.

[111] Karlo, C, et al. CT- and MRI-based volumetry of resected liver specimen: comparison to intraoperative volume and weight measurements and calculation of conversion factors. Eur J Radiol, (2010). , e107-e111.

[112] Dello, S. A, et al. Liver volumetry plug and play: do it yourself with ImageJ. World J Surg, (2007). , 2215-2221.

[113] Van Der Vorst, J. R, et al. Virtual liver resection and volumetric analysis of the future liver remnant using open source image processing software. World J Surg, (2010). , 2426-2433.

[114] Dello, S. A, et al. Prospective volumetric assessment of the liver on a personal computer by nonradiologists prior to partial hepatectomy. World J Surg, (2010). , 386-392.

[115] Dubois, F, Berthelot, G, & Levard, H. Laparoscopic cholecystectomy: historic perspective and personal experience. Surg Laparosc Endosc, (1991). , 52-57.

[116] Dagher, I, et al. Laparoscopic liver resection: results for 70 patients. Surg Endosc, (2007). , 619-624.

[117] Descottes, B, et al. Laparoscopic liver resection of benign liver tumors. Surg Endosc, (2003). , 23-30.

[118] Farges, O, et al. Prospective assessment of the safety and benefit of laparoscopic liver resections. J Hepatobiliary Pancreat Surg, (2002). , 242-248.

[119] Morino, M, et al. Laparoscopic vs open hepatic resection: a comparative study. Surg Endosc, (2003). , 1914-1918.

[120] Simillis, C, et al. Laparoscopic versus open hepatic resections for benign and malignant neoplasms--a meta-analysis. Surgery, (2007). , 203-211.

[121] Kaneko, H. Laparoscopic hepatectomy: indications and outcomes. J Hepatobiliary Pancreat Surg, (2005). , 438-443.

[122] Kelling, G. Ueber Oesophagoskopie, Gastroskopie und Kölioskopie. Münch Med Wochenschr, (1902). , 21-24.

[123] Jacobeus, H. Ueber die Möglichkeit die Zystoskopie bei Untersuchung seröser Höhlungen anzuwenden. Münch Med Wochenschr, (1910). , 2090-2092.

[124] Jacobeus, H. Kurze Uebersichtüber meine Erfahrungen mit der Laparo-thoraskopie. Münch Med Wochenschr, (1911). , 2017-2019.

[125] Cuesta, M. A, et al. Limited laparoscopic liver resection of benign tumors guided by laparoscopic ultrasonography: report of two cases. Surg Laparosc Endosc, (1995). , 396-401.

[126] Azagra, J. S, et al. Laparoscopic anatomical (hepatic) left lateral segmentectomy-technical aspects. Surg Endosc, (1996). , 758-761.

[127] Cherqui, D, et al. Laparoscopic liver resections: a feasibility study in 30 patients. Ann Surg, (2000). , 753-762.

[128] Cherqui, D. Laparoscopic liver resection. Br J Surg, (2003). , 644-646.

[129] Dagher, I, et al. Laparoscopic hepatectomy for hepatocellular carcinoma: a European experience. J Am Coll Surg, (2010). , 16-23.

[130] Gigot, J. F, et al. Laparoscopic liver resection for malignant liver tumors: preliminary results of a multicenter European study. Ann Surg, (2002). , 90-97.

[131] Buell, J. F, et al. An initial experience and evolution of laparoscopic hepatic resectional surgery. Surgery, (2004). , 804-811.

[132] Chang, S, et al. Laparoscopy as a routine approach for left lateral sectionectomy. Br J Surg, (2007). , 58-63.

[133] Buell, J. F, et al. The international position on laparoscopic liver surgery: The Louisville Statement, 2008. Ann Surg, (2009). , 825-830.

[134] Nguyen, K. T, et al. Minimally invasive liver resection for metastatic colorectal cancer: a multi-institutional, international report of safety, feasibility, and early outcomes. Ann Surg, (2009). , 842-848.

[135] Mirnezami, R, et al. Short- and long-term outcomes after laparoscopic and open hepatic resection: systematic review and meta-analysis. HPB (Oxford), (2011). , 295-308.

[136] Van Dam, R. M, et al. Initial experience with a multimodal enhanced recovery programme in patients undergoing liver resection. Br J Surg, (2008). , 969-975.

[137] Wind, J, et al. Systematic review of enhanced recovery programmes in colonic surgery. Br J Surg, (2006). , 800-809.

[138] Fearon, K. C, et al. Enhanced recovery after surgery: a consensus review of clinical care for patients undergoing colonic resection. Clin Nutr, (2005). , 466-477.

[139] Kehlet, H, & Wilmore, D. W. Multimodal strategies to improve surgical outcome. Am J Surg, (2002). , 630-641.

[140] Wilmore, D. W, & Kehlet, H. Management of patients in fast track surgery. Bmj, (2001). , 473-476.

[141] Basse, L, Madsen, J. L, & Kehlet, H. Normal gastrointestinal transit after colonic resection using epidural analgesia, enforced oral nutrition and laxative. Br J Surg, (2001). , 1498-1500.

[142] Basse, L, et al. Accelerated postoperative recovery programme after colonic resection improves physical performance, pulmonary function and body composition. Br J Surg, (2002). , 446-453.

[143] Delaney, C. P, et al. Prospective, randomized, controlled trial between a pathway of controlled rehabilitation with early ambulation and diet and traditional postoperative care after laparotomy and intestinal resection. Dis Colon Rectum, (2003). , 851-859.

[144] Zutshi, M, et al. Randomized controlled trial comparing the controlled rehabilitation with early ambulation and diet pathway versus the controlled rehabilitation with early ambulation and diet with preemptive epidural anesthesia/analgesia after laparotomy and intestinal resection. Am J Surg, (2005). , 268-272.

[145] Podore, P. C, & Throop, E. B. Infrarenal aortic surgery with a 3-day hospital stay: A report on success with a clinical pathway. J Vasc Surg, (1999). , 787-792.

[146] Trondsen, E, et al. Day-case laparoscopic fundoplication for gastro-oesophageal reflux disease. Br J Surg, (2000). , 1708-1711.

[147] Basse, L, et al. A clinical pathway to accelerate recovery after colonic resection. Ann Surg, (2000). , 51-57.

[148] Schoetz, D. J, et al. Ideal" length of stay after colectomy: whose ideal? Dis Colon Rectum, (1997). , 806-810.

[149] Kehlet, H. Multimodal approach to control postoperative pathophysiology and rehabilitation. Br J Anaesth, (1997). , 606-617.

[150] Kehlet, H, & Dahl, J. B. Anaesthesia, surgery, and challenges in postoperative recovery. Lancet, (2003). , 1921-1928.

[151] Nygren, J, et al. A comparison in five European Centres of case mix, clinical management and outcomes following either conventional or fast-track perioperative care in colorectal surgery. Clin Nutr, (2005). , 455-461.

[152] Spelt, L, et al. Fast-track programmes for hepatopancreatic resections: where do we stand? HPB (Oxford), (2011). , 833-838.

[153] Stoot, J. H, et al. The effect of a multimodal fast-track programme on outcomes in laparoscopic liver surgery: a multicentre pilot study. HPB (Oxford), (2009). , 140-144.

[154] Adam, R. Chemotherapy and surgery: new perspectives on the treatment of unresectable liver metastases. Ann Oncol, (2003). Suppl 2: , ii13-ii16.

[155] Abdalla, E. K, et al. Improving resectability of hepatic colorectal metastases: expert consensus statement. Ann Surg Oncol, (2006). , 1271-1280.

[156] Adam, R, et al. Rescue surgery for unresectable colorectal liver metastases downstaged by chemotherapy: a model to predict long-term survival. Ann Surg, (2004). discussion 657-8., 644-657.

[157] Yeo, S. G, et al. Whole-liver radiotherapy for end-stage colorectal cancer patients with massive liver metastases and advanced hepatic dysfunction. Radiat Oncol, (2010). , 97.

[158] Krishnan, S, et al. Conformal radiotherapy of the dominant liver metastasis: a viable strategy for treatment of unresectable chemotherapy refractory colorectal cancer liver metastases. Am J Clin Oncol, (2006). , 562-567.

[159] Schefter, T. E, & Kavanagh, B. D. Radiation therapy for liver metastases. Semin Radiat Oncol, (2011). , 264-270.

[160] Andolino, D. L, et al. Stereotactic body radiotherapy for primary hepatocellular carcinoma. Int J Radiat Oncol Biol Phys, (2011). , e447-e453.

[161] Minn, A. Y, Koong, A. C, & Chang, D. T. Stereotactic body radiation therapy for gastrointestinal malignancies. Front Radiat Ther Oncol, (2011). , 412-427.

[162] Chang, D. T, et al. Stereotactic body radiotherapy for colorectal liver metastases: a pooled analysis. Cancer, (2011) . 4060-4069.

[163] Saxena, A, et al. Factors predicting response and survival after yttrium-90 radioembolization of unresectable neuroendocrine tumor liver metastases: a critical appraisal of 48 cases. Ann Surg, (2010). , 910-916.

[164] Evans, K. A, et al. Survival outcomes of a salvage patient population after radioembolization of hepatic metastases with yttrium-90 microspheres. J Vasc Interv Radiol, (2010). , 1521-1526.

[165] Jakobs, T. F, et al. Hepatic yttrium-90 radioembolization of chemotherapy-refractory colorectal cancer liver metastases. J Vasc Interv Radiol, (2008). , 1187-1195.

[166] Smits, M. L, et al. Holmium-166 radioembolization for the treatment of patients with liver metastases: design of the phase I HEPAR trial. J Exp Clin Cancer Res, (2010). , 70.

[167] Mayo, S. C, & Pawlik, T. M. Thermal ablative therapies for secondary hepatic malignancies. Cancer J, (2010). , 111-117.

[168] Jiao, D, et al. Microwave ablation treatment of liver cancer with 2,450-MHz cooled-shaft antenna: an experimental and clinical study. J Cancer Res Clin Oncol, (2010). , 1507-1516.

[169] Bhardwaj, N, et al. Microwave ablation for unresectable hepatic tumours: clinical results using a novel microwave probe and generator. Eur J Surg Oncol, (2009). , 264-268.

[170] Martin, R. C, Scoggins, C. R, & Mcmasters, K. M. Safety and efficacy of microwave ablation of hepatic tumors: a prospective review of a 5-year experience. Ann Surg Oncol, (2009). , 171-178.

[171] Kobayashi, S, et al. A single-incision laparoscopic hepatectomy for hepatocellular carcinoma: initial experience in a Japanese patient. Minim Invasive Ther Allied Technol, (2010). , 367-371.

[172] Gaujoux, S, et al. Single-incision laparoscopic liver resection. Surg Endosc, (2010). , 1489-1494.

[173] Patel, A. G, et al. Video. Single-incision laparoscopic left lateral segmentectomy of colorectal liver metastasis. Surg Endosc, (2010). , 649-650.

[174] Giulianotti, P. C, et al. Robotic liver surgery: results for 70 resections. Surgery, (2010). , 29-39.

[175] Jain, G, et al. Stretching the limits of laparoscopic surgery": two-stage laparoscopic liver resection. J Laparoendosc Adv Surg Tech A, (2010). , 51-54.

[176] Machado, M. A, et al. Two-stage laparoscopic liver resection for bilateral colorectal liver metastasis. Surg Endosc, (2010). , 2044-2047.

[177] Alkari, B, Owera, A, & Ammori, B. J. Laparoscopic liver resection: preliminary results from a UK centre. Surg Endosc, (2008). , 2201-2207.

[178] Starzl, T. E, & Fung, J. J. Themes of liver transplantation. Hepatology, (2010). , 1869-1884.

[179] Calne, R. Y. Early days of liver transplantation. Am J Transplant, (2008). , 1775-1778.

[180] Starlz, T. E, et al. Reconstructive problems in canine liver homotransplantation with special reference to the postoperative role of hepatic venous flow. Surg Gynecol Obstet, (1960). , 733-743.

[181] Moore, F. D, et al. Experimental whole-organ transplantation of the liver and of the spleen. Ann Surg, (1960). , 374-387.

[182] Starzl, T. E, et al. HOMOTRANSPLANTATION OF THE LIVER IN HUMANS. Surg Gynecol Obstet, (1963). , 659-676.

[183] Starzl, T. E, Marchioro, T. L, & Waddell, W. R. THE REVERSAL OF REJECTION IN HUMAN RENAL HOMOGRAFTS WITH SUBSEQUENT DEVELOPMENT OF HOMOGRAFT TOLERANCE. Surg Gynecol Obstet, (1963). , 385-395.

[184] Starzl, T. E, et al. Orthotopic homotransplantation of the human liver. Ann Surg, (1968). , 392-415.

[185] Schroter, G. P, et al. Infections complicating orthotopic liver transplantation: a study emphasizing graft-related septicemia. Arch Surg, (1976). , 1337-1347.

[186] Calne, R. Y, et al. Liver transplantation in man. II. A report of two orthotopic liver transplants in adult recipients. Br Med J, (1968). , 541-546.

[187] GooszenLeerboek chirurgie. Bohn Stafleu van Loghum, (2006). , 425-426.

[188] Levi, D. M, et al. Liver transplantation with preservation of the inferior vena cava: lessons learned through 2,000 cases. J Am Coll Surg, (2012). discussion 698-9., 691-698.

[189] Abbasoglu, O. Liver transplantation: yesterday, today and tomorrow. World J Gastro-enterol, (2008). , 3117-3122.

[190] Groth, C. G, et al. Historic landmarks in clinical transplantation: conclusions from the consensus conference at the University of California, Los Angeles. World J Surg, (2000). , 834-843.

[191] Calne, R. Y, et al. Cyclosporin A initially as the only immunosuppressant in 34 recipients of cadaveric organs: 32 kidneys, 2 pancreases, and 2 livers. Lancet, (1979). , 1033-1036.

[192] Starzl, T. E, et al. The use of cyclosporin A and prednisone in cadaver kidney transplantation. Surg Gynecol Obstet, (1980). , 17-26.

[193] Starzl, T. E, et al. Liver transplantation with use of cyclosporin a and prednisone. N Engl J Med, (1981). , 266-269.

[194] Starzl, T. E, et al. FK 506 for liver, kidney, and pancreas transplantation. Lancet, (1989). , 1000-1004.

[195] Todo, S, et al. Liver, kidney, and thoracic organ transplantation under FK 506. Ann Surg, (1990). discussion 306-7., 295-305.

[196] Grady, O, et al. Tacrolimus versus microemulsified ciclosporin in liver transplantation: the TMC randomised controlled trial. Lancet, (2002). , 1119-1125.

[197] Haddad, E. M, et al. Cyclosporin versus tacrolimus for liver transplanted patients. Cochrane Database Syst Rev, (2006). , CD005161.

[198] Bismuth, H, & Houssin, D. Reduced-sized orthotopic liver graft in hepatic transplantation in children. Surgery, (1984). , 367-370.

[199] Pichlmayr, R, et al. Transplantation of a donor liver to 2 recipients (splitting transplantation)--a new method in the further development of segmental liver transplantation]. Langenbecks Arch Chir, (1988). , 127-130.

[200] Ng, K. K, & Lo, C. M. Liver transplantation in Asia: past, present and future. Ann Acad Med Singapore, (2009). , 322-310.

[201] Chen, C. L, & De Villa, V. H. Split liver transplantation. Asian J Surg, (2002). , 285-290.

[202] Adam, R, et al. Evolution of liver transplantation in Europe: report of the European Liver Transplant Registry. Liver Transpl, (2003). , 1231-1243.

[203] Raia, S, Nery, J. R, & Mies, S. Liver transplantation from live donors. Lancet, (1989). , 497.

[204] Strong, R. W, et al. Successful liver transplantation from a living donor to her son. N Engl J Med, (1990). , 1505-1507.

[205] Sugawara, Y, & Makuuchi, M. Living donor liver transplantation: present status and recent advances. Br Med Bull, (2005). , 15-28.

[206] Broering, D. C, et al. Is there still a need for living-related liver transplantation in children? Ann Surg, (2001). discussion 721-2., 713-721.

[207] Lo, C. M, et al. Adult-to-adult living donor liver transplantation using extended right lobe grafts. Ann Surg, (1997). discussion 269-70., 261-269.

[208] Shimada, M, et al. Living-donor liver transplantation: present status and future perspective. J Med Invest, (2005). , 22-32.

[209] Brown, R. S, et al. A survey of liver transplantation from living adult donors in the United States. N Engl J Med, (2003). , 818-825.

[210] Malago, M, Burdelski, M, & Broelsch, C. E. Present and future challenges in living related liver transplantation. Transplant Proc, (1999). , 1777-1781.

[211] Dutkowski, P, et al. Current and future trends in liver transplantation in Europe. Gastroenterology, (2010). e1-4., 802-809.

[212] Busuttil, R. W, & Tanaka, K. The utility of marginal donors in liver transplantation. Liver Transpl, (2003). , 651-663.

[213] White, S. A, & Prasad, K. R. Liver transplantation from non-heart beating donors. BMJ, (2006). , 376-377.

[214] Ikeda, T, et al. Ischemic injury in liver transplantation: difference in injury sites between warm and cold ischemia in rats. Hepatology, (1992). , 454-461.

[215] Abt, P, et al. Liver transplantation from controlled non-heart-beating donors: an increased incidence of biliary complications. Transplantation, (2003). , 1659-1663.

[216] Abt, P. L, et al. Survival following liver transplantation from non-heart-beating donors. Ann Surg, (2004). , 87-92.

Chapter 2

Anesthetic Considerations for Patients with Liver Disease

Aparna Dalal and John D. Jr. Lang

Additional information is available at the end of the chapter

1. Introduction

The liver is the largest gland in the body. The average human liver weighs approximately 1.5-1.7 kg, and holds a blood volume of approximately 500 ml. It receives approximately 25% of the cardiac output, of which 75% is supplied by the portal vein and the other 25% by the hepatic artery. Its venous drainage is to the inferior vena cava via the hepatic veins. The hepatic ductal system produces the bile which is then stored in the gall bladder.

The liver synthesizes most proteins, with the exception of gamma globulins and factor VIII. It is also responsible for protein degradation, glucose homeostasis, fatty acid β-oxidation, bilirubin production and excretion. Hepatocytes are embryologically less differentiated; hence the liver is the only organ capable of regeneration after surgical resection or trauma.

Hepatic blood flow is predominantly dependent upon systemic blood flow and pressure-based on pressure flow regulation and hepatic arterial buffer response. There is also central nervous system control of the hepatic blood flow via the thoracic sympathetic fibers. Sympathetic stimulation may cause the blood volume which is present in the liver to be expelled into the circulation, thus providing additional circulatory volume if needed.

Hepatic blood flow is reduced by all anesthetic agents and techniques via reductions in hepatic blood flow and hepatic oxygen uptake. The volatile agents, desflurane and sevoflurane have the least significant effect on total hepatic blood flow. Other perioperative causes of a reduction of hepatic blood flow include mechanical ventilation, hypercarbia, positive end-expiratory pressure, hypotension, hemorrhage, hypoxemia and surgery. A significant decrease in hepatic blood flow can result in parenchymal centrilobular necrosis when extreme resulting in further worsening of perioperative liver dysfunction.

In liver disease, anesthetic drug distribution, metabolism and elimination may be altered. Uptake and onset of anesthetic drug action is usually unaffected. Hepatic clearance of an agent is dependent upon volume of distribution, functional hepatic blood flow, hepatic extraction ratio and hepatic microsomal activity. As a result, opioids may accumulate and the pharmacological actions of drugs such as benzodiazepines maybe prolonged. In extreme situations, actions of non-depolarizing muscle relaxants such as vecuronium and rocuronium maybe also be prolonged.

The liver plays a critical role in coagulation as it is the principal site of synthesis for the majority of clotting factors: II, V, VII, IX, X, XI, and XII. All coagulation factors except for VIII, which is mainly produced by the endothelium, are markedly reduced in patients with liver disease. Patients with chronic liver disease may also develop thrombocytopenia secondary to splenomegaly caused by prolonged portal hypertension. Additionally, reduced levels of thrombopoietin, which regulates platelet production in the liver, may also further contribute to platelet counts in more advanced disease. Also, antithrombin-III (AT-III) levels fall due to reduced synthesis and/or increased consumption due to fibrinolysis. All of the proteins involved in fibrinolysis except for tissue plasminogen activator (tPA) and plasminogen activator inhibitor (PAI-1) are synthesized in the liver. However, tPA levels can be increased due to decreased clearance by the liver predisposing patients to further risks of intra- and perioperative hemorrhage. Hemostatic changes associated with surgical bleeding are thrombocytopenia, platelet function defects, inhibition of platelet aggregation and adhesion by nitric oxide and prostacyclin, decreased levels of coagulation factors: II, V, VII, IX, X, XI, quantitative and qualitative abnormalities of fibrinogen, low levels of α2-antiplasmin, Factor XIII and thrombin activatable fibrinolysis inhibitor, and elevated tPA. Hemostatic changes associated with thrombosis are elevated vWF, decreased levels of ADAMTS-13 (a vWF cleaving protease), and decreased levels of anti-coagulants: ATIII, Protein C and S, α2 macroglobulin, elevated levels of heparin cofactor II, elevated VIII, decreased levels of plasminogen, normal or increased PAI-1. Hypercoagulability can occur in patients with liver disease, especially those with cholestatic disease.

In the setting of acute liver failure (ALF), the coagulopathy encountered can be much more severe. Plasma concentrations of coagulation factors with the shortest half-life fall first; factors V and VII (12 hrs and 4-6hrs respectively) and factors II,VII and X subsequently. In a review of over 1000 patients with ALF by the US Acute Liver Failure Study Group, the mean international normalization ratio (INR) in ALF was 3.8 +/- 4.0 (range 1.5 - >10) with most having a moderately prolonged INR (1.5 to 5) and only 19% with an INR >5. Moreover, thrombocytopenia is common with 40% of patients having platelet counts < 90,000 on admission. [1]

2. Pathophysiology of End Stage Liver Disease

Liver disease can be acute or chronic. Common causes of chronic liver disease are viral hepatitis (B & C), autoimmune hepatitis, non-alcoholic steatohepatitis (NASH), Laennec's cirrhosis, cryptogenic cirrhosis, and metabolic diseases such as hemachromatosis and Wilson's disease. Cholestatic causes of liver disease include primary biliary cirrhosis and primary sclerosing cholangitis.

Predominant pathophysiological manifestation of liver disease is portal hypertension. There is increased resistance to portal blood flow due to hepatic parenchymal scarring and fibrosis, and splanchnic hyperemic resulting in hypersplenism, thrombocytopenia and the progression formation of varices. Normal portal pressures are usually in the range of 5-12 mmHg. Portal hypertension is generally defined when any 2 of the following 3 criteria are met: splenomegaly, ascites or bleeding esophageal varices. Portal pressures at this time are usually > 20 mmHg.

The combination of decreased production of albumin and portal hypertension results in the accumulation of ascites. It also occurs due to renal retention of sodium and water, and localization of this excess fluid in the peritoneal cavity. Tense ascites may decrease functional residual capacity (FRC), adversely affect pulmonary gas exchange and increase risk of aspiration. Hydrothorax or pleural effusions may produce atelectasis. Secondary hyperaldosteronism may manifest as hypokalemic metabolic alkalosis. Additionally, there is intra- and extra-pulmonary shunting, elevated mixed venous oxygen saturation (SvO2), altered lactate metabolism. The hyperdynamic circulation is a result of decreased systemic vascular resistance (SVR) and compensatory increased cardiac output to maintain tissue perfusion. Inadequate synthesis of coagulation factors produces coagulopathy. There is delayed gastric emptying creating putting the patient at-risk for aspiration. Increased ammonia levels (hyperammonemia) can result in hepatic encephalopathy.

3. Other clinically relevant associations with patients with liver disease includes

Portopulmonary hypertension (POPH) is a pulmonary hypertension syndrome with vascular obstruction and increased resistance to pulmonary arterial flow due to varying degrees of pulmonary endothelial/smooth muscle proliferation, vasoconstriction and in-situ thrombosis. The development of POPH has not been demonstrated to correlate with the severity of liver disease.

Hepatopulmonary syndrome (HPS) is characterized by arterial hypoxemia caused by intra-pulmonary vascular dilatations. The clinical triad of 1) portal hypertension; 2) hypoxemia; and 3) pulmonary vascular dilatations characterizes the clinical presentation of HPS [2].

Hepatorenal syndrome is a form of pre-renal acute kidney injury that occurs in decompensated cirrhosis. The syndrome is classified into two types: Type 1 is characterized by a doubling of the serum creatinine level to greater than 2.5 mg/dl in less than 2 weeks while Type 2 is characterized by a stable or slower progressive course of renal failure [3].

Hepatic encephalopathy occurs due to accumulation of circulating neurotoxins such as unmetabolized ammonia, gamma aminobutyric acid, gut-derived false neurotransmitters leading to altered neurotransmission by glutamate or altered cerebral energy homeostatsis. [4] Clinically, it is manifested by neuropsychiatric abnormalities and generalized clonus on clinical examination.

4. Assessing perioperative risk

Patient operative risk is dictated by severity of liver disease, co-existing medical diseases and type of surgery (i.e., upper abdominal, emergent, cardiac etc.) It may also be dependent on s on the anesthetic conducted and ability to maintain of hepatic blood flow.

An important measure for assessing mortality risk is the Child-Pugh Classification. Though this was first used to stratify risk for surgical correction of portal hypertension, it is also found to be predictive of survival in cirrhosis. The score is assigned based upon bilirubin, albumin, prothrombin time (PT), ascites and encephalopathy. One point is given for each of the following: albumin > 3.5 g/dl, INR < 1.7, bilirubin <2mg/dl, no ascites, no encephalopathy. 2 points are given for each of the following: Albumin 1.8- 3.5 g/dl, INR between1.7-2.3, bilirubin 2-3 mg/dl, slight to moderate ascites, grade 1-2 encephalopathy. 3 points are given for each of the following: albumin < 1.8 g/dl, INR >2.3, bilirubin > 3 mg/dl, tense ascites, grade 3-4 encephalopathy. Class A = 5-6 points, Class B = 7-9 points, Class C = 10-15 points. [5] Child Pugh A, B, C predicts a perioperative mortality risk of 10, 30 and 80 % respectively. [6]

Other measures for predicting mortality include ascites, increased serum creatinine, preoperative GI bleed, high ASA physical status score and previous abdominal surgery. Steatosis and steatohepatitis may also be considered as risk factors for postoperative complications, especially after abdominal procedures. The Model of End Liver Disease (MELD) score predicts severity based upon serum creatinine, total bilirubin, and PT INR. It is used to estimate long term survival, as well as list patients for liver transplantation with the United network of Organ Sharing (UNOS). (need a reference here)

Elective surgery is contraindicated when the patient has acute viral hepatitis, alcoholic hepatitis, fulminant hepatic failure, severe chronic hepatitis, is a Child Pugh C patient or has other manifestations of end stage liver disease.

Patients with advanced liver disease should be effectively managed so that hepatic perfusion and hepatic oxygen delivery are maximized l and sequelae of their liver disease such as hepatic encephalopathy, cerebral edema, coagulopathy, hepatopulmonary syndrome, portopulmonary hypertension and portal hypertension has been identified and treated accordingly if possible.

5. Preoperative evaluation of patients with liver disease

Assessment of hepatic function includes evaluating risks for aggravating underlying liver disease, extra-hepatic complications, alterations of hepatic synthetic function and altered drug disposition.

Liver function tests do not measure hepatic function. They represent release of damaged or dead hepatocyte intracellular contents into the systemic circulation, hence provide a snapshot at that point in time only. Actual liver function is represented by albumin, prothrombin time and pseudocholinesterase concentrations. Obtaining liver function tests in healthy patients is not recommended as abnormal liver function tests (LFTs) exist in about 1 in 700 patients, and a vast majority of these patients do not have advanced liver disease. Thus, patients with asymptomatic elevations in serum transaminase levels (less than two times normal values) may undergo anesthesia and surgery with good outcomes.

Patients with chronic hepatitis should be screened prior to elective surgery even if they are asymptomatic. The INR is the most sensitive indicator of hepatocellular dysfunction. At present, though it is accepted that abnormal hemostasis is a result of liver disease, it is debatable whether the abnormal tests really predict bleeding risk [7]. Moreover, the relationship of coagulation profiles to the risk of bleeding with chronic as well as acute liver disease is uncertain [8]. Low platelet count may not be solely responsible for an increased risk of bleeding as the platelet function is also important. Bleeding time is no longer recommended as a test of platelet function. The current consensus is for a pre-procedure platelet count > 50,000, since it appears that a platelet count above 50,000 is likely to be adequate based on previous studies [9].

It is also important to assess the patient for extra-hepatic pathophysiology related to liver disease. The diagnostic criteria for POPH include a mean pulmonary artery pressure (mPAP) greater than 25 mmHg at rest and a pulmonary vascular resistance (PVR) of > 240 dynes.s.cm^{-5} [10]. A better measure is a transpulmonary gradient > 12 mmHg (mPAP-PAOP) as this reflects the obstruction to flow (PVR) and also distinguishes the contribution of intravascular volume and flow to the mPAP [11].

The European Respiratory Society (ERS)/European Association for Study of the Liver (EASL) Task Force have certain set diagnostic criteria for hepatopulmonary syndrome (HPS). These include diagnosis of liver disease, an A-a oxygen gradient > 15 mmHg, pulmonary vascular dilatation documented by "positive" delayed, contrast-enhanced echocardiography with left heart, detection of microbubbles for > 4 cardiac cycles after right heart opacification of microbubbles and brain uptake > 6% following 99mTc macroaggregated albumin (MAA) lung perfusion scanning. HPS can be diagnosed when there is a cirrhosis with ascites, serum creatinine of >1.5 mg/dL, no improvement of serum creatinine after at least 2 days with diuretic withdrawal and volume expansion with albumin, absence of shock, no current or recent treatment with nephrotoxic drugs and absence of parenchymal kidney disease as indicated by proteinuria > 500 mg/day, microhematuria, and/or abnormal renal ultrasonography. [12]

6. Cardiac assessment of End Stage Liver Disease (ESLD) patients

Cirrhotic patients with ESLD may suffer from cirrhotic cardiomyopathy. This is comprised of increased cardiac output and compromised ventricular response to stress. This entity is likely mediated by decreased beta-agonist transduction, increased circulating inflammatory mediators resulting in cardiac depression, and accompanying repolarization abnormalities [13-18]. Low systemic vascular resistance and bradycardia are also commonly seen in ESLD. Patients with ESLD may also demonstrate diastolic dysfunction. [19]. The electrophysiologic abnormalities found in cirrhotic cardiomyopathy include QT-interval prolongation, electrical and mechanical dyssynchrony and chronotropic incompetence [20-22]. Carvedilol administered to patients with ESLD has been demonstrated to reduce portal pressures by decreasing net splanchnic blood flow. [23].

Additionally, ESLD are also at risk for the development of coronary artery disease (CAD), however the liver itself has not been implicated. Approximately 25 % of these patients have at least one moderate or severe coronary artery with critical stenosis. Obstructive CAD was most common among patients with 2 traditional cardiac risk factors such as smoking, diabetes mellitus (DM),and/or hyperlipidemia [24]. Left ventricular hypertrophy and hyperdynamic systolic function in ESLD may result in hemodynamically significant left ventricular outflow tract obstruction (LVOTO). One retrospective review of 106 transplant recipients found inducible LVOTO on pre-operative dobutamine stress echocardiography (DSE) in 40% of patients [25]. In this study, an outflow gradient of 36 mm Hg was significantly associated with intraoperative hypotension. Many ESLD patients also have prolonged corrected QT interval (QTc) on an electrocardiogram which can be associated with an increased risk of ventricular arrhythmias. Though it is not a contraindication to surgery and anesthesia, one should look for electrolyte disturbances or the use of QT interval-prolonging drugs. All patients with ESLD should undergo a preoperative echocardiography to assess ventricular function, ventricular size, valvular function, pulmonary artery pressure, and to exclude the presence of a significant LVOTO or pericardial effusion. Pre-operative echocardiography is useful to calculate pulmonary artery systolic pressure. Pulmonary artery systolic pressures (PASP) values of 45-50 mmHg and /or right ventricular dysfunction are usually used for screening POPH. Right heart catheterization should be performed to gauge the mean pulmonary artery pressure (PAP), pulmonary capillary wedge pressure (PCWP) and transpulmonary gradient (TPG) as 5% to 10% of ESLD candidates have POPH [26],. A preoperative mPAP of 35 to 50 mm Hg has been associated with a 50% risk of mortality after liver transplantation in patients with POPH [26], and mortality approached 100% among patients with POPH and mPAP ≥50 mm Hg [27]. Thus, POPH warrants perioperative treatment with vasodilators such as epoprosterenol, sildenafil or nitric oxide. Stress testing of ESLD patients can be done to detect CAD. Dobutamine stress echocardiography has been found to have a negative predictive value in ESLD patients to be 85%.[28,29]. The predictive value of nuclear single-photon emission computed tomography (SPECT) stress imaging is limited by the chronic vasodilatory state exhibited by patients with ESLD [30]. The specificity of abnormal SPECT findings for obstructive CAD by coronary angiography is only 61% [31]. Coronary angiography is the gold standard for detecting CAD. When possible, it is important make an assessment of CAD risk in the ESLD patient before revascularization becomes contraindicated (usually an excessive bleeding risk due to coagulopathy and/or thrombocytopenia). Transesophageal echocardiography (TEE) and/or pulmonary artery catheterization may be used intraoperatively to allow for real-time hemodynamic monitoring and volume management..

7. Anesthetic agents

All volatile anesthetics decrease the mean arterial pressure and portal blood flow. Halothane has consistently the most dramatic effect in reducing hepatic arterial blood flow. [32,33]. On the other hand, sevoflurane, desflurane and isoflurane have been consistently shown to better preserve hepatic blood flow and function. Intravenous anesthetics have a modest impact on hepatic blood flow, and no meaningful adverse impact on postoperative liver function if the mean arterial pressure is adequately maintained throughout the time anesthetized. Induction agents such as etomidate and thiopental decrease hepatic blood flow, either from increased hepatic arterial vascular resistance or from reduced cardiac output and/or blood pressure. [34]. Ketamine has little impact on hepatic blood flow. [35] Propofol increases total hepatic blood flow in both hepatic arterial and portal venous circulation, suggesting a significant vasodilator effect. [36,37].

Opioids such as morphine have significantly reduced metabolism in patients with advanced cirrhosis. The elimination half-life of morphine is prolonged, potentially exaggerating sedative and respiratory depressant effects. Fentanyl is highly lipid soluble with a short duration of action, which is also metabolized in the liver. Fentanyl elimination is not appreciably altered in patients with cirrhosis. [38,39]. However, unlike fentanyl, the half-life of alfentanil is almost doubled in patients with cirrhosis. [40]. Remifentanil is a synthetic opioid with an ester linkage that allows for rapid hydrolysis by blood and tissue esterases. It elimination is unaltered in patients with severe liver disease. [41].

Thiopental has a small hepatic extraction ratio. However, its elimination half-life is unchanged in cirrhotics, as it has a large volume of distribution. The clearance of etomidate is unchanged in cirrhotic patients, but its clinical recovery time maybe unpredictable due to increased volumes of distribution [42]. The elimination kinetic profile of propofol is similar in cirrhotic patients as well as normal patients, but the mean clinical recovery times maybe longer after discontinuation of infusions. [43]. The half-life of midazolam is prolonged due to reduced clearance, reduced protein binding, resulting in a prolonged duration of action and an enhanced sedative effect, especially after multiple doses or prolonged infusions. [44] Dexmedetomidine, an α2-adrenergic agonist, with sedative and analgesic properties, is primarily metabolized in the liver. Dose adjustments are therefore indicated when used in patients with significant hepatic dysfunction. [45].

Vecuronium and rocuronium are steroidal muscle relaxants which undergo hepatic metabolism, hence have decreased clearance, prolonged half-lives, and prolonged neuromuscular blockade in patients with cirrhosis. [46 47]. Atracurium and cisatracurium which undergo Hofmann elimination and ester hydrolysis respectively, have clinical duration of actions similar to those in normal patients. [48,49]

8. Intraoperative considerations

For liver surgery where major bleeding is anticipated, it is prudent to secure intravenous access using large bore peripheral catheters as well as central venous access catheters. Rapid sequence induction is recommended in patients with tense ascites to minimize the risk of aspiration. Circulatory collapse should be prevented by concomitant administration of intravenous colloid solutions because intravascular volume re-equilibrium occurs 6 to 8 hrs after removal of larger volumes of ascitic fluid. [50]. Large volumes of colloids and crystalloids maybe given within a few minutes with the assistance of commercially available rapid infusion devices. Red cell salvage should be facilitated with use of Cell savers with/without leukocyte filters. Blood administration may be associated with hyperkalemia and hypocalcemia.

Bleeding during liver surgery could be either surgical, due to previous or acquired coagulation disturbances, or both. The preoperative INR has no predictive value in relation to intraoperative blood loss and the value of fresh frozen plasma (FFP) administration to correct abnormal INR values is debatable and may even increase bleeding due to the volume load [51]. Intraoperative hemostasis panels consisting of INR, fibrinogen and platelet count, and platelet function assays for both platelet count and function, may help to differentiate between the above. A very useful intraoperative test for coagulation is the thromboelastograph (TEG). This test denotes the net effect of pro and anti-coagulants and pro and anti-fibrinolytic factors and the resulting clot tensile strength. It provides information on the rate and strength of clot formation and also clot stability/fibrinolysis. (Table 1)

Parameter	Interpretation	Preferred therapy for abnormal values
R	R is the time of latency from the time that the blood was placed in the TEG® analyzer until the initial fibrin formation.	FFP
α	The α-value measures the rapidity (kinetics) of fibrin build-up and cross-linking and the speed of clot strengthening.	Cryoprecipitate
K	K time is a measure of the rapidity to reach a certain level of clot strength.	FFP
MA	MA, or Maximum Amplitude, is a direct function of the maximum dynamic properties of fibrin and platelet bonding and represents the ultimate strength of the fibrin clot.	Platelet

Table 1. TEG Parameters

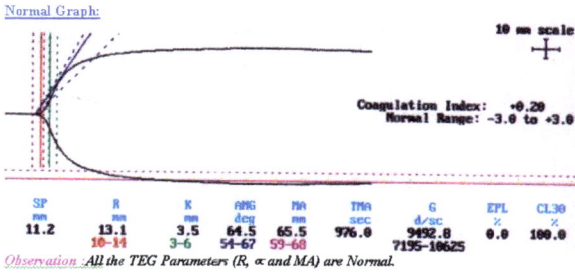

Figure 1. The Normal TEG Graph

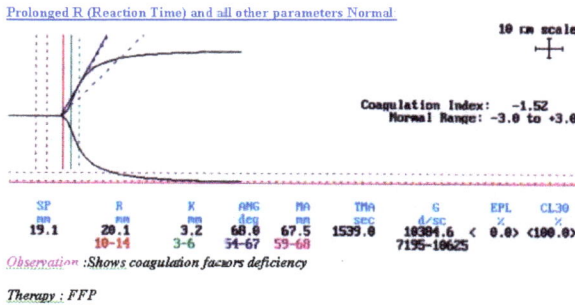

Figure 2. Prolonged Reaction Time

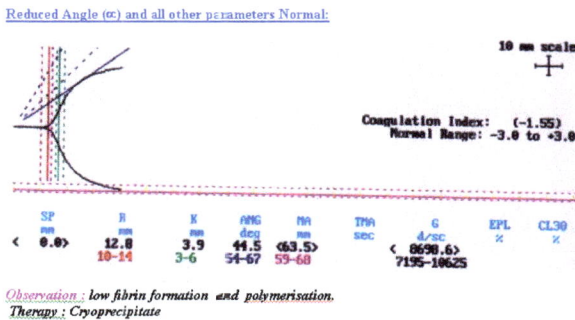

Figure 3. Reduced Angle

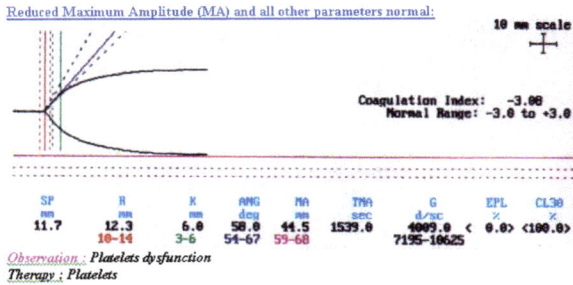

Figure 4. Reduced Maximum Amplitude.

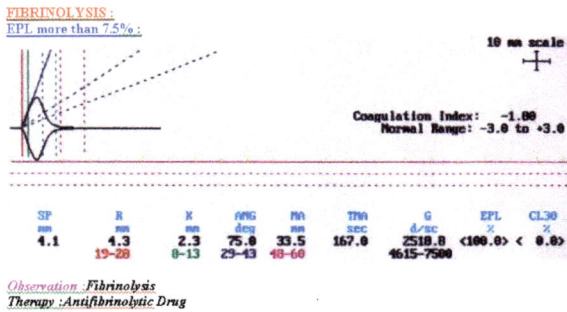

Figure 5. Fibrinolysis

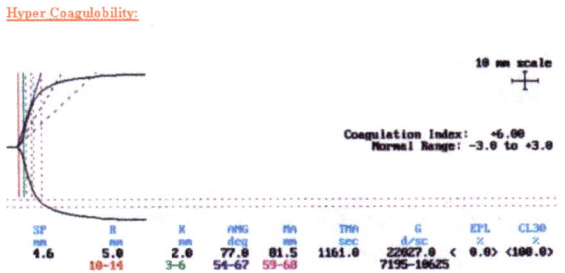

Figure 6. Hypercoagubility

In addition, it is possible to detect heparin-like activity and to measure functional fibrinogen.(Figure 1-5,) Moreover, the only way to currently detect intraoperative hypercoagubility is via TEG. (Figure 6) Thus, TEG may act to facilitate specific goal directed therapy. If fibrinolysis is diagnosed on the TEG and it is causing clinically significant microvascular ooze, small doses of epsilon aminocaproic acid (EACA) or tranexamic acid (TA) are suitable antifibrinolytics. Factor VII has been used to control massive bleeding during liver surgery; however, it has not proved to be consistently effective to control bleeding and is associated with significant side effects. [52]

Transesophageal echocardiography (TEE) is a very useful cardiac monitoring tool to monitor function of the ventricles and assess intraoperative regional wall motion abnormalities (RWMAs), especially in patients with CAD. The monitoring of right heart systolic function is essential in patients with POPH. Moreover, it can be used effectively to assess volume status and guide fluid therapy.

9. Post-operative considerations

Surgery and anesthesia can further worsen hepatic function. Moreover, undiagnosed preexisting liver disease is often the cause of hepatic dysfunction postoperatively. Depending upon the surgical procedure, one may observe continued "third space" losses.. Potential for renal dysfunction or failure as a result of surgery is exacerbated with preexisting liver disease. As well, preoperative or intraoperative coagulopathy can continue postoperatively or can develop during first 24-48 hrs after surgery secondary to worsening hepatic dysfunction.

Postoperative jaundice occurs as a result of overproduction and under excretion of bilirubin, direct hepatocellular injury, or extra-hepatic obstruction. [53] Multiple blood transfusions can increase the levels of unconjugated bilirubin because approximately 10 % of stored whole blood undergoes hemolysis within 24 hours of transfusion. Each 0.5 – 1 unit of blood stored in CPDA-1 yields 7.5 g of hemoglobin, which is then converted to approximately 250 mg of bilirubin. [54] This may overwhelm the liver's ability to conjugate and excrete bilirubin. Immediate postoperative jaundice (< 3wks) can also occur due for multiple reasons including but not exclusive to hemolysis, anesthesia, hypotension, hypovolemia, drugs, infection, sepsis, bleeding, resorption of hematoma, bile duct ligation or injury, hepatic artery ligation, retained common bile duct stone, postoperative pancreatitis, Gilbert's syndrome, Dubin-Johnson Syndrome, inflammatory bowel syndrome, heart failure. [53] Delayed postoperative jaundice (>3 wks) can be a result of drugs, blood transfusion, postintestinal bypass status and total parenteral nutrition. [53]

10. Postoperative pain relief role of epidural analgesia

Thoracic epidural analgesia provides excellent analgesia for liver resections. [55] The catheter is usually inserted at the T6-T9 space. Ropivacaine or bupivacaine are common local anesthetics used with or without the addition of small amounts of opioids such as fentanyl, sufentanil, hydromorphone or morphine. It also reduces the gastrointestinal paralysis compared with systemic opioids. [56]. There is benefit of using combined general and epidural anesthesia in patients with high-risk surgery, but this has not been extensively studied in hepatic surgery. The reasons are probably associated with the concerns with coagulation issues in this group. Additional concerns maybe harbored as neuroaxial blocks themselves are associated with risks. Estimated risk of having serious neurological injury may be as high as 0.08 %.[57, 58]. Moreover, direct spinal cord injury can occur without paraesthesias, whereas pain is more common in lesions affecting nerve roots. [59]. The incidence of persistent neurological deficit has been reported as 0.005-0.07 %. [60,61]. At our institution, we follow a practice where time from anticoagulant drug administration to epidural catheter placement is 3-5 days for warfarin, INR < 1.5, 4 hrs for heparin low dose subcutaneously, 12 hrs for low molecular weight heparin (LMWH), 5 days for clopidogrel and zero for aspirin. The time from epidural catheter removal to anticoagulant drug administration is at least 24 hrs for warfarin, 2 hrs for low dose heparin and 6-8 hrs for LMWH.

It is essential to understand that the degree of underlying parenchymal disease is not the only factor which is responsible for perioperative coagulopathy. Other important factors include amount of blood loss, dilution coagulopathy, amount and quality of residual liver parenchyma, its exposure to ischemia to name a few. [62-64]. Persistent pain or transient coagulopathy may cause delayed epidural catheter removal in patients undergoing partial hepatectomy [65]. The risk of meningitis or epidural abscess is in the range of 0.0004-0.05% [66,67].

11. Liver – specific surgical procedures

Transjugular Intrahepatic Portosystemic Shunt Procedure (TIPS)

TIPS is a procedure used in patients with end stage liver disease to decrease portal pressure and attenuate complications related to portal hypertension. It is usually done in the interventional radiology suite. The goal of this procedure is diversion of portal blood flow into the hepatic vein. The stent is passed through the internal jugular vein over a wire into the hepatic vein, which is located using fluoroscopic guidance. This stent is then advanced through the hepatic parenchyma into the portal vein. This will decompress the portal circulation. Usually, general anesthesia is requested for this procedure, as the radiologists prefer that the patients do not move during this procedure and it may be prolonged. Sedation is usually not preferred as there maybe potential respiratory depression in cirrhotic patients with underlying pulmonary dysfunction or hypoxemia from hepatopulmonary syndrome. Additionally, the presence of ascites may produce risk of aspiration. For this procedure, the central venous pressure (CVP) is monitored. After the stent is placed, the portal pressures are measured. Reduction of the difference between the two reflects the effectiveness of TIPS. Potential complications of this procedure include pneumothorax with internal jugular vein (IJV) cannulation, hematoma formation, inadvertent carotid puncture, cardiac arrhythmia with intracardiac catheter passage, acute life threatening hemorrhage with hepatic artery puncture, hepatic capsular tear, extrahepatic portal venous puncture, development of pulmonary edema and congestive cardiac failure.

12. Radiofrequency Ablation (RFA) of hepatic tumors

Radiofrequency ablation of tumors up to 3 cm in size is currently used to treat non-resectable malignant tumors. During this procedure, a high-frequency, alternating current is delivered through a needle-like probe into the tumor, which induces coagulative necrosis of the tumor and surrounding tissue.[68,69]. RFA is done either percutaneously or laparoscopically. In a study which analyzed nationwide RFAs, it was found that procedure-specific complications were frequent (18.2 %), with transfusion requirements (10.7 %), intraoperative bleeding (4.3 %), and hepatic failure (2.3 %) being the most common. Postoperative complications were also common (12.0 %), with arrhythmias, heart failure, coagulopathy, and open surgical approach acting as significant predictors. [70]

Transarterial Chemoembolization (TACE)

Usually, an adequate amount of emulsion containing oil-based contrast agent Lipiodol and anticancer agents is injected through a catheter then the selected arteries are embolized by embolic agents. Superselective TACE is generally used to minimize damage to non-tumorous areas by using a microcatheter to embolize only the cancerous subsegment.[71-73] Epirubicin and cisplatin are commonly used as anticancer agents, and miriplatin, a new platinum drug, came into use in 2010.[74,75]. Indications for TACE are wide-ranging, and the procedure is generally performed in patients with hypervascular hepatocellular cancer (HCC) who are not indicated for surgery or local therapy for reasons such as multiple bilobar HCC, liver dysfunction, old age or co-morbidity, and in whom the first branch from the main portal vein is not occluded. In practice, this technique is commonly indicated for patients who are Child–Pugh class A or B with multiple tumors with a diameter of 3 cm or more or with four or more HCC. [76,77]. When TACE is combined with RFA, there may be several advantages. For example, TACE decreases the blood flow which in turn reduces the heat loss, thus increasing the size of the RFA ablative zone. In addition, the inclusion of TACE makes the evaluation of ablative margins easier, and enhances the control of satellite lesions.

Hepatic Resections

Liver resections can be done either open or robotic/laparoscopic. Hepatic resection procedures include partial resection, subsegmental resection, segmental resection, two segment resection, extended two-segment resection or three-segment resections. Pre-operative assessment should include the evaluation of the risk assessment using the CTP or MELD score, hepatic parenchymal function, and correction of severe anemia or coagulopathy, management of severe esophageal varices. The choice of anesthetic drugs as well as their doses should be based on the above assessment. There is a risk of significant blood loss. Therefore, it may be prudent to secure large bore intravenous access and be prepared for rapid infusion of colloids and crystalloids. Blood and blood products should be made available for perioperative use. Control of bleeding during resection is usually done with pressure, coagulation and hilar clamping or via the Pringle maneuver. Hilar occlusion produces a minimal increase in systemic arterial pressure, increase in systemic vascular resistance and a minimal decrease in cardiac index. There may be risk of air embolism with extensive resection and disruption of hepatic veins. Most surgeons request a low central venous pressure to facilitate dissection and minimize blood loss from the hepatic vessels and vena cava. Postoperative concerns are similar to those in major abdominal surgery. Central neuroaxial analgesia is not recommended if there is risk of coagulopathy which may result in hematoma formation in the epidural or spinal space.

Donor Liver Hepatectomy

One method of expanding donor pool for liver transplantation is the use of living donor grafts. Adult-to-adult living donor liver transplantation (LDLT) is a complex procedure that poses serious health risks to and provides no direct health benefit for the donor. Because of this uneven risk-benefit ratio, ensuring donor autonomy through informed consent is critical. However, informed consent for LDLT is sub-optimal as donors do not adequately appreciate disclosed information during the informed consent process, despite United Network for Organ Sharing/CMS regulations requiring formal psychological evaluation of donor candidates. [78] Types of donor liver grafts can be left lobe, left lobe and caudate, right lobe, extended right lobe and right lateral sector. After preoperative evaluation and screening, a virtual resection and volume analysis is done using contrast enhanced computed tomography (CT). These not only estimate SLV but can also determine segmental volume, delineate surgical planes, define anatomical landmarks of hepatic vasculature and biliary structures and calculate anticipated graft and remnant liver volumes post resection. It is essential that the minimal donor remnant volume be at least 30% of the original volume. Additionally, when right-lobe LDLT is planned, whether the middle hepatic vein (MHV) should remain in the donor or be resected is controversial. The MHV primarily provides various drainage of the right anterior lobe and segment IV. Most transplant surgeons prefer to leave the MHV in the donor to avoid congestion of segment IV and reduce the risk of liver failure in the donor.[79] The anesthesia management is similar to that of hepatectomy. In donors, several complications have been reported. In one study, right hepatectomy (resection of segments 5–8) was done in 101 donors, left lobectomy (resection of segments 2–3) in 11 donors, and left hepatectomy (resection of segments 2–4) in one donor. Minor anesthetic complications were shoulder pain, pruritus and urinary retention related to epidural morphine, and major morbidity included central venous catheter-induced thrombosis of the brachial and subclavian vein, neuropraxia, foot drop and prolonged postdural puncture headache. One of 113 donors died from pulmonary embolism on the 11th postoperative day. [80]. It was also observed that donor patients experienced significant postoperative pain despite the use of thoracic patient-controlled epidural analgesia (PCEA) infusion catheters as compared to patients who had undergone major hepatic resection. This was attributed to the longer surgical duration for donor hepatectomy and neuroplasticity which may play a role in exaggerated postoperative pain perception along with various psychological factors.[81]. It is also interesting to note that approximately 10% of donors had a platelet count < 150,000 x 10⁹/liter, 2 to 3 years post-donation. [82]

13. Conclusion

Patients with liver disease are at increased risk for both perioperative morbidity and mortality. They require delineation of the degree of liver dysfunction present prior to undergoing surgery and have outcomes that are primarily dictated by the degree of hepatic dysfunction and type of surgery performed. They can certainly pose significant challenges for perioperative care.

Author details

Aparna Dalal* and John D. Jr. Lang

*Address all correspondence to: dalala1@uw.edu

The University of Washington School of Medicine, Department of Anesthesiology & Pain Medicine, NE Pacific, Seattle, WA, USA

References

[1] Munoz S, Reddy R, Lee W et al Coagulopathy in acute liver failure. Neurocrit Care 2008;9:103-7

[2] Rodriguez-Roisin, R, Krowka MJ, Hepatopulmonary syndrome: a liver–induced lung vascular disorder. N Eng J Med 2008; 358: 2378-2387.

[3] Gines P, Schrier RW. Renal failure in cirrhosis. N Engl J Med, 2009; 361: 1279-90.

[4] Riordan SM, Williams R: Treatment of hepatic encephalopathy. N Engl J Med 1997; 337:473-479.

[5] Pugh RHN, Murray-Lyon IM, Dawson JL, et al. Transection of oesophagus for bleeding of oesophageal varices. Br J Surg 60:646-649,1973.

[6] Mansour A, Watson W, Shayani V, Pickelman J: Abdominal operations in patients with cirrhosis: Still a major surgical challenge. Surgery 1997; 22:730-736.

[7] Agarwal B, Shaw S, Hari MS et al. Continuous renal replacement therapy in patients with liver disease. J. Hepatol 2009;51:504-9

[8]   Rockey DC, Caldwell SH, Goodman ZD et al. Liver Biopsy (AASLD Position Paper) Hepatology, 2009;3:1017- 1044.

[9]   Tripodi A, Primignsni M, Chantarangkul V et al. Thrombin generation in patients with cirrhosis: the role of platelets. Hepatology 2006;44:440-445.

[10] Rodriguez-Roisin R, Krowka M, Hervé P, Fallon M. Pulmonary-hepatic vascular disorders (PHD). Eur Respir J 2004; 24:861-880.

[11] Ramsay M. Portopulmonary Hypertension and Right Heart failure in Patients with Cirrhosis. Curr Opin Anaesthesiol 2010;

[12] Salerno F, Gerbes A, Gines P et al. Diagnosis, prevention and treatment of hepatorenal syndrome in cirrhosis. Gut, 2007; 56: 1310-8.

[13] Alqahtani SA, Fouad TR, Lee SS. Cirrhotic cardiomyopathy. Semin Liver Dis 2008; 28:59–69.

[14] Baik SK, Fouad TR, Lee SS. Cirrhotic cardiomyopathy. Orphanet J Rare Dis 2007; 2:15.

[15] Gaskari SA, Honar H, Lee SS. Therapy insight: cirrhotic cardiomyopathy. Nat Clin Pract Gastroenterol Hepatol 2006;3:329 –37.

[16] Liu H, Song D, Lee SS. Cirrhotic cardiomyopathy. Gastroenterol Clin Biol 2002; 26:842–7.

[17] Moller S, Henriksen JH. Cirrhotic cardiomyopathy: a pathophysiological review of circulatory dysfunction in liver disease. Heart 2002;87:9 –15.

[18] Myers RP, Lee SS. Cirrhotic cardiomyopathy and liver transplantation. Liver Transpl 2000;6 Suppl 1:44 –52.

[19] Ward CA, Liu H, Lee SS. Altered cellular calcium regulatory systems in a rat model of cirrhotic cardiomyopathy. Gastroenterology 2001;121:1209–8.

[20] Bernardi M, Calandra S, Colantoni A, et al. Q-T interval prolongation in cirrhosis: prevalence, relationship with severity, and etiology of the disease and possible pathogenetic factors. Hepatology 1998;27: 28–34.

[21] Henriksen JH, Fuglsang S, Bendtsen F, Christensen E, Moller S. Dyssynchronous electrical and mechanical systole in patients with cirrhosis. J Hepatol 2002;36:513–20.

[22] Kelbaek H, Rabol A, Brynjolf I, et al. Haemodynamic response to exercise in patients with alcoholic liver cirrhosis. Clin Physiol 1987;7:35– 41.

[23] Tripathi D, Hayes PC. The role of carvedilol in the management of portal hypertension. Eur J Gastroenterol Hepatol 2010;22:905–11.

[24] Tiukinhoy-Laing SD, Rossi JS, Bayram M, et al. Cardiac hemodynamic and coronary angiographic characteristics of patients being evaluated for liver transplantation. Am J Cardiol 2006;98:178–81

[25] Maraj S, Jacobs LE, Maraj R, et al. Inducible left ventricular outflow tract gradient during dobutamine stress echocardiography: an association with intraoperative hypotension but not a contraindication to liver transplantation. Echocardiography 2004;21:681–5.

[26] Swanson KL, Wiesner RH, Nyberg SL, Rosen CB, Krowka MJ. Survival in portopulmonary hypertension: Mayo Clinic experience categorized by treatment subgroups. Am J Transplant 2008;8: 2445–53.

[27] Martinez-Palli G, Taura P, Balust J, Beltran J, Zavala E, Garcia- Valdecasas JC. Liver transplantation in high-risk patients: hepatopulmonary syndrome and portopulmonary hypertension. Transplant Proc 2005;37:3861– 4.

[28] Williams K, Lewis JF, Davis G, Geiser EA. Dobutamine stress echocardiography in patients undergoing liver transplantation evaluation. Transplantation 2000;69:2354–6.

[29] Donovan CL, Marcovitz PA, Punch JD, et al. Two-dimensional and dobutamine stress echocardiography in the preoperative assessment of patients with end-stage liver disease prior to orthotopic liver transplantation. Transplantation 1996;61:1180–8.

[30] Davidson CJ, Gheorghiade M, Flaherty JD, et al. Predictive value of stress myocardial perfusion imaging in liver transplant candidates. Am J Cardiol 2002;89:359–60.

[31] Aydinalp A, Bal U, Atar I, et al. Value of stress myocardial perfusion scanning in diagnosis of severe coronary artery disease in liver transplantation candidates. Transplant Proc 2009;41:3757– 60.

[32] Gatacel C, Losser MR, Payen D: The postoperative effects of halothane versus isoflurane on hepatic artery and portal vein blood flow in humans. Anesth Analg 2003; 96:740-745.

[33] Grundmann U, Zizzis A, Bauer C, Bauer M: In vivo effects of halothane, enflurane, and isoflurane on hepatic sinusoidal microcirculation. Acta Anaesthiol Scand 1997; 41:760-765.

[34] Thomson IA, Fitch W, Hughes RL, et al: Effects of certain I.V. anaesthetics on liver blood flow and hepatic oxygen consumption in the greyhound. Br J Anaesth 1986; 58:69-80.

[35] Thomson IA, Fitch W, Campbell D, et al: Effects of ketamine on liver blood flow and hepatic oxygen consumption: Studies in the anaesthetized greyhound. Acta Anaesthiol Scand 1988; 32:10-14.

[36] Carmichael FJ, Crawford MW, Khayyam N: Effect of propofol infusion on splanchnic hemodynamics and liver oxygen consumption in the rat. Anesthesiology 1993; 79:1051-1060.

[37] Wouters PF, Van de Velde MA, Marcus MAE, et al: Hemodynamic changes during induction of anesthesia with eltanolone and propofol in dogs. Anesth Analg 1995; 81:125-131.

[38] Tegeder I, Lötsch J, Geisslinger G: Pharmacokinetics of opioids in liver disease. Clin Pharmacokinet 1999; 37:17-40.

[39] Haberer JP, Schoeffler P, Couderc E, et al: Fentanyl pharmacokinetics in anaesthetized patients with cirrhosis. Br J Anaesth 1982; 54:1267-1270.

[40] Ferrier C, Marty J, Bouffard Y, et al: Alfentanil pharmacokinetics in patients with cirrhosis. Anesthesiology 1985; 62:480-484.

[41] Dershwitz M, Hoke JF, Rosow CE, et al: Pharmacokinetics and pharmacodynamics of remifentanil in volunteer subjects with severe liver disease. Anesthesiology 1996; 84:812-820

[42] Van Beem H, Manger FW, Van Boxtel C, et al: Etomidate anaesthesia in patients with cirrhosis of the liver: Pharmacokinetic data. Anaesthesia 1983; 38:61-62

[43] Servin F, Cockshott ID, Farinotti R, et al: Pharmacokinetics of propofol infusions in patients with cirrhosis. Br J Anaesth 1990; 65:177-183.

[44] Trouvin JH, Farinotti R, Haberer JP, et al: Pharmacokinetics of midazolam in anaesthetized cirrhotic patients. Br J Anaesth 1988; 60:762-767.

[45] Baughman VL, Cunningham FE, Layden T: Pharmacokinetic/pharmacodynamic effects of dexmedetomidine in patients with hepatic failure. Anesth Analg 2000; 90(Suppl):S391.

[46] Arden JR, Lynam DP, Castagnoli KP, et al: Vecuronium in alcoholic liver disease: A pharmacokinetic and pharmacodynamic analysis. Anesthesiology 1988; 68:771-776.

[47] Magorian T, Wood P, Caldwell J, et al: The pharmacokinetics and neuromuscular effects of rocuronium bromide in patients with liver disease. Anesth Analg 1995; 80:754-759.

[48] De Wolf AM, Freeman JA, Scott VL, et al: Pharmacokinetics and pharmacodynamics of cisatracurium in patients with end-stage liver disease undergoing liver transplantation. Br J Anaesth 1996; 76:624-628.

[49] Ward S, Neill EA: Pharmacokinetics of atracurium in acute hepatic failure (with acute renal failure). Br J Anaesth 1983; 55:1169-1172.

[50] Menon KVN, Kamath PS: Managing the complications of cirrhosis. Mayo Clin Proc 2000; 75:501-509.

[51] Massicotte L, Capitanio U, Beaulieu D et al. Independent validation of a model predicting the need for RBC transfusion in liver transplantation. Transplantation 2009;88:386-91.

[52] Shami VM, Caldwell SH, Hespenheidee E. Recombinant factor VIIa for coagulopathy in fulminant hepatic failure compared to conventional therapy. Liver Transplant 2003.9:138-143.

[53] Nyberg LM, Pockros PJ: Postoperative jaundice. In Schiff ER, Sorrell MF, Maddrey WC, ed. Schiff's Diseases of the Liver, 8th ed. Philadelphia: Lippincott-Raven; 1999:599-605.

[54] Zuck TF, Basinger TA, Peck CC, et al: The in vivo survival of red blood cells stored in modified CDP with adenine: Report of a multi-institutional cooperative effort. Transfusion 1972; 17:374-382.

[55] Werawatganon T, Charuluxanun S. Patient controlled intravenous opioid analgesia versus continuous epidural analgesia for pain after intra-abdominal surgery. The Cochrane Database of Systematic Reviews 2005, Issue 1. Art. No.: CD004088.pub2. DOI: 10.1002/14651858.CD004088.pub2.

[56] Jørgensen H, Wetterslev J, Mø.niche S, Dahl JB. Epidural local anaesthetics versus opioid-based analgesic regimers for postoperative gastrointestinal paralysis, PONV and pain after abdominal surgery. The Cochrane Database of Systematic Reviews 2001, Issue 1. Art. No.: CD001893. DOI: 10.1002/14651858.CD00001893.

[57] Horlocker TT, Abel MD, Messick JM Jr, Schroeder DR. Small risk of serious neurologic complications related to lumbar epidural catheter placement in anesthetized patients. Anesth Analg 2003;96:1547-52

[53] Rosenquist RW, Birnbach DJ. Editorial Epidural insertion in anesthetized adults: will your patients thank you? Anesth Analg 2003;96:1545-6.

[59] Tsui BC, Armstrong K. Can direct spinal cord injury occur without paresthesia? A report of delayed spinal cord injury after epidural placement in an awake patient. A nesth Analg 2005;101:1212-4.

[60] Wheatley RG, Schug SA, Watson D. Safety and efficacy of postoperative epidural analgesia. Br J Anaesth 2001;87:47-61.

[61] Horlocker TT, Wedel DJ. Neurologic complications os spinal and epidural anesthesia. Reg Anesth Pain Med. 2000;25(1):83-98. Review.

[62] Borromeo CJ, Stix MS, Lally A, Pomfret EA. Epidural catheter and increased prothrombin time after right lobe hepatectomy for living donor transplantation. Anesth Analg. 2000 Nov;91(5):1139-41.

[63] Schumann R, Zabala L, Angelis M, Bonney I, Tighiouart H, Carr DB. Altered hematologic profiles following donor right hepatectomy and implications for perioperative analgesic management. Liver Transpl. 2004 Mar;10(3):363-8.

[64] Siniscalchi A, Begliomini B, De Fietri L, Braglia V, Gazzi M, Masetti M, Di Benedetto F, Pinna AD, Miller CM, Pasetto A. Increased prothrombin time and platelet counts

in living donor right hepatectomy: implications for epidural anesthesia. Liver Transpl. 2004 Sep;10(9):1144-9.

[65] Tsui S L, Young B H, NG KFJ, et al.Delayed epidural catheter removal: the impact of postoperative coagulopathy. Anaesth Intensive Care 2004;32:630-6

[66] Moen V, Dahlgren N, Irestedt L. Severe neurological complications after central neuraxial blockades in Sweden 1990-1999. Anesthesiology 2004;101:950-9.

[67] Wang LP, Hauerberg J, Schmidt JF. Incidence of spinal epidural abscess after epidural analgesia. Anesthesiology 1999;91:1928-36.

[68] Curley SA, Marra P, Beaty K, Ellis LM, Vauthey JN, Abdalla EK, et al. Early and late complications after radiofrequency ablation of malignant liver tumors in 608 patients. Ann Surg. ;239:450–8.

[69] Krishnamurthy VN, Casillas J, Latorre L. Radiofrequency ablation of hepatic lesions: A review. Appl Radiol. 2003:32:11–26.

[70] Justin P. Fox, MD 1, Joshua Gustafson, MD2, Mayur M. Desai, PhD MPH1,3, Minia Hellan, MD4, Thav Thambi-Pillai, MD5, and James Ouellette, DO4. Short-Term Outcomes of Ablation Therapy for Hepatic Tumors: Evidence from the 2006–2009 Nationwide Inpatient Sample Ann Surg Oncol DOI 10.1245/s10434-012-2397-0.

[71] Matsui O, Kadoya M, Yoshikawa J et al. Small hepatocellular carcinoma: treatment with subsegmental transcatheter arterial embolization. Radiology 1993; 188: 79–83.

[72] Matsui O, Kadoya M, Yoshikawa J, Gabata T, Takashima T,Demachi H. Subsegmental transcatheter arterial embolization for small hepatocellular carcinomas: local therapeutic effect and 5-year survival rate. Cancer Chemother Pharmacol 1994; 33 (Suppl): S84–8.35

[73] Takayasu K, Arii S, Kudo M et al. Superselective transarterial chemoembolization for hepatocellular carcinoma. Validation of treatment algorithm proposed by Japanese guidelines. J Hepatol 2012; 56: 886–92.

[74] Okabe K, Beppu T, Haraoka K et al. Safety and short-term therapeutic effects of miriplatin-lipiodol suspension in transarterial chemoembolization (TACE) for hepatocellular carcinoma. Anticancer Res 2011; 31: 2983–8.

[75] Okusaka T, Kasugai H, Ishii H et al. A randomized phase II trial of intra-arterial chemotherapy using SM-11355 (Miriplatin) for hepatocellular carcinoma. Invest New Drugs 2011; doi. 10.1007/s10637-011-9776-4.

[76] Kudo M, Izumi N, Kokudo N et al. Management of hepatocellular carcinoma in Japan: Consensus-Based Clinical Practice Guidelines proposed by the Japan Society of Hepatology (JSH) 2010 updated version. Dig Dis 2011; 29: 339–64.

[77] Clinical Practice Guidelines for hepatocellular carcinoma – The Japan Society of Hepatology 2009 update. Hepatol Res 2010; 40 (Suppl 1): 2–144.

[78] Elisa J. Gordon,1,2,5 Amna Daud,et al. Informed Consent and Decision-Making About Adult-to-Adult Living Donor Liver Transplantation: A Systematic Review of Empirical Research (Transplantation 2011;92: 1285–1296)

[79] Hertl M, Cosimi AB: Living donor liver transplantation: how can we better protect the donors? Transplantation 83:263, 2007

[80] S. Ozkardeslera, D. Ozzeybeka, et al. Anesthesia-Related Complications in Living Liver Donors: The Experience from One Center and the Reporting of One Death American Journal of Transplantation 2008; 8: 2106–2110

[81] Jacek B. Cywinski, MD, Brian M. Parker, MD, Meng Xu, Samuel A. Irefin, MD. A Comparison of Postoperative Pain Control in Patients After Right Lobe Donor Hepatectomy and Major Hepatic Resection for Tumor. Anesth Analg 2004;99:1747–52.

[82] James F. Trotter, et al. Laboratory Test Results After Living Liver Donation in the Adult-to-Adult Living Donor Liver Transplantation Cohort Study LIVER TRANSPLANTATION 17:409-417, 2011

Chapter 3

Critical Care Issues After Major Hepatic Surgery

Ashok Thorat and Wei-Chen Lee

Additional information is available at the end of the chapter

1. Introduction

Major hepatic resections have become the routine aspect of managing certain liver conditions such as primary liver malignancies and certain secondaries. Five-year survival is negligible in un-treated patients compared with around 30% in those receiving hepatic resection [1]. Patients with liver disease who require surgery are at greater risk for surgical and anesthesia related complications than those with a healthy liver [2, 3, 4]. The magnitude of the risk depends upon the type of liver disease and its severity, the surgical procedure, and the type of anesthesia.

The first few days after major hepatic surgery are critical to successful outcome of the procedure. Metabolic and functional changes after hepatic resection are unique and cause significant challenges in management. A multidisciplinary approach is required along with effective communication among all caregivers. With attentive, anticipatory care, many potential problems can be averted and new problems can be detected early and treated appropriately. Contemporary critical care management after major hepatic surgery doesn't differ from standard intensive care which includes invasive hemodynamic monitoring, mechanical ventilation, vital parameter monitoring, strict antisepsis measures, metabolic control with due attention to the glycemic control and nutritional aspect which more or less always affected in the patients with cirrhosis.

The post-operative management after hepatic surgery is greatly influenced by hemodynamic monitoring intraoperatively. Patient's intra-operative course, blood loss, requirement of blood products during surgery largely defines the outcome in post-operative period along with patient's nutritional status, liver functions and associated comorbidities. Hence close co-operation with the anesthesiologist and surgeon is necessary.

Majority of postoperative management issues after liver resection are unique and require a thorough understanding of liver metabolism and the pathophysiology of liver disease. The purpose of this review is to elaborate on specific early postoperative management issues after liver resection, examine current evidence and present the management options.

2. Hepatic resections and general considerations

Through the recent surgical advances, hepatic resection could be carried out under the condition of liver cirrhosis or obstructive jaundice, but there are many complications and associated mortality in these cases. Hepatic cirrhosis limits the ability of the liver to regenerate. Fortunately, it appears that most of the advanced cirrhotic livers can tolerate even major resections, and the presence of cirrhosis should not preclude potentially curative or life-prolonging surgery [5]. Careful patient selection based on preoperative Child-Pugh score and ICG test, resections can be limited leaving behind enough liver parenchyma to avoid postoperative liver dysfunction. But such patients are more vulnerable to perioperative insults secondary to ischemia and hypoperfusion, which is reflected in perioperative morbidity and mortality [6]. The Child-Pugh clinical scoring system has been used as a reliable, validated prognostic tool for patients with chronic liver disease undergoing general or porto-caval shunt surgery and has gained widespread use in hepato-biliary surgery. It has recently been suggested that patients with scores of B or C should not receive liver resection surgery [7].

The associated cirrhosis greatly increases the risk for partial hepatectomy. In normal liver. even up to 70% of resection of liver is well tolerated. With underlying liver cirrhosis, partial hepatectomy is only offered to patients who are Pugh-Child's A and the most favorable class B patients [8]. While in Child C patients even minor hepatic surgery or even locoregional therapy can cause hepatic dysfunction. Post-operative outcome and level of post-operative care largely influenced by the underlying cirrhosis and post cirrhotic complications present at the time of surgery. Hence, even enucleation of hepatocellular carcinoma in Child C patients is a major surgery and procedure related mortality is present in one-third of patients [9].

3. Post-operative care

Variables such as severity of underlying cirrhosis, degree of debility before surgery, associated co-morbid diseases and operative complexity appear to have a significant influence on the rapidity at which patients progress through their early postoperative recovery phase.

Attributed to regenerating capacity of the liver, most of the major liver resections are well tolerated and seldom patients have significant biochemical abnormalities. Patients with compensated liver cirrhosis and its complications are more prone for intraoperative blood loss causing deterioration of organ functions and loss of reserve capacity to withstand even minor stress causing life-threatening complications. The disturbances in cardio-respiratory function should be carefully monitored in high Dependency unit. The complications are more in elderly patients. The condition of older patients can change rapidly and therapy may need to be adjusted every few hours if optimum cardio-respiratory function is to be maintained.

• Planning of intensive monitoring for high risk patients with associated co-morbidities should be done during surgery and in postoperative wards

• Diagnose and treat complications quickly

• Institute invasive monitoring and elective ventilation when required

• Continue postoperative care to increase the rate of recovery.

3.1. Immediate post operative

Initial postoperative assessment begins in operating room. Most patients with pre-operative normal liver functions and child A patients recover without any systemic effects. Such patients may not need intensive care unit and can directly be transferred to inpatient wards after an appropriate period of extremely close observation in recovery unit. Many centers usually monitor the patients in ICU for 24 hours before being transferred to inpatient setting after major liver resections.

Most of the patients are awakened in operating room after surgery, and if extubation criteria are fulfilled, the patient is extubated [10, 11]. It is advised that not all patients are candidates of early extubation and each case should be judged on its own merits. But prolonged intubation and mechanical ventilation in postoperative period associated with more pulmonary complications that further prolongs patient's recovery and increases the mortality & morbidity [12]. In addition, Mandell et al. demonstrated that immediately extubated patients experienced a shorter stay in the ICU, resulting in a significant reduction in ICU services and associated costs for extubated patients [13].

After arrival in ICU, initial vital assessment should be done. Most centers follow more or less same protocol. Fluid management is strictly based on patient's present hemodynamic conditions and blood products are administered as per the present condition requires. Input and output fluid charts are maintained with due attention to hourly urine output which should be minimum 0.5 ml/kg/min. Any renal dysfunction in the form of oliguria should be treated immediately because optimum renal function is of paramount importance as a determinant of good outcome [14].

Routine blood investigations, coagulation profile and organ specific tests are ordered. Patients still on ventilatory support, baseline arterial blood gas estimation is done at the arrival. Serum lactate level is determined as it depicts the imbalance between tissue oxygen supply and consumption, thus an indirect measure of tissue perfusion and cardiac output [15]. Postoperative aminotransferase and alanine aminotransferase and total bilirubin levels are not routinely measured after trauma-related surgery. However, in postoperative liver resection and living donor hepatectomy, these values are to be followed to ensure recovery of liver function [16]. A transient early increase in serum hepatic transaminase and alkaline phosphatase levels as a result of hepatocellular damage is common, but a persisting elevation suggests ongoing hepatic ischemia.

Hypothermia in postoperative period is prevented and core body temperature is maintained above 37°C. Hypothermia can cause vaso-constriction and coagulopathy. Core temperature should be monitored and normothermia maintained using warmed fluids and forced warm air blankets. The abdominal drains are examined for the color and content as postoperative hemorrhage is not uncommon after major liver resections and may require re-exploration. In liver transplant setting, due to underlying coagulopathy, ongoing hemorrhage must be detected at earliest. Gross blood stained drain fluid with acute fall in hemoglobin level should alarm surgeon and patient should be re-explored at earliest.

All patients receive broad spectrum antibiotics. The choice of antibiotics is usually center dependent. In our center, we usually administer single broad spectrum antibiotic, mostly third generation cephalosporin in stable patients with Child A score. But in high risk patients, defined by Child score & nutritional status, and patients who are on ventilator support postoperatively, we prefer to use combination broad spectrum antibiotics. In presence of fever, blood culture and antibiotics sensitivity defines the course of antibiotics administered.

3.2. Monitoring of vital parameters

Monitoring the vital parameters like pulse, blood pressure, respiratory rate, ECG, oxygen saturation and the urine output and immediate intervention are instituted to prevent postoperative complications. Vital organ functioning is monitored as follow:

1. Blood pressure, temperature, pulse, respiratory rate.

2. Electrolytes, glycemic control, liver and renal functions

3. Fluid balance and urine output

4. Drain and wound status and appropriate care

5. Medication for pain relief

6. Neurological and cardiac functions

7. Good nutritional intake and bowel movement

3.2.1. Cardiac Monitoring

Central venous line, arterial blood pressure monitoring, continuous record of pulse rate and heart rate are routine standards for monitoring the patients after major hepatic surgery. Arterial blood pressure monitoring accurately measures blood pressure even in presence of hypotension and hypovolemia. In addition, repeated blood sampling can be obtained for routine laboratory investigations and arterial blood gas monitoring. Patients are usually tachycardic postoperatively. But heart rate >100/min should be thoroughly checked for ongoing insults such as persistent hypovolemia, pain, ongoing hemorrhage (drain fluid & falling HB level are indicators) or cardiac arrhythmias. Sinus tachycardia is common after major surgery and should revert without any complications. If tachycardia increases, persistent infection, hypovolemia, pain or presence of cardiac arrhythmia are detected and treated promptly.

At least two large-bore intravenous cannulas are inserted. Although rapid infusion devices are seldom needed, they are available and primed in the ICU at all times. Pulmonary artery catheterization is reserved for patients with known preoperative left-ventricular dysfunction. This allows continuous measurement of cardiac output and instantaneous calculation of systemic vascular resistance. Real-time ECG monitoring is carried out routinely on most critically ill patients. Changes in rate, rhythm, and character can be identified rapidly by physicians and nurses and acted on immediately.

Monitoring of central venous pressure (CVP) is an important aspect in patients after major liver resections. Measurement of CVP acts as guide for fluid management and hemodynamic manipulation. Liver resections usually carried out under low CVP, usually between 2-5 mm of Hg, to prevent blood loss. This especially an important strategy in patients with underlying liver cirrhosis with child score B & C. CVP is usually kept in same range after surgery and excess fluid administration is restricted. If patient is normotensive and urine output is adequate (>0.5 mL/kg/hr), any attempt to administer extra fluid to elevate CVP is avoided especially in first 48 hours. But after major liver resection, a hyperdynamic state with increased cardiac index and augmented splanchnic blood flow persists for at least 3 days postoperatively [17]. This increased blood supply to the residual liver parenchyma ensures rapid growth.

Signs and symptoms of the heart failure can easily be overlooked as they mimic those of cirrhosis and liver failure. Transthoracic echocardiography is a useful modality in such patients which can measure right ventricular systolic pressure and also shows the cardiac changes.

3.2.2. Pulmonary monitoring

Pulmonary functions are assessed by continuous pulse oximetry, intermittent arterial blood gas analysis, respiratory rate and if patient is on ventilator support, patients are observed via end-tidal carbon dioxide monitoring in addition to the standard ventilatory monitoring and alarm systems.

The course of extubated patients is fairly predictable and most of them recover without any complications. However, after major resections pulmonary complications such as pleural effusion, right sub-diaphragmatic collection causing right lung collapse and pulmonary edema are frequent. edema. These complications range from 50% to over 80% according to literature [18, 19]. Atelectasis is most common amongst these. Atelectasis can be reduced by early mobilization, aggressive chest physiotherapy, adequate pain control and incentive spirometry. Extubated patients should be given chest physiotherapy and incentive spirometry exercises as early as 8 hours post operatively. This will help the expansion of lung and prevent accumulation of the secretions causing atelectasis. Nebulisation with saline with or without anti-cholinergics is given daily 2-3 times and continued till patients are ambulatory.

If the patient is admitted to the ICU while intubated after reversal of the paralytic agents, the ventilatory settings are adjusted according to the patient's respiratory status and arterial blood gases. Patients with good cough and gag reflex, respiratory rate <30 breaths per minute, tidal volume >5ml/kg and aterial PO2 >70mmHg can be extubated. But in presence of pulmonary complications (described later), in very ill, malnourished patients weaning is not possible and may require prolonged ventilation. Metabolic abnormalities such as hypophosphatemia, hypomagnesemia, hypocalcemia and hypokalemia may lead to respiratory muscle dysfunction and inability to wean from ventilator [20]. Such patients in whom prolonged mechanical ventilation is needed for more than 1 week, tracheostomy should be considered to clear airway secretions and reduce the resistance that accompanies the use of standard long endotracheal tubes.

Extubated patients with postoperative hypoxemia are benefited by continuous positive airway pressure (CPAP) that increases the lung expansion and improves fair gas exchange across alveolar capillary membrane Appropriate analgesia is essential to prevent pulmonary complications, but oversedation needs to be carefully avoided. In absence of coagulopathy and other contraindications, epidural analgesia should be considered and it has been shown to reduce the pulmonary complications [21]. Deep vein thrombosis prophylaxis is strongly encouraged after major liver surgery to prevent any thromboembolic complications.

However, in absence of complications in relatively stable postoperative patients, recovery is smooth and extubation is possible within 12 hours.

3.2.3. Renal function monitoring

Maintenance of effective renal function is a critical factor after major hepatic surgery including liver transplantation [22]. 3% of patients experience permanent and 10% transient renal dysfunction following major liver surgery [23]. Hence every attempt must be made to prevent and control renal failure in perioperative period.

Renal autoregulation effectively ceases below renal perfusion pressures of 70 mmHg to 75 mmHg, below which flow becomes pressure dependent. In cirrhotic patients, the concomitant sympathetic activation results in a rightward shift of the autoregulation curve; thus these patients have even less tolerance of reductions in renal perfusion pressure [24]. Adequate fluid management is imperative for both adequate renal perfusion pressure and flow throughout the entire post-operative period to prevent renal impairment.

Hourly monitoring of urine output and laboratory values such as blood urea and serum creatinine are good measures of adequate renal functioning. Urine output is monitored with as indwelling catheter and urine output is maintained at more than 1-2 ml/kg. Any decrease in urinary output should be assessed for the intravascular volume and hypovolemia if any should be corrected.

In presence of normal blood pressure and satisfactory intravascular volume, diuretics are used to improve the urine output. 1 to 2 mg/kg furosemide is given intravenously as bolus followed by a furosemide infusion of 0.2-0.4 mg/kg/hr titrated to maintain adequate urine flow. Continuous infusion results in increased urine output without much alteration in volume status often seen with intermittent bolus therapy.

Intraoperative hemodynamic instability and clamping of major vessels during major liver resections are the main causes of postoperative renal failure. Intraoperative blood loss can lead to renal perfusion problems leading to acute tubular necrosis (ATN) especially in cirrhotic patients with marginal renal functions from the outset. Drug induced nephrotoxicity is another cause of post-operative renal insufficiency.

Renal insufficiency, probably the most ominous perioperative complication in patients with liver disease, is usually a predictor of markedly reduced survival and a sign that hepatorenal syndrome may have developed.

3.2.4. Neurological assessment

Postoperative drowsiness and confusion are commonly caused by neuraxial or systemic opioid administration, which responds to simple changes in administration. However, these patients should be carefully assessed for more serious pathology. Most of the patients show normal neurological recovery. The patients who are extubated immediately after surgery, neurological recovery is complete and not associated with any morbidity. The intubated patients who require mechanical ventilation are usually sedated and neurological assessment in such patients is difficult and usually misleading.

Assessment of the patient's neurological status is done by Glasgow coma scale (GCS) scoring system that records the conscious state of the patient. Patients with GCS score 12 or more are fully conscious and if with endotracheal tube, can be extubated if other pulmonary criteria for extubation are met. Mechanically ventilated patients with sedation and under effect of paralyzing drugs are difficult to assess neurologically and assessment should be performed after wearing of effects of these drugs.

In patients undergoing liver transplantation, the marginal metabolism of anesthetic agents can cause delayed emergence from surgery, as well as residual hepatic encephalopathy [9]. Many patients usually resolve without any neurological aftereffects after major hepatic resections, but prolonged ICU stay due to postoperative complications can result in neurological dysfunction that range from anxiety, depression and sleep deprivation to frank hallucinations and delusional states. ICU psychosis is not uncommon. Patients developing postoperative hepatic dysfunction may develop hepatic encephalopathy which reflects a spectrum of neuropsychiatric abnormalities seen in patients with altered liver functions after exclusion of other known central nervous system disorders [25]. Drugs such as narcotics and sedatives should be avoided in patients with postoperative impairment of liver functions and used cautiously with underlying liver cirrhosis as they may cause prolonged depression of consciousness and precipitate hepatic encephalopathy [4]. Encephalopathy must be considered in a patient with deteriorating liver function and unexplained neurological symptoms. Measurement of blood ammonia may be useful if the diagnosis is unclear. Encephalopathy is treated with cardio-respiratory optimization, further lactulose and may require invasive ventilation.

3.3. Fluid and electrolyte management

Optimizing perioperative fluid management is essential in reducing the risk of postoperative complications and mortality as the cirrhotic patients tend to have limited physiologic reserve. Adequate fluid administration may reduce the stress response to surgical trauma and support recovery [26].

The immediate postoperative period after hepatic resection is characterized by fluid and electrolyte imbalances that are further accentuated by derangements of liver function. Maintenance of adequate fluid balance and normal renal function is critical. Cirrhotics are prone to fluid shifts, vasodilation and resultant hypotension. In this setting, colloids rather than crystalloids should be administered to restore intravascular volume. 50% of patients will also develop significant but self-limiting ascites during the first 48 h, which can cause hypovolemia. Management with sodium restriction and judicious use of diuretic therapy is recommended. Paracentesis may be necessary to prevent tense ascites [27].

At present, no widely accepted recommendations are available for the optimal peri-operative fluid regimen to be used in major non-thoracic surgery. The exact balance of fluid transfusion will be determined by the size of resection, plasma electrolytes and glucose measurements, and volaemic status of the patient. In liver transplantation, fluid overload has been shown to be a predictor of poor graft function and increased postoperative morbidity [28]. In liver resection it has been shown repeatedly that keeping the CVP low results in reduced blood loss and blood transfusion requirements [29-33].

Crystalloids mainly, 0.9% saline and lactated ringer, usually are used postoperatively as replacement and maintenance fluid. Colloids act as plasma expander and can be added as maintenance fluid, but should not be used as resuscitation fluid in case of shock.

Electrolyte abnormalities are common after major hepatic resections, especially beyond Child A patients. Hyponatremia is often seen in patients with cirrhosis and ascites. However, asymptomatic patients treated with normal saline and serum sodium is monitored. Sodium deficit is corrected gradually. In symptomatic patients, a goal increase of sodium with 1.5-2 mEq/L/hr for 3-4 hours until symptoms resolve appears to be safe. But it should not exceed 10 mEq/L in first 24 hours [34]. Rapid correction in any patients is avoided as it may result in central pontine myelinosis.

Hyperlactemia and hypophosphatemia are common derangements in patients undergoing liver resection. Due to the additive effects of lactate-containing intravenous solution, non-lactate containing solutions are recommended for postoperative use [35]. Hypophosphatemia is encountered in nearly all patients after major hepatic resection is believed to be due to increased phosphate uptake by regenerating hepatocytes. It may cause impaired energy metabolism in many organs and may lead to respiratory failure, cardiac arrhythmias, hematologic dysfunction, insulin resistance, and neuromuscular dysfunction [36, 37]. Standard liver resection management includes adequate replacement of phosphate with supplementation of maintenance fluids with potassium phosphate and oral/parenteral replacement.

Correction of potassium is an ongoing process after major liver resections. Patients with high urine output may have hypokalemia which should be corrected. In most cases supplementation is administered by the intravenous route, but it can also be given orally via nasogastric tube. Patients who have received multiple transfusions tend to have hyperkalemia. Before potassium correction underlying metabolic acidosis must be treated first. Severe hyperkalemia in patients with renal dysfunction or failure requires urgent treatment with pharmacological agents or early dialysis. In presence electrocardiographic changes, intravenous calcium to stabilize the cardiac membrane, intravenous insulin and glucose can be given to decrease serum potassium levels. However associated hypomagnesemia should be corrected as it is commonly seen in association with hypokalemia and hypercalcemia.

3.4. Glycemic control

Strict control of blood glucose in surgical patients admitted to intensive care unit has been shown to reduce morbidity and mortality [38]. Hyperglycemia may be induced by surgical stress causing dysregulation of liver metabolism and immune function, resulting in adverse postoperative outcomes [39]. Insulin therapy is particularly important and blood glucose levels are monitored serially to keep glucose levels in target range of 90-120 mg/dl. But development of insulin resistance after the liver resection makes adequate blood glucose control challenging. Some centers use insulin-sliding scale to keep blood glucose in target range, in which blood glucose levels are monitored at regular intervals and doses of insulin changed accordingly while some centers use continuous insulin infusion to control glucose levels. The doses of insulin are to be modified depending on the blood sugar levels.

Okabayashi et al. [40] examined the safety and effectiveness of closed loop insulin administration system, a type of artificial pancreas (STG-22, Nikkiso, Tokyo, Japan) in patients undergoing hepatic resection, but the mean sugar level was above the target levels 90-120 mg/dl. Hypoglycemia after insulin therapy is not uncommon. Hypoglycaemia may as well occur in postoperative due to result of impaired hepatic mobilization of glucose is in high-risk patients or large resections and may necessitate glucose infusion. Dextrose solutions are used to restore normal sugar level. If patients can take orally, or no contraindications for enteral feeding, oral or nasogastric feeding is always preferred.

3.5. Nutrition

Malnutrition is common in patients with liver disease and it may increase risk of postoperative complications after major liver surgeries [41].

The post-hepatic resection period the high demand of the regenerating liver is characterized by a catabolic state and often has glucose and electrolyte imbalances. Nutritional support during this critical period is of paramount importance to ensure adequate hepatic regeneration and postoperative-recovery. Non-cirrhotic patients with adequate preoperative nutritional status may not require any special intervention and should be started on early oral/enteral diet.

But patients who have poor nutritional intake, with or without compromised liver functions (cirrhosis or steatosis), after major liver resections the short-term outcome in such patients may be improved with the use of supplemental enteral nutrition. This may as well improve the child class of patients and reduce the mortality in patients with cirrhosis and malnutrition. If oral feeding can be tolerated, enteral feeding is always preferred over parenteral as it also maintains the intestinal integrity.

Richter, et al. [42] evaluated five randomized controlled studies that compared enteral versus parenteral nutrition in the post-hepatic resection patients [43-45] and concluded that the postoperative complications were significantly low in patients with enteral feeding. In addition, supplementation of branched chain amino acids has got immunomodulating role. Liver disease alters the metabolism of amino acids resulting in low levels of branched chain amino acids such as leucine, isoleucine and valine. Branched chain amino acids (BCAA) supplementation in patients with advanced cirrhosis is associated with improved nutritional status and decreased frequency of complications of cirrhosis. Okabayashi et al. showed improved quality of life in patients supplemented with BCAA after they underwent major hepatic resections [46]. Ishikawa et al. demonstrated increased levels of erythropoietin after short term supplementation with BCAA in non-hepatitis patients undergoing curative resection [47]. Erythropoietin has got protective effects on liver cells from ischemic injury.

Thus, adequate perioperative nutritional support and institution of early enteral nutrition are crucial. Protein restriction is advised only in presence of neurological complications like encephalopathy.

3.6. Correction of coagulopathy

Derangements in conventional markers of coagulation such as prothrombin time/ international ratio (PT/INR), partial thromboplastin time (PTT) and platelet count are common post hepatectomy and correlates with the extent of resection. Postoperative coagulopathy peaks 2-5 days post surgery. Decreased synthetic functions of the liver remnant and consumption of coagulation factors postoperatively can cause increase in INR postoperatively between 1 to 5 days with corresponding decrease in platelets and fibrinogen [48, 49].

Prolongation of PT/INR is often self-limited and usually resolves without the need for transfusion of fresh frozen plasma (FFP) in non-cirrhotics. In patients with cirrhosis, decreased hepatic protein synthesis contributes to a prolonged prothrombin time and partial thromboplastin time, both of which are prolonged usually in direct proportion to the impairment of hepatic reserve. Administration of fresh frozen plasma provides all necessary clotting factors and can correct underlying coagulopathy.

Patients having preoperative obstructive jaundice should receive vitamin K injection both before and after surgery. Sometimes determination of the precise cause of coagulopathy may be difficult in some patients with advanced liver disease, both vitamin K and fresh frozen plasma given together in such patients. In case of postoperative drop in hemoglobin and hematocrit, fresh whole blood transfusion is ideal replacement. A platelet count of 50,000/µl is acceptable. Administration of platelets in the absence of bleeding often results in platelet antibodies, even if type-specific platelets are used. Thrombocytopenia should be treated with platelet transfusion only if platelet count is less than 10,000/ µl or between 10,000-30,000/ µl in presence of active bleeding.

Currently, there is no consensus regarding the criteria for prophylactic FFP transfusion after hepatic resection. Cirrhotics are at increased risk of bleeding after resection. A combination of FFP transfusions, vitamin K, octreotide and human r-FVIIa may be utilized to correct coagulopathy and prevent bleeding.

4. Pain management

Postoperative pain following liver surgery is significant, and adequate analgesia remains a challenge for the caregivers. It helps in early mobilization, improves respiratory functions, permits smooth extubation and decreases systemic blood pressure [50]. Opioids are mainstay of postoperative pain management, morphine and fentanyl being most commonly used analgesics. However, opioids can certainly cause sedation, respiratory depression and exacerbation of hepatic encephalopathy. Due to decreased metabolism of opioids in cirrhotic patients, the bioavailability of these drugs is increased. Size of liver resection has been correlated with impaired opioid metabolism, larger volume resections result in greater impairment of opioid metabolism [51]. Hence, patients should be closely monitored for any signs of respiratory depression. In presence of renal dysfunction, fentanyl is better choice as it is less affected by renal impairment [52].

Epidural analgesia has emerged as an important pain management option in major surgeries and with adjunct to intravenous analgesics provides better pain control & less sedation. But many patients presenting for hepatic surgery have a coagulopathy or thrombocytopenia that makes them ineligible for an epidural or intrathecal therapy. The prolonged prothrombin time potentially predisposes these patients to spinal hematoma and cord compression. In our institute we use epidural analgesia only in patients with normal coagulation profile and good hepatic functions. Intrathecal morphine in doses of 0.5 mg to 0.7 mg can be used as an alternative in patients without coagulopathy. This significantly reduces systemic morphine requirements postoperatively.

Patient controlled analgesia (PCA) is newly emerged concept of self administration of analgesics in controlled doses by patient himself with a pump. This is preferred mode of administrating opioids for moderate to severe pain. Randomized controlled trials have shown the effectiveness of PCA over conventional parenteral analgesia in providing better pain control and increased patient satisfaction [53].

The use of NSAIDs is not recommended post hepatectomy in cirrhotic patients and in renal insufficiency due to risk of hemorrhage and hepatorenal syndrome. However, intravenous acetaminophen can be used in doses not exceeding more than 2 g/day in patients with liver impairment [54].

5. Postoperative complications

Approximately 20% of otherwise healthy patients may experience postoperative complications after elective liver resections [6]. Postoperative complications included surgical complications (bleeding from the surgical site and bile leak), hepatic dysfunction, cardiovascular, respiratory, and renal system dysfunction, and infection. Preoperative American Society of Anesthesiologists (ASA) classification [55], presence of steatosis, extent of resection, simultaneous extrahepatic resection, and perioperative blood transfusion [56] have been found to be independent predictors for the development of postoperative complications.

5.1. Infections

Infection after hepatic resection is a major contributor of postoperative morbidity and mortality and might be predictive of long-term outcomes [57]. Obesity, preoperative biliary drainage, extent of hepatic resection, operative blood loss, comorbid conditions and postoperative bile leak are the risk factors predictive of postoperative infectious complications [58, 59]. Standard measures to reduce the incidence of postoperative infectious complications such as early mobilization, strict antiseptic measures during patient care, changing or removing the urinary catheters within 10 days, removal of central venous catheters earliest possible and aggressive chest physiotherapy should be routine in the postoperative period.

Most frequent complications are pulmonary infection and intra-abdominal infections with abscess formation. Both of these complications are well responsive to the antibiotics. Intra-abdominal collections either biloma or frank abscesses should be drained under radiologic guidance. Septic shock is rare and associated mortality is high if develops. Early recognition of postoperative infection, prompt institution of broad-spectrum antibiotics and aggressive source control is of utmost importance.

Early enteral feeding has protective role in maintaining gut mucosal barrier function. Disruption of this barrier results in translocation of intestinal organisms that is the source of postoperative infections especially in malnourished patients. Strategies such as early enteral nutrition are aimed to protect the gut-barrier function and reduce infectious complication.

5.2. Post operative hemorrhage

Less frequent complications include post-operative hemorrhage that is associated with increased mortality. Underlying coagulopathy is the main reason. Patients with cirrhosis, steatosis, and after chemotherapy are at especially increased risk of coagulopathy and bleeding. Postoperative coagulopathy is at its peak 2-5 days post surgery may act as another contributory factor. Immediate re-exploration and hemostasis is the treatment. This may necessitate the blood transfusion.

5.3. Pulmonary complications

Pulmonary complications are not uncommon after major hepatic resections. Pulmonary complications are a major cause of morbidity and mortality during the postoperative period [60]. Common pulmonary complications occurring in the postoperative period include pulmonary atelectasis, pleural effusion, pulmonary edema and pneumonia.

5.3.1. Atelectasis

Atelectasis is one of the most common postoperative pulmonary complications, particularly following abdominal and thoraco-abdominal procedures [(61). Postoperative atelectasis is usually caused by decreased compliance of lung tissue, impaired regional ventilation, retained airway secretions, and/or postoperative pain that interferes with spontaneous deep breathing and coughing [62]. After major hepatic resections right sub-diaphragmatic collections and postoperative pain are the major causes. Continuous positive airway pressure (CPAP) is beneficial to patients who develop hypoxemia and/or increased respiratory effort due to postoperative atelectasis in the setting of few secretions. Patients with abundant respiratory secretions receive frequent chest physiotherapy such as postural drainage & percussion and oral suctioning. Flexible bronchoscopy should be performed for the patients who are unresponsive to chest physiotherapy and oral suctioning.

Any accumulation in right sub-diaphragmatic space should be drained under radiologic guidance. Atelectasis can be reduced by early mobilization, incentive spirometry, aggressive chest physiotherapy and adequate postoperative analgesia.

5.3.2. Pneumonia

Pneumonia is uncommon complication but may prove life threatening. It usually tends to occur within first five postoperative days [63]. It presents with fever, leukocytosis, increased secretions, and pulmonary infiltrates on chest radiographs. Patients develop hypoxemia and eventually respiratory distress. Postoperative pneumonia should be suspected in presence of fever, leukocytosis and development of new pulmonary infiltrates on chest radiographs. Empiric antibiotic treatment must be started and tailored as per the microbiological analysis of sputum samples.

5.3.3. Pleural effusion

Pleural effusion occurs mostly on right side and related to surgical manipulation or hepatic hydrothorax. Minimal pleural effusion is common during the immediate postoperative period and disappears within few days. However, larger collections and persistent pleural effusion affecting respiratory functions must be drained.

Subphrenic abscess is a complication of surgery that may induce pleural effusions; however, the effusions associated with a subphrenic abscess are distinct from the usual postoperative pleural effusion in that they usually become apparent about 10 days after surgery and are typically associated with signs and symptoms of systemic infection [64]. Subphrenic abscess must be drained and appropriate antibiotic treatment should be started.

5.3.4. Pulmonary edema

Extravascular lung-water accumulation, indicating mild to moderate pulmonary edema following liver resection, has been reported; however, this does not appear to affect oxygenation significantly in the postoperative period [65]. Early onset may be related to transfusion-related acute lung injury or overzealous fluid administration. It is due to increased permeability across alveolar capillary membrane [66]. Other causes include sepsis and acute respiratory distress syndrome. Treatment of pulmonary edema includes fluid restriction, diuretics and continuous positive airway pressure. Most cases resolve spontaneously in a relatively short period of time with no long-term sequelae [67]

5.3.5. Hepatic dysfunction

Postoperative hepatic failure remains a significant challenge. Liver dysfunction is common after liver surgery and anesthesia. It can range from mild enzyme elevations to fulminant hepatic failure. The abnormalities of liver functions noted postoperatively are mostly due to surgery itself or anaesthetic agents used. Although increased serum bilirubin is common postoperatively especially in cirrhotic patients (upto 20%), jaundice is infrequent (<1%) and its presence should prompt a thorough evaluation of the cause.

Although low residual liver volume was found to be associated with postoperative liver failure, the regenerative ability of the liver is remarkable, and the residual, otherwise healthy liver is expected to double in size within the first week following the resection. Increase in hepatic parenchymal mass does not necessarily result in full restoration of functional ability. Pre-existing cirrhosis or positive virus carrier status limits liver regeneration, and these patients are more susceptible to developing postoperative hepatic failure. Liver regenerating is also reduced in diabetic patients predisposing them for the liver failure after major resections [68].

But liver dysfunction can also occur in absence of any pre-existing liver disease. The hepatocellular dysfunction may occur due to drugs including anaesthetic agents, ischemia, shock, iatrogenic injury or viral hepatitis. Known causes of cholestatic dysfunction include sepsis, prolonged blood transfusions, drugs, biliary tract injury, choledocholithiasis and total parenteral nutrition [4]. Even if abnormalities are not noted on computed tomography or ultrasonography, choalngiographic studies are warranted in presence of strong suspicion of biliary obstruction.

Most cases of benign postoperative jaundice (without any obvious cause) eventually resolve spontaneously with supportive treatment only. Usually all cases of hepatic dysfunction are managed in ICU and liver functions are monitored serially along with the coagulation parameters. Hepatic failure is a life threatening complications. Presence of hepatic encephalopathy increases mortality. Increased ammonia due to underlying hepatic failure is a key element in the pathogenesis of encephalopathy. Coagulation parameters are often deranged with underlying liver failure and should be corrected with blood transfusion and fresh frozen plasma transfusion. If patient doesn't respond to the supportive medical management, liver transplantation must be considered. However, hepatic failure is rare complication after major resection and presence of underlying liver dysfunction should prompt specialized management of underlying cause to prevent progression of liver failure.

5.3.6. Other complications

In-hospital mortality following liver resection has been associated with perioperative myocardial infarction, sepsis with multiple organ failure and pulmonary embolism. After major abdominal surgeries, the risk of deep venous thrombosis and pulmonary embolism is 15-40% that increases the mortality, morbidity and length of hospital stay significantly [69].

Early mobilization, intermittent pneumatic compression devices and pharmacologic agents have important role in prevention of venous thromboembolism (VTE). While pharmacologic thromboprophylaxis is widely accepted for most general surgery procedures, the fear of bleeding after major hepatectomy has limited its use. But venous thromboembolism can still occur even in presence of deranged coagulation parameters (prolonged INR & aPTT [70]. A higher incidence of VTE has been noted in patients not receiving thromboprophylaxis and should be administered starting the day of surgery unless high risk of bleeding exists.

6. Summary

The expansion of major liver surgery as a treatment option for various liver tumours has presented new challenges to surgeons and physicians in terms of the assessment and management of postoperative complications, particularly those involving hepatic insuffi- ciency and susceptibility to infection. Understanding of hepatic pathophysiology is impor- tant for optimal perioperative care. Multiple factors contribute to increased mortality in patients with underlying liver disease. But due to advances in surgery, anesthesia and im- proved critical care management, there is progressive improvement in survival even in complex situations. Patient selection with evaluation of the risk factors in various liver conditions is needed. Reduction in mortality in patients with liver disease undergoing re- section depends on close attention to coagulation, intravascular volume, renal function, electrolyte levels, cardiovascular status and nutrition. patient selection, appropriate moni- toring, and multidisciplinary postoperative management are the key elements in im- proved survival among patients undergoing liver resections.

Author details

Ashok Thorat and Wei-Chen Lee*

*Address all correspondence to: weichen@cgmh.org.tw

Division of Liver and Transplantation Surgery, Department of General Surgery, Chang- Gung Memorial Hospital at Linkou, Chang-Gung University College of Medicine, Taiwan

References

[1] Simmonds, P. C., Primrose, N. J., Colquitt, J. L., Garden, O. J., Poston, G. J., & Rees, M. (2006). Surgical resection of hepatic metastases from colorectal cancer: a systemat- ic review of published studies. *Br J Cancer*, 94, 982-99.

[2] O'Leary, J. G., Yachimski, P. S., & Friedman, L. S. (2009). Surgery in the patient with liver disease. *Clin Liver Dis*, 13(2), 211-231.

[3] Friedman, L. S. (1999). The risk of surgery in patients with liver disease. *Hepatology*.

[4] Patel, T. (1999). Surgery in the patient with liver disease. *Mayo Clin Proc*, 593-9.

[5] Redai, I., Emond, J., & Brentjers, T. (2004). Anesthetic considerations during liver surgery. *Surg Clin N Am*, 84, 401-411.

[6] Belghiti, J. , Hiramatsu, K., Benoist, S., Massault, P. P., Sauvanet, A., & Farges, O. (2000). Seven hundred forty-seven hepatectomies in the 1990s: an update to evaluate the actual risk of liver resection. *J Am Coll Surg*, 191(1), 38-46.

[7] Clavien, P. A., Petrowsky, H., De Olveira, M. L., & Graf, R. (2007). Strategies for safer liver surgery and partial liver transplantation. *N Engl J Med*, 356, 1545-59.

[8] Franco, D., & Borgonovo, G. (1994). Liver resection in cirrhosis of the liver. In: Blumgart LH (ed). *Surgery of the Liver and Biliary Tract, 1st ed. Edinburgh: Churchill Livingstone*, 1539-1555.

[9] Bismuth, H., Chiche, L., Adam, R., et al. (1993). Liver resection versus transplantation for hepatocellular carcinoma in cirrhotic patients. *Ann Surg*, 218(2), 145-151.

[10] Mandell, S., Lockrem, J., & Kelley, S. (1997). Immediate tracheal extubation after liver transplantation: Experience of two transplant centers. *Anesth Analg*, 84, 249-253.

[11] Wong, D., Cheng, D., Kustra, R., et al. (1999). Risk factors of delayed extubation, prolonged length of stay in intensive care unit, and mortality in patients undergoing coronary artery bypass graft with fast-track cardiac anesthesia: A new cardiac risk score. *Anesthesiology*, 91, 936-950.

[12] Krowka, M. J., & Cortese, D. A. (1985). Pulmonary aspects of chronic liver disease and liver transplantation. *Mayo Clin Proc*, 60, 407-418.

[13] Neelakanta, G., Sopher, M., Chan, S., Pregler, J., Steadman, R., Braunfeld, M., et al. (1997). Early tracheal extubation after liver transplantation. *J Cardiothorac Vasc Anesth*, 11, 165-167.

[14] Bilbao, I., Armadans, L., Lazaro, J. L , Hidalgo, E., Castells, L., & Margarit, C. (2003). Predictive factors for early mortality following liver transplantation. *Clin Transplant*, 17, 401-11.

[15] Basaran, M., Sever, K., Ugurlucan, M., et al. (2006). Serum Lactate Level Has Prognostic Significance After Pediatric Cardiac Surgery. 20(1), 43-47.

[16] Imamura, H., Kokudo, N., Sugawara, Y., Sano, K., Kaneko, J., & Takayama, T. (2004). Pringle's manoeuvre and selective inflow occlusion in living donor liver hepatectomy. *Liver Transpl*, 10(6), 771-8.

[17] Thasler, W. E., Bein, T., & Jauch. K. H. (2002). Perioperative effects of hepatic resection surgery on hemodynamics, pulmonary fluid balance, and indocyanine green clearance. *Langenbecks Arch Surg*, 387(2), 271-5.

[18] Pirate, A., Ozgur, S., Torgay, A., Arslan, G., et al. (2004). Risk factors for postoperative respiratory complications in adult liver transplant recipients. *Transplant Proc, 36,* 218-220.

[19] Duran, F. G., Piqueras, B., Romero, M., Clemente, G., et al. (1998). Pulmonary complications following orthotopic liver transplant. *Transplant Int,* 11(1), 255-259.

[20] Aubeir, M., Murciano, D., Lecocguic, Y., et al. (1985). Effect of hypophosphatemia on diaphragmatic contractibility in patients with acute respiratory failure. *N Engl J Med,* 313-420.

[21] Lawrence, V. A., Cornell, J. E., & Smetana, G. W. (2006). Strategies to reduce postoperative pulmonary complications after noncardiothoracic surgery: systematic review for the American College of Physicians. *Ann Int Med,* 144, 596-608.

[22] Nair, S., Verma, S., & Thuluvath, P. J. (2002). Pretransplant renal function predicts survival in patients undergoing orthotopic liver transplantation. *Hepatology, 35,* 1179-1185.

[23] Melendez, J. A., Arslan, V., Fisher, M. E., Wuest, D., Jarnagin, W. R., Fong, Y., et al. (1998). Perioperative outcomes of major hepatic resections under low central venous pressure anesthesia: blood loss, blood transfusion, and the risk of postoperative renal dysfunction. *J Am Coll Surg,* 187(6), 620-5.

[24] Dagher, L., & Moore, K. (2001). The hepatorenal syndrome. *Gut,* 49(5), 729-737.

[25] Ferenci, P., Lockwood, A. , Mullen, K. , et al. (2002, 1998). Hepatic encephalopathy-definition, nomenclature, diagnosis and qualification: final report of the working party at the 11th World Congresses of Gastroenterology. Vienna. *Hepatology, 35,* 716-721.

[26] Hamilton, M. A. (2009). Perioperative fluid management: Progress despite lingering controversies. *Cleveland clinic journal of Medicine,* 28-31.

[27] Wrighton, L. J., O'Bosky, K. R., Namm, J. P., & Senthil, M. (2012). Postoperative management after hepatic resection. *J Gastrointest Oncol,* 3, 41-47.

[28] Bennett-Guerrero, E., Feierman, D. E., Winfree, W. J., et al. (2001). Preoperative and intraoperative predictors of postoperative morbidity, poor graft function, and early rejection in 190 patients undergoing liver transplantation. *Arch Surg,* 136, 1177-83.

[29] Furrer, K., Deoliveira, M. L., Graf, R., & Clavien, P. A. (2007). Improving outcome in patients undergoing liver surgery. *Liver Int,* 27, 26-39.

[30] Vassilios, S., Georgia, K., Kassiani, T., Dimitrios, T., & Contis, J. C. (2004). The role of central venous pressure and type of vascular control in blood loss during major liver resections. *Am J Surg,* 187, 398-402.

[31] Jones, R., Moulton, C. E., & Hardy, K. J. (1998). Central venous pressure and its effect on blood loss during liver resection. *Br J Surg,* 85, 1058-60.

[32] Melendez, J. A., Arslan, V., Blumgart, L. H., et al. (1998). Perioperative outcomes of major hepatic resections under low central venous pressure anesthesia: blood loss, blood transfusion, and the risk of postoperative renal dysfunction. *Am Coll Surg*, 187, 620-5.

[33] Wang, W. D., Liang, L. J., Huang, X. Q., & Yin, X. Y. (2006). Low central venous pressure reduces blood loss in hepatectomy. *World J Gastroenterol*, 12, 935-9.

[34] Sterns, R., Cappuccino, J., Silver, S., et al. (1994). Neurologic sequelae after treatment of severe Hyponatremia: a multicenter prospective. *J Am Soc Nephrol*, 4, 1522.

[35] Watanabe, I., Mayumi, T., Arishima, T., Nakao, A., et al. (2007). Hyperlactemia can predict the prognosis of liver resection. *Shock*, 28, 35-8.

[36] Geerse, D. A., Bindels, A. J., Kuiper, M. A., Roos, A. N., Spronk, P. E., & Schultz, M. J. (2010). Treatment of hypophosphatemia in the intensive care unit: a review. *Crit Care*, 14, R147.

[37] Shor, R., Halabe, A., Rishver, S., Tilis, Y., Matas, Z., Fux, A., et al. (2006). Severe hypophosphatemia in sepsis as a mortality predictor. *Ann Clin Lab Sci*, 36, 67-72.

[38] Van den Berghe, G., Wouters, P., Weekers, F., Verwaest, C., Bruyninckx, F., Schetz, M., et al. (2001). Intensive insulin therapy in the critically ill patients. *N Engl J Med*, 345, 1359-67.

[39] Huo, T. I., Lui, W. Y., Huang, Y. H., Chau, G. Y., Wu, J. C., Lee, P. C., et al. (2003). Diabetes mellitus is a risk factor for hepatic decompensation in patients with hepatocellular carcinoma undergoing resection: a longitudinal study. *Am J Gastroenterol*, 98, 2293-8.

[40] Okabayashi, T., Hnazaki, K., Nishimori, I., Sugimoto, T., Maeda, H., Yatabe, T., et al. (2008). Continuous post-operative blood glucose monitoring and control using a closed-loop system in patients undergoing hepatic resection. *Dig Dis Sci*, 53, 1405-10.

[41] Dicecco, S. R., Wieners, E. J., Weisner, R. H., et al. (1989). Assessment of nutritional status of patients with end-stage liver disease undergoing liver transplantation. *Mayo Clin Proc*, 64, 95-102.

[42] Richter, B., Schmandra, T. C., Golling, M., & Bechstein, W. O. (2006). Nutritional support after open liver resection: a systematic review. *Dig Surg*, 23, 139-45.

[43] Shirabe, K., Matsumata, T., Shimada, M., Takenaka, K., Kawahara, N., Yamamoto, K., et al. (1997). A comparison of parenteral hyperalimentation and early enteral feeding regarding systemic immunity after major hepatic resection--the results of a randomized prospective study. *Hepatogastroenterology*, 44, 205-9.

[44] Mochizuki, H., Togo, S., Tanaka, K., Endo, I., & Shimada, H. (2000). Early enteral nutrition after hepatectomy to prevent postoperative infection. *hepatogastroenterology*, 47, 1407-10.

[45] Hu, Q. G., & Zheng, Q. C. (2003). The influence of Enteral Nutrition in postoperative patients with poor liver function. *World J Gastroenterol*, 9, 843-6.

[46] Okabayashi, T., Iyoki, M., Sugimoto, T., Kobayashi, M., & Hanazaki, K. (2011). Oral supplementation with carbohydrate- and branched-chain amino acidenriched nutrients improves postoperative quality of life in patients undergoing hepatic resection. *Amino Acids*, 40, 1213-20.

[47] Ishikawa, Y., Yoshida, H., Mamada, Y., Taniai, N., Matsumoto, S., Bando, K., et al. (2010). Prospective randomized controlled study of short-term perioperative oral nutrition with branched chain amino acids in patients undergoing liver surgery. *Hepatogastroenterology*, 57, 583-90.

[48] De Pietri, L., Montalti, R., Begliomini, B., Scaglioni, G., Marconi, G., Reggiani, A., et al. (2010). Thromboelastographic changes in liver and pancreatic cancer surgery: hypercoagulability, hypocoagulability or normocoagulability? *Eur J Anaesthesiol*, 27, 608-16.

[49] Shontz, R., Karuparthy, V., Temple, R., & Brennan, T. J. (2009). Prevalence and risk factors predisposing to coagulopathy in patients receiving epidural analgesia for hepatic surgery. *Reg Anesth Pain Med*, 34, 308-11.

[50] Recart, A., Duchene, D., White, P. F., Thomas, T., Johnson, D. B., & Cadeddu, J. A. (2005). Efficacy and safety of fast-track recovery strategy for patients undergoing laparoscopic nephrectomy. *J Endourol*, 19(10), 1165.

[51] Rudin, A., Lundberg, J. F., Hammarlund-Udenaes, M., Flisberg, P., & Werner, M. U. (2007). Morphine metabolism after major liver surgery. *Anesth Analg*, 104, 1409-14.

[52] Chandok, N., & Watt, K. D. (2010). Pain management in the cirrhotic patient: the clinical challenge. *Mayo Clin Proc*, 85, 451-8.

[53] Hudcova, J., Mc Nicol, E., Quah, C., Lau, J., & Carr, D. B. (2006). Patient controlled opioid analgesia versus conventional opioid analgesia for postoperative pain. *Cochrane Database Syst Rev*.

[54] Mimoz, O., Incagnoli, P., Josse, C., Gillon, M. C., Kuhlman, L., Mirand, A., et al. (2001). Analgesic efficacy and safety of nefopam vs. propacetamol following hepatic resection. *Anaesthesia*, 56, 520-5.

[55] Wolters, U., Wolf, T., Stutzer, H., & Schroder, T. (1996). ASA classification and perioperative variables as predictors of postoperative outcome. *British Journal of Anaesthesia*, 77, 217-222.

[56] Melendez, J., Ferri, E., Zwillman, M., Fischer, M., De Matteo, R., Leung, D., et al. (2001). Extended hepatic resection: A 6-year retrospective study of risk factors for perioperative mortality. *J Am Coll Surg*, 192(1), 47-53.

[57] Neal, C. P., Mann, C. D., Garcea, G., Briggs, C. D., Dennison, A. R., & Berry, D. P. (2011). Preoperative systemic inflammation and infectious complications after resection of colorectal liver metastases. *Arch Surg*, 146, 471-8.

[58] Kaibori, M., Ishizaki, M., Matsui, K., & Kwon, A. H. (2011). Postoperative infectious and non-infectious complications after hepatectomy for hepatocellular carcinoma. *Hepatogastroenterology*, 58, 1747-56.

[59] Okabayashi, T., Nishimori, I., Yamashita, K., Sugimoto, T., Yatabe, T., Maeda, H., et al. (2009). Risk factors and predictors for surgical site infection after hepatic resection. *J Hosp Infect*, 73, 47-53.

[60] Lawrence, V. A., Hilsenbeck, S. G., Mulrow, C. D., Dhanda, R., Sapp, J., & Page, C. P. (1995). Incidence and hospital stay for cardiac and pulmonary complications after abdominal surgery. *J Gen Intern Med*, 10(12), 671.

[61] Xue, F. S., Li, B. W., Zhang, G. S., et al. (1999). The influence of surgical sites on early postoperative hypoxemia in adults undergoing elective surgery. *Anesth Analg*, 88, 213.

[62] Platell, C., & Hall, J. C. (1997). Atelectasis after abdominal surgery. *J Am Coll Surg*, 185, 584.

[63] Montravers, P., Veber, B., Auboyer, C., et al. (2002). Diagnostic and therapeutic management of nosocomial pneumonia in surgical patients: results of the Eole study. *Crit Care Med*, 30, 368.

[64] Goodman, L. R. (1980). Postoperative chest radiograph: I. Alterations after abdominal surgery. *AJR Am J Roentgenol*, 134, 533.

[65] Thasler, W. E., Bein, T., & Jauch, K. H. (2002). Perioperative effects of hepatic resection surgery on hemodynamics, pulmonary fluid balance, and indocyanine green clearance. *Langenbecks Arch Surg*, 387(2), 271-5.

[66] Barrett, N. A., & Kam, P. C. (2006). Transfusion-related acute lung injury: a literature review. *Anaesthesia*, 61, 777-785.

[67] Mulkey, Z., Yarbrough, S., Guerra, D., et al. (2008). Postextubation pulmonary edema: a case series and review. *Respir Med*, 102, 1659.

[68] Shirabe, K., Shimada, M., Gion, T., Hasegawa, H., Takenaka, K., Utsunomiya, T., et al. (1999). Postoperative liver failure after major hepatic resection for hepatocellular carcinoma in the modern era with special reference to remnant liver volume. *J Am Coll Surg*, 188(3), 304-7.

[69] Geerts, W. H., Bergqvist, D., Pineo, G. F., Heit, J. A., Samama, C. M., Lassen, M. R., et al. (2008). Prevention of venous thromboembolism: American College of Chest Physicians Evidence-Based Clinical Practice Guidelines. 8th Edition, *Chest*, 133, 381S-453S.

[70] Lesmana, C. R., Inggriani, S., Cahyadinata, L., & Lesmana, L. A. (2010). Deep vein thrombosis in patients with advanced liver cirrhosis: a rare condition? *Hepatol Int*, 4, 433-8.

Chapter 4

Essential Functional Hepatic and Biliary Anatomy for the Surgeon

Ronald S. Chamberlain

Additional information is available at the end of the chapter

1. Introduction

That every surgeon will experience complications is a certainty. Indeed, it has been said that if one has no complications, one does not do enough surgery. Yet, major surgical complications are often avoidable and frequently the result of three tragic surgical errors. These errors are: 1) a failure to possess sufficient knowledge of normal anatomy and function, 2) a failure to recognize anatomic variants when they present, and 3) a failure to ask for help when uncertain or unsure. All but the last of these errors are remediable with study and effort. In regard to the last error, most surgeons learn humility through their failures and at the expense of their patients, while some never learn.

The importance of a precise knowledge of parenchymal structure, blood supply, lymphatic drainage, and variant anatomy on outcome is perhaps nowhere more apparent than in hepatobiliary surgery. Though the liver was historically an area where few brave men dared to tread, and even less returned a second time, recent advances in anesthetic technique and perioperative care now permit hepatic surgery to be performed with low morbidity and mortality in both academic and community hospitals. That said, surgeons are duly cautioned to inventory their own skills and knowledge before venturing forward into the right upper quadrant. This chapter will review functional biliary and hepatic anatomy necessary for the conduct of safe and successful hepatic operations.

2. The liver

2.1. Surface anatomy

The liver is situated primarily in the right upper quadrant, and usually benefits from complete protection by the lower ribs. Most of the liver substance resides on the right side, although it is not uncommon for the left lateral segment to arch over the spleen. The superior surface of the liver is molded to, and abuts the undersurface of the diaphragm on both the right and left side. During normal inspiration, the liver may rise as high as the 4th or 5th intercostal space on the right.

The liver itself is completely invested with a peritoneal layer except on the posterior surface where it reflects onto the undersurface of the diaphragm to form the right and left triangular ligaments. The liver is attached to the diaphragm and anterior abdominal wall by three separate ligamentous attachments, namely the falciform, round, and right and left triangular ligaments. (Figure 1) The falciform ligament, which is situated on the anterior surface of the liver, arises from the anterior leaflets of the right and left triangular ligaments and terminates inferiorly where the ligamentum teres enters the umbilical fissure. The gallbladder is normally attached to the undersurface of the right lobe and directed towards the umbilical fissure. At the base of the gallbladder fossa, is the hilar transverse fissure through which the main portal structures to the right lobe course. Additional important landmarks on the posterior liver surface include a deep vertical groove in which the inferior vena cava is situated, and a large bare area (i.e. no peritoneal coating) that is normally in contact with the right hemidiaphragm and right adrenal gland. The left lateral segment of the liver arches over the caudate lobe that is situated to the left of the vena cava. The caudate lobe is demarcated on the left by a fissure containing the ligamentum venosum (a remnant of the umbilical vein). Additional left-sided important surface features include the gastrohepatic omentum that is located between the left lateral segment and the stomach. The gastrohepatic omentum may contain replaced or accessory hepatic arteries. Finally, there is usually a thick fibrous band that envelops the vena cava high on the right side and runs posteriorly towards the lumbar vertebrae. This band, which is sometimes referred to as the vena caval ligament, must be divided to allow proper visualization of the suprahepatic cava and right hepatic veins.

2.2. Parenchyma (the liver substance)

The liver is comprised of two main lobes, a large right lobe, and a smaller left lobe. Although the falciform ligament is often thought to divide the liver into a right and left lobe, the true "anatomic" or "surgical" right and left lobes of the liver are defined by the course of the middle hepatic vein that runs through the main scissura of the liver. Although various descriptions of the internal anatomy of the liver have been proffered over the last century, Couinaud's (1957) segmental anatomy of the liver is the most useful for the surgeon.

Couinaud's classification system divides the liver into four unique sectors based upon the course of the three major hepatic veins. Each sector receives its blood supply from a separate portal pedicle. Within the *main scissura* lies the middle hepatic vein that courses from the left side of the suprahepatic vena cava to the middle of the gallbladder fossa. Functionally, the main scissura divides the liver into separate right and left lobes which have independent portal inflow, and biliary architecture. (Figures 2 and 3) An artificial line that divides the liver into right and left hemilivers is known as Cantlie's line. The right hepatic veins runs within the right segmental scissura and divides the right lobe into a right posterior and anterior sector,

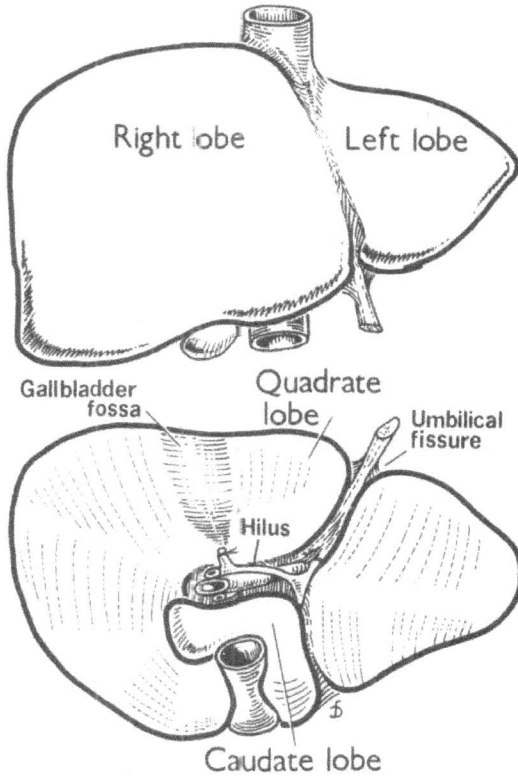

Figure 1. Surface anatomy of the liver. (A) Anterior surface, (B) Inferior surface of the liver. Reprinted with permission from Hahn and Blumgary, Functional Hepatic and Raciologic Anatomy in Surgery of the Liver and Biliary Tract (3rd Edition), Blumgart LH, Fong Y and WH Jarnigan (Ecs.) Lippincott Williams, London, UK (2000).

while the left hepatic veins follows the path of the falciform ligament and divides the left lobe into a medial and lateral segment.

The right and left lobes of the liver are further divided into 8 segments based upon the distribution of the *portal scissurae*. At the hilus, the right portal vein pursues a very short course (1 – 1.5 cm) before entering the liver. Once entering the hepatic parenchyma, the portal vein divides into a right anterior sectoral branch that arches vertical in the frontal plane of the liver, and a posterior sectoral branch that follows a more posterolateral course. The right portal vein supplies the anterior (or anteriomedial) and posterior (or posterolateral) sectors of the right lobe. The branching pattern of these sectoral portal veins subdivides the right liver into 4 segments -- segments V (anterior and inferior) and VIII (anterior and superior) form the anterior sector, and segments VI (posterior and inferior) and VII (posterior and superior) form the posterior sector.

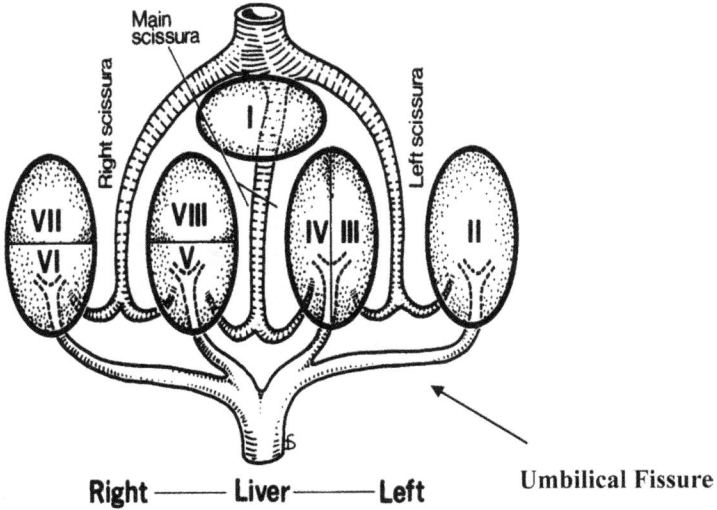

Figure 2. Segmental and sectoral anatomy of the liver. The liver is divided into three main scissura by the right, middle, and left hepatic vein branches. The middle hepatic courses through the main scissura (or Cantlie's line) and divides the liver into right and left lobes. The right hepatic vein divides the right liver into anterior (segments V and VIII) and posterior (segment VI and VII) sectors, while the left hepatic vein divides the left lobe into medial (segments IV A and B) and lateral segments (segments II and III). The intrahepatic branching of the right and left hepatic ducts, arteries and portal veins (shown) in the horizontal plane of the liver divides the liver into eight separate segments. The caudate lobe (segment I) is neither part or left lobe. Rather the caudate lobe receives venous and arterial branches from both the right and left side of the liver, and drains directly into the inferior vena cava.

In contrast to the right portal vein, the left portal vein has a long extrahepatic length (3 – 4 cm) coursing beneath the inferior portion of the quadrate lobe (segment 4B) enveloped in a peritoneal sheath (the hilar plate.) Upon reaching the umbilical fissure, the left portal vein runs anteriorly and superiorly within the liver substances, and gives off horizontal branches to the quadrate lobe medially (segments IV A (superior) and B (inferior)) and to the left lateral segment (segments III (inferior) and II (superior)) (Figure 3).

The caudate lobe (segment I) is neither part of the left nor right lobes, though it lies mostly on the left side (Figure 4). More precisely, it is the most dorsal portion of the liver situated behind the left lobe and embracing the retrohepatic vena cava from the hilum to the diaphragm. The portion of the caudate lobe that is within the right liver is usually quite small, and lies posterior to segment 4B. Figure 3 illustrates the location of the caudate lobe which lies between the left portal vein and vena cava on the far left, and the middle hepatic vein and vena cava within the right liver. The caudate lobe receives blood vessels and biliary tributaries from both the right and left hemilivers. The right side of the caudate lobe, and the caudate process, receives its blood supply from branches of the right or main portal vein, while the left side of the caudate receives a separate vessel from the left portal vein.

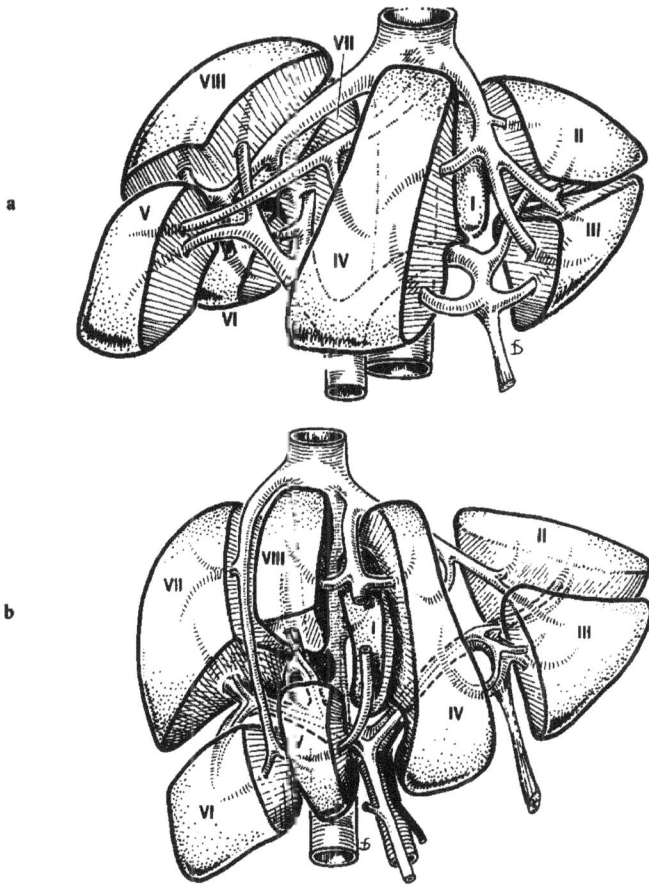

Figure 3. Couninaud's segmental anatomy of the l ver. (a) *in vivo* appearance; (b) *ex vivo* appearance.

Aberrant segmental anatomy of the liver is uncommon. The presence of a diminutive left lobe is the most common anomaly reported, and is important only because it may serve as a limitation to the performance of extended right hepatectomies. Although reports of "accessory" hepatic lobes are not uncommon, these do not represent separate segments with independent intrahepatic vascular supply, but rather elongated tongues of normal liver tissue. Riedel's lobe is the most common of these "accessory" lobes, and is reality, an extended piece of liver tissue hanging inferiorly off segments 5 and 6.

3. Hepatic veins (Outflow)

The three major hepatic veins (the right, middle and left) comprise the main outflow tract for the liver, although additional veins (5 – 20) of varying size are always present as direct communications between the vena cava and the posterior surface of the right lobe. Uniquely, the caudate lobe (segment I) drains principally through direct communications with the retrohepatic cava.

The hepatic veins lie within the three major scissura of the liver dividing the parenchyma into the right anterior and posterior sectors, and the right and left lobes. (Fig 2 and 3) The right hepatic vein lies within the right scissura (or segmental fissure) and divides the right lobe into a posterior (segments VI and VII) and anterior (segments V and VIII) sector. The middle hepatic veins lies within the main hepatic scissura (or main lobar fissure) separating the right anterior sector (segments V and VIII) from the quadrate lobe (segment IV). Anatomically, the main scissura separates the liver into right and left lobes. The left hepatic vein lies within the left scissura (or the left segmental fissure) in line with or just to the right of the falciform ligament. The right hepatic vein drains directly into the suprahepatic cava, while the middle and left hepatic vein coalesce to form a short common trunk prior to entry. The umbilical vein represents an additional alternative site of venous efflux. It is located beneath the falciform ligament and eventually terminates in the left hepatic vein, or less commonly in the confluence of the middle and left hepatic veins.

4. Hepatic venous anomalies

Although the outline above should suffice as cursory knowledge of hepatic venous anatomy, it is far from exhaustive. For example, large accessory right hepatic veins are commonly found, and an appreciation of these structures on axial imaging can be important to operative planning. If a large accessory right hepatic vein is present, it may be possible to divide all three major hepatic veins in the performance of an extended left hepatectomy. Most importantly, the surgeon embarking on hepatic resection should have a thorough knowledge of the internal course of the hepatic veins, as the danger posed by hepatic venous bleeding cannot be overestimated.

5. Hepatic arteries (Inflow)

5.1. Extrahepatic arterial anatomy

"Normal" hepatic arterial anatomy is anything but normal. Indeed standard celiac arterial anatomy as described in most major anatomic treatise is found in only 60% of cases. An *accessory* hepatic artery refers to a vessel that supplies a segment of liver that also receives blood supply from a normal hepatic artery. An aberrant hepatic artery is called a *replaced* hepatic artery as it represents the only blood supply to a specific hepatic segment. Precise knowledge of normal hepatic arterial anatomy is necessary to appreciate abnormal anatomy and will be the focus of this section.

The celiac artery arises from the aorta shortly after it emerges through the diaphragmatic hiatus. The celiac trunk itself is typically very short and divides into the left gastric, splenic, and common hepatic artery shortly after its origin. (Figure 5). The common hepatic artery typically passes forward for a short distance in the retroperitoneum where it them emerges at the superior border of the pancreas and left side of the common hepatic duct. The common hepatic artery supplies 25% of the liver's blood supply, with the portal vein supplying the remaining 75%.

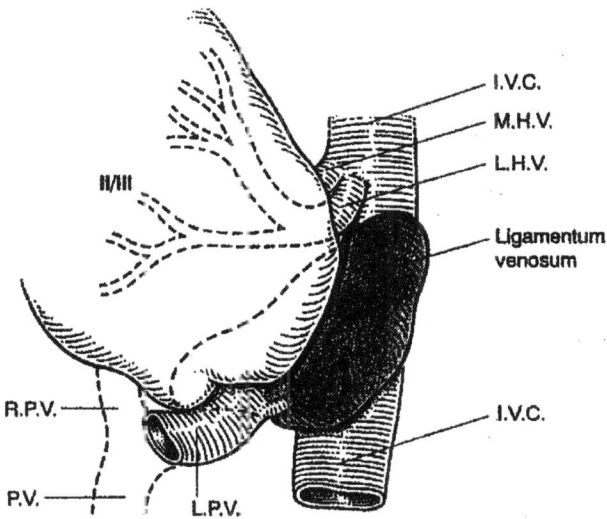

Figure 4. Caudate lobe anatomy. The caudate lobe is situated to the left of the inferior vena cava (I.V.C). Superiorly the caudate lobe is covered by segments II and I I which are reflected laterally in this diagram. The ligamentum venousm, a remnant of the fetal umbilical vein, courses across the anterior surface of the caudate lobe to enter the left hepatic vein. The caudate lobe runs along the retrohepatic vena cava from the common trunk of the middle and left hepatic veins (M.H.V., L.H.V.) to the portal vein (P.V.) inferiorly. (Left (L.P.V.) and right portal vein (R.P.V.)). Small venous tributaries drain the caudate lobe directly into to the I.V.C. On its medial surface, the caudate lobe is attached to the right iver by the caudate process.

After arising from the celiac axis, the common hepatic artery turns upward and runs lateral and adjacent to the common bile duct. The gastroduodenal artery that supplies the proximal duodenum and pancreas is typically the first branch of the common hepatic artery. The right gastric artery takes off shortly thereafter and continues within the lesser omentum along the lesser curve of the stomach. At this point the common hepatic artery is referred to as the proper hepatic artery. The proper hepatic artery courses towards the hilum, and soon divides into the right and left hepatic arteries. Prior to the bifurcation, a small cystic artery branches off to

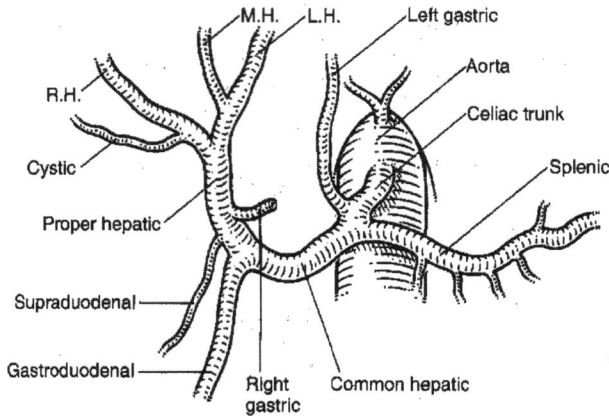

Figure 5. Normal celiac axis anatomy. The presence of the right hepatic (R.H.), middle hepatic (M.H.) to segment IV, and left hepatic (L.H.) artery are demonstrated.

provide blood supply to the gallbladder. While coursing through the hepatoduodenal ligament, the proper hepatic artery, common bile duct, and portal vein are enveloped in a peritoneal sheath within the hepatoduodenal ligament. The proper hepatic artery bifurcates earlier than the common bile duct and portal vein. In 80% of cases the right hepatic artery courses posterior to the common hepatic duct before entering the hepatic parenchyma. In 20% of cases, the right hepatic artery may lie anterior to the common hepatic duct. Upon reaching the hepatic parenchyma, the right hepatic artery branches into right anterior (Segments V and VIII), and right posterior sectoral branches (Segments VI and VII). The posterior sectoral branch initially runs horizontally through the hilar transverse fissure (of Gunz), normally present at the base of Segment V and adjacent to the caudate process. The left hepatic artery runs vertically towards the umbilical fissure where it gives off a small branch (often called the middle hepatic artery) to segment IV, before continuing on to supply Segments II and III. Additional small branches of the left hepatic artery supply the caudate lobe (segment I), although caudate arterial branches may also arise from the right hepatic artery. The sectoral and segmental bile ducts and portal veins follow the course of the hepatic artery branches. Intrahepatic branching of these structures will be discussed in more detail below.

The blood supply to the common bile duct is varied and multiple. Branches of the common hepatic, gastroduodenal, and pancreaticoduodenal arteries have all been shown to provide arterial supply at various levels.

5.2. Hepatic arterial anomalies

Variations in the arterial blood supply to the liver are common. Although the hepatic artery typically arises from the celiac axis, complete replacement of the main hepatic artery or its' branches occur with variable frequency. Similarly, duplication or accessory hepatic arterial branches, particularly an accessory left hepatic artery, may be more the norm than an anomaly. The most common hepatic arterial anomaly involving a replaced vessel is a replaced right hepatic artery (25%). In this situation, the replaced right hepatic artery usually arises from the superior mesenteric artery and runs lateral and posterior to the portal vein within the hepatoduodenal ligament. (Figure 6). In rare instances, the entire common hepatic artery, or its' individual branches may arise directly off the celiac trunk or aorta.

6. Portal venous anatomy

The portal vein is formed by a union of the superior mesenteric vein (SMV) and splenic vein behind the neck and body of the pancreas. In up to one third of all individuals, the inferior mesenteric vein may also join this confluence. Venous tributaries from the pancreas may also drain directly into the portal vein, and generally correspond to the arterial supply. More precisely, there are anterior, posterior, superior and inferior pancreatic vessels. In addition, the left gastric vein and inferior mesenteric vein typically drain into the splenic vein, but in rare instances these vessels may enter the portal vein directly. Surgical dogma states that there are no venous branches on the anterior surface of the portal vein and, for the most part this is true – most veins enter the portal vein tangentially from the side. However, having paid homage to surgical dogma, the reality is that small anterior venous branches may exist, and any manipulation posterior to the pancreatic neck and anterior to the portal vein should be performed with maximum operative exposure and care.

Access to the portal vein is typically obtained by identifying the superior mesenteric vein on the inferior surface of the pancreas. In some circumstances it is necessary to first locate the middle colic vein within the transverse mesocolon and follow it inferiorly to the SMV. The length of the SMV is highly variable, and may range from only a few millimeters up to 4 cm. In many circumstances the SMV is made up of 2 to 4 venous branches that coalesce shortly before joining the portal vein rather than a single dominant vein. The inferior pancreaticoduodenal vein, which can be quite prominent, is the only vein that normally enters the SMV directly. Proper identification of this vein is necessary to avoid injury (and often substantial blood loss). All other pancreatic venous tributaries enter the portal vein, rather than the SMV.

In the performance of a pancreaticoduodenal resection, early division of the common bile duct (CBD) provides great exposure to the right lateral side of the portal vein, and facilitates the creation of a "tunnel" above the portal vein, and beneath the pancreas. Once a determination has been made regarding the resectability of the pancreatic lesion, we favor early transection of the common bile duct. If the tumor later proves unresectable, a palliative end to side bilioenteric bypass can be performed.

In addition to those variants described above, there are additional (but rare) congenital anomalies of the portal vein with which the surgeon should be aware. The two most common are an anterior portal vein that lies above the pancreas and duodenum, and a direct entry of the portal vein into the inferior vena cava-- a congenital "portocaval" shunt. The importance

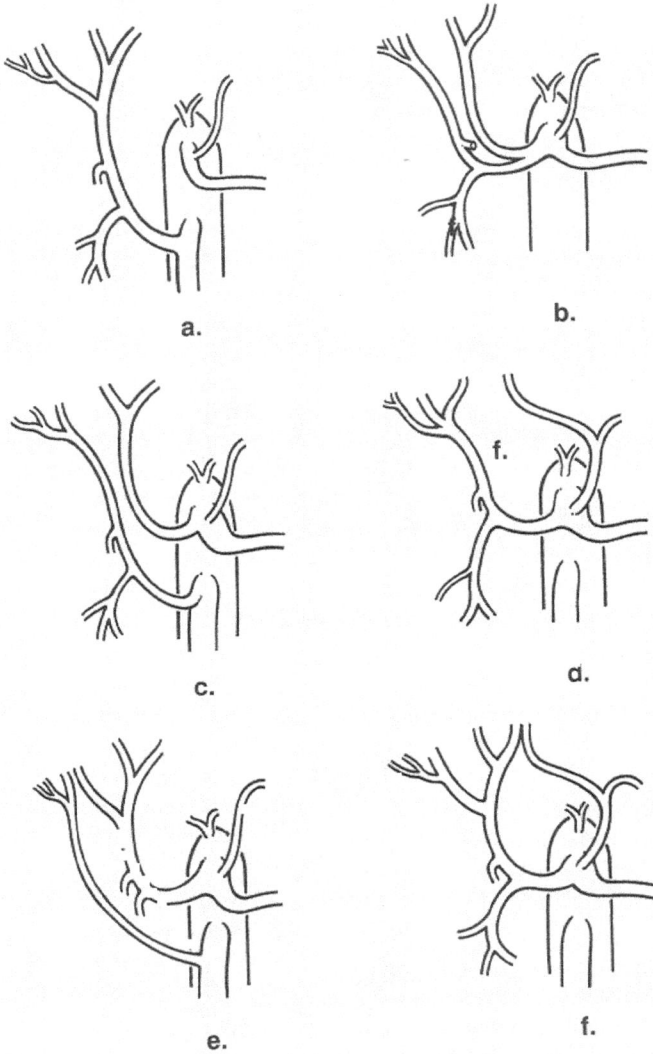

Figure 6. Hepatic arterial anomalies. (a) Replaced main hepatic artery arising from the superior mesenteric artery (SMA), (b) Independent origin of the right and left hepatic artery from the celiac axis, (c) Replaced right hepatic artery arising from the SMA, (d) Replaced left hepatic artery arising from the left gastric artery (LGA), (e) Accessory right hepatic artery arising from the SMA, (f) Accessory left hepatic artery arising from the LGA.

of careful dissection around the portal vein cannot be overemphasized. Inadvertent injury or transection of the portal vein or a main tributary is difficult to correct, and remains among the most lethal of surgical errors.

7. Intrahepatic arterial and portal venous anatomy

Throughout the course of the liver, the sectoral and segmental bile ducts, hepatic arteries and portal venous branches run together. (Figure 7) Whereas knowledge of precise intrahepatic biliary anatomy is of most practical value to the operating surgeon, further detail about intrahepatic anatomy will be discussed in that section below.

8. The biliary tract

Extrahepatic hepatic biliary anatomy

The extrahepatic biliary system consists of the extrahepatic portions of the right and left bile ducts that join to form a single biliary channel coursing through the posterior head of the pancreas to enter the medial wall of the second portion of the duodenum. The gallbladder and cystic duct form an additional portion of this extrahepatic biliary system that typically joins with the terminal portion of the common hepatic duct to form the common bile duct. In most instances, the confluence of the right and left bile ducts lies to the right of the umbilical fissure and anterior to the right branch of the portal vein. The right hepatic duct is typically short (< 1cm) and branches into a right posterior sectoral duct (segments VI/VII) and a right anterior sectoral duct (segments V/VIII) shortly after entering the hepatic parenchyma. In contrast, the left hepatic duct has a relatively long extrahepatic course (2- 3 cm) along the base of the quadrate lobe (segment IV) and enters the hepatic parenchyma at the umbilical fissure. Lowering the hilar plate (i.e., connective tissue enclosing the left hepatic elements and Glisson's capsule) at the base of the quadrate lobe provides great exposure to both the biliary hilum and the extrahepatic portion of the left hepatic duct. (Figure 8)

9. The common bile duct

By convention, the entry point of the cystic duct divides the main extrahepatic biliary channel into the common hepatic duct (above) and the common bile duct (below). The common bile duct continues inferiorly positioned anterior to the portal vein, and lateral to the common hepatic artery. If the hepatic artery bifurcates early, the right hepatic artery may be seen coursing below (80% of the time) the common bile duct (see details above). At the junction of the 1st and 2nd portion of the duodenum, the common bile duct ducks behind the duodenum posterior to the pancreatic head, in order to enter the medial wall of the duodenum (2nd portion) at the sphincter of Oddi.

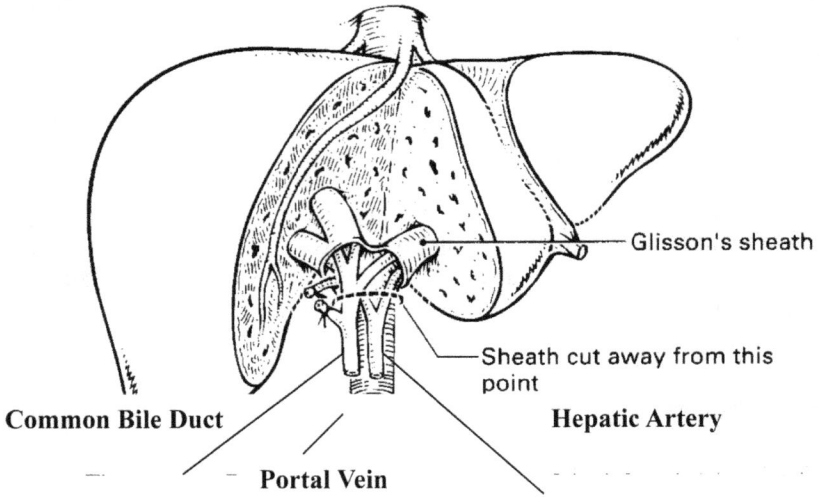

Figure 7. Portal pedicles. This cutaway view of the right and left portal pedicles demonstrate the course of the right and left portal veins, hepatic ducts, and hepatic arteries as they enter the hepatic parenchyma

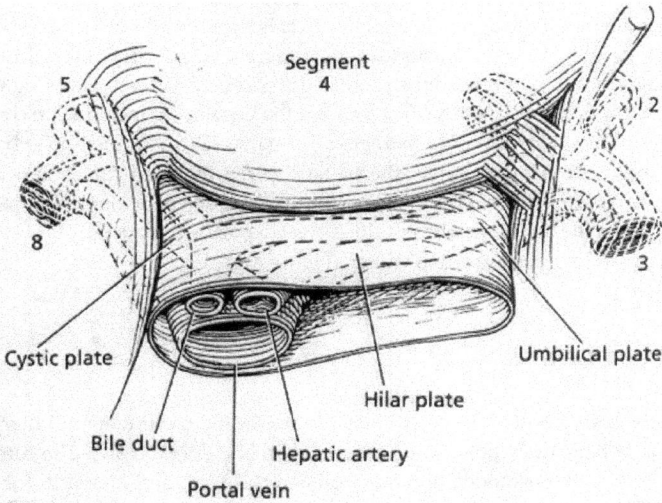

Figure 8. Lowering of the hilar plate and exposure of the left hepatic duct. The left hepatic duct runs at the base of the quadrate lobe (segment 4) and is covered by the hilar plate (a layer of connective tissue running between the hepatoduodenal ligament and the Glissonian capsule of the liver. Dividing this layer demonstrates the extrahepatic portion of the left hepatic duct arising from the umbilical fissure. (Numbers 2,3,4 and refer to segmental liver anatomy).

10. Gallbladder and cystic duct

The gallbladder is situated on the undersurface of the anterior inferior sector (segment V) of the right lobe of the liver. Though often densely adherent, it is separated from the liver parenchyma by the cystic plate, a layer of connective tissue arising from Glisson's capsule and in continuity with the hilar plate at the base of segment IV. In rare instances, the gallbladder is only loosely attached to the undersurface of the liver by a thinly veiled mesentery and may be prone to volvulus. Variations in gallbladder anatomy are rare. These variations include (a) bilobed or double gallbladders, (b) septated gallbladders, or (c) gallbladder diverticulums.

The cystic duct arises from the infindibulum of the gallbladder and runs medial and inferior to join the common hepatic duct. The cystic duct is typically 1-3 mm in diameter, and can range from 1 mm to 6 cm in length depending upon its union with the common hepatic duct. Spiral mucosal folds, referred to as valves of Heister, are present in the mucosa of the cystic duct. Cystic duct abnormalities are uncommon and include (a) double cystic ducts (very rare), (b) aberrant cystic duct entry sites, and (c) aberrant cystic duct union with the common hepatic duct. Aberrant entry points for the cystic duct include a low entry into the common hepatic duct retroduodenal or retropancreatic, and anomalous entry into the main right hepatic duct or sectoral duct. Aberrant union of the cystic duct and common hepatic duct can take multiple forms including (a) absence of a cystic duct (< 1%), (b) parallel course of the cystic duct and common hepatic artery with a shared septum (20%), and (c) an anomalous passage of the cystic duct posterior to the common hepatic duct with entry on the medial wall (5%). (Figure 9)

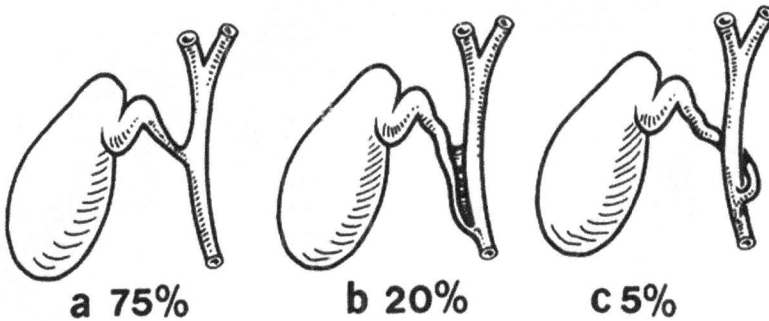

a 75%   **b 20%**   **c 5%**

Figure 9. Variations in cystic ductal anatomy.

Typically, the cystic artery is a single vessel that courses lateral and posterior to the cystic duct. However, variations in the anatomy of the cystic artery are common. (Figure 10) Multiple cystic arteries, origin of the cystic artery from a segmental or lobar hepatic artery, aberrant course of the cystic artery over the cystic duct, and various other anomalies have been reported. A careful intra-operative determination of cystic artery anatomy is important to prevent unnecessary hemorrhage during cholecystectomy.

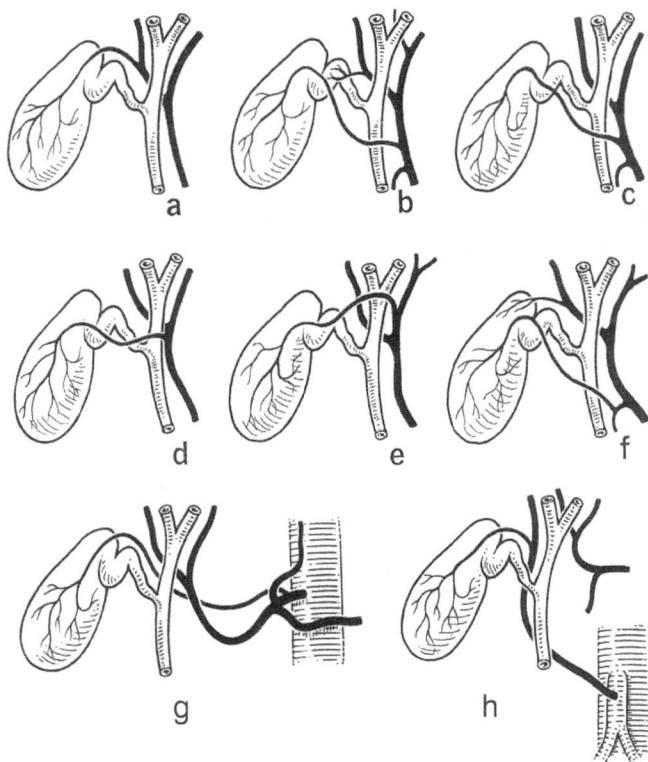

Figure 10. Cystic artery anomalies. (A) Typical course, (B) Double cystic artery, (C) cystic artery crossing anterior to the main bile duct, (D) cystic artery originating from the right branch of the hepatic artery and crossing the common hepatic duct anteriorly, (E) cystic artery originating from the left branch of the hepatic artery, (F) cystic artery originating from the gastroduodenal artery, (G) the cystic artery may arise from the celiac axis, (H) cystic artery originating from a replaced right hepatic artery.

10. Intrahepatic bile duct anatomy

An understanding of intrahepatic ductal anatomy is obviously important and vital to the performance of a high biliary anastomoses for cholangiocarcinoma (Klatskin tumors), an intrahepatic bilioenteric bypass, and complex hepatic resections such as caudate lobectomy, and left and right trisegmentectomy. The right and left lobes of the liver are drained separately by the right and left hepatic ducts. In contrast, 1 – 4 smaller ducts from either the right or left hepatic ducts drain the caudate lobe. Within the liver parenchyma, the intrahepatic biliary radicals parallel the major portal triad tributaries directed toward each hepatic segment of the liver. More specifically, bile ducts are usually situated superior to its complementary portal vein branch, while the hepatic artery lies inferiorly.

The left hepatic duct drains all 3 segments of the left liver. (Segment II, III, and IV). In some textbooks, segment IV, the quadrate lobe, is futher sub-divided into sub-segments (4A, superior, and 4B, inferior). So conceptually, both the right and left hepatic ducts each drains 4 segments. Although the left hepatic duct originates within the liver and terminates in the common hepatic duct, it is easier to describe its' path in reverse since the extrahepatic areas are readily visible to the operating surgeon. After the bifurcation into the right and left hepatic ducts, the left duct courses towards the umbilical fissure along the under surface of segment IVB above and behind the left branch of the portal vein. Access to this area can be gained by lowering the hilar plate (described above). Several small branches from the quadrate lobe (Segment 4) and the caudate lobe (Segment 1) may enter the left duct at this location. The left hepatic duct is formed within the umbilical fissure by the segment III (lateral), and segment IVB (medial) ducts. Following the course of the umbilical fissure vertically towards the falciform ligament, the segment II (lateral), and segment IVA (medial) branches are formed. Although a careful and tedious dissection is required to access the segmental biliary ducts for anastomoses, (e.g., a segment III bypass), control of the segmental portal triads to all areas of left lobe is readily achievable within the umbilical fissure. (see Figure 11)

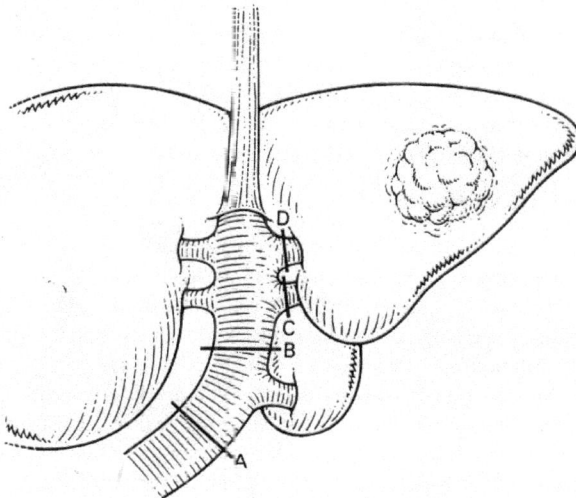

Figure 11. Left portal vein pedicle. The union of the segment IV, II, and III portal veins within the umbilical fissure forms the left portal vein. A separate segment I portal vein also enters the left portal vein before it coalesces with the right portal vein at the hilus. Lines A, B, C, D demonstrate various lines of portal vein transection which are required to complete various hepatic resections. Line A is the line of transection for completion of a left hepectecomy and caudate lobectomy. Line B is the line of transection for completion of a left hepatectomy. Line C is the line of transection for a segment II resection. Line D is the line of transection for a segment III resection.

The right hepatic duct emerges from the liver at the base of segment V just to right of the caudate process. This duct drains segments V, VI, VII, and VIII and originates at the junction of the right posterior (segments VI and VII), and anterior (segments V and VIII) sectoral ducts. The right posterior sectoral duct follows an almost horizontal course at the base of segments V and VI that can often been seen lying within a transverse fissure on the superficial surface of the liver. Segmental biliary branches from segments VI (inferior) and VII (superior) converge to form the main right posterior sectoral duct. Segmental branches from segments V and VIII form the right anterior sectoral duct. While the right posterior sectoral duct follows a horizontal course, the right anterior sectoral duct runs almost vertical within segment V, and receives branches from both segment V (inferior) and VIII (superior).

Biliary drainage of the caudate lobe is less predictable. Conceptually, the caudate lobe has three distinct areas -- a right part, a left part, and the caudate process. In some instances three separate bile ducts may be present. The caudate process represents a narrow bridge of tissue that connects the caudate to the right lobe (segment V). In more than 75% of cases the caudate drains into both the right and left hepatic ductal system, but isolated drainage into the right (< 10%), or left hepatic duct (~15%) can occur.

11. Anomalous biliary drainage

Normal intra- and extrahepatic biliary anatomy is present in approximately 75 percent of cases. (Figure 12) Every effort should be made to define existing intrahepatic anatomy based on pre-operative imaging, since failure to do so may result in devastating complications. Anomalies in both sectoral and segmental anatomy may exist together or separately. The more common type of each of the anomalies will be described in more detail below.

Anomalous sectoral biliary anatomy

Although the union of the right and left hepatic duct typically occurs at the hilum, a triple confluence of the right posterior and anterior sectoral ducts with the left hepatic duct may, exist in up to ~15% of cases. (Figure 12) In 20% of cases, one of the right sectoral ducts, more commonly the anterior sectoral duct, may enter the common hepatic duct distal to the confluence. If this situation is not recognized it can be very dangerous, and represents a common cause of injury during laparoscopic cholecystectomy. Less commonly (~5%), the right posterior sectoral duct (and rarely the right anterior sectoral duct) may cross to enter the intrahepatic portion of the left hepatic duct. Failure to appreciate this anomaly prior to right or left hepatectomy, can lead to significant post-operative problems. Note some authorities believe that this anomaly represents the most common intrahepatic biliary variations.

Anomalous segmental biliary anatomy

A large number of segmental biliary anomalies have been reported. Most are unimportant to the surgeon and of anatomical interest only. Figure 13 illustrates the more common anomalies that have been reported within the right lobe and the medial segment of the left lobe.

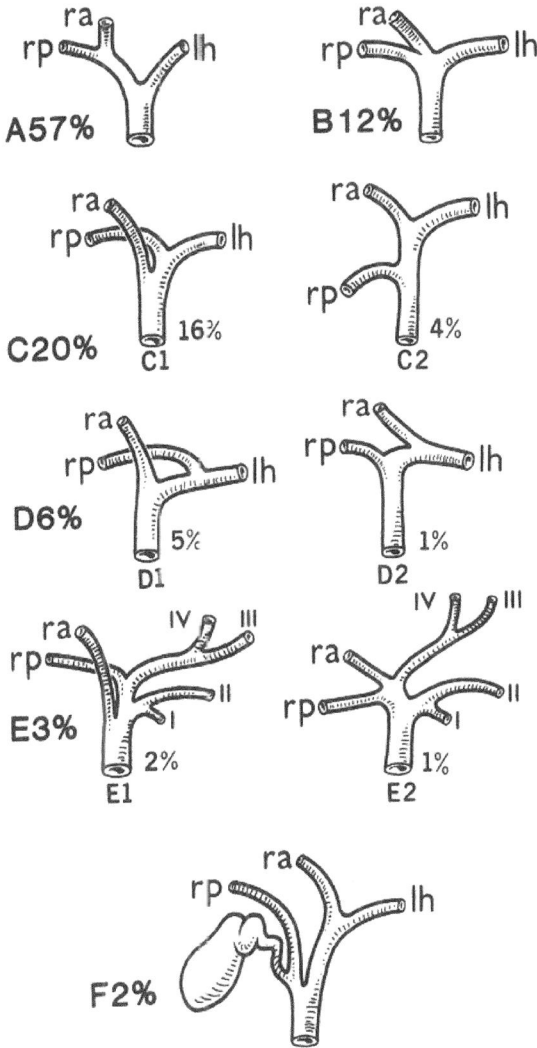

Figure 12. Normal and aberrant sectoral ductal anatomy. (A) Typical ductal anatomy, (B) triple confluence, (C) Ectopic drainage of a right sectoral duct into the common hepatic duct (C1, right anterior duct draining into the common hepatic duct; C2, right posterior duct draining into the common hepatic duct), (D) ectopic drainage of a right sectoral duct into the left hepatic ductal system (D1, right posterior sectoral duct draining into the left hepatic ductal system; D2, right anterior sectoral duct draining into the left hepatic ductal system, (E) absence of the hepatic duct confluence, (F) absence of right hepatic duct and ectopic drainage of the right posterior duct into the cystic duct.

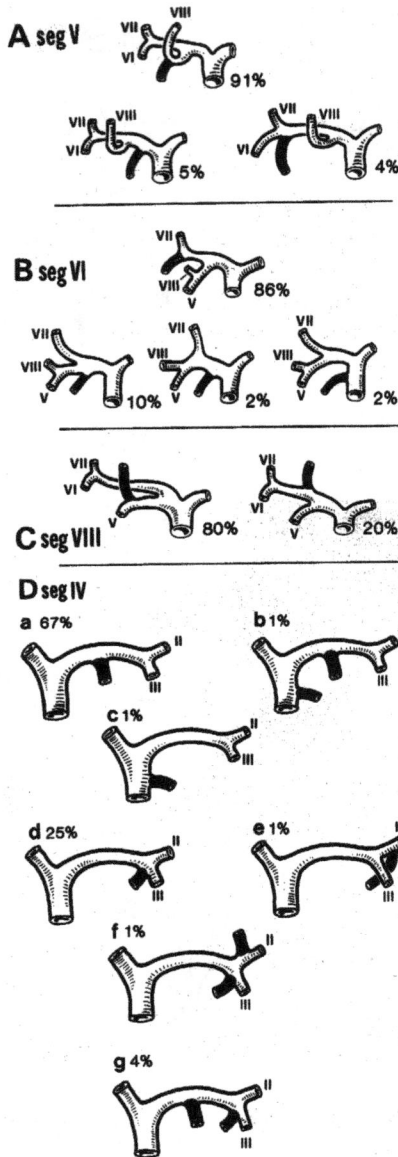

Figure 13. Normal and aberrant segmental ductal anatomy. (A), variations of segment V, (B) variations of segment VI, (C) variations of segment VIII, (D) variations of segment IV. Note there is no variation of drainage of segments II, III, and VII.

12. Summary

A comprehensive understanding of normal and aberrant anatomy is the cornerstone of surgery. The truth of this statement is nowhere more apparent than in the performance of complex hepatobiliary surgery. Mastery of the segmental anatomy of the liver, as well as a comprehensive understanding of both normal and anomalous arterial, venous and biliary anatomy, are the *sine qua non* for performing safe hepatic resections. Recent advances in perioperative management of patients with hepatobiliary diseases (detailed elsewhere in this book), permit the surgeon to perform increasingly radical hepatic procedures (upon sicker patients.) Although the expertise offered by our radiology and anesthesiology colleagues is important, it is incumbent upon every surgeon who performs liver resection to be well prepared. An age-old surgical axiom states "98% of the surgical outcome is determined in the operating room." A good outcome in the performance of hepatic resections requires one to become a student of the game.

Author details

Ronald S. Chamberlain[1,2,3]

1 Department of Surgery, Saint Barnabas Medical Center, Livingston, NJ, USA

2 Department of Surgery, University of Medicine and Dentistry of New Jersey, Newark, NJ, USA

3 Saint George's University School of Medicine, Grenada, West Indies

References

[1] Abdalla, E. K, Vauthey, J. N, & Couinaud, C. The caudate lobe of the liver: implications of embryology and anatomy for surgery. Surg Oncol Clin N Am (2002). , 11, 835-48.

[2] Bismuth, H. Surgical anatomy and anatomical surgery of the liver. World J Surg (1982). , 6, 3-9.

[3] Bismuth, H. Surgical anatomy and anatomical surgery of the liver. In: Blumgart LH, editor. Surgery of the liver and biliary tract. Edinburgh (UK): Churchill Livingstone; (1988). , 3-10.

[4] Couinaud, C. Lobes et segments hepatiques: note sur l'architecture anatomique et chirurgicale du foie. Presse Med (1954).

[5] Ger, R. Surgical anatomy of the liver. Surg Clin N Am (1989). , 69, 179-93.

[6] Goldsmith, N. A, & Woodburne, R. T. Surgical anatomy pertaining to liver resection. Surg Gynecol Obstet (1957).

[7] Healey JE JrSchroy PC. Anatomy of the biliary ducts within the human liver: analysis of the prevailing pattern of branchings and the major variations of the biliary ducts. Arch Surg (1953).

[8] Healey JE JrVascular anatomy of the liver. Ann N Y Acad Sci (1970).

[9] Healey JE JrClinical anatomic aspects of radical hepatic surgery. J Int Coll Surg (1954).

[10] Hjortsjo, C. H. The topography of the intrahepatic duct system. Acta Anat (1951)., 11, 599-615.

[11] Longmire, W. P. Historic landmarks in biliary surgery. South Med J (1982)., 75, 1548-50.

[12] Meyers, W. C, Ricciardi, R, & Chiari, R. S. Liver. Anatomy and development. In: Townsend CM, editor. Sabiston textbook of surgery. 16th edition. Philadelphia: WB Saunders; (2001)., 997-1034.

[13] Mizumoto, R, & Suzuki, H. Surgical anatomy of the hepatic hilum with special reference to the caudate lobes. World J Surg (1988)., 12, 2-10.

[14] Nakamura, S, & Tsuzuki, T. Surgical anatomy of the hepatic veins and the inferior vena cava. Surg Gynecol Obstet (1981)., 152, 43-50.

[15] Skandalakis, L. J, Colborn, G. L, Gray, S. W, et al. Surgical anatomy of the liver and extrahepatic biliary tract. In: Nyhus LM, Baker RJ, editors. Mastery of surgery. 2nd edition. Boston: Little, Brown and Co; (1992)., 775-805.

[16] Skandalaki, J. E, Skandalakis, L. J, Skandalakis, P. N, & Mirilas, P. Hepatic Anatomy. Surg Clinc N Am 84 ((2004).

[17] Smith, R. In: Suzuki T, Nakayusu A, Kauabe K, et al, editors. Surgical significance of anatomic variations of the hepatic artery. Am J Surg (1971)., 122, 505-12.

Chapter 5

Experimental Models in Liver Surgery

M.B. Jiménez-Castro, M. Elias-Miró,
A. Casillas-Ramírez and C. Peralta

Additional information is available at the end of the chapter

1. Introduction

Ischemia-Reperfusion (I/R) injury is an important cause of liver damage occurring during surgical procedures including hepatic resections and liver transplantation (LT) [1-3]. The shortage of organs has led centers to expand their criteria for the acceptance of marginal grafts that exhibit poor tolerance to I/R [4]. Some of these include the use of organs from older donors and grafts such as small-for-size or steatotic livers. However, I/R injury is the underlying cause of graft dysfunction in marginal organs [4]. Indeed, the use of steatotic livers for transplantation is associated with an increased risk of primary nonfunction or dysfunction after surgery [5]. In addition, the occurrence of postoperative liver failure after hepatic resection in a steatotic liver exposed to normothermic ischemia has been reported [6]. A large number of factors and mediators play a part in liver I/R injury. The relationships between the signalling pathways involved are highly complex and it is not yet possible to describe, with absolute certainty, the events that occur between the beginning of reperfusion and the final outcome of either poor function or a non-functional liver graft. We will show that the mechanisms responsible for hepatic I/R injury depends on the experimental model used, who are valuable tool for understanding the physiopathology of hepatic I/R injury and discovering novel therapeutic targets and drugs. Several strategies to protect the liver from I/R injury have been developed in animal models and, some of these, might find their way into clinical practice. The species used for experimental investigation of hepatic I/R injury range from mice to pigs. The book chapter will discuss the numerous experimental models used to study the complexity of hepatic I/R injury, data reported in choice of the animal model, when selecting an animal species, the age, the sex, the degree of steatosis…etc. Thus, the different strengths and limitations of the different experimental models will be discussed. Also the standardized experimental conditions, such as anesthetic and analgesic procedures will be described. We also attempt to highlight the fact that the types of ischemia (cold and warm ischemia) play an important role in experimental liver surgery. The most existing reviews concerning about mechanisms responsible of I/R does not make a distinction between cold and warm ischemia. We will discuss the different experimental models of normothermic ischemia including global hepatic ischemia with portocaval decompression, global liver ischemia with spleen transposition and partial liver ischemia. Among the different experimental models of cold hepatic I/R injury, we will described the different experimental models used, including a section on orthotopic liver transplantation (OLT) because it is a common yet and complex microsurgical technique. In an attempt to expand the size of the donor pool, the different surgical techniques including reduced-size liver transplantation (RSLT), split liver transplantation (SLT) and living donor liver transplantation (LDLT) will be mentioned in the book chapter. In line with this, the optimization of graft function and survival through the static organ preservation and machine perfusion will also discused. Static organ preservation was a breakthrough and remains the conventional method of preservation. The machine perfusion has emerged as a suitable strategy for preserving liver grafts with promising data over the past decade, especially when marginal organs such as steatotic liver are used for transplantation. The strengths and disadvantages of the different types of machine perfusion (normothermic, hypothermic and subnormothermic machine perfusion) will be discussed. Furthermore some factors, including the duration and extent of hepatic ischemia, starvation, graft, age, and steatosis-which must be considered before the selection of an experimental model of hepatic I/R-will be mentioned. All of these factors contribute to enhancing liver susceptibility to I/R injury. In line with this, we will focused on the negative effects of ischemia on liver regeneration in both normal and marginal livers when they are subjected to liver surgery associated with hepatic resections or LT. The different experimental models of hepatic I/R in which both conditions-ischemia and resection- are present will be described.

2. Hepatic ischemia-reperfusion injury

Due to the complexity of hepatic I/R injury, the present review summarizes the established basic concepts of the mechanisms and cell types involved in this process (Fig. 1). The imbalance between nitric oxide (NO) and endothelin production, contributes to microcirculatory diseases associated with I/R. Concomitantly, the activation of Kupffer cells (KC) releases reactive oxygen species (ROS) and proinflammatory cytokines, including tumour necrosis factor-α (TNF-α) and interleukin-1 (IL-1) [7-9]. ROS can also derive from mitochondria and the xanthine dehydrogenase/xanthine oxidase (XDH/XOD) pathway in activated SEC and hepatocytes. Cytokines promote neutrophil activation and accumulation, thereby contributing to the progression of parenchymal injury by releasing ROS and proteases [7,10]. Capillary narrowing also contributes to hepatic neutrophil accumulation [11]. Besides, IL-1 and TNF-α recruit and activate CD4+ T-lymphocytes, which produce granulocyte-macrophage colony-stimulating factor (GM-CSF), interferon gamma (INF-γ) and TNF-β. These cytokines amplify KC activation and TNF-α and IL-1 secretion and promote neutrophil recruitment and adherente into the liver sinusoids [12]. Platelet activating factor can prime neutrophils for ROS generation, whereas leukotriene B4 (LTB4) contributes to the amplification of the neutrophil response [7,10]. In addition, I/R initiates protein misfolding in the endoplasmic reticulum (ER), which can activate a highly conserved unfolded protein response (UPR) signal transduction pathway. The UPR is characterized by coordinated activation of three ER transmembrane proteins, inositol-requiring enzyme 1 (IRE1), PKR-like ER kinase (PERK) and activating transcription factor (ATF)-6. If the damage is so severe that homeostasis cannot be restored, ER stress signal transduction pathways ultimately initiate apoptosis and necrosis [9]. In addition to the high ROS level–generating system found in liver grafts shows low levels of antioxidants such as glutathione (GSH) and superoxide dismutase (SOD) [1,9]. Alterations in the renin-angiotensin system (RAS), retinol binding protein 4 (RBP4), adiponectin and peroxisome proliferator activated receptor gamma (PPARγ) contribute to oxidative stress. Toll like receptor (TLR4) signaling pathway is also responsible for the hepatic I/R damage. Myeloid differentiation primary response gene 88 (MyD88) and TIR-domain-containing adapter-inducing interferon-β (TRIF) activate intracellular signaling cascades that ultimately trigger an inflammatory response [9,13].

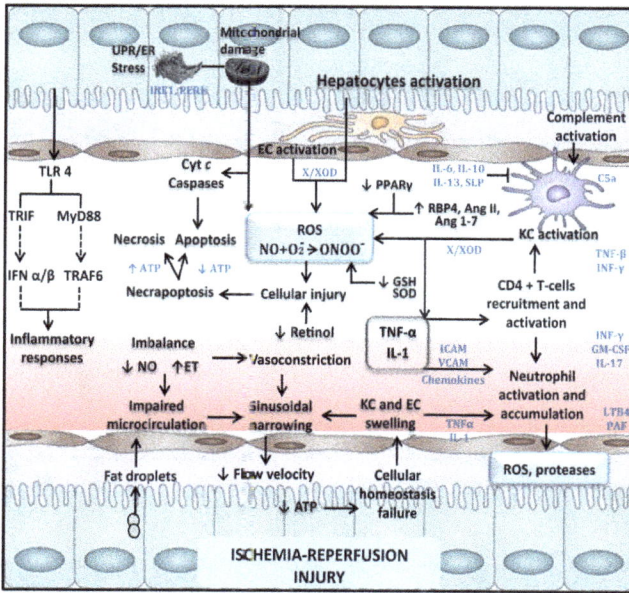

Figure 1. *Mechanisms involved in hepatic ischemia-reperfusion injury.* ATP, adenosine triphosphate; Cyt c: cytochrome c; EC, endothelial cell; ET, endothelin; GM-CSF, granulocyte-macrophage colony-stimulating factor; GSH, glutathione; ICAM, intracellular cell adhesion molecule; IFN α/β, interferon α/β; IL, interleukin; INF, interferon; IRE1, inositol-requiring enzyme 1; KC, kupffer cell; LTB4, leucotriene B4; MyD88, myeloid differentiation primary response gene 88; NO, nitric oxide; ONOO⁻, peroxynitrite; PAF, platelet activating factor; PERK, protein kinase-like endoplasmic reticulum kinase; PPARγ, peroxisome proliferator-activated receptor γ; RBP4, retinol binding protein 4; Renin-Angiotensin system (RAS): Ang II and Ang 1-7, angiotensin; ROS, reactive oxygen species; SLP, secretory leukocyte protease inhibitor; SOD, superoxide dismutase; TLR4, toll-like receptor 4; TNF, tumor necrosis factor; TRAF6, TNF receptor-associated factor 6; TRIF, TIR-domain-containing adapter-inducing interferon- β; UPR/ER, unfolded protein response/endoplasmic reticulum; VCAM, vascular cell adhesion molecule; X/XOD, xanthine/xanthine oxidase

3. Experimental models

Experimental surgery is an activity within the scientific development, offering a wide range of possibilities for the progress of medicine. As a discipline can be accessed from various branches of science and allows testing and development of surgical procedures and learning the scientific method, so that, working with laboratory animals has been and is required prelude to innovation and development of advances in clinical surgery. The reproduction and validation of experimental models has facilitated the extrapolation of the knowledge acquired to Medicine [16]. The animals used in research models have been divided into four groups: spontaneous, induced, negative and orphans. 1) The spontaneous or non-manipulated models are obtained by selection of inbred animals that express a variable or among populations in which a large number of animals that express variable; 2) Induced or manipulated models are obtained by an experimental challenge that can be classified into five groups: A. Administration of biologically active substances, eg., induction of steatosis after alcohol ingestion. B. Surgical manipulation, such as partial hepatectomy (PH) for the study of liver regeneration. C. Administration of modified diets, lack or surplus components, e.g., in the study of hyperlipidemia. D. Genetic manipulation and transgenic animals which produce special models that are being helpful in understanding mechanisms of pathogenesis and therapy. 3) The negative patterns are those in which a given variable does not develop. The interest is in studying the mechanisms that provide resistance. 4) Orphan models are those expressing an unknown variable in humans [16].

The speed of human studies is slow, the majority of human tissues are not routinely accessible for research purposes, and there is a very limited opportunity for interventional studies. Although scientific research has always relied on the use of cell cultures, information that is obtained through *in vitro* studies can be extrapolated to biomedical research only when analyzed within a complex organism with metabolic functioning. Therefore, one avenue holding tremendous potential in the search for therapies against I/R damage is the use of intact living systems, in which complex biological processes can be examined. There are many advantages of animal studies: large numbers of animals (especially rodents) can be bred and studied, interventional studies can be performed, and established and emerging tools for targeted manipulation of gene expression levels provide insight into the function of mediators in hepatic I/R injury.

Comparison of the results of animal studies and their extrapolation to human beings is feasible, but with limitations. Among the primary obstacles are differences in hypothermia and ischemia tolerance, differences in the anatomy of the livers of various species and subspecies, differences between and within the experimental models used, and differences in the modes of administration, dosage, and metabolic breakdown of the drugs under investigation. Thus, it is very important to choose the animal species and the experimental model and to standardize the protocol according to the clinical question under study.

Small and large animals have their own advantages and disadvantages but the ultimate choice of animal species depends essentially on the scientific problema in question. Small animals such as mice and rats are exceptionally useful because they are easy to manage, present minimal logistical, financial, or ethical problems, and provide the potential for genetic alterations (e.g., transgenic and knockout animals). However, an important drawback is that the results of studies performed in small animals are of limited applicability to human beings due to their varying size and anatomy of the liver and their faster metabolism [17]. Large animals such as pigs, sheep. and dogs exhibit greater similarity in their anatomy and physiology to human beings. Thus, they are more suited for the study of problems of direct clinical relevance. However, their use is restricted by serious logistical and financial difficulties and often by ethical concerns. Furthermore, the technical possibilities of blood and tissue processing are extremely restricted because of the limited availability of immunological tools for use in large animal species [17].

Extensive data exist on liver anatomy in various species of animals, but a few examples of species variations will suffice to prove that caution is warranted in the extrapolation of this data to humans. Mice and rats each have 4 liver lobes: median (or middle), left, right, and caudate and all, except the left, are further subdivided into 2 or more parts. Human liver lobes can be subdivided into 9 segments based on the vascular and ductal branching patterns to the right and lefts sides. The hepatic lobes of the rat appear to have similar fundamental portal and hepatic venous systems, and thus segments, comparable to that of human liver. The vascular systems to or from lobes show individual variations in humans as well as in rats. In humans and other mammals, sinusoids drain only into the terminal hepatic veins whereas in the rat sinusoids enter the hepatic venous system at all levels of the hepatic venous tree. In rats, unlike humans, the sinusoids are supplied not only by the terminal portal venules but also directly from larger venous branches. In addition, rat livers lack the septal vein branches, which are present both in humans and pigs [18]. The presence of arterio-portal anastomosis is very frecuent in rats but not in hamsters and humans. The rat is unique in possessing a perihilar biliary plexus, which is present from the large hiliar portal tracts to smaller portal tracts. An equivalent, less developed structure exists in humans only in large portal tracts. The biliary system in pigs lacks this plexus altogether, but contains numerous side pouches throughout the course of the bile duct [18,19]. Mice and humans have a gall bladder, but not the rat. Significant difference is present among the species with respect to the extent of hepatic parenchymal innervation and the human has the most abundant supply of autonomic nerves in the intraparenchymal region [20]. Differences in hepatic cell types have been reported depending of species evaluated. For example, regarding to endothelial cells, rats have relatively higher fenestrae compared to some other species. Defenestration is though to play a role in some liver diseases [18]. Intrinsic biochemical differences between the hepatocytes of the various species have been also reported. Rats and mice are extremely sensitive to the response of peroxisome proliferators, hamsters show a less marked response while primates and humans are insensitive or non-responsive [21]. There are two principle hypotheses to explain species differences in response to PPs: quantity of PPARa and/or the quality of the PPARα-mediated response [22].

When selecting an animal species, the age and sex of the animals should be considered. Depending on the duration of ischemia, young (35–50 g) and older rats (250–400 g) exhibit significant differences in their hepatic microcirculation [23]. A mature rat weighing more than 250 g (14–16 weeks old) is the most suitable because younger rats can present technical problems, whereas older rats are more prone to respiratory infections and fat accumulation. Sex selection also affects experimental results, as hormone levels in female animals are dependent on the estrous cycle, which certainly affects the ischemia tolerance of the liver. For instance, a study demonstrated that after normothermic liver ischemia, male rats were less sensitive to reperfusion injury than female rats.

Considering the relevancy of hepatic steatosis in surgery, experimental models of hepatic I/R injury in the presence of steatosis have been developed. However, the mechanisms involved in hepatic I/R injury, as it will be described in following sections, are different depending on the method used to induce steatosis. The different models of steatosis include 1) induced genetic models; 2) animals fed diets with high levels of saturated fat and/or carbohydrates and/or proteins; 3) animals fed diets deficient in methyl groups (choline, methionine, folates); and 4) animals fed modified high-fat diets (lower methionine and choline and higher-fat content).

The induction of I/R injury must be performed under standardized experimental conditions. Of primary importance are the conditions under which the animals are kept such as adequate acclimatization time, maintenance under climatized conditions with 12 hours light / 12 hours darkness, and standardized diets. The anesthetic method and postoperative analgesic regimen must also be standardized. When choosing the anesthetic and analgesic procedures, possible interactions with liver metabolism must be considered. Attention must be paid to adequate monitoring of blood pressure, heart rate, and body temperature.

4. Normothermic hepatic ischemia

4.1. Global hepatic ischemia with portocaval decompression

The model of global liver ischemia with portal decompression ideally simulates the clinical situation of warm ischemia after the Pringle maneuver for liver resection and LT. The first successful shunt operation in humans was performed by Vidal in 1903 [24]. Blakemore was one of the first workers to report successful portal-systemic anastomosis in rats working principally with endothelium-lined tubes [25]. Burnett et al., modified this technique to form a portocaval shunt [26]. In 1959 Bernstein and Cheiker developed the portosystemic shunt that conducted the portal blood after functional hepatectomy into one of the iliac veins [27]. In small animals, in addition to many other shunt techniques such as the portofemoral shunt and the mesentericocaval shunt via the jugular vein, in 1995, Spiegel et al., developed the splenocaval shunt [28] (Figure 2).

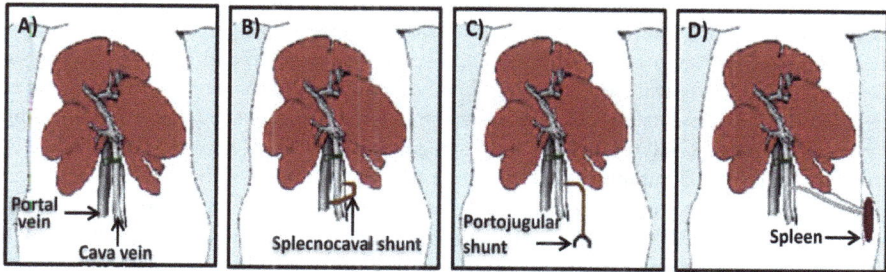

Figure 2. Models of global normotermic liver ischemia. A) Pringle-maneuver. B) Splecnocaval shunt. C) Portojugular shunt. D) Spleen transposition.

4.2. Global liver ischemia with spleen transposition

Bengmark et al., developed this model in 1970 for the surgical treatment of portal hypertension [29]. In 1981 Meredith and Wade presented a rat model that by transposition of the spleen produced a portosystemic shunt in the anhepatic rat [30]. A small incision is made in the left hypochondrium. After transposition of the spleen into a subcutaneous pouch, adequate portosystemic anastomoses arise after two to three weeks (Figure 2). Reversal of blood flow in the splenic vein, induced by the transposition, stimulates angiogenesis. In the second step 2 weeks later, the surgeon performs a median laparotomy and temporary occlusion of the hepatoduodenal ligament. This decompression by spleen transposition does not require microsurgical technique and is therefore easy to perform. Two-to-three weeks postoperatively, the spleen will have been encapsulated without any signs of bleeding or inflammation. One disadvantage of this model is the long time lapse (3 weeks) until the formation of adequate portosystemic collaterals. Not until this point in time are the collaterals sufficiently large to take over portal vein flow completely. Furthermore, it is uncertain how the changes in hepatic inflow will react upon the collaterals [31].

4.3. Partial liver ischemia and liver regeneration

In 1982, Yamauchi et al., described a model of hepatic ischemia [32]. In this technique, ischemia is induced by occlusion of the hepatic artery, the portal vein, and the bile duct of the left and median lobes. An extracorporeal shunt is not necessary because blood flow continues through the right and caudal liver lobes. This model of 70% partial ischemia has been widely used in experimental studies of hepatic I/R [13,33]. Additionally, an experimental model of 30% partial liver ischemia has been used in which blood supply to the right lobe of the liver is interrupted by occlusion at the level of the hepatic artery and portal vein [34]. It is known that, in clinical situations, PH under I/R is usually performed to control bleeding during parenchymal dissection. *In vitro* studies, although they have proved helpful in disclosing the signal transmission pathways of various hepatocyte mitogens, need to be supplemented by *in vivo* studies with experimental animals so as to simulate the interactions between the various cell populations of the liver. Different strategies have been adopted for the experimental induction of liver regeneration as follow below [35]. On the other hand, the use of an experimental model including both hepatic regeneration and I/R injury is advisable to simulate the clinical situation of selective or hemihepatic vascular occlusion for liver resections. In experimental model, after resection of left hepatic lobe, a microvascular clamp is placed across the portal triad supplying the median lobe (30%). Congestion of the bowel is avoided during the clamping period by preserving the portal flow through the right and caudate lobes. At the end of ischemia time, the right lobe and caudate lobes are resected, and reperfusion of the median lobe is achieved by releasing the clamp. This model of hepatic resection does not require any portal decompression and also fulfills certain important criteria such as reversibility, good reproducibility, and simple performance [36].

4.4. Other experimental models of liver regeneration – Regeneration after liver injury

There are large numbers of toxins that can cause liver damage and cell death in the liver parenchyma followed by liver regeneration. Carbon tetrachloride, d-galactosamine, ethanol, thioacetamide and acetaminophen are the hepatotoxins that have been most frequently employed to induce experimental liver regeneration in the hope of answering various questions [35]. In contrast to PH, these so-called hepatotoxic models of liver regeneration are easier to perform and of greater clinical relevance. Whereas PH leaves all the remaining hepatic acini intact, hepatotoxins can be used selectively to induce centrilobular or periportal necrotic lesions and can thus better simulate certain liver diseases. One serious weakness of toxin-induced liver regeneration is the por reproducibility and standardability of the models, because the local and systemic effects of the toxin depend on the dose, the mode of administration, the species of animals, their age and nutritional status and other factors, and the extent of the liver injury and the regeneration can vary accordingly. The regenerative response of the liver is often determined by the dose and mode of administration. Furthermore, the toxins can directly interfere with the cellular and molecular mechanisms of liver regeneration, e.g., by damaging membranes (interruption of the interaction between growth factors and membrane receptors), impairment of gene expression and protein synthesis, inflammatory reactions (increased production of cytokines and oxygen radicals) or activation of nonparenchymal cells [37]. Finally, in these toxic models the processes of liver injury and repair are closely interwoven, a fact that adds to the difficulties of investigating liver regeneration. It is therefore difficult to predict the extent of liver damage and liver regeneration and to avoid significant variability between individual experiments [35].

5. Liver transplantation

The development and implementation of different surgical techniques in LT have been based upon animal experimental studies. LT in larger laboratory animals such as dogs and pigs is technically easier. However, the rat has become the most important subject for experimental LT because of, among other factors, the availability of genetically defined animals [38]. The first experimental liver replacement with OLT was reported by Cannon in 1956, but none of those dogs survived [39]. Surgical techniques for experimental OLT on pigs were started by Garnier et al., in 1965 [40]. OLT in mice is technically very difficult, even without reconstruction of the hepatic artery. By contrast, OLT in rats is technically accessible, producing more clinically relevant and reliable data [41]. The development of clinically relevant OLT models in rats [41] has advanced clinical knowledge in LT. These experimental models facilitate the study of new preservation methods, tolerance induction, rejection mechanisms, and novel immunosuppressor therapies [42].

The first model of OLT in the rat was described by Lee et al., in 1973 using hand-suture techniques [43]. This technique includes standard microvascular suture technique for venous anastomoses and a miniaturized extracorporeal portal-tojugular shunt ("microsuture model"). Rearterialization of the graft is performed by anastomosing the donor aorta end-to-side to the host aorta, and the donor bile duct is implanted into the duodenum [43]. Two years later, in 1975, Lee reported a modified model without hepatic artery reconstruction and temporal shunt of the portojugular venovenous bypass [44]. However, these models were not widely used due to the prolonged surgical time and technical demand. In 1979, Zimmermann introduced a microsuture model [45] that is similar to the simplified model of Lee [44]. He developed a new technique for bile duct reconstruction that preserves the sphincter of ampulla "splint technique". In the same year, Kamada and Calne [46] developed a cuff technique for anastomoses of portal vein and bile duct to simplify Lee's model and especially to shorten the anhepatic time and reduce biliary complications. With the cuff method being introduced by Kamada and Calne [46], OLT in rats without hepatic artery reconstruction became globally accepted [41]. Other models introduced by later investigators contain for the most part only a few modifications. In 1980 Miyata introduced the "three-cuff model" [47] with cuff technique for the three venous anastomoses. Bile duct anastomosis is performed by using the splint technique first described by Zimmermann [45], in which reestablishment of hepatic blood flow is not carried out. Anastomosis of the portal vein is done by the method of Kamada and Calne [46]. For connecting the bile duct, splint technique was used [47]. In 1982 Engemann [48] devised a microsuture model that corresponds closely to the model of Lee [43]. During the anhepatic time he dispensed with portosystemic bypass and used an aortic-celiac segment for rearterialization. This had been already prepared in the donor operation, and anastomosed end-toside to the infrarenal aorta of the recipient. Bile duct anastomosis is performed using the splint technique [48]. Portal vein clamping causes a rise of endotoxin in the portal vein, which could lead to disturbances in hepatic microcirculation. Lee was the first to use a portosystemic shunt, but in further models it has not been established because the acceleration of the transplantation procedure by improved anastomotic techniques was expected to preclude the need for this complicated operative procedure [38]. Kitakado completed the "two-cuff model" in 1992 by developing a bioabsorbable material (synthesis of D, L-lactic acid and glycolic acid). Its *in vivo* degradation time is about 4 months when used for cuff anastomosis of portal vein and infrahepatic vein cava [49]. He established a longterm model in OLT in rat. This surgical procedure is usually performed according to the procedure described by Kamada and Calne [46]. After arterial and portal perfusion, the suprahepatic vena cava is dissected free from the diaphragmatic ring, and the intrathoracic vena cava is transected. The aorta is cut around the celiac axis to form the aortic patch. Finally, the inferior vena cava, the portal vein, and the bile duct are cut, and the graft is placed in a cold preservation solution (Figure 3). OLT is then performed by suture or mechanical microvascular anastomoses. Sutured vascular anastomosis reduces the incidence of thrombosis but takes a long time to perform. Suprahepatic vena cava anastomosis is performed by the continuous suturing technique. Then, portal vein and infrahepatic vena cava anastomosis is performed in the same manner. Hepatic artery reconstruction in rat LT can prevent bile duct ischemia and preserve the structure of the liver [50]. Several techniques of rearterialization by suture have been proposed [50], the best being the aortic segment anastomosis technique. After rearterialization, the common bile duct is anastomosed. OLT by hand-sewn microanastomosis is a very useful method because this technique comes closest to the techniques used in human transplantation surgery. Alternatively, livers can be satisfactorily allografted in rats by using the rapid cuff-ligature technique for anastomosis [46]. In the simplified technique, the donor hepatic artery can be ligated because it will not be anastomosed [42].

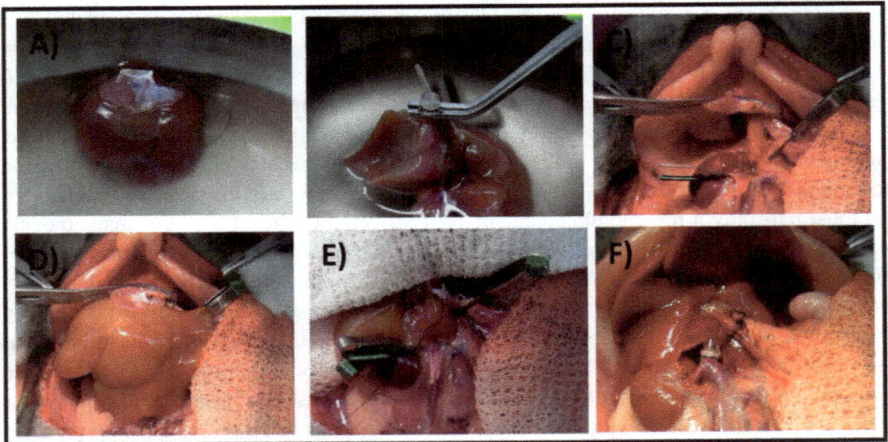

Figure 3. Liver transplantation procedure. A) Suprahepatic cava vein prepared for the anastomosis. B) Inferior vein cava cuff attachment. C) Anhepatic phase in the recipient rat. D) Anastomosis of suprahepatic cava vein by continuous suture. E) Portal vein anastomosis trhough the cuff method. F) Anastomosis of the bile duct.

6. Strategies to expand the size of the donor pool

In an attempt to expand the size of the donor pool, a number of surgical techniques have been developed over the past 15 years, including reduced-size liver transplantation (RSLT), split liver transplantation (SLT) and living donor liver transplantation (LDLT) [51]. For children and small adult recipients, RSLT has been developed to maximize the use of donor organs. Bismuth and Houssin in 1984, transplanted the left lateral segment of the left liver lobe from a cadaveric donor into a small child and discarded the remainder of the donor liver [52]. Couinaud's anatomical classification permits the creation of partial liver allografts from either deceased or living donors. Couinaud's classification divides the liver into eight independent segments, each of which has its own vascular inflow, outflow, and biliary drainage [53]. Segments IV to VIII are used for adults, whereas left lateral lobes (Segments II and III) or left lobes (Segments II, III, and IV) are used for pediatric recipients. Bleeding, bilomas, and portal vein thrombosis are complications related to the procedure itself, which are associated with an increased number of re-operation. SLT, first performed in 1988, allows the division of the adult donor liver, together with its vascular and biliary structures, into two or more functional grafts, which can be transplanted into two or more recipients [54]. Liver splitting is performed either *ex situ* or *in situ*. So far, there is no consensus on which technique is superior because both techniques demonstrate similar patient and graft survival rates compared with whole liver grafting [54]. Biliary complications occur in 22% of recipients. In 1990, Broelsch et al., reported the first 20 series of LDLT in the USA [55]. In 1996, Lo et al., [56] performed the first successful LDLT using an extended right lobe from a living donor for an adult recipient. One of the benefits of reduced-size grafts from living donors is a graft of good quality with a short ischemic time, this latter being possible because live donor procurements can be electively timed with the recipient procedure. Conversely, the major concern over the application of LDLT for adults is graft-size disparity. Small grafts require posterior regeneration to restore the liver/body ratio. A small graft may result in malfunction or the small for size syndrome in which the recipient fails to sustain adequate metabolic function. It is well known that I/R significantly reduce liver regeneration after hepatectomy. Thus, the identification and subsequent modulation of mechanism that are involved in liver injury and regeneration might favor the recovery and functioning of the transplanted organ.

To mimic some of the pathophysiological events that occur during such clinical situations, several experimental models of RSLT have been developed. For example, OLT with the implantation of liver grafts that approximated 30%–70% of the normal mass of a rat liver has been performed. Graft size is important for normal liver function and host survival [51]. It has been reported that 100% of recipient rats that were implanted with 40%, 50%, 60%, or 70% of the liver survived regardless of the duration of preservation. This suggests that graft sizes of 40% or greater are sufficient to meet the metabolic demands of the recipients. The transplantation of a graft of 30% of the normal liver mass provides an extreme model of hepatic reduction that presumably stimulated a maximal regenerative response [51]. Three possibilities exist with respect to the timing of the graft reduction: in the donor before perfusion, in the container (ex situ), or in the recipient after reperfusion. If the reduction is done *in vivo* prior to the removal of the donor liver, then two concerns exist: 1) excessive bleeding might stimulate systemic responses that could alter the liver and 2) the immediate phase of the regeneration response could be initiated in the donor animal. The second choice, ex situ reduction, can be done without the risk of damaging the graft by manipulation or affecting anastomosis after reperfusion. Finally, resection of the graft after implantation in the recipient adds surgical stress and the risk of bleeding.

7. Modes of organ preservation and optimizing the graft

The ideal method of organ preservation should: 1) Reverse injury sustained during donor death and organ procurement; 2) Provide viability testing; 3) Prolong safe preservation time and 4) Improve the graft quality [57]. There are currently 2 modes of preservation methods for livers: static and dynamic (Figure 4). Simple cold storage is the main method for static storage while hypothermic machine perfusion (HMP) and normothermic machine perfusion (NMP) comprise some of the methods for dynamic preservation. Of these methods, only simple cold store is roved clinically for livers. The remaining methods are in various stages of pre-clinical and early clinical studies. Dynamic preservation methods require some dynamic movement of either fluid or gas to facilitate preservation. The advantage of these methods over simple cold storage is that they all have been shown to improve recovery of donor after cardiac death organs. These organs have the potential to increase the donor pool by 20–40%.

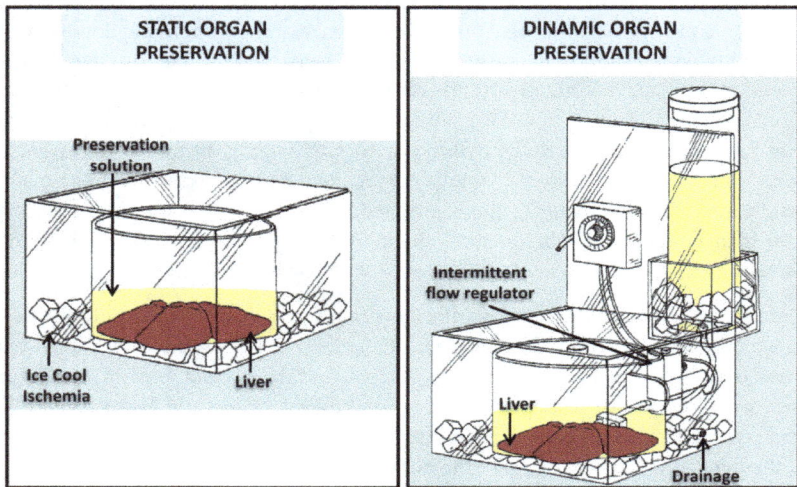

Figure 4. Illustrative modes of organ preservation. Static or dynamic organ preservation.

7.1. Static organ preservation

Static cold storage (SCS) is the most commonly used preservation method used for all organs. The principles underlying cold preservation are the slowing of metabolism (by cooling) and the reduction of cell swelling due to the composition of preservation solutions. The introduction of the University of Wisconsin (UW) solution by Belzer and Southard for SCS was a breakthrough and remains the conventional method of preservation. Reduction of metabolic activity (by cooling) is the major principle of organ preservation [57,58]. At the moment the flow of oxygenated blood is terminated, the supply of oxygen, cofactors and nutrients stops and the accumulation of metabolic waste products begins. Although metabolism is slowed 1.5- to 2-fold for every 10°C drop in temperature, anaerobic metabolism continues, which leads to depletion of energy stores and concomitant build up of an acidotic milieu. Depletion of ATP causes loss of transcellular electrolyte gradients, influx of free calcium and the subsequent activation of phospholipases, and therefore is the main contributor for cell swelling and lysis. Ischaemia creates the basis for the subsequent production of toxic molecules after reperfusion, particularly reactive oxygen intermediates, the basis of the cascade of events that characterize the I/R injury. Even with the most effective preservation solutions, cold storage aggravates graft injury at the time of transplantation. This situation is due to two processes, one proportional to the duration of ischemia and the other specifically related to cooling [57]. Using this preservation method, however, organs undergo injury at several consecutive stages: warm ischemia prior to preservation, cold preservation injury, ischemic rewarming during surgical implantation and reperfusion injury. With the extension of criteria to include expanded criteria donor and donation after cardiac death organs, static preservation is associated with increased delayed graft function and graft loss. In organs retrieved from non-heart-beating donors (NHBD) -with an inevitable period of oxygen deprivation between cardiac arrest and organ perfusion – the deleterious effects of cold ischaemia are superimposed on the injury sustained during warm ischaemia [57]. Only a few studies have demonstrated the optimization of graft function and survival with modification of static preservation. It is doubtful that considerable improvements in organ preservation and especially in the rescue of marginal organs will be possible as long as the strategy is based on static principles [58]. In 1990s, Minor et al., developed a new method, called venous systemic oxygen persufflation (VSOP) to supply gaseous oxygen to livers during SCS preservation [59]. The oxygen was introduced into hepatic vasculature via the suprahepatic vena cava. This technique was employed on steatotic rat livers for 24 h, and resulting in improved preservation of mitochondria and sinusoidal endothelial linings, less KC activation and reduced hepatocellular enzyme release compared to SCS preservation. Recently, by assessing the enzyme release, energy storage, bile production, and cell death during isolated reperfusion, it was demonstrated that application of VSOP for 90 minutes may rescue the steatotic livers after extended (18 h) SCS preservation [60].

7.2. Machine perfusion

Machine liver perfusion is an alternative preservation method to SCS which can be further categorized based on the temperature employed and has emerged with promising data over the past decade because it has significant potential in graft preservation and optimization when the use of marginal organs is the objective. Machine perfusion involves pulsatile perfusion of the liver using a machine as opposed to SCS. This can be performed by perfusing the liver with a hypothermic perfusate or with a normothermic perfusate. There is experimental evidence in animal models that machine perfusion protects against liver I/R injury [61]. The safety and efficacy of machine perfusion compared to SCS to decrease liver I/R injury is yet to be assessed in humans by randomized controlled trials [61,62].

Compared with simple cold storage, machine perfusion confers many anticipated advantages such as the following: 1) provision of continuous circulation and better preservation of the microcirculation; 2) continuous nutrient and oxygen delivery; 3) removal of metabolic waste products and toxins; 4) opportunity to assess organ viability; 5) improved clinical outcomes via improved immediate graft function rates; 6) prolonged preservation time without increased preservation damage; 7) administration of cytoprotective and immunomodulating substances; and 8) lower graft dysfunction incidence, shorter hospital stays, and better graft survival rates [62].

7.3. Normothermic machine perfusion

In the first half of the 20th century, Alexis Carrel perfused different organs with normothermic, oxygenated serum and demonstrated viability for several days [63]. Actually, the first successful human LT carried out by Starzl [64], were transplanted after liver graft pretreatment by machine perfusion with diluted, hyperbaric oxygenated blood. Most perfusion circuits were assembled from standard cardiopulmonary bypass components. Principle constituents are a centrifugal pump, a membrane oxygenator and a heat exchanger. Other critical components of the perfusate include nutrition (glucose, insulin, aminoacids), drugs to prevent thrombosis or microcirculatory failure (heparin, prostacyclin) and agents to reduce cellular oedema, cholestasis and free radical injury [57]. Normothermic machine perfusion (NMP) provides a physiologically-relevant environment to the isolated donor organ, the quality of liver grafts can be manipulated more efficiently than those simply stored in an ice-box during SCS, because NMP maintains and mimics normal *in vivo* liver conditions and function during the entire period of preservation, thus avoiding hypothermia and hypoxia and minimizing preservation injury [58,62]. In contrast to cold storage preservation the concept of normothermic preservation is to maintain cellular metabolism. The underlying principle is the combination of continuous circulation of metabolic substrates for ATP regeneration and removal of waste products. There is accumulating evidence for the superiority of the more physiological approach of normothermia in association with an oxygenated blood-based perfusion solution [57].

Schön et al., [65] studied NMP to preserve pig livers for transplantation and to rescue them from warm ischemia in a model of donor after cardiac death. Short (5 h) or prolonged (20 h) NMP preservation is superior to SCS for normal and ischemically damaged livers, respectively [62]. The longest preservation of steatotic livers was the NMP preservation for 48 hours in a pig model by Jamisson et al., who employed blood containing additional insulin and vasodilators as perfusate, and observed a mild reduction of steatosis from 28% to 15%. The NMP circuit dually perfuses 1.5 L of autologous heparinized blood at physiological pressures, which allows hepatic blood flow autoregulation. Prostacyclin, taurocholic acid, and essential amino acids are infused continuously. Apart from logistics, one potential drawback of NMP is the mandatory use of oxygen carriers if blood is not available [62]. Perhaps the only weakness is that SCS prior to NMP revokes its beneficial effect. Therefore, immediately after cardiac asystole, normothermic perfusion in the donor should be installed, as described by Fondevila et al., [66] for the preservation of livers from uncontrolled donation after cardiac death. The use of

NHBDs as a source of liver grafts for transplantation has long been debated. The concept of normothermic recirculation in the context of NHBDs was first developed by Garcia-Valdecasas et al., [67]. With 4 h of NMP, hepatic damage incurred during 90 minutes of cardiac arrest can be reverted, achieving 100% graft survival after 5 days of postransplant follow-up. These results offer the hope that NMP will be able to increase the clinical applicability of NHBD LT over that offered by traditional cold storage [67].

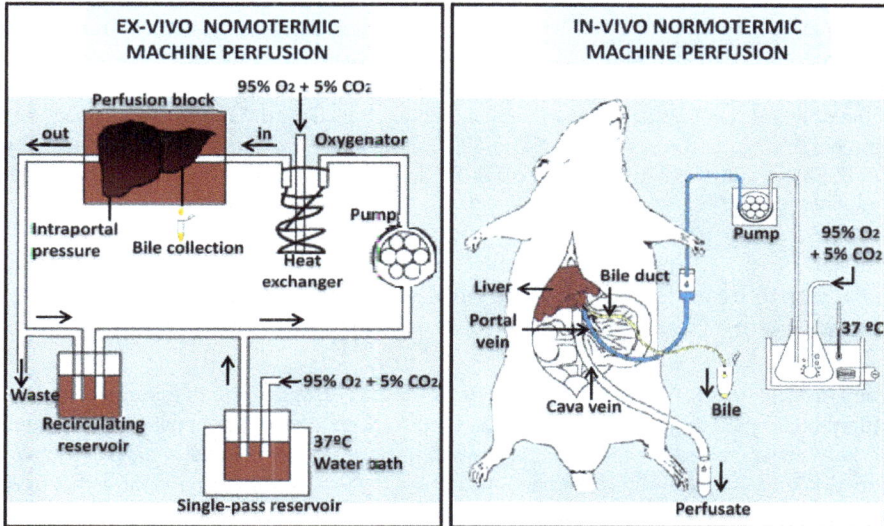

Figure 5. Esquematic illustration for ex-vivo and *in-vivo* normotermic machine perfusion

7.4. Hypothermic machine perfusion

For decades, cooling down organs to cold temperatures allowed successful organ transplantation within a limited period. The first and most prominent difference between SCS and (oxygenated) hypothermic machine perfusion (HMP) is the restoration of the tissue's energy charge and glycogen content while preventing ATP depletion [62]. In 1990, Pienaar et al., [68] reported that seven of eight dogs survived after LT with HMP preservation for 72 h and a similar outcome after 48 h of SCS. HMP is increasingly being used as an alternative method to SCS for the preservation of grafts obtained from nonoptimal donors. Indeed, several studies have reported a greater reduction in delayed graft function after HMP preservation than after SCS. Bessems et al., employed HMP preservation with UW-gluconate solution on steatotic rat livers for 24 h and alleviated I/R compared to SCS [69]. There is a substantial body of research, predominantly in rodents, demonstrating improved preservation by providing oxygen to livers [70]. Nevertheless, clear guidelines towards target values/ranges for oxygen levels regarding the optimal duration of oxygenation during HMP are lacking. HMP can also be applied at the end of the cold storage period, which is attractive for logistical reasons. The disadvantage here is the time-dependent increase in vascular resistance, bearing the risk of damage to the sinusoidal endothelium [58].

7.5. Subnormotermic machine perfusion

Subnormothermic machine perfusion (SNMP) preservation lies between HMP and NMP, but it remained relatively unexplored until recently despite holding promising applications [71]. In an isolated rat liver perfusion model, SNMP enhanced the functional integrity of steatotic livers compared with SCS findings. Organ protecting properties mediated by decreasing the temperature to a 20–28ºC have been observed previously. SNMP avoids some of the downsides of hypothermia while maintaining mitochondrial function and it may circumvent the logistical rest raints of NMP [62]. Vairetti et al., preserved steatotic rat livers by SNMP (20ºC) with Kreb-Henseleit solution for 6 hours and obtained reduced I/R damage compared to SCS [71].

8. Factors to be considered before the selection of an experimental model of hepatic I/R

Many investigators have used rodent models of warm (*in situ*) liver I/R to mimic some of the pathophysiological events that occur during LT. Although a great deal of useful information has been generated from these studies, an overriding question remains: Are the mechanisms responsible for transplant-mediated liver injury and dysfunction the same as those that have been reported for warm liver I/R injury? The answer is yes and no; that is, some of the mechanisms are similar, but many are dissimilar. It is important to make a distinction between the different types of ischemia, because there already is some controversy regarding the pathophysiological mechanisms depending on the type of ischemia (cold or normothermic), and it should be considered that the type of ischemia, the extent and time of ischemia, the type of liver submitted to I/R, and the presence of liver regeneration, all lead to differences in the pathophysiological mechanisms of hepatic I/R. These are discussed below to provide the reader with a guide to select the appropriate experimental model of hepatic I/R depending on the aims being pursued.

8.1. Relevance of the type of surgical procedure

The mechanisms responsible for hepatic I/R injury as well as the effects of pharmacological treatments are dependently of the liver surgical procedure. There is a range of potentially conflicting results with regard to the mechanisms responsible for ROS generation in liver I/R injury depending of the liver surgical procedure evaluated. XDH/XOD system is the main ROS generator in hepatocytes and LT-related lung damage [72]. However, results obtained in experimental models of the isolated perfused liver have underestimated the importance of the XDH/XOD system, and suggest that mitochondria could be the main source of ROS

[9]. In addition, studies by Metzger et al., in experimental models of normothermic hepatic ischemia showed that the increased vascular oxidant stress after 30 and 60 minutes of ischemia was attenuated by inactivation of KC but not by high dose of allopurinol in experimental models of normothermic hepatic ischemia [73].

It should be considered that the effectiveness of drugs on hepatic regeneration and damage could be different depending on the surgical conditions evaluated. Thus, gadolinium chloride treatment protected against hepatic damage in conditions of I/R without hepatectomy and improved liver regeneration after PH without I/R [74]. However, the same drug had injurious effects on hepatic damage and impaired liver regeneration in conditions of PH under I/R [75]. It should be also considered that the effectiveness of RAS blockers on hepatic regeneration and damage could be different depending on the surgical conditions evaluated. In conditions of PH under I/R, the AT1R antagonist for nonsteatotic livers and the AT1R and AT2R antagonists for steatotic ones improved regeneration in the remnant liver. The combination of AT1R and AT2R antagonists in steatotic livers showed stronger liver regeneration than either antagonist used separately and also provided the same protection against damage as that afforded by AT1R antagonist alone. However, the loss of protection of Ang II receptor antagonists against damage in conditions of PH under I/R (only AT1R antagonist protected steatotic liver against damage) compared with the study of I/R without hepatectomy (in which both Ang-II receptor antagonists reduced damage in both liver types) could be explained by the different surgical conditions. In the model of I/R without hepatectomy [33], the blood supply to the left and median liver lobes (70% hepatic mass) was interrupted, and the other hepatic lobes remained intact. However, in the conditions evaluated herein, only blood supply to the remnant liver (30% hepatic mass) was interrupted and the other hepatic lobes were excised. Compared with the study of I/R without hepatectomy [33], in PH under I/R, there are two main differences, the percentage of hepatic mass that is deprived of blood supply and hepatic resection. It is well known that the mechanisms of hepatic damage are different depending on the percentage of hepatic mass that is deprived of blood supply [76,77]. In addition, the inherent mechanisms of hepatic damage derived from the massive removal of hepatic mass should be considered. This may explain, at least partially, why the same drug, such as an Ang II receptor antagonist, may show differential effect on hepatic injury depending on surgical conditions [36]. In line with this, clinical and experimental studies revealed the injurious effects of NO on damage in the remnant liver in conditions of PH under I/R [36]. However NO protect against hepatic damage in an experimental model of I/R without PH [11]. In PH under I/R, Ang-II is an appropriate therapeutic target to protect steatotic livers against hepatic damage and regenerative failure. However, this target could be not appropriate in steatotic LT, since the results indicate a novel target for therapeutic interventions in LT within the RAS cascade, based on Ang 1-7, which could be specific for this type of liver. Indeed, Ang 1-7 receptor antagonist reduced necrotic cell death and increased survival in recipients transplanted with steatotic liver grafts [15].

The results, based on isolated perfused liver, indicated that the addition of epidermal growth factor (EGF) and isulin-like growth factor 1 (IGF-I) separately or in combination to UW reduced hepatic injury and improved function in both liver types. EGF increased IGF-I, and both additives up-regulated AKT in both liver types. This was associated with glycogen synthase kinase-3β (GSK3β) inhibition in non-steatotic livers and PPARγ over-expression in steatotic livers [78]. The benefits of EGF and IGF-I as additives in UW solution were also clearly seen in an experimental model of normothermic hepatic ischemia. However, the relationship between EGF and IGF-I was different dependently of the surgical procedure. Indeed, under these conditions, IGF-I increased EGF, thus protecting steatotic and non-steatotic livers against I/R damage. The beneficial role of EGF on hepatic I/R damage may be attributable to p38 inhibition in non-steatotic livers and to PPARγ overexpression in steatotic livers [79].

PPARα agonists as well as ischemic preconditioning (IP), through PPARα, inhibited mitogen-activated protein kinase expression following I/R in steatotic livers undergoing normothermic hepatic ischemia. This in turn inhibited the accumulation of adiponectin in steatotic livers and reduced its negative effects on oxidative stress and hepatic injury [13]. In line with this, adiponectin silent small interfering RNA (siRNA) treatment decreased oxidative stress and hepatic injury in steatotic livers. However, another study by Man et al., 2006 [80] in small fatty grafts, adiponectin treatment exerted anti-inflammatory effects that down-regulated TNFα mRNA and vasoregulatory effects that improved the microcirculation. Adiponectin anti-inflammatory effects also include the activation of cell survival signaling via the phosphorylation of Akt and the stimulation of NO production. Additionally, the studies by Man et al., [80] showed the anti-obesity and proliferative properties of adiponectin in small fatty transplants. Taken together, the aforementioned data indicate that the action mechanisms of adiponectin depend on the surgical conditions. Thus, on the basis of the different results reported to date in hepatic I/R, it is difficult to discern whether we should aim to inhibit adiponectin, or administer adiponectin to protect steatotic livers against cold ischemia associated with transplantation. Moreover, the adiponectin data reported for these experimental models of hepatic I/R [13,80] should not be extrapolated to cadaveric organ transplantation. For small liver grafts (which are relatively common) and under conditions of warm ischemia, the periods of ischemia range from 40 to 60 minutes; this range may not be accurate for cadaveric donor LT.

RBP4 is an adipokine synthesized by the liver, whose known function is to transport retinol in circulation. However, the role of RBP4 in hepatic I/R could depend on the liver surgical procedure. Steatotic liver grafts were found to be more vulnerable to the down-regulation of RBP4. RBP4 treatment-through AMP-activated protein kinase (AMPK) induction- reduced PPARγ over-expression, thus protecting steatotic liver grafts against I/R injury associated with transplantation. In terms of clinical application, therapies based on RBP4 treatment and PPARγ antagonists might open new avenues for steatotic LT and improve the initial conditions of donor livers with low steatosis that are available for transplantation [81]. On the other hand, the effects of RBP4 could depend on the surgical conditions. Indeed, RBP4 administration not only failed to protect both liver types from damage and regenerative failure, it exacerbated the negative consequences of liver surgery in PH under I/R [82]. Under these conditions, RBP4 affected the mobilization of retinol from steatotic livers, revealing actions of RBP4 independent of simple retinol transport. The injurious effects of RBP4 were not due to changes in retinol levels. Thus, strategies based on modulating RBP4 could be ineffective and possibly even harmful in both liver types in PH under I/R or surgical conditions including small-for-size LT.

8.2. Relevance of the duration of hepatic ischemia

The severity of hepatocyte damage depends on duration of ischemia. Depending on the objectives of the research, it is important to consider a specific ischemia duration. In other words, if you want to study the mechanisms involved in hepatic I/R injury or the protective mechanisms of a drug, it is more appropriate to use a duration of ischemia associated with high survival. If the purpose is to study the relevance of a drug in hepatic I/R injury, then it is advisable to assess survival, and, therefore, it is more adequate to use experimental models in which the ischemic period is associated with low survival. These observations are based on the following data reported in the literature. It appears that short periods (60 minutes) of warm ischemia result in reversible cell injury, in which liver oxygen consumption returns to control levels when oxygen is resupplied after ischemia. Reperfusion after more prolonged periods of warm ischemia (120-180 minutes) results in irreversible cell damage. These observations agree with a previous report on rat liver subjected to I/R, indicating a cellular endpoint for hepatocytes after 90 minutes of ischemia [83]. In human LT, a long ischemic period is a predicting factor for posttransplantation graft dysfunction, and some transplantation groups hesitate to transplant liver grafts preserved for more than 10 h. Some studies in experimental models of LT indicate that cold ischemia for 24 h induces low survival. However, LT, following shorter ischemic periods, may also result in primary organ dysfunction [72].

It is important to distinguish between the types of Ischemia (warm and cold) because there is already some controversy about the pathophysiological mechanisms of cold ischemia, which may depend, for example, on the time. The mechanisms of hepatic I/R injury are also different depending on the duration of hepatic ischemia. Along these lines, in the same experimental model of LT, XDH/XOD plays a crucial role in hepatic I/R injury only in conditions under which significant conversion of XDH to XOD occurs (80–90% of XOD) such as 16 h of cold ischemia. However, this ROS generation system does not appear to be crucial for shorter ischemic periods such as 6 h of cold ischemia [72]. Similarly, it should also be noted that oxidative stress in hepatocytes and the stimulatory state of KCs after I/R depend on the duration of ischemia and may also differ between ischemia at 4°C and that at 37°C, which probably leads to different developmental mechanisms of liver damage.

Our previous results indicate that PPARα does not play a crucial role in I/R injury in non-steatotic livers. This contrasts with a study published by Okaya and Lentsch [84], in which the authors reported the benefits of PPARα agonists on postischemic liver injury. Although the dose and pretreatment time of the PPARα agonist WY-14643 were similar in both studies, Okaya and Lentsch reported an ischemic period of 90 minutes; ours was 60 minutes, which is the ischemic period currently used in liver surgery [3]. Thus, 60 minutes of ischemia seems to be insufficient to induce changes in PPARα in nonsteatotic livers [13].

8.3. Relevance of the extent of hepatic ischemia

Another factor to consider before selecting the experimental model of hepatic I/R is the percentage of hepatic ischemia applied. The extent of hepatic injury as well as the hepatic I/R mechanisms, including the recovery of blood flow and energy charge during hepatic reperfusion is dependent on the extent of ischemia-whether total or partial (70%) hepatic ischemia is applied [36]. This fact could be explained by the stealing phenomenon. In contrast to 100% hepatic ischemia, during ischemia in the left and median lobes, the flow is shunted via the right lobes and following the release of the occlusion of the left and median lobes, a significant amount of shunting via the right lobes will continue during reperfusion until vascular resistance in the postischemic lobes decreases. This occurs because blood flows through the path of least resistance. The reasons for this may be cellular swelling endothelial, stasis, or other changes. Thus, the recovery of blood flow and hepatic perfusion of the preischemic lobe is later in the case of 70% hepatic ischemia than in 100% hepatic ischemia [76]. In line with these observations, the benefits of some drugs such as ATP-MgCl2 were dependent on the extent of hepatic ischemia used [32,77].

8.4. Relevance of the type of liver submitted to I/R

A variety of clinical factors including starvation, graft age, and steatosis have been studied in different experimental models of hepatic I/R because of the relevance of these factors in clinical practice. These factors enhance liver susceptibility to I/R injury, further increasing the patient risks related to reperfusion injury.

8.4.1. Starvation

The pre-existent nutritional status is a major determinant of the hepatocyte injury associated with I/R. In clinical LT, starvation of the donor, due to prolonged intensive care unit hospitalization or the lack of adequate nutritional support, increases the incidence of hepatocellular injury and primary nonfunction [85]. Based on the nutritional state status, several experimental and clinical studies support the hypothesis that the availability of glycolytic substrates is important for maintenance of hepatic ATP levels during I/R. Fasting exacerbates I/R injury because the low content of glycogen stores results in more rapid ATP depletion during ischemia. In addition, fasting causes alterations in tissue antioxidant defenses, accelerates the conversion of XDH to XOD during hypoxia and induces mitochondrial alterations [85]. Caraceni et al., [86] have shown that mitochondrial damage is greatly enhanced by fasting which decreases the hepatic content of antioxiants and therefore sensitizes the mitochondrial to the injurious effects of ROS. Considering these observations, an artificial nutritional support may represent a new approach for the prevention of reperfusion injury in fasted livers. On the contrary, fasting has been reported to improve organ viability and survival [87], as it reduces phagocytosis and the generation of TNF-α [87]. To understand these apparent contradictory results, it is important to consider the different experimental conditions in these investigations. A beneficial effect of high glycogen content can mainly be expected under conditions of long preservation times and long periods of warm ischemia. Under these conditions, high metabolic reserves of the liver may attenuate ischemic cell injury and preserve defense functions against cytotoxic mediators of KCs. Conversely, short ischemic periods require lower metabolic reserves, and the extent of KC activation can be the dominant factor in early graft injury.

8.4.2. Age

A number of distinct age-related alterations have been identified in the hepatic inflammatory response to hepatic I/R [88]. Under warm hepatic ischemia, mature adult mice had greatly increased neutrophil function, increased intracellular oxidant levels, and decreased mitochondrial function compared with the findings in young adult mice. These alterations contributed to the increased liver injury after I/R observed in mature adult mice compared with that in young adult mice. The results obtained in an experimental model of isolated perfused liver indicate that during reperfusion, livers obtained from old rats generate a lower amount of oxyradicals than livers from young rats. This fact could be explained by the lower KC activity, the reduction of liver blood flow, and the impaired functions and structural alterations observed in the livers of old rats. In fact, in hepatocytes from mature adult mice, delayed activation of nuclear factor kappa B (NFκB) in response to TNF-α and virtually no production of macrophage inflammatory protein 2 have been detected, which may be due to an agerelated defect in hepatocytes [88].

8.4.3. Steatosis

The first step to minimize the adverse effects of I/R in steatotic livers is a full understanding of the mechanisms involved in I/R injury in these marginal organs. This can be achieved only with the selection of an appropriate method to induce steatosis in livers undergoing I/R. It is well known that the mechanisms involved in hepatic I/R injury are different depending on the type of liver (nonsteatotic versus steatotic livers). In addition to the impairment of microcirculation, mitochondrial ROS generation dramatically increases during reperfusion in steatotic livers [9,86]. Results obtained under warm hepatic ischemia indicate that apoptosis is the predominant form of hepatocyte death in the ischemic nonsteatotic liver, whereas the steatotic livers develop massive necrosis after an ischemic insult [9]. Steatotic livers differed from nonsteatotic livers in their response to the UPR and ER stress since IRE1 and PERK were weaker in the presence of steatosis [89]. Decreased ATP production and dysfunction of regulators of apoptosis, such that Bcl-2, Bcl-xL and Bax have been proposed to explain the failure of apoptosis in steatotic livers. Differences were also observed when we analyzed the role of the RAS, as the nonsteatotic grafts exhibited higher Ang-II levels than steatotic grafts whereas steatotic grafts exhibited higher Ang 1-7 levels [15]. In the context of I/R injury associated with LT, the axis ACE-Ang II-ATR and ACE2-Ang 1-7-Mas play a major role in nonsteatotic and steatotic grafts, respectively. From the point of view of clinical application, these findings may open up new possibilities for therapeutic interventions in LT within the RAS cascade, based on Ang 1-7 for steatotic livers and Ang II for non-steatotic ones [15]. Moreover, reduced RBP4 and increased PPARγ levels were observed in steatotic livers compared to non-steatotic livers [81]. The vulnerability of steatotic livers subjected to warm ischemia is also associated with increased adiponectin, oxidative stress, and IL-1 levels and a reduced ability to generate IL-10 and PPARα [13,90].

It should be considered that there are differences in the mechanisms involved in hepatic I/R injury depending on the method used to induce steatosis. In contrast with other experimental models of steatosis, both dietary high fat and alcohol exposure induced the production of SOD/catalase-insensitive ROS, which may be involved in the mechanism of steatotic liver failure after OLT [9]. Neutrophils have been involved in the increased vulnerability of steatotic livers to I/R injury, especially in alcoholic steatotic livers. However, neutrophils do not account for the differentially greater injury in non-alcoholic steatotic livers during the early or late hours of reperfusion. Similarly, the role of TNF in the vulnerability of steatotic livers to I/R injury may be dependent on the type of steatosis [1,9].

8.5. Relevance of regeneration in experimental models of hepatic I/R

It is known that different experimental models trigger different responses when a common mechanism or the same drug is investigated. This situation is witnessed when analyzing liver injury in models of I/R with or without hepatectomy. This situation is illustrated by Ramalho et al., [36] regarding the loss of protection of Ang-II receptor antagonists against liver damage in conditions of PH under I/R compared with the study of I/R without hepatectomy, in which Ang-II receptor antagonists reduced hepatic damage. These different results could not be explained by differences in the dose or frequency of drug administration but rather by differences in surgical conditions (percentage of hepatic ischemia and the presence or absence of hepatectomy). In the model of I/R without hepatectomy [33], the blood supply to the left and median liver lobes (70% hepatic mass) was interrupted, and the other hepatic lobes remained intact. However, in PH under I/R, only blood supply to the remnant liver (30% hepatic mass) was interrupted and the other hepatic lobes were excised [36].

According to the cell type and experimental or pathologic conditions, TNF-α may stimulate cell death or it may induce hepatoprotective effects mediated by antioxidant, antiapoptotic, and other anti-stress mediators coupled with a pro-proliferative biologic response. For example, although the deleterious effect of the TNF-α in local and systemic damage associated with hepatic I/R in experimental models of normothermic hepatic ischemia is well established [91], this mediator is also a key factor in hepatic regeneration [92], an important process in RSLT and PH associated with hepatic resections [93]. These differential effects observed for TNF-α can also be extrapolated to transcription factors. It is well known that NFκB can regulate various downstream pathways and thus has the potential to be both pro- and antiapoptotic [8]. Currently it is not clear whether the beneficial effects of NFκB activation in protection against apoptosis or its detrimental proinflammatory role predominate in liver I/R [8]. Hepatic neutrophil recruitment and hepatocellular injury are significantly NFκB activation is suppressed in mice following partial hepatic I/R. However, NFκB activation is essential for hepatic regeneration after rat LT, and reduces apoptosis and hepatic I/R injury [94].

9. Strategies applied in experimental models of hepatic I/R

9.1. Pharmacological treatment and additives in preservation solution

Numerous experimental studies have focused on the developing *in vivo* pharmacological strategies aimed at inhibiting the harmful effects of I/R [9,72,89,90,95-99]. Some of these studies are summarized in Table 1. However, none of these treatments has managed to prevent hepatic I/R injury. A large number of ingredients-which have been introduced into UW solution in experimental models of hepatic cold ischemia [9,95,100-102] (Table 1). However, none of these modifications to the UW solution composition have found their way into routine clinical practice. Further studies will be required to elucidate whether the use of perfluorochemicals (PFC) in preservation solutions might improve the viability of liver grafts undergoing transplantation. PFC are hydrocarbons with high capacity for dissolving respiratory and other nonpolar gases. A negligible O_2-binding constant of PFC allows them to release O_2 more effectively than hemoglobin into the surrounding tissue (acts as an oxygen-supplying agent). PFC differs from hemoglobin preparations in that it is a totally synthetic compound formed on a liquid hydrocarbon base. Unlike hemoglobin, acidosis, alkalosis, and temperature seem to have no or little effect on the oxygen delivery of PFC, allowing this compound to be used effectively during cold storage of organs [103]. A recently study, used Oxycyte, a PFC added to UW solution can be beneficial after cardiac death liver graft preservation in a rat model [103]. However, their effects on reperfusion injury were not evaluated in that study. In fact, the possibility that preoxygenated PFC exacerbates the ROS during reperfusion should not be discarded since the use of gaseous oxygen applied to the livers during the storage period was only effective in improving hepatic viability upon reperfusion when antioxidants were added to the UW rinse solution [104].

It should be also considered that the inclusion of some components in the UW solution has been both advocated and criticized. Indeed, simplified variants of the UW solution in which some additive were omitted were demonstrated to have similar or even higher protective potential during cold liver storage. Another limitation of the UW solution is that some of its constituent compounds, including allopurinol do not offer very good protection because they are not present at a suitable concentration and encounter problems in reaching their site of action [9]. The possible side effects of some drugs may frequently limit their use in human LT. For example, idiosyncratic liver injury in humans is documented for chlorpromazine, pernicious systemic effects have been described for NO donors, allopurinol therapy can cause hematological changes and gadolinium can induce coagulation disorders. Some case reports of acute hepatotoxicity attributed to rosiglitazone have been published [105]. The development of therapeutic strategies that utilize the protective effect of heme oxygenase-1 induction is hampered by the fact that most pharmacological inducers of this enzyme perturb organ function by themselves [106].

Pharmacological treatment-derived difficulties must also be considered. In this regard, SOD and GSH exhibit inadequate delivery to intracellular sites of ROS action [9]. The administration of anti-TNF antibodies does not effectively protect against hepatic I/R injury, and this finding has been related to the failure of complete TNF-α neutralization locally [11]. Although this also occurs in non-steatotic livers, modulating I/R injury in steatotic livers poses a greater problem. Differences in the action mechanisms between steatotic and non-steatotic livers mean that therapies that are effective in non-steatotic livers may prove useless in the presence of steatosis, and the effective drug dose may differ between the two liver types. Findings such as these must be considered when applying pharmacological strategies in the same manner to steatotic and non-steatotic livers because the effects may be very different. For example, caspase inhibition, a highly protective strategy in non-steatotic livers, had no effect on hepatocyte injury in steatotic livers [9]. Moreover, whereas in an LT experimental model, an NO donor reduced oxidative stress in non-steatotic livers, the same dose increased the vulnerability of steatotic grafts to I/R injury. Furthermore, there may be drugs that would only be effective in steatotic livers. This was the case of compounds such as cerulenin, which reduce UCP-2 expression in steatotic livers and carnitine [9].

Pharmacological Therapy – Warm Ischemia			
Species	Drug	Ischemic Time	Effect
Mice	Cerulenin (Fatty acid synthase inhibitor)	15 min	↓ UPC2, ↑ ATP
	Catalase and derivatives	30 min	↓ Oxidative stress
	Apocynin (NAPH oxidase inhibitor)		↓ Oxidative stress
	TBC-1269 (Pan-selectin antagonist)	90 min	↓ Inflammatory response, ERK ½
Rat Rat	Lisinopril (ACE inhibitor)	30 min	↓ Oxidative stress
	Ascorbate (ROS scavenger)		↓ Apoptosis
	Allopurinol (XOD inhibitor)	30 60 min	↓ Oxidative stress
	Melatonin (Hormone)	40 min	↓ IKK, JNK pathways
	SOD (antioxidant)	45 min	↓ Microcirculatory disturbances, leukocyte acumulation
	L-arginine (NO precursor)		↑ NO, ATP ↓ Neutrophil accumulation
	Tocopherol (Antioxidante)	45 90 min	↓ Microcirculatory disturbances, Lipid peroxidation, SEC damage
	IL-10	60 min	↓ IL-1, Oxidative stress
	Anti-ICAM-1		↓ Adherence of leukocytes in postsinusoidal venules
	Gabexate mesilate (Protease inhibitor)		↓ TNF-α, Leukocyte activation
	OP-2507 (Analogue of prostacyclin)		↓ Microcirculatory disturbance
	WY-14643 (PPARα agonist)		↓ Oxidative stress, Inflammatory cytokines

	n-3 PUFA		↓ Liver injury, Oxidative stress
	Glutathione (Antioxidant)		↓ Microcirculatory disturbances ↑ Detoxification of ROS
	Spermine NONOate (NO donor)	60 90 min	↓ IL-1α, Oxidative stress
	FK506 (Immunosupressant)		↓ TNF
	Rosiglitazone (PPARα agonist)		↑ Autophagy ↓ Cytokines
	AMPK activators		↑ NO, ATP
	Adenosine	90 min	↑ NO
	Anti-TNF antiserum		↓ TNF, Leukocyte accumulation
	α-Lipoic acid (Antioxidant)		↑ Liver regeneration, ↓ Apoptosis

Pharmacological Therapy – Warm Ischemia with Hepatectomy

Species	Drug	Ischemic Time	Effect
Rat	Tauroursodeoxycholate (Bile acid)	60 min	↓ Endoplasmic reticulum stress
	Sirolimus (Immunossupressant)		↓ Linfocytes
	IL-1ra (IL-1 receptor antagonist)	90 min	↓ TNF, Oxidative stress
Dog	FK 3311 (Cox-2 inhibitor)	60 min	↓ Neutrophil infiltration, Cox-2

Pharmacological Therapy – Liver Trasplantation

Species	Drug	Ischemic Time	Effect
Mice	Cerulenin (fatty acid synthase inhibitor)	80 min	↓ UPC2, ↑ ATP
Rat	FK 409 (NO donor)	80 min	↑ HSP, IL-10, ↓ SEC damage, IL-1
	CS1 peptides (FN-α4β1 interac blocker)	4 h	↓ Neutrophil and lymphocyte T infiltration, TNF-α, iNOS
	Tocopherol (antioxidante)	5 h	↓ Lipid peroxidation, SEC damage, Microcirculatory disturbance
	Hemin (HO-1 inducer)		↑ Bcl-2
	Cobalt-protoporphyrin IX (HO-1 inducer)	6 h	↓ Macrophages infiltration and T cells
	PSGL-1 (P-selectin blocker)		↓ Neutrophil infiltration, TNF-α, INFγ, iNOS
	Anti-TNF antiserum	6, 24 h	↓ TNF, Leukocyte accumulation
	SOD (antioxidant)	8 h	↓ Microcirculatory disturbance, Leukocyte acumulation
	Tauroursodeoxycholate (Bile acid)		↓ Endoplasmic reticulum stress
	Allopurinol (XOD inhibitor)	8, 16 h	↓ Oxidative stress

	Drug		Effect
	Z-DEVD-FMK *(caspase 3 and 7 inhibitor)*	16 h	↑ Microvascular perfusión, Bcl-2 ↓ Apoptosis
	L-arginine *(NO precursor)*	18 h	↑ NO, ATP, ↓ Neutrophil accumulation
	Treprostinil *(Prostacyclin analogue)*		↓ Liver injury, Platelet deposition, microcirculatory disturbance
	ANP *(vasodilating peptide)*	24 h	↑ PI3K/Akt, ↓ Apoptosis
	Bucillamine *(antioxidant)*		↓ Oxidative stress
	Chlorpromazine *(Ca^2 + channel antagonist)*		↑ ATP ↓ Mitocondrial dysfunction, Alterations in lipid metabolism
	sCR1 *(complement inhibitor)*		↓ Microcirculatory disturbance, Leukocyte adhesion
	Glutathione *(antioxidant)*		↓ Microcirculatory disturbance ↑ Detoxification of ROS
	N-acetylcysteine *(glutathione precursor)*		↓ Microcirculatory disturbance
	Anti-ICAM-1		↓ Adherence of leukocytes in postsinusoidal venules
	Glycine *(Kupfer cell modulator)*		↓ Neutrophil accumulation, TNF-α
	GdCl3 *(Kupffer cell blocker)*		↓ Neutrophil accumulation, TNF-α
	Cbz-Val-Phe methyl ester *(calpain inhibitor)*	24, 40h	↓ Calpain activation, SEC apoptotic
	EHNA *(adenosine deaminase inhibitor)*	24, 44 H	↑ Interstitial adenosine ↓ Microcirculatory disturbance, Leukocytes rolling
	CGS-21680 *(adenosine A2 receptor agonist)*	30 h	↑ cAMP, ↓ SEC Killing
	Sotrastaurin *(PKC Inhibitor)*		↓ Apoptosis, macrophage/neutrophil accumulation
	FR167653 *(IL-1β and TNF-α supressor)*	48 h	↓ TNF-α, IL-1α, Kupffer cell activation
	Doxorubicin *(Heat shock proteins inducer)*		↓ TNF-α, MIP-2, NKκB
Pig	Sodium ozagrel *(Thromboxane synthase inhibitor)*	8 h	↓ ET-1

Additives to UW solution – Liver Trasplantation

Species	Drug	Ischemic Time	Effect
Mouse	Erythropoietin *(EPO)*	24 h	↓ Liver injury
Rat	Meloxicam *(COX-2 Inhibitor)*	1 h	↓ Apoptosis, Liver injury, Oxidative stress
	Simvastatin *(KLF2-inducer)*	1, 6, 16 h	↓ Inflammation, Liver injury, Oxidative stress,
	Tauroursodeoxycholate *(Bile acid)*	2 h	↓ Endoplasmic reticulum stress
	S-nitroso-N-acetylcysteine	2, 4, 6 h	↓ Liver injury

	LY294002 *(PI3K inhibitor)*	7, 9, 24 h	↓ Apoptosis
	8br-cAMP, 8br-cGMP *(nucleotide analogs)*	24 h	↓ TNF-α and neutrophil accumulation
	Ruthenium red *(mitochondrial Ca2+ uniporter inhibitor)*		↓ Mitocondrial dysfunction
	Melatonin *(Hormone)*		↓ Oxidative stress, Liver injury
	OP-4183 *(PGI2 analogue)*		↓ Oxidative stress
	SAM *(ATP precursor)*		↓ Oxidative stress
	IDN-1965 *(caspase inhibitor)*	24, 30 h	↓ Apoptosis
	Pifithrin-alpha *(p53 inhibitor)*	24, 48 h	↓ Apoptosis
	Sodium nitroprusside *(NO donor)*		↓ Microcirculatory dysturbances
	FR167653 *(p38 inhibitor)*	30 h	↓ Microcirculatory dysturbances
	GSNO *(NO donor)*	48 h	↓ SEC damage
Dog	Trifluoperazine *(calmodulin inhibitor)*	24 h	↓ Microcirculatory dysturbances
Pig	E5880 *(PAF antagonist)*	8 h	↓ Microcirculatory dysturbances
	EGF, IGF-1, NGF-α	18 h	↑ ATP

Table 1. In *vivo* pharmacological therapy and adcitives in preservation solution in experimental models of warm hepatic ischemia (with or whithout hepatectomy) and liver transplantation

9.2. Gene therapy

Advances in molecular biology provide new opportunities to reduce liver I/R injury by using gene therapy. Genome manipulation can be achieved by: A) germ line manipulation (oocyte injections); B) stem cell transformation and reintroduction into embryos, and C) targeting specific cells or organs with vectors or viruses (gene transfer). The first 2 approaches include germ-line alterations and are neither feasible nor accepted by society. The third approach would lend to the treatment of individual patients with either acquired or congenital diseases [12]. In the last years, significant advances in gene therapy vectors have occurred. Gene transfer can be accomplished by direct injection of DNA into a target organ or tissue, transduction by recombinant viral vectors carrying a specific gene of interest, e.g., adenovirus (Ad) or retrovirus, transfection of cells by chemical methods (e.g., cationic liposomes), or stem cell transduction and reintroduction of genetically-altered cells back into embryos [107] (Table 2). Currently, researchers in gene transfer have focused efforts toward targeting vectors to specific cells or organs without loss of transduction ability [108,109], allowing high level gene transduction of the liver without affecting other organs [12,107].

		Genetic material	Packaging capacity	Duration of experiment	Integration into genome	Transduction of postmitotic cells
Recombinant viruses	Oncoretrovirus	RNA	9 kb	Long	Yes	Low
	Lentivirus	RNA	10 kb	Long	Yes	Low
	Foamy	RNA	12 kb	Long	No	High
	Herpes virus	DNA	"/>30 kb	Transient	No	High
	Adenovirus	DNA	30 kb	Transient	Rarely	Moderate
	AAV	DNA	4.6 kb	Long postmitotic tissues	Rarely	Moderate
	Oncoretrovirus	RNA	9 kb	Long	Yes	Low
	Lentivirus	RNA	10 kb	Long	Yes	Low
Non-viral methods	siRNA	RNA	No limitation	Transient	No	Zero
	DNA injection	DNA	No limitation	Transient	No	Zero
	Cationic liposomes	DNA	No limitation	Transient	No	Zero
	Stem cell transduction	DNA	No limitation	Transient	No	Zero

Table 2. Summary of gene therapy vectors commonly used.

Antiapoptotic Strategies (Bcl-2/Bcl-Xl, Bag-1 and caspases): Bcl-2 blocks apoptosis and necrosis and has been implicated in the prolongation of cell survival [110]. Given its functional importance in the cell death cascade, it constitutes one of the key targets for cytoprotective therapeutic manipulation for the regulation of apoptosis [110,111]. As demostrated by Bilbao et al., [111] in a mouse hepatic I/R model, overexpression of Ad-mediated Bcl-2 gene significantly decreased hepatocyte apoptosis and necrosis, improved hepatic function, and prolonged survival as compared with controls. In addition, Bag-1 is a Bcl-2 binding protein resulting in a prolonged and stabilized antiapoptotic activity [112]. In addition, Bag-1 appears to exert an indirect silencing effect on TNF receptor R1 and hence suppresses the death receptor signal. A recent study by Sawitzki et al., [113] has demonstrated the cytoprotective effect of Ad-mediated Bag-1 gene transfer in rat liver I/R. Using a model of cold ischemia and OLT, Ad-Bag-1 transfer improved portal venous blood flow, increased bile production, and improved hepatic function with decreased neutrophil accumulation in the graft. Furthermore, Ad-mediated Bag-1 expression preserved hepatic architecture and reduced inflammation. The activation of T cells infiltrating the graft was inhibited, since decreased expression of TNF-α, CD25, IL-2, and IFNγ [107]. Caspase-8 is presumed to be the apex of the death-mediated apoptosis pathway, whereas caspase-3 belongs to the "effector" proteases in the apoptosis cascade. Contreras et al., demonstrated that inhibition of caspase-8 and caspase-3 by siRNA provided significant protection against warm hepatic I/R injury and decreased animal mortality. In addition, animals given siRNA caspase-8, or more significantly siRNA caspase-3, presented lower neutrophil infiltration and better histologic profiles [114].

Antioxidant therapy (SOD, HO-1, Ferritin): Oxidative stress can activate NF-κB and the AP-1 pathway and induce expression of proinflammatory genes including cytokines, adhesion molecules, and chemokines leading to neutrophil-mediated inflammation [115-117]. To inhibit the burst of ROS or its effect on hepatocytes, several oxygen stress inhibitory proteins have been studied, e.g., SOD and catalase have been transfected by either adenovirus, liposomes or polyethylene-glycol [8,12,118]. Using partial hepatic I/R models, Ad-mediated MnSOD administration reduced liver tissue damage and activation of both NF-κB and AP1 [119,120] when compared with lacZ-transduced controls. In another study, He et al., [121] demonstrated that SOD or catalase gene delivery by polylipid nanoparticles injected via the portal vein 1 day prior to the warm I/R procedure resulted in high levels of the transgene enzyme activity in the liver, and markedly attenuated hepatic I/R injury [121]. However, results with NFκB activation have been conflicting. Takahashi et al. reported that overexpression of IκB, an NFκB inhibitor (mediated by Ad-IκB) resulted in partial protection in hepatic I/R injury [122]. Heme oxygenase 1 (HO-1) is a stress responsive protein and can be induced by various conditions such as hypoxia [12,107]. Several studies have shown that HO-1 exhibits potent cytoprotective effects after hepatic I/R [123,124]. In a cold ex-vivo rat liver perfusion model and a syngeneic liver transplant OLT model, treatment of genetically obese Zucker rats with Ad-HO-1 improved portal venous blood flow, increased bile production, and decreased hepatocyte injury [123]. Unlike in untreated rats, upregulation of HO-1 correlated with preserved hepatic architecture, improved liver function, and depressed infiltration by T cells and macrophages. Ad-mediated HO-1 gene overexpression increased survival of recipients from 40% to 80% [12,107]. Ad-HO-1 gene transfer decreased macrophage infiltration in the portal areas and inducible nitric oxide synthetase (iNOs) expression; it also increased the expression of antiapoptotic genes Bcl-2/Bcl-xl and Bag-1, as compared with controls [107]. Iron chelation is another approach to ameliorate the I/R injury cascade. Free iron has been shown to play a role in the formation of the free radicals through the Fenton reaction; these contribute to endothelial cell damage. Ferritin induction is a result of the action of HO-1 on the heme porphyrin causing the release of Fe2+. Ferritin can reduce the availability of intracellular free Fe2+, which can participate in free radical generation [125]. Studies by Ke et al., [107] demostrated that overexpression of Ad vector carrying the ferritin heavy chain (H-ferritin) gene protects rat livers from I/R injury [126]. In these studies, the protective effect of H-ferritin was associated with the inhibition of endothelial cell and hepatocyte apoptosis. Evidence suggested that H-ferritin exerts an antiapoptotic role and may be used as a therapeutic measure to prevent I/R [107].

Immunoregulatory cytokines (IL-10 and IL-13) and IL-1 receptor antagonist (IL-1R): IL-13 regulates liver inflammatory I/R injury via the signal transducer and activator of transcription 6 (STAT6) pathway [127]. IL-10 induces antioxidant HO-1 gene expression in murine macrophages and exerts anti-inflammatory effects [128]. In recent studies, Ad-IL-13 gene transfer in cold ischemia models has shown powerful cytoprotective effects [129]. Gene transfer of IL-13 improved hepatic function, upregulated HO-1, and prevented hepatic apoptosis through the upregulation of Bcl-2/Bcl-xl [107]. The beneficial effects of IL-13 correlated with *in vivo* cross talk between innate TLR4 and adaptive Stat6 immunity [130]. In fact, using an experimental model of warm hepatic ischemia, Stat6-deficient mice with Ad-IL-13 failed to improve hepatic function and hepatic histological features. Transfer of Ad-IL-13 increased anti-oxidant HO-1 expression and inhibited TLR4 activation in WT mice, whereas low HO-1 and enhanced TLR4 expression was shown in Stat6-deficient mice [107]. It has been demonstrated that the pro-inflammatory cytokine IL-1 plays a critical role in the pathophysiological response to I/R. Experimental results have shown that blockade of the IL-1R reduced TNF production and liver damage [131]. In a partial hepatic I/R model, gene transfer of Ad-mediated IL-1R antagonist prolonged animal survival and improved hepatic function while preserving the histological architecture. In addition, a marked decrease in production of proinflammatory cytokines such as IL-1, TNF-α, and IL-6 was present [107].

T-cell co-stimulation blockade: CD40-CD154. A number of studies have shown that CD4+ T lymphocytes play an important role as key cellular mediators in I/R injury mediated inflammatory responses. The CD40–CD154 co-stimulation pathway provides the essential second signal in the initiation and maintenance of T-cell-dependent immune esponses [132]. Recent studies have demonstrated that CD40-CD154 is required for the mechanism of hepatic warm I/R injury [133]. In OLT, prolonged *in vivo* blockade of the CD40-CD154 interaction following pretreatment of liver isografts with Ad-CD40Ig exerted potent cytoprotection against I/R injury. Apoptosis was prevented and neutrophil accumulation was reduced. Evidence also demonstrated prevention of Th1-type cytokine (interferon γ (IFN-γ) and IL-2) upregulation and the local expression of antioxidant HO-1 and antiapoptotic Bcl-2/Bcl-xl genes were triggered [107].

Adipocytokine, sphyngolipid and TLR4 regulation: Massip-Salcedo et al., [13] demostrated though the systemic delivery of adiponectin in livers treated with adiponectin siRNA that steatotic livers by themselves can generate adiponectin as a consequence of I/R. This study reports evidence of the injurious effects of adiponectin in stetatotic livers under warm ischemic conditions, and results suggest the clinical potential of gene therapy for I/R damage in steatotic livers by siRNA-mediated adiponectin gene silencing [13]. Products of sphingolipid metabolism are important second messengers that regulate a variety of cell processes including cell death, proliferation, and inflammation. Using a mice warm hepatic I/R model, Shi et al., demonstrated that SK2 knockdown by siRNA effectively prevented hepatocyte death [134]. Jiang et al., [135] reported a hepatocyte-specific delivery system for the treatment of liver I/R, using galactose-conjugated liposome nanoparticles (Gal-LipoNP). Heptocyte-specific targeting was validated by selective *in vivo* delivery as observed by increased Gal-LipoNP accumulation and gene silencing in the liver. Gal-LipoNP TLR4 siRNA treatment reduced hepatic damage, neutrophil accumulation and the inflammatory cytokines IL-1 and TNF-α [135].

Advances in molecular biology have provided new opportunities to reduce liver I/R injury using gene therapy [9,12,13,96,114] (Table 3). However, the experimental data indicate that there are a number of problems inherent in gene therapy, such as vector toxicity, difficulties in increasing transfection efficiencies and protein expression at the appropriate time and site, and the problem of obtaining adequate mutants (in the case of NFκB) due to the controversy regarding NFκB activation [136]. Although non-viral vectors (such as naked DNA and liposomes) are likely to present fewer toxic or immunological problems, they suffer from inefficient gene transfer [136]. In addition, LT is an emergency procedure in most cases, which leaves very little time to pre-treat the donor with genetic approaches. Efforts to reduce the time between gene therapy and LT might open new venues for preventative gene therapy [12]. Currently, viral vectors hydrodynamic injection and cationic liposomes are the main methods for delivering siRNA *in vivo*. While viral vectors are associated with severe side effects, other methods require large volume and high injection speed, which are not clinically applicable [135]. Systemic administration of small interfering RNA (siRNA) may cause globally nonspecific targeting of all tissues, which impedes clinical use.

9.3. Cell therapy – Hepatocyte transplantation

The liver was among the first organs considered for strategies based on the transplantation of isolated cells. The first hepatocyte transplant was performed to treat the Gunn rat, the animal model for Crigler-Najjar syndrome, which is congenitally unable to conjugate bilirubin and consequently exhibits life long hyperbilirubinemia. The transplant resulted in a decreased plasma bilirubin concentration. Later, isolated hepatocytes were transplanted into rats with liver failure induced by dimethylnitrosamine. These experiments demonstrated that hepatocyte transplantation could potentially be used for the treatment of liver failure and innate defects of liver-based metabolism. More than 30 years later, these models are still used in work to improve hepatocyte engraftment and/or function [137].

Many studies have shown that hepatocytes transplanted into rodents via the spleen or the portal vasculature enter through portal vein branches and are entrapped in proximal hepatic sinusoids; consequently, the hepatocytes are distributed predominantly in periportal regions of the hepatic lobules. Transplanted hepatocytes cause both portal hypertension and transient I/R injury. The portal hypertension, in experimental animals at least, usually resolves within 2 to 3 hours with no obvious long-term detrimental effects, and microcirculatory abnormalities disappear within 12 hours. Numerous hepatocytes (up to 70% of transplanted cells) remain trapped in the portal spaces, and most of them are destroyed by the phagocytic responses of KC, which are activated shortly after deposition of hepatocytes in liver sinusoids [138]. The remaining cells translocate from sinusoids into the liver plates through a process involving disruption of the sinusoidal endothelium and release of vascular endothelial growth factor by both host and transplanted cells. In rodents, hepatic remodeling is complete within 3 to 7 days, and the engrafted cells become histologically indistinguishable from host cells. Transplantation of 2×10^7 hepatocytes in rats has led to the engraftment of about 0.5% of the transplanted cells in the recipient livers [139]. Only hepatocytes harboring a selective advantage for survival/proliferation can efficiently repopulate a recipient liver, and as a result, many repopulation strategies have been developed using approaches involving the induction of acute or chronic liver injury [137]. Despite decades of research, the processes and factors underlying cell engraftment and *in situ* proliferation are only partially understood, and a good understanding of these mechanisms is essential for the development of new and efficient treatments of human liver diseases. The prevention of early loss of transplanted cells would undoubtedly improve hepatocyte transplantation. First, it has been recently shown that cell-cell interactions between transplanted hepatocytes and hepatic stellate cells modulate hepatocyte engraftment in rat livers. After cell transplantation, soluble signals activating hepatic stellate cells are rapidly induced along with early up-regulated expression of matrix metalloproteinases and their inhibitors [140]. Second, the interaction between integrin receptors and the extracellular matrix plays a role in cell engraftment. Third, hepatocytes express soluble and membrane-bound forms of tissue factor–dependent activation of coagulation and exert tissue factor–dependent hepatocyte-related procoagulant activity [137].

Gene	Specie	Ischemia	Vector	Effect
Bcl-2	Mouse	Warm ischemia	Adenovirus	↓ Apoptosis and Necrosis ↑ Survival
eNOS	Mouse	Warm ischemia	Adenovirus	↓ Liver injury
SOD	Mouse/Rat	Warm ischemia	Adenovirus	↓ Liver injury
IL-13	Mouse/Rat	Cold ischemia	Adenovirus	↓ Liver injury, Neutrophil infiltration, TLR4 activation, Apoptosis ↑ HO-1 expression, Survival
Bag-1	Rat	Cold ischemia	Adenovirus	↓ Liver injury, Neutrophil infiltration
CD40Ig	Rat	Cold ischemia	Adenovirus	↓ Liver injury, Neutrophil accumulation, Apoptosis and Necrosis
IkB	Rat	Cold ischemia	Adenovirus	↓ Liver injury
HO-1	Rat	Cold ischemia	Adenovirus	↓ Liver injury, Macrophage infiltration, iNOS ↑ Survival
Ferritin	Rat	Cold ischemia	Adenovirus	↓ Liver injury, Apoptosis
IL-1R antagonist	Rat	Warm ischemia	Cationic liposomes	↓ Liver injury ↑ Survival
SOD	Mouse	Warm ischemia	Polyplexes	↓ Liver injury ↑ Antioxidative enzyme activity
Catalase	Mouse	Warm ischemia	Polyplexes	↓ Liver injury ↑ Antioxidative enzyme activity
SK2	Mouse	Warm ischemia	siRNA	↓ Liver injury, Apoptosis ↑ survival
Caspase-3	Mouse	Warm ischemia	siRNA	↓ Liver injury, Neutrophil infiltration
Caspase-8	Mouse	Warm ischemia	siRNA	↓ Liver injury, Neutrophil infiltration
TLR4	Mouse	Warm ischemia	siRNA	↓ Liver injury, Neutrophil infiltration, ROS, Inflammation
Adiponectin	Rat	Warm ischemia	siRNA	↓ Liver injury

Table 3. Summary of gene therapy using specific target genes in hepatic ischemia-reperfusion

In recent years, the development of different animal models has allowed significant progress in hepatocyte transplantation. In rats, the occlusion of portal branches of the two anterior liver lobes results in a regeneration response in the remaining nonoccluded lobes leading to their hypertrophy. This procedure, portal branch ligation, favors efficient retroviral transduction of hepatocytes *in vivo*. Furthermore, hepatic tissue engineering using primary hepa-tocytes is an emerging therapeutic approach to liver diseases. Two recent studies reported engraftment of functional hepatocytes in a neovascularized subcutaneous cavity in mice. A method to manipulate uniform sheets of hepatic tissue allowing the formation, *in vivo*, of a 3-dimensional miniature liver system that maintained its biological function for several months has been also described [137,139]. In the view of clinical practice, treatment of fulmi-nant hepatic failure patients by hepatocyte transplantation has been attempted by a number of investigators [141]. In one report, patients who received a hepatocyte transplant, one pa-tient fully recovered and three were successfully bridged to OLT [141]. In a prospective study of five patients who were transplanted with cryopreserved human hepatocytes, three patients were successfully bridged to OLT [142]. Other reports have described clinical im-provement and relatively longer survival in hepatocyte- transplanted patients [143] but poor final outcome has also been reported, possibly related to immunosuppression, inadequate number of transplanted cells, and limited engraftment time [137].

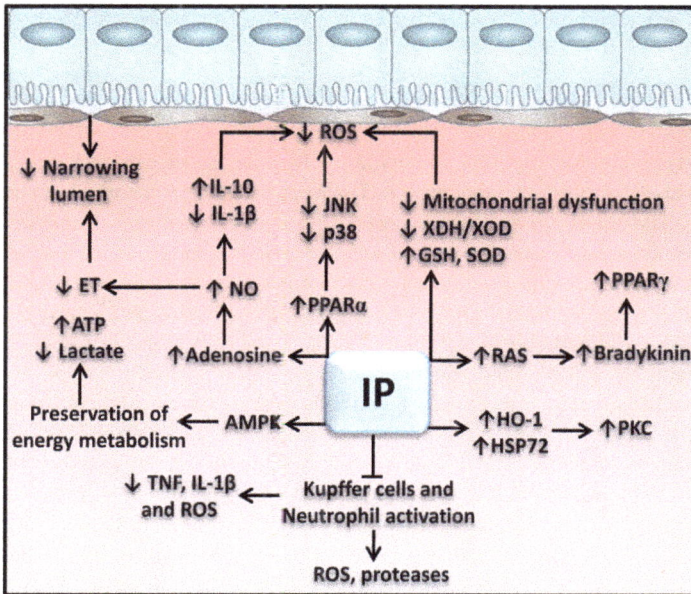

Figure 6. *Mechanisms of Ischemic preconditioning in hepatic ischemia-reperfusion injury.* AMPK, AMP-activated pro-tein kinase; ATP, adenosine triphosphate; ET, endothelin; GSH, glutathione; HO-1, heme oxygenase 1; HSP72, heat shock protein 72; IL, interleukin; JNK, c-Jun N-terminal kinase; NO, nitric oxide; PKC, protein kinase C; PPAR, peroxi-some proliferator-activated receptor; RAS, renin-angiotensin system; ROS, reactive oxygen species; SOD, superoxide dismutase; TNF, tumor necrosis factor; XDH/XOD, xanthine/xanthine oxidase

9.4. Surgical strategies

The response of hepatocyte to ischemia never ceases to surprise. In fact, contrary to what might be expected, the induction of consecutive periods of ischemia in the liver does not induce an additive effect in terms of hepatocyte lesions. Ischemic preconditioning (IP) based on brief periods of ischemia followed by a short interval of reperfusion prior to a prolonged ischemic stress protects the liver against I/R injury by regulating different cell types and multiple mechanisms such as energy metabolism, microcirculatory disturbances, leukocyte adhesion, KC activation, proinflammatory cytokine release, oxidative stress, apoptosis and necrosis [96] (Figure 6). This is an advantage in relation with the use of drugs that exerts its action on a specific mechanism. The benefits of IP observed in experimental models of hepatic warm and cold ischemia [96] prompted human trials of IP. To date, IP has been successfully applied in human liver resections in both steatotic and non-steatotic livers but unfortunately, it proved ineffective in elderly patients [144]. Preliminary clinical studies have reported the benefits of IP in LT [145,146]. IP may also have a role in the transplantation of small grafts whose pathophysiology overlaps with I/R injury. Additional randomized clinical studies are necessary to confirm whether this surgical strategy can be commonly used in clinical liver surgery.

10. Conclusion and perspectives

From the data obtained in experimental models of hepatic I/R, we can state that I/R injury is a multifaceted and intriguing phenomenon. The increasing use of marginal donors in major liver surgery and the fact that these organs are more susceptible to ischemia highlight the need for further research directed at the mechanisms of I/R injury. Machine perfusion has been criticized for its complicated logistics and for possibly damaging the organ and vital structures such as the endothelium. On the contrary, NMP fulfils all ideal organ preservation criteria by avoiding hypoxia and hypothermia. Responses to the strategies aimed at reducing hepatic I/R injury might depend on the surgical procedure, type of liver and percentage of hepatic ischemia. Further research is required to elucidate whether the pharmacological approaches presented in this review can be translated into liver surgery associated with hepatic resections and LT. Advances in molecular biology have provided new opportunities to reduce liver I/R injury using gene therapy. However, there are a number of problems inherent in gene therapy, such as vector toxicity and difficulties in increasing transfection. Liver-cell transplantation is at an early stage. Numerous approaches to isolating stem cells of hepatic or extrahepatic origin, including embryonic stem cells, are being developed. However, extensive work is still required to assess the number of cells that need to be expanded and differentiated, and the functionality of the different cell types needs to be carefully addressed in animal models. Surgical strategies such as IP affect multiple aspects of I/R injury, whereas pharmacological approaches often affect only a few mediators and might have systemic side effects.

Acknowledgments

Jiménez-Castro M.B. and Elias-Miró M., contributed equally to this work. Jiménez-Castro M.B., is in receipt of a fellowship from SETH Foundation (Sociedad Española de Transplante Hepatico) Spain.

Author details

M.B. Jiménez-Castro[1], M. Elias-Miró[1], A. Casillas-Ramírez[1] and C. Peralta[1,2*]

*Address all correspondence to: cperalta@clinic.ub.es

1 August Pi i Sunyer Biomedical Research Institute, Barcelona, Spain

2 Networked Biomedical Research Center of Hepatic and Digestive Diseases, Barcelona, Spain

References

[1] Serafin A., Rosello-Catafau J., Prats N., Xaus C., Gelpi E., Peralta C. Ischemic preconditioning increases the tolerance of fatty liver to hepatic ischemia-reperfusion injury in the rat. *American Journal of Pathology* 2002;161(2) 587–601.

[2] Clavien P., Harvey P., Strasberg S. Preservation and reperfusion injuries in liver allografts. An overview and synthesis of current studies. *Transplantation* 1992;53(5) 957-978.

[3] Huguet C., Gavelli A., Chieco P., Bona S., Harb J., Joseph J., et al. Liver ischemia for hepatic resection: where is the limit? *Surgery* 1992;111(3) 251-259.

[4] Busuttil R., Tanaka K. The utility of marginal donors in liver transplantation. *Liver Transplantation* 2003;9(7) 651–663.

[5] Ploeg R., D'Alessandro A., Knechtle S., Stegall M., Pirsch J., Hoffmann R., et al. Risk factors for primary dysfunction after liver transplantation-a multivariate analysis. *Transplantation* 1993;55(4) 807-813.

[6] Behrns K., Tsiotos G., DeSouza N., Krishna M., Ludwig J., Nagorney D. Hepatic steatosis as a potential risk factor for major hepatic resection. *Journal of Gastrointestinal Surgery* 1998;2(3) 292-298.

[7] Jaeschke H. Molecular mechanisms of hepatic ischemia-reperfusion injury and preconditioning. *American Journal of Physiology Gastrointestinal and Liver Physiology* 2003;284(1) G15–G26.

[8]   Fan C., Zwacka R., Engelhardt J. Therapeutic approaches for ischemia/reperfusion in-
      jury in the liver. *Journal of Molecular Medicine* 1999;77(8) 577-592.

[9]   Elias-Miro M., Massip-Salcedo M., Jiménez-Castro MB., Peralta C. Does adiponectin
      benefit steatotic liver transplantation?. *Liver Transplantation* 2011;17(1) 993-1004.

[10] Jaeschke H. Mechanisms of reperfusion injury after warm ischemia of the liver. *Jour-
      nal ofHepatobiliary and Pancreatic Surgery* 1998;5(4) 402–408.

[11] Peralta C., Fernandez L., Panes J., Prats N., Sans M., Pique J., et al. Preconditioning
      protects against systemic disorders associated with hepatic ischemia-reperfusion
      through blockade of tumor necrosis factor-induced P-selectin up-regulation in the
      rat. *Hepatology* 2001;33(1) 100–113.

[12] Selzner N., Rudiger H., Graf R., ClavienP. Protective strategies against ischemic in-
      jury of the liver. Gastroenterology 2003;125(3) 917–936.

[13] Massip-Salcedo M., Zaouali M., Padrissa-Altés S., Casillas-Ramírez A., Rodés J.,
      Roselló-Catafau J., Peralta C. Activation of peroxisome proliferator-activated recep-
      tor-alpha inhibits the injurious effects of adiponectin in rat steatotic liver undergoing
      ischemia-reperfusion *Hepatology* 2008;47(2) 461-472.

[14] Bader M., Peters J., Baltatu O., Müller D., Luft FC., Ganten D. Tissue rennin-angioten-
      sin systems: New insights from experimental animal models in hypertension re-
      search. *Journal of Molecular Medicine* 2001;79 76-102.

[15] Alfany-Fernández I., Casillas-Ramírez A., Bintanel-Morcillo M., Brosnihan K., Ferrar-
      io C., Serafin A., et al. Therapeutic targets in liver transplantation: angiotensin II in
      nonsteatotic grafts and angiotensin-(1-7) in steatotic grafts. *American Journal of Trans-
      plantation* 2009;9(3) 439-451.

[16] Quijano-Collazo Y. Trasplante hepatico experimental. Brasil:Atheneu Hispánica;
      2006.

[17] Abdo E., Cunha J., Deluca P., Coelho A., Bacchella T., Machado M. Protective effect
      of N2-mercaptopropionylglycine on rats and dogs liver during ischemia/reperfusion
      process. *Arquivos de Gastroenterologia* 2003;40(3) 177–180.

[18] Malarkey D., Johnson K., Ryan L., Boorman G., Maronpot R. New insights into func-
      tional aspects of liver morphology. *Toxicologic Pathology* 2005;33 27-34.

[19] Saxena R., Theise N., Crawford J. Microanatomy of the human liver – exploring the
      hidden interfaces. Hepatology 1999; 30(6) 1339-1346.

[20] Lin Y., Nosaka S., Amakata Y., Maeda T. Comparative study of the mammalian liver
      innervations: an immunohistochemical study of protein gene product 9.5, dopamine
      β-hydroxylase and tyrosine hydroxylase. *Comparative Biochemistry and Physiology*
      1995;110A(4) 289-298

[21] Bentley P., Calder I., Elcombe C., Grasso P., Stringer D., Wiegand H. Hepatic peroxisome proliferator in rodents and its significance for humans. *Food and Chemical Toxicology* 1993:31(11) 857-907.

[22] Hasmall S., James N., Macdonald N., Soames A., Roberts R. Species differences in response to diethylhexylphthalate:suppression of apoptosis, induction of DNA synthesis and peroxisome proliferator activated receptor alpha-mediated gene expression. *Archives of Toxicology* 2000;74 85-91.

[23] Yahanda A., Paidas C., Clemens M. Susceptibility of hepatic microcirculation to reperfusion injury: A comparison of adult and suckling rats. *Journal of Pediatric Surgery* 1990;25(2) 208–213.

[24] Vidal M. Traitement chirurgical des ascites. *La Presse Médicale* 1903;11 747–749.

[25] Blakemore A., Lord J. The technic of using vitallium tubes in establishing portacaval shunts for portal hypertension. *Annals of Surgery* 1945;122(4) 449–475.

[26] Burnett W., Rosemond G., Weston J., Tyson R. Studies of hepatic response to changes in blood supply. *Surgical Forum* 1951;94 147–153.

[27] Bernstein D. Cheiker S. Simple technique for porto-caval shunt in the rat. *Journal of Applied Physiology* 1959;14(3) 467–470.

[28] Spiegel H., Bremer C., Boin C., Langer M. Reduction of hepatic injury by indomethacin-mediated vasoconstriction: a rat model with temporary splenocaval shunt. *Journal of Investigative Surgery* 1995;8(5) 363-369.

[29] Bengmark S., Börjesson B., Olin T., Sakuma S., Vosmic J. Subcutaneous transposition of the spleen: An experimental study in the rat. *Scandinavian Journal of Gastroenterology* 1970;7 175–179.

[30] Meredith C., Wade D. A model of portal-systemic shunting in the rat. *Clinical and Experimental Pharmacology and Physiology* 1981;8 651–652.

[31] Suzuki S., Nakamura S., Sakaguchi T., Mitsuoka H., Tsuchiya Y., Kojima Y., et al. Pathophysiological appraisal of a rat model of total hepatic ischemia with an extracorporeal portosystemic shunt. *Journal of Surgical Research* 1998;80(1) 22–27.

[32] Hasselgren P., Jennische E., Fornander J., Hellman A. No beneficial affect of ATP-MgCl2 on impaired transmembrane potential and protein synthesis in liver ischemia. *Acta Chirurgica Scandinavica* 1982;148(7) 601–607.

[33] Casillas-Ramírez A., Amine-Zacuali M., Massip-Salcedo M., Padrissa-Altés S., Bintanel-Morcillo M., Ramalho F., et al. Inhibition of angiotensin II action protects rat steatotic livers against ischemia-reperfusion injury. *Critical Care Medicine* 2008;36(4) 1256-1266.

[34] Peralta C., Bartrons R., Riera L., Manzano A., Xaus C., Gelpí E., Roselló-Catafau J. Hepatic preconditioning preserves energy metabolism during sustained ischemia.

American Journal of Physiology -Gastrointestinal and Liver Physiology 2000(279)1 G163–G171.

[35] Palmes D., Spiegel H. Animal models of liver regeneration. *Biomaterials* 2004;25 1601-1611.

[36] Ramalho F., Alfany-Fernandez I., Casillas-Ramírez A., Massip-Salcedo M., Serafín A., Rimola A., et al. Are angiotensin II receptor antagonists useful strategies in steatotic and nonsteatotic livers in conditions of partial hepatectomy under ischemia-reperfusion?. *Journal of Pharmacology and Experimental Therapeutics* 2009;329(1) 130-140.

[37] Czaja MJ. Liver regeneration following hepatic injury. London: Chapman and Hall; 1998.

[38] Spiegel H. Palmes D. Surgical techniques of orthotopic rat liver transplantation. *Journal of Investigative Surgery* 1998;11(2) 83-96.

[39] Cannon J. Organs. *Transplantation Bulletin* 1956;3 7. En: Cordier G., Garnier H., Clot J., Camplez P., Gorin J., Clot P. Orthotopic liver graft in pigs. 1st results. *Mémories de l'Académie Nationale de Chirurgie* 1966;92(27) 799-807.

[40] Garnier H., Clot J., Bertrand M., Camplez P., Kunlin A., Gorin J., et al. Liver transplantation in the pig: surgical approach. *CR Hebd Seances Academic Science* 1965;260(21) 5621-5623.

[41] Hori T., Nguyen J., Zhao X., Ogura T., Hata T., Yagi S., et al. Comprehensive and innovative techniques for liver transplantation in rats: A surgical guide. *World Journal of Gastroenterology* 2010;16 (25) 3120-3132.

[42] Aller M., Mendez M., Nava M., Lopez L., Arias J., Arias J. The value of microsurgery in liver research. *Liver International* 2009;29(8) 1132-1140.

[43] Lee S., Charters A., Chandler J., Orloff M. A technique for orthotopic liver transplantation in the rat. *Transplantation* 1973;16(6) 664-669.

[44] Lee S., Charters A., Orloff M. Simplified technic for orthotopic liver transplantation in the rat. *American Journal of Surgery* 1975;130(1) 38-40.

[45] Zimmermann F., Butcher G., Davies H., Brons G., Kamada N., Turel 0. Techniques for orthotopic liver transplantation in the rat and some studies of the immunologic responses to fully allogeneic liver grafts. *Transplantation Proceedings* 1979;11 571-577.

[46] Kamada N.. Calne R. Orthotopic liver transplantation in the rat. Technique using cuff for portal vein anastomosis and biliary drainage. *Transplantation* 1979;28(1) 47-50.

[47] Miyata M., Fischer J., Fuhs M., lsselhard W., Kasai Y. A simple method for orthotopic liver trasplantation in the rat. Cuff technique for three vascular anastomoses. *Transplantation* 1980;30 335-338.

[48] Engemann R., Ulrichs K., Thiede A., Muller-Ruchholtz W., Hamelmann H. Value of a physiological liver transplant model in rats. Induction of specific graft tolerance in a fully allogeneic strain combination. *Transplantation* 1982;33 566-568.

[49] Kitakado Y., Tanaka K., Asonuma K., Uemoto S., Matsuoka S. A new bioabsorbable material for rat vascular cuff anastomosis: Establishment for the long-term orthotopic liver transplantation model. *Archives Ipn Chirurgie* 1992;61 445-453.

[50] Ma Y., Wang G., Guo Z., Guo Z., He X., Chen C. Surgical techniques of arterialized orthotopic liver transplantation in rats. *Chinese Medical Journal* 2007;120(21) 1914-1917.

[51] Urakami H., Abe Y., Grisham M. Role of reactive metabolites of oxygen and nitrogen in partial liver transplantation: lessons learned from reduced-size liver ischemia and reperfusion injury. *Clinical and Experimental Pharmacology and Physiology* 2007;34(9) 912-919.

[52] Bismuth H., Houssin D. Reduced-sized orthotopic liver graft in hepatic transplantation in children. *Surgery* 1984;95(3) 367–370.

[53] Couinaud L. Le foie; études Anatomiques et Chirurgicales. France:Masson; 1957

[54] Gong N., Chen X. Partial liver transplantation. *Frontier Medical* 2011;5(1) 1-7.

[55] Broelsch C., Emond J., Whitington P., Thistlethwaite J., Baker A., Lichtor J. Application of reduced-size liver transplants as split grafts, auxiliary orthotopic grafts, and living related segmental transplants. *Annals of Surgery* 1990;212(3) 368-375.

[56] Lo C., Fan S., Liu C., Lo R., Lau G., Wei W., et al. Extending the limit on the size of adult recipient in living donor liver transplantation using extended right lobe graft. *Transplantation* 1997;63(10) 1524-1528.

[57] Vogel T., Brockmann J., Friend P. Ex-vivo normothermic liver perfusion: an update. *Current Opinion in Organ Transplantation* 2010;15(2) 167-172.

[58] Rougemont O., Lehmann K., Clavien P. Preconditioning, organ preservation, and postconditioning to prevent ischemia-reperfusion injury to the liver. *Liver Transplantation* 2009;15(10) 1172-1182.

[59] Minor T., Saad S., Nagelschmidt M., Kotting M., Fu Z., Paul A., et al. Successful transplantation of porcine livers after warm ischemic insult in situ and cold preservation including postconditioning with gaseous oxygen. *Transplantation* 1998;65(9): 1262-1264.

[60] Minor T., Stegemann J., Hirner A., Koetting M. Impaired autophagic clearance after cold preservation of fatty livers correlates with tissue necrosis upon reperfusion and is reversed by hypothermic reconditioning. *Liver Transplantation* 2009;15(7) 798-805.

[61] Gurusamy K., Gonzalez H., Davidson B. Current protective strategies in liver surgery. *World Journal of Gastroenterology* 2010;16(48) 6098-6103.

[62] Monbaliu D., Brassil J. Machine perfusion of the liver: past, present and future. *Current Opinion in Organ Transplantation* 2010;15 160-166.

[63] Carrel A. The culture of whole organs. *Science* 1935;14 621–623.

[64] Starzl T., Groth C., Brettschneider L., Moon J., Fulginiti V., Cotton E., Porter K. Extended survival in 3 cases of orthotopic homotransplantation of the human liver. *Surgery* 1968;63 549–563.

[65] Schön M., Kollmar O., Wolf S., Schrem H., Matthes M., Akkoc N., et al. Liver transplantation after organ preservation with normothermic extracorporal perfusion. *Annals of Surgery* 2001;2338(1) 114–123.

[66] Fondevila C., Hessheimer A., Ruiz A., Calatayud D., Ferrer J., Charco R., et al. Liver transplant using donors after unexpected cardiac death: novel preservation protocol and acceptance criteria. *American Journal of Transplantion* 2007;7(7) 1849-1855.

[67] García-Valdecasas J., Fondevila C. In-vivo normothermic recirculation: an update. *Current Opinion in Organ Transplantation* 2010;15(2) 173-176.

[68] Pienaar B., Lindell S., Van Gulik T., Southard J., Belzer F. Seventy-two-hour preservation of the canine liver by machine perfusion. *Transplantation* 1990;49(2) 258–260.

[69] Bessems M., Doorschodt B., van Marle J., Vreeling H., Meijer A., van Gulik T. Improved machine perfusion preservation of the non-heart-beating donor rat liver using Polysol: a new machine perfusion preservation solution. *Liver Transplantation* 2005;11 1379–1388.

[70] Vekemans K., Liu Q., Brassil J., Komuta M., Pirenne J., Monbaliu D. Influence of flow and addition of oxygen during porcine liver hypothermic machine perfusion. *Transplantation Proceedings* 2007;39(8) 2647-2651.

[71] Vairetti M., Ferrigno A., Carlucci F., Tabucchi A., Rizzo V., Boncompagni E., et al. Subnormothermic machine perfusion protects steatotic livers against preservation injury: a potential for donor pool increase?. *Liver Transplantation* 2009;15(1) 20–29.

[72] Fernández L., Heredia N., Grande L., Gómez G., Rimola A., Marco A., et al. Preconditioning protects liver and lung damage in rat liver transplantation: role of xanthine/xanthine oxidase. *Hepatology* 2002;36(3) 562–572.

[73] Metzger J., Dore S. Lauterburg B. Oxidant stress during reperfusion of ischemic liver: no evidence for a role of xanthine oxidase. *Hepatology* 1998;8(3) 580–584.

[74] Rai R., Yang S., McClain C., Karp C., Klein A., Diehl A. Kupffer cell depletion by gadolinium chloride enhances liver regeneration after partial hepatectomy in rats. *American Journal of Physiology* 1996;270 G909–G918.

[75] Watanabe M., Chijiiwa K., Kameoka N., Yamaguchi K., Kuroki S., Tanaka M. Gadolinium pretreatment decreases survival and impairs liver regeneration after partial hepatectomy under ischemia/reperfusion in rats. *Surgery* 2000;127 456–463.

[76] Hayashi H., Chaudry I., Clemens M., Baue A. Hepatic ischemia models for determining the effects of ATP-MgCl$_2$ treatment. *Journal of Surgical Research* 1986;40(2) 167–175.

[77] Chaudry I., Clemens M., Ohkawa M., Schleck S., Baue A. Restoration of hepatocellular function and blood flow following hepatic ischemia with ATP–MgCl$_2$. *Advances in Shock Research* 1982;8 177–186.

[78] Zaouali M., Padrissa-Altés S., Ben Mosbah I., Alfany-Fernandez I., Massip-Salcedo M., Casillas Ramirez A., et al. Improved rat steatotic and nonsteatotic liver preservation by the addition of epidermal growth factor and insulin-like growth factor-I to University of Wisconsin solution. *Liver Transplantation* 2010;16(9) 1098-111.

[79] Casillas-Ramírez A., Zaouali A., Padrissa-Altés S., Ben Mosbah I., Pertosa A., Alfany-Fernández I., et al. Insulin-like growth factor and epidermal growth factor treatment: new approaches to protecting steatotic livers against ischemia-reperfusion injury. *Endocrinology* 2009;150(7):3153-3161.

[80] Man K., Zhao Y., Xu A., Lo C., Lam K., Ng K., et al. Fat-derived hormone adiponectin combined with FTY720 significantly improves small-for-size fatty liver graft survival. *American Journal of Transplantation* 2006;6(3) 467-476.

[81] Casillas-Ramírez A., Alfany-Fernández I., Massip-Salcedo M., Juan M., Planas J., Serafin A., et al. Retinol-Binding protein 4 and peroxisome proliferator-activated receptor-γ in steatotic liver transplantation. *Journal of Pharmacology and Experimental Therapeutics* 2011;338(1) 143-153.

[82] Elias-Miró M., Massip-Salcedo M., Raila J., Schweigert F., Mendes-Braz M., Ramalho F, et al. Retinol binding protein 4 and retinol in rat steatotic and non-steatotic livers in partial hepatectomy under ischemia-reperfusion. *Liver Transplantation* 2012; doi: 10.1002/lt.23489.

[83] Gonzalez-Flecha B., Cutrin J., Boveris A. Time course and mechanism of oxidative stress and tissue damage in rat liver subjected to in vivo ischemia-reperfusion. *Journal of Clinical Investigation* 1993;91(2) 456-464.

[84] Okaya T., Lentsh A. Peroxisome proliferator-activated receptor-alpha regulates postischemic liver injury. *American Journal of Physiology - Gastrointestinal and Liver Physiology* 1994;286 G606-G612.

[85] Stadler M., Nuyens V., Seidel L., Albert A., Boogaerts J. Effect of nutritional status on oxidative stress in an ex vivo perfused rat liver. *Anesthesiology* 2005;103(5) 978–986.

[86] Caraceni P., Domenicali M., Vendemiale G., Grattagliano I., Pertosa A., Nardo B., et al. The reduced tolerance of rat fatty liver to ischemia reperfusion is associated with mitochondrial oxidative injury. *Journal of Surgical Research* 2005;124(2) 160–168.

[87] Sankary H., Chong A., Foster P., Brown E., Shen J., Kimura R. et al. Inactivation of Kupffer cells after prolonged donor fasting improves viability of transplanted hepatic allografts. *Hepatology* 1995;22(4) 1236–1242.

[88] Okaya T., Blanchard J., Schuster R., Kuboki S., Husted T., Caldwell C., et al. Age-dependent responses to hepatic ischemia/reperfusion injury. *Shock* 2005;24(5) 421–427.

[89] Ben Mosbah I., Alfany-Fernández I., Martel C., Zaouali M., Bintanel-Morcillo M., Rimola A., et al. Endoplasmic reticulum stress inhibition protects steatotic and nonsteatotic livers in partial hepatectomy under ischemia-reperfusion. *Cell Death and Disease* 2010;1 e52(1-12).

[90] Serafin A., Rosello-Catafau J., Prats N., Gelpi E., Rodes J., Peralta C. Ischemic preconditioning affects interleukin release in fatty livers of rats undergoing ischemia/reperfusion. *Hepatology* 2004;39(3) 688–698.

[91] Peralta C., Leon O., Xaus C., Prats N., Jalil E., Planell E., et al. Protective effect of ozone treatment on the injury associated with hepatic ischemia-reperfusion: antioxidant-prooxidant balance. *Free Radical Research* 1999;31(3) 191–196.

[92] Teoh N., Leclercq I., Pena A., Farell G. Low-dose TNF-alpha protects against hepatic ischemia-reperfusion injury in mice: implications for preconditioning. *Hepatology* 2003;37(1) 118–128.

[93] Tian Y., Jochum W., Georgiev P., Moritz W., Graf R., Clavien P. Kupffer cell-dependent TNF-alpha signaling mediates injury in the arterialized small-for-size liver transplantation in the mouse. *Proceedings of the National Academy of Sciences* 2006;103 4598-4603.

[94] Bradham C., Schemmer P., Stachlewitz R., Thurman R., Brenner D. Activation of nuclear factor-kappaB during orthotopic liver transplantation in rats is protective and does not require Kupffer cells. Liver *Transplantation and Surgery* 1999;5(4) 282–293.

[95] Casillas-Ramírez A., Ben Mosbah I., Ramalho F., Rosello-Catafau J., Peralta C. Past and future approaches to ischemia-reperfusion lesion associated with liver transplantation. *Life Science* 2006;79 1881–1894.

[96] Bahde R., Spiegel H. Hepatic ischaemia-reperfusion injury from bench to bedside. *British Journal of Surgery* 2010;97(10) 1461-1475.

[97] Zúñiga J., Cancino M., Medina F., Varela P., Vargas R., Tapia G., et al. N-3 PUFA supplementation triggers PPAR-α activation and PPAR-α/NF-κB interaction: anti-inflammatory implications in liver ischemia-reperfusion injury. *PLoS One* 2011;6(12) e28502.

[98] Ghonem N., Yoshida J., Stolz D., Humar A., Starzl T., Murase N., Venkataramanan R. Treprostinil, a prostacyclin analog, ameliorates ischemia-reperfusion injury in rat orthotopic liver transplantation. *American Journal of Transplantation* 2011;11(11) 2508-2516.

[99] Kamo N., Shen X., Ke B., Busuttil R., Kupiec-Weglinski J. Sotrastaurin, a protein kinase C inhibitor, ameliorates ischemia and reperfusion injury in rat orthotopic liver transplantation. *American Journal of Transplantation* 2011;11(11) 2499-2507.

[100] Stoffels B., Yonezawa K., Yamamoto Y., Schäfer N., Overhaus M., Klinge U., et al. Meloxicam a COX-2 inhibitor, ameliorates ischemia/reperfusión injury in non-heart-beating donor livers. *European Surgical Research* 2011;47(3) 109-117.

[101] Li W., Meng Z., Liu Y., Patel R., Lang J. The hepatoprotective effect of sodium nitrite on cold ischemia-reperfusion injury. *Journal of Transplantation* 2012; 635179.

[102] Eipel C., Hübschmann U., Abshagen K., Wagner K., Menger M., Vollmar B. Erythro-poietin as additive of HTK preservation solution in cold ischemia/reperfusion injury of steatotic livers. *Journal of Surgical Research* 2012;173(1) 171-179.

[103] Bezinover D., Ramamoorthy S., Uemura T., Kadry Z., McQuillan P., Mets B., et al. Use of a third-generation perfluorocarbon for preservation of rat DCD liver grafts. *Journal of Surgical Research* 2011; 1-7.

[104] Minor T., Kötting M. Gaseous oxygen for hypothermic preservation of predamaged liver grafts: fuel to cellular homeostasis or radical tissue alteration?. *Cryobiology* 2000;40 182-186.

[105] Reynaert H., Geerts A., Henrion J. Review article: the treatment of non-alcoholic stea-tohepatitis with thiazolidinediones. *Alimentary Pharmacology and Therapeutics* 2005;22(10) 897-905.

[106] Schmidt R. Hepatic organ protection: from basic science to clinical practice. *World Journal of Gastroenterology* 2010;16(48) 6044-6045.

[107] Ke B., Lipshutz G., Kupiec-Weglinski J. Gene therapy in liver ischemia and reperfu-sion injury. *Current Pharmacology* 2006;12 2969-2975.

[108] Mizuguchi H., Hayakawa T. Targeted adenovirus vectors. *Human Gene Therapy* 2004;15(11) 1034-1044.

[109] Drazan K., Csete M., Da Shen X., Bullington D., Cottle G., Busuttil R., Shaked A. Hepatic function is preserved following liver-directed, adenovirus-mediated gene transfer. *Journal of Surgical Research* 1995;59(2) 299-304.

[110] Kroemer G. The proto-oncogene Bcl-2 and its role in regulating apoptosis. *Nature Medicine* 1997;3: 614-20.

[111] Bilbao G., Contreras J., Eckhoff D., Mikheeva G., Krasnykh V., Douglas J., et al. Re-duction of ischemia-reperfusion injury of the liver by in vivo adenovirus-mediated gene transfer of the antiapoptotic Bcl-2 gene. *Annals of Surgery* 1999;230 185-93.

[112] Takayama S., Sato T., Krajewski S., Kochel K., Irie S., Millan J., et al. Cloning and functional analysis of BAG-1: a novel Bcl-2-bonding protein with anti-cell death ac-tivity. *Cell* 1995;80 279-84.

[113] Sawitzki B., Amersi F., Ritter T., Fisser M., Shen X., Ke B., et al. Upregulation of Bag-1 by ex vivo gene transfer protects rat livers from ischemia/reperfusion injury. *Human Gene Therapy* 2002:13 1495-504.

[114]　Contreras J., Vilatoba M., Eckstein C., Bilbao G., Anthony J., Eckhoff D. Caspase-8 and caspase-3 small interfering RNA decreases ischemia/reperfusion injury to the liver in mice. *Surgery* 2004;136(2) 390-400.

[115]　Palmer H., Paulson K. Reactive oxygen species and antioxidants in signal transduction and gene expression. *Nutrition Reviews* 1997;55 353-361.

[116]　Zwacka R., Zhang Y., Zhou W., Halldorson J., Engelhardt J. Ischemia/reperfusion injury in the liver of BALB/c mice activates AP-1 and nuclear factor kappaB independently of IkappaB degradation. *Hepatology* 1998;28 1022-30.

[117]　Baeuerle P., Henkel T. Function and activation of NF-kappa B in the immune system. *Annual Review of Immunology* 1994;12 141-79.

[118]　Okaya T., Lentsch A. Hepatic expression of S32A/S36A Ikappa B alpha does not reduce postischemic liver injury. *Journal of Surgical Research* 2005;124(2) 244-249.

[119]　Zwacka R., Zhou W., Zhang Y., Darby C., Dudus L., Halldorson J., et al. Redox gene therapy for ischemia/reperfusion injury of the liver redices AP1 and NF-κB activation. *Nature Medicine* 1998;4 698-704.

[120]　Wheeler M., Katuna M., Smutney O., Froh M., Dikalova A., Mason R., et al. Comparison of the effect of adenoviral delivery of three superoxide dismutase genes against hepatic ischemia-reperfusion injury. *Human Gene Therapy* 2001;12 2167-2177.

[121]　He S., Zhang Y., Venugopal S., Dicus C., Perez R., Ramsamooi R., et al. Delivery of antioxidative enzyme genes protects against ischemia7reperfusion-induced liver injury in mice. *Liver Transplantation* 2006;12(21) 1869-1879.

[122]　Takahashi Y., Ganster R., Ishikawa T., Okuda T., Gambotto A., Shao L., et al. Protective role of NF-kappaB in liver cold ischemia/reperfusion injury: effects of IkappaB gene therapy. *Transplantation Proccedings* 2001;33(1) 602.

[123]　Amersi F., Buelow R., Kato H., Ke B., Coito A., Shen X., et al. Upregulation of heme oxygenase-1 protects genetically fat Zucker rat livers from ischemia/reperfusion injury. *Journal of Clininical Investigation* 1999;104(11) 1631-1639.

[124]　Tsuchihashi S., Fondevila C., Kupiec-Weglinski J. Heme oxygenase system in ischemia and reperfusion injury. *Ann Transplant* 2004;9(1) 84-87.

[125]　Halliwell B., Gutteridge J. Biologically relevant metal ion-dependent hydroxyl radical generation. An update. *FEBS Letter* 1992;307(1) 108-112.

[126]　Berberat P., Katori M., Kaczmarek E., Anselmo D., Lassman C., Ke B., et al. Heavy chain ferritin acts as an antiapoptotic gene that protects livers from ischemia reperfusion injury. *FASEB Journal* 2003;17 1724-1726.

[127]　Kato A., Yoshidome H., Edwards M., Lentsch A. Regulation of liver inflammatory injury by signal transducer and activator of transcription-6. *American Journal of Pathology* 2000;157 297-302.

[128] Lee T., Chau L. Heme oxygenase-1 mediates the anti-inflammatory effect of interleukin-10 in mice. *Nature Medicine* 2002;8 240-246.

[129] Ke B., Shen X., Lassman C., Gao F., Busuttil R., Kupiec-Weglinski J. Cytoprotective and antiapoptotic effects of IL-13 in hepatic cold ischemia/reperfusion injury are heme oxygenase-1 dependent. *American Journal of Transplantation* 2003;3 1076-1082.

[130] Ke B., Shen X., Gao F., Busuttil R., Kupiec-Weglinski J. Interleukin 13 gene transfer in liver ischemia and reperfusion injury: role of Stat6 and TLR4 pathways in cytoprotection. *Human Gene Therapy* 2004;15 691-698.

[131] Harada H., Wakabayashi G., Takayanagi A., Shimazu M., Matsumoto K., Obara H., et al. Transfer of the interleukin-1 receptor antagonist gene into rat liver abrogates hepatic ischemia-reperfusion injury. *Transplantation* 2002;74 1434-1441.

[132] Ke B., Shen X., Gao F., Busuttil R., Lowenstein P., Castro M., et al. Gene therapy for liver transplantation using adenoviral vectors: CD40-CD154 blockade by gene transfer of CD40Ig protects rat livers from cold ischemia and reperfusion injury. *Molecular Therapy* 2004;9 38-45.

[133] Shen X., Ke B., Zhai Y., Amersi F., Gao F., Anselmo D., et al. CD154-CD40 T cell costimulation pathway is required in themechanism of hepatic ischemia/reperfusion injury, and its blockade facilitates and dependents on heme oxgenase-1 mediated cytoprotection. *Transplantation* 2002;74 315-319.

[134] Shi Y., Rehman H., Ramshesh V., Schwartz J., Liu Q., Krishnasamy Y., et al. Sphingosine kinase-2 inhibition improves mitochondrial function and survival after hepatic ischemia-reperfusion. *Journal of Hepatology* 2012;56(1) 137-145.

[135] Jiang N., Zhang X., Zheng X., Chen D., Zhang Y., Siu L., et al. Targeted gene silencing of TLR4 using liposomal nanoparticles for preventing liver ischemia reperfusion injury. *American Journal of Transplantation* 2011;11(9) 1835-1844.

[136] Somia N., Verma I. Gene therapy: trials and tribulations. *Nature Reviews: Genetics* 2000;1(2) 91–99.

[137] Weber A., Groyer-Picard M., Franco D., Dagher I. Hepatocyte transplantation in animal models. *Liver Transplantation* 2009;15 7-14.

[138] Joseph B., Malhi H., Bhargava K., Palestro C., McCuskey R., Gupta S. Kupffer cells participate in early clearance of syngeneic hepatocytes transplanted in the rat liver. *Gastroenterology* 2002;123 1677-1685.

[139] Allen K., Soriano H. Liver cell transplantation: the road to clinical application. *Journal of Laboratory and Clinical Medicine* 2001;138 298-312.

[140] Benten D., Kumaran V., Joseph B., Schattenberg J., Popov Y., Schuppan D., et al. Hepatocyte transplantation activates hepatic stellate cells with beneficial modulation of cell engraftment in the rat. *Hepatology* 2005;42 1072-1081.

[141]   Soriano H. Liver cell transplantation: human applications in adults and children. London: Kluwer Academic Publishers; 2002.

[142]   Strom S., Fisher R., Thompson M., Sanyal A., Cole P., Ham J., et al. Hepatocyte transplantation as a bridge to orthotopic liver transplantation in terminal liver failure. *Transplantation* 1997;63(4) 559–569.

[143]   Bilir B., Guinette D., Karrer F., Kumpe D., Krysl J., Stephens J., et al. Hepatocyte transplantation in acute liver failure. *Liver Transplantation* 2000;6(1) 32–40.

[144]   Clavien P., Selzner M., Rudiger H., Graft R., Kadry Z., Rousson V., et al. A prospective randomized study in 100 consecutive patients undergoing major liver resection with versus without ischemic preconditioning. *Annals of surgery* 2003;238(6) 843–850.

[145]   Azoulay D., Del Gaudio M., Andreani P., Ichai P., Sebag M., Adam R., et al. Effects of 10 minutes of ischemic preconditioning of the cadaveric liver on the graft's preservation and function: the ying and the yang. *Annals of Surgery* 2005;242(1) 133–139.

[146]   Amador A., Grande L., Martí J., Deulofeu R., Miquel R., Solá A., et al. Ischemic preconditioning in deceased donor liver transplantation: a prospective randomized clinical trial. *American Journal of Transplantation* 2007;7(9) 2180-2189.

Strategies to Decrease Morbidity After Hepatectomy for Hepatocellular Carcinoma

Hiroshi Sadamori, Takahito Yagi and
Toshiyoshi Fujiwara

Additional information is available at the end of the chapter

1. Introduction

In-hospital mortality rates after hepatectomy for HCC have been greatly improved due to advances in surgical techniques and perioperative management [1-4]. However, relatively high morbidity rates remain problematic, and bile leakage and organ/space surgical site infection (SSI) are still common causes of major morbidity after hepatectomy for HCC [5-13].

Various types of hepatectomy in many centres have recently been performed based on the degree of hepatic functional reserve and the location of the HCC. Anatomic hepatectomy for HCC, including subsegmentectomy, reportedly contributes to the prognosis for patients with HCC [14-16]. In addition, the rate of repeat hepatectomy for recurrent HCC has recently increased from 10% to 31% as the prognosis for patients with HCC has improved [17-22].

In our institution, anatomic and repeat hepatectomies for HCC have been performed aggressively [12, 16, 22]. We investigated risk factors for bile leakage and organ/space SSI following hepatectomies for HCC in the present series, which included a large number of patients with a high proportion of anatomic or repeat hepatectomy. Furthermore, causes, management and outcomes of intractable bile leakage and organ/space SSI were investigated and strategies to reduce major morbidity were considered.

2. Methods

2.1. Patients

Medical records of 359 patients who underwent hepatectomy without biliary reconstruction for HCC in our department between January 1, 2001 and March 31, 2010 were studied retrospectively. Patients comprised 292 men and 67 women, with a mean age of 65 years (range, 32-89 years). The aetiology of liver disease was hepatitis C virus in 163 patients, hepatitis B virus in 122 patients, both hepatitis C virus and hepatitis B virus in 31 patients, and alcoholic liver disease in 16 patients. Child-Pugh class was A in 332 patients and B in 27 patients. A total of 296 patients (82.5%) underwent anatomic hepatectomy including subsegmentectomy. Repeat hepatectomy was performed for 59 patients (16.4%). Repeat hepatectomy was indicated when all tumours detected on preoperative imaging could be resected within the hepatic functional reserve. When recurrent HCC tumours were 2 cm in maximum diameter and 3 were present, percutaneous ablation therapies were selected despite the feasibility of repeat hepatectomy, depending on tumour location in the liver.

2.2. Surgical procedure

Laparotomy was performed through a J incision in 287 patients, a Mercedes incision in 33 patients, a midline incision in 23 patients, and a thoraco-abdominal incision in 16 patients. Preoperative cholangiography was not usually performed. Intraoperative ultrasonography was performed to determine the extent of HCC and the line of parenchymal transection. Parenchymal transection was performed using an ultrasonic dissector (Sonop 5000; Aloka, Tokyo, Japan) combined with bipolar electrocautery. Glisson's pedicles in livers dissected by the ultrasonic dissector were ligated and small pedicles were resected using metallic surgical clips. For hemihepatectomies or extended operations, hilar dissection was performed to divide the ipsilateral branches of the hepatic artery and portal vein. The hepatic duct was exposed inside the liver during parenchymal transection and was ligated or oversewn using fine non-absorbable sutures. Parenchymal transection in hemihepatectomy or extended operations was performed largely without occlusion of vascular inflow. For segmentectomies or subsegmentectomies, Glisson's pedicle was transected at the hepatic hilus and an intermittent Pringle manoeuvre was applied during parenchymal transection.

Intraoperative cholangiography was undertaken for selected patients when the integrity of the bile duct was in doubt. A bile leakage test using a cholangiography catheter was also performed for selected patients when many Glisson's pedicles were exposed in the plane of hepatic resection. In principle, two abdominal drainage tubes were systematically positioned and the method of placing the drainage tubes was changed according to the type of hepatectomy. In hemihepatectomy, one drainage tube was placed on the cut surface of the liver and another was positioned at the Winslow hiatus. In subsegmentectomy and segmentectomy, one drainage tube was placed on the cut surface of the liver and another was positioned in the right subphrenic space. From 2001 to 2005, an open drainage system was employed using 12-mm silicone Penrose drains (Kaneka, Osaka, Japan). From 2006 to 2010, a closed drainage system was used with 24-Fr BLAKE silicone drains (Johnson & Johnson, Somerville, NJ, USA). Drains were removed when the drainage was serous and contained no bile, usually around postoperative day (POD) 5.

2.3. Definition of bile leakage

Postoperative bile leakage was defined as the drainage of macroscopic bile from surgical drains for more than 7 days after surgery. Major bile leakage was defined as macroscopic bile discharge >100 ml/day that did not decrease from one day to the next. Minor bile leakage was defined as bile leakage that did not fulfil the definition for major bile leakage. Intractable bile leakage was defined as bile leakage requiring endoscopic retrograde biliary drainage (ERBD) or percutaneous transhepatic biliary drainage (PTBD) during postoperative management.

2.4. Definition of SSIs

SSIs were defined according to the National Infections Surveillance system [23]. Using these criteria, SSIs are classified as either incisional (superficial or deep) or organ/space. Criteria for superficial incisional SSI included infection occurring at the incision site within 30 days after surgery that involved only the skin and subcutaneous tissue and at least one of the following: 1) pus discharge from the incision; 2) bacteria isolated from a sample culture from the superficial incision; 3) localized pain, tenderness, swelling, redness, or heat; and 4) wound dehiscence. Criteria for deep incisional SSI included infection of the fascia or muscle related to the surgical procedure occurring within 30 days after surgery and at least one of the following: 1) pus discharge from the deep incision; 2) spontaneous dehiscence of the incision; or 3) deliberate opening of the incision when the patient displayed the previously described signs and symptoms of infection. The definition of organ/space SSI was based on postoperative findings of at least one of the following: 1) purulent drainage from a drain without macroscopic bile discharge; or 2) intra-abdominal collection of purulent fluid confirmed at the time of reoperation or percutaneous drainage. If intra-abdominal collection at the time of reoperation or percutaneous drainage contained macroscopic bile discharge, bile leakage was considered present. If purulent fluid was drained first and macroscopic bile leakage subsequently became apparent, this was defined as bile leakage. In contrast, if drainage of purulent fluid was still observed after the cessation of macroscopic bile leakage, this was defined as organ/space SSI.

2.5. Antimicrobial prophylaxis

Prophylactic antibiotics regimens were as follows. With initial hepatectomy, a first-generation cephalosporin was injected intravenously within 30 min prior to skin incision. In patients who underwent operations lasting longer than 3 h, additional antimicrobial agents were injected intravenously every 3 h as recommended by the Center for Disease Control guidelines [23]. These agents were also administered up to POD 2. In repeat hepatectomy, second-generation cephalosporin was injected intravenously in the same manner as in the initial hepatectomy and continued until POD 3.

2.6. Intervention for methicillin-resistant *Staphylococcus aureus* (MRSA)

With the exception of two emergency cases, all patients underwent preoperative evaluation for MRSA, including nasal culture. As a result, 9 of the 359 patients (2.5%) showed colonisation with MRSA on admission to our institution. In those 9 patients with detection of MRSA colonisation from preoperative nasal cultures, decolonisation was performed using intranasal mupirocin therapy (administered twice daily for 3-5 days preoperatively). Prophylactic intravenous infusion of vancomycin was not applied in the 9 patients with intranasal MRSA colonisation.

2.7. Analysis of risk factors for bile leakage and SSIs

Patient demographics, operative and tumour factors, and preoperative liver function were evaluated to determine impacts on the occurrence of bile leakage and organ/space SSI. Preoperative factors included patient age, sex, aetiology of liver disease, Child-Pugh classification, indocyanine green dye retention rate at 15 min (ICG-R15), serum albumin, history of diabetes mellitus, previous radiofrequency ablation (RFA) and previous transarterial chemoembolisation (TACE). The cut-off level for ICG-R15 was set at 20%, because ICG-R15 <20% has been reported as the safe range for bisegmentectomy [3,5,9]. Surgical factors were evaluated for the type of skin incision, type of hepatectomy, number of hepatectomies, blood loss, operative time, blood transfusion, and method of abdominal drainage. With regard to the type of hepatectomy, anterior segmentectomies and medial (S4) segmentectomies were sub-grouped for analysis. The cut-off point for operative time was determined by an analysis of the receiver operating characteristics curve for bile leakage. The optimal cut-off for operative time was 306 min; sensitivity and specificity were 0.696 and 0.728, respectively. We thus set 300 min as the cut-off level for operative time. Tumour factors included the number of HCC lesions and the maximum diameter of HCC. Cut-off level for HCC diameter was determined according to results from previous reports that analysed risk factors for morbidity after hepatectomy for HCC [3,5,9,12].

2.8. Investigation of intractable bile leakage

Management and outcomes were investigated for 46 patients with postoperative bile leakage. Indications for ERBD to treat postoperative bile leakage were based on postoperative findings of at least one of the following: 1) amount of macroscopic bile discharge from surgical drains >200 ml/day at 2 weeks after surgery; 2) amount of macroscopic bile discharge from surgical drains >100 ml/day at 4 weeks after surgery; or 3) macroscopic bile discharge from surgical drains still continuing at 6 weeks after surgery. PTBD was indicated when postoperative cholangiography and biliary drainage by ERBD were considered impractical. Intractable bile leakage necessitating ERBD or PTBD was encountered in 8 patients. The operative procedure, number of hepatectomies, timing of biliary procedures, sites of bile leakage and possible causes of bile leakage were evaluated in these 8 patients with intractable bile leakage.

2.9. Investigation of characteristics in organ/space SSI

Organ/space SSI was classified according to the modified Clavien system [24]: grade I, minor risk events not requiring special treatment; grade II, potentially life-threatening complications requiring pharmacological treatment; grade III, complications requiring surgical, endoscopic or radiological intervention, either with (III-b) or without (III-a) general anaesthesia; grade IV, life-threatening complications involving dysfunction of one (IV-a) or multiple (IV-b) major organs; and grade V, complications resulting in the death of the patient. Management and outcomes were investigated for 31 patients with organ/space SSI. In addition, the causative bacterium was identified for both incisional and organ/space SSIs. Furthermore, pre- and intraoperative parameters, causative bacteria and hospitalisation were compared between groups classified by the number of hepatectomies in patients with organ/space SSI.

2.10. Statistical analysis

Operative time, blood loss and postoperative hospital stay are presented as mean ± standard error of the mean. Differences in qualitative variables were assessed using Fisher's exact test or the ^{2} test, while differences in quantitative variables were analysed using the Mann-Whitney test. Uni- and multivariate logistic regression analyses were used to identify risk factors for bile leakage and organ/space SSI based on the 18 above-mentioned clinical factors. Relative risk was described by the estimated odds ratio (OR) with a 95% confidence interval. Two-sided P-values were computed and an effect was considered significant at the level of P 0.05. All statistical analyses were performed using SPSS II statistical software (SPSS, Tokyo, Japan).

3. Results

3.1. Risk factors for bile leakage (Tables 1, 3)

Univariate logistic regression analysis revealed several factors associated with increased risk of developing bile leakage. Repeat hepatectomy influenced the risk of developing bile leakage, with an OR of 3.78 compared to the initial hepatectomy. In contrast, neither previous RFA nor TACE had any significant impact on the occurrence of bile leakage. Operative time 300 min was associated with increased risk (OR, 5.32; P< 0.001), as was blood loss 2 000 ml (OR, 4.12; P< 0.001). Multivariate analysis regarding bile leakage confirmed operative time 300 min as an independent risk factor.

Variable	OR	95%CI	P
Bile leakage			
Operative time (<300 min vs. ≥ 300 min)	5.32	2.71–10.4	<0.001
Blood loss (<2000 ml vs. ≥ 2000 ml)	4.12	2.07–3.20	<0.001
Number of hepatectomies (initial vs. repeat)	3.78	1.91–7.48	<0.001

Table 1. Univariate analysis of risk factors for bile leakage.

3.2. Risk factors for SSIs (Tables 2, 3)

SSIs developed in 14.5% of patients (n=52), and 3 patients showed both incisional and organ/space SSIs. Univariate logistic regression analysis revealed several factors associated with increased risk of developing SSIs. Repeat hepatectomy influenced the risk of developing SSIs, with an OR of 8.27 for initial hepatectomy. Operative time 300 min was associated with increased risk (OR, 4.46; P<0.001). The presence of blood transfusion influenced the risk of developing SSIs. Presence of bile leakage was associated with increased risk of SSIs (OR, 6.40; P=0.002). Multivariate analysis regarding SSIs confirmed both repeat hepatectomy and operative time 300 min as independent risk factors.

3.3. Risk factor for incisional SSI (Tables 2, 3)

Incidence of incisional SSI was 6.7% (n=24). Univariate logistic regression analysis revealed that the presence of blood transfusion was associated with increased risk of developing incisional SSI. Type of skin incision classified according to the presence or absence of transverse incision showed no significant influence on the occurrence of incisional SSI in this series. Multivariate analysis regarding incisional SSI confirmed the presence of blood transfusion as an independent risk factor.

3.4. Risk factors for organ/space SSI (Tables 2, 3)

Organ/space SSI developed in 8.6% of patients (n = 31). Univariate logistic regression analysis revealed several factors associated with increased risk of developing organ/space SSI. Repeat hepatectomy influenced the risk of developing organ/space SSI, with an OR of 4.29 compared to initial hepatectomy. In contrast, neither previous RFA nor TACE exerted any significant impact on occurrence of organ/space SSI.

Variable	OR	95%CI	P
SSIs			
Operative time (<300 min vs. ≥ 300 min)	4.46	1.64–5.46	<0.001
Number of hepatectomies (initial vs. repeat)	8.27	2.24–8.24	<0.001
Bile leakage (absence vs. presence)	6.40	1.55–6.46	0.002
Blood transfusion (absence vs. presence)	2.05	1.37–4.55	0.003
Incisional SSI			
Blood transfusion (absence vs. presence)	4.38	1.85–10.4	<0.001
Organ/space SSI			
Number of hepatectomies (initial vs. repeat)	4.29	3.79–18.0	<0.001
Bile leakage (absence vs. presence)	3.16	2.90–14.3	<0.001
Operative time (<300 min vs. ≥ 300 min)	2.99	2.03–9.81	<0.001
Blood loss (<2000 ml vs. ≥ 2000 ml)	2.63	0.73–6.59	0.010

Table 2. Univariate analysis of risk factors for SSIs.

The method of abdominal drainage (open Penrose drains or closed suction drains) showed no significant influence. Operative time 300 min was associated with increased risk of organ/space SSI (OR, 2.99; P< 0.001). Presence of bile leakage was likewise associated with increased risk (OR, 3.16; P = 0.01). Blood loss 2 000 ml was associated with increased risk (OR, 2.63; P< 0.001). Multivariate analysis confirmed both repeat hepatectomies and presence of bile leakage as independent risk factors for organ/space SSI.

Variable	OR	95%CI	P
Bile leakage			
Operative time (<300 min vs. ≧ 300 min)	25.9	2.28 – 29.4	0.009
SSIs			
Number of hepatectomies (initial vs. repeat)	3.43	1.73 – 6.80	<0.001
Operative time (<280 min vs. ≧ 280 min)	2.32	1.22 – 4.43	0.011
Incisional SSI			
Blood transfusion (absence vs. presence)	7.56	2.58 – 22.1	<0.001
Organ/space SSI			
Number of hepatectomies (initial vs. repeat)	6.15	2.69 – 14.1	<0.001
Bile leakage (absence vs. presence)	3.01	1.20 – 7.56	0.018

Table 3. Multivariate analysis of risk factors for bile leakage and SSIs.

3.5. Management and outcomes of bile leakage (Figure 1)

Management and outcomes of the 46 patients with bile leakage are shown in Figure 1.

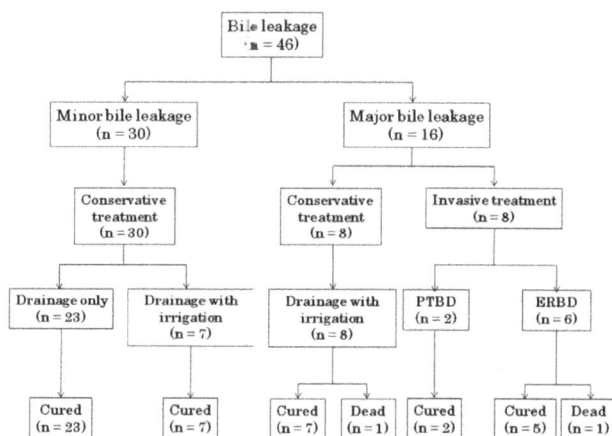

Figure 1. Medical management and outcomes for patients with postoperative bile leakage.

Minor bile leakage in 30 patients (65%) was controllable and cured by conservative therapies comprising drainage alone in 23 patients and drainage with irrigation in 7 patients. Sixteen patients (35%) showed complications of major bile leakage. In 8 of these patients, the major bile leakage was treated using drainage with irrigation. One patient died due to subsequent intractable ascites and liver failure during drainage with irrigation, while the other 7 patients healed. The remaining 8 patients with major bile leakage needed either ERBD or PTBD.

3.6. Characteristics of 8 patients with intractable bile leakage (Table 4)

We investigated the characteristics of the 8 patients who needed either ERBD or PTBD for bile leakage. High-risk surgical procedures were performed in most of these cases and repeat hepatectomy was performed in 6 of the 8 patients. The median timing of biliary procedures was POD 21.5 (range, POD 2-45). Bile leakage sites identified on postoperative cholangiography included the hepatic duct in 2 patients and the raw surface of the liver in 6 patients. Possible causes of bile leakage as assessed by postoperative cholangiography were as follows: stricture of the hepatic duct that existed preoperatively, possibly due to previous treatments for HCC in 4 patients (2 patients due to previous hepatectomies, 1 patient due to previous TACE, 1 patient due to previous RFA), stricture of the hepato-jejunostomy from previous pancreatoduodenectomy in 1 patient, dyskinesis of the papilla of Vater in 1 patient and intraoperative injury of the left hepatic duct related to repeat hepatectomy in 2 patients. Three of these 8 patients subsequently showed complications of intractable ascites. In 2 patients, both bile leakage and intractable ascites were cured without intra-abdominal septic complications. The other patient with stricture and injury of the left hepatic duct caused by a previous RFA died due to intractable ascites, uncontrollable biliary infection and liver failure. Bile leakage in the other 5 patients healed after either ERBD or PTBD, with no other major morbidities.

Age/Sex	Operative procedure	Number of hepatectomies	Biliary procedure (Timing)	Site of bile leakage	Cause of bile leakage	Outcome
62/M	Caudate lob.	Repeat	Endoscopic (14POD)	Lt. hepatic duct	Stricture and injury of lt. hepatic duct due to previous RFA	Died (Biliary infection and liver failure)
53/M	Anterior seg.	Repeat	Endoscopic (10POD)	Lt. hepatic duct	Intra-operative injury of lt. hepatic duct	Cured
72/M	S5 subseg.	Initial	PTBD (43POD)	Raw surface of liver	Stricture of hepato-jejunostomy of previous pancreatoduodenectomy	Cured
60/M	S6 subseg.	Repeat	Endosopic (30POD)	Raw surface of liver	Stricture of rt. hepatic duct due to previous hepatectomy	Cured
62/M	S8 partial hep.	Initial	Endosopic (2POD)	Raw surface of liver	Dyskinesis of the papilla of Vater	Cured
67/F	Central biseg.	Repeat	PTBD (45POD)	Raw surface of liver	Stricture of rt. hepatic duct due to intra-operative injury	Cured
50/M	Posterior seg.	Repeat	Endoscopic (15POD)	Raw surface of liver	Stricture of rt. hepatic duct due to previous TACE	Cured
70/F	S4 seg.	Repeat	Endoscopic (28POD)	Raw surface of liver	Stricture of lt. hepatic duct due to previous hepatectomy	Cured

lob = lobectomy; seg = segmentectomy; hep = hepatectomy; PTBD = percutaneous transhepatic biliary drainage; RFA = radiofrequency ablation; TACE = transcatheter arterial chemoembolization; POD = postoperative day

Table 4. Characteristic and management of 8 patients with intractable bile leakage.

3.7. Management and outcome of organ/space SSI

Organ/space SSI in 31 patients was classified as follows: abscess on the cut surface of the liver in 26 patients; right subphrenic abscess in 4 patients; and liver abscess in 1 patient. One of the 31 patients with organ/space SSI was treated by reoperation due to right subphrenic abscess, but died due to myocardial infarction. Eleven patients needed percutaneous drainage of organ/space SSI and all of them were cured. Organ/space SSI in 19 patients healed with irrigation of the pre-existing drain. As a result, 31 patients with organ/space SSI were stratified according to the modified Clavien system as follows: grade I, 0 patients; II, 13 patients; III-a, 15 patients; III-b, 2 patients; IV-a, 1 patient; IV-b, 0 patients; and V, 0 patients. No mortality was associated with organ/space SSI in this series, but the postoperative hospital stay was significantly longer for patients with organ/space SSI (53 7.2 days) than for patients without organ/space SSI (27 0.9 days, P = 0.001).

3.8. Bacteria causing incisional and organ/space SSI (Table 5)

Causative bacteria for incisional and organ/space SSI comprised gram-positive cocci in 17 patients (70.8%) and 19 patients (61.3%), and gram-negative rods in 6 patients (25.0%) and 9 patients (29.0%), respectively, indicating similar proportions of gram-positive cocci and gram-negative rods in both incisional and organ/space SSI. MRSA was the causative bacteria in 12 of 19 patients with organ/space SSI caused by gram-positive cocci.

Causative bacteria	Incisional (n = 24)	Organ/space (n = 31)
Gram-positive cocci		
MRSA	8	12
MSSA	0	1
S. epidermidis	4	2
Enterococcus sp.	4	4
Steptococcus sp.	1	0
Total	17 (70.8%)	19 (61.3%)
Gram-negative bacilli		
Escherichia coli	1	2
Klebsiella sp.	2	1
Pseudomonas sp.	2	4
Enterobacter sp.	1	1
Bacteroides sp.	0	1
Total	6 (25.0%)	9 (29.0%)
Negative	1	3

MRSA: Methicillin-resistant Staphylococcus aureus
MSSA: Methicillin-sensitive Staphylococcus aureus

Table 5. Causative bacteria of incisional and organ/space SSI.

3.9. Comparison between initial and repeat hepatectomies in patients with organ/space SSI (Table 6)

We compared clinical parameters between initial and repeat hepatectomies in patients with organ/space SSI (Table 6). HCC diameter was significantly larger in patients with organ/space SSI who underwent initial hepatectomy than in patients who underwent repeat hepatectomy. No significant differences were seen between groups in any other preoperative parameters, including patient demographics and preoperative liver function. No significant differences were identified between groups in operative parameters, including blood loss, operative time and blood transfusion. Rates of bile leakage were similar between groups. In contrast, in terms of bacteria causing organ/space SSI, detection of MRSA was significantly more frequent in the repeat hepatectomy group than in the initial group.

	Number of hepatectomies		
	Initial (n=14)	Repeat (n=17)	*P value*
Age	59.6 ± 3.2	62.2 ± 2.5	0.523
Etiology of liver disease			
HCV-related	5	5	0.713
HBV-related	6	11	0.231
HCV+HBV	2	0	0.113
Child-Pugh class			
A/B	14/0	15/2	0.192
ICG R15(%)	12.7 ± 1.7	18.5 ± 3.4	0.149
Albumin (g/dl)	4.0 ± 0.1	3.9 ± 0.1	0.610
Diabetes mellitus			
Negative/Positive	12/2	13/4	0.524
Number of HCC lesions			
1/>1	11/3	8/9	0.078
Diameter of HCC (cm)	4.5 ± 0.9	2.4 ± 0.3	<0.05
Type of hepatectomy			
Partial hepatectomy	0	3	0.104
Subsegmentectomy	3	5	0.619
Segmentectomy	4	5	0.960
Hemihepatectomy	7	4	0.132
Trisegmentectomy	0	0	
Blood loss (ml)	1833 ± 511	1697 ± 307	0.822
Operative time (min)	333 ± 11	343 ± 29	0.767
Blood transfusion			
Absence/Presence	9/5	8/9	0.345
Bile leakage			
Absence/Presence	9/5	10/7	0.531
MRSA			
Negative/Positive	12/2	7/10	<0.05
Hospital stay (days)	41 ± 7	63 ± 11	0.111

HCV hepatitis C virus, HBV hepatitis B virus, ICG R-15 indocyanine green dye retension rate at 15 min, MRSA Methicillin-resistant Staphylococcus aureus

Table 6. Comparison between initial nad repeat hepatectomies in patients with organ/space SSI.

4. Discussion

In-hospital mortality rates after hepatectomy for HCC have been greatly improved due to advances in surgical techniques and perioperative management [1-4]. However, relatively high morbidity rates remain problematic. The overall morbidity rates after hepatectomy for liver tumors have been reported to be 22.6 – 47.7%, and bile leakage and organ/space surgical site infection (SSI) are still common causes of major morbidity after hepatectomy for HCC [5-13]. Various types of hepatectomy in many centres have recently been performed based on the degree of hepatic functional reserve and the location of the HCC. In addition, the rate of repeat hepatectomy for recurrent HCC has recently increased from 10% to 31% as the prognosis for patients with HCC has improved [17-22]. The characteristic of our study is that this series consisted of a large number and percentage of both anatomic and repeat hepatectomies for HCC.

Rates of bile leakage after hepatectomy for liver tumours and benign lesions have been reported as 3.6%-12.0%, varying widely among different studies [6, 7, 11, 12, 25-30]. However, no standardised definition of bile leakage after hepatectomy has been established. In previous reports [6, 8, 11, 13, 30], the definition based on the drainage of macroscopic bile has been adopted. Several studies has proposed the definition on quantitative basis using the bilirubin concentration within the drain [26, 28], but these cut-off values varied. Currently, the International Study Group of Liver Surgery has proposed a consensus definition of bile leakage based on the postoperative course of bilirubin concentration in serum and drainage fluid [31]. Application of a uniform definition of bile leakage is indispensable to enabling standardised comparison of the results of different clinical reports and to facilitating objective evaluation of therapeutic modalities in the field of hepatectomies.

In the present study, prolonged operative time was identified as an independent risk factor for bile leakage and the type of hepatectomy had no significant impact on the rate of bile leakage. Several groups have reported that hepatectomies in which the cut surface exposed the major Glisson's sheath (i.e., central bisegmentectomy, S4 segmentectomy, and S8 subsegmentectomy) were independent risk factors for bile leakage [8, 28-30]. However, our results indicate that the standard types of hepatectomy were not risk factors for bile leakage, even if a wide cut surface with an exposed major Glisson's sheath was necessary, when assessment of liver function was appropriate and surgical procedures were performed carefully during transection of the liver parenchyma. We assume that the prolongation of operative time in this study was related to the extended duration of liver parenchymal transection and/or resection for severe intra-abdominal adhesions around the liver.

Our results revealed latent stricture of the biliary anatomy and intraoperative injury of the hepatic duct related to repeat hepatectomy as the main causes of intractable bile leakage requiring invasive treatment. Preoperative assessment of the biliary anatomy should therefore be considered for selected patients at high risk of intractable bile leakage. Various measures could also be applied during surgery to diminish the incidence of major and intractable bile leakage. First, intraoperative cholangiography should be used, particularly in repeat hepatectomies and in patients who have been treated with RFA or TACE for HCC located in the hepatic hilar region, as the identification of bile duct injury or stricture could allow immediate correction. Second, T-tube drainage or trans-cystic duct drainage of the common bile duct could be indicated in patients needing decompression of the biliary tree, such as patients with dyskinesis of the papilla of Vater. Third, particularly in repeat systematised hepatectomies, division of the bile ducts could be performed inside the liver during parenchymal transection, as this procedure could decrease the risk of injury to the bile ducts compared to division of the bile ducts at the liver hilum.

In the 1980s and 1990s, organ/space SSI formation after hepatectomy was reported as a fatal complication causing liver failure and death [32-34]. Although rates of organ/space SSI after hepatectomy have been reported as 4.7%-25% [35-42], hospital mortality rates caused by organ/space SSI have declined [7-10, 36, 40]. Several groups have reported high patient age and presence of diabetes mellitus as independent risk factors for organ/space SSI [36, 39]. However, these variables were not identified as independent risk factors for organ/space SSI in the present study. Our key result was the identification of repeat hepatectomy as an independent risk factor for organ/space SSI, suggesting that treatment strategies need to be established to reduce the high rate of organ/space SSI after repeat hepatectomy.

Repeat hepatectomy was identified as an independent risk factor for SSI and organ/space SSI, but previous RFA and TACE were not. Repeat hepatectomy for recurrent HCC is useful in establishing the good long-term outcomes. Cumulative 5-year survival rates after second hepatectomy have been reported as 41-69% [17-22]. RFA has recently been confirmed as a safe and promising therapy for recurrent HCC after hepatectomy. However, sufficient evidence does not exist to confirm whether RFA actually improves long-term outcomes. Cumulative 5-year survival rates after RFA for recurrent HCC after hepatectomy have been reported as 18-51.6% [43-45]. RFA is sometimes ineffective for HCC on the liver surface or near large vessels. In addition, postoperative adhesions between the remnant liver and gastrointestinal tract may prevent safe percutaneous RFA in patients with recurrent HCC.

In this study, MRSA was detected more frequently in organ/space SSI after repeat hepatectomy compared with after initial hepatectomy. We assume that most organ/space SSIs with MRSA after repeat hepatectomy develop as a result of contamination when the surgical procedure comes into contact with intra-abdominal colonisation or micro-abscesses of MRSA that had formed after the initial hepatectomy. This assumption might be partially supported by our result that the method of abdominal drainage (open or closed) had no significant influence on the occurrence of organ/space SSI. If this assumption is valid, preoperative interventions for MRSA, consisting of nasal culture and decolonisation of nasal MRSA, will not greatly reduce the occurrence of organ/space SSI involving MRSA after repeat hepatectomy. Walsh et al. recently reported that an MRSA intervention program, in which all patients received intranasal mupirocin and those patients colonised with MRSA received prophylactic intravenous infusion of vancomycin, resulted in near-complete and sustained elimination of MRSA SSIs after cardiac surgery [46]. Regarding patients who undergo repeat hepatectomies, preoperative detection of intra-abdominal colonisation or micro-abscess containing MRSA is difficult. MRSA intervention programs thus need to be improved, particularly for patients who undergo repeat hepatectomies, by considering the prophylactic intravenous administration of vancomycin.

In conclusion, our results reveal prolonged operative time as an independent risk factor for bile leakage, and latent stricture of the biliary anatomy and intraoperative injury of the hepatic duct related to repeat hepatectomy as the main causes of intractable bile leakage necessitating invasive treatment. Repeat hepatectomy was also identified as an independent risk factor for organ/space SSI, with MRSA as the main causative bacteria in organ/space SSI after repeat hepatectomy for HCC. Establishment of treatment strategies is thus important for reducing the high rate of organ/space SSI after repeat hepatectomy. In addition, preoperative assessment of the biliary anatomy and surgical procedures to decrease the incidence of major bile leakage should be considered for selected patients at high risk of intractable bile leakage.

Author details

Hiroshi Sadamori*, Takahito Yagi and Toshiyoshi Fujiwara

*Address all correspondence to: sada@md.okayama-u.ac.jp

Department of Gastroenterological Surgery, Okayama University Graduate School of Medicine, Dentistry and Pharmaceutical Sciences, Okayama, Japan

References

[1] Fan, S. T., Lo, C. M., Liu, C. L., et al. (1999). Hepatectomy for hepatocellular carcinoma: toward zero hospital deaths. *Ann Surg*, 229, 323-330.

[2] Fong, Y., Sun, R. L., Jarnagin, W., & Blumgart, L. H. (1999). An analysis of 412 cases of hepatocellular carcinoma at a Western center. *Ann Surg*, 229, 790-800.

[3] Torzilli, G., Makuuchi, M., Inoue, K., et al. (2007). No mortality liver resection for hepatocellular carcinoma in cirrhotic and noncirrhotic patients. *Arch Surg*, 134, 984-992.

[4] Sadamori, H., Yagi, T., Matsuda, H., et al. (2010). Risk factors for major morbidity after hepatectomy for hepatocellular carcinoma in 293 recent cases. *J Hepatobiliary Pancreat Sci*, 17, 709-718.

[5] Shimada, M., Takenaka, K., Fujiwara, Y., et al. (1998). Risk factors linked to postoperative morbidity in patients with hepatocellular carcinoma. *Br J Surg*, 85, 195-198.

[6] Lo, C. M., Fan, S. T., Liu, C. L., Lai, E. C. S., & Wong, J. (1998). Biliary complication after hepatic resection-Risk factors, management, and outcome. *Arch Surg*, 133, 156-161.

[7] Belghiti, J., Hiramatsu, K., Benoist, S., Massault, P., Sauvanet, A., & Farges, O. (2000). Seven hundred forty-seven hepatectomies in the 1990s: an update to evaluate the actual risk of liver resection. *J Am Coll Surg*, 191, 38-46.

[8] Yamashita, Y., Hamatsu, T., Rikimaru, T., et al. (2001). Bile leakage after hepatic resection. *Ann Surg*, 233, 45-50.

[9] Capussotti, L., Muratore, A., Amisano, M., Polastri, R., Bouzari, H., & Massucco, P. (2005). Liver resection for hepatocellular carcinoma on cirrhosis: analysis of mortality, morbidity and survival-a European single center experience. *Eur J Surg Oncol*, 31, 986-993.

[10] Taketomi, A., Kitagawa, D., Itoh, S., et al. (2007). Trends in morbidity and mortality after hepatic resection for hepatocellular carcinoma: An institute's experience with 625 patients. *J Am Coll Surg*, 204, 580-587.

[11] Virani, S., Michaelson, J., Hutter, M., et al. (2007). Morbidity and mortality after liver resection: Results of the patient safety in surgery study. *J Am Coll Surg*, 204, 1284-1292.

[12] Sadamori, H., Yagi, T., Shinoura, S., et al. (2012). Risk factors of organ/space surgical site infection after hepatectomy for hepatocellular carcinoma in 359 recent cases. *J Hepatobiliary Pancreat Sci*, Jan 25. [Epub ahead of print]. PMID: 22273719.

[13] Sadamori, H., Yagi, T., Shinoura, S., et al. (2012). Intractable bile leakage after hepatectomy for hepatocellular carcinoma in 359 recent cases. *Dig Surg*, 29, 149-156.

[14] Hasegawa, K., Kokudo, N., Imamura, H., et al. (2005). Prognostic impact of anatomic resection for hepatocellular carcinoma. *Ann Surg*, 242, 252-259.

[15] Eguchi, S., Kanematsu, T., Arii, S., et al. (2008). Liver Cancer Study Group of Japan. Comparison of the outcomes between an anatomical subsegmentectomy and a non-anatomical minor hepatectomy for single hepatocellular carcinomas based on a Japanese nationwide survey. *Surgery*, 143, 469-475.

[16] Sadamori, H., Matsuda, H., Shinoura, S., et al. (2009). Anatomical subsegmentectomy in the lateral segment for hepatocellular carcinoma. *Hepatogastroenterology*, 56, 971-977.

[17] Farges, O., Regimbeau, J. M., & Belghiti, J. (1998). Aggressive management of recurrence following surgical resection of hepatocellular carcinoma. *Hepatogastroenterology*, 45, 1275-1280.

[18] Shimada, M., Takenaka, K., Taguchi, K., et al. (1998). Prognostic factors after repeat hepatectomy for recurrent hepatocellular carcinoma. *Ann Surg*, 227, 80-85.

[19] Poon, R. T., Fan, S. T., Lo, C. M., Liu, C. L., & Wong, J. (1999). Intrahepatic recurrence after curative resection of hepatocellular carcinoma: long-term results of treatment and prognostic factors. *Ann Surg*, 229, 216-222.

[20] Minagawa, M., Makuuchi, M., Takayama, T., & Kokudo, N. (2003). Selection criteria for repeat hepatectomy in patients with recurrence hepatocellular carcinoma. *Ann Surg*, 238, 703-710.

[21] Itamoto, T., Nakahara, H., Amano, H., et al. (2007). Repeat hepatectomy for recurrent hepatocellular carcinoma. *Surgery*, 141, 589-597.

[22] Umeda, Y., Matsuda, H., Sadamori, H., Matsukawa, H., Yagi, T., & Fujiwara, T. (2011). A prognostic model and treatment strategy for intrahepatic recurrence of hepatocellular carcinoma after curative resection. *World J Surg*, 35, 170-177.

[23] CDC NNIS System. (2004). National Infections Surveillance (NNIS) system report, data summary from January 1992 to June 2004, issued October 2004. *Am J Infect Control*, 32, 470-485.

[24] Dindo, D., Demartines, N., & Clavien, P. A. (2004). Classification of surgical complications: a new proposal with evaluation in a cohort of 6336 patients and results of a survey. *Ann Surg*, 240, 205-213.

[25] Benzoni, E., Cojutti, A., Lorenzin, D., et al. (2007). Liver resective surgery: a multivariate analysis of postoperative outcome and complication. *Langenbecks Arch Surg*, 392, 45-54.

[26] Tanaka, S., Hirohashi, K., Tanaka, H., et al. (2002). Incidence and management of bile leakage after hepatic resection for malignant hepatic tumors. *J Am Coll Surg*, 195, 484-489.

[27] Reed, D. N., Jr Vitale, G. C., Wrightson, W. R., Edwards, M., & Mc Masters, K. (2003). Decreasing mortality of bile leaks after elective hepatic surgery. *Am J Surg*, 185, 316-318.

[28] Nagano, Y., Togo, S., Tanaka, K., et al. (2003). Risk factors and management of bile leakage after hepatic resection. *World J Surg*, 27, 695-698.

[29] Lee, C. C., Chau, G. Y., Lui, W. Y., et al. (2005). Risk factors associated with bile leakage after hepatic resection for hepatocellular carcinoma. *Hepatogastroenterology*, 52, 1168-1171.

[30] Capussotti, L., Ferrero, A., Vigano, L., Sgotto, E., Muratore, A., & Polastri, R. (2006). Bile leakage and liver resection: Where in the risk? *Langenbecks Arch Surg*, 141, 690-694.

[31] Koch, M., Garden, O. J., Padbury, R., et al. (2011). Bile leakage after hepatobiliary and pancreatic surgery: A definition and grading of severity by the International Study Group of Liver Surgery. *Surgery*, 149, 680-688.

[32] Yanaga, K., Kanematsu, T., Takenaka, K., & Sugimachi, K. (1986). Intraperitoneal septic complications after hepatectomy. *Ann Surg*, 203, 148-152.

[33] Anderson, R., Saarela, A., Tranberg, K. G., & Bengmark, S. (1990). Intraabdominal abscess formation after major liver resection. *Acta Chir Scand*, 156, 707-710.

[34] Nagasue, N., Kohno, H., Tachibana, M., Yamanoi, A., Ohmori, H., & El-Assai, O. (1999). Prognostic factors after hepatic resection for hepatocellular carcinoma associated with Child-Turcotte class B and C cirrhosis. *Ann Surg*, 229, 84-90.

[35] Wu, C. C., Yeh, D. C., Lin, M. C., Liu, T. J., & P'eng, F. K. (1998). Prospective randomized trial of systemic antibiotics in patients undergoing liver resection. *Br J Surg*, 85, 489-493.

[36] Togo, S., Matsuo, K., Tanaka, K., et al. (2007). Perioperative infection control and its effectiveness in hepatectomy. *J Gastroenterol Hepatol*, 22, 1942-1948.

[37] Shiba, H., Ishii, Y., Ishida, Y., et al. (2009). Assessment of blood-products use as predictor of pulmonary complications and surgical infection after hepatectomy for hepatocellular carcinoma. *J Hepatobiliary Pancreat Surg*, 16, 69-74.

[38] Okabayashi, T., Nishimori, I., Yamashita, K., et al. (2009). Risk factors and predictors for surgical site infection after hepatic resection. *J Hospital Infect*, 73, 47-53.

[39] Kobayashi, S., Gotohda, N., Nakagohri, T., Takahashi, S., Konishi, M., & Kinoshita, T. (2009). Risk factors of surgical site infection after hepatectomy for liver cancers. *World J Surg*, 33, 312-317.

[40] Uchiyama, K., Ueno, M., Ozawa, S., et al. (2011). Risk factors for postoperative infectious complications after hepatectomy. *J Hepatobiliary Pancreat Sci*, 18, 67-73.

[41] Togo, S., Kubota, T., Takahashi, T., et al. (2008). Usefulness of absorbable sutures in preventing surgical site infection in hepatectomy. *J Gastrointest Surg*, 12, 1041-1046.

[42] Arikawa, T., Kurokawa, T., Ohwa, Y., et al. (2011). Risk factors for surgical site infection after hepatectomy for hepatocellular carcinoma. *Hepatogastroenterology*, 58, 143-146.

[43] Lau, W. Y., & Lai, E. C. (2009). The current role of radiofrequency ablation in the management of hepatocellular carcinoma: a systemic review. *Ann Surg*, 249, 20-5.

[44] Choi, D., Lim, H. K., Rhim, H., Kim, Y. S., Yoo, B. C., Paik, S. W., et al. (2007). Percutaneous radiofrequency ablation for recurrent hepatocellular carcinoma after hepatectomy: long-term results and prognostic factors. *Ann Surg Oncol*, 14, 2319-9.

[45] Taura, K., Ikai, I., Hatano, E., Fujii, H., Uyama, N., & Shimahara, Y. (2006). Implication of frequent local ablation therapy for intrahepatic recurrence in prolonged survival of patients with hepatocellular carcinoma undergoing hepatic resection: an analysis of 610 patients over 16 years old. *Ann Surg*, 244, 265-73.

[46] Walsh, E. E., Greene, L., & Kirshner, R. (2011). Sustained reduction in methicillin-resistant Staphylococcus aureus wound infections after cardiothoracic surgery. *Arch Intern Med*, 171, 68-73.

Chapter 7

The Aim of Technology During Liver Resection — A Strategy to Minimize Blood Loss During Liver Surgery

Fabrizio Romano, Mattia Garancini, Fabio Uggeri,
Luca Gianotti, Luca Nespoli, Angelo Nespoli and
Franco Uggeri

Additional information is available at the end of the chapter

1. Introduction

Liver resection is considered the treatment of choice for liver tumours. Despite standardized techniques and technological advancing for liver resections, an intra-operative haemorrhage rate ranging from 700 and 1200 ml is reported with a post-operative morbidity rate ranging from 23 and 46% and a surgical death rate ranging from 4 and 5% [1],[2],[3],[4],[5],[6].

The parameter "**Blood loss**" has a central role in liver surgery and different strategies to minimize it are a key to improve these results. Bleeding has to be considered a major concern for the hepatic surgeon because of several reasons. At first it is certainly the major intra-operative surgical complication and cause of death and historically one of the major postoperative complication together with bile leaks and hepatic failure [5],[6],[7],[8],[9].

Besides a high intra-operative blood loss is associated with higher rate of post-operative complication and shorter long-term survival [10],[11],[12],[13]. Furthermore it is associated with an extensive use of vessel occlusion techniques, directly correlated with higher risk of post-operative hepatic failure. Last, a higher value of intra-operative blood loss is associated with a higher rate of peri-operative transfusions; host immunosuppression associated with transfusions with a dose-related relationship is correlated with a higher rate of complication (in particular infections) and recurrence of malignancies in neoplastic patients [11],[12],[14],[15], [16],[17],[18],[19],[20],[21]. In order to reduce transfusions hepatic surgeon has also not to misinterpret post-operative fluctuations of blood parameter: Torzilli at al. demonstrated that haemoglobin rate and haematocrit after liver resection show a steady and significant decrease until the third post-operative day and then an increase; so this situation has to be explained as physiological and does not justifies blood administration [22].Although the mechanism of bleeding in surgical interventions is multifactorial, technical factors may be responsible for a significant amount of intraoperative and early postoperative bleeding. The main progress in reducing perioperative blood loss has been made through improved surgical and anesthetic techniques and through better understanding of hemostatic disorders in patients who have liver disease. developments in surgical, anesthesiologic, and pharmacologic strategies that have contributed to a reduction of blood loss during liver surgery in cirrhotic and noncirrhotic patients. The clinical relevance of different types of strategies may vary, depending on the stage of the operation. For example, topical hemostatic agents have a role in reducing blood loss from the hepatic resection surface after partial liver resection, whereas surgical techniques play a more important role during transection of the liver parenchyma (Fig. 1).

2. How can we reduce bleeding in liver surgery?

Figure 1 shows the amount of blood loss during the different phases of liver surgery. It is clear that the higher risk for bleeding and the greater amount of blood loss occur during the parenchymal transection phase of the procedure.

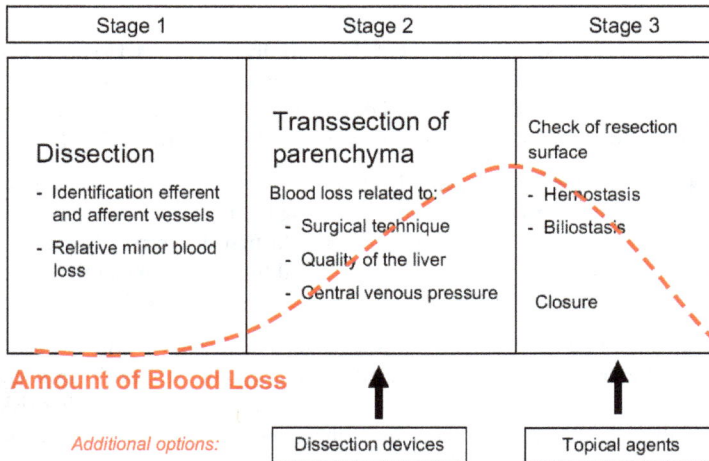

Figure 1. The mechanisms of bleeding and the relative amount of blood loss (dotted line) during the three surgical stages of partial liver resections. In general, most bleeding can be encountered during transsection of the liver parenchyma. In this stage of the operation, blood loss is mainly caused by bleeding from the resection surface of the liver.

The aim of the study is to investigate the principal solutions to the problem of high blood loss in hepatic resection, considering the role of surgeons and anestesiologists. Table 1 resume all the methods to prevent or reduce bleeding during liver surgery. Moreover we focused our attention on the technological aspects of liver parenchima transection. We will describe each technology and instrument discussing the principle of functioning, the technical characteristics and analysed the advantages (**A**) and the disadvantages (**D**) correlated to their employment during liver transection. We divided the instruments taking into account the energy employed for their functioning.

Surgical
Vascular clamping techniques
Inflow occlusion
Continuous Pringle maneuver
Continuous Pringle maneuver after ischemic preconditioning
Intermittent Pringle maneuver
Total vascular occlusion
Dissection devices for transsection of liver parenchyma
Classic methods
Scalpel
''Finger-fracture'' method
Clamp crushing
Ultrasonic dissection
Hydro-jet dissection
Electro coagulation (monopolar, bipolar, Argon coagulation)
Radiofrequency ablation-based devices
Staplers
Topic hemostatic agents
Anesthesiologic
Maintaining low central venous pressure by using
Volume contraction
Phlebotomy
Vasodilatation
If needed, forced diuresis
Blood products
Use of pharmacologic agents
Ant fibrinolytics
Recombinant factor VIIa

Table 1. Surgical and anesthesiologic methods used to reduce blood loss in liver surgery

Moreover we tried to compare the different instruments and technologies basing on literature data to identify the best instruments for each type of liver resection (open surgery, laparoscopic surgery, resective surgery, oncologic surgery, liver transplantation).

3. The role of the surgeon

Most blood loss during liver resection occurs during parenchymal transection. Hepatic surgeon has different ways to control bleeding:

Vessel occlusion techniques: Those technique are based on the idea that to limit the blood flow through the liver during parenchymal transection can reduce the haemorrhage. Although various forms and modified techniques of vascular control have been practiced, there are basically two main strategies; inflow vascular occlusion and total vascular exclusion23-24 Inflow vascular occlusions are techniques that limit anterograde blood flow with the clamping of all the triad of the hepato-duodenal ligament (*Pringle 's manoeuvre, PM*), only of the vascular pedicles (selective clamping of the portal vein and the hepatic artery or *Bismuth technique*) or *intravascularportal clamping*. During Pringle's manouvre the hepatoduodenal ligament is encircled with a tape, and then a vascular clamp or tourniquet is applied until the pulse in the hepatic artery disappears distally. The PM has relatively little general haemodynamic effect and no specific anaesthetic management is required. However, bleeding can still occur from the backflow from the hepatic veins and from the liver transection plane during unclamping. The other concern is the ischaemic-reperfusion injury to the liver parenchyma, especially in patients with underlying liver diseases25. The continuous Pringle manoeuvre (CPM) can be safely applied to the normal liver under normothermic conditions for up to 60 minutes and up to 30 minutes in pathological (fatty or cirrhotic) livers, although much longer durations of continuous clamping 127 minutes in normal livers and 100 minutes in pathological livers have been reported to be safe26-27. One way to extend the duration of clamping and to reduce ischaemia to the remnant liver is by the intermittent Pringle manoeuvre (IPM). It involves periods of inflow clamping that last for 15-20 minutes followed by periods of unclamping for five minutes (mode 15/5 or 20/5), or five minutes clamping followed by one minute unclamping (mode 5/1)28-29 IPM permits a doubling of the ischaemia time, when compared with CPM and the total clamping time can be extended to 120 minutes in normal livers and 60 minutes in pathological livers. The disadvantage of IPM is that bleeding occurs from the liver transaction surface during the unclamping period and, thus, the overall transection time is prolonged as more time is spent in achieving haemostasis. Belghiti et al (1999) revealed that there was no significant difference in total blood loss or volume of blood transfusion between CPM and IPM (mode 15/5). However, they noticed that pathological livers tolerated CPM poorly.

A newer perspective on inflow occlusion comes from the concept of ischaemic preconditioning (IP). It refers to an endogenous self-protective mechanism by which a short period of ischaemia followed by a brief period of reperfusion produces a state of protection against subsequent sustained ischaemia-reperfusion injury [30]-[31]. The IP is performed with ten minutes of ischaemia followed by ten minutes of reperfusion before liver transaction with CPM [32].

Hemihepatic clamping (half-Pringle manoeuvre) interrupts the arterial and portal inflow selectively to the right or left liver lobe that is to be resected [33]-[34]. It can be performed with or without prior hilar dissection. It can also be combined with simultaneous occlusion of the ipsilateral major hepatic vein. The advantage of this technique is that it avoids ischaemia in the remnant liver, avoids visceral congestion and allows clear demarcation of the resection margin. The disadvantage is that bleeding from the parenchymal cut surface can occur from the nonoccluded liver lobe.

Segmental vascular clamping entails the occlusion of the ipsilateral hepatic artery branch and balloon occlusion of the portal branch of a particular segment. The portal branch is identified by intra-operative ultrasound and puncture with a cholangiography needle through which a guide wire and balloon catheter is passed [35],[36].

Total vascular exclusion (TVE) combines total inflow and outflow vascular occlusion of the liver, isolating it completely from the systemic circulation. It is done with complete mobilisation of the liver, encircling of the suprahepatic and infrahepatic IVC, application of the Pringle manoeuvre, and then clamping the infrahepatic IVC followed by clamping of the suprahepatic IVC. TVE is associated with significant haemodynamic changes and warrants close invasive and anaesthetic monitoring. Occlusion of the IVC leads to marked reduction of venous return and cardiac output, with a compensatory 80% increase in systemic vascular resistance and 50% increase in heart rate and, thus, not every patient can tolerate it. TVE can be applied to a normal liver for up to 60 minutes and for 30 minutes in a diseased liver. The ischaemic time can be extended when combined with hypothermic perfusion of the liver [37]-[38]. Apart from the unpredictable haemodynamic intolerance, post-operative abdominal collections or abscesses and pulmonary complications are more common in TVE, when compared with CPM.

Inflow occlusion with extraparenchymal control of hepatic veins is a modified way of performing TVE. The main and any accessory right hepatic vein, the common trunk of the middle and left hepatic veins, or the separate trunks of the middle and left hepatic veins (15% of cases) are first dissected free and looped. It has been reported that the trunks of the major hepatic veins can be safely looped in 90% of patients [39]-[40]. The loops can then be tightened or the vessels clamped after inflow occlusion is applied, so that the liver lobe is isolated from the systemic circulation without interrupting the caval flow. It can be applied in a continuous or intermittent manner. The maximal ischaemia time is up to 58 minutes under continuous occlusion. This technique is more demanding than TVE, but it can avoid the haemodynamic drawbacks of TVE while at the same time provide almost a bloodless field for liver transection.

Instruments and technique for resections: Although a large part of improvements of these last decades in liver surgery can be correlated to a better knowledge of the surgical hepatic anatomy (Couinaud's segmentation of liver [41]), better monitoring during anaesthesia and introduction of intra-operative ultrasonography and of other imaging techniques, the choice of surgical technique for sectioning the liver has surely important repercussions on the intervention's outcome. Furthermore in the last two decades improvements in technology allowed the development of a large number of instruments with the aim to reduce blood loss during surgical procedure. The main part of these tools have been developed or applied to liver surgery. The rationale in liver transection is to employ an instrument that can selectively eliminate parenchyma leaving vital structures intact. In other words, a resistance modulated device, able to fragment low-resistance tissue (hepatic parenchyma) preserving fibrous (high-resistance) components such as vessels and biliary ducts, successively ligated by the surgeon. To date, no single instrument has been designed to adequately satisfy both of these tasks.

There are two techniques we could define traditional: the *finger fracture method* and the *clamp crushing method*. These are the oldest techniques for hepatic transection and are still employed especially by long experienced surgeons. Techniques of liver transection gained marked attention since the introduction of the clamp-crushing technique in the 1970s.10,11 As a refinement of the finger fracture method, it has served as the reference technique for liver transection ever since.

The use of traditional techniques to isolate bile ducts and vascular pedicles from the surrounding parenchyma provides for employment of clips or sutures for sealing bile ducts and vascular vessels and for other haemostasis techniques to stop haemorrhage from the resection's surface. There are several studies those sustain that traditional methods are still competitive with new technique based on utilization of special devices [1],[42],[96] In a recent Metanalysis Rahbari and coll concluded that the clamp-crushing technique could be still reccomended as the reference method for the transection of the parenchyma during liver surgery [12], [4].

Introduction of new devices for liver dissection surely have an important role, in particular for reduction of intra-operative blood loss. Actually the most important devices useful for liver resection are the followings, presented as they are from a technical point of view and analysed to find the advantages (A) and the disadvantages (D) correlated to their employment and divede according to the source of energy employed. There are two types of transection devices: those mainly used for dissection (e.g. the haemostatic clamps or ultrasonic dissector) and mainly used for haemostasis and coagulation (e.g. sutures, endo-staplers, sealers, etc.) (table 2). Moreover the water-jet, the ultrasonic aspirator (CUSA®) and the blunt dissection can be categorised under selective dissection techniques. Non-selective techniques cannot discriminate between duct structures and parenchyma. To mention are finger fracture and mechanical instruments as the scalpel, the scissors and with reservation the linear stapler as well as thermal instruments as the high-frequency electrocoagulator, the laser, the bipolar forceps or the scissors of the UltraCision®

Preparation	Transection
Finger fracture	ligation
crush/clamp	clips
suction knife	electrocoagulation (mono/bipolar)
CUSA	Microwave tissue coagulation
Water Jet	Ultracision
Jet-Cutter	Ligasure
Tissuelink	Gyrus
Aquamantys	

Table 2. Surgical techniques for preparation and tissue transection of the liver

Furthermore most attempts have involved use of radiofrequency ablation-based instruments in a "precoagulation strategy" in which the energy device is used to burn and seal the parenchyma before sharp dissection. In the second strategy, ultrasonic-activated instruments cut through the liver while sealing the vessels. Both method suffer from the fact that large vessels are poorly visualized and can bleed on transection. In addition, blood or biliary vessels from adjacent parts of the liver meant to be salvage can be inadvertently injured by this "blind" coagulation.

3.1. Tools based on ultrasound technology

Harmonic Scalpel, HS (Johnson and Johnson Medical, Ethicon, Cincinnati, USA): Also known as "Ultrasonically Activated Scalpel" or "Ultrasonic Coagulation Shears", this instrument was introduced in the early 1990s. The ultrasound scissors System includes a generator with a foot switch, the reusable handle for the scalpel and the cutting device with scissors. The scissors are composed by a moveable blade and by a fixed longitudinal blade that vibrates with a ultrasonic frequency of 55,5 kHz (55.500 vibrations per second). HS can simultaneously cut and coagulate causing protein denaturation by destroying the hydrogen bonds in proteins and by generation of heat in vibrating tissue. This generated heat denatures proteins and forms a sticky coagulum that covers the edges of dissection. Although the heat produces no smoke and thermal injury is limited, the depth of marginal necrosis is greater than incurred by either the water jet or CUSA The lateral spread of the energy is 500 micrometers.

A: HS is the only instrument that can simultaneously cut and coagulate (it can coagulate vessel until 2-3 mm of diameter [43]); it's useful on cirrhotic liver [44]; no electricity passes through the patient and there's no smoke production (especially useful in laparoscopic surgery); it can be used in laparoscopic and laparotomic surgery. **D**: The instrument results in a continuous bleeding risk related to the blind tissue penetration to coagulate vessels hidden into the hepatic parenchyma. Studies demonstrate that HS is not capable to reduce blood loss and operating time compared to traditional techniques [45]-[46], cannot coagulate vessel over 2-3 mm of diameter which have to be clipped, legated or sealed with other instruments; HS is not easy to use as a blunt dissector and have substantially demonstrated its usefulness only during the resection of the superficial part of liver (2, 3 cm) free from large vessels and bile ducts; besides some studies have demonstrated that HS increases the rate of post-operative bile leaks [47]-[48] raising concern that HS may not be effective in sealing bile ducts. this postoperative bile leakage occurred because Glisson's sheath was not completely sealed when the HA is used blindly in the deep liver parenchymal layer. It was difficult to seal the sheath precisely in the deep liver parenchymal layer.

The use of HS in liver cirrhosis is controversial. The greatest concern with the use of the harmonic scalpel is the risk of shearing [49]. Slight errors of movement can shear parenchyma without completely coagulating vessels and/or ducts. Moreover it's expensive (the generator costs US$ 20.000 and the handle USS 250). An evolution fo the harmonic Scalpel is the Harmonic FOCUS. Using this device the liver parenchyma is crushed by the nonactivated HF, which blades are similar to Kelly forceps, and the tiny areas of residual tissue are checked and completely sealed with the activated HF without changing to forceps. This device allow, after accurate exposure, a sealing "under view" of tiny vasculatures and biliary structures and this seems to reduce bleeding and postoperative bile leakages. [125-126] This new technique has been called "fusion technique".The attempt to accomplish both the task of division and of hemostasis is provided by a recently introduced device, which intends to crush liver parenchyma simul taneously sealing the vessels without the need to change the instrument, the so called focus-clysis or 'fusion technique'

Functionally, the instrument should be compared to a Kelly, in which the surgeon can adjust the precision and depth of cutting by modulating blade pressure; parenchyma crushing exposes the tiny vessels that can be coagulated employing the harmonic technology provided in high power (1–2 mm vessels) and low power (up to 5 mm). Vessels larger than 5 mm in diameter should be divided and ligated in a traditional fashion. It seems that the 'fusion technique' could reduce blood loss and the incidence of biliary fistula, with a cost comparable to other technologies.

Cavitron Ultrasonic Surgical Aspirator, CUSA (Valleylab) (Fig 2): The use in liver surgery of this instrument, also known as Ultrasonic Dissector, was described for the first time in literature in 1979 by Hodgson [50]. CUSA is a surgical system in which a pencil-grip surgical hand piece contains a transducer that oscillates longitudinally at 23 kHz and to which a hollow conical titanium tip is attached. The vibrating tip of the instrument causes explosion of cells with a high water content (just like hepatocytes) and fragmentation of parenchyma sparing blood and bile vessel because of their walls prevalently composed by connective cells poor of water but rich of intracellular bonds. This device (together with hydrojet dissector) should be considered among that tools able to selectively divide parenchyma from vessels according to their different mechanical resistance (in which hepatocytes contain less fibrous tissue than the vessel, thus offering less resistance to crushing during parenchymal division), the so called selective dissection technique.

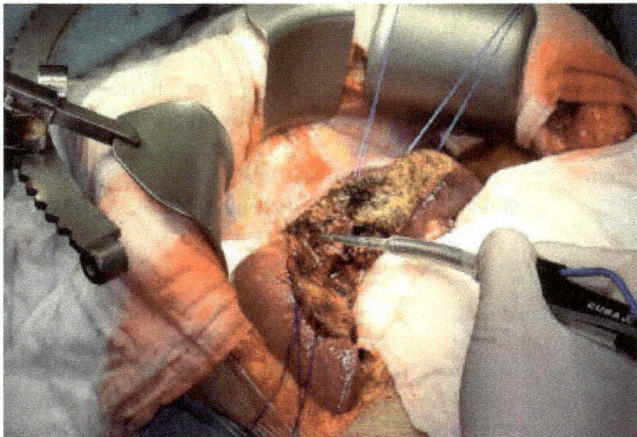

Figure 2. Parenchima transection using CUSA

The device is equipped by a saline solution irrigation system that cools the hand piece and wash the transection plane and by a constant suction system that removes fragmented bits of tissue and permits excellent visualization. A: CUSA is capable to dissect offering excellent visualization resulting useful in particular during non-anatomical resections and approaching the deeper portion of the transection plane [51]-[52]. The instrument allows surgeons to see clearly blood and biliary vessels as they dissect through the liver [53]. (2) use of the instrument allows them to avoid prolonged extrahepatic vascular control, and (3) the operation actually takes less time because the vessels are continuously controlled during the dissection and there is little need for a prolonged search for bleeding or biliary vessels after the specimen has been removed.

A previous retrospective study from Fan showed that the ultrasonic dissector resulted in lower blood loss, lower morbidity, and lower mortality compared with the clamp crushing technique [54] Furthermore, ultrasonic dissection resulted in a wider tumor-free margin because of a more precise transection plane.

D: CUSA can't coagulate or realize haemostasis so it need to be used in couple with an other instruments to achieve hemostasis and biliostasis. Even if some studies sustain it to be capable to reduce intra-operative blood loss, operating time and duration of vessel occlusion [55], important studies demonstrate that CUSA can't offer these advantages if compared with traditional techniques; a prospective trial by Rau et al. showed no statistical difference in reduction of blood loss with the use of CUSA as compared to conventional methods [56]; and another trial by Takayama et al. [52], in fact, noted a greater median blood loss. CUSA causes more frequent tumour exposure at the surgical margin than traditional techhiques[1] and it's less useful for cirrhotic livers because the associated fibrosis prevents easy removal of hepatocytes [57]; besides some authors found using CUSA method (compared to clamp crushing method) an increase of venous air embolism without evidence of hemodynamic compromise but with increased risk of paradoxical embolism in cirrhotic patients [58]. Moreover CUSA should be used in association with other devices which are able to perform hemostasis. The instrument seems cumbersome and complicated to inexperienced operating room personnel. Therefore, it is easy for the instrument to malfunction. The fact that the instrument works by removing a margin of liver tissue makes it, by nature, less attractive for harvesting liver for living-donor transplantation.

3.2. Tools based on radiofrequency technology

Tissuelink Monopolar Floating Ball, TMFB (Floating Ball, TissueLink medical, Dover, NH, USA) (Fig 3): This new instrument put on the market in 2002 is a linear device that employs Radiofrequency energy focused at the tip to coagulate target tissue. The tip is provided with a low volume (4-6 ml/min) saline solution irrigation that makes easier the conduction of RF in surrounding tissue and cools the tip itself avoiding formation of chars. TMFB can seal vascular and bile structures up to 3 mm in diameter by collagen fusion. These qualities makes this device an excellent instrument for achieving haemostasis and in particular for pre- coagulating (with a painting movement) parenchyma and vessels prior to transection, preventing blood loss.

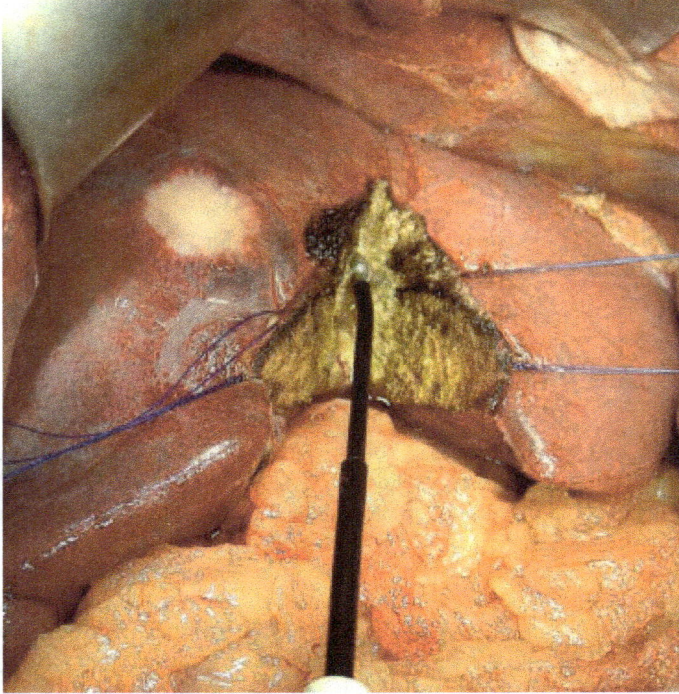

Figure 3. The Tissue Link working performing a liver resection

Otherwise continuously heating tissue underneath a cool layer, however, causes a build up of steam that can result in tissue destruction. The latter phenomenon is known as steam popping [59].

There are two models on the market, the DS3.0 with blunt tip that simply coagulates and the DS3.5-C Dissecting Sealer that is provided with sharp tip that can also dissect. **A**: The instrument is, in a sense, "friendlier" to most surgeons. In other words, surgeons, who are usually adept at using cautery, can easily understand this mechanism of action and use it accordingly. TMFB can coagulate (and the Dissecting Sealer can also cut) tissues and seals blood and bile ducts up to 3 mm in diameter, is able to reduce blood loss and the recourse to vessel occlusion techniques if compared to traditional technique s [60],[61],[62], offers good results also in cirrhotic livers and cystopericystectomy [63] and has a saline irrigation that avoids production of smoke, chars and sticky coagulum to which the device could stick causing new bleeding when it's moved away. TMFB, used on the cut liver surface after dissection, destroys eventual additional cancer cells at the margin of resection; in order to assure sterile margins, extra tissue destruction at the margins of resection may be desirable for tumor excisions. Otherwise this could be a disadvantage in case of living donor liver transplantation. It's available for both laparotomic and laparoscopic surgery and it's quite cheap and compatible with most electro-surgical generator currently available.

D: TMFB is not able to coagulate vessel over 2-3 mm of diameter which have to be clipped, legated or sealed with other instruments [64]. So the instrument should be used in combination with other instruments or clips or ties. Moreover studies do not demonstrate it's efficacy to reduce operating time if compared with traditional techniques [65].

Bipolar Vessel Sealing Device, BVSD (LigaSure, Valleylab Inc. Boulder, Colorado, USA) (fig 4): The use in liver surgery of this instrument was described for the first time in literature in 2001 by Horgan [67]. The LigaSure System includes a generator with a foot switch and a clamp-form hand piece that can be used for parenchymal fragmentation and isolation of blood and bile structures just like in clamp crushing technique before application of energy; it employs RF to realize permanent occlusion of vessels or tissue bundle. The LigaSure generator has a Valleylab's Instant Response technology, a feedback-controlled response system that diagnoses the tissue type in the instrument jaws and delivers the appropriate amount of energy to effectively seal the vessel: when the seal cycle is complete, a generator tone sound, and output to the handset is automatically discontinued. BVSD is capable to obliterate the lumen of veins and arteries up to 7 mm in diameter by the fusion of elastin and collagen proteins of the vessel walls; that makes BVSB the only safe and real alternative to sutures and clips for sealing vessel [68],[69],[70].

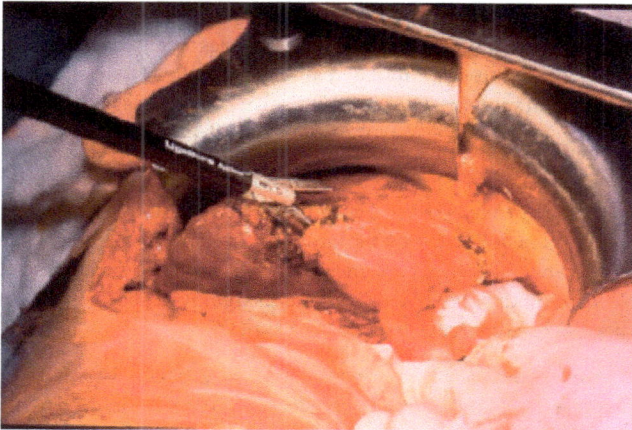

Figure 4. The Ligasure Atlas during parenchyma transection

A: BVSD coagulates sealing vessels up to 7 mm in diameter with minimal charring, thermal spread or smoke, it's capable to reduce blood loss and the need for vessel occlusion techniques if compared to traditional techniques [8],[71],[72], A recently published randomized controlled trial demonstrated that the use of Ligasure in combination with a clamp crushing technique resulted in lower blood loss and faster transaction speed in minor liver resections compared with the conventional technique of electric cautery or ligature for controlling vessels in the transection plane [73]. Otherwise a more recent randomized trial from the same team was not able to show a real difference between the traditional techniques and the Ligasure vessel sealing system [74]. The instrument is available for both laparotomic and laparoscopic surgery [75]. Furthermore the use of Ligasure System is not correlated with an increase of the rate of postoperative bile leaks and in some study bile leakage was nihill [76]-[127] and that proves his effectiveness in obliterate also bile vessel. **D**: after the application the coagulated tissue often sticks to the instrument's jaws causing new bleeding when the device is moved away; BVSD seems to be less effective in presence of cirrhosis for two reasons: first the portal hypertension correlated with cirrhosis causes thinning of the dilate portal vein's walls and makes their obliteration less effective; second cirrhosis makes crushing technique difficult and the hepatic tissue between the blades may disperse the power applied causing vessel to bleed [128]; moreover it seems to be ineffective in cystopericystectomy [77] (even if some surgeons sustain his effectiveness in this surgery [78]). Ligasure vessels sealing system has been widely use during liver transection in a "blind" way [70]-[71], achieving parenchymal fracture and vessel sealing in the same time without identification of tiny vasculatures and bile ducts. This could be considered a limits of this tools which do not allow the surgeon to clearly check the structures which are going to be sealed. To overcome this limit a technique similar to the "fusion technique" used with Harmonic FOCUS has been developed for the Ligasure vessel sealing system [130]., using the Ligasure precise. With this technique using LigaSure itself, the hepatic parenchyma was widely and gently crushed and confirmed that the remnant vessels and tiny vessels (2mm in diameter) were divided by the LigaSure under direct vision. This allow to coagulate only vessels appropriate for sealing with this instrument and imprtant vascular pedicles to adjacent segments can be visualized and protected. Larger vessels (3mm in diameter) were tied by absorbable braid. This approach seems to reduce transection time and is the so called "postcoagulation technique" [138].

Habib's technique: This technique, invented by Habib in 2002, is also known as Bloodless Hepatectomy Technique [10],[88]. Resection is conducted using cooled tip RadioFrequency probe those contain a 3 cm exposed tip to coagulate liver resection margins. Once a 2 cm-wide coagulative necrosis zone is created by multiple applications of the probes in adjacent zones and at different depths, the division of the parenchyma with a surgical scalpel is possible without any bleeding. Both the remnant liver and the removed specimen have on the margin of resection a portion of necrotic coagulated liver l cm thick.

A: The primary problem with each of the previous devices is that whilst small vessels can be coagulated during transection, larger vessels are often left patent and injured, which can result in considerable blood loss requiring tedious clipping and suturing in order to achieve haemostasis.

Habib's Technique allows hepatic resections with marginal blood loss, without any vessel occlusion technique or intra or post-operative transfusions, coagulating each vessel encountered in the field of energy application; In a preliminary study of 15 cases of mainly segmental or wedge resection reported by Weber et al., the mean blood loss was only 30±10 ml, and no complications such as bile leakage were observed [88]. Another group also reported low blood loss using this technique in liver resection [89]. Haemostasis is obtained only by RF thermal energy: no additional devices like stitches, knots, clips or fibrin glue are needed [10],[88],[90],[91]; it's effective also in the cirrhotic liver and the l-cm-thick of burned coagulated surface assures margins free from tumour. The technique has the advantage of simplicity compared with the aforementioned transection techniques As the RF assisted technique allows parenchymal sparing during the first resection, this in turn results in more repeat liver resections being possible for recurrences. It also enables nonanatomical resections during these repeat resections.

D: Habib's technique cannot be applied near the hilum or the cava vein for fear of damaging this structures and because the blood flow of large vessels subtracts RF energy and involves an incomplete coaugulative necrosis [92],[93] (up to now the technique has been experienced only for segmental resection); the l-cm-thick of burned coagulated layer in the surface involves the loss of part of healthy parenchyma and a higher rate of postoperative abdominal abscesses [91],[94]. Moreover one potential disadvantage of this technique is the sacrifice of parenchymal tissue in the liver remnant, with a 1 cm wide necrotic tissue at the transection margin, which may be critical in cirrhotic patients who require major liver resection or in case of liver resection for living donor liver transplantation. An evolution of the Habib probe is the Habib 4X [92]which adress the problem of time consuming and the risk of skin burns from the grounding pad related to previous device.The device was introduced perpendicularly into the liver, abutting the transection line (Figure 5). The generator was programmed to produce an alert signal when energy delivery had been automatically stopped, thus avoiding over coagulation and carbonation. The probe was gently moved to and fro in its vertical axis for 3e5 mm throughout the coagulation process to avoid adherence of the probe to the liver parenchyma. The probe was then reintroduced adjacent to the last coagulated area in a serial fashion, until the area to be transected was fully ablated. The number of applications required to create a complete zone of desiccation was related to the size of the cut surface of the resection margin.

1. A second line of ablation, parallel to the first line and closer to the tumour edge, was then done to ensure complete tissue coagulation and perfect haemostasis prior to transection

2. The Habib 4X was then applied perpendicularly to the previous two lines of ablation, so as to ensure complete coagulation of any residual normal liver parenchyma. This allowed a margin of coagulated liver parenchyma to remain; ensuring vessels and bile ducts remained sealed. For deeper tumours the device was applied at an angle of 45 degrees to the surface. This technique allow to achieve a very low rate of blood transfusion in a very large series [88]

Gyrus plasmakinetic pulsed bipolar coagulation device: Gyrus /Gyrus medical inc., Maple Groves, Mn, USA) is a bipolar cautery device which seals the hepatic parenchima using a combination of pressure and energy that results in the fusion of collagen and elastin in the walls of the hepatic vasculature and bile ducts [98]. The device can reliably seals vessels up to 7 mm in diameter minimizing the amount of blood loss during the transection of the liver. Thermal spread and sticking to tissues is reduced by a cooling period after each pulse as the impedance of the coagulated tissue increased. This instrument has been previously widely used in gynaecological procedures and it's use in liver surgery is relatively new.

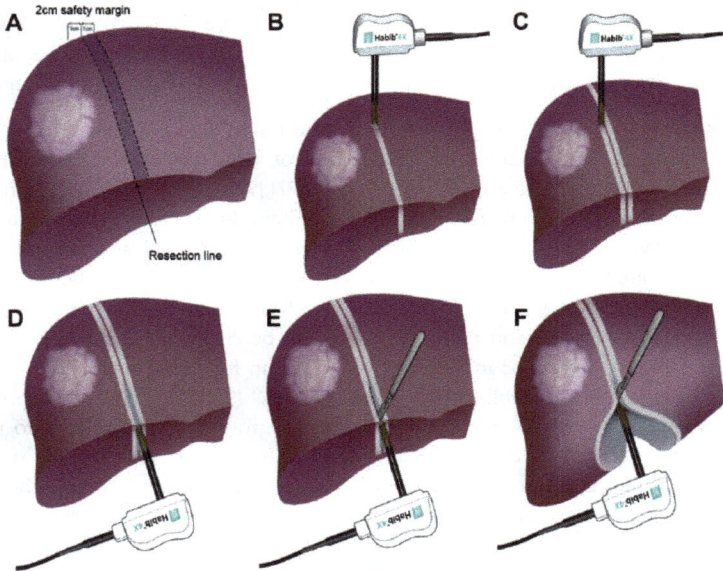

Figure 5. Habib technique for liver resection

A: It could be used in a similar manner to the clamp-crush technique to transect hepatic parenchyma. After incising the hepatic capsule with bovie the instrument is inserted into the liver in an open manner and bipolar energy is applied as the forceps are slowly closed over the parenchyma. The cauterized liver is subsequently transected with Metzenbaum scissors. The device was used for the entire hepatic parenchymal transection; only named vascular and biliary structures required additional attention and were stapled or suture ligated. The device exhibits a minimal thermal spread of 2–3 mm and was frequently used for parenchymal transection abutting the hepatic hilum. With the exception of large, named vascular and biliary structures which were routinely stapled or ligated, excellent haemostasis and biliary duct fusion were achieved uniformly.

In a recent series median blood loss rate compare favourably with those in several large series using the traditional clamp-crush technique [99]. Moreover blood loss and transfusion rates were comparable with those cited in recent report of alternative parenchymal transection, as showed by results of Tan et Al [100]. In this study Gyrus compared favourably with Harmonic scalpel in term of Bile leakage and the author underlined the concorrential cost of the device. Moreover it seems to be useful even in case of cirrhotic patients. Corvera et al. [98] have also reported the use of the Gyrus device in cirrhotic livers comparing it to the clamp and crush technique. They evaluated five patients in each group showing similar results between the two groups in terms of operating time, blood loss and major post-operative complications.

D: as the ligasure vessel sealing device one of the limit of this device is the "blind" use without clear identification of vascular and biliary structures before sealing

The Aquamantis System: The Aquamantys System employs Transcollation ® technology (fig 6) to simultaneously deliver RF (radiofrequency) energy and saline for haemostatic sealing and coagulation of soft tissue and bone at the surgical site. Transcollation technology is used in a wide variety of surgical procedures, including orthopaedic joint replacement, spinal surgery, orthopaedic trauma and surgical oncology.Transcollation technology simultaneously integrates RF (radiofrequency) energy and saline to deliver controlled thermal energy to the tissue. This allows the tissue temperature to stay at or below 100°C, the boiling point of water. Unlike conventional electrosurgical devices which operate at high temperatures, Transcollation technology does not result in smoke or char formation when put in contact with tissue. Blood vessels contain Type I and Type III collagen within their walls. Heating these collagen fibers causes radial compression, resulting in a decrease in vessel lumen diameter. Using the Aquamantys generator with patented bipolar and monopolar sealers, surgeons can achieve broad tissue-surface haemostasis by applying Transcollation technology in a painting motion, or it can be used to spot-treat bleeding vessels. This is capable of sealing structures 3–6 mm in diameter without producing high temperature or excessive charring and eschar. Structures more than 6 mm in diameter should be divided in conventional manner with clips or ties. Constant suction is required to clear the saline used for irrigation.

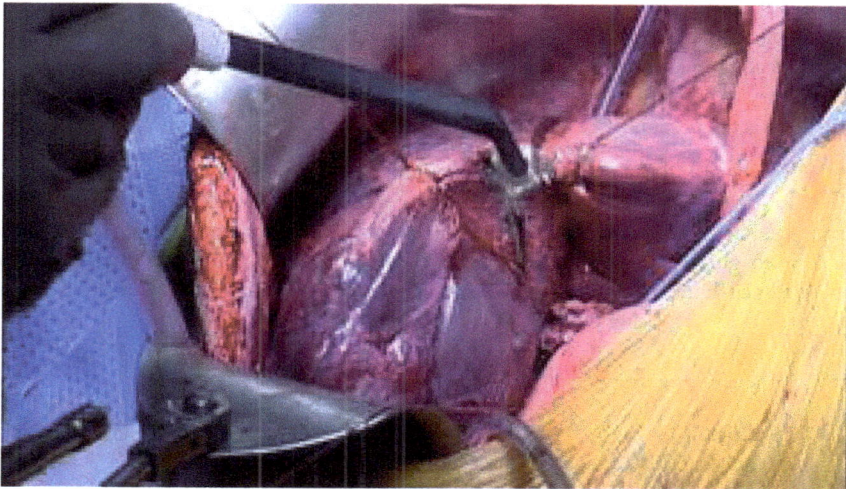

Figure 6. Aquamantys transcollation technology performing liver resection

A: it's use is "friendlier" to most surgeons, easy to learn most surgeons are comfortable after 5–6 procedures. It seals blood and bile ducts up to 6 mm in diameter, is able to reduce blood loss and the recourse to vessel occlusion techniques. Moreover it offers good results also in cirrhotic livers [66] and destroys eventual additional cancer cells at the margin of resection.

D: it is expensive and pace of liver transection could be low. Moreover there is a lack of data reported in literature due to the relative novelty of this device.

Coolinside: The new Coolinside® device (Apeiron Medical, Valencia, Spain) is a hand-held device which simultaneously coagulates (using RF) and cuts (by means of a cold scalpel) the liver. This device and its manipulation is built for both laparotomic and laparoscopic procedures.Coagulation is performed by a blunttip metallic electrode positioned at the distal edge, which is electrically connected to a Cosman CC-1 coagulator system (Radionics, Burlington, MA, USA) operating at a maximum power of 90W. The liver tissue is cut using a thin blade at the distal edge. Inside it the active electrode has a closed hydraulic circuit containing saline solution at a temperature of 0 ºC, which is propelled to the distal edge by a Radionics continuous perfusion pump (Burlington, MA, USA) at a speed of approximately 130 mL/min. The cold liquid keeps the surface of the tissue below 100 ºC by refrigerating the active electrode. The feedback system for the warm saline solution means that it can never come into contact with the patient (as in the case of the Tissuelink® device).

A: The key to the performance of the device is in the fact that the depth of hepatic parenchymal transection is adapted to the coagulation effect achieved by the proximal edge of the active electrode, that part which first comes into contact with the tissue. In this way, every time the surgeon moves the device over the surface of the liver, the parenchyma is cut and coagulated simultaneously [132]. In this study 11 hepatic resection were performed entirely with coolinside without the need for ligature or clips or pringle manouver, with no bile leak complication and high transection speed. This device combines coagulation and transection capacity and it does not need to be combined with other devices (not even stitches or clips). Moreover, it is not necessary to perform vascular occlusion, parenchymal coagulation is homogeneous and, lastly, there is the possibility of using it in laparoscopic surgery.

D:As with other RF devices, tissue pre-coagulation can change structures so that it can be difficult to identify the main hepatic vessels or conduits. Moreover, the amount of hepatic tissue that is sacrificed may be greater than in the case of other techniques, given that with this device the coagulated area may be up to 5 mm, which might limit but not contraindicate this technique in cirrhotic patients. Moreover this could be considered a disadvantages in case of liver resection during living donor liver transplantation

3.3. Others source of energy

Water Jet Scalpel, WJS: The WJS was introduced in 1982 by Papachristou [79]. This tools could achieve, as well as CUSA, a selective dissection.

The dissection modalities which take advantage of the anatomic conditions are called selective. The water-jet effects hereby like an intelligent knife and separates the more resistant duct- and vessel structures automatically from the parenchyma which thus become visible. When visible they can be closed easily under controlled conditions.

The device consists of a pressure generating pump and a flexible hose connected to the hard piece. The liquid (saline solution) flows at a steady stream and is projected through the nozzle at the tip of the hand piece. The jet hits the liver at the desired line of transection and washes away the parenchyma, leaving the intra-hepatic ducts and vessel undamaged; then the vascular and bile structures can be legated and the transection plane coagulated. The tip is reinforced by a suction tube which removes excess fluid; besides splashing is avoided by covering the area of dissection with a transparent sheet or a Petri dish. Compared to the CUSA, the water jet leaves a smoother cut surface and little hepatic degeneration or necrosis at the borders.

A: WJS can dissect offering excellent visualization and is effective also in the cirrhotic liver. In the only available prospective randomized trial of water jet in the literature, in which 31 patents underwent liver resection using water jet and another 30 patients underwent liver resection using CUSA, water jet transection reduced blood loss, blood transfusion, and transection time compared with CUSA [80]. water jet techniques is quite good for dissecting out major hepatic veins when tumors are in proximity. This allows for delineation of hepatic veins, particularly at the junction with the inferior vena cava, and prevents positive margin. It allow the so called selective dissection technique.

D: WJS can't coagulate or realize haemostasis and some study demonstrate that it cannot achieve a reduction of intra-operative blood loss and operating time if compared with traditional techniques [81],[82]; using this technique is possible cancerous seeding of the healthy abdominal organs and infection of the operators by hepatic viruses. Moreover in literature some cases of gas embolism are described using this device [83]. Furthermore the instrument may be more effective than the CUSA with respect to operating in the presence of cirrhosis. Papachristou and Barters [79] initially reported that the water jet was likely to be ineffective when there is increased fibrotic tissue. Later papers, however, describe successful resections with cirrhosis by using higher jet pressures. Une et al. [80] report that one does not need to use higher water jet pressures to dissect cirrhotic tissue effectively; instead, the same pressures as for normal parenchyma just need to be applied longer. The major concern of surgeons using the water jet is the associated splash. The latter effect is caused by solution bouncing off tissues. Besides the obvious infectious concerns of the possibility of contaminating operating room personnel, the splash brings up the notion of the possibility of cancerous seeding. This possibility must be considered in operations for malignancy and one needs to take additional care not to expose the gross tumor during the dissection.

Staplers (fig 7): Since the nineties vascular staplers to divide hepatic veins and portal branches during hemihepatectomy are considered an achievement that aids in minimizing blood loss and thereby reduces the need for inflow occlusion. Further, staplers seem to be advantageous in the unroofing of hepatic cysts since any inadvertently injured bile duct or blood vessel is sealed [84].

Staplers can be used in liver surgery for control of inflow and outflow vessels, or to divide liver parenchyma [84],[85]. The stapler is rarely used as the principal instrument in hepatic resection. The device can add speed to the operation in open or laparoscopic surgery. Its primary use is for achieving control of hepatic vasculature, particularly the hepatic veins.The use of vascular staplers to divide hepatic veins and portal branches is considered an achievement that has aided in minimizing blood loss and thereby reduced the need for inflow occlusion. Recent publications reporting a number of techniques using stapling devices in liver surgery showed them to be extraordinarily useful in the safe ligation of inflow and outflow vessels.

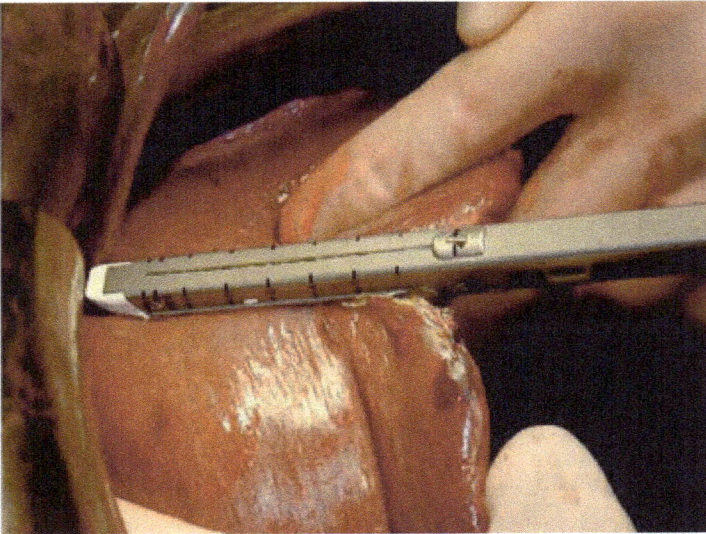

Figure 7. Parenchima transection performed using a Stapler

Biliary radicals can be incorporated efficiently into the staple line. Division of the hepatic veins with a stapler as opposed to direct ligation proffers several advantages. First, it eliminates the risk of dissecting the hepatic veins and minimize the risk of slipped ligature. Furthermore the stapler simultaneously divides multiple venous branches, especially on the right side, that are too short to allow for a safe and rapid more traditional ligation.

A: It is particularly useful in dividing the major trunk of hepatic veins or the middle hepatic vein deep in the transaction. Vascular staplers also can be used to divide the hepatic duct pedicle in right or left hepatectomy [7]. The procedure starts by dividing the liver capsule by diathermy the use of a stapler for transection of the liver parenchyma following by fracturing the liver tissue with a vascular clamp in a stepwise manner and subsequently divided with an EndoGIA vascular stapler. In a large series of 300 stapler hepatectomies, including 193 major hepatectomies, mortality of 4% and morbidity of 33% were reported which is comparable with conventional liver resection techniques. Vascular control was necessary in only 10% of the series, with an overall median blood loss of 700 mL [86]. The rate of biliary leakage seems to be very low, with a 8% reported in the largest series [86]. Moreover the trasection speed is the highest among all the techniques employed. Most recently, an ultrasound-directed transparenchymal application of vascular staplers to selectively divide major intrahepatic blood vessels before the parenchymal phase of liver resection has been shown to minimize blood loss, warm ischemia time, and operative time [131]

More to the point, in cases of difficult parenchymal transection with ongoing bleeding, the stapler device offers faster specimen removal giving the surgeon the opportunity to control the loss of blood from the raw liver surface

D:Although the technique appears attractive, the financial cost is a serious drawback. One problem associated with the use of a stapler for liver transection is increased risk of bile leak, since the stapler is not very effective in sealing small bile ducts [87]. Otherwise other studies report a very low rate of biliary injury and leakage. Moreover the surgeon must also be selective in the use of a stapler for the treatment of tumors particularly near the hilum in order to obtain sufficient margin. In case of stapler malfunction the surgeon should be ready with a back up technique to achieve vein control in case of sudden hemorrage. Serious blood loss can theoretically occur when the stapler has sealed only half the diameter of the vessel or after misfire of the device.

Chang's needle technique: This technique presented by Chang in 2001 [95] is based on the utilization of a special instrument equipped with a 18 cm straight inner needle with an hook near its top; Chang needle can be applied repeatedly to make overlapping interlocking mattress sutures with N° l silks along the inner side of the division line. After this phase liver parenchyma can be divided directly by scissors, electrocautery or traditional resection methods applying new suture only for tubular structures of significant size.

A: Chang's needle technique can be performed without application of any vessel occlusion techniques, without any other haemostatic technique and reducing blood transfusions; this method seems to be capable to reduce both intra-operative blood loss and resection time; besides it's surely cheap and is reported to be simple too [96].

D: It can't be applied if the lesion is too close to inferior cava vein [97]

3.4. Combined techniques

In the last decades a combined use of the devices previously analyzed has been reported in literature to increase the efficacy of each device, based on consideration that we have 2 different kind of instruments (as shown in table 3): those that allow a preparation of vascular structures achieving a selective dissection and those that allow a non-selective dissection (with a blind coagulation of the vasculature and biliary structures). Efficient and safe liver parenchymal transection is dependent on the ability to simultaneously address 2 tasks: parenchymal division and hemostasis. Because no single instrument has been developed that is adequate for both of these tasks, most hepatic parenchymal transections are performed using a combination of instruments and techniques.

Aloia e coll developed a 2-surgeons technique which combine saline-linked cautery and ultrasonic dissection [133]. This techniques allowed a reduction in the operative time when compared to ultrasonic dissection alone. Moreover blood loss and lenght of operation seems to be reduced.

Reference	Patients	Technique	Blood loss/ transfused patients	Operative time, min	Transection speed, cm²/s
Takayama et al. [8]	132 (66 vs. 66)	Clamp-crush technique Cavitron ultrasonic surgical aspirator	452[a]/NA 515[a]/NA	54[b] 61[b]	1.0 1.1
Rau et al. [9]	61	Hydrojet dissector Cavitron ultrasonic surgical aspirator	NA/1.5 NA/2.5	28[b] 46[b]	NA
Koo et al. [22]	50 (25 vs. 25)	Clamp-crush technique Cavitron ultrasonic surgical aspirator	792[a]/NA 875[a]/NA	119 139	NA
Lesurtel et al. [16]	100 (4 groups, 25 each)	Clamp-crush technique Cavitron ultrasonic surgical aspirator Hydrojet dissector Radiofrequency dissecting sealer	1.5[c]NA 4[c]/NA 3.5[c]/NA 3.4[c]/NA	NA	3.9 2.3 2.4 2.5
Arita et al. [21]	80 (40 vs. 40)	Clamp-crush technique Radiofrequency dissecting sealer	733[a]/0 665[a]/2	80 79	0.89 0.99
Smyrniotis et al. [20]	82 (41 vs. 41)	Clamp-crush technique Sharp transection	460[a]/15 500[a]/13	211 205	NA
Lupo et al. [23]	50 (26 vs. 24)	Clamp-crush technique Radiofrequency dissecting sealer	NA/8 NA/13	292 278	NA

Only randomized trials are reported. NA = Not available in the study.
[a] Blood loss is expressed in ml.
[b] Value refers only to transection time.
[c] Blood loss is expressed in ml/cm². The number of patients transfused is expressed as a mean only in the trial by Rau et al. [9].

[a] Blood loss is expressed in ml.

[b] Value refers only to transection time.

[c] Blood loss is expressed in ml/cm². The number of patients transfused is expressed as a mean only in the trial Rau et al. [9].

Table 3. Only randomized trials are reported. NA= Not available in the study.

In January 2004, Sakamoto et al retrospectively compared their experience with 16 liver resections in which SLC was used in combination with a bipolar vessel-sealing device and a matched set of 16 patients undergoing liver resections in which a crush-clamp technique was used.[134] They found that fewer patients in the SLC group required inflow occlusion and that blood loss was reduced. Differences in total operative time were not reported, but liver transection time was prolonged in the SLC group. Aldrighetti et al. [135] published a relatively larger series comparing clamp-crushing with ultrasonic plus harmonic scalpel dissection. The latter resulted in longer operative time, but with a reduced blood loss (and consequently a lower transfusion rate) and with a lower rate of biliary fistula. However, the retrospective method of the study, and the relatively long period of inclusion may have biased these results against the clamp-crush technique. Lesurtel and Tanai combined ultrasonic dissection with bipolar coagulation [136-137]. They concluded that UD associated with efficient bipolar forceps cautery is probably one of the safest and the most efficient device for liver transection, even if its superiority over the clamp crushing technique has not been well established. In a recent paper Yokoo et coll [139] combined the use of ultrasonically activated scalpel with a saline linked radiofrequency dissecting sealer versus bipolar cautery with a saline-irrigation system and ultrasonically activated. Scalpel. The first technique resulted in shorter operative time and lower postoperative complication rate. Moreover Gruttadauria and coll developed a combination of utrasonic surgical aspirator in association with a monopolar floating ball in elderly patients. This new technique reduced length of stay, procedure length, and use of perioperative blood in a cohort of patients [140]. Nagano and coll evaluated the efficacy of combination of CUSA plus argon beam colagulator in comparison with CUSA plus bipolar coagulation, and showed that the first approach allowed to a shorter transection time and lower blood loss [141]

Haemostasis techniques: Coagulation of vessels over 1 mm of diameter can be achieved positioning clips or sutures before division, or using devices like LigaSure, TMFB or HS for their target vessels or staplers for the largest veins. Clips and sutures are used especially during transection through traditional techniques.

During and after liver's transaction haemostasis of the vascular structures under 1 mm of diameter is another important concern of the surgeon: first because the continuous bleeding from the little vessels in the parenchyma represents a considerable part of intra-operative blood loss, and second because it makes hard for the surgeon the visualization of the surgical field. The stop of tearing small vessels that causes oozing from the cut surface can be achieved with normal monopolar or bipolar electrocoagulator, better if equipped with saline irrigation that makes them less traumatic and avoids formation of sticky coagulum An alternative is represented by employment of Argon Beam Coagulator or TMFB that probably is the best device for stopping tearing of small vessels on the cut surface of the liver.

After the resection other two precautions can be taken: application of mattress sutures for providing to a mechanical compression of the bare surface and application of biological glue for realizing complete haemostasis through a chemical/biological action.

Choice of surgical strategy: The choice of surgical strategy is based on the pre-operative evaluation and on the now indispensable Intra-Operative Ultrasonography (IOUS); in fact several studies have demonstrated that the IOUS is capable to change surgical strategy in over 40% of cases finding new lesions or diagnosing as inoperable lesions those were thought operable at the previous evaluation [101-104].The kind of surgical strategy chosen for the intervention on the base of affects strongly influences the operative outcome and the amount of operative blood loss. The most considerable aspect is the amplitude of the resection: a large resection like a right hemi-hepatectomy (or another typical resection) involves a higher bleeding and risk of complications. From this point of view the choice of segmental or wedge limited resections, when they are possible in respect of radical oncology standards, has to be consedered the best option [105,106]. Usual surgical margins for removal of liver tumours are 1 cm of healthy parenchyma surrounding the lesion. Kokudo et al. in 2002 demonstrated that for colorectal metastases the surgical margin can be, in particular situations, lowered to 2 mm with increase of the pathology recurrence rate from O% for 5 mm margin to 6% for 2 mm margin [107].

This finding, combined with a contrast-enhanced IOUS during the resection, could be a rationale incentive for practising limited resections [108-110], and the possibility of an accurate investigation of the remnant liver through the IOUS

Drug administration for reducing intra-operative blood loss: Liver resection may cause a variable degree of hyperfibrinolytic states; this phenomenon occurs in the days immediately after hepatectomy and is more pronounced in patients with a diseased liver or in patients who have undergone to a wider hepatectomy extent [111-116]So some authors propose the utilization of drugs with antifibrinolytic effect like Aprotinin that is reported to be capable to reduce intra-operative blood loss (especially during liver resection time) and transfusions [117-119]. Other authors propose utilization of the cheaper Tranexamic acid reporting similar results [120]. Although a theoretical risk of thromboembolic complications is present, no adverse drug effects like deep venous thrombosis, pulmonary embolism or other circulatory disturbances were detected in both these studies.

3.5. Comparison of different liver transection techniques

The choice of transection techniques is currently a matter of preference of surgeons, as there are few data from prospective randomized trials that compared different techniques. It has been shown in small prospective randomized trials that clamp crushing or water jet may be preferable to CUSA in terms of quality of transection or speed of transection [1],[122]. Moreover Water-jet dissection.

Seems to be considerably faster than CUSA® or blunt dissection and Pringle-time and blood loss can be reduced by using this device [83]. However, the results of these trials remain to be validated by larger-scale trials. CUSA dissection is still a widely used technique worldwide.

Several studies have been addressed to clarify these critical points, underlining the advantages and the drawbacks of each device. One of the first randomized studies [52] comparing the ultrasonic dissector versus the clamp-crush technique showed that the ultrasonic dissector is more frequently associated with tumor exposure at the resection margin and with incomplete appearance of landmark hepatic veins on the cut surface. The authors did not find any difference in postoperative morbidity and blood loss, concluding that clamp-crushing technique resulted in a higher quality of hepatectomy, thus being the option of choice.

Aldrighetti et al. [135] published a relatively larger series comparing clamp-crushing with ultrasonic plus harmonic scalpel dissection. The latter resulted in longer operative time, but with a reduced blood loss (and consequently a lower transfusion rate) and with a lower rate of biliary fistula. However, the retrospective method of the study, and the relatively long period of inclusion may have biased these results against the clamp-crush technique. The study performed by Takayama and colleagues found no difference in transection speed between the crush/clamp technique and ultrasonic dissection. This same study also demonstrated that the crush/clamp technique resulted in increased precision and improved quality of hepatectomy according to a grading system considering such factors as positive surgical margins, appearance of landmark hepatic veins on the cut hepatic surface, and postoperative morbidity. Koo and colleagues also demonstrated that no difference existed with blood loss, transfusion requirements, speed of resection, or total operative time between crush/clamp and the ultrasonic dissector

A randomized study [73] comparing LigaSure with the conventional method, demonstrated no statistical difference (p = 0.185) in blood loss and mortality rate between the two groups. But, LigaSure was slightly superior in terms of transection speed, number of ties per cm 2 and hemostasis time. The resulting total operating time decreased by 27 min, and hospital stay was shortened by 2 days in the LigaSure group. The authors performed also a cost analysis which found a highly cost-effective ratio in favor of LigaSure due to shorter operative time, hospital stay and low capital cost of the disposable device. They considered 3 mm as the range of maximal effectiveness in sealing portal triads (without increasing the rate of biliary fistula). A more recent randomized study [74] did not demonstrate this difference in blood loss, operating time and hospital stay, failing to find a superiority of one technique over the other. In this particular situation, the cost-effectiveness of LigaSure in the clamp-crush method was not confirmed, favoring once again the latter. Radiofrequency-assisted hepatic transection has also been studied in a randomized, controlled fashion. The results of this study indicated that postoperative morbidity, including abscesses and biliary complications, was significantly higher with the use of radiofrequency-assisted resection compared to crush/clamp.

As recently described in non-randomized settings [85]-[86], liver transection could be also performed with the stapling technique. As reported, the technique appears to be safe and quicker. Commonly, staplers are considered to be expensive tools, but they increase only the total material cost. However, owing to decreased blood loss, transfusion rate, shorter operative time and in-hospital stay, the global cost for a hepatectomy (especially for the major ones) has considerably decreased especially in high-volume centers. It should also be noticed that the stapling technique [142] can reduce the time of vascular control (i.e. Pringle). This fact turns out to be relevant when the resection is conducted in injured parenchyma due to prolonged chemotherapy (hepatic steatosis, sinusoidal obstruction syndrome, steatohepatitis, etc.). Cataldo et al [143] comparing stapler, crush/clamp and dissecting sealer demostrate that liver trasnection with stapler was quicker, but mean blood loss and oncological margin were similar for the three techniques. A recent study of clearly demonstrate that there is no benefit of any alternative method that has so far been compared with the clamp-crushing technique within a RCT regarding morbidity, mortality, and transfusion rates. Moreover, available RCTs failed to show an advantage of these novel devices to reduce blood loss, parenchymal injury, operation time, and hospital stay.Recently, a randomized trial compared four methods of liver transection, namely clamp crushing, CUSA, Hydrojet, and dissecting sealer, with 25 patients in each group [121]. In that study, clamp crushing was associated with the fastest transection speed, lowest blood loss, and lowest blood transfusion requirement. Furthermore, clamp crushing was the most cost-effective technique. However, in that study, clamp crushing was performed with the Pringle maneuver, whereas the other techniques were performed without the Pringle maneuver. This might have resulted in bias in favor of clamp crushing. An other recent comparative study between clamp crushing technique (CRUSH), ultrasonic dissection

(CUSA) or bipolar device (LigaSure), failed to show any difference between the three techniques in terms of intraoperative blood loss, blood transfusion, postoperative complications and mortality [72]. Further prospective randomized studies are needed to determine which transection technique is the best. Moreover a recent review of the Cochrane conclude that Clamp-crush technique is advocated as the method of choice in liver parenchymal transection because it avoids special equipment, whereas the newer methods do not seem to offer any benefit in decreasing the morbidity or transfusion requirement. Otherwise in the comparison of different techniques, apart from the efficacy in transaction with low blood loss, the relative speed of transection and the potential complications are other parameters to be considered. [122] Furthermore, the use of special instruments for transection is costly, especially when two instruments are used in combination for transection and hemostasis. It is difficult to compare the relative cost of different transection instruments because some are reusable whereas others are designed for single use, and the cost of the same instrument varies substantially in different countries. The clamp–crush and sharp dissection techniques do not involve any additional instruments. A cost comparison between the clamp–crush technique and other techniques revealed that clamp–crush is two to six times cheaper than other methods, depending on the number of surgeries performed each year. Nonetheless, the cost of these various techniques should play a part in the surgeon's decision as to whether to use them or not.

Besides reduction of blood loss and perioperative complications, radical resection with tumor-free margins is a major goal in surgery for malignant hepatic lesions. Disease-positive resection margins are a strong prognostic factor for local tumor recurrence and overall survival. Unfortunately, pathohistological data on resection margins were only available for two trials.Takayama et al. demonstrated comparable resection margins in their comparison of the clamp-crushing to the ultrasonic dissector technique.[52] However, Smyrniotis et al. reported far greater length of the narrowest tumor-free margin in their sharp transection group. [144] The question of whether any alternative transection technique provides a benefit in longterm survival of cancer patients needs further evaluation within clinical trials.

4. The role of the anaesthesiologist

Patients those are subjected to liver surgery are usually pre and intra-operatorially treated with infusion of liquids, plasma expanders and blood products: normally hepatic resections are in fact conduced in condition of euvolaemia or hypervolaemia to protect patients from the risk of consistent haemorrhage and haemodynamic's instability.

Despite this idea several studies have demonstrated that a condition of Low Central Venous Pressure (LCVP) can reduce bleeding, recourse to vessel occlusion techniques and transfusions during resection [111,112,113]. It has been scientifically demonstrated that intra-operative blood loss is correlated with inferior retro-hepatic vena cava pressure [114].

Mendelez obtained very low blood loss results in major hepatic resections managed keeping theCVP under 5 mmHg: this is possible with abstention from practising any infusion but intra-operative liquid infusion at the low speed of 75 ml/h and without any drug administration but employing hypotensive effects of normal anaesthetics (like Isoflurane, morphine and Fenta‐lyn). It's obvious that LCVP technique needs a strict monitoring of several parameters: in particular systolic arterial pressure has constantly to be kept over 90 mmHg and diuresis over 25 ml/h. After the specimen is removed and after the realization of complete haemostasis starts the infusion of liquids, and if necessary of plasma expanders and blood products until euvolaemia is obtained and haemoglobin value is over 8-10 g/dl [115].

LCVP has to be abandoned in case of uncontrollable haemorrhage (over 25% of total blood volume) or application of total vascular exclusion technique. Mendelez using LCVP reports a 0,4% rate of gas embolism [116]. This illustrates the importance of collaboration between surgeons and anaesthetists for a successful hepatectomy.

5. Conclusions

Improvement in the techniques of liver transection is one of the most important factors for improved safety of hepatectomy in recent years. The use of intraoperative ultrasound aids delineation of the proper transection plane and allow to transect tumor close to main vessels without bleeding. Clamp crushing and ultrasonic dissection are currently the two most popular techniques of liver transection. The role of new instruments such as ultrasonic shear and RFA devices in liver transection remains unclear, with few data available in the literature.

The role of vascular exclusion including Pringle's manouver seems to be decreasing with improved transection technique. However, it remains a useful technique in reducing bleeding from inflow vessels, especially for surgeons with less experience in liver resection, and recent results show safety of this technique even for prolonged total time of ischemia. Maintenance of low central venous pressure remains an important adjunctive measure to reduce blood loss in liver transection.

As clear data for comparison of various liver transection techniques are lacking, currently the choice of technique is often based on the individual surgeon's preference. However, certain general recommendations can be made based on existing data and the author's experience. Clamp crushing is a lowcost technique but it requires substantial experience to be used effectively for liver transection, especially in the cirrhotic liver. CUSA can be used in both cirrhotic and non-cirrhotic liver, is associated with low blood loss and it has a well established safety record, with low risk of bile leak. It is particularly useful in major hepatic resections when dissection of the major branches of the hepatic veins is required, or in cases where the tumor is in close proximity to a major hepatic vein, as it allows clear dissection of the hepatic vein from the tumor. This could be the preferred

5.1. Technique in oncological resection

5.1.1. The main disadvantage of the CUSA technique is slow transection

Newer instruments such as the Harmonic Scalpel, Ligasure and TissueLink Dissector enhance the capability of hemostasis and allow faster transection. However, they lack the preciseness of CUSA in dissection of major hepatic veins, and, HS more than others may be associated with increased risk of bile leak. Moreover they are particularly useful in laparoscopic liver resection. They can also be used in combination with CUSA for sealing of vessels, but this increases the cost substantially. RFA-assisted transection is probably the most speedy liver transaction technique. However, the risk of thermal injury to major bile duct is a serious concern and its use is probably restricted to minor resection Gyrus and Aquamantis are relatively new instrument and literature do not allow to draw any conclusion about their efficacy and safety.

The experience of the surgeon in practising hepatic surgery, whatever is the method to perform it, is still a factor of primary importance. In spite of that, the advent of new diagnostic instruments, new devices for resection and coagulation, a better knowledge of the liver's anatomy and pathology and a closer collaboration with the anaesthetist make the hepatic surgery a kind of surgery more defined and rational. From this point of view new studies based on the use of different surgical strategies, association of different devices and employment of different diagnostic and anaesthetic techniques is desirable.

5.2. Summary of advantages and disadvantages of the parenchymal-division instruments (table 4)

Table.4lists the primary advantages and disadvantages of five instruments used for parenchymal division during liver resection. The CUSA has the principal advantage of precise identification of both vascular and biliary vessels so that they may be controlled by ligature or other methods. In addition, the CUSA provides some haptic feedback to the surgeon so that dissection planes may remain clear. The principal disadvantages of the CUSA are threefold: (1) While the instrument permits removal of a large margin around tumors, the proof of adequate margins ends up in the suction container; (2) due to its mechanism of action, the CUSA is not very good for dissection through the fibrotic tissues found in cirrhotic livers; (3) without considerable education of the operating room personnel, the complexity of the mechanism may be cause for delays or malfunctions during procedures. The water jet affords many of the same advantages as the CUSA. Additionally, it produces minimal marginal necrosis, making it an ideal instrument in certain scenarios. The most important concern with this instrument, however, is the splash, for reasons described above. The harmonic scalpel's primary advantage is its ability to simultaneously cut and coagulate. The associated coagulum, however, may cause delayed complications. Originally devised for laparoscopic use, the harmonic scalpel's design is not particularly advantageous for open cases. Used as an adjunctive instrument, the stapler provides the possibility for speedier dissections. On the other hand, the stapler is a relatively imprecise instrument that also has the potential to malfunction during procedures. The floating ball is a surgeon-friendly instrument, particularly for the novice liver resectionist. Its mode of action may be particularly helpful in cirrhosis. The instrument acts by "controlled" burning and therefore is, by nature, an imprecise instrument; plus, there are concerns both for delayed complications related to the coagulum and for steam popping.

5.3. Ranking the clinical usefulness of the five instruments (table 5)

Table 2 subjectively ranks the five instruments according to perceived usefulness in various clinical scenarios. For resection of malignancies, we rank the CUSA number one because of its ability to stay within tissue planes during resections while preserving vessels for ligature. The water jet was second due to concerns about the splash. Third on the list is the floating ball because of its user friendliness. The harmonic scalpel lands fourth on our list because we expect laparoscopic liver resections to increase. We find the water jet to be the most useful instrument for living-donor resections because of the minimal necrotic margin. After the water jet, we advocate the more traditional, fine instrument (e.g., mosquito clamp) dissections. We rank the CUSA third because with experience, the surgeon may minimize the disadvantage of tissue removal.

Instrument	Advantages	Disadvantages
CUSA	permits identification of vessels; tactile feedback	pathologic confusion, use difficult in cirrhosis mechanically complicated do not coagulate, need a combined technique
Water Jet	selective dissection minimal marginal necrosis	splash; possible electolyte imbalances
Harmonic scalpel	cut and coagulate simultaneosly	coagulum precoagulation technique; blind dissection
Ligasure	cut and coagulate simultaneosly	coagulum precoagulation technique; blind dissection
Gyrus	ut and coagulate simultaneosly	coagulum precoagulation technique; blind dissection
TissueLink	friendliness to novices	imprecision, steam popping precoagulation technique
Aquamantys	Friendliness	precoagulation technique
Stapler	speed	imprecision, malfunction
Habib technique	coagulate large vessels	speed

Table 4. Advantages and disadvantages of most common devices

The harmonic scalpel tops the instruments for laparoscopic surgery, primarily because the scalpel is designed for laparoscopic surgery. Another reason the scalpel is particularly useful here is that the principal tumors being removed now via the laparoscope are small benign ones.

Therefore, the imprecision of this instrument is not so much of a disadvantage. The CUSA comes in second primarily because its suction competes with insufflation. Staplers are number three because of their ability to gain quick control over vessels during laparoscopic dissections. Finally, because laparoscopic hepatic surgery is rapidly evolving, we believe there will soon be new uses for old instruments or development of new instruments that will be particularly useful for this approach. For cirrhotic livers, we rank the floating ball number one due to its effective burning of fibrotic tissue. The harmonic scalpel may also be effective. Because of their relative precision, we rank the water jet and CUSA lower than the other two. Staplers do also have a role here.

Scenario	Instrument ranking
resection of malignancies	Cusa
	Water Jet
	Habib
	Tissuelink
	HS and Ligasure
Living donor resections	Tissuelink and Aquamantis
	Water jet
	CUSA
	HS and Ligasure and Gyrus
	Habib
Laparoscopic procedures	HS and Ligasure
	CUSA
	Stapler
Cirrhosis	Tissuelink and Aquamantis
	Habib
	HS and Gyrus
	Water Jet
	CUSA

Table 5. Instrument ranking in various clinical scenarios based on perceived usefulness

Author details

Fabrizio Romano, Mattia Garancini, Fabio Uggeri, Luca Gianotti, Luca Nespoli, Angelo Nespoli and Franco Uggeri

Department of Surgery, University of Milan Bicocca, San Gerardo Hospital Monza, Milan, Italy

References

[1] Poon RT, Fan ST, Lo CM, et al. Improving perioperative outcome expands the role of hepatectomy in management of benign and malignant hepatobiliary diseases: analysis of 1222 consecutive patients from a prospective database. Ann Surg. 2004;240:698 –708

[2]   Rees M, Plant G, Wells J et al; One hundred and fifty hepatic resections: evolution of tecnique towards bloodless surgery. British Joumal of Surgery 1996; 83:1526-1529

[3]   Doci R, Gennari L, Bignami P et al. Morbidity and Mortality after Hepatic Resection of Metastases from Colorectal Cancer. Br L Surg 1995;377-381

[4]   Belghiti J, Hiramatsu K, Benois S, et al. Seven Hundered Hepatectomies in tehe 1990s: an update to evaluate the actual risk of Liver Resection. Journal of American Surgeon 2000,19138-46

[5]   Gozzetti G, Mazziotti A, Grazi L et al: Liver Resection without Blood Transfusion. Br J Surg 1995;82:1105-1110

[6]   Cunningham JD, Fong Y, Shriver C et al: One Hundred consecutive Hepatic Resections: Blood Loss, Transfusion and Operative Technique. Archives of Surgery 1994;129:1050-1056

[7]   Descottes B, Lachachi F, Durand-Fontanier S et al: Right hepatectomies without vascular clamping: report of 87 cases. Journal of Hepatobiliary Pancreatic Surgery 2003; 10:90-94

[8]   Romano F, Franciosi C, Capretti R, Uggeri F, Uggeri F. Hepatic surgery using the Ligasure Vessel System. World Journal of Surgery 2005; 29:110-112

[9]   Jarnagin WR, Gonen M, Fong Y, et al: Improvement in Perioperative Outcome after Hepatic Resection: Analysis of 1803 consecutive cases over the past decade. Annals of Surgery 2002;236:397-406

[10] Navarra G, Spalding D, Zacharoulis D, Nicholls JP, Kirby S, Costa 1, Habib NA. Bloodlcss Hcpatectomy Technique. HPB Surg 2002;4:95-97

[11] Rosen CB, Nagomey DM, 1`aswell HF, I-Iegelson S, Ilstrup D, Van Heerden JA. Perioperative blood trasfusion and determinants of survival after liver resection for metastatic colorectal carcinama. Annals of Surgery 1992: 216:493-505

[12] Stephenson KR, Steinberg SM, Hughes KS, Vetto JT, Sugarbaker PH, Chang AE. Perioperative blood trasfusions are associated with decreased time to recurrence and decreased survival after resection for colorectal liver metastases. Annals of Surgery 1988; 208: 679-687

[13] Torzilli G, Makuuchi M, Midorikawa Y et al: Liver Resection Without Total Vascular Exclusion: Hazardous or Bencfuical? An analysis of our Experience. Annals of Surgery 2001; 233:167-175

[14] Kooby DA, Stockman J, Ben-Pcrat L, Gonen M, Jarnagin WR, Dematteo RP, Tuorto S, Wuest D, Blumgart LH, Fong Y. Influence fo Trasfusions on Perioperative and Long-Term Outcome in Patients Following Hepatic Resection for Colorectal Metastases. Annals of Surgery 2003; 237:860-870

[15] Fujimoto J, Okamoto E, Yamanaka N et al: Adverse Effect of Perioperative Blood Transfusions on Survival after hepatic Resection for Hepatocellular Carcinoma. Hepato- Gastroenterlogy 1997; 44:1390-1396

[16] Ohio M, Contini P, Mazzei C et al; Soluble HLA class I, HLA class II and FAS Ligand in Blood Components. A possible key to explain the Immunomodulatory Effects of Allogenic Blood Transfusionsl Blood 1999; 93:1770-1777

[17] Tait BD, d'Apice AJF, Morrow L, Kennedy L. Changes in suppressor cell activity in renal dialysis patients after blood transfusion. Transplant Proc 1984; 16:995-997

[18] Kaplan J, Samaik S, Levy J. Transfusion-induced immunologic abnormalities not related to the AIDS virus. N Engl J Med 1985; 313:1227

[19] Donnelly PK, Shenton BK, Alomran AM, Francis DM, Proud G, Taylor RM. A new mechanism of humoral immuno-depression in chronic renal failure and its importance to dialysis and transplantation. Proceedings of the European Dialysis and transplant Association 1983; 20:297-304

[20] Lenliard V, Gemsa D, Opelz G. Transfusion-induced release of prostaglandin E2 and its role in the activation ofT suppressor cells. Transplant Proc 1985; 17:2380-2382

[21] Lawrence RJ, Cooper AJ, Lozidou M, Alexander P, Taylor 1. Blood transfusion and recurrence of colorectal cancer: the role of platelet-derived growth factors. British Journal of Surgery 1990; 77:1106-1 109

[22] Torzilli G, Gambetti A, Del Fabbro D, Leoni P, Olivari N, Donadon M, Montorsi M, Makuuch M. _Techniques for Hepatectomies Without Blood Transfusion, Focusing on Interpretation of Postoperative Anemia. Archives of Surgery 2004; 139:1061-1065

[23] Abdalla EK, Noun R, Belghiti J. Hepatic vascular occlusion: which technique? Surg Clin North Am 2004; 84; 563-85.

[24] Smyrniotis V, Farantos C, Kostopanagiotou G, Arkadopoulos N. Vascular control during hepatectomy: Review of methods and results. World J Surg 2005; 29: 1384-96.

[25] Kim YI. Ischemia-reperfusion injury of the human liver during hepatic resection. J Hepatobiliary Pancreat Surg 2003; 10: 195-9.

[26] Smyrniotis VE, Kostopanagiotou GG, Contis JC, Farantos CI, Voros DC, Kannas DC, Koskinas JS. Selective hepatic vascular exclusion (SHVE) versus Pringle manoeuvre in major liver resections: A prospective study. World J Surg 2003; 27: 765-9.

[27] torzilli

[28] Belghiti J, Noun R, Malafosse R, Jagot P, Sauvanet A, Pierangeli F, et al. Continuous versus intermittent portal triad clamping for liver resection: a controlled study. Ann Surg 1999; 229: 369-75.

[29] Capussotti L, Muratore A, Ferrero A, Massucco P, Ribero D, Polastri R. Randomized clinical trial of liver resection with and without hepatic pedicle clamping. Br J Surg 2006; 93:685-689

[30] Clavien PA, Yadav S, Sindram D, Bentley RC. Protective effects of ischaemic preconditioning for liver resection performed under inflow occlusion in humans. Ann Surg 2000; 232: 155-62

[31] Nuzzo G, Giuliante F, Vellone M, De Cosmo G, Ardito F, Murazio M, et al. Pedicle clamping with ischemic preconditioning in liver resection. Liver transpl 2004; 10: S53-S57.

[32] Clavien PA, Selzner M, Rudiger HA, Graf R, Kadry Z, Rousson V, Jochum W. A prospective randomized study in 100 consecutive patients undergoing major liver resection with versus without ischemic preconditioning. Ann Surg 2003; 238: 843-52.

[33] Makuuchi M, Mori T, Gunven P, Yamazaki S, Hasegawa H. Safety of hemihepatic vascular occlusion during resection of the liver. Surg Gynecol Obstet 1987; 164: 155-8.

[34] Horgan PG, Leen E. A simple technique for vascular control during hepatectomy: The half-Pringle. Am J Surg 2001; 182: 265-7.

[35] Castaing D, Garden OJ, Bismuth H. Segmental liver resection using ultrasound-guided selective portal venous occlusion. Ann Surg 1989; 210: 20-23.

[36] Goseki N, Kato S, Takamatsu S, Dobashi Y, Hara Y, Teramoto K, et al. Hepatic resection under the intermittent selective portal branch occlusion by balloon catheter. J Am Coll Surg 1994; 179: 673-8.

[37] Huguet C, Addario-Chieco P, Gavelli A, Arrigo E, Harb J, Clement RR. Technique of hepatic vascular exclusion for extensive liver resection. Am J Surg 1992; 163: 602-05.

[38] Eyraud D, Richard O, Borie DC. Schaup B, Carayon A, Vezinet C, et al. Hemodynamic and hormonal responses to the sudden interruption of caval flow: Insights from a prospective study of hepatic vascular exclusion during major liver resections. Anesth Analg 2002; 95: 1173-8.

[39] Torzilli G, Makuuchi M, Midorikawa Y, Sano K, Inoue K, Takayama T, Kubota K. Liver resection without total vascular exclusion: hazardous or beneficial? An analysis of our experience. Ann Surg 2001; 233: 161-75.

[40] Elias D, Dube P, Bonvalot S, Lebanne B, Plaud B, Lasser P. Intermittent complete vascular exclusion of the liver during hepatectomy: Technique and indications. Hepatogastroenterology 1998; 45: 389-95.

[41] Couinaud C; Le foie: etudes anatomique et chirurgicales. Paris: Masson, 1957

[42] Meyers WC, Shekherdimian S. Owen SM, Ringe BH, Brooks AD. Sorting through methods of dividing the liver. European Surgery 2004; 36:289-295

[43] Schmidbauer S, Hallfeldt KK et al: Experience with Ultrasound Scissors and Blades (UltraCision) in open and laparoscopic liver resection. Annals of Surgery 2002; 235(1): 27-30

[44] H Sugo,Y Mikami, F Matsumoto et al: Hepatic resection using Harmonic Scalpel. Surgery Today 2000; 30:959-962

[45] Kim J, Ahamad SA, Lowy AM et al: Increased biliary fistulas after liver resection with the Harmonic Scalpel. The American Surgeon 2003; 69(9):815-819

[46] Okamoto T, Nakasato Y, Yanagisawa S et al: Hepatectomy using the Coaugulating Shears type of Ultrasonically Activated Scalpel. Digestive Surgery 2001; 18(6):427- 430

[47] Fun ST, Lai ECS, Lo CM et al. Hepatectomy with an Ultrasonic Dissector for hepato-cellular carcinoma. British Journal of Surgery 1996; 83:117-120

[48] Nakayama H, Masuda H, Shibata M, Amano S, Fukuzawa M. Incidence of bile leakage after three types of hepatic parenchymal transection. Hepatogastroenterology 2003; 50:1517-1520

[49] W. Schweiger, A. El-Shabrawi, G. Werkgartner, H. Bacher, H. Cerwenka, M. Thalham-mer and H. J. Mischinger Impact of parenchymal transection by Ultracision® harmonic scalpel in elective liver surgery. Eur Surg 2004;36:285-288

[50] Hodgson WJB, Aufses A Jr. Surgical ultrasonic dissection of liver. Surgical Rounds 1979; 2:68

[51] Fusulo F, Giori A, Fissi S et al:-Cavitron Ultrasonic Surgical Aspirator'(CUSA) in liver resection. International Surgery 1992; 77:64-66

[52] Takayama T, Makuuchi M, Kubota K, Harihara Y, Hui AM, Sano K, et al. Randomized comparison of ultrasonic vs clamp transection of the liver. Arch Surg 2001; 136: 922-8.

[53] E Felekouras, E Prassas, M Kontos, I Papaconstantinou, E Pikoulis, A Giannopoulos, C Tsigris, M Tzivras, C Bakogiannis, M Safioleas, E Papalambros, E Bastounis.Liver Tissue Dissection: Ultrasonic or RFA Energy? World J Surg 2006;30:2210-2216

[54] Fan ST, Lai EC, Lo CM, Chu KM, Liu CL, Wong J.Hepatectomy with an ultrasonic dissector for hepatocellular carcinoma Br J Surg. 1996:117-20.

[55] Yamamoto Y, Ikai I, Kume M et al: New simple technique for hepatic parenchymal resection using a Cavitron Ultrasonic Surgical Aspirator and Bipolar Cautery Equipped with a Chamnel for Water Dripping. World Journal of Surgery 1999; 23:1032-1037

[56] Rau HG, Wichmann MW, Schinkel S, Buttler E, PickelmannS, Schauer R, et al. Surgical techniques in hepatic resections: Ultrasonic aspirator versus Jet-Cutter. A prospective randomized clinical trial. Zentralbl Chir 2001;/126:/586_90.

[57] Wrightson WR, Edwards MJ, McMasters KM. The role of the ultrasonically activated shears and vascular cutting stapler in hepatic resection. Am Surg 2000;66:1037–1040.

[58] Koo BN, Kil HC, Choi JS, Kim JY, Chun DH, Hong YW. Hepatic resection by the Cavitron Ultrasonic Surgical Aspirator increases the incidence and severity of venous air embolism. Anesth Analg 2005; 101.966-970

[59] Topp SA, McClurken M, Lipson D, Upadhya GA, Ritter JH, Linehan D, Strasberg SM (2004) Saline-linked surface radiofrequency ablation: factors affecting steam popping and depth of injury in the pig liver. Ann Surg 239: 518–527

[60] Sakamoto Y, Yamamoto J et al: Bloodless liver resection using the Monopolar Floating Ball plus Ligasure Diathermy: preliminary results of 16 liver resections. World Journal of Surgery

[61] Di Carlo I, Barbagallo F, Toro A et al. Hepatic resection using a water-cooled, high-density, Monopolar Device: a new technology for safer surgery. Journal of gastrointestinal surgery 2004; 5 596-600

[62] Aloia TA, Zorzi D, Abdalla EK, Vauthey JN. Two surgeon technique for hepatic parenchymal transection of the non-cirrhotic liver using a salin-linked cautery and ultrasonic dissection. Ann Surg 2005;242;172-177

[63] Torzilli G, Donadon M, Marconi M, Procopio F, Palmisano A, Del Fabbro D, Botea F, Spinelli A, Montorsi M. Monopolar floating ball versus bipolar forceps for hepatic resection: a prospective trial. J Gastrointest Surg. 2008 Nov;12(11):1961-6

[64] Arita J, Hasegawa K, Kokudo N. Randomized clinical trial of the effect of a saline-linked radiofrequency coagulator on blood loss during hepatic resection. Br J Surg. 2005;92:954–959.

[65] Sandonato L, Soresi M, Cipolla C, Bartolotta TV, Giannitrapani L, Antonucci M, Galia M, Latteri MA. Minor hepatic resectio for hepatocellular carcinoma in cirrhotic patients: kelly clamp crushing resection versus heat coagulative necrosis with bipolar radiofrequency devices Am Surg. 2011;1490-5

[66] Geller DA, Tsung A, Maheshwari V, et al. Hepatic resection in 170 patients using saline-cooled radiofrequency coagulation. HPB 2005;7:208.

[67] Horgan PG: A novel technique for parenchymal division during hepatectomy. The American Journal of Surgery 2001; 181: 236-237

[68] Strasberg SM, Drebin JA, Linehan D. Use of Bipolar Vessel-Sealing Device for Parenchymal Transaction During Liver Surgery. Journal of Gastrointestinal Surgery 2002,6:569-574

[69] Nanashima A, Tobinaga S, Abo T, Nonaka T, Sawai T, Nagayasu T. Usefulness of the combination procedure of crash clumping and vessel sealing for hepatic resection. J Surg Oncol. 2010 Aug 1;102:179-83

[70] Tepetes K, Christodoulidis G, Spryridakis EM. Tissue Preserving Hepatectomy by a Vessel Sealing Device Journal of Surgical Oncology 2008;97:165–168

[71] Patrlj L, Tuorto S, Fong Y. Combined blunt-clump dissection and Ligasure ligation for hepatic parenchyma dissection: postcoagulation technique. J Am Coll Surg. 2010;210:39-44

[72] Doklestić K, Karamarković A, Stefanović B, Stefanović B, Milić N, Gregorić P, Djukić V, Bajec D. The Efficacy of Three Transection Techniques of the Liver Resection: A Randomized Clinical Trial. Hepatogastroenterology. 2011 ;59:117-121.

[73] Saiura A, Yamamoto J, Koga R, Sakamoto Y, Kokudo N, Seki M, et al. Usefulness of LigaSure for liver resection: analysis by randomized clinical trial. Am J Surg 2006;/192:/41-45.

[74] M Ikeda, K Hasegawa, K Sano, H Imamura, Y Beck, Y Sugawara,, N Kokudo, M Makuuchi. The Vessel Sealing System (LigaSure) in Hepatic Resection. A Randomized Controlled Trial Ann Surg 2009;250:199-203

[75] Slakey DP. Laparoscopic liver resection using a bipolar sealing device: Ligasure. HPB 2008;10:253-5.

[76] S Evrard, Y Bécouarn, R Brunet, M Fonck, C Larrue, S Mathoulin-Pélissier. Could bipolar vessel sealers prevent bile leaks after hepatectomy? Langenbecks Arch Surg; 392: 41–44

[77] Andoh H, Sato Y, Yasui O et al: Laparoscopic right hemihepatectomy for a case of polycystic disease with right predominance. Journal of Hepatobiliary Pancreatic Surgery 2004; 11:1l6-118

[78] Garancini M, Gianotti L, Mattavelli I, Romano F, Degrate L, Caprotti R, Nespoli A, Uggeri F. Bipolar vessel sealing system vs. clamp crushing technique for liver paren-chyma transection. Hepatogastroenterology. 2011 Jan-Feb;58:127-32

[79] Papachristou DN, Barters R: Resection of the liver with a waterjet. British journal of Surgery 1982; 69:93-94

[80] Une Y, Uchino J, Shimamura T et al: Water Jet Scalpel for liver resection in Hepatocel-lular Carcinoma with or without Cirrhosis. International Surgery 1996; 81:45-48

[81] Izumi R, Yabushita K, Yagi M et al: Hepatic resection using a water jet dissector. Surgery today 1993; 23:31-35

[82] Rau HG, Wichmann MW, Schinkel S, Buttler E, Pickelmann S, Schauer R, Schildberg FW. Surgical techniques in hepatic resections: Ultrasonic aspirator versus Jet-Cutter. A prospective randomized clinical trial]. Zentralbl Chir. 2001 Aug;126:586-90..

[83] Rau HG, Duessel AP, Wurzbacher S. The use of water-jet dissection in open and laparoscopic liver resection. HPB 2008;10:275-80

[84] Fong Y, Blumgart LH. Useful stapling techniques in liver surgery. J Am Coll Surg 1997;/185:/93 -100.

[85]   Kaneko H, Otsuka Y, Takagi S, Tsuchiya M, Tamura A, Shiba T. Hepatic resection using stapling devices. Am J Surg 2004;/ 187:/280 4.

[86]   P Schemmer, H Friess, U Hinz, A Mehrabi, T W. Kraus, K Z'graggen, J Schmidt, W Uhl, MW. Bu chler. Stapler hepatectomy is a safe dissection technique. Analysis of 300 patients. World J Surg 2006;30;419-430

[87]   Wang WX, Fan ST. Use of the Endo-GIA vascular stapler for hepatic resection. Asian J Surg 2003;/26:/193 6.

[88]   Weber JC, Navarra C, Jiao NR, Nicholls JP, Jensen SL, Habib NA. New technique for liver resection using heat coagulative necrosis. Annals of Surgery 2002,236: 1-4

[89]   Stella M, Percivale A et al: Radiofrequency-assisted liver resection. Journal of Gastro-intestinal Surgery 2003; 7:797-801

[90]   Haghighi KS, Wang F, King J, Daniel S, Morris DL. In-line radiofrequency ablation to minimize blood loss in hepatic parenchymal transection. Am J Surg 2005;/190:/43 7.

[91]   Pai M, Frampton AE. Mikhail S, Resende V, Kornasiewicz O, Spalding DR, Jiao LR, Habib NA. Radiofrequency assisted liver resection: analysis of 604 consecutive cases.. Eur J Surg Oncol. 2012;38:274-80

[92]   Pai M, Jiao LR, Khorsandi S, et al. Liver resection with bipolar radiofrequency device: Habibtrade mark 4X. HPB 2008;10: 256–60.

[93]   A Ayav, L Jiao, R Dickinson, J Nicholls, M Milicevic, R Pellicci, P Bachellier,N Habib. Liver Resection With a New Multiprobe Bipolar Radiofrequency Device. Arch Surg 2008;143:396-401

[94]   Ayav A, Bachellier P, Habib NA, et al. Impact of radiofrequency assisted hepatectomy for reduction of transfusion requirements. Am J Surg. 2007;193:143-148.

[95]   Chang YC, Nagasue N, Lin XZ et al: Easier hepatic resection with a straight needle. American Journal of Surgery 2001; 182:260-264

[96]   Chang YC, Nagasue N, Chen CS, Lin XZ. Simplified hepatic resection with the use on Chang's Needle. Annals of Surgery 2006; 243:169-172

[97]   Y. C. Chang, N. Nagasue. Blocking intrahepatic inflow and backflow using Chang's needle during hepatic resection: Chang's maneuver. HPB 2008;10:244-248

[98]   CU Corvera, SA Dada, JG Kirkland, BS Ryan, D Garrett, BA Lawrence, W Way, L Stewart. Bipolar Pulse Coagulation for Resection of the Cirrhotic Liver. Journal of Surgical Research 2006;136, 182-186

[99]   MR Porembka, MB Majella Doyle, NA Hamilton, PO Simon, SM Strasberg, DC Linehan, WG Hawkins. Utility of the Gyrus open forceps in hepatic parenchymal transection. Hpb 2009;11:258-263

[100] J Tan, A Hunt, R Wiesuriya, L Delriviere, A Mitchell. Gyrus PlasmaKinetic bipolar coagulation device for liver resection ANZ J Surg 2010;80:182-185

[101] Shukla PJ, Pandey D, Rao PP, Shrinkhande SV, Thakur MH, Arya S, Ramani S, Mehta S, Mohandas KM. Impact of intra-operative ultrasonography in liver surgery. Indian J oumal of Gastroenterology 2005; 24(2):62-65

[102] Bismuth H, Castaing D, Garden OJ. The use of operative ultrasound in surgery of primary liver tumors. World Journal of Surgery 1987;11:610-614

[103] Staren ED, Gambla M, Deziel DJ et al: Intraoperative ultrasound in the management of liver neoplasm. American Surgeon 1997;63:591-596

[104] Parker GA, Lawrence W Jr, Florsley JS et al. Intraoperative ultrasound of the liver affects operative decision making. Annals of Surgery 1989;209:569-577

[105] DeMatteo RP. Anatomic segmental hepatic resection is superior to wedge resection as an oncologic operation for colorectal liver metastases. Journal of Gastrointestinal Surgery 2000; 4:178-184

[106] Kokudo N. Anatomical Major resection versus nonanatomical limited resection for liver metastases from colorectal carcinoma. American Journal of Surgery; 181:153-159

[107] Kokudo N, Miki Y, Sugai S, Yanagisawa A, Kato Y, Sakamoto Y, Yamamoto J, Yama-guchi T, Muto T, Makuuchi M. Genetic and histological assessment of surgical margins in resected liver metastases from colorectal carcinoma: minimum surgical margins for successful resection. Archives of Surgery 2002; 137:833-840

[108] Torzilli G, Del Fabbro D, Olivari N, Calliada F, Montorsi M, Makuuchi M. Contrast-enhanced ultrasonography during liver surgery. British Journal of Surgery 2004; 91:1165-1167

[109] Torzilli G, Olivari N, Del Fabbro D, Gambetti A, Leoni P, Montorsi M, Makuuchi M. Contrast-enhanced intraoperative ultrasonography in surgery for hepatocellular carcinoma in cirrhosis, Liver Transplantation 2004; 10:534-38

[110] Torzilli G, Del Fabbro D, Palmisano A, Donadon M, Bianchi P, Roncalli M, Balzarini L, Montorsi M. Contrast-enhanced intraoperative ultrasonography during hepatectomies for colorectal cancer liver metastases. Journal of Gastrointestinal Surgery 2005; 9:1148-1153

[111] Melendez JA, Arslan V, Fischer ME, Wuest D, Jarnagin WR, Fong Y, Blumgart LH. Perioperative Outcomes of Major Hepatic Resection under Low Central Venous Pressure Anesthesia: Blood Loss, Blood Trasfusion, and the Risk of Postoperative Renal Dysfunction. Journal of American College of Surgeons 1998; 187:620-625

[112] Terai C, Anada H, Matsushima S et al: Effect of mild Trendelemberg on Central Hemodynamics and Intemal Jugular velocity, cross sectional area, and Flow. American Journal of Emergency Medicine 1995; 13:255-258

[113] Hughson RL, Maillet A, Gauquelin G, et al: Investigation of hormonal effects during 10-h head-down tilt on heart rate and blood pressure variability. Journal of Applicated Physiology 1995; 78:583-596

[114] Smymiotis V, Kostopanagiotou G, Theodoraki K, Tsantoulas D, Contis JC. The role of central venous pressure and type of vascular control in blood loss during major liver resections. American Journal of Surgery 2004; 187:398-402

[115] Chen H, Merchant NB, Didolkar MS. Hepatic resection using intermittent vascular inflow occlusion and low central venous pressure anesthesia improves morbidity and mortality. Journal of Gastrointestinal Surgery 2000; 4:162-167

[116] Johnson M, Mannar R, Wu AVO. Correlation between Blood Loss and Interior Vena Cava Pressure during Liver Resection. British Journal of Surgery 1998; 85:188-190

[117] Paputheodoridis GV, Burroughs AK. Hemostasis in hepatic and biliary disorders. In: Blumgart LH, Fong Y, eds. Surgery of the liver and biliary tract, 3rd ed. London: Saunders, 2000:199-213

[118] Oguro A, Taniguchi H, Daidoh T et al. FActers relating to coagulation, fibrinolysis and hepatic damage after liver resection. Hepatobiliary Pancreatic Surgery 1993; 7:43-49

[119] Lentschener C, Benhamou D, Mercier FJ Boyer-Neumann C, Naveau S, Smadja C, Wolf M, Franco D. Aprotinin reduces blood loss in patients undergoing elective liver resection. Anesth Analg 1997; 84:875-881

[120] Wu CC, Ho WM, Cheng SB, Yeh DC, Wen MC, Liu TJ, P'eng FK. Perioperative parenteral Tranexamic Acid in liver tumour resection: a prospective randomized trial toward a "blood transfusion"-free hepatectomy. Annals of Surgery 2006; 243:173- 180

[121] Lersutel M, Selzner M, Petrowsky S, McCormack L, Clavien PA. How should transection of the liver be performed? a prospective randomized study in 100 consecutive patients: comparing four different transection strategies. Ann Surg 2005;/242:/814_22.

[122] KS Gurusamy, V Pamecha, D Sharma, BR Davidson. Techniques for liver parenchymal transection in liver resection. Cochrane Library Copyright © 2009 The Cochrane Collaboration. Published by JohnWiley & Sons, Ltd.

[123] Torzilli G, Procopio F, Donadon M, Del Fabbro D, Cimino M, Montorsi M. Safety of intermittent Pringle maneuver cumulative time exceeding 120 minutes in liver resection: a further step in favor of the "radical but conservative" policy.Ann Surg. 2012 Feb;255:270-80

[124] Rahbari NN, Koch M, Schmidt T, Motschall E, Bruckner T, Weidmann K, Mehrabi A, Büchler MW, Weitz J. Meta-analysis of the clamp-crushing technique for transection of the parenchyma in elective hepatic resection: back to where we started? Ann Surg Oncol 2009;16:630-639

[125] Jagannath P,ChhabraDG, Sutariya KR et al. (2010)Fusion technique for liver transection with Kelly-clysis and harmonic technology. World J Surg 34:101–105.

[126] N Gotohda, M Konishi, S Takahashi, T Kinoshita, Y Kato, T Kinoshita. Surgical Outcome of Liver Transection by the Crush-Clamping Technique Combined with Harmonic FOCUS. World J Surg (2012) 36:2156–2160

[127] Romano F, Garancini M, Caprotti R, Bovo G, Conti M, Perego E, Uggeri F. Hepatic resection using a bipolar vessel sealing device: technical and histological analysis.HPB 2007;9:339-44

[128] Romano F, Franciosi C, Caprotti R, Uggeri F, Uggeri F. Hepatic surgery using the Ligasure vessel sealing system. World J Surg. 2005 Jan;29:110-2

[129] Pai M, Jiao LR, Khorsandi S, et al. Liver resection with bipolar radiofrequency device: Habibtrade mark 4X. HPB 2008;10: 256–60.

[130] A Nanashima,S Tobinaga, T Abo, T Nonaka, T Sawai, T Nagayasu. Usefulness of the Combination Procedure of Crash Clamping and Vessel Sealing for Hepatic Resection. Journal of Surgical Oncology 2010;102:179–183

[131] Smith DL, Arens JF, Barnett CC, et al. A prospective evaluation of ultrasound directed transparenchymal vascular control with linear cutting staplers in major hepatic resections. Am J Surg 2005;190:23–29.

[132] MÁ Martínez-Serrano, L Grande, F Burdío, E Berjano, I Poves, R Quesada. Sutureless hepatic transection using a new radiofrequency assisted device. Theoretical model, experimental study and clinic trial. CIR ESP. 2011;89:145–151

[133] Aloia TA, Zorzi D, Abdalla EK, Vauthey JN. Two-surgeon technique for hepatic transection of the noncirrhotic liver using salinelinked cautery and ultrasonic dissection. Ann Surg 2005; 242: 172–177.

[134] Sakamoto Y, Yamamoto J, Kokudo N, et al. Bloodless liver resection using the monopolar floating ball plus ligasure diathermy: preliminary results of 16 liver resections. World J Surg. 2004;28:166–172.

[135] Aldrighetti L, Pulitano C, Arru M, Catena M, Finazzi R, Ferla G. "Technological" approach versus clamp crushing technique for hepatic parenchymal transection: a comparative study. J Gastrointest Surg. 2006;10:974–9.

[136] Lesurtel M, Belghiti J. Open hepatic parenchymal transection using ultrasonic dissection and bipolar coagulation. HPB, 2008; 10: 265- 270

[137] Taniai N, Onda M, Tajiri T, Akimaru K, Yoshida H, Mamada Y. Hepatic parenchymal resection using an ultrasonic surgical aspirator with electrosurgical coagulation. Hepatogastroenterology 2002;/49:/1649 -1651

[138] Patrlj L, Tuorto S, Fong Y. Combined blunt clamp dissection and ligasure ligation for hepatic parenchyma dissection: postcoagulation technique. J Am Coll Surg 2010;210:39-44

[139] Yokoo H, Kamiyama T, Nakanishi K, Tahara M, Fukumori D, Kamachi H, Matsushita M, Todo S. Effectiveness of using ultrasonically activated scalpel in combination with radiofrequency dissecting sealer or irrigation bipolar for hepatic resection. Hepatogastroenterology. 2012 ;59:831-5

[140] Gruttadauria S, Doria C, Vitale CH, Cintorino D, Foglieni CS, Fung JJ, Marino IR. Preliminary report on surgical technique in hepatic parenchymal transection for liver tumors in the elderly: a lesson learned from living-related liver transplantation. J Surg Oncol. 2004 Dec 15;88:229-33

[141] Nagano Y, Matsuo K, Kunisaki C, Ike H, Imada T, Tanaka K, Togo S, Shimada H. Practical usefulness of ultrasonic surgical aspirator with argon beam coagulation for hepatic parenchymal transection. World J Surg. 2005 Jul;29:899-902

[142] Schemmer P, Bruns H, Weitz J, Schmidt J, Büchler MW. Liver transection using vascular stapler: a review. HPB (Oxford) 2008; 10:249-252

[143] Cataldo ET, Earl TM, Chari RS, Gorden DL, Merchant NB, Wright JK, Feurer ID, Wright Pinson C. A clinica comparative analisys of crush/clamp, stapler and dissections sealer hepatic transection method. HPB 2008;10:321-326

[144] Smyrniotis V, Arkadopoulos N, Kostopanagiotou G, Farantos C, Vassiliou J, Contis J, et al. Sharp liver transection versus clamp rushing technique in liver resections: a prospective study. Surgery. 2005;137:306–11

Chapter 8

The Role of Ultrasound in Hepatic Surgery

Mattia Garancini, Luca Gianotti, Fabrizio Romano,
Vittorio Giardini, Franco Uggeri and Guido Torzilli

Additional information is available at the end of the chapter

1. Introduction

The first experiences of ultrasonography (US) during surgical operations dated at the first years of the sixties, when some surgeons employed ultrasound in order to identify urinary or biliary stones [1,2]. These experiences gave birth to 2 important areas of application of ultrasound in the surgical field: intra-operative ultrasonography (IOUS) and interventional ultrasonography.

The first reports concerning the usage of IOUS in liver surgery dated at 1980-81 [3,4].

Hepatic surgery became the most important field of development of IOUS and nowadays the ultrasounds are employed for several goals: the precise localization of lesions and their relationship with surrounding biliary and vascular structures, the examination of the liver anatomy in order to plan the surgical strategy in respect of the oncologic principles, the intra-operative re-staging with identification of new nodules.

In 1968 Gramiak and Shah firstly introduced the ultrasound contrast agents (USCA); later, the introduction of ultrasound contrast agents for the study of the liver in 1999 [5], and then the intra-operative contrast enhanced ultrasound (CEIOUS) [6] offered further development to this important technique.

Nowadays the IOUS is considered an invaluable tool for hepatic surgery and its usage should be considered mandatory. CEIOUS demonstrates great potentialities but its role has not been established yet, even considering recent developments of multi-slice computerized tomography and magnetic resonance with liver-specific contrast agents.

2. IOUS and CEIOUS: Technical aspects

If compared to the trans-abdominal conventional US, IOUS offers several advantages. First, the higher resolution of the ultrasonographyc images, because the probe is in direct contact with the liver avoiding the absorption of acoustic waves by the abdominal wall. Second, during conventional US the liver has to be "spied" within the acoustic windows (es: transcostal), meanwhile during IOUS the proper intra-operative probes [Fig 1] can be placed in contact with the anterior, superior, inferior or posterior liver surface and a lesion can be studied from different point of view; consequently, IOUS performed after liver mobilization offers much more information. Third, during IOUS, information obtained with the ultrasound study and information gained by inspection and palpation can complement each other.

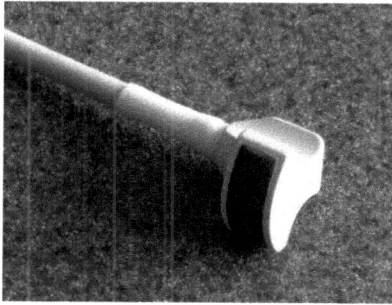

Figure 1. Intra-operative ultrasound probe

The non-panoramic nature of the study represents the main limitation of every US examination, included the IOUS. A great attention should be paid to examine the whole liver parenchyma, avoiding to leave some portion of the liver unexplored. For this reason, information gained by IOUS should always be integrated with the ones obtained from the pre-operative and panoramic study like computerized tomography (CT) and magnetic resonance (MR).

Before IOUS of the liver, partial hepatic mobilization with section of the round and falciform ligaments is always suggested.

Firstly, the liver should be explored using a standard convex (frequencies: 3.75-10 MHz) or micro-convex probe (frequencies: 3.75-10 MHz), in order to obtain a wide ultrasonographyc imaging. The probe should be initially placed between segment 4a and 4b to visualize the hepatic hilum, and then moved on the liver surface evaluating presence of eventual anatomical abnormalities of portal, arterial or biliary pedicles and of sub-hepatic venous system. Then the liver should be explored and mapped searching for focal lesions; precise localization of the lesions detected at pre-operative staging must be confirmed and new lesions must be mapped. A standardized sequential study of each segment is suggested for that, avoiding to leave unexplored portion of liver.

Afterwards, when indicated, the CEIOUS can be performed; main goals of the CEIOUS are characterization of lesions of uncertain nature and detection of new lesions not previously visualized.

The contrast agent (example: 4.8 ml of Sulfur Hexafluoride) has to be injected in a peripheral vein (cannula of 21 gauge or larger) and the arterial, portal and late phases are monitored (CEIOUS phases are reported in Table 1 and Figure 2); if necessary the USCA can be repeated twice.

Phase	Time (seconds)
Injection	0
Arterial phase	10-45
Portal phase	45-90
Late phase	90-240

Table 1. CEUS and CEIOUS vascular phases

The main advantage of the trans-abdominal contrast enhanced ultrasound (CEUS) and of the CEIUOS is that they allow a continuous real-time imaging; consequently they offer much more information for characterization of nodules than contrast-enhanced CT and MR, whose main limitation is that they are non-continuous techniques.

3. CEIOUS: The contrast agents

The acoustic difference between the intra-vascular gas microbubbles and the surrounding blood and tissues represents the basis for use of ultrasound contrast agents (USCA). The gas content of first-generation USCA (eg: Levovist, Schering AG, Berlin, Germany) is air, and the outward diffusion of air results in a relatively rapid decrease in the acoustic reflection and hence limited clinical utility. The stability of newer USCA like Optison (GE Healthcare, Amersham, Buckinghamshire, England), SonoVue (Bracco, Milan, Italy), and Definity (Bristol-Myers Squibb, Billerica, MA) is achieved by use of highmolecular-weight gases, and the slower outward diffusion of these gases makes such second generation USCAs more effective and long lived in the vascular system.

In recent years microbubbles taken up by Kupffer cells, thus possessing a "post-vascular" phase, were registered as a new second-generation USCA in Japan (Sonazoid, GE Healthcare). During the post-vascular Kupffer-phase, the tumour appears as a contrast defect image due to the lack of Kupffer cells and can consequently be characterized.

The usage of some USCAs is not approved in Italy, and authors' experience hare reported is limited to the Sulfur Hexafluoride (SonoVue, Bracco, Milan, Italy).

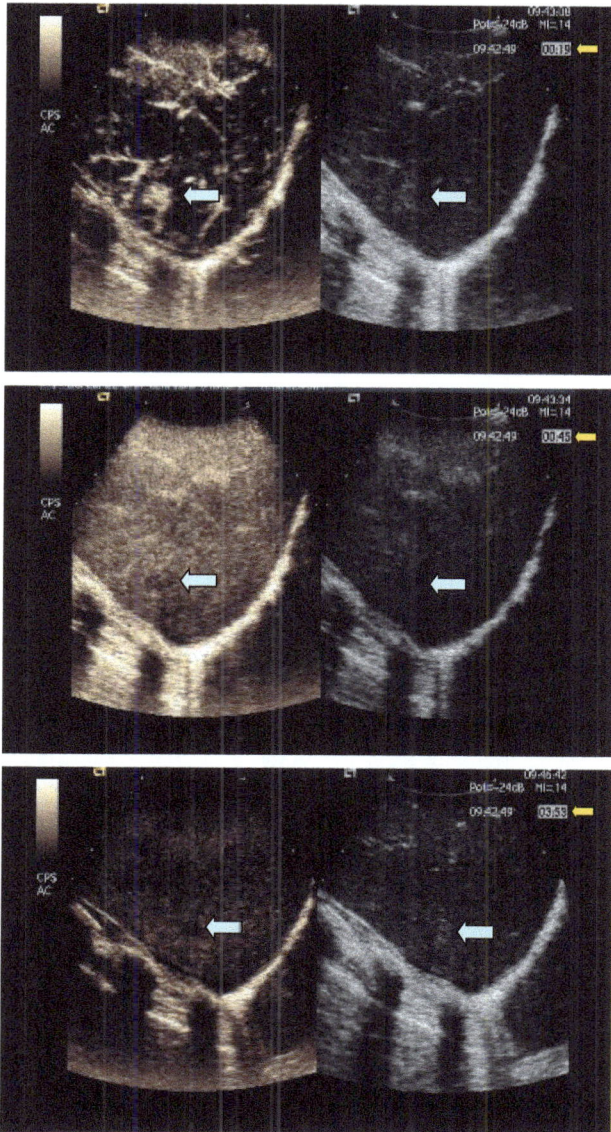

Figure 2. This picture shows a colorectal liver metastases (indicated by azure-blu arrows) in the arterial, portal and tardive phases (time passes form the injection of contrast agents is indicated by yellow arrows) during an contrast-enhanced intra-operative ultrasound study.

4. IOUS and CEIOUS: The intra-operative re-staging

Nowadays, IOUS is still considered the most accurate diagnostic technique for detecting focal liver lesions [7,8]. Nevertheless, it's remarkable that recent technical ameliorations in radiology allowed better outcomes in terms of sensitivity and specificity regarding the detection of primitive and metastatic liver lesions. In particular, the recent availability of multidetector-row Computerized Tomography (CT) with more than 64 channels and of liver specific contrast agents in Magnetic Resonance (MR) represent a great improvement in the diagnostic accuracy of liver tumours.

Hepatocellular carcinoma (HCC) and colo-rectal liver metastases (CRLM) represent the most common malignant liver lesions and the most common indication to liver resection worldwide, and consequently in this chapter the attention will be focused to the staging of these tumours.

4.1. Colorectal liver metastases

Concerning the detection of synchronous liver metastases, the contrast enhanced CT is reported to have a sensitivity and the specificity of respectively 64-72% and 64-72% [9,10], although some recent studies conducted on smaller populations showed values of sensibility of 71.7–92% [11-14]. On the other hand, concerning the detection of metachronous liver metastases, the efficacy of contrast enhanced CT revealed to be unimpressive in term of sensibility [15].

MR showed sensibility ranging from 42 and 100%; the sensibility of MR with liver-specific non-superparamagnetic contrast agents ranges between 64 and 98%, resulting generally superior when compared to the sensibility of CT scan, and a specificity of 75-79% [16-20].

In literature the data regarding the sensibility of the Positron Emission Tomography (PET) – CT appear contrasting, meanwhile the specificity is considered higher than the ones obtained by contrast enhanced CT and MR [21-23].

The trans abdominal CEUS showed high values of sensibility (80-98%) and specificity (66%-98%) for the detection of CRLM; for lesions larger than 20 mm, when sulphur-hexafluoride microbubbles (SonoVue®, Bracco, Milan, Italy) SonoVue is employed as contrast agent, the sensibility is 100% and consequently superior to conventional ultrasound and comparable to contrast enhanced CT [24-29]. It's remarkable that the studies included in these reviews regard mostly comparisons between different radiologic techniques without the anatomo-pathologic or follow up data.

Several studies demonstrated that IOUS of the liver is useful for the intra-operative re-staging of patients undergoing to liver resection for CRLM [7,8] and of patients undergoing to colorectal resection of the primitive neoplasm even in absence of liver lesions detected during pre-operative work-up [30-31]. The superiority of the IOUS compared to pre-operative studies in terms of sensibility leads to a modification of the surgical strategy [32]. The main limitation of the IOUS is the difficult characterization of the nodules; it has been ridden out after the introduction of the USCAs.

Several studies demonstrated that the CEIOUS is the most accurate diagnostic technique for detection and characterization of liver nodules; both the sensibility and specificity of CEIOUS in studies based on the comparison with other diagnostic techniques like CT and MR rises up to 100% and downsize the diagnostic accuracy of the pre-operative staging and even of the IOUS [33-37].

Preliminary results of a prospective study [59] based on the comparison among CEIOUS, CEUS, CT and RM with liver-specific contrast agent for the detection of liver metastases in patients submitted to colorectal resection for cancer, showed that CEIOUS has higher sensibility and specificity when singularly compared to any other pre-operative technique [Table 2]. These results are consistent with the ones previously published in similar setting of patients. In this survey, when the pre-operative work up is analysed on the whole (CT + RM + CEUS), CEIOUS did not offer an amelioration in terms of sensibility but showed an increased value of specificity for better characterization of liver lesions. Moreover, the CEIOUS modified the surgical strategy in 44.4% of patients even when the pre-operative work up is analysed on the whole (CT + RM + CEUS).

	TC	RM	CEUS	CEIOUS
Sensibility	80%	90%	80%	100%
Specificity	93%	79%	100%	100%

Table 2. Sensibility and specificity of CT, RM, CEUS and CEIOUS

Consequently, all patients undergoing liver resection for CRLM and all patients undergoing colorectal resection for cancer should be submitted to IOUS of the liver; moreover, among these patients, all the ones studied during the pre-operative staging with only one radiologic technique, all the ones studied with more than one technique reporting contrasting results and all the ones with new hepatic nodules during IOUS should be submitted to CEIOUS.

4.2. Hepatocellular carcinoma

Studies regarding the accuracy of pre-operative radiologic examinations for HCC assessed values of sensitivity and specificity of 60-93% and 50-95% for CT and of 52-100% and 42-97% for MR respectively; regarding the RM the best results have been obtained employing liver-specific contrast agents [39-40]. On the other side, it's remarkable that the studies included in these reviews regard mostly comparisons between different radiologic techniques without the anatomo-pathologic or follow up data.

Several studies reported that IOUS detects additional nodules in 33-41% of patients undergoing liver resection for HCC [41-43]. In cirrhotic patients with HCC, IOUS is a useful tool to detect new nodules but cannot differentiate malignant lesions from other liver nodules which account for 70–80% [44]. In fact, the risk nowadays is to overestimate the tumour stage with IOUS or laparoscopic ultrasonography considering that, except for those nodules with mosaic ultrasonographic pattern which are malignant in 84% of cases, only 24–30% of hypoechoic nodules, and 0–18% of those hyperechoic are malignant [45]. To overcome this problem even biopsy seems not to be adequate. When sulphur-hexafluoride microbubbles (SonoVue®, Bracco, Milan, Italy), the CEIOUS analysis of nodules vascularization may provide crucial information for their differentiation.

In this sense, Torzilli et al proposed in 2007 [41] a classification for the patterns of enhancement during CEIOUS in 4 categories: A1 (full enhancement in the arterial phase and washout in the delayed phases), A2 (intralesional signs of neovascularization during all phases), A3 (no nodular enhancement but detectability during the liver enhancement), and B (undetectability during the liver enhancement). Following this classification, resection is recommended for A1-3 nodules for high risk of malignancy and no treatment is recommended for B nodules.

With its intra-operative re-staging, CEIOUS shows sensibility of 100%, specificity of 69-100% and can modify the surgical strategy up to 79% of patients [41-43]. All patients undergoing liver resection for HCC should be submitted to IOUS; moreover, all the patients carriers of liver tumours of uncertain differentiation at pre-operative work up and all patients with new nodules at IOUS should be submitted to CEIOUS.

5. Echo-guided liver resection

The modern liver surgery is based on two concepts: a liver resection has to be radical following the oncologic principles and has to be conservative in a parenchyma sparing policy [46]. Consequently the exact resection plane should be carefully planned before the resection using the invaluable ultrasound guidance.

An ultrasound probe is placed on the liver surface and the target lesion to be removed has to be visualized. The surgeon draws on the liver surface the resection plane that includes the tumour; this procedure is simplified by the usage of a linear probe, because if the acoustic waves are parallel, to define the projection of the lesion or of the resection's area on the liver surface is easier (Fig). After that, the parenchymal transection can start, but during the resection the echo-guidance should be used to check if the resection plane is correct or has to be modified.

It's remarkable that the oncologic principles those have to be respected can vary depending on the type of liver tumour.

In presence of CRLM, the most important aspect regards the tumour margin. Positive hepatectomy margin has been indicated as an independent negative prognostic factor for carriers of CRLM [47], but the minimum safe width of free margin has to be established yet. Data regarding the presence of micro-metastases around CRLM are contrasting, reporting rates of micro-metastases ranging from 2% to 58% of patients; consequently, these authors suggested different widths of free margin ranging from 2 to 10 mm [48,49]. If the presence of micro-metastases around a CRLM could be related to the cytoreduction after some type chemotherapy has to be clarified yet. The rate of cut edge recurrence is reported to be up to

13.3% for a margin inferior to 2 mm, but if the surgical margin could represents a prognostic factor for patients survival is still debated [48,49]. Anyway, all the authors agree that micrometastases are confined to a short distance from the tumour (mostly less than 5-10 mm) and that a tumour margin of 10 mm is safe without risk of cut-edge recurrence. The more reasonable approach for carriers of CRLM should be to guarantee a 10 mm margin when possible, so the surgeon during the echo-guided definition of the resection plane should consider this margin. Anyway, because liver resection plus chemotheraphy provides the best chance of cure for carriers of CRLM, complete removal of the tumour with a minimum margin (even less than 2 mm) is justified when technically unavoidable for tumours size, location or number. This aspect is of paramount importance in presence of tumours next to or in contact with major vessels; in these cases, in absence of clear signs of vascular invasion at the IOUS, the vessel resection and consequent major liver resection should be avoided, offering with a parenchyma sparing policy lower post-operative morbidity and mortality. Moreover the avoidance of major hepatectomy allows the possibility of further repeated hepatectomies in patients with disease recurrence, those have shown similar morbidity and mortality compared to first hepatectomy [50].

In presence of HCC, the most important aspect regards the type of surgical resection to be performed, anatomic or non-anatomic. Anatomic resection should be considered the gold standard approach for liver resection in patients with HCC, meanwhile non-anatomic resection should be indicated only in selected patients with HCC set on cirrhosis with poor liver function. Indeed, tumour dissemination from the main lesion through the portal branches demands an anatomic approach with removal of at least the portal area which includes the lesion. The surgical margin per se does not represent a main aspect, because an anatomic resection (segmentectomy or sub-segmentectomy) can be considered adequate even in presence of a narrow margin, while a non-anatomic resection of a nodule with a 10 or even 20 mm margin could be inadequate if the portal branch feeding the nodule has not be removed. HCCs are usually associated with liver cirrhosis, and several series reported that liver resection in cirrhotic patients is related to not negligible postoperative mortality and morbidity [51,52]. The main problem to overcome when planning a surgical approach is to find a balance between the liver volume to be resected, which should be drastically reduced, and the need to perform, if possible, an anatomic resection. The use of IOUS as guidance is indispensable in this sense, but there are several methods up to now available for this procedure. The most diffused technique is the puncture technique proposed by Makuuchi et al in 1981 [53,54]. With this technique, the portal branch feeding the tumour to be resected is punctured under IOUS-guidance, through a free-hand technique or with a proper device, and then dye (usually indigo-carmine) is injected into the vessel while the hepatic artery at the hepatic hilum is clamped. The stained area becomes evident on the liver surface, it is marked with the electrocautery, and hepatic artery clamping is released. The main disadvantage of this technique, other than the quite high skill in puncturing millimetric vessels, is the fact that if the ink regurgitates or is injected into the wrong portal branch, it could be difficult to identify the proper area to be removed. Furthermore, clamping of the hepatic artery is recommended but not always feasible without the need for a hilar dissection to tape the vessel to be clamped. Other methods have been proposed such as a balloon catheter inserted transhepatically to occlude the feeding portal branch [55], or, more recently, through the mesenteric vein [56]. Mazziotti et al. proposed for segment 8 resection the division of the liver along the main portal fissure, and subsequently to approach the segment 8 glissonian pedicle intraparenchymally [57]. Santambrogio et al. have even recently suggested ablation of the feeding portal and arterial branches [58].

More recently Torzilli et al. proposed the ultrasound-guided finger compression technique, consisting in the demarcation of the resection area (segmental either subsegmental) by IOUS-guided finger compression of the vascular pedicle feeding the tumor at the level closest to the tumour but oncologically suitable. This maneuver is constantly monitored in real-time by simply using the same IOUS probe and it is maintained until the surface of the targeted liver area begins to discolor and can be easily marked with the electrocautery [59]. Torzilli's technique offers several advantages, including the non-invasiveness (no intravascular catheter) and rapid reversibility, and consequently can be repeated if necessary.

In general, any other type of primitive or metastatic liver tumours, when a surgical treatment is indicated, can be managed by means of a surgical resection with adequate margins, but in literature data concerning other specific tumours are still lacking.

One more and recent application of IOUS in hepatic surgery concerns the management of liver tumours involving an hepatic vein (HV) next to the caval confluence. These lesions traditionally require a major hepatectomy, with resection of the involved vein and the portion of parenchyma drained by that vein. Nevertheless, as previously reported, morbidity and mortality after major hepatic resections are not negligible, especially in cirrhotic patients [51,52]. A careful intra-operative study of the liver anatomy can offer alternatives to major hepatectomy. In 1987 Makuuchi M et al. introduced a new hepatectomy procedure for resection of the right hepatic vein (when invaded by a tumour) and preservation of the inferior right hepatic vein, an accessory hepatic vein draining segment VI present in 20-25% of patients [60]. Then, in 2010 Torzilli et al. suggested a set of criteria to be met for a parenchyma-sparing liver resection in presence of liver tumours invading any HV at its caval confluence [61]. The criteria are based on the direct or indirect signs of presence of venous anastomoses connecting adjacent HV, those had been previously highlighted in 1958 by Couinaud C et al. during studies performed on liver specimens [62] and can now be detected intra-operatively during IOUS [63].

A segment of a HV can be resected while avoiding the removal of the complete portion of the liver drained by that vein when, during HV finger compression at the hepatocaval confluence, at least one of these criteria is satisfied:

1. Reversal flow direction in the peripheral portion of the hepatic vein to be removed, which suggests drainage through collateral circulation in adjacent HV or inferior cava vein (IVC)

2. Hepatopetal flow in the portal branch feeding the areas to be spared

3. Detectable connecting veins with adjacent HV or IVC

It is remarkable how every surgical procedure performed on the liver is strictly dependent from the knowledge of the liver anatomy and from the ultrasounds; definitely in liver surgery the ultrasounds represent the link between the surgical anatomy and the surgical intervention.

6. Laparoscopic ultrasound

Due to improvements in technologies and increasing surgeon's experiences, the number of hepatectomy performed laparoscopically increased exponentially around the world in the recent years, and consequently the usage of laparoscopic ultrasound (LUS) of the liver [64]. Main goals of LUS are the same of ones presented in open liver surgery; anyway LUS has a few theoretical drawbacks if compared to traditional IOUS, including the difficulty in the ultrasound study of the superior and posterior segments and the limited diffusion of laparoscopic probe equipped for the contrast enhanced study.

Other indications to LUS include the re-staging before laparotomic liver surgery or before laparoscopic resection of gastrointestinal cancer (more frequently of colorectal cancer). Diagnostic laparoscopy combined with LUS is considered an adequate staging modality for primary liver malignancies and permits to avoid unnecessary laparotomies [65]. Nevertheless, the LUS seems to play a limited role in staging patients with potentially resectable CRLM candidates for open liver resection; this is owing mainly to the low sensitivity rate of 59% [66]. Consequently there may be a role for laparoscopy for diagnosing suspected peritoneal disease, but LUS should not be used routinely in patients with CRLM candidates for open liver resection.

The LUS of the liver at the time of primary resection of colorectal cancer is reported to yield more lesions than preoperative contrast-enhanced computerized tomography and could be considered for routine use during laparoscopic oncologic colorectal surgery [67].

One further indication for LUS is laparoscopic radiofrequency in patients carriers of HCC and not amenable to liver resection or percutaneous ablation; in these patients, LUS is an invaluable tool, either in the pre-treatment imaging to re-stage the patient, evaluate the relationship of the tumour with the surrounding structures and to guide the insertion of the electrode into the tumour, either for the post-treatment imaging evaluation [68].

7. Conclusions

Ultrasonography is an invaluable tool in hepatic surgery, either for the intra-operative re-staging, either for the guidance during the surgical procedure. The only major drawback of this IOUS-guided liver surgery is the need for hepatic surgeons to be trained in the use of ultrasound. Indeed, to be fully profitable, IOUS and CEIOUS should be carried out by the surgeon himself who can then use the information obtained by the ultrasound exploration in a surgical perspective. Organization of a training program for liver surgeons is far from being carried out worldwide, but it should be considered a main goal for hepato-biliary surgeons, because the liver surgeons must be equipped with ultrasound skills as like as with surgical technical skills.

Author details

Mattia Garancini[1], Luca Gianotti[1], Fabrizio Romano[1], Vittorio Giardini[1], Franco Uggeri[1] and Guido Torzilli[2]

*Address all correspondence to: mattia_garancini@yahoo.it

1 Department of General Surgery, Ospedale San Gerardo, Monza, Italy

2 Third Department of Surgery, University of Milan School of Medicine, IRCCS Istituto Clinico Humanitas, Rozzano, Milan, Italy

References

[1] Eiseman B, Greenlaw RH, Gallagher JQ. Localization of common duct stones by ultrasound. Arch Surg 1965;91:195

[2] Schliegel TU, Diggdon P, Cuellar J. The use of ultrasound for localizing renal calculi. J Urol 1961;86:367

[3] Sigel B, Coelho JCU, Spigos DG, et al. Real-time ultrasonography during biliary surgery. Radiology 1980;137:531

[4] Makuuchi M, Hasegawa H, Yamazaki S. Intraoperative ultrasonic examination for hepatectomy. Jap J Clin Oncol 1981;11:367

[5] Blomley MJK, Albrecht T, Cosgrove DO, et al. Improved detection of liver metastases with stimulated acoustic emission in the late phase of enhancement with the US contrast agent SH U 508A: early experience. Radiology 1999;210:409-416

[6] Torzilli G, Olivari N, Moroni E, Del Fabbro D, Gambetti A, Leoni P, Montorsi M, Makuuchi M. Contrast-enhanced intraoperative ultrasonography in surgery for hepatocellular carcinoma in cirrhosis. Liver Transpl 2004;10(2 Suppl 1):S34-38

[7] Sahani DV, Kalva SP, Tanabe KK, et al. Intraoperative US in patients undergoing surgery for liver neoplasms: comparison with MR imaging. Radiology 2004; 232:810–814.

[8] Torzilli G, Makuuchi M. Intraoperative ultrasonography in liver cancer. Surg Oncol Clin N Am 2003;12:91–103.

[9] Bipat S, van Leeuwen MS, Comans EFI, et al. Colorectal liver metastases: CT, MR imaging, and PET for diagnosis—meta-analysis. Radiology 2005;237(1):123-131

[10] Kinkel K, Lu Y, Both M, Warren RS, Thoeni RF. Detection of hepatic metastases from cancers of the gastrointestinal tract by using noninvasive imaging methods (US, CT, MR imaging, PET): a meta-analysis. Radiology 2002;224(3):748-756

[11] Ashraf K, Ashraf O, Haider Z, Rafique Z. Colorectal carcinoma, preoperative evaluation by spiral computed tomography. J Pak Med Assoc 2006;56(4):149-153

[12] Scott DJ, Guthrie JA, Arnold P. et al. Dual phase helical CT versus portal venous phase CT for the detection of colorectal liver metastases: correlation with intra-operative sonography, surgical and pathological findings. Clin Radiol 2001;56(3):235-242

[13] Soyer P, Poccard M, Boudiaf M. et al. Detection of hypovascular hepatic metastases at triple-phase helical CT: sensitivity of phases and comparison with surgical and histopathologic findings. Radiology 2004;231(2):413-420.

[14] Wicherts DA, de Haas RJ, van Kessel CS, et al. Incremental value of arterial and equilibrium phase compared to hepatic venous phase CT in the preoperative staging of colorectal liver metastases: An evaluation with different reference standards. Eur J Radiol 2011;77(2):305-311

[15] Glover C, Douse P, Kane P, et al. Accuracy of investigations for asymptomatic colorectal liver metastases. Dis Colon Rectum 2002;45(4):476-484

[16] Bartolozzi C, Donati F, Cioni D, et al. Detection of colorectal liver metastases: a prospective multicenter trial comparing unenhanced MRI, MnDPDP-enhanced MRI, and spiral CT. Eur Radiol 2004;14(1):14-20

[17] Balci NC, Befeler AS, Leiva P, Pilgram TK, Havlioglu N. Imaging of liver disease: comparison between quadruple-phase multidetector computed tomography and magnetic resonance imaging. J Gastroenterol Hepatol 2008;23(10):1520-1527

[18] Regge D, Campanella D, Anselmetti GC, et al. Diagnostic accuracy of portal-phase CT and MRI with mangafodipir trisodium in detecting liver metastases from colorectal carcinoma. Clin Radiol 2006;61:338-347

[19] Rappeport ED, Loft A, Berthelsen AK, et al. Contrast-enhanced FDG-PET/CT vs. SPIO-enhanced MRI vs. FDG-PET vs. CT in patients with liver metastases from colorectal cancer: a prospective study with intraoperative confirmation. Acta Radiol 2007;48(4):369-378

[20] Koh DM, Brown G, Riddell AM. et al. Detection of colorectal hepatic metastases using MnDPDP MR imaging and diffusion-weighted imaging (DWI) alone and in combination. Eur Radiol 2008;18(5):903-910

[21] Selzner M, Hany TF, Wildbrett P, et al. Does the novel PET/CT imaging modality impact on the treatment of patients with metastatic colorectal cancer of the liver? Ann Surg 2004;240(6):1027-1034; discussion 1035-1036

[22] D'souza MM, Sharma R, Mondal A, et al. Prospective evaluation of CECT and 18F-FDG-PET/CT in detection of hepatic metastases. Nucl Med Commun 2009;30(2): 117-125

[23] Kong G, Jackson C, Koh DM, et al. The use of 18F-FDG PET/CT in colorectal liver metastases-comparison with CT and liver MRI. Eur J Nucl Med Mol Imaging 2008;35(7):1323-1329

[24] Albrecht T, Blomley MJK, Burns PN, et al. Improved detection of hepatic metastases with pulse-inversion US during the liver-specific phase of SHU 508A: multicenter study. Radiology 2003;227(2):361-370

[25] Larsen LPS, Rosenkilde M, Christensen H, et al. The value of contrast enhanced ultrasonography in detection of liver metastases from colorectal cancer: a prospective double-blinded study. Eur J Radiol 2007;62(2):302-307

[26] Konopke R, Kersting S, Bergert H, et al. Contrast-enhanced ultrasonography to detect liver metastases : a prospective trial to compare transcutaneous unenhanced and contrast-enhanced ultrasonography in patients undergoing laparotomy. Int J Colorectal Dis 2007;22(2):201-207

[27] Gültekin S, Yücel C, Ozdemir H, et al. The role of late-phase pulse inversion harmonic imaging in the detection of occult hepatic metastases. J Ultrasound Med 2006;25(9): 1139-1145

[28] Oldenburg A, Hohmann J, Foert E, et al. Detection of hepatic metastases with low MI real time contrast enhanced sonography and SonoVue. Ultraschall Med 2005;26(4): 277-284

[29] Rappeport ED, Loft A, Berthelsen AK, et al. Contrast-enhanced FDG-PET/CT vs. SPIO-enhanced MRI vs. FDG-PET vs. CT in patients with liver metastases from colorectal cancer: a prospective study with intraoperative confirmation. Acta Radiol 2007;48(4):369-378

[30] Milsom JW, Jerby BL, Kessler H, et al. Prospective, blinded comparison of laparoscopic ultrasonography vs. contrast-enhanced computerized tomography for liver assessment in patients undergoing colorectal carcinoma surgery. Dis Colon Rectum 2000;43(1):44-49

[31] Stone MD, Kane R, Bothe A, et al. Intraoperative ultrasound imaging of the liver at the time of colorectal cancer resection. Arch Surg 1994;129(4):431-435; discussion 435-436

[32] Agrawal N, Fowler AL, Thomas MG. The routine use of intra-operative ultrasound in patients with colorectal cancer improves the detection of hepatic metastases. Colorectal Dis 2006;8(3):192-194

[33] Torzilli G, Del Fabbro D, Palmisano A, et al. Contrast-enhanced intraoperative ultrasonography during hepatectomies for colorectal cancer liver metastases. J Gastrointest Surg 2005;9(8):1148-1153; discussion 1153-1154

[34] Leen E, Ceccotti P, Moug SJ, et al. Potential value of contrast-enhanced intraoperative ultrasonography during partial hepatectomy for metastases: an essential investigation before resection? Ann Surg 2006;243(2):236-240

[35] Fioole B, de Haas RJ, Wicherts DA, et al. Additional value of contrast enhanced intraoperative ultrasound for colorectal liver metastases. Eur J Radiol 2008;67(1):169-176

[36] Conlon R, Jacobs M, Dasgupta D, Lodge JPA. The value of intraoperative ultrasound during hepatic resection compared with improved preoperative magnetic resonance imaging. Eur J Ultrasound 2003;16(3):211-216

[37] Torzilli G. Contrast-enhanced intraoperative ultrasonography in surgery for liver tumors. Eur J Radiol 2004;51 Suppl:S25-29

[38] Garancini M. Contrast-enhanced intra-operative ultrasound vs pre-operative imaging for the detection of liver metastases in patients with colo-rectal cancer: a prospective study. Specialization thesis. University of Milano-Bicocca; 2011. (available at: http://www.slc.livermeta.net/index.php/congressi) [accessed 21/09/2012]

[39] Bolog N, Andreisek G, Oancea I, Mangrau A. CT and MR imaging of hepatocellular carcinoma. J Gastrointestin Liver Dis 2011;20:181-189.

[40] Willatt JM, Hussain HK, Adusumilli S, Marrero JA. MR Imaging of hepatocellular carcinoma in the cirrhotic liver: challenges and controversies. Radiology 2008;247(2): 311-330

[41] Torzilli G, Palmisano A, Del Fabbro D, Marconi M, Donadon M, Spinelli A, Bianchi PP, Montorsi M. Contrast-Enhanced Intraoperative Ultrasonography Durino Surgery for Hepatocellular Carcinoma in Liver Cirrhosis: Is It Useful or Useless? A Prospective Cohort Study of Our experience. Ann Surg Oncol 2007;14:1347–1355

[42] Lu Q, Luo Y, Yuan CX, Zeng Y, Wu H, Lei Z, Zhong Y, Fan YT, Wang HH, Luo Y. Value of contrast-enhanced intraoperative ultrasound for cirrhotic patients with hepatocellular carcinoma: A report of 20 cases. World J Gastroenterol 2008;14:4005–4010.

[43] Wu H, Lu Q, Luo, He XL, Zeng Y. Application of contrast-enhanced intraoperative ultrasonography in the decision-making about hepatocellular carcinoma operation. World J Gastroenterol 2010;16:508–512.

[44] Takigawa Y, Sugawara Y, Yamamoto J et al. New lesions detected by intraoperative ultrasound during liver resection for hepatocellular carcinoma. Ultrasound Med Biol 2001;27:151–156

[45] Kokudo N, Bandai Y, Imanishi H, et al. Management of new hepatic nodules detected by intraoperative ultrasonography during hepatic resection for hepatocellular carcinoma. Surgery 1996;119:634–640

[46] Torzilli G, Montorsi M, Donadon M, Palmisano A, Del Fabbro D, Gambetti A, Olivari N, Makuuchi M. "Radical but conservative" is the main goal for ultrasonography-

guided liver resection: prospective validation of this approach. J Am Coll Surg 2005;201(4):517-528.

[47] Hughes KS, Rosenstein RB, Songhorabodi S, Adson MA, Ilstrup DM, Fortner JG, Maclean BJ, Foster JH, Daly JM, Fitzherbert D, et al. Resection of the liver for colorectal carcinoma metastases. A multi-institutional study of long-term survivors. Dis Colon Rectum 1988;31:1-4.

[48] Kokudo N, Miki Y, Sugai S, Yanagisawa A, Kato Y, Sakamoto Y, Yamamoto J, Yamaguchi T, Muto T, Makuuchi M. Genetic and histological assessment of surgical margins in resected liver metastases from colorectal carcinoma: minimum surgical margins for successful resection. Arch Surg 2002;137(7):833-40.

[49] Wakai T, Shirai Y, Sakata J, Valera VA, Korita PV, Akazawa K, Ajioka Y, Hatakeyama K. Appraisal of 1 cm hepatectomy margins for intrahepatic micrometastases in patients with colorectal carcinoma liver metastasis. Ann Surg Oncol 2008;15(9): 2472-2481.

[50] Lopez P, Marzano E, Piardi T, Pessaux P. Repeat hepatectomy for liver metastases from colorectal primary cancer: a review of the literature. J Visc Surg 2012;149:97-103

[51] Poon RT, Fan ST, Lo CM, et al (2002) Extended hepatic resection for hepatocellular carcinoma in patients with cirrhosis: is it justified? Ann Surg 236:602–611

[52] Schroeder RA, Marroquin CE, Bute BP, et al. Predictive indices of morbidity and mortality after liver resection. Ann Surg 2006;243:373–379.

[53] Makuuchi M, Hasegawa H, Yamazaki S. Intraoperative ultrasonic examination for hepatectomy. Jpn J Oncol 1981;11:367–390.

[54] Makuuchi M, Hasegawa H, YamazakiS, et al. Ultrasonically guided systematic subsegmentectomy. Surg Gynecol Obstet 1985;161:346-350

[55] Shimamura Y, Gunve'n P, Takenaka Y, et al. Selective portal branch occlusion by balloon catheter during liver resection. Surgery 1986;100:938–941.

[56] Ou JR, Chen W, Lau WY. A new technique of hepatic segmentectomy by selective portal venous occlusion using a balloon catheter through a branch of the superior mesenteric vein. World J Surg 2007;31:1240 –1242.

[57] Mazziotti A, Maeda A, Ercolani G, et al. Isolated resection of segment 8 for liver tumors: a new approach for anatomical segmentectomy. Arch Surg 2000;135:1224 – 1229.

[58] Santambrogio R, Costa M, Barabino M, et al. Laparoscopic radiofrequency of hepatocellular carcinoma using ultrasound-guided selective intrahepatic vascular occlusion. Surg Endosc 2008;22:2051–2055.

[59] Torzilli G, Procopio F, Cimino M, Del Fabbro D, Palmisano A, Donadon M, Montorsi M. Anatomical segmental and subsegmental resection of the liver for hepatocellular

carcinoma: a new approach by means of ultrasound-guided vessel compression. Ann Surg 2010;251(2):229-235

[60] Makuuchi M, Hasegawa H, Yamazaki S, et al. Four new hepatectomy procedures for resection of the right hepatic vein and preservation of the inferior right hepatic vein. Surg Gynecol Obstet 1987;164:68–72.

[61] Torzilli G, Palmisano A, Procopio F, Cimino M, Botea F, Donadon M, Del Fabbro D, Montorsi M. A new systematic small for size resection for liver tumors invading the middle hepatic vein at its caval confluence: mini-mesohepatectomy. Ann Surg 2010;251(1):33-39

[62] Couinaud C, Nogueira C. Les veines sus-hepatique chez l'homme. Acta Anat (Basel) 1958;34:84-110

[63] Torzilli G, Garancini M, Donadon M, Cimino M, Procopio F, Montorsi M. Intraoperative ultrasonographic detection of communicating veins between adjacent hepatic veins during hepatectomy for tumours at the hepatocaval confluence. Br J Surg 2010;97(12):1867-1873

[64] Nguyen KT, Nguyen KT, Gamblin TC, Geller DA (2009) World review of laparoscopic liver resection: 2804 patients. Ann Surg 250:831–841

[65] de Castro SM, Tilleman EH, Busch OR, van Delden OM, Laméris JS, van Gulik TM, Obertop H, Gouma DJ. Diagnostic laparoscopy for primary and secondary liver malignancies: impact of improved imaging and changed criteria for resection. Ann Surg Oncol 2004;11:522-529.

[66] Hariharan D, Constantinides V, Kocher HM, Tekkis PP. The role of laparoscopy and laparoscopic ultrasound in the preoperative staging of patients with resectable colorectal liver metastases: a meta-analysis. Am J Surg 2012;204:84-92.

[67] Milsom JW, Jerby BL, Kessler H. Hale JC, Herts BR, O'Malley CM. Prospective, blinded comparison of laparoscopic ultrasonography vs. contrast-enhanced computerized tomography for liver assessment in patients undergoing colorectal carcinoma surgery. Dis Colon Rectum 2000;43(1):44-49.

[68] Santambrogio R, Opocher E, Costa M, Cappellani A, Montorsi M. Survival and intrahepatic recurrences after laparoscopic radiofrequency of hepatocellular carcinoma in patients with liver cirrhosis. J Surg Oncol 2005;89:218-225; discussion 225-226.

Chapter 9

Two-Step Hanging Maneuver for an Isolated Resection of the Dorsal Sector of the Liver

Hideaki Uchiyama, Shinji Itoh and Kenji Takenaka

Additional information is available at the end of the chapter

1. Introduction

Resection of malignant lesions arising in the dorsal sector of the liver is a challenging procedure because the sector is located deep in the abdominal cavity and surrounded by the inferior vena cava (IVC) and the major hepatic veins [1 – 9]. A hanging maneuver is an innovative procedure in hepatic surgeries, in which the liver parenchyma is hung by a tape, thereby making a straight cutting line [10 – 14]. This technique was applied in two patients who had a hepatocellular carcinoma (HCC) in the dorsal sector. Patient 1 was a 46-year-old female, who was found to have an HCC, approximately 3 cm in diameter, located just above the IVC. The patient had a large inferior right hepatic vein (IRHV). The superior right hepatic vein (SRHV) and the IRHV were individually controlled with a tape after dividing several short hepatic veins from the right side of the IVC. A cotton tape was introduced from the groove between the SRHV and the middle hepatic vein (MHV) to the right and left Glisson sheaths via the space just next to the left side of the IRHV. The liver was split into the right and left hemilivers by pulling the tape upwards. Next, the tape was introduced from the space behind the confluence of the MHV and the left hepatic vein (LHV) to the space behind the left Glisson sheath via the fissure of the ligamentum venosum after dividing a few small Glisson branches into the caudate lobe from the left Glisson sheath. The liver parenchyma was divided between the medial sector and the dorsal sector by pulling the tape medially, Finally, the dorsal sector including the tumor was resected by dividing the short hepatic veins from the left side of the IVC. Patient 2 was a 59-year-old male, who was found to have an HCC, approximately 3 cm in diameter, located in the Spiegel lobe (a part of the dorsal sector) during a follow-up for chronic hepatitis B. The tumor compressed the left side of the IVC and protruded inferomedially. Cotton tape was introduced from the groove between the MHV and the LHV to the groove between the right and left Glisson sheaths via the posterior surface of the liver after dividing all the short hepatic veins from the right side of the IVC. The liver was split into the right and left hemilivers by pulling the tape upwards. The liver parenchyma was divided between the medial sector and the dorsal sector as in Patient 1. The operation time was 623 and 435 minutes and the intraoperative blood loss was 834 and 1320 grams, respectively. No complications occurred in the two patients. The application of hanging maneuvers enables surgeons to safely resect tumors located deep in the dorsal sector of the liver.

This surgical technique requires a lot of indispensable procedures for hepatic surgeries. This chapter presents the step-by-step surgical procedures regarding hanging maneuvers for an isolated resection of the dorsal sector.

2. Patients

The patients' characteristics and preoperative laboratory data are summarized in Table 1. Patient 1 had a cirrhotic liver caused by hepatitis B and had undergone laparoscopic splenectomy approximately two months before hepatectomy to control intractable ascites caused by splenomegaly accompanied with cirrhosis. Patient 2 had a fibrotic liver caused by chronic hepatitis B. Both patients had a solitary HCC in the dorsal sector.

	Patient 1	Patient 2
age	46	59
gender	female	male
native liver disease	cirrhosis caused by hepatitis B	chronic hepatitis B
white blood cell (/μl)	4900	5400
hemoglobin (g/dl)	7.9	14.7
platelet (× 10³ /μl)	235	171
total bilirubin (mg/dl)	0.49	0.42
albumin (g/dl)	3.2	4.6
prothrombin time – international normalized ratio	1.05	0.95
indocyanine green dye retention at 15 minutes (%)	27	13
tumor diameter (cm)	3	3

Table 1. Patient characteristics and preoperative laboratory data

3. Surgical procedures in patient 1

The HCC, approximately 3 cm in diameter, was located just above the IVC (Figure 1). A limited hepatectomy was selected because the patient had a relatively advanced cirrhotic liver and the preoperative evaluations predicted that an extended hepatectomy would have led to postoperative liver failure.

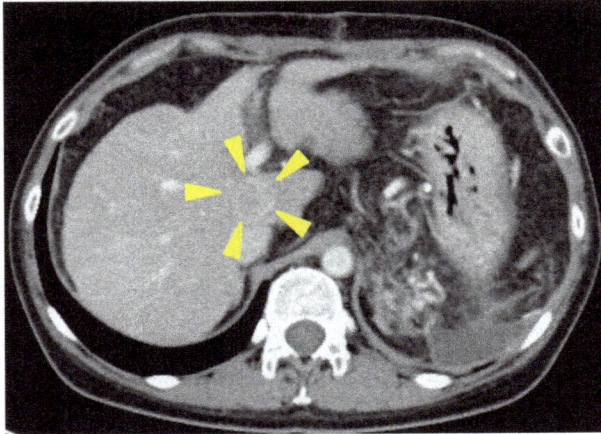

Figure 1. Hepatocellular carcinoma in Patient 1 located just above the inferior vena cava

Figure 2 shows a schematic diagram of the surgical procedure. Patient 1 had a relatively large IRHV. This vein was kept intact because its division could have caused congestion of the posterior sector. The liver was split into the right and left hemilivers by dividing the liver parenchyma along the right side of the middle hepatic vein using a hanging maneuver with a cotton tape introduced into the space between the posterior surface of the liver and the anterior surface of the IVC. The liver parenchyma was divided between the medial sector and the dorsal sector using a hanging maneuver with a cotton tape placed in the fissure of the ligamentum venosum.

The patient was placed in the supine position. The abdomen was opened by bilateral subcostal incisions with an upper midline extension. There was a small amount of ascites and the liver had a cirrhotic appearance. Cholecystectomy was performed and a tube was inserted into the cystic duct for cholangiography. The right lobe was mobilized clockwise by dividing the right triangular ligament. The IVC ligament was divided, and the SRHV and the IRHV were individually encircled with a tape. A thin cotton tape was introduced from the groove between the SRHV and the confluence of the MHV and the LHV to the left-side space of the IRHV (Figure 3).

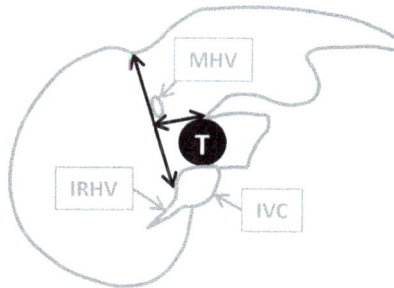

Figure 2. Schematic diagram of the hanging maneuvers for the isolated resection of the dorsal sector used in Patient 1 IRHV, the inferior right hepatic vein; IVC, the inferior vena cava; MHV, the middle hepatic vein; T, tumor

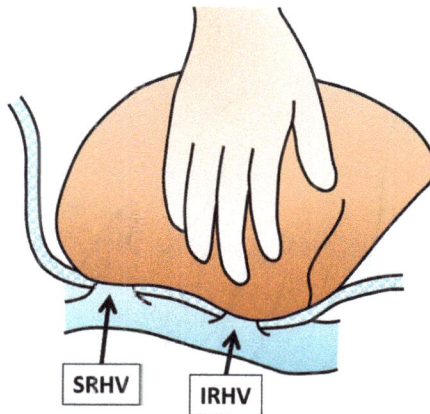

Figure 3. Introducing a cotton tape along the left-side spaces of the superior and the inferior hepatic veins SRHV, the superior right hepatic vein; IRHV, the inferior right hepatic vein

The procedure moved on to the hepatic hilum. The right Glisson sheath was encircled with a tape. A small notch was made on the lowest part of the dividing plane as a hook for the hanging tape (Figure 4).

The left lateral lobe was mobilized counterclockwise by dividing the left triangular ligament. The ligamentum venosum was divided near the LHV (Figure 5). Thereafter, the confluence of the MHV and the LHV was encircled with a tape.

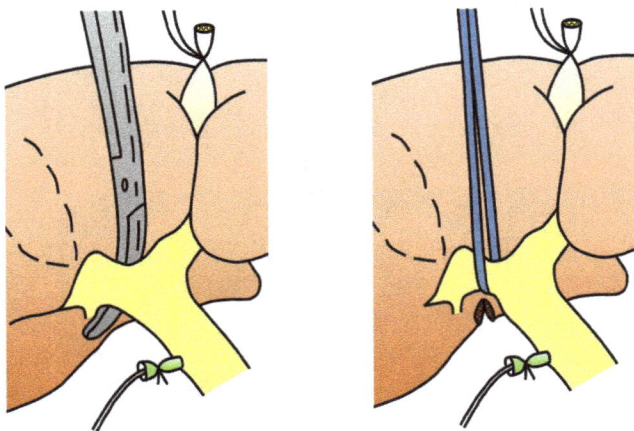

Figure 4. Taping of the right Glisson sheath

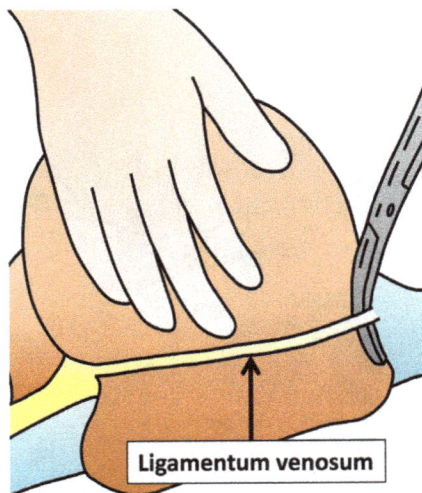

Ligamentum venosum

Figure 5. Division of the ligamentum venosum

The tail of the cotton tape was introduced into the groove between the right and the left Glisson sheath. The liver was split into the right and the left hemilivers by pulling up the cotton tape upwards (Figure 6, 7).

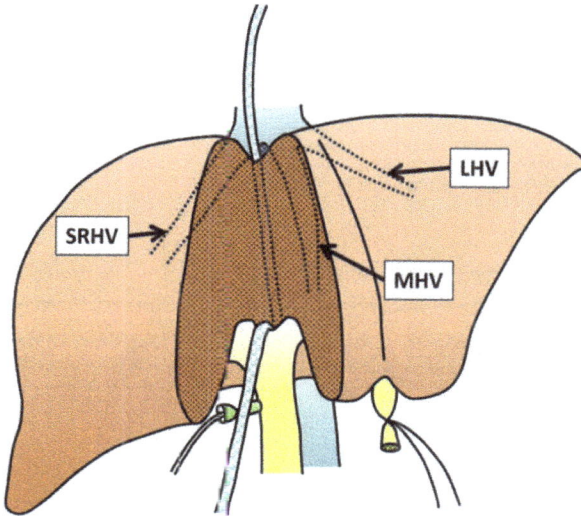

Figure 6. Splitting of the liver into the right and left hemilivers using a hanging maneuver (schematic diagram) LHV, the left hepatic vein; MHV, the middle hepatic vein ; SRHV, the superior right hepatic vein

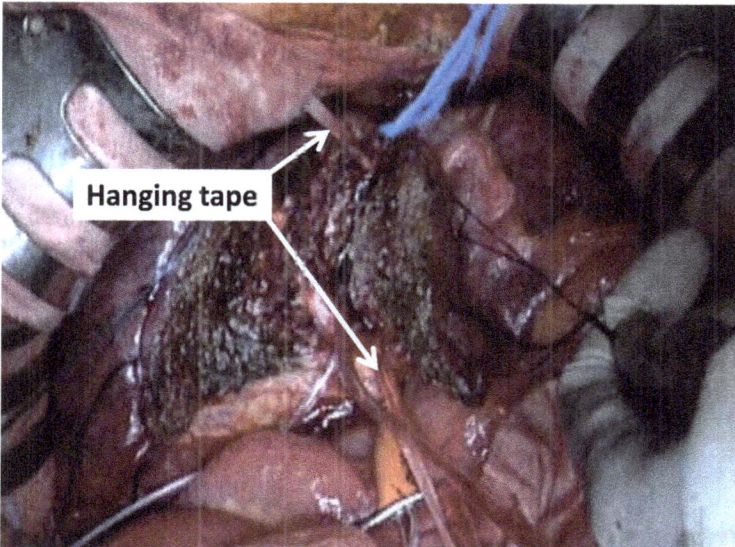

Figure 7. Splitting of the liver into the right and left hemilivers using a hanging maneuver (photograph

Splitting the liver into the two hemilivers revealed a few caudate branches from the left Glisson sheath (Figure 8). These branches were divided to make a space behind the left Glisson sheath (Figure 9). A cotton tape was introduced from the space behind the confluence of the MHV and the LHV to the space behind the left Glisson sheath via the fissure of the ligamentum venosum. The liver parenchyma was transected between the medial sector and the dorsal sector by medially lifting the cotton tape (Figure 10).

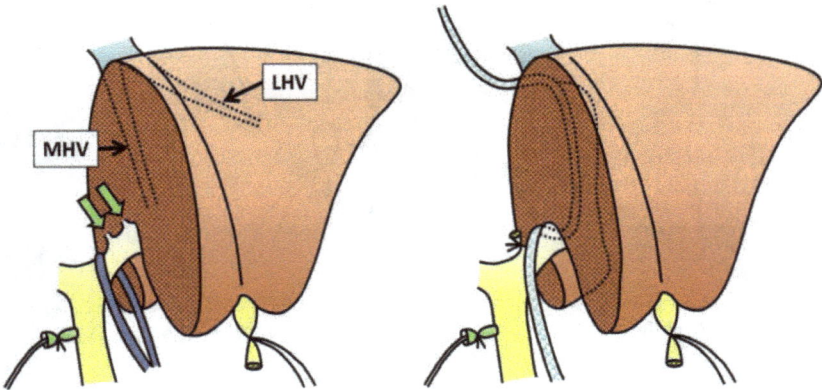

Figure 8. Division of the caudate branch from the left Glisson sheath (left) and a hanging maneuver for transecting the liver parenchyma between the medial sector and the dorsal sector (right) Green arrows indicate the caudate branch from the left Glisson sheath. LHV, the left hepatic vein; MHV, the middle hepatic vein

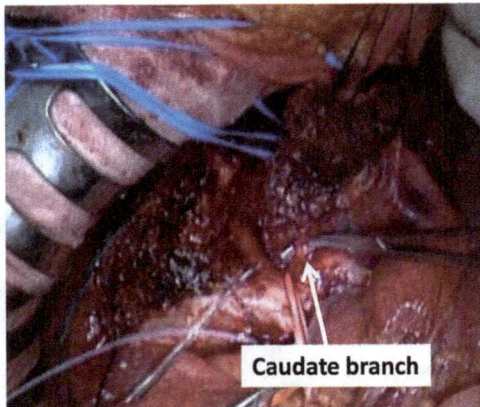

Figure 9. Division of the caudate branch from the left Glisson sheath

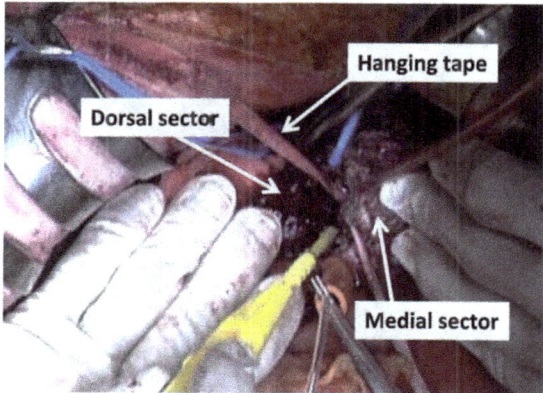

Figure 10. A hanging maneuver for transecting the liver parenchyma between the medial sector and the dorsal sector

All the short hepatic veins from the dorsal sector were divided from the left side of the IVC (Figure 11, 12). The IVC ligament was divided, and the dorsal sector including the tumor was retrieved from the surgical field (Figure 13).

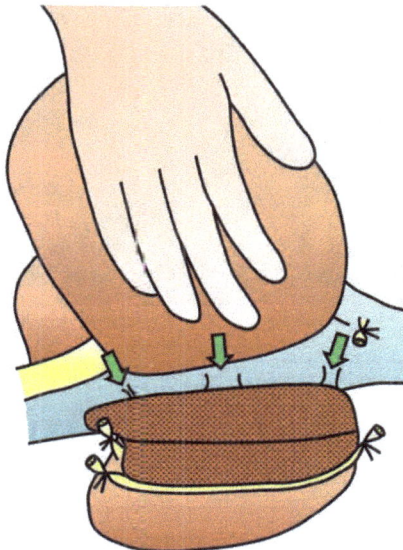

Figure 11. Division of the short hepatic veins from the left side of the inferior vena cava (schematic diagram) Green arrows indicate the short hepatic veins to be divided

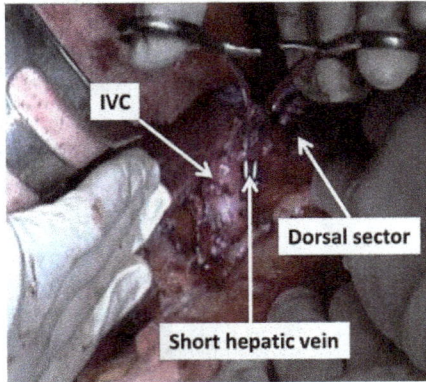

Figure 12. Division of the short hepatic veins from the left side of the inferior vena cava (photograph) IVC, the inferior vena cava

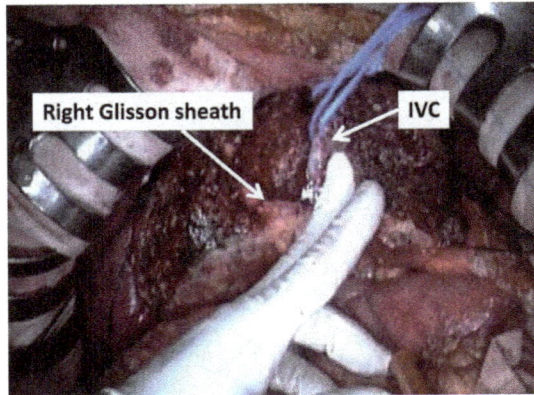

Figure 13. Completion of the isolated resection of the dorsal sector IVC, the inferior vena cava

4. Surgical procedures in patient 2

The surgical procedures in Patient 2 were reported previously [15]. The procedures differed in two points from the procedures used in Patient 1: All the short hepatic veins were divided from the right side of the IVC and the liver was split into hemilivers along the left side of the MHV by introducing cotton tape through the groove between the MHV and the LHV.

5. Surgical results

The surgical results are summarized in Table 2. Patient 1 required transfusion of two units of red blood cell because of pre-existing anemia. The resected specimens had an acceptable tumor-free surgical margin. Kinetics of the laboratory data are shown in Figure 14 and 15. Both patients exhibited rapid recovery of laboratory data. Follow-up CT after the surgeries demonstrated that there were no perfusion abnormalities in the livers (Figure 16).

	Patient 1	Patient 2
operation time (minutes)	623	435
intraoperative blood loss (grams)	834	1320
blood transfusion	two units of concentrated red blood cell	none
length of postoperative hospital stay (days)	13	15
complications	none	none

Table 2. Surgical results

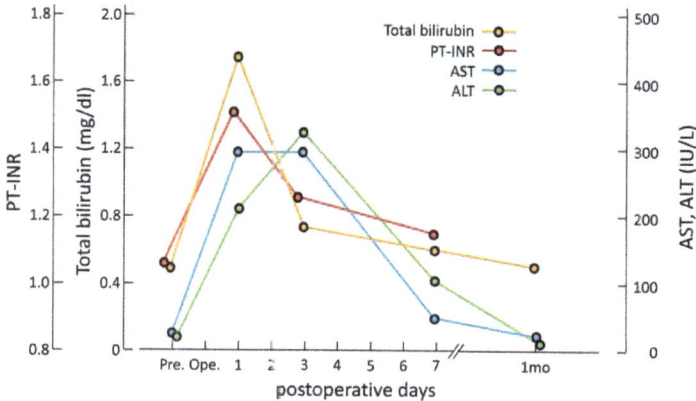

Figure 14. Kinetics of laboratory data in Patient 1 ALT, alanine aminotransferase; AST, aspartate aminotransferase; PT-INR, prothrombin time – international normalized ratio

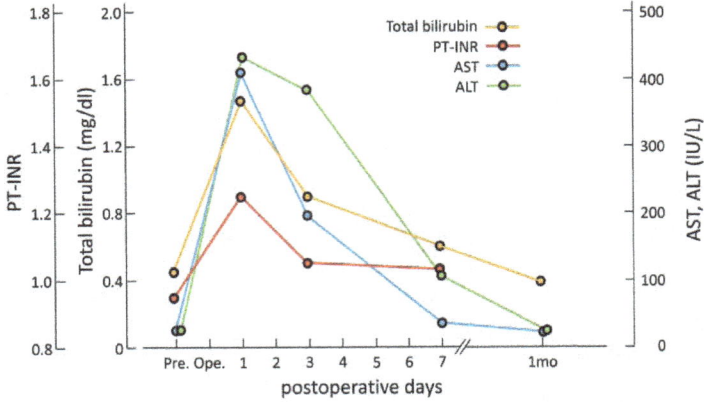

Figure 15. Kinetics of laboratory data in Patient 2 ALT, alanine aminotransferase; AST, aspartate aminotransferase; PT-INR, prothrombin time – international normalized ratio

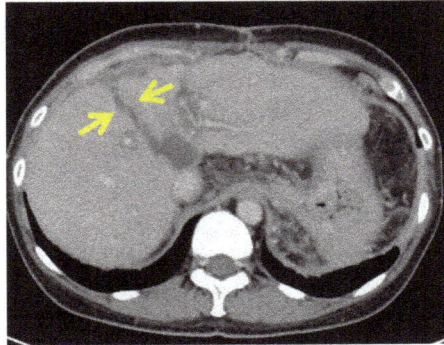

Figure 16. Follow-up CT of Patient 1 two months after the surgery Yellow arrows indicate the dividing plane between the right and left hemilivers.

6. Conclusion

Livers with malignant lesions to be resected are often cirrhotic. Parenchymal transection of cirrhotic liver from the dorsal direction may cause uncontrollable bleeding. The application of hanging maneuvers to an isolated resection of the dorsal sector enables surgeons to safely transect the liver parenchyma only via an anterior approach.

Author details

Hideaki Uchiyama*, Shinji Itoh and Kenji Takenaka
*Address all correspondence to: huchi@surg2.med.kyushu-u.ac.jp
Department of Surgery, Fukuoka City Hospital, Japan

References

[1] Abdalla EK, Vauthey JN, Couinaud, C. The caudate lobe of the liver: implications of embryology and anatomy for surgery. Surg Oncol Clin N Am 2002; 11(4): 835-848.

[2] Asahara T, Dohi K, Hino H, Nakahara H, Katayama K, Itamoto T, Ono E, Moriwaki K, Yuge O, Nakanishi T, Kitamoto M. Isolated caudate lobectomy by anterior approach for hepatocellular carcinoma originating in the paracaval portion of the caudate lobe. J Hepatobiliary Pancreat Surg 1998; 5(4): 416-421.

[3] Chaib E, Ribeiro MA Jr, Souza YE, D'Albuquerque LA. Anterior hepatic transection for caudate lobectomy. Clinics 2009; 64(11): 1121-1125.

[4] Kosuge T, Yamamoto J, Takayama T, Shimada K, Yamasaki S, Makuuchi M, Hasegawa H. An isolated, complete resection of the caudate lobe, including the paracaval portion, for hepatocellular carcinoma. Arch Surg 1994; 129(3): 280-284.

[5] Takayama T, Tanaka T, Higaki T, Katou K, Teshima Y, Makuuchi M. High dorsal resection of the liver. J Am Coll Surg 1994; 179(1): 72-75.

[6] Yanaga K, Matsumata T, Hayashi H, Shimada M, Urata K, Sugimachi K. Isolated hepatic caudate lobectomy. Surgery 1994; 115(6): 757-761.

[7] Utsunomiya T, Okamoto M, Tsujita E, Ohta M, Tagawa T, Matsuyama A, Okazaki J, Yamamoto M, Tsutsui S, Ishida T. High dorsal resection for recurrent hepatocellular carcinoma originating in the caudate lobe. Surg Today 2009;39(9): 829-832.

[8] Yamamoto J, Kosuge T, Shimada K, Yamasaki S, Takayama T, Makuuchi M. Anterior transhepatic approach for isolated resection of the caudate lobe of the liver. World J Surg 1999;23(1): 97-101.

[9] Yamamoto T, Kubo S, Shuto T, Ichikawa T, Ogawa M, Hai S, Sakabe K, Tanaka S, Uenishi T, Ikebe T, Tanaka H, Kaneda K, Hirohashi K. Surgical strategy for hepatocellular carcinoma originating in the caudate lobe. Surgery 2004;135(6): 595-603.

[10] Belghiti J, Guevara OA, Noun R. Saldinger PF, Kianmanesh R. Liver hanging maneuver: a safe approach to right hepatectomy without liver mobilization. J Am Coll Surg 2001; 193(1): 109-111.

[11] Kim SH, Park SJ, Lee SA, Lee WJ, Park JW, Hong EK, Kim CM. Various liver resections using hanging maneuver by three Glisson's pedicles and three hepatic veins. Ann Surg 2007; 245(2): 201-205.

[12] Kim SH, Park SJ, Lee SA, Lee WJ, Park JW, Kim CM. Isolated caudate lobectomy using the hanging maneuver. Surgery 2006; 139(6): 847-850.

[13] López-Andújar R, Montalvá E, Bruna M, Jiménez-Fuertes M, Moya A, Pareja E, Mir J. Step-by-step isolated resection of segment 1 of the liver using the hanging maneuver. Am J Surg 2009; 198(3): e42-48.

[14] Ogata S, Belghiti J, Varma D, Sommacale D, Maeda A, Dondero F, Sauvanet A. Two hundred liver hanging maneuvers for major hepatectomy: a single-center experience. Ann Surg 2007; 245(1): 31-35.

[15] Uchiyama H, Itoh S, Higashi T, Korenaga D, Takenaka K. A two-step hanging maneuver for a complete resection of Couinaud's segment I. Dig Surg 2012; 29(3): 202-205.

Chapter 10

Segmental Oriented Liver Surgery

O. Al-Jiffry Bilal and Khayat H. Samah

Additional information is available at the end of the chapter

1. Introduction

Understanding the vascular and biliary anatomy of the liver is mandatory for a successful anatomical liver resection. It is also extremely important in complex liver operations, althaugh it might not be in cases of simple wedge resection for benign disease. As the presence of HCC is usually in the background of liver cirrhosis, the importance of anatomical resection to be able to clear the tumour and have sufficient amount of liver to avoid post-operative liver failure. In this chapter we will try to illustrate the importance of anatomical liver resection and give an idea of the latest liver anatomy with a demonstration on how to identify and resect each part of the liver.

2. Why anatomical liver resection

As many general surgeons might like to do wedge non anatomical liver resections because it is less complicated and gets the tumour out. There are several reasons to perform anatomical resection:

1. In Hepato-Cellular carcinoma (HCC) which is the most common reason to perform liver resections, were it is the first line of treatment nowadays [1,2]. As the HCC are able to invade the portal veins and disseminate through its inter segmental branches [3] (cough reflux), segmentectomy is preferable. Intrahepatic metastasis [4,5] and invasion to the portal and hepatic venous system will affect the post operative prognosis. To improve the post surgical outcome the segmental liver resection is indicated. It involves the removal of the whole segment containing the tumor with its vasculature which might be affected by the tumor invasion [1,4 - 6]. Satellite micro metastasis will also be removed as their feeding vessel for that segment [3].

Anatomic liver resection is superior to non anatomic from the oncologic and anatomic aspects [7]. Anatomically based hepatectomy is the best means of achieving a negative margin[8].The recurrence rate within 2 years associated with aggressive tumor biology such as high tumor grade, satellite lesions and microvascular invasion [7], is higher in non anatomical resection.

In small HCC <4cm anatomic resection achieves better disease-free survival than limited resection without increasing the postoperative risk [9-10].

The overall survival and the disease-free survival rates were significantly better in the anatomic resection compared to the non anatomic resection group [1,11-12],as well as the recurrence disease free survival [10].

A meta-regression analysis was done and published in June 2012 that was conducted on 9036 patients from 1990-2011 and demonstrated that the 5 years disease free survival and the 5 year survival was significantly better in the anatomic resection group than the non anatomic resection group with no effect on the post operative mortality and morbidity[13].

2. Less bleeding with almost no need for transfusion in the intra-operative period as there is no transaction of the vessels. Also there is few vessels present in the inter segmental planes. Relatively the inter segmental area is a non-vascular plane, so segmental identification, control of the feeding vessels and the vascular pedicle will decrease the blood loss. This is one of the direct causes of decreased post operative morbidities and mortalities [3,14 - 16].

3. Segmental resection will preserve as much of the liver parenchyma [3] and will enable sufficient liver volume especially in cirrhotic patients [16] and in patients with multiple liver lesions [17] or in patients who will need another resection in the future. Also it will decrease the post operative liver insufficiency from small liver remnant in cirrhotic patients [3,14,15,16].

4. In colorectal metastasis segmental resection is superior to non anatomical resection as it results in better tumour clearance and free margins. Multiple studies demonstrated that it did affect the disease free survival, and the control of micro-metastasis through segmental portal branches. Segmentectomy offered disease-free and overall survival rates similar to those after major resection. [3,14]

For metastasis it has been found that with wedge resection the recurrence rate and positive margins were higher compared to the segmental resection. This resulted in inadequate tumour resection especially in deep lesions where the incidence of inadvertently cutting into the tumour is higher. Also the bleeding rate is high due to the difficult control of the venous branches that will obscure the resection plane.

Wedge resections are usually inadequate and potentially dangerous, especially for large tumours, and are often associated with greater blood loss and a greater incidence of positive histological margins.[8,11]. Liver failure due to parynchymal necrosis or small liver remnant are observed in non anatomical (wedge) liver resection. It also results in higher incidence of biliary fistula and infection because of the remnant devitalized liver tissue [18].

Non-anatomical liver resection (Wedge) can be done in certain circumstances; in resections where the tumour is small (<3cm) and located peripherally at the edge of a cirrhotic liver or when the tumour is situated at the border of several segments and its resection requires the removal of large volume which is not possible due to the liver status.

Also, in cases of benign liver resection were no safety margin is required and the surgeon would like to preserve as much liver volume as possible, so the lesion can be enucleated. However, care should be taken not to injure nearby vessels or bile ducts.

3. Segmental liver anatomy

3.1. The history

The understanding of liver segments was first established in 1953 by Healy [19] and was further reinforced by Couinaud in 1957 [20]. When trying to understand their description it might be somewhat confusing, however we will try to make it as simple as possible.

They both used the new division by Cantlie who disapproved the old terminology of the right and left liver which was divided by the falciform ligament and used his description of the right and left liver divided by the midline which is oblique and extended from the gallbladder bed to the right side of the inferior vena cava. Healy then divided the liver using the arteriobiliary segmentation. This lead to the division of the right liver into the two segments, the right anterior and the right posterior segments (called now sections). The left side was divided by the falciform into the left medial and left lateral segments (called now sections). However, Couinaud used the hepatic veins and divided the right liver into right anterior and right posterior sectors. The left side was divided by the left hepatic vein into the left medial and left lateral sectors, and the middle hepatic vein was running in the midplane of the liver (Cantile line). Then recently the terminology of segments that was described by Healy was changed to sections leading the way to the word section and sector that you see in all papers involving the liver anatomy. They both divided these sectors or sections to segments according to the portal vein anatomy and we reached to our 8 segments that we know today. **Figure.1**

When looking at these two description you will find that both agreed on the anatomy of the right liver cause there was no difference between the right anterior (segment 5&8) and the right posterior (segments 6&7) section or sector. However, on the left side there was a difference, because of the anatomical variations and we believe this is what led to this misunderstanding. The left medial section (segment 4) is not the same as the left medial sector (segment 3&4), and the left lateral section (segment 2&3) is not the same as the left lateral sector (segment 2). They also both agreed on the separation of segment 1 (caudate Lobe) as it has its own blood supply and drains directly to the inferior vena cava.

Another thing when looking to the terminology is the word "Lobe". Some authors use the term left lobectomy to describe the resection of segments 2&3, which is the functional left lateral section. Also the right lobe as segments 4 to 8 were it is an extended right trisectionectomy.

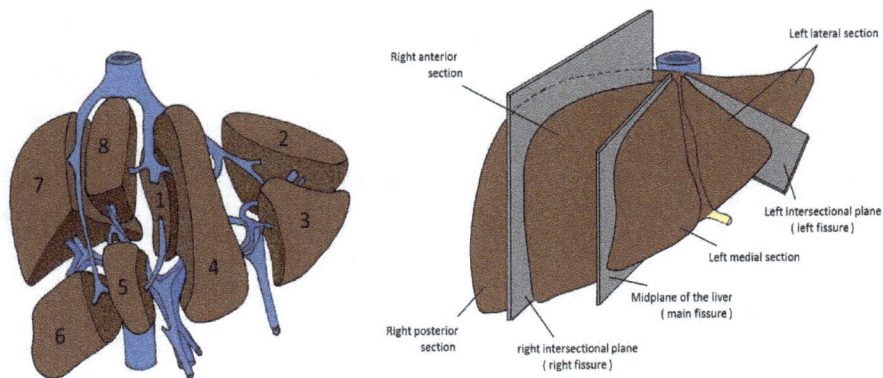

Figure 1. Liver sections, plane and segments

This description was based on the anatomical land mark of the liver using the falciform ligament and not the functioning liver segments as described above.

3.2. The new terminology

The use of many different terminologies and difficulty in understanding the description described above, were the European societies adopted Couinaud's description and the American societies used the Healy's description. So the scientific committee of the International Hepato-Pancreato-Biliary Association with experts around the world came up with the Brisbane 2000 Terminoloy of Liver Anatomy and resection which we have been using and will use for our description in this chapter [21].

To understand this terminology, first the liver is divided into two parts, the main liver and the caudate lobe (called the dorsal sector by Cauinaud). Then the main liver is divided into the right and left liver.

This part is called the first order division, where the liver is divided into the right liver or right hemiliver, and the left liver or the left hemiliver. Notice the word lobe has been removed completely for the confusion we mentioned above, so the resection of the right side is called right hepatectomy or right hemihepatectomy (segments 5 to 8). The left side is called; left hepatectomy or left hemihepatectomy (segments 2 to 4). **Figure.2**

The second order division, where the right liver is divided into two parts. The right anterior section giving the right anterior sectionectomy (segment 5&8), the right posterior section leading to the resection of the right posterior sectionectomy (segments 6&7). On the left side there will be the left medial section giving the left medial sectionectomy (segment 4), and the left lateral section leading to the resection of the left lateral sectionectomy (segment 2&3). **Figure.3**

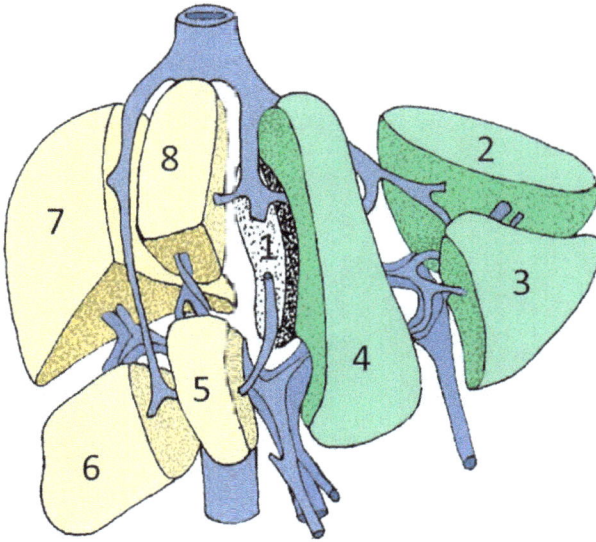

Figure 2. Right (yellow) and left liver (green)

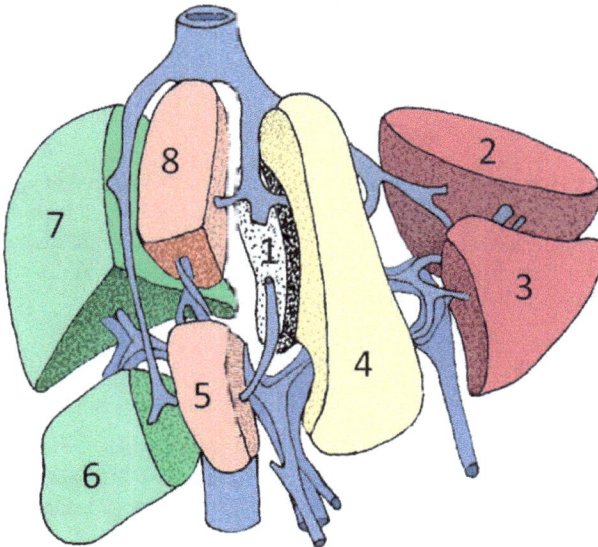

Figure 3. Sections Green: right posterior, orange: right anterior Yellow: left medial, red: left latera

The third order division, is the division of each of these sections into segments as we mentioned above. The resection of any of these segments is called a segmentectomy and if two or more segments were resected that are not related as described in the second order division it is called bisegmentectomy or trisegmentectomy. This should not be coinfused with the trisectionectomy of the right or left side were we resect three sections and not segments. **Figure.4**

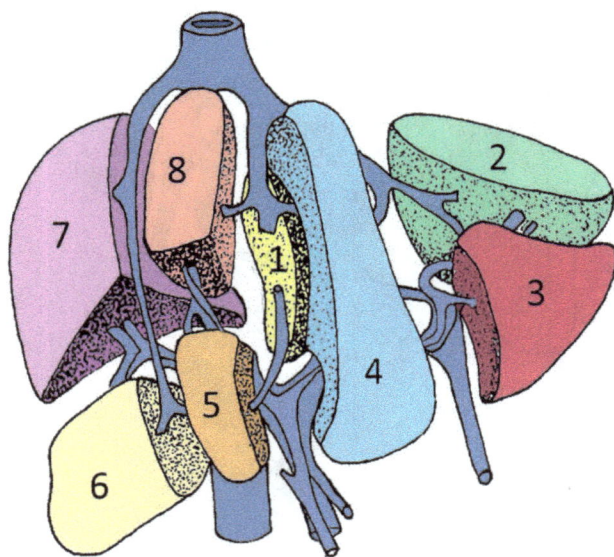

Figure 4. Segments, each with a different color

An addition was added also if the word sector were to be used instead of section. This is the same on the right side and on the left we had a left medial sector with a left medial secterectomy (segment 3&4), and the left lateral sector giving rise to the resection of the left lateral secterectomy (segment 2). So the term section or sector has to be used very cautiously on the left side to describe exactly what you mean.

3.3. Clinical applications

1. As each liver segment can be resected separately, liver resection can be segment based

2. Segement 4, is divided into 4A and 4B. This was made because of multiple indications were segment 4A is rsected without the resection of segment 4B like in cases of gallbladder cancer. Also the resection of segment 4A is counted as the most difficult liver resection as it lies between the middle and the left hepatic vein.

3. This terminology has gained wide acceptance and has removed most of the confusion that use to exist in the past.

3.4. Intrahepatic glissonian triads

The extra hepatic portal triad is consisted of the portal vein, the hepatic artery and the common hepatic duct. These structures are enclosed in a connective tissue and peritoneum up to the hepatic hilum. The term Glissonian sheath is reserved for the part that extendeds into the intrahepatic portion of the liver beyond the hilum. This sheath surrounds the portal triad structure before they enter into each section, giving rise to the resection of each segment (liver unit) separately without affecting the other segments [22]. This gives rise to the aberrance of the central segments 4, 5 and 8 ramifications like a bush and fan shaped. Consequently, a single segment resection will require several Glissonian sheath at various depth and is much more difficult. Were the priphral segments 6, 7, 2 and 3 have long branches that travels a distance reaching to these segments giving the appearance of tree like making their resection less complicated and usually requiring a single Glissonian sheath ligation [23].

3.5. Portal vein and liver resection

On the right side the portal vein is similar to the arteriobiliary segmentation. On the left side they differ from each other. The left portal vein consists of a transverse and an umbilical portion. The transverse portion only sends small branches to segment 4 and one or two branches to segment 1. All the larger branches arise beyond the attachment of the ligamentum venosum (umbilical portion of the left portal vein). **Figure.5**. This part of the vein gives right branches to segment 4 and on the left side it gives one branch to segment 2 and more than one to segment 3. The portal vein terminates where it joins the ligamentum teres at the edge of the liver. This unique structure explains the duael function of the left portal vein during in-utero and then in-adult life.

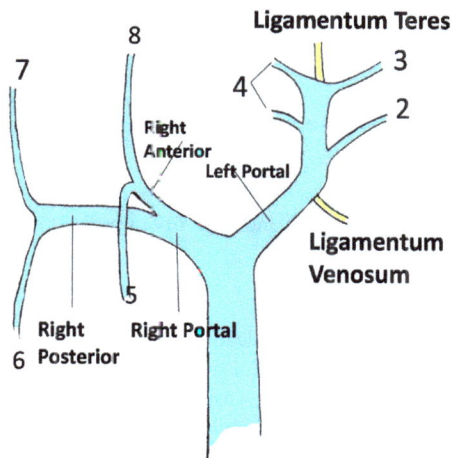

Figure 5. Portal vein with its divisions

On the right side the portal vein is usually very short and gives rise to the right anterior and right posterior branches. Each of these branches gives rise to two main segmental divisions. The right anterior gives both segment 5 and 8, where the right posterior gives segment 6 and 7. **Figure.6**.

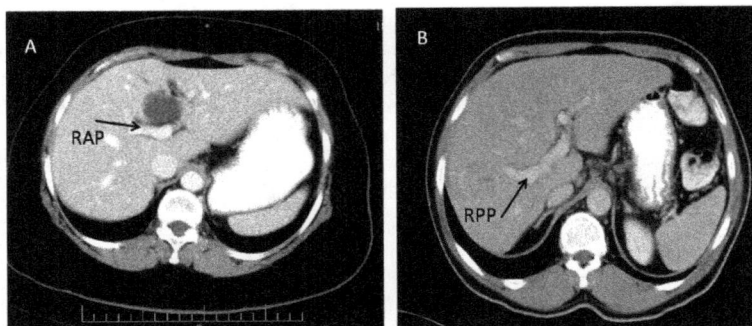

Figure 6. A) right anterior portal branch (RAP) B) right posterior portal branch (RPP)

Usually there are very little variations in the portal vein. The commonest one is where the right anterior branch joins the left portal vein. This is very important to recognise especially when doing a left hepatectomy causes injury could happen to the right anterior section leading to the loss of segments 5 and 8. Another common anomaly is the absence of the main right portal vein giving rise to a trifurcation at the hilum of the portal vein to the left main, right anterior and the right posterior branches. This is important when doing a right hepatectomy to transect each branch separately not to injure the left portal vein [24-25].

4. Clinical identification of the liver segments

For the clinical description of this part we will try to simulate what happens in clinical practice by dividing it to pre-operative radiology and intra-operative by intra-operative ultrasound.

4.1. Pre–operative

To try and make this part as simple as possible for the reader we will try to identify land marks that you should look for in the ultrasound, CT or MRI. The ultrasound is the usual screening tool used to see the whole liver and identify cystic from solid lesions. Then most centres will request a Triphasic CT scan of the liver in the hope to identify the nature of the lesion and the location. A physician should not comment on any lesion seen until full examination of all three phases (arterial, venous and delayed) are examined and the lesion is seen on all three phases to give the best chance of reaching the right diagnosis.

As we described the anatomy of the liver by the first order division and its landmark the middle hepatic vein, it is the same here. The middle hepatic vein can be seen on any of the above mentioned x-ray investigation. This will lead to the division of the liver to the right and left liver and identifying the lesion in which liver it lies. **Figure.7**.

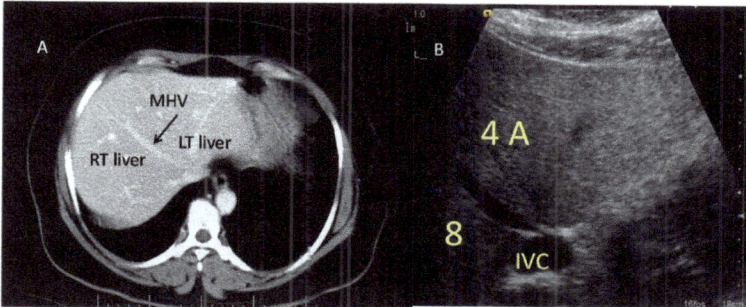

Figure 7. Middle hepatic vein (MHV) A) CT B) Ultrasound

The next step is to identify the falciform ligament and the right hepatic vein. This will divide the left liver to the medial and lateral sections and the right liver to the anterior and posterior sections alternatively. By this any lesion will be clearly seen in each section of the hemi-liver. **Figure.8**.

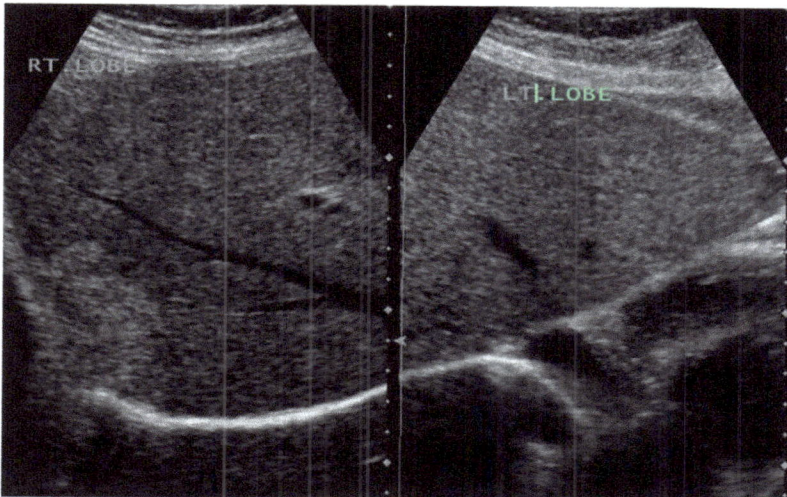

Figure 8. A) right hepatic vein (RHV) – with a lesion in seg 7 B) falciform (FL)

The last step is to identify the main portal vein and follow it till you reach to the bifurcation of the right and left branches which corresponds to the line that divides the liver into the upper and lower segments. This will give rise to the division of each section to its corresponded segments as described before in the anatomy part **Figure.9**.

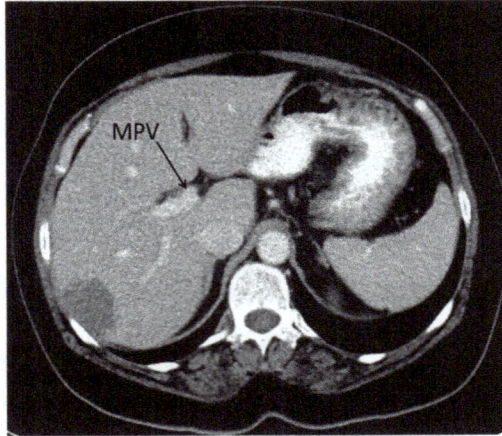

Figure 9. Main portal vein (MPV), with a lesion seen in the right posterior lower segment (segment6)

If this simple technique is adopted a full idea of the lesions identity and location could be achieved with a high degree of certainty making the surgical planning much more feasible. **Figure.10**.

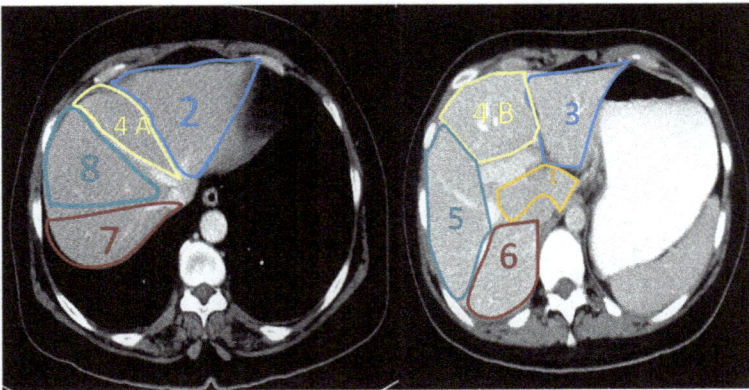

Figure 10. All segments identified on CT pre-opretive

4.2. Intra–operative

This is usually carried out by the intra-operative ultrasound [26-30], which we believe no liver resection should be done without mastering its use especially in malignant liver lesions. There are six simple steps that should be followed to get the best results of the ultrasound. 1) General inspection the whole liver as CT is not the ideal tool to identify superficial liver lesions. 2) A systemic recognition of all three hepatic veins and the main portal veins with its branches to identify all the liver segments. 3) Localize the tumour and determine which segments are involved. 4) Determine which segments needs to be resected to achieve good margins and balance it with the state of the liver trying at all times to go thru the anatomical lines to get an anatomical liver resection when possible to achieve the advantages mentioned before. 5) Mark the liver resection line on the liver surface. 6) Redetermine the distance from the tumour and the resection lines to be certain not to be close or even worse go thru the lesion.

To identify the segments the same method that was done pre-operative on CT is adopted by the localisation of the middle hepatic vein and drawing a line on it to get the right and left livers. **Figure.11**.

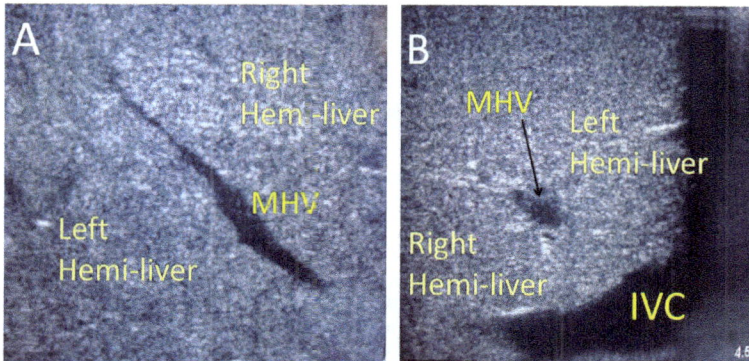

Figure 11. Intra operative ultrasound middle hepatic vein. A) longitudinal B) sagetal

The falciform ligament which divides the left liver to the medial and lateral sections can be seen on the surface. The left hepatic vein that divides segment 2 and 3 can be identified. On the right side the right hepatic vein is seen and a line is made to divide the right liver to the anterior and posterior sections. **Figure.12**.

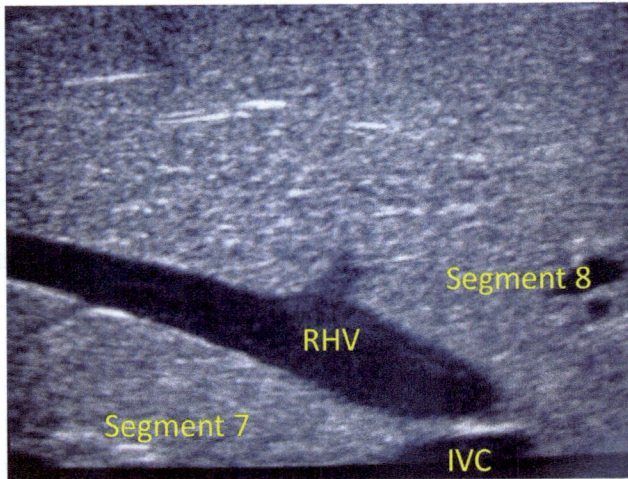

Figure 12. Intra operative ultrasound right hepatic vein.

The portal vein is then identified and followed to get all its branches and a line is made horizontally to get the upper and lower segments of the liver. **Figure.13**. After connecting all these lines the liver segments will be seen on the surface with the exception of segment 1 which is separate as we indicated before and can be seen over the IVC as the caudate lobe [31]. **Figure.14**

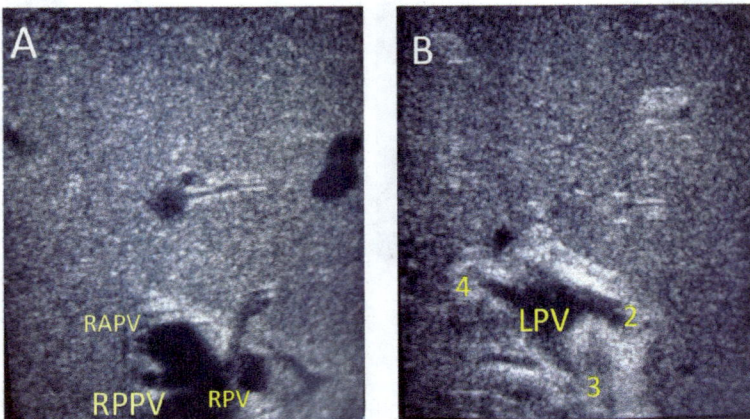

Figure 13. A) Right Portal veins (RPV) and its bifurcation to right anterior (RAPV) and right posterior (RPPV). B) Left portal vein (LPV) with its segmental branches

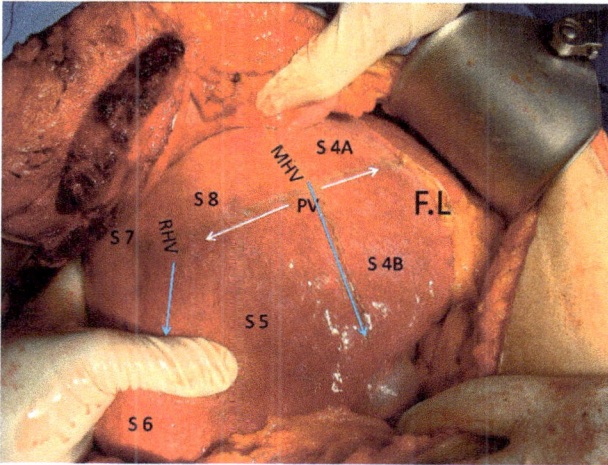

Figure 14. Liver segments on the liver intra operatively Portal vein (PV) in weight, middle hepatic vein (MHV), falciform ligament (FL), right hepatic vein (RHV)

5. Liver resection

A full pre-operative evaluation is necessary before embarking on a liver resection especially that most of the patients with HCC are also cirrhotic. There are multiple models to evaluate these patients and the most widely used one is the Child-Pugh score. This model stratifies patients into stage A, B, and C. Also the size of the tumour and the patient's physiological function are very important. Therefor most recent staging systems for HCC has included three important factors to evaluate the patient before any liver resection, the tumour, the liver status and the patient factor. Although chronic liver disease is not an absolute contraindication to liver resection, the morbidity and mortality increases prohibitively with increasing hepatic dysfunction. Childs class C or late B patients are generally excluded from major resections whereas Childs A or early B patients may be candidates [8,31].

As we mentioned above radiological studies are important in determining the presence of portal hypertension, ascitis, tumour localization, feasibility of the resection, tumour extension, distance from the pedicles and segments necessary to be resected as well as extra-hepatic metastasis [8].

5.1. Position and skin incision

The patient is usually in supine position, with the arms extended 90° when possible [8]. To minimize risk of air embolism from disrupted hepatic veins[8] and to minimize blood loss from the resected raw liver surface[3]. The resection is performed with the patient in the

Trendelenburg position and as recommended by all liver surgeon with a low central venous pressure of 0-5 mmHg(15°).**Figure.15.**

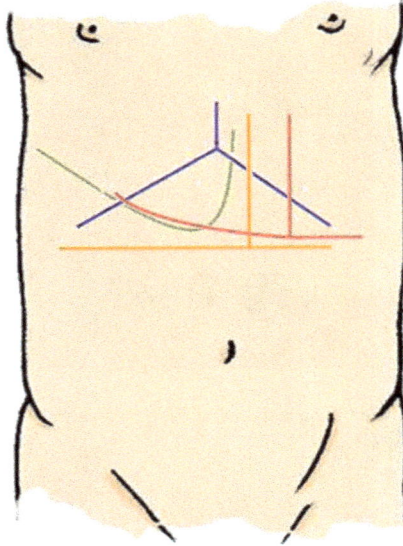

Figure 15. Skin incisions for liver surgery

Preparation of the operative field includes the area from the lower abdomen up to and including the chest, extending from axillary line to axillary line [8]. The majority of liver resections are performed with either a right subcostal incision with upper midline extension (inverted hockey stick) or a chevron (Mercedes) incision [8]. Intra-operative ultrasound is done as described above and the necessary ligaments are released according to the segments of the liver that needs to bee resected. Usually the falciform ligament is released to allow free mobilization of the liver and a better access for the ultrasound.

5.2. Approaches to liver resection

A liver surgeon should be familiar with all the techniques of liver resection because each has advantages and disadvantages making different resections more feasible.

5.2.1 Anterior approach

This technique is started by dissection of the portal triad and the hilar plate, where the right and left portal veins are identified.Figure.16. This makes the ligation of each portal branch more feasible. Then the vascular line of demarcation is seen and with the aid of intra-operative ultrasound to identify the rest of the vascular structures and the tumour. The liver is then mobilized according to the part being resected. Parynchymal transaction is then carried out followed by ligation of the hepatic veins. This type is usually applied in patients with less liver fibrosis and a right or left liver resection is needed.

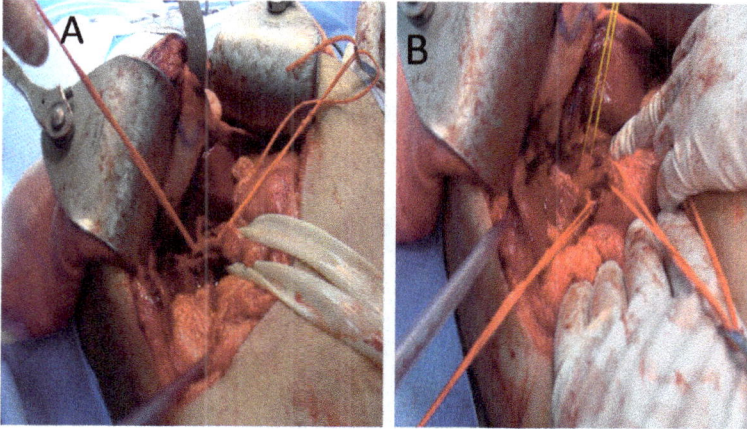

Figure 16. Anterior Approach. A) the tape is around the main and right hepatic artery. B) The yellow tape is around the left portal vein

5.2.2 Posterior approach

The liver is mobilized according to the part being resected. This will give access to the right or left hepatic vein which is usually circled and controlled. Then two ways can be done, were some surgeons transect the vein followed by Pringle and transect the liver parenchyma by the fast technique in about 10-15 min. This is usually fast and has less bleeding and can be done in patients with right, left and both left lateral and right posterior (peripheral sections) liver resections specially if the patient has liver fibrosis because of the time and bleeding. However, this technique requires the excellent use of ultrasound to avoid injury to the main vascular structures, and prevent a long Pringle time for the unresected part of the liver.

The other way is to start with the liver transaction. This will not require the routine use of Pringle, however it can be associated with more blood lose, and longer transection time to control the bleeding. This is usually done in non cirrhotic patients specially in living related liver transplant.

By using also the posterior approach the portal pedicle will be transacted at the end in the liver. This will decrease the injury or the narrowing of the unresected pedicle.

5.2.3 Hanging technique

This approach was adopted recently and was mainly applied in the right liver donors for living related liver transplant. This technique usually relies on the principle of keeping the liver well vascularised till the last minute to keep the liver viable.

The approach is done by using the avascular plane on the anterior part of the inferior vena cava and the window between the right and middle hepatic vein. This makes the passage of a tape from the inferior part of the liver to the superior part over the inferior vena cava. **Figure. 17**. The live is then transacted over the tape slowly while maintaining good haemostasis. Then the right hepatic vein and the right pedicle are transacted.

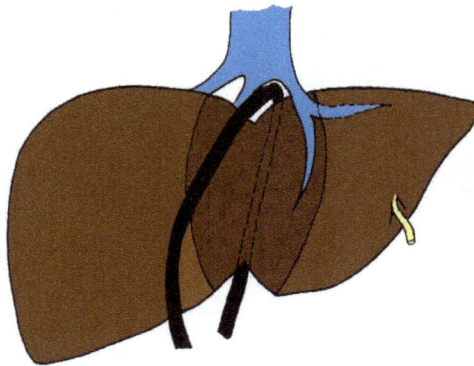

Figure 17. Hanging technique

This method is used mainly in right liver resection, and the tape can be moved in any plane wanted with the aid of the ultrasound. It also has a non touch like technique, were the liver is not mobilized till the vascular inflow and outflow are transacted. However, it requires time and very experienced surgeon not to injure the inferior vena cava during insertion of the tape. It's also time consuming and not applied in cirrhotic liver because bleeding will be more.

5.2.4. Hilar plate dissection

This technique is started by hilar plate dissection and reaching to each sectional branch or even to each segmental branch. Control of the inflow is done first followed by mobilization of the part intended to be resected. The liver resection is then carried out and the outflow is then transacted. **Figure.18**

This method is best for central liver resection, however the hilar dissection requires experience and cannot be carried out in cirrhotic livers as bleeding will be difficult to control. Intra-operative ultrasound is very important to locate the portal branches and the outflow veins to decrease its injury, also the tumor localization is important not to cut through it.

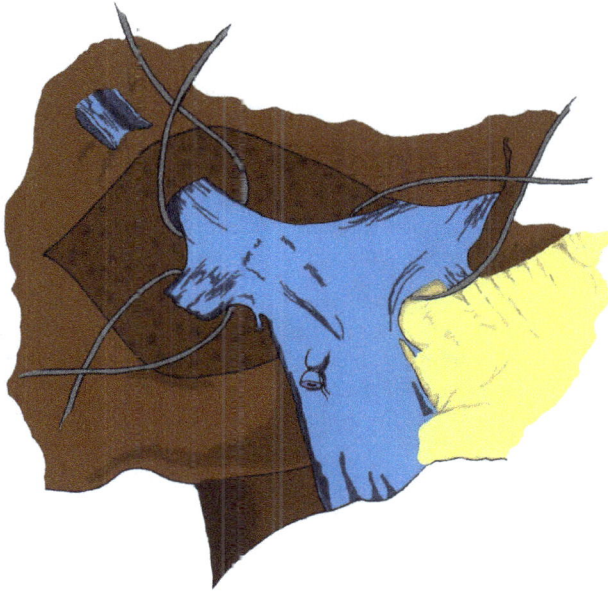

Figure 18. Hillar dissection. Tapes around the sectional portal branches on the right and the main left portal vein on the left

5.2.5. Intra–hepatic ligation

Peripheral and non anatomical liver resections are usually done by this approach. Intra-operative ultrasound is done to see the tumour and its blood supply. Mobilization followed by parenchymal transaction, were the inflow and outflow vessels are transacted in the liver.

5.2.6. Radio–frequency assisted liver resection

The first description of RFA-assisted liver resection was published by Habib's group [32]. This technique showed a major improvement of liver surgery with low/no morbidity and mortality observed [33]. It also showed decrease in the anesthetic time, operative time, hospital stay, and blood loss. Liver resection became a comparatively safer procedure [34].

Liver resection utilizing radiofrequency-induced resection plane coagulation as a safe alternative to the established resection techniques. The residual zone of coagulation necrosis remains basically unchanged during a follow up of three years, with a safety margins of 0.5-3.5 cm and Histopathological proof [35].

The RadioFrequency Assisted liver resection has 5 steps [32, 36]:

Step 1: First or inner line is made on the liver capsule with argon diathermy to mark the periphery of the tumor. This is done by bimanual palpation and intraoperative ultrasound.

Step 2: Second or outer line, again using argon diathermy, is made on the liver capsule 2 cm outside (away from) the inner line to mark the site where the probe is positioned to achieve coagulative necrosis.

Step 3: Coagulative necrosis is produced along a line that follows the second or outer line. The cooled-tip RF probe and a 500-kHz RF Generator, which produces 100 W of power and allows measurements of the generator output, tissue impedance, and electrode tip temperature. The probe contains a 3-cm exposed electrode, a thermocouple on the tip to monitor temperature and impedance. Two coaxial cannulae through which chilled saline is circulated during RF energy application to prevent tissue boiling and cavitation immediately adjacent to the needle.

Step 4: Further probe applications are deployed to obtain a zone of necrosis according to the depth of the liver parenchyma to be resected. Application of the RF energy should begin with the area deepest and farthest from the upper surface of the liver. Once the deepest 3 cm of tissue is coagulated, the probe is withdrawn by 3 cm to coagulate the next cylinder of tissue, and so on until the upper surface of the liver is reached. Each application requires about 60 seconds of RF energy.

Before each probe removal, the saline infusion is stopped to increase the temperature close to the electrode. This results in coagulation of the needle tract during withdrawal and reduces the possibility of bleeding from the probe tract and the liver capsule.

Step 5: The liver parenchyma is divided using the scalpel. The plane of division should be situated midway between the first and second line so as to leave a 1-cm resection margin away from the tumor and leave in situ 1 cm of burned coagulated surface.

5.2.7. Total hepatic vascular exclusion

This method combines total inflow and outflow vascular occlusion of the liver, isolating it completely from the systemic circulation. It is achieved after complete liver mobilization, application of inflow occlusion by Pringle manoeuvre, and then placing a clamp across the infra-hepatic IVC above the renal veins and the right adrenal vein followed by a supra-hepatic IVC clamp above the opening of the major hepatic veins. After the parenchymal transection and hemostasis, the clamps are removed in the reverse order[37]. **Figure 19.**

This results in a significant haemodynamic instability, with a substantial reduction in cardiac output, though blood pressure is usually maintained [38]. Around 10% of patients cannot tolerate it haemodynamically[39].

The ischaemic limit is 60-90 mins for patients with normal liver function [40]. In patients with cirrhosis, the maximal ischaemic time is halved and, in addition, the liver function before surgery must be at the better end of the spectrum[41]

However this technique is not done nowadays with the advanced surgical techniques except in rare conditions like tumour thrombus reaching the IVC or the atrium **Figure 20.** It also prevents intra-operative thrombus migration, and allows major hepatic veins or IVC reconstruction [37].

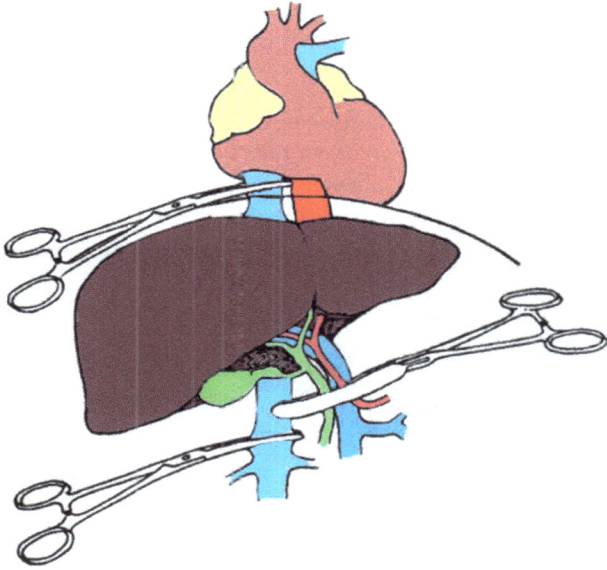

Figure 19. Total hepatic vascular occlusion

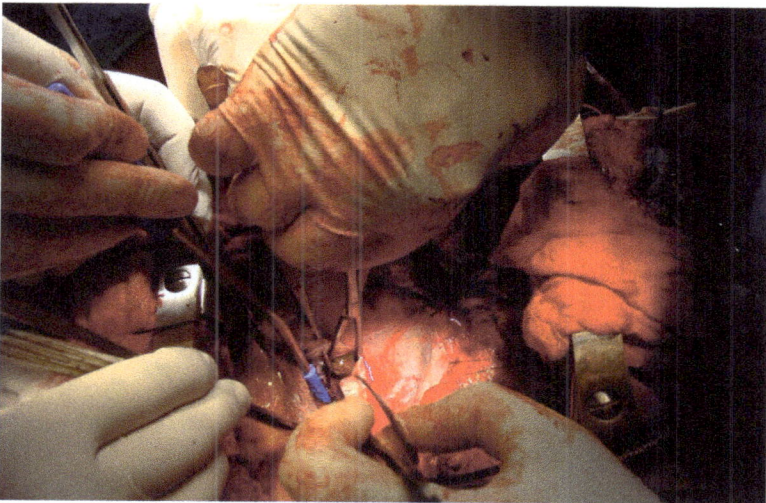

Figure 20. A case of HCC with atrial thrombus with total vascular occlusion. The thrombus is being removed from the right atrium.

However, with the development of liver surgery there has been use of some part of vascular occlusion done selectively or in compinations:

1. In-flow control:

2. Pringle manoeuvre; This is done by occluding the total inflow to the liver. This is usually in cases of central liver resection or majour resections where a large volume of blood is suspected to be lost. **Figure 21**

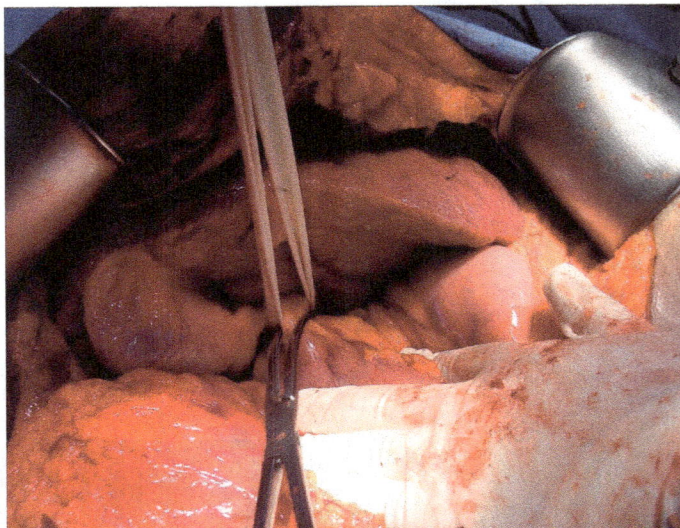

Figure 21. Pringles manoeuvre

1. Hemi-Hepatic control; This is done as described before in the right or left hemi-hepatectomy by controlling the right or left pedicle at the glissonian sheath. **Figure 16**

2. Sectional control; This is done by isolating and controlling the sectional branches as described in the (Hilar Plate Disection) as described above to be able to isolate each section without affecting other parts of the liver. **Figure 17**

3. Out-Flow Control:

4. Total hepatic control; this is achieved by either clamping of the IVC above and below or clamping the hepatic veins without affecting the flow of the IVC as nowadays done in piggy-back liver transplant.

5. Isolated hepatic vein control; this is done as described in the posterior approach where full mobilization of the liver is done and the right or left hepatic vein is isolated and clamped with-out affecting the IVC or the other hepatic veins

These all can be done separately or combined to achieve a bloodless liver resection and maintain patient stability.

5.3. Parynchymal transaction

Meyer-May described the use of Kocher-like clamps to crush liver parenchyma in 1939 [12,42] and haemostatic clamps such as Kelly clamps [43] are still used to crush small areas of the parenchyma, leaving the vessels intact.

Lortat-Jacob used the handle of a scalpel[9] and Lin described the use of finger fracture to remove parenchyma under inflow occlusion to isolate vessels and bile ducts for ligation[44,45].

Ultrasonic dissection has been developed using the CUSA (Cavitron Ultrasonic Surgical Aspirator)[42], this allows for delineation of the hepatic veins, particularly at the junction with the inferior vena cava, and prevents positive margin [45]. It has been shown to be very effective for division of the parenchyma with low blood losse [46,47].

Water-jet dissection [48-49] reduced blood loss, blood transfusion, and transaction time compared with CUSA, but there is increased risk of venous air embolism [45].

Harmonic Scalpel allows sealing of small vessels during the transaction of liver parenchyma. It can be used alone or in combination with clamp crushing or CUSA. It also have been adopted for laparoscopic resections [50,51] with limitation in the dissection around the main trunk of the hepatic veins [52]. **Figure.22**

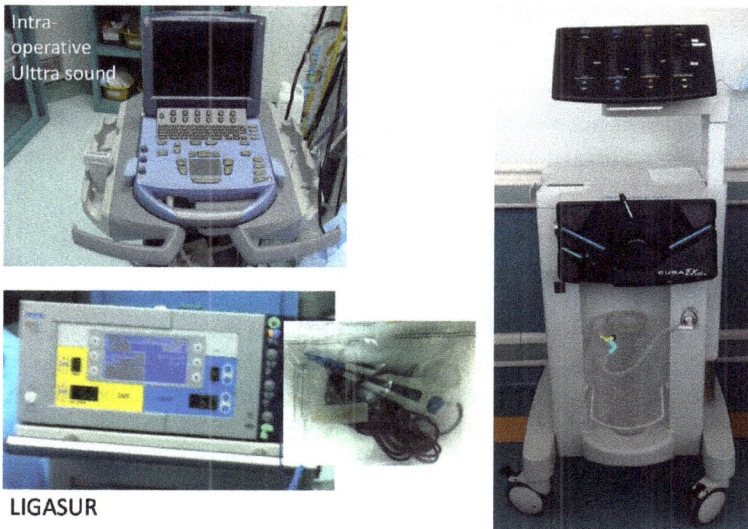

Figure 22. Instruments used for liver resection

Ligasure designed to seal small vessels by a combination of Ultrasonic dissection of liver parenchyma using compression pressure and bipolar radiofrequency (RF) energy [45], it was found to be more useful in laparoscopic resection than open.

Tissue Link dissecting sealer, where saline runs to the tip of the electrode to couple RF energy to the liver surface and achieve coagulation [45].

All these instruments have been used and according to many authors each has been claimed to be better than the other. Our believe is that a surgeon should be familiar with all techniques and instruments as each hospital has its own and when instrument malfunction occurs he will have the ability to adopt and rise up to the situation.

5.4. Specific liver resections

5.4.1. Right hepatectomy

Resection of the right hemiliver (segments 5, 6, 7 & 8) is one of the most common types of liver resection. It involves removing all hepatic parenchyma to the right of the middle hepatic vein [8]. This can be done by the Anterior, Posterior or the hanging techniques described above. However, it is important to see which approach will be better for each patient taking into consideration the tumour and the status of the liver.

This starts with mobilization of the right liver by division of the falciform, coronary and right triangular ligaments. Then vascular inflow and outflow control should take place. Three general approaches have been described for achieving vascular inflow control: 1) extrahepatic dissection within the porta hepatis, with division of the right hepatic artery and right portal vein prior to division of the parenchyma (anterior approach) 2) intrahepatic control of the main right pedicle within the substance of the liver prior to parenchymal transection (Intra-Hepatic ligation); and 3) intrahepatic control of the pedicle after parenchymal transaction (hanging technique or posterior approach) [8].

Then the right hepatic artery, right portal vein and the right hepatic duct are lighted and divided extrahepatic. The right liver is then dissected from the inferior vena cava either before or after according to which approach is being adopted. The short hepatic veins that drain from the right hemi liver to the inferior vena cava should be ligated and divided as well as the Hepato-caval ligament. The right hepatic vein is then dissected extrahepatic and ligated. After this step a clear line of demarcation will appear as the right hemi liver will became darker and ischemic. Liver parenchyma transaction will be done on the right border of the middle hepatic vein. Some vascular anomalies can cause the demarcation line of a right hepatectomy to be along the left border of the middle hepatic vein so care must tacked to preserve segment 4 branches or it will become congested. Blood loss control can be achieved by pringle's maneuver, using of low central pressure or extrahepatic clamping of the middle and left hepatic veins.**Figure.23**

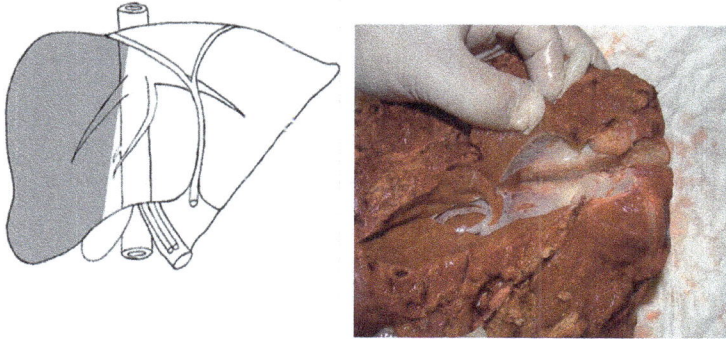

Figure 23. Right Hepatectomy; a right liver specimen with tumor invasion in the right hepatic vein

5.4.2. Extended right hepatectomy (Right trisectionectomy)

Right hepatectomy + extrahepatic ligation and division of the branches of the hepatic artery, portal vein and bile ducts to segment 4 with the division of the right and middle hepatic veins leaving the left hepatic vein and portal triad supplying the left lateral section intact [18].

The left triangular ligament may be preserved to prevent liver rotation and venous outflow occlusion post resection [42].**Figure.24**

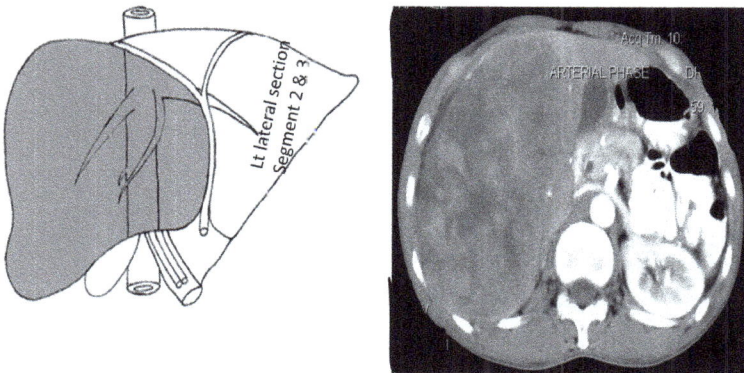

Figure 24. Right extended tri-sectionectomy: a CT scan of a liver tumor that was resected as shown in the drawing

5.4.3. Left hepatectomy

This can be done in the same manner as the right liver resection, however it will require the identification of the left portal triad. Starting with mobilization of the left liver by division of the falciform and the left triangular ligaments. Extrahepatic division of the extrahepatic branches of the left hepatic artery, left portal vein and left hepatic duct. Isolation of the trunk of the middle and left hepatic vein. Parynchymal transaction done along the plane demarcated by the ischemic left liver along a plan on the left side of the middle hepatic vein. The same should be considered as the line of demarcation can be on the right of the middle hepatic vein. The left hepatic vein is ligated intrahepaticly. Blood Loss can be reduced by using Pringle's maneuver plus either low central venous pressure or selective hepatic vascular occlusion by clamping the right hepatic vein. **Figure 25**

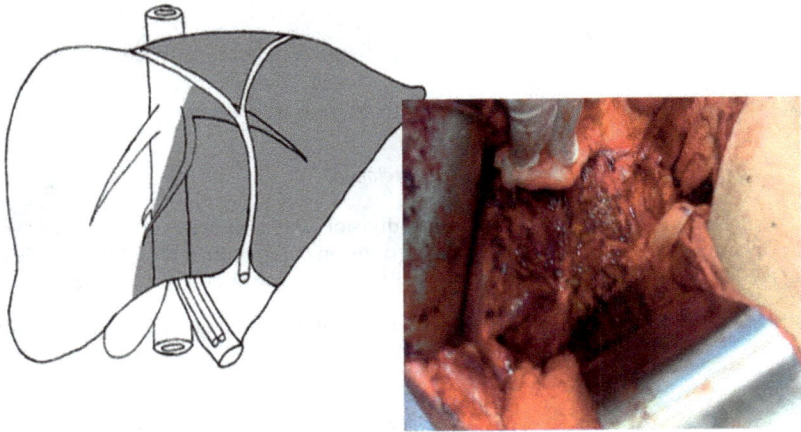

Figure 25. A case of Left Liver resection; the middle hepatic vein seen in the remnant liver with a schematic demonstration

5.4.4. Extended left hepatectomy (Left trisectionectomy)

Similar to left hepatectomy in addition of the right anterior section. Care should be done to preserve the hepatic arterial, portal venous and bile duct branches to the right posterior section and the right hepatic vein. If the right inferior hepatic vein is large it should be preserved so the venous drainage to segment 6 will not be affected [18].

5.4.5. Left lateral sectionectomy

Isolated segment II or III resection is uncommonly performed because of the ease of combined segment II and III (left lateral section) and the small volume of each segment. In the presence of cirrhosis or when multiple segmental resections are performed, isolated resection may be necessary. The left hepatic vein is identified extrahepaticaly and the left lateral sectional portal triad is ligated at the umbilical vein and the falciform ligament.**Figure.26**. Then the hepatic transaction is carried out with very minimal blood lose.

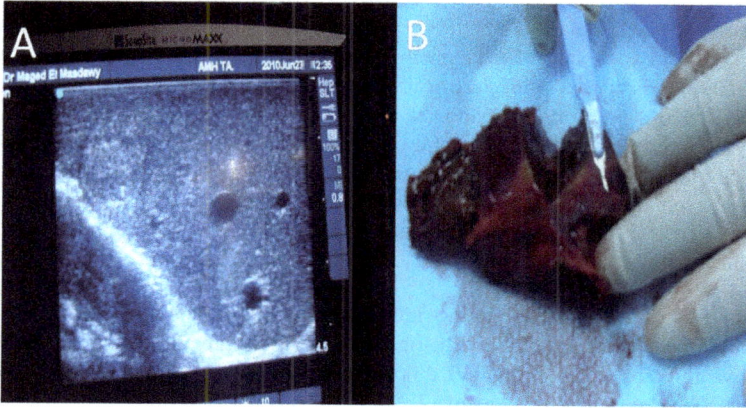

Figure 26. A) left lateral section as seen on intra-operaive ultrasound. B) the specimen with tumour to check for margins

5.4.5. Right posterior sectionectomy (Segment VI and VII)

This can be achieved by most techniques described above depending on the tumour size and the status of the Liver. Full mobilization of the right liver with division of the posterior draining veins. The right portal pedicle is exposed, and the anterior and posterior branches are identified (Hilar plate approach). The posterior pedicle is clamped, and the line of demarcation is evident. The pedicle may be divided, and parenchymal dissection may be performed in standard fashion. The line of transection is horizontal and posterior to the right hepatic vein. However, the right vein may be sacrificed during this procedure since the anterior section will be adequately drained by the middle hepatic vein[8]. If the liver is cirrhotic we would advise the use of an extrahepatic approach like the posterior approach to minimize the blood lose and injury to the right anterior portal triad.

5.4.7. Right anterior sectionectomy (Segment V and VIII)

This is extreamly rare and very difficult because of its location between both the right and middle hepatic veins with the importance of not injuring any of them. This is why if it is done it is combined with segment IV (Central liver resection) to remove the middle hepatic vein and have a safe distance from the right hepatic vein. The approach is similar to the right posterior sectionectomy were the right anterior portal triad is seen and ligated to stop the inflow and get the line of demarcation.

5.4.8. Isolated segment II or III resection

For removal of either segment II or III, the inflow pedicle is ligated, but the main left hepatic vein is preserved because it provides the only venous drainage to the remaining segment. The inflow pedicles to segments II and III branches directly from the umbilical portion of the left main portal vein. To isolate these pedicles, the left lateral section is shifted cephalad using traction on the divided falciform ligament. If present, the parenchymal bridge between segment III and IV is divided with electrocautery. Dissection of the umbilical fissure to the left of the portal vein is performed. Ligation of either segment II or III pedicles demarcates the boundary between them. The left hepatic vein may be clamped to reduce blood loss, but clamping is generally unnecessary if the central venous pressure is low. Liver transection then proceeds in an oblique antero-cranial plane with attention to preserve the left hepatic vein [3].

5.4.9. Isolated segment VII

To expose this segment dissection of the right triangular ligament is necessary. The vascular pedicle of segment VII originate from the right lateral glissoian pedicle and enters the parenchyma in a common trunk at segment VI, this will run deep and divide to two branches anterior to segment VI and posterior to segment VII.

After mobilizing the infindibulum of the gall bladder and dividing the lateral peritoneum of the hepato-duodenal ligament the lateral pedicle can be easily freed as well as the artery. Once this is identified with the bile duct, the right branch of the portal vein is freed. The bile ducts will never be dissected outside the parenchyma but only transparenchymaly at the end of the resection to prevent damage to the adjacent hepatic ducts. Clamping of the arterial branch will lead to blanching of the entire right anterior section. The fissure of the right hepatic vein will indicate the upper resection margin. The vein could be left in place or removed in case of neoplasm infiltration, also isolated resection of segment VII with ligation of the right hepatic vein can be safely performed, venous out flow of segment VI should be insured by preserving the accessory hepatic veins and the right inferior hepatic vein(present in 25%) to prevent the transitory venous congestion of segment VI with hemorrhage from the resection margins after isolated removal of segment VII. After clamping of the lateral glissonian pedicle the trunk of the right hepatic vein will be clamped and divided. Parenchymal dissection will follow the appearing ischemic demarcation line and the dissection plane will start from the top downward between segment VII and VIII. The pedicle will be exposed with the dissection once it have been divided the arterial and portal branches at the hilum can be unclamped, segment VI returns to its normal color and the inferior demarcation line will become evident.

5.4.10. Isolated segment VI

Similar to segment VII, after mobilization of the right liver, ligation of the inferior or accessory suprahepatic vein if present and clamping of the arterial and portal brances which will produce the ischemic demarcation line.

The parenchyma is divided starting from the lower margin of the liver proceeding along an oblique plane from the right to the left and from the front to back. Deep in the parenchyma the lateral pedicle is ligated. The glissonian pedicle is then unclamped at the hilum and segment VII will return to the normal colour, the upper dissection margin will follow the ischemic line between the two segments VI and VII.

5.4.11. Isolated segment IV

Segment IV is divided into two subsegments, IVA and IVB, based on the inflow pedicles. Isolated resection of IVB is usually done in a intra-hepatic ligation method and most often with segment V in cases of gallbladder carcinoma. Were outflow control for segment IV resection is usually not obtained until the liver is divided. After dissection of the hepatoduodenal ligament the left branches of the hepatic artery are identified and then the middle branch is ligated and divided. Dissection will be carried out along the gall bladder-inferior vena cava plane. Glissonian capsule divided above the hilar plate. The portal branch is usually seen with the hilar plate and dissection with control by Bull-dog clamps to see the line of demarcation. At this point segment IV will only be attached to the Middle Hepatic vein which will be transfixed.

5.4.12. Isolated segment I ''Caudate lobe''

This is the least popular liver resection as all the other segments can be done in an intra-hepatic ligation method or in a non-anatomical approach. However, the Caudate liver resection has its own unique location above the inferior vena cava and its own blood supply giving it the excellent challenge for any liver surgeon. There are 5 approaches:

1. Bilateral approach: For isolated caudate lobectomy, the caudate lobe is approached from both right and left side after complete mobilization of the liver with controll of the suprahepatic and intrahepatic inferior vena cava as well as the right hepatic vein and the common trunk of the middle and left hepatic veins. Then the caudate lobe is detached from the inferior vena cava along the anterior surface of the retro-haptic IVC and the short hepatic veins are identified and divided. The hepatogastric ligament is detached from the undersurface of the liver and the fibrous hepatocaval ligament need to be divided to free the spieglian lobe from the IVC and the diaphragm. All short hepatic veins are ligated and divided. So the caudate lobe is free from the inferior vera cava. The branches of the to the para caval portion of the caudate lobe from the right portal vein, right hepatic artery and duct, branches to the spiegelian lobe from the left portal vein, left hepatic artery and duct are ligated and divided. By carefull dissection the liver is detached from the surroundings and the right, middle and left hepatic veins. In this step; 2 important land marks for this dissection : A) the angle between the right hepatic vein and the inferior vans cava i.e the top of the caudate lobe. B) the meeting point between the caudate process and the right liver. An imaginary line joining these two points is considered as the caudate boundary for the liver transection. Meticulous care should be applied not to injure the major vessels or induce bleeding which will be difficult to control.

2. Left sided approach: Similar to the bilateral approach whit the exception that the dissection is mainly from the left side of the liver. In small tumours <3cm, if an isolated partial caudate lobecetomy or left hepatectomy combined with complete caudate lobecetomy is carried out. **Figure 27**

3. Right sided approach: Similar to the bilateral approach whit the exception that the dissection is mainly from the right side of the liver. In thin patients with right hepatectomy combined with caudate lobecetomy.

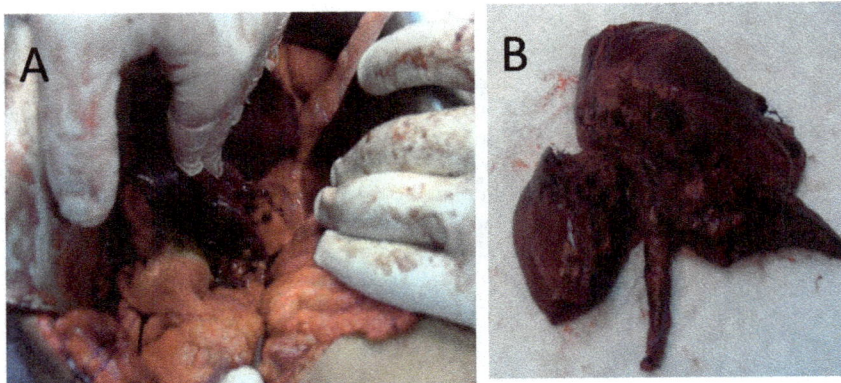

Figure 27. Caudate liver resection, A) the lobe is removed from the IVC and lifted up (left approach). B) The specimen of the caudate with the left liver and the CBD for cholangiocarcinoma

4. Anterior approach: This approach provides a better operative field by opening the mid plane of the liver widely so the major hepatic veins and the Hilar plate will be exposed to direct vision thus will facilitate tumour resection from the main vessels. For tumours >4cm especially when the tumour is located in the paracaval portion or in close contact with the major hepatic veins. With tha same technique of the bilateral approach. After freeing the caudate lobe from the reto-hepatic inferior vena cava, pringle's meneuver is then applied. The liver is transacted through the mid plane starting from the point between the root of the right and middle hepatic veins to the fossa of the gall bladder. This is better done using the hanging technique. When the transection reaches the Hilar plate at the hilum, the portal triade of the caudate lobe is isolated and divided. The caudate lobes then separated from the major hepatic veins in one block with the tumour. After removal of the specimen all bleeding points and bile leak should be controlled individually.

5. Retrograde caudate lobectommy: Used if the tumour is closely adherent to or infiltrating the inferior vena cava, or if the tumour is too large in size to be turned from one side to the other. Mobilization of the liver by the division of all the ligaments, control of the hepatoduodenal ligament, suprahepatic and intrahepatic inferior vena cava for possible occlusion if necessary. The liver is transected along the mid plane 1cm from the tumour, the hepatic veins are exposed under direct vision and carefully dissected from the specimen, ligation and division of the caudate portal triad from the right/left hepatic arteries and veins. In combined Left/right hepatectomy with caudate lobecetomy the hepatic pedicel can be transected accordingly. The specimen will be attached only to the inferior vena cava. The last step here will be the division of the short hepatic veins, and if the tumour is attached to the IVC part of it could be resected with the tumour and then it'll be repaired or reconstructed.

5.4.13. Central liver resection

Segments IV, V, and VIII (also known as mesohepatectomy) is rarely performed. This resection involves ligation of inflow vessels from both the right and left portal pedicles. The resection is performed by combining the techniques of segment IV resection and right anterior sectionectomy. Dissection begins at the hilum and the umbilical fissure with the goal of inflow control. The right anterior sectional pedicle is isolated, as are the segment IV pedicles. The division of the liver parenchyma begins to the right of the umbilical fissure (or within it if the tumor is nearby). **Figure.28.**Care should be given to avoid ligating the left main portal umbilical branch. Dissection is continued upward to the main trunk of middle hepatic vein. The right anterior sectional pedicle is ligated to demarcate the boundary of the liver resection on the right side. Liver transection proceeds in the plane of the right hepatic vein until it meets the left resection plane. At this point, one should be cautious with handling the freely dangling central lobe. Excessive traction may tear the thin-walled middle hepatic vein, resulting in massive hemorrhage. Gently hold the lobe and divide the base of the middle hepatic vein. This procedure removes the gallbladder, central lobe, and middle hepatic vein en bloc, leaving the caudate, right posterior section, and left lateral section intact. The raw liver surface may be covered with a flap of omentum.

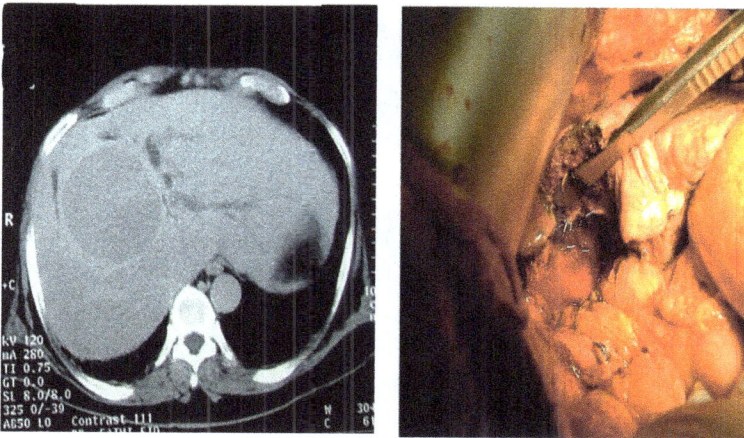

Figure 28. Central liver lesion as seen on CT scan and the same patient intra-operatively after resection

5.5. Control of bleeding

To minimize blood loss from the resected raw liver surface the patient is placed 15 degres in the Trendelenburg position [3]. Low venous pressure is maintained by minimizing fluid infusion and restricting intraoperative blood transfusion unless more than 25% of the blood volume is lost [53,54]. Systolic blood pressure is kept above 90 mm Hg, and intraoperative urine output is maintained at about 25 mL/hour [3].

Dissection and control of the hepatic veins performed prior to parenchymal transaction.[8].

Venous outflow draining is divided after dividing the inflow vessels[8], unless the posterior approach is adopted with a Pringles manoeuvre to prevent liver congestion.

Control of the suprahepatic and intrahepatic inferior vena cava [18], pringle's maneuver[3,18], and mobilization with parenchymal transection performed with a low central venous pressure < 5 mm Hg [3,11] can decrease the bleeding amount significantly.

Author details

O. Al-Jiffry Bilal[1,2] and Khayat H. Samah[2]

1 Surgery, Taif University, Taif, Saudi Arabia

2 Surgery, AlHada Military Hospital, Taif, Saudi Arabia

References

[1] Prognostic impact of anatomical resection for hepatocellular carcinomaKiyoshi Hasegawa, Norihiro Kokudo, Hiroshi Imamura, Yutaka Matsuyama, Masami Minagawa, Keiji Sano, Yasuhiko Sugawara, Tadatoshi Takayama, Masatoshi Makuuchi. Ann surg. (2005). , 242, 252-259.

[2] Eltawil et alDifferentiating the impact of anatomic and non-anatomic liver resection on early recurrence in patients with Hepatocellular Carcinoma. World journal of surgical oncology (2010).

[3] Segment-Oriented approach to liver resectionK.H. Liau, L.H. Blumgart, R.P. dematteo. Surg clin N Am (2004). , 84(2004), 543-561.

[4] Ultrasonically guided subsegmentectomyMakuuchi M, Hasegawa H, Yamaxaki S. Surg Gynecol Obstet. (1986). , 161, 346-359.

[5] Segmental liver resection using ultrasound-guided selective portal vein occlusion-Casting D, Garden J, Bismuth H. Ann Surg. (1989). , 210, 20-23.

[6] Segment-Oriented hepatic resection in the management of Malignant neoplasms Billigsley KGJarnagin WR, Fong Y, Blumgart LH. of the liver. J Am Coll Surg (1998). Nov; , 187(5), 471-81.

[7] Anatomic versus limited nonanatomic Resection for solitary hepatocellular carcinomaTanaka K, Shimada H, Matsumoto K, Nagano Y, Endo I, Togo SSurgery (2008).

[8] Techniques of Hepatic ResectionHoward m. Karpoff, William r. Jarnagin, José melen-dez, Yuman fong, Leslie h. Blumgart. Hepatobiliary Cancer.

[9] Ueno, S, et al. (2008). Efficacy of anatomic resection vs nonanatomic resection for small nodular hepatocellular carcinoma based on gross classification. J Hepatobiliary Pancreat Surg , 15(5), 493-500.

[10] Extent of liver resection influences the outcome in patients with cirrhosis and small hepatocellular carcinomaRegimbeau JM, Kianmanesh R, Farges O, Dondero F, Sauvanet A, Belghiti J. Surgery (2002). Mar;, 131(3), 311-7.

[11] Dematteo, R. P, Palese, C, Jarnagin, W. R, et al. Anatomic segmental hepatic resection is superior to wedge resection as an oncologic operation for colorectal liver metastases. J Gastrointest Surg (2000). , 4, 178-84.

[12] Imamura, H, Matsuyama, Y, Miyagawa, Y, et al. Prognostic significance of anatomical resection and des-_-carboxy prothrombin in patients with hepatocellular carcinoma. *Br J Surg.* (1999). , 86, 1032-1038.

[13] Cucchetti, A, et al. Acomprehensive meta-analysis on outcome of anatomic resection versus nonanatomic resection for hepatocellular carcinoma. Annal of Surgical oncology 27 June (2012).

[14] Segmental liver resection for Colorectal metastasis, Daniel V. Kosov, Georgi L. Kabakov. Gastrointestin Liver Dis.December (2009). , 18(4)

[15] Anatomic segmental resection compared to major hepatectomy in the treatment of liver neoplasmsThomas S. Helling, Benoit Blondeau. HPB,(2005). , 7, 222-225.

[16] Postoperative liver dysfunction and future remnant liver: where is the limit? Results of a prospective studyFerrero A. Vigano L, Polastri R, et al. World J Surg (2007). , 31, 1643-1651.

[17] Expanding criteria for resectability of colorectal liver metastasesPawlik TM, Schulick RD, Choti MA. Oncologist (2008). , 13, 51-64.

[18] Applied anatomy in liver resection and liver transplantW.Y. Lau 978-7-11712-875-9R-12876

[19] Healey JE JrSchriy PC. Anatomy of the biliary ducts within the human liver: analysis of the prevailing pattern of branching and the major variations of the biliary ducts. Arch Surg (1953). , 66, 599-616.

[20] Couinaud, C. Lefoie. Etudes Anatomiques et Chirugicales. Paris:Masson & Cie, (1957).

[21] The brisbane (2000). Terminology of Liver Anatomy and ResectionTerminology committee of the IHPBA, HPB 2000, . 333-9.

[22] The importance of Glisson's capsule and its sheaths in the intrahepatic approach to resection of the liverLaunois B, Jamieson G. Surg Gynecol Obstet (1992). , 174, 7-10.

[23] Kida, H, Uchimura, H, & Okamoto, K. Intrahepatic architecture of bile and portal vein. J Biliary tract and pancreas (1987). , 8, 1-7.

[24] Jamieson, G, & Launois, B. Liver resection and liver transplantation: the anatomy of the liver and associated structures. In :The anatomy of general surgical operations. Ed. Jamieson GG, Ilsevier Edinburgh, (2006). Chapter , 2, 8-23.

[25] Strsberg, S. M. liver terminology and Anatomy. In Hepatobiliary Carcinoma. Editor: W.Y. Lau World scintific Singapore (2007). chapter 2, , 25-50.

[26] The impact of intraoperative ultrasonography on surgery for liver neoplasmsKane R, Hughes L, Qcua E. J Ultrasound Med (1994).

[27] Laparoscopic staging and intraoperative ultrasonography for liver tumor management Ravikumar T. Surg Oncol Clin North Am (1996).

[28] Liver resection by ultrasonic dissection and intraoperative ultrasonographyHanna SS, Nam R, Leonhardt C. HPB Surg (1996). , 9, 121-8.

[29] Intraoperative ultrasonography and other techniques for segmental resectionsTakayama T, Makuuchi M. Surg Oncol Clin North Am (1996). , 5, 261-9.

[30] The use of operative ultrasound as an aid to liver resection in patients with hepatocellular-carcinomaMakuuchi M, Hasegawa H, Yamazaki S, Takayasu K, Moriyama N. World J Surg (1987). , 11, 615-21.

[31] Operative risks of major hepatic resectionsCapussotti L, Polastri R. Hepatogastroenterology (1998). , 45, 184-90.

[32] Weber, J. C, Navarra, G, Jiao, L. R, Nicholls, J. P, Jensen, S. L, & Habib, N. A. New technique for liver resection using heat coagulative necrosis. Ann Surg (2002).

[33] Navarra, G, Lorenzini, C, Curro, G, Basaglia, E, & Habib, N. H. Early results after radiofrequency-assisted liver resection. Tumori (2004).

[34] Tepel, J, Klomp, H. J, Habib, N, Fandrich, F, & Kremer, B. Modification of the liver resection technique with radiofrequency coagulation. Chirurg (2004).

[35] Radiomorphology of the Habib Sealer-Induced Resection Plane during Long-Time Followup: A Longitudinal Single Centere Experience after 64 Radiofrequency-Assisted Liver ResectionsRobert Kleinert, RogerWahba, Christoph Bangard, Klaus Prenzel,Arnulf H. H°olscher,1, 2 and Dirk Stippel.

[36] Radiofrequency ablation-assisted liver resection: review of the literature and our experiencePeng Yao & David L. Morris. HPB, (2006).

[37] Methods of vascular control technique during liver resection: a comprehensive review, Wan-Yee Lau, Eric C. H. Lai and Stephanie H. Y. Lau. Hepatobiliary Pancreat Dis Int,October 15,(2010). (5)

[38] Delva, E, Barberousse, J. P, Nordlinger, B, Ollivier, J. M, Vacher, B, Guilmet, C, et al. Hemodynamic and biochemical monitoring during major liver resection with use of hepatic vascular exclusion. Surgery (1984).

[39] Belghiti, J, Noun, R, Zante, E, Ballet, T, & Sauvanet, A. Portal triad clamping or hepatic vascular exclusion for major liver resection. A controlled study. Ann Surg (1996). , 224, 155-61.

[40] Huguet, C, Gavelli, A, Chieco, P. A, Bona, S, Harb, J, Joseph, J. M, et al. Liver ischemia for hepatic resection: where is the limit? Surgery (1992). , 111, 251-9.

[41] Emond, J, Wachs, M. E, Renz, J. F, Kelley, S, Harris, H, Roberts, J. P, et al. Total vascular exclusion for major hepatectomy in patients with abnormal liver parenchyma. Arch Surg (1995). discussion 830-1., 130, 824-30.

[42] A review of techniques for liver resection. AG Heriot, ND Karanjia. Ann R Coll Surg Engl 2002; 84: 371-380.

[43] One hundred consecutive hepatic resectionsBlood loss, transfusion, and operative technique. Cunningham JD, Fong Y, Shriver C, Melendez J, Marx WL, Blumgart LH. Arch Surg (1994). , 129, 1050-6.

[44] A simplified technique for hepatic resection. Lin T. Ann Surg 1974; 180: 225-9.

[45] Current techniques of liver transactionRONNIE T.P. POON. HPB, (2007).

[46] Cavitron ultrasonic surgical aspirator (CUSA) in liver resectionFasulo F, Giori A, Fissi S, Bozzetti F, Doci R, Gennari L. Int Surg (1992). , 77, 64-6.

[47] Resection of colorectal liver metastasesScheele J, Stang R, Altendorf-Hofmann A, Paul M. World J Surg (1995). , 19, 59-71.

[48] New water-jet dissector: initial experience in hepatic surgery Baer HUMaddern GJ, Blumgart LH. [published erratum appears in Br J Surg 1994; 81: 1103]. Br J Surg (1991). , 78, 502-3.

[49] A comparison of different techniques for liver resection: blunt dissection, ultrasonic aspirator and jet-cutter. Rau HG, Schardey FM, Buttler E, Reuter C, Cohnert TU, Schildberg FW.Eur J Surg Oncol 1995; 21: 183-7.

[50] Experience with ultrasound scissors and blades (UltraCision) in open and laparoscopic liver resectionSchmidbauer S, Hallfeldt KK, Sitzmann G, Kantelhardt T, Trupka A Ann Surg (2002).

[51] Cherqui, D, Husson, E, Hammoud, R, Malassagne, B, Stephan, F, Bensaid, S, et al. Laparoscopic liver resections: a feasibility study in 30 patients. Ann Surg (2000).

[52] Hepatic resection using the harmonic scalpelSugo H, Mikami Y, Matsumoto F, Tsu-mura H, Watanabe Y, Kojima K, et al. Surg Today (2000).

[53] Perioperative outcomes of major hepatic resections under low central venous pres-sure anesthesia: blood transfusionand the risk of postoperative renal dysfunction. Melendez JA, Arslan V, Fischer ME, Wuest D, Jarnagin W, Fong Y, et al. JAmColl Surg (1998). , 187, 620-5.

[54] Recent advances in hepatic resectionDeMatteo RP, Fong YM, Jarnagin WR, Blumgart LH. Semin Surg Oncol (2000). , 19, 200-7.

Chapter 11

Right Anterior Sectionectomy for Hepatocellular Carcinoma

Hiromichi Ishii, Shimpei Ogino, Koki Ikemoto,
Kenichi Takemoto, Atsushi Toma,
Kenji Nakamura and Tsuyoshi Itoh

Additional information is available at the end of the chapter

1. Introduction

Hepatectomy is an established first-line therapeutic option for hepatocellular carcinoma. Because there is high likelihood of cancer cells from hepatocellular carcinoma spreading throughout the portal venous system, anatomical hepatectomy is effective for eradication of the intrahepatic metastases of hepatocellular carcinoma [1, 2].

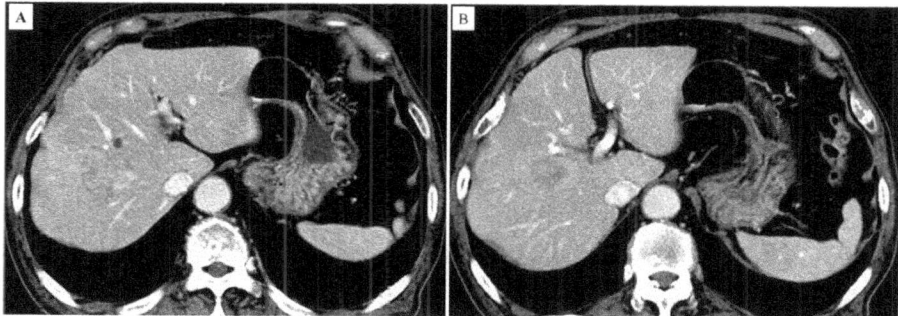

Figure 1. The hepatocellular carcinoma is located in segment 8 of the liver (A) and close to the root of the right anterior Glissonean pedicle (B).

For patients with hepatocellular carcinomalocated in the right anterior section or close to the root of the right anterior Glissonean pedicle (Figure 1A, 1B), right anterior sectionectomy has an important advantage, i.e., preservation of nontumorous parenchyma, over conventional hemihepatectomy. Although right anterior sectionectomy is a difficult hepatic resection because of the danger of intraoperative bleeding from the middle and right hepatic veins and risk factor of postoperative bile leakage [3, 4], this surgical procedure is safe and effective in selected patients [5, 6]. Laparoscopic mesohepatectomy is performed at limited institutions [7, 8]; however, the use of this procedure is limited and controversial to date because of the high degree of proficiency required. Herein, we describe techniques of right anterior sectionectomy using theGlissonean pedicle transection method via a conventional open laparotomy approach. The Brisbane 2000 terminology of liver anatomy andresections is used in this manuscript.

2. Surgical technique

Laparotomy is performed through an upper midline incision with right lateral subcostal extension (reversed L-shaped incision). The xiphoid process is excised, the round ligament is ligated and divided, and the falciform ligament is divided along the surface of the liver. We routinely conduct an intraoperative ultrasonography for hepatectomy to define the tumor location and vessels to be manipulated for resection.The right hemiliver is mobilized by dividing the coronary and right triangular ligaments; however, the right adrenal gland is not dissected from the right hemiliver. The ventral surfaces of the root ofthe right and middle hepatic veins are exposed. A cholecystectomy is performed and a 4-Fr. biliary tube is inserted through the cystic duct for a bile leakage test after removing the specimen.

The hepatoduodenal ligament is encircled and taped. The peritoneum of the hepatoduodenal ligament is dissected at the ventral and dorsal sides of the hepatic hilum, the hilar plate is detached blindly and bluntly from the liver parenchyma, and then, the right Glissonean pedicle is encircled extrahepatically using Kelly forceps.To avoid injury to the elements of the caudate lobe, the right Glissonean pedicle should be encircled on the right side of the caudate process branch. After the cystic plate is dissected, the right anterior Glissonean pedicle is identified and encircled extrahepatically[9, 10] (Figure 2). If a large liver tumor is located near the root of the right Glissonean pedicle, it is difficult to approach the Glissonean pedicle extrahepatically; therefore, the anterior branches of the right hepatic artery and right portal vein are encircled separately [11].

After the right anterior Glissonean pedicle is clamped,discoloration of the right anterior section is confirmed, and the demarcation line is then marked by electrocautery (Figure 3).

Using the Pringle maneuver, a parenchymal dissection between the left medial and right anterior sections is performed along the demarcation line from the caudal towardthe cranial direction using an ultrasonic surgical aspirator and the right side of the middle hepatic vein is exposed on the raw surface of the liver. The branches of the middle hepatic vein originating from the anterior section are ligated and divided, and the thick branches should be clamped with vascular clamp forceps, divided and sewn with a continuous suture (Figure 4). At the cranial and caudal ends of the parenchymal dissection, the right side of the middle hepatic vein root and the left side of the right anterior Glissonean pedicle are identified, respectively. The dorsal end point of the parenchymal dissection is the line which connects the root of the middle hepatic vein and the hilar plate.

Using right hemihepatic vascular occlusion [12], a parenchymal dissection between the right anterior and posterior sections is performed along the demarcation line from the caudal toward the cranial direction using anultrasonic surgical aspirator and the left side of the right hepatic vein is exposed on the raw surface of the liver. After the parenchymal dissection is progressed toward the right anterior Glissonean pedicle, the anterior Glissonean pedicle is exposed as distally as possible to avoid biliary injury of the right posterior section (Figure 5) and divided using the stapler or double transfixing sutures (Figure 6). At the cranial end of parenchymal dissection, the left side of the right hepatic vein root is identified.

By retracting the anterior section upward, the parenchymal dissection between the right anterior section and caudate lobe is advanced from the caudal to the cranial direction (Figure 7). Then, the right anterior section is removed (Figure 8).

Hemostasis of the raw surface of the liver is confirmed and the bile leakage test performed. Then, the biliary tube is extracted, and the stump of the cystic duct is ligated.

A closed drain is placed in the raw surface of the liver.

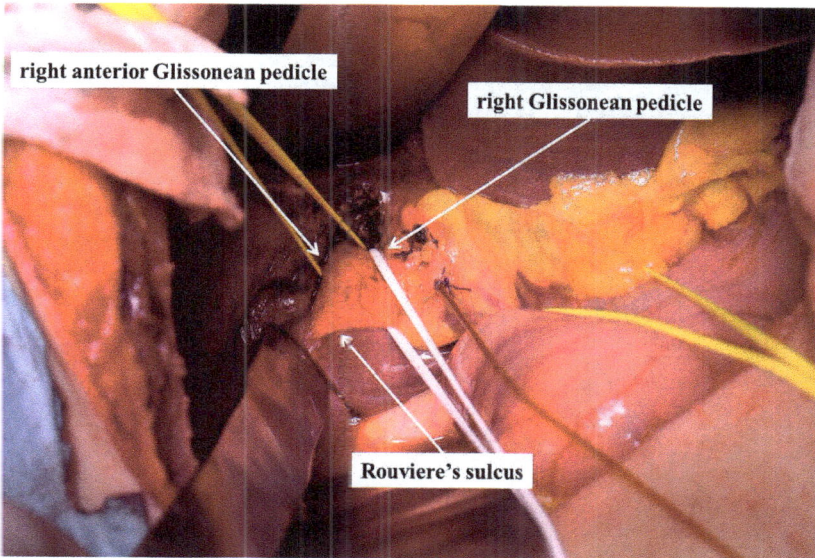

Figure 2. The right and right anterior Glissonean pedicles are encircled extrahepatically.

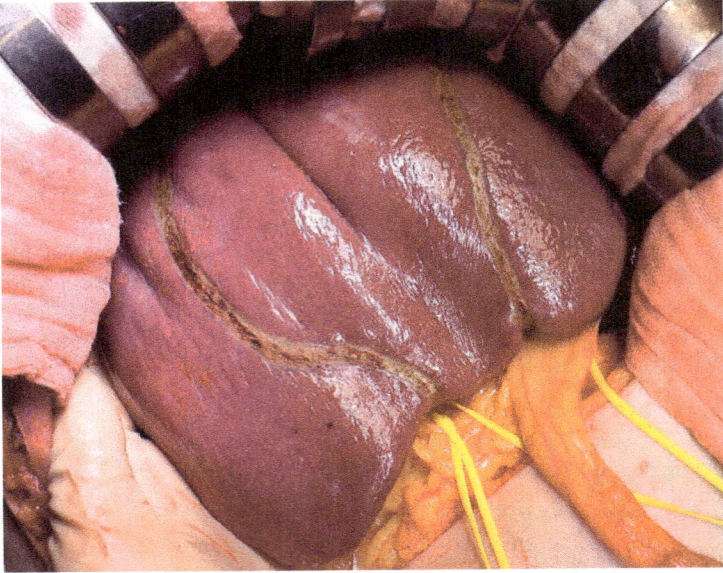

Figure 3. The right anterior section is marked by electrocautery.

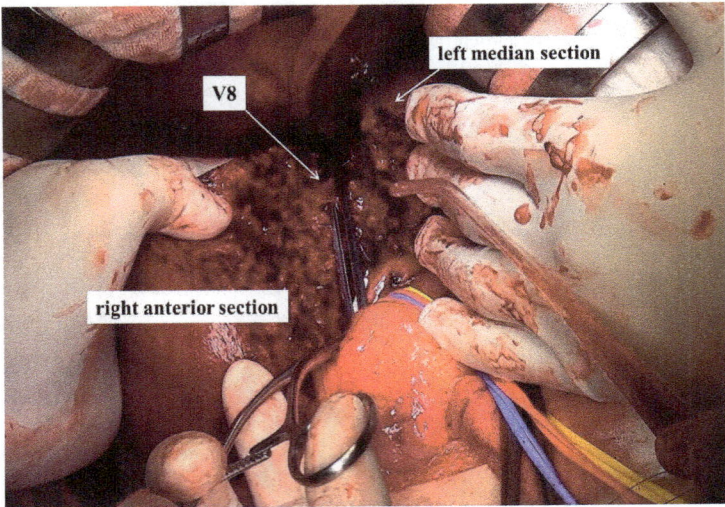

Figure 4. The branch of the middle hepatic vein originating from the anterior section (V8) is clamped with vascular clamp forceps.

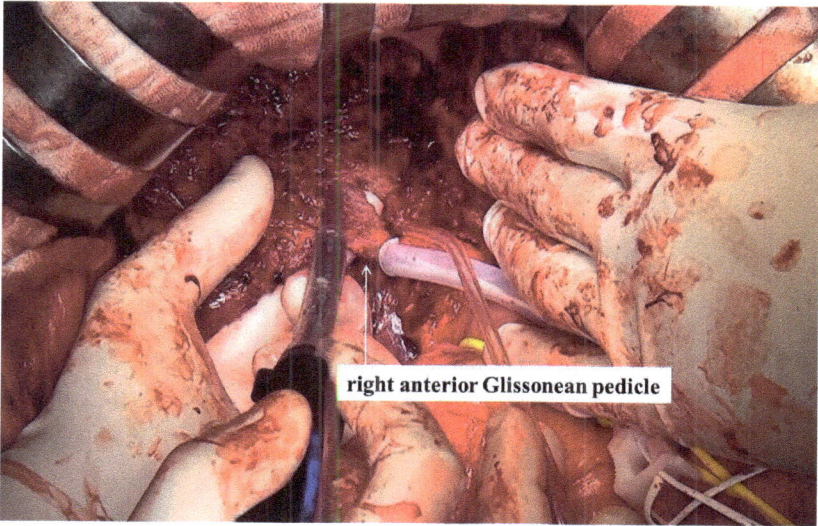

Figure 5. The anterior Glissonean pedicle is exposed as distally as possible.

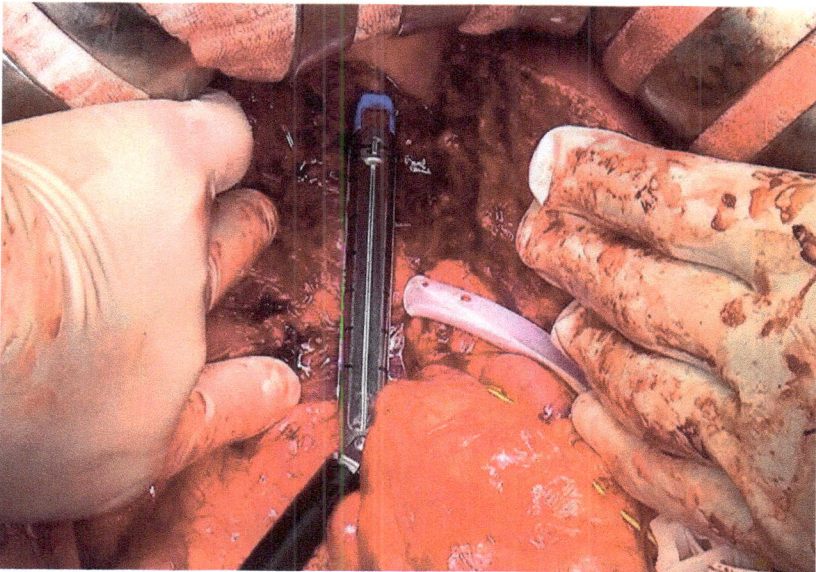

Figure 6. The anterior Glissonean pedicle is divided using the stapler.

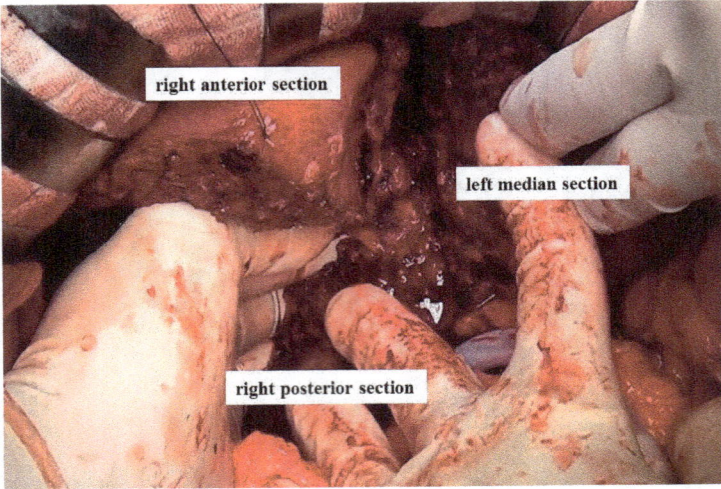

Figure 7. By retracting the anterior section upward, the parenchymal dissection between the right anterior section and caudate lobe is advanced from the caudal to the cranial direction.

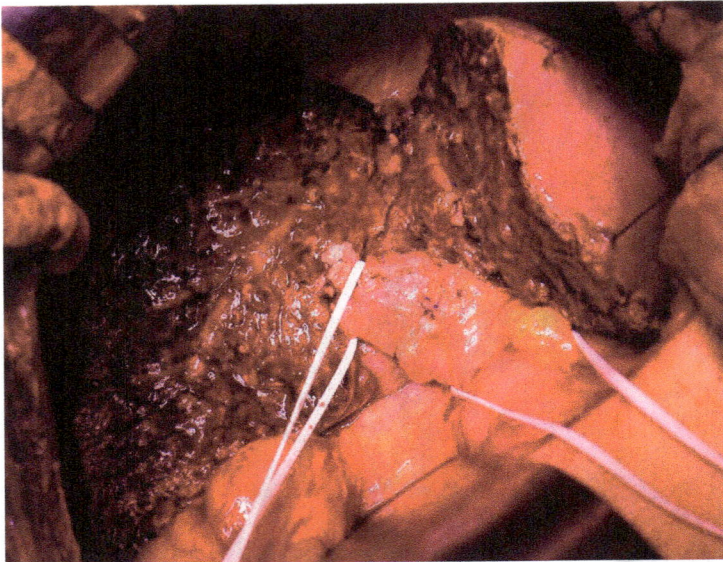

Figure 8. After the right anterior section is removed, the right side of the middle hepatic vein and the left side of the right hepatic vein are exposed on the raw surface of the liver.

3. Comments

Between April 2010 and May 2012, 8 patients underwent a right anterior sectionectomyusing the Glissonean pedicle transection method for hepatocellular carcinoma at our institution. The median surgical time was 323 minutes (range: 227-468 minutes)and the median surgical blood loss was 830.5 ml (range: 180-2009 ml). There was one postoperative complication, i.e., bile leakage, and no mortality.

TheextrahepaticGlissonean pedicle approach is preferable to avoid postoperative lymphatic leakage than separately dividing the arterial and portal branches of the right anterior section. It is important to divide the right anterior Glissonean pedicle as distally as possibleto avoid biliary injury of the right posterior section.

Author details

Hiromichi Ishii[1*], Shimpei Ogino[1], Koki Ikemoto[1], Kenichi Takemoto[1], Atsushi Toma[1], Kenji Nakamura[1] and Tsuyoshi Itoh[1]

*Address all correspondence to: ishii0512h@yahoo.co.jp

1 Division of Surgery, Kyoto Prefectural Yosanoumi Hospital, Japan

References

[1] Hasegawa, K., Kokudo, N., Imamura, H., Matsuyama, Y., Aoki, T., Minagawa, M., Sano, K., Sugawara, Y., Takayama, T., & Makuuchi, M. (2005). Prognostic impact of anatomical resection for hepatocellular carcinoma. *Ann Surg*, 242(2), 252-9.

[2] Arii, S., Tanaka, S., Mitsunori, Y., Nakamura, N., Kudo, A., Noguchi, N., & Irie, T. (2010). Surgical strategies for hepatocellular carcinoma with special reference to anatomical hepatic resection and intraoperative contrast-enhanced ultrasonography. *Oncology*, 78(1), 125-30.

[3] Yamashita, Y., Hamatsu, T., Rikimaru, T., Tanaka, S., Shirabe, K., Shimada, M., & Sugimachi, K. (2001). Bile leakage after hepatic resection. *Ann Surg*, 233(1), 45-50.

[4] Hayashi, M., Hirokawa, F., Miyamoto, Y., Asakuma, M., Shimizu, T., Komeda, K., Inoue, Y., Arisaka, Y., Masuda, D., & Tanigawa, N. (2010). Clinical risk factors for postoperative bile leakage after liver resection. *IntSurg*, 95(3), 232-8.

[5] Hu, R. H., Lee, P. H., Chang, Y. C., Ho, M. C., & Yu, S. C. (2003). Treatment of centrally located hepatocellular carcinoma with central hepatectomy. *Surgery*, 133(3), 251-6.

[6] Kim, K. H., Kim, H. S., Lee, Y. J., Park, K. M., Hwang, S., Ahn, C. S., Moon, D. B., Ha, T. Y., Kim, Y. D., Kim, K. K., Song, K. W., Choi, S. T., Kim, D. S., Jung, D. H., & Lee, S. G. (2006). Clinical analysis of right anterior segmentectomy for hepatic malignancy. *Hepatogastroenterology*, 53(72), 836-9.

[7] Nitta, H., Sasaki, A., Fujita, T., Itabashi, H., Hoshikawa, K., Takahara, T., Takahashi, M., Nishizuka, S., & Wakabayashi, G. (2010). Laparoscopy-assisted major liver resections employing a hanging technique: the original procedure. *Ann Surg*, 251(3), 450-3.

[8] Machado, M. A., & Kalil, A. N. (2011). Glissonian approach for laparoscopic mesohepatectomy. *SurgEndosc*, 25(6), 2020-2.

[9] Couinaud, C. (1985). A simplified method for controlled left hepatectomy. *Surgery*, 97(3), 358-61.

[10] Takasaki, K. (1998). Glissonean pedicle transection method for hepatic resection: a new concept of liver segmentation. *J HepatobiliaryPancreatSurg*, 5(3), 286-91.

[11] Makuuchi, M., Hashikura, Y., Kawasaki, S., Tan, D., Kosuge, T., & Takayama, T. (1993). Personal experience of right anterior segmentectomy (segments V and VIII) for hepatic malignancies. *Surgery*, 114(1), 52-8.

[12] Makuuchi, M., Mori, T., Gunven, P., Yamazaki, S., & Hasegawa, H. (1987). Safety of hemihepatic vascular occlusion during resection of the liver. *SurgGynecolObstet*, 164(2), 155-8.

Chapter 12

Benign Hepatic Neoplasms

Ronald S. Chamberlain and Kim Oelhafen

Additional information is available at the end of the chapter

1. Introduction

Historically benign liver tumors were encountered incidentally during laparotomy or more recently during laparoscopy at which time definitive histological diagnosis can be established. However, with the utilization of advanced imaging modalities hepatic neoplasms have been increasingly identified, with a prevalence rate of up to 50% reported among the general population [1]. Among these incidental lesions, 83% were characterized as benign neoplasms, as outlined in Table 1 [1-3]. Benign hepatic neoplasms represent a diverse group of tumors that develop from either epithelial or mesenchymal cell lines (Table 2), and while the frequency of such lesions is not well documented, more than 50% are classified as hemangiomas [1]. Focal nodular hyperplasia (FNH) and hepatic adenomas represent the next most frequently diagnosed benign tumors. A variety of additional exceedingly rare benign lesions have also been described most of which are sufficiently infrequent enough to be classified as "fascinomas" [1].

Neoplasm	Relative frequency
Hemangioma	52%
Focal nodular hyperplasia	11%
Metastatic tumor (T_xN_xM1)	11%
Hepatocellular adenoma	8%
Focal fatty infiltration	8%
Hepatocellular carcinoma	6%
Extrahepatic process (eg., abscess, adrenal tumor)	3%
Other benign hepatic process	1%

Table 1. Diagnostic frequency of incidentally identified solid liver neoplasms[1,2,9]

Cell of origin	Tumors
Epithelial	
Hepatocellular	Focal nodular hyperplasia (FNH)
	Hepatocellular adenoma (HA)
	Regenerative nodule
Cholangiocellular	Biliary adenoma
	Biliary cystadenoma
Other	Epitheliod leiomyoma
Mesenchymal	
Endothelial	Hemangioma
	Cavernous
	Capillary
	Hemangioendothelioma
	Adult
	Infantile
Mesothelial	Solitary fibrous tumor
	Benign mesothelioma
	Fibroma
Adipocyte	Lipoma
	Myelolipoma
	Angiomyelipoma
Miscellaneous	
Tumors	Biliary hamartoma

Table 2. Benign solid liver neoplasms[1,9]

Most benign tumors are asymptomatic which makes standardizing the work-up difficult. The evaluation of incidental solid hepatic tumors should be individualized based upon the patient's age, sex, past medical history, medications, and associated clinical signs. Although physical examination of the abdomen is typically unremarkable it may rarely reveal localized tenderness and/or a palpable mass. Liver function tests are indicated though are seldom abnormal in asymptomatic patients. Additional laboratory testing such as alpha-fetoprotein (AFP), carcinoembryonic antigen (CEA), carbohydrate antigen (CA) 19-9 and, lactate dehydrogenases may also be ordered depending on the clinical scenario.

Substantial advancements and the widespread availability and use of modern imaging modalities to diagnose and treat abdominal pain, has led to a marked increase in the identification of benign liver tumors. A full discussion of the advantages and disadvantages of individual imaging techniques is beyond the scope of this chapter but is outlined in Table 3. Briefly, B-mode ultrasonography (US) can effectively differentiate cystic and solid neoplasms and is usually the initial study of choice [4,5]. Contrast-enhanced computed tomography (CT) provides greater sensitivity than US for determination of lesion number, size, and location [5, 6]. Magnetic resonance imaging (MRI) represents the most sensitive and specific study to discriminate between various benign liver lesions, particularly when contrast agents are used [5-7]. Finally, fluorodeoxyglucose positron emission tomography ([18]FDG-PET) can aid in the differentiation of benign versus malignant tumors based on the metabolic activity of the lesion [8]. Although modern imaging techniques can precisely diagnose the vast majority of incidental benign tumors, laparoscopic or open biopsy is necessary to exclude malignancy when precise diagnosis remains elusive.

Tumor	US	CT	MRI	Tc 99 RBC scan	Tc 99 SC scan
Hemangioma	Hyperechoic Well-demarcated Increased vascular flow Central venous pooling	Highly sensitive **Non-contrast** Isodense **Contrast:** Hypoense Irregular peripheral enhancement with delayed central filling	Highly sensitive Isodense on T1 Hyperdense on T2 Gadolinium enhanced scan shows similar findings to contrast CT	Blood pooling of radionucleotide	Not indicated
Focal Nodular Hyperplasia (FNH)	Non-specific Hyperechoic	Highly specific **Non-contrast** Isodense **Contrast:** isodense and well-demarcated with central scar	Highly specific Isodense of T1 & T2 Early hyperdense after gadolinium	Not indicated	Takes up Tc99 Tc99 SC contains bile ducts and Kupffer cells
Hepatic Adenoma	Non-specific Hyperechoic Increased blood flow on duplex scanning	Non-specific **Non-contrast** hypo to isodense **Contrast:** isodense with peripheral enhancement with subsequent centripetal flow	Non-specific Iso- hpointense T1 & T2 Uniform enhancement after gadolinium	Not indicated	Generally does not take Tc99 SC because of the lack of bile ducts and Kupffer cells

US = ultrasonography; CT = computed tomography; MRI = magnetic resonance imaging; T1 = T1-weighted MRI; T2 = T2-weighted MRI; Tc-99m RBC = technetium-99m-labeled red blood cell; Tc-99m SC = technetium-99m sulfur colloid.

Tab e 3. Radiographic appearance of benign liver neoplasms[1,9]

Accurate diagnosis is essential to the appropriate management of hepatic neoplasms. Although patients may require surgical intervention for diagnostic purposes, few benign tumors require surgical management for symptomatic relief. As such, surgical intervention for benign tumors is primarily indicated (1) for definitive diagnosis when imaging is inconclusive, (2) to prevent malignant transformation, such as in the case of hepatic adenoma, (3) to reduce the risk of rupture and, (4) for the treatment of rare life-threatening complications as a result of rupture or haemorrhage [9].

2. Hemangioma

Hemangioma is the most common benign mesenchymal neoplasm of the liver and occurs in two variants, capillary and cavernous. Hepatic hemangiomas are identified in 0.4% to 20% of all imaging studies preformed [10-14]. Hemangiomas are frequently discovered incidentally on autopsy studies with 60%-80% identified in individuals in their 4th- 6th decade of life [12-16].The precise etiology of hemangiomas is poorly understood but they are generally considered to be benign congenital hamartomas composed of disorganized venous vasculature separated by intervening fibrous tissue [17]. Hemangiomas vary greatly in size from a few millimeters to over 50 cm, with the majority (up to 80%) less than 4 cm [1,12,18]. Although most commonly solitary, up to 40% of patients with hemangiomas have multiple tumors [19].

Capillary hemangiomas are more prevalent than are cavernous hemangiomas [1,20]. However, these hypervascular lesions are typically small (2 cm) and are rarely clinically significant [1]. As such, the management of capillary hemangiomas requires the exclusion of malignancy and patient reassurance that routine surveillance is not necessary in the absence of symptoms [9].

Cavernous hemangiomas are far more often clinically relevant than capillary hemangiomas. The incidence of cavernous hemangiomas is 3 times greater among women than men, with a mean age of 45 years [12,16]. Whether this reflects a true increase in incidence or a result of more frequent imaging amongst females remains unclear as evident by one autopsy series in which there was a nearly equal sex incidence [1,21]. Although no link between oral contraceptive pill (OCP) use and hemangioma incidence has been established, early studies suggest a link between OCP use and increased hemangioma size at initial presentation [18].

3. Clinical presentation

The most frequently reported symptoms of liver hemangiomas include abdominal pain, nausea, vomiting, early satiety, and prolonged fever [1,22]. Most symptoms of hepatic hemangioma are attributable to rapid expansion, thrombosis, or infarction, resulting in inflammation or stretching of Glisson's capsule [1]. Large hemangiomas (> 10 cm) may occasionally present as a non-tender palpable mass in the right upper quadrant, however physical exam more often reveals only vague abdominal tenderness without a mass [1,23]. Occasionally, a bruit maybe detected over the liver. Evidence of intratumoral or intraperitoneal rupture may be reflected by hemoperitoneum and subsequent shock, which requires emergent surgical intervention. Rarely biliary colic, obstructive jaundice, gastric obstruction, torsion of a pedunculated lesion, pulmonary embolism, spontaneous intraperitoneal hemorrhage, and consumptive coagulopathy have been reported [22,24,25]. Kasabach-Merritt syndrome, which was originally used to describe thrombocytopenia and afibrinogenemia associated with hemangiomas on the skin and spleen of infants, is frequently used to define hepatic hemangioma patients with severe thrombocytopenia and concomitant consumptive coagulopathy [26].

4. Pathology

Hemangiomas are typically well demarcated from surrounding hepatic tissue, which often permits surgical enucleation [27]. In tumors not well demarcated, the tumor-parenchymal interface defines the ease with which enucleation versus formal resection is required. Four interface variants between the hemangioma and hepatic parenchyma have been described. The "fibrolamellar" interface is characterized by a capsule-like fibrous ring of various thickness and is the most common [9]. The involved veins parallel the periphery of the hemangioma or traverse the fibrous lamella. The healthy hepatic parenchyma is often atrophic and a plane between the hemangioma and uninvolved liver tissue is well defined. A second variant, the "compression" interface consists of a hemangioma in which the periphery of the neoplasm is well demarcated despite the absence of a fibrous lamella [1]. An "interdigiting" pattern lacks a fibrous lamella and instead is replaced by an ill-defined plane between the vascular channels of the hemangioma and uninvolved hepatic parenchyma [1]. Finally, an "irregular" or "spongy" interface occur when the hemangioma appears to intercalate into the surrounding hepatic parenchyma [1]. Despite the invasive appearance of this variant, hemangiomas do not possess any malignant potential.

The diagnosis of cavernous hemangioma is generally easy to establish with modern imaging techniques. However, in some instances atypical hemangiomas may be confused for other pathology, including but not limited to, hemorrhagic telangiectasia (Osler-Rendu-Weber), hemangioendothelioma, and peliosis hepatis [9]. When diagnosis remains unclear, indeterminate lesions should be managed surgically as percutaneous biopsy may result in uncontrollable hemorrhage [1].

5. Radiographic evaluation

Accurate radiographic diagnosis of hepatic hemangioma is essential since once definitive diagnosis is established no additional intervention is typically required [9]. Radiographic evaluation is largely dictated by clinical presentation as most hemangiomas are discovered incidentally on imaging studies completed for unrelated symptomology and/or pathology. Depending on the initial degree of diagnostic certainty additional imaging maybe superfluous.

B-mode ultrasonography is typically the initial imaging study performed [1]. On US hemangiomas appear as a homogenous hyperechoic mass that is well demarcated from surrounding liver parenchyma [1,28,29]. The addition of duplex US provides additional information regarding peripheral blood flow and central pooling of venous blood [1,28]. As malignant lesions may demonstrate similar acoustic patterns, additional imaging modalities are often required for definitive confirmation. On contrast enhanced compute tomography (CE-CT) hemangiomas initially appear as hypodense masses with a pattern of irregular peripheral nodular enhancement following initial injection of contrast [30,31]. Delayed venous images subsequently demonstrate characteristic central venous filling of the hypodense mass [30,31]. Magnetic resonance imagining (MRI), though rarely needed for diagnosis of most hemangiomas, is the most sensitive and specific modality for the detection and diagnosis of hemangioma [6,32]. T-1 weighted images reveal a smooth well-demarcated homogenous isodense mass, whereas T-2 weighted studies demonstrate a hyperdense pattern [33,34]. The administration of intravenous gadolinium diethylenetriaminepentaacetic acid (Gd-DTPA) contrast results in the pathognomonic pattern of peripheral nodular enhancement with central filling on delayed images [1,35,36]. This enhancement pattern is typical of most hemangiomas > 2 cm [37]. Hemangiomas < 2 cm may demonstrate rapid uniform enhancement which is indistinguishable from hypervascular hepatocellular carcinoma (HCC) [37]. ^{18}F-FDG PET scan may be useful for differentiation between benign and malignant hepatic tumors [38]. Studies have shown that the activity of both glucose-6-phosphatase and glucose transporters are increased in HCC resulting in decreased uptake of ^{18}F-FDG in hemangiomas as compared to HCC [8]. Historically, technetium-99 labeled red blood cells scintigram (Tc-99 RBC scan) was the gold standard for the diagnostic evaluation of hemangiomas, but technological advancements in axial imaging has led to a decline in the reliance on RBC scintigraphy [31,39]. Finally, selective hepatic angiography typically yields a characteristic neovascular "corkscrewing" appearance with rapid central filling from the neovascular periphery described as "cottonwool" [1]. Despite these characteristic findings, the high diagnostic yield of less invasive modalities makes arteriography rarely necessary.

6. Diagnosis & treatment

The majority of hemangiomas are asymptomatic, particularly those lesions < 1.5 cm in size [1]. Although hemangiomas can grow to great sizes, they generally do not compromise liver function and as such liver function tests are often normal. In rare instances thrombosis or intraparenchymal hemorrhage may occur acutely affecting liver function tests. Spontaneous rupture of hepatic hemangiomas is an exceptionally rare event with a review of the literature revealing less than 30 cases of spontaneous rupture since 1898. Given the low yet significant risk of bleeding, fine needle aspiration (FNA) should be avoided [1]. As a rule, biopsy is only indicated if a histologic diagnosis is unclear or will alter planned treatment, thus in the absence of clinical symptoms the most appropriate treatment strategy is careful observation [1].

Surgical resection should be considered in patients with disabling pressure or pain suggestive of extrinsic compression of adjacent structures, in those experiencing acute symptoms related to rupture, or when malignancy cannot be ruled out [22,40]. In general clinical symptoms increase concurrently with tumor size, with most symptomatic tumors having a mean size of 10 ± 8 cm as compared with 6.8 ± 5.8 cm for asymptomatic lesions [41].

Surgical intervention should be approached no differently than for treatment of other hepatic tumors. It is essential that surgeons possess an extensive knowledge of the anatomy and vascular supply of the liver. The extent of hepatic resection required is directly related to the anatomic location of the lesion and its proximity to surrounding vasculature. Thus, the location of the lesion will largely dictate the operative approach hence a full evaluation of the tumor's extent is critical. Large central lesions which border the inferior vena cava, hepatic outflow tract, or the portal vein, may pose an exorbitant surgical risk and as such may not allow for resection [1].

While enucleation is often indicated, formal resection is required in certain instances. Recall it is the histological features of the tumor-parenchymal interface which defines how easily a parenchymal-sparing technique may be utilized. Unlike malignant lesions, resection of hemangiomas does not necessitate removal of a margin of normal tissue with the tumor. Enucleation is carried out by careful dissection within the proper plane between the hepatic parenchyma and tumor. Division and ligation of the principal hepatic artery should be completed early in the operation as this often results in significant tumor decompression thereby facilitating resection [1]. The majority of hemangiomas are contained within a tough fibrous capsule which can be clamped and used for retraction purposes [1]. As hepatic venous branches are encountered extending from the lesion they should be controlled with clips or ties [1]. Presently, mortality outcomes for resection and enucleation are comparable [42].

Hepatic artery ligation for treatment of hemangioma has also been described anecdotally [9]. Although its benefits are likely transient, hepatic artery embolization and/or ligation play a pivotal role only in the temporary management of uncontrolled hemorrhage from rupture [43,44]. Finally, radiation therapy for symptomatic hemangiomas has also been reported. Though data validating the use of radiotherapy is limited, it seems a reasonable approach for symptomatic hemangioma where surgical intervention is clearly contraindicated.

7. Special issue: Hemangioma in children

Hepatic hemangiomas of infancy and childhood differ substantially in their appearance, presentation, and progression than those in adults [1]. These lesions are frequently large and symptomatic. In contrast to adult hemangiomas, the risk of spontaneous rupture in infancy is greater [1]. Similarly, Kasabach-Merritt syndrome occurs more frequently and results more often in death among affected infants. As a result of the numerous venous lakes within these lesions, which serve as siphons for a large proportion of the total cardiac output, severe congestive heart failure and death may result. Initial treatment of high output cardiac failure in children includes oxygen, diuretics, digitalis, corticosteroids, hepatic artery ligation, and radiation therapy [2, 45-48]. Contrary to the conservative management of adult hemangiomas, hemangiomas of infancy and childhood more frequently require life-saving surgical intervention.

8. Focal nodular hyperplasia

Focal nodular hyperplasia (FNH) is the second most common benign hepatic lesion [20]. FNH is found predominately in women (in a ratio of 8-9:1) between the ages of 20-50 years, and has a prevalence of 4 - 8% in the general population [49,50]. Similar to hemangiomas, the prevalence of FNH has markedly increased over the past several decades, which likely reflects the proficiency and widespread use of advanced imaging modalities [1].

Although Klatskin (1977) and Vana (1979) each reported an association between OCP use and the development of FNH, the high frequency of FNH in the absence of OCP use suggests no causal relationship [32,51]. However, enlargement of FNH lesions has been described in the setting of pregnancy and long-term OCP use [52]. While the etiology of these lesions has not yet been clearly delineated, it has been suggested that FNH is a hyperplastic polyclonal response of normal hepatic parenchyma to localized areas of increased arterial perfusion [53]. Expectantly, FNH has been found in association with vascular disorders and malformations including hereditary hemorrhagic telangiectasia, hemihypertrophy Klippel-Trenaunay-Weber syndrome, and congenital absence of the portal vein [49,54-57].

While typically small (< 5 cm), FNH lesions have been reported as large as 19 cm [48,50]. The majority of FNH lesions are solitary in nature (80%-95%), although up to 20% of individuals are reported to have multiple lesions [1, 48, 50]. When multifocal, FNH often occurs in conjuncture with other benign hepatic lesions including hemangiomas [58].

9. Clinical presentation

FNH is frequently asymptomatic with up to 75% of lesions discovered incidentally during radiologic workup, laparotomy, or laparoscopy for unrelated pathology [59]. Similar to hepatic hemangiomas, spontaneous rupture is extremely rare as illustrated by Chamberlain et. al (2003) management of 33 patients with FNH where no ruptures were evident [9]. Large, peripheral, pedunculated lesions may result in a palpable mass associated with abdominal pain and/or fullness, but acute symptoms associated with rupture, necrosis, or infarction are a rarity.

10. Pathology

Macroscopically FNH is a firm pale to red colored lesion with sharp margins. Lesions are typically small, pedunculated, and peripherally located. Unlike hemangiomas and hepatic adenomas, FNH lack a capsule. Histologically FNH appears as regenerative nodules making histopathological differentiation from cirrhosis difficult. Lesions contain normal hepatic elements with a haphazard arrangement of cords and sinusoids [5]. Proliferating bile ducts, fibrous septae, Kupffer cells, and sinusoids are typically present in FNH, and are characteristically absent in hepatocellular adenomas [13,50,59]. Generally FNH contain a large artery with multiple branches radiating through disorganized fibrous septa to the periphery. This radiating arterial pattern produces a spoke and wheel image on angiography and is responsible for the central scar appearance on radiographic imaging studies [60,61].

11. Radiographic imaging

Definitive diagnosis of FNH can be challenging. FNH lesions are well visualized on US but are highly variable and exhibit no distinct characteristic features. Helical CE-CT reveals a well-demarcated lesion that is often isodense [29]. However, during the portal venous phase the pathognomonic central scar may be appreciated. Distinguishing FNH on standard MRI can prove challenging as the lesion is composed of the similar elements as the normal liver parenchyma. FNH may appear isointense with a central scar on T-1 and T-2 weighted imaging [62]. MRI with Gd-DTPA demonstrates a hyperintense lesion early, which becomes isointense with central scar enhancement on delayed imaging [63-65]. The use of reticuloendothelial agents including Ferridex, which is taken up selectively by Kupffer cells, increases the specificity of both CT and MRI imaging [1]. Technetium-99-labeled sulfur colloid scintigraphy may prove helpful in demonstrating the presence of Kupffer cells within the FNH lesion, however this finding is not specific enough for definitive diagnosis [1,66,67]. Angiography, though rarely indicated for the diagnosis of FNH, usually demonstrates a hypervascular mass with a single central artery and enlarged peripheral vessels in a "spoken wheel" appearance [66-68]. Finally, ^{18}F-FDG PET can aid in the differentiation between benign and malignant lesions, but it is neither sensitive nor specific enough for diagnosis of FNH [8,38].

12. Diagnosis & treatment

The natural course of an FNH lesion is generally indolent with minimal risk of rupture or complication. Laboratory testing generally reveals normal liver function tests and alpha-fetoprotein levels, although minor elevations in aspartate and alanine aminotransferase, alkaline phosphatase, and gamma glutamyl transpeptidase may occasionally be seen. Definitive diagnosis of FNH in an asymptomatic patient warrants conservative management and includes close observation with repeat imaging every four to six months [9]. When radiology is equivocal, most surgeons still choose close observation with follow-up studies preformed every three to four months. Biopsy is generally not indicated, as results are seldom diagnostic [69].

Although it may be impossible to distinguish FNH from a well-differentiated HCC without surgical excision, FNH tumors do not undergo malignant transformation. Thus indications for surgical intervention should be limited to those situations where there is a change in the size or number of lesion(s), a change in the intensity of symptoms, or where classic imaging characteristics are absent and diagnostic dilemma remains [70]. Hence, the role of the surgeon is typically limited to patient reassurance and close observation [9].

13. Hepatic adenoma

Hepatic adenomas are identified predominately in women of reproductive age [49]. The estimated prevalence of hepatic adenomas within the general population on postmortem exams is approximately 1% [10]. Etiologically, hepatic adenomas are of epithelial origin. Unlike hepatic hemangiomas and FNH, a clear association between the use of OCPs and hepatic adenomas has been established. First described in 1973, multiple studies have documented a reciprocal relationship between OCP use and adenoma incidence based on estrogen dose and exposure time [71-75]. Approximately 90% of individuals with adenomas have previous OCP exposure [1]. The prevalence of hepatic adenomas is estimated at 1 per 1,000,000 among women who have never used OCP as compared with 30-40 per 1,000,000 amongst long-term OCP users [72,76]. OCPs also affect the course of disease progression as lesions are generally larger, more numerous, and more likely to bleed than tumors in OCP-naïve individuals [32,75,77,78]. Adenoma regression has been observed in patients after discontinuation of OCP with recurrence ensuing during pregnancy and/or OCP re-administration [72,79,-82]. Despite these findings, the mechanism by which estrogen therapy affects the development and course of hepatic adenomas has yet to be clearly elucidated.

Hepatic adenomas are typically small (< 5 cm), soft, solitary lesions but may be multiple in up to 30% of cases [9]. Of note, hepatic adenomatosis disease, defined as the presence of >10 lesions, is a distinct disease entity from that of hepatic adenoma and as such will not be described in further detail [83]. Hepatic adenomas have been associated with type I glycogen storage disease, galactosemia, Klienfelter's syndrome, and Turner's syndrome as well as with androgen, domiphene, danazol and growth hormone use [1,84-86]. Although hepatic adenomas are benign, these lesions have been associated with spontaneous hemorrhage, rupture, and malignant transformation, making prognosis more grave than that of other benign hepatic tumors [5,87].

14. Clinical presentation

Since adenoma and FNH both present in women of reproductive age and have similar radiographic appearances they are frequently confused. Differential diagnosis is critical given that the recommended treatment of each respective lesions differs. Hepatic adenomas are most often diagnosed as a result of imaging done for unrelated pathology or following workup of a palpable abdominal mass (30% patients) [88]. Occasionally episodic pain may be evident as a result of an enlarged liver, intratumoral bleed, or tumor necrosis [9]. Up to 33% of patients with hepatic adenomas present with acute rupture and concomitant intraperitoneal bleeding [1]. The development of acute severe pain associated with hypotension reflects spontaneous rupture and carries a 20% mortality rate if not appropriately identified and treated [32,89-91].

15. Pathology

Grossly hepatic adenomas appear as smooth, soft, and pale yellow tumor on cut surface [1]. These lesions often contain prominent blood vessels that have a high potential for rupture and hemorrhage [1]. As adenomas lack a fibrous capsule intraparenchymal bleeding may occur, which frequently results in a variegated appearance.

Microscopically hepatic adenomas appear as well circumscribed lesions composed of monotonous sheets of hepatocytes laden with glycogen and lipids [5]. These lesions lack normal hepatic architecture and demonstrate thickened trabeculae interspersed with sinusoids and prominent thin walled vessels [1,5]. Biliary ducts and portal tracts are distinctly absent from adenomas.

While the malignant potential of adenomas remains controversial, several authors have reported a low (5%) yet consistent risk of transformation [87]. Histological differentiation between well differentiated HCC and adenoma can be difficult, especially in the presence of fibrolamellar HCC which is also more common in women of reproductive age. This issue is further explained in situations in which HCC and hepatic adenoma have been found adjacent to one another [61,50,89,92,93].

16. Radiological imaging

Although radiographic evaluation is important for complete workup of hepatic adenoma radiographic features are often nonspecific [94]. As such, despite the use of multiple imaging techniques, diagnosis often remains equivocal. Ultrasound exhibits a mixed echogenic pattern with an overall heterogeneous appearance [1,29]. Lesions appear hyperechoic as a result of their high lipid content with a heterogeneous pattern reflecting intratumoral hemorrhage and necrosis [95]. CE-CT imaging is frequently utilized for adenoma visualization and typically demonstrates a hypo- to isodense lesion as a result of low attenuation on non-contrast phase [1]. A variegated appearance with peripheral enhancement during the early contrast phase with subsequent centripetal flow during the venous phase may be apparent, however CT can demonstrate a spectrum of disparate findings [96]. MRI findings for hepatic adenoma are similar to those on CT. Due to the high fat and glycogen content, adenomas are usually well demarcated on MRI imaging [29]. While most adenomas appear iso- to hyperintense on both T-1 and T-2 weighted images, findings are highly variable [1,97]. The administration of contrast agents including gadolinium or gabodenate dimeglumine (Gd-BOPTA) results in early markedly uniform enhancement on arterial phase, which subsequently becomes isodense on the portal venous phase [98]. The use of [18]FDG PET scan may also aid in the differentiation of benign versus malignant disease in which where adenomas demonstrate poor uptake of [18]FDG as compared to HCC [8,38].

Additional imaging modalities infrequently used include technetium-99 sulfur colloid scanning. This imaging modality is particularly useful in differentiating between hepatic adenoma and FNH, as hepatic adenomas lack bile duct components and frequently appear as a "cold nodules" on imaging [99]. Occasionally however, a minority of lesions do take up the sulfur colloid, rendering them indistinguishable from FNH [99]. Although rarely utilized, angiography typically reveals hypervascular lesions with areas of hemorrhage and necrosis [1,28].

17. Diagnosis & treatment

In the absence of acute hemorrhage, serological tests rarely assist in diagnosis. Liver function tests and tumor makers including CEA, alpha-fetoprotein, and CA 19-9 are invariably normal.

Hepatic adenomas pose a greater risk for rupture (33%) and malignant transformation (5%) than do other benign hepatic lesions [9,87]. As such all patients with suspected or confirmed hepatic adenoma > 3 cm should undergo enucleation or surgical resection [1,100]. The approach to surgical excision should be as previously described. Since all adenomas are suspected to harbor malignancy an adequate margin of normal parenchyma should be taken [1]. When surgical exploration is not feasible angiographic embolization or ligation can provide temporary yet life saving relief.

As a result of the relationship between OCP and adenoma incidence, it is recommended that all individuals suspected of having an adenoma discontinue the use of OCP immediately and indefinitely [1,61]. Patients should also be advised against pregnancy until after adenoma resection, as the growth and rupture risk of hepatic adenomas is highly unpredictable during gestation [101]. Yearly follow-up with imaging is advised among all patients where a causal link between OCP use and adenoma is absent [9]. As a result of improved safety of hepatic resection and the use of minimally invasive techniques in hepatectomy it is suggested that all hepatic adenomas > 3 cm be resected [1,100]. In patients with significant contraindications to surgical intervention, OCP should be discontinued and the patient enrolled in an ongoing surveillance program [9].

18. Additional liver tumors

18.1. Epithelial tumor

Biliary hamartomas

Bile duct adenomas and hamartomas are common tumors. Bile duct adenomas appear as small, white, solitary, subcapsular masses [1]. They are defined histologically by narrow lumen bile ducts surrounded by fibrosis. Hamartomas appear as small gray-white nodules that lie just beneath the capsule of the liver [102]. Biliary hamartomas are frequently multifocal and are characterized microscopically by the presence of dilated mature bile ducts surrounded by fibrous tissue [1]. These lesions are especially important as they are frequently misinterpreted as metastatic tumor by the operating surgeon. This notion heightens the importance of confirmatory diagnosis to rule out malignancy for all hepatic lesions. Precise diagnosis is most important in situations in which the presence of a metastatic liver disease will alter the proceedings of a planned operation.

18.2. Mesenchymal tumors

Solitary fibrous tumor (other names include benign mesothelioma or fibroma)

Solitary fibrous tumors (SFT) are rare mesenchymal tumors that are frequently mistaken for metastatic lesions as a result of their radiographic and intra-operative appearance. Grossly SFT's appear as white-to-gray lesions and can vary greatly in size ranging from 2 – 20 cm in diameter [1]. Despite their large size, most SFT's remain asymptomatic. Histologically, most have a classic short storiform pattern and display an absence of cellular atypia, mitoses, and/or necrosis [1]. However when malignant, SFTs frequently possess a high mitotic rate and marked cellular atypia. Immunohistochemically SFTs display a strong positive staining for vimentin and CD-34 [1]. Since definitive histologic examination is required for diagnosis of either a benign or malignant SFT, surgical resection is indicated in nearly all circumstances.

Lipoma, myelolipoma, or angiomyelipoma

Similar to several other benign hepatic lesions, most benign fatty hepatic tumors are identified at the time of autopsy with only isolated reports of histological diagnosis following operative resection [13]. Multiple variants including angiolipoma, myelolipoma, and angiomyolipoma have been described [13,103]. Additionally, "pseudolipomas" have been described as lesions in which there is an extracapsular fatty tumor with involutional changes. It is probable that this lesion results when a free-floating piece of fat becomes entrapped between diaphragm and liver surface [1,10]. In most situations definitive diagnosis requires surgical resection to exclude malignancy.

Mesenchymal Hamartomas

Mesenchymal hamartomas are exceedingly rare congenital liver tumors which occur most frequently in infants under 1 year of age [9,104]. Microscopically these lesions demonstrate a myxoid background of highly cellular embryonal mesenchyme with haphazard groupings of bile ducts, cysts, and hepatic cells [105]. Generally, the cystic element is the most prominent feature resulting in a characteristic "honeycomb" appearance [106]. In contrast to biliary hamartomas, which are clinically insignificant, mesenchymal hamartomas can significantly impair hepatic function as a result of their large size [106]. Although benign, these lesions can result in death due to mass effect and/or hepatic insufficiency [1]. Thus, all suspected mesenchymal hamartomas should be completely excised when possible. If complete surgical excision cannot be achieved surgical debulking may be sufficient as there have been no reports of recurrence after an incomplete surgical resection to date [107].

Myxoma

Myxomas are exceptionally uncommon benign lesions of the liver. To date fewer than five cases have been reported [9,58,108]. These lesions arise from primitive connective tissue. Histologically myxomas demonstrate a myxoid matrix with scattered proliferation of connective tissue cells [108]. Similar to other types of hepatic tumors described above, surgical resection is generally indicated to exclude malignancy.

Teratoma

Primary teratomas are remarkably rare benign hepatic lesions. A review of the literature revealed only 7 reports to date, with the majority of lesions occurring in children [109]. Secondary hepatic teratomas have been observed following systemic chemotherapy administration for treatment of testicular cancer [1]. Teratomas arise from pluripotent cells and frequently contain components from all three germ layers. Teratomas are typically encapsulated cystic lesions that are easily resectable [1,110]. Imaging characteristics reflect tissue heterogeneity and are often non-specific [110]. Surgical resection of hepatic teratomas is indicated to exclude malignancy.

19. Conclusion

A thorough understanding of the natural history and accurate histologic diagnosis are fundamental to appropriate management of patients with benign liver tumors. Although advancements in imaging have drastically improved the detection and characterization of both benign and malignant liver neoplasms, the ultimate burden of responsibility for diagnosis and treatment remains that of the surgeon. Ongoing improvements in perioperative care and surgical techniques, coupled with increased surgical experience presently permit hepatic resection to be performed with a high level of safety. Despite these developments, a conservative approach including close observation with serial examination and imaging seems most appropriate for asymptomatic patients in which malignancy is not suspected.

Symptomatic patients without medical or anatomic contraindication to a major hepatic resection, as well as patients in whom a malignancy cannot be excluded (including individuals with adenomas > 3 cm), should be considered for surgical intervention. Preoperative needle biopsy is frequently contraindicated due to a high risk of rupture and hemorrhage, and therefore should only be considered after exclusion of hemangioma. Additionally, it is important to note that distinguishing particular lesions (especially adenoma and FNH) on needle biopsy is exceedingly difficult. As such caution should exercised when using this information to make clinical evaluations. Excisional biopsy of small and peripheral lesions and adequate wedge incision biopsy of large lesions should permit the pathologist to make an accurate histologic diagnosis and exclude a malignancy. If doubt remains, formal hepatic resection is indicated.

Author details

Ronald S. Chamberlain[1,2,3] and Kim Oelhafen[3]

1 Department of Surgery, Saint Barnabas Medical Center, Livingston, NJ, USA

2 Department of Surgery, University of Medicine and Dentistry of New Jersey, Newark, NJ, USA

3 Saint George's University School of Medicine, Grenada, West Indies, Grenada

References

[1] Chamberlain RS, DeCorato D, Jarnagin W. Benign liver lesions. In Blumgart L, Fong Y, & Jarnagin W. (ed.) American Cancer Society Atlas of Clinical Oncology Hepatobiliary Cancer. British Colombia: Decker Inc; 2001. p1-30.

[2] Little JM, Kenny J, Hollands MJ. Hepatic incidentaloma: a modern problem. World J Surg 1990;14(4): 448-51.

[3] Little JM, Richardson A, Tait N. Hepatic dyschoma: a five-year experience. HPB Surg 1991;4(4): 291-8.

[4] Izzo F, Cremona F, Ruffolo F, Palaia R, Parisi V, Curley SA. Outcome of 67 patients with hepatocellular cancer detected during screening of 1125 patients with chronic hepatitis. Ann Surg 1998;227(4): 513-8.

[5] Sonnenday C, Welling T, Pelletier S. Hepatic Neoplasms. In Mulholland M & Lillemoe K, et al (5th ed.) Greenfield's Surgery: Scientific Principles & Practice. Philadelphia: Wolter Kluwer/Lippincott Williams & Wilkins; 2011. p934-94.

[6] Yoon SS, Charny CK, Fong Y, Jarnagun WR, Schwartz LH, Blumgart LH et al. Diagnosis, management, and outcomes of 115 patients with hepatic hemangioma. J Am Coll Surg 2003;197(3): 392-402.

[7] Balci NC, Befeler AS, Leiva P, Pilgram TK, Havlioglu N. Imaging of liver disease: comparison between quadruple-phase multidetector computed tomography and magnetic resonance imaging. J Gastroenterol Hepatol 2008;23(10): 1520-7.

[8] Sacks A, Peller P, Surasi D, Chatburn L, Mercier G, Subramaniam RM. Value of PET/CT in the Management of Primary Hepatobiliary Tumors, Part 2. AJR Am J Roentgenol 2011;197(2): W260-5.

[9] Chamberlain RS. Benign Tumors of the Liver: a surgical perspective. In Chamberlain RS & Blumgart LH (ed.) Hepatobiliary Surgery. Texas: Landes Bioscience; 2003. p81-99.

[10] Karhunen PJ. Benign hepatic tumours and tumour like conditions in men. J Clin Pathol 1986;39(2): 183-88.

[11] Lam KY. Autopsy findings in diabetic patients: a 27-yr clinicopathologic study with emphasis on opportunistic infections and cancers. Endocr Pathol 2002;13(1): 39-45.

[12] Gandolfi L, Leo P, Solmi L, Vitelli E, Verros G, Colecchia A. Natural history of hepatic hemangiomas: clinical and ultrasound study. Gut 1991;32(6): 677-80.

[13] Ishak KG, Rabin L. Benign tumors of the lover. Med Clin North Am 1975;59(4): 995-1013.

[14] Gilon D, Slater PE, Benbassat J. Can decisions analysis help in the management of giant hemangioma of the liver? J Clin Gastroenterol 1991;13(3): 255-8.

[15] Edmondson HA. Tumors of the liver and intrahepatic bile duct. In: Atlas of tumor pathology. Section VII, fascicle 25. Washington DC: Armed Forces Institute of Pathology; 1958.

[16] Farges O, Daradkeh S, Bismuth H. Cavernous hemangioma of the liver: are there any indications for resection? World J Surg 1995;19(1): 19-24.

[17] Sewell JH, Weiss K. Spontaneous rupture of hemangioma of the liver. A review of the literature and presentation of illustrative case. Arch Surg 1961;83: 729-33.

[18] Glinkova V, Shevah O, Boaz M, Levine A, Shirin H. Hepatic haemangiomas: possible association with female sex hormones. Gut 2004;53(9): 1352-5.

[19] Little JM. Benign tumors of the liver. In: Terblanche J (ed). Hepatobiliary malignancies: its multidisciplinary management. London: Edward Arnold; 1994. p325-49.

[20] Tait N, Richardson AJ, Muguti G, Little JM. Hepatic cavernous hemangioma: a 10-year review. Aust N Z J Surg 1992;62(7): 521-4.

[21] Dockerty MB, Gray HK, Henson SW. Benign tumors of the liver. II. Hemangiomas. Surg Gynecol Obstet 1965;103(3): 327-31.

[22] Shumacker HB. Hemangioma of the liver: discussion of symptomatology and report of patient treated by operation. Surgery 1942;11: 209-22.

[23] Griecco MB, Miscall BG. Giant hemangioma of the liver. Surg Gyncecol Obstet 1978;147(5): 783-7.

[24] Ochsner JL, Halpert B. Cavernous hemangioma of the liver. Surgery 1958;43(4): 577-82.

[25] Dennis M. Fatal pulmonary embolism due to thrombosis of a hepatic cavernous hemangioma. Med Law 1980;20(4): 287-8.

[26] Hall GW. Kasabach-Merritt syndrome: pathogenesis and management. Br J Haematol 2001;112(4): 851-62.

[27] Baer HU, Dennsion AR, Mouton W, Stain SC, Zimmermann A, Blumgart LH. Enucleation of giant hemangiomas of the liver. Technical and pathologic aspects of a neglected procedure. Ann Surg 1992;216(6): 673-6.

[28] Assy N, Nasser G, Djibre A, Beniashvilli Z, Zidan J. Characteristics of common solid liver lesions and recommendations for diagnostic workup. World J Gastroenterol 2009;15(26): 3217-27.

[29] Madrazo BL. Use of imaging studies to aid in the diagnosis of benign liver tumors. Gastroenterol Hepatol (N Y) 2011;7(10): 683-5.

[30] Trastek VF, van Heerden JA, Sheedy PF II, Adson MA. Cavernous hemangiomas of the liver: resect of observe? Am J Surg 1983;145(1): 49-53.

[31] Foster JH. Evaluation of asymptomatic solitary hepatic lesions. Annu Rev Med 1988;39: 85-93.

[32] Klatskin G. Hepatic tumors: possible relationship to use of oral contraceptives. Gastroenterology 1977;73(2): 386-94.

[33] McFarland EG, Mayo-Smith WW, Saini S, Hahn PF, Goldberg MA, Lee MJ. Hepatic hemangiomas and malignant tumors: improved differentiation with heavily T2-weighted conventional spin-echo MR imaging. Radiology 1994;193(1): 43-7.

[34] Goshima S, Kanematsu M, Kondo H, Yokoyama R, Kajita K, Tsuge Y, et al. Hepatic hemangiomas: a multi-institutional study of appearance on T2-weighted MR findings and apparent diffusion coefficients. Eur J Radiol 2009;70(2): 325-30.

[35] Adam A, Dixon AK, Grainger RG, et al. A Textbook of Medical Imaging. 5th ed. Philadelphia, PA: Churchill Livingston/Elsevier; 2008.

[36] Fulcher AS, Sterling RK. Hepatic neoplasms: computed tomography and magnetic resonance features. J Clin Gastroenterology 2002;34(4): 463-71.

[37] Kim T, Federle MP, Baron RL, Peterson MS, Kawamori Y. Discrimination of small hepatic hemangiomas from hypervascular malignant tumors smaller than 3cm with three-phase helical CT. Radiology 2001;219(3): 699-706.

[38] Kurtaran A, Becherer A, Pfeffel F, Muller C, Traub T, Schmalijohann J, et al. 18F-fluorodeoxyglucose (FDG)-PET features of focal nodular hyperplasia (FNH) of the liver. Liver 2000;20(6): 487-90.

[39] Farlow DC, Chapman RP, Gruenewald SM, Antico VF, Farrell GC, Little JM. Investigation of focal hepatic lesions: is tomographic red blood cell imaging useful? World J Surg 1990;14(4): 463-7.

[40] Alper A, Ariogul O, Emre A, Uras A, Okten A. Treatment of liver hemangiomas by enucleation. Arch Surg 1988;123(5): 660-1.

[41] Charny CK, Jarnagin WR, Schwartz LH, Frommeyer HS, DeMatteo RP, Fong Y, et al. Benign liver tumors: radiologic and surgical management. Br J Surg 2000;88(6): 808-13.

[42] Giuliante F, Ardito F, Vellone M, Giordano M, Ranucci G, Piccoli M, et al. Reappraisal of surgical indications and approach for liver hemangioma: a single center experience on 74 patients. Am J Surg 2011;201(6): 741-8.

[43] Nishida O, Satoh N, Alam AS, Uchino J. The effect of hepatic artery ligation for irresectable cavernous hemangioma of the liver. Am Surg 1988;54(8): 483-6.

[44] DeLorimier AA, Simpson BB, Braum RS, Carlsson E. Hepatic-artery ligation for hepatic hemangiomatosis. N Engl J Med 1967;277(7): 333-7.

[45] Park WC, Phillips R. The role of radiation therapy in the management of hemangiomas of the liver. JAMA 1970;212(9): 1496-8.

[46] Dehner LP, Ishak KG. Vascular tumors of the liver in infants and children. A study of 20 cases and review of the literature. Arch Pathol 1971;92(2): 101-11.

[47] Clatworthy HW, Boles ET, Newton WA. Primary tumors of the liver in infants and children. Arch Dis Child 1960;35: 22–8.

[48] Nguyen L, Shandling B, Ein S, Stephens C. Hepatic hemangioma in childhood: medical management or surgical resection? J Pediatr Surg 1982; 17(5):576-9.

[49] Wanless IR, Mawdsley C, Adams R. On the pathogenesis of focal nodular hyperplasia of the liver. Hepatology 1985;5(6): 1194-200.

[50] Craig J, Peters R, Edmundson H. Tumors of the liver and intrahepatic bile ducts, Fascicle 26. (2nd ed.). Washington DC: DC Armed Forces Institute of Pathology; 1989. p6.

[51] Vana J, Murphy GP, Aronoff BL, Baker HW. Survey of primary liver tumors and oral contraceptive use. J Toxicol Environ Health 1979;5(2-3): 255-73.

[52] Scott LD, Katz AR, Duke JH, Cowan DF, Maklad NF. Oral contraceptives, pregnancy, and focal nodular hyperplasia of the liver. JAMA 1984;251(11): 1461-3.

[53] Poon RT, Fan ST. Assessment of hepatic reserve for indication of hepatic resection: how I do it. J Hepatobiliary Pancreat Surg 2005;12(1): 31-7.

[54] Rebouissou S, Bioulac-Sage P, Zucman-Rossi J. Molecular pathogenesis of focal nodular hyperplasia and hepatocellular adenoma. J Hepatol 2008;48(1): 163-70.

[55] Altavilla G, Guariso G. Focal nodular hyperplasia of the liver associated with portal vein agenesis: a morphological and immunohitsochemical study of one case and review of the literature. Adv Clin Path 1999;3(4): 139-45.

[56] Buscarini E, Danesino C, Plauchu H, de Fazio C, Olivieri C, Brambilla G, et al. High prevalence of hepatic focal nodular hyperplasia in subjects with hereditary hemorrhagic telangiectasia. Ultrasound Med Biol 2004;30(9): 1089-97.

[57] De Gaetano AM, Gui B, Macis G, Manfredi R, Di Stasi C. Congenital absence of the portal vein associated with focal nodular hyperplasia in the liver in a adult woman: imaging and review of the literature. Abdom Imaging 2004;29(4): 455-9.

[58] Mathieu D, Zafrani ES, Anglade MC, Dhumeaux D. Association of focal nodular hyperplasia and hepatic hemangioma. Gastroenterology 1989;97(1): 154-7.

[59] Goodman, ZD. Benign Tumors of the Liver. In: Okuda K, Ihak KD. (ed.) Neoplasms of the Liver. Tokyo: Springer; 1987. p105.

[60] Whelan Jr, Baugh JH, Chandon S. Focal nodular hyperplasia of the liver. Ann Surgery 1973;177(2): 150–8.

[61] Kerlin P, Davis GL, McGill DB, Weiland LH, Adson MA, Sheedy PF 2nd . Hepatic adenoma and focal nodular hyperplasia: clinical, pathologic and radiologic features. Gastroenterology 1983;8(5 Pt 1): 994-1002.

[62] Mattison GR, Glazer GM, Quint LE, Francis IR, Bree RL, Ensminger WD. MR imaging of hepatic focal nodular hyperplasia: characterization and distinction from primary malignant hepatic tumors. ARJ 1987;148(4): 711-5.

[63] Irie H, Honda H, Kaneko K, Kuroiwa T, Fukuya T, Yoshimitsu K, et al. MR imaging of focal nodular hyperplasia of the liver: value of contrast-enhanced dynamic study. Radiat Med 1997;15(1): 29-35.

[64] Mahfouz AE, Hamm B, Taupitz M, Wolf KJ. Hypervascular liver lesions: differentiation of focal nodular hyperplasia from malignant tumors with dynamic gadolinium-enhanced MR imaging. Radiology 1993;186(1): 133-8.

[65] Rummeny E, Weissleder R, Sironi S, Stark DD, Comptom CC, Hahn PF, et al. Central scars in primary liver tumors: MR features, specificity, and pathologic correlation. Radiology 1989;171(2): 323-6.

[66] Mergo PJ, Ros PR. Benign Lesions of the Liver. In The Radiologic Clinics of North America 2nd ed. Philadelphia: WB Saunders; 1998. p319.

[67] Rogers JV, Mack LA, Freeny PC. Johnson ML, Sones PJ. Hepatic focal nodular hyperplasia: angiography, CT, sonography, and scintigraphy. ARJ Am J Roentgenol 1981;137(5): 983-90.

[68] Welch TJ, Sheedy PF 2nd, Johnson CM, Stephens DH, Charboneau JW, Brown ML, et al. Focal nodular hyperplasia and hepatic adenoma: comparisons of the angiography, CT, US, and scintigraphy. Radiology 1985;156(5): 593-5.

[69] Fabre A, Audet P, Vilgrain V, Nguyen BN, Valla D, Belghiti J, et al. Histological scoring of liver biopsy in focal nodular hyperplasia with atypical presentation. Hepatology 2002;35(2): 414-20.

[70] Bonney GK, Gomez D, Al-Mukhtar A, Toogood GJ, Lodge JP, Prasad R. Indication for treatment and long-term outcome of focal nodular hyperplasia. HPB (Oxford) 2007;9(5): 368-72.

[71] Baum JK, Bookstein JJ, Holtz F, Klein EW. Possible association between benign hepatomas and oral contraceptives. Lancet 1973;2(7835): 926-9.

[72] Rooks JB, Ory HW, Ishak KG, Strauss LT, Greenspan JR, Hill AP, et al. Epidemiology of hepatocellular adenoma. The role of oral contraceptive use. JAMA 1979;242(7): 644-8.

[73] Nime F, Pickren JW, Vana J, Aronoff BL, Baker HW, Murphy GP. The histology of liver tumors in oral contraceptive users observed during a national survey by the American College of Surgeons Commission on Cancer. Cancer 1979;44(4): 1481-9.

[74] Rosenberg L. The risk of liver neoplasia in relation to combined oral contraceptive use. Contraception 1991;43(6): 643-52.

[75] Søe KL, Søe M, Gluud C. Liver pathology associated with the use of anabolic-androgenic steroids. Liver 1992;12(2): 73-9.

[76] Reddy KR, Schiff ER. Approach to a liver mass. Semin Liver Dis 1993;13(4): 423-35.

[77] Shortell CK, Schwartz SI. Hepatic adenoma and focal nodular hyperplasia. Surg Gynecol Obstet 1991;173(5): 426-31.

[78] Meissner K. Hemorrhage cause by ruptured liver cell adenoma following long term oral contraceptives: a case report. Hepatogastroenterolog 1998;45(19): 224-5.

[79] Edmondson HA, Reynolds TB, Henderson B , et al. Regression of liver cell adenoma associated with oral contraceptives. Ann Intern Med 1977:86(2): 180-2.

[80] Kawakatsu M, Vilgrain V, Erlinger S, Nahum H. Disappearance of liver cell adenoma: CT and MR imaging. Abdom Imaging 1997;22(3): 274-6.

[81] Aseni P, Sansalone CV, Sammartino C, Benedetto FD, Carrafiello G, Giacomoni A, et al. Rapid disappearance of hepatic adenoma after contraceptive withdrawal. J Clin Gastroenterology 2001;33(3): 234-6.

[82] Norris, S. Drug- and Toxin-Induced Liver Injury. In: Comprehensive Clinical Hepatology, O'Grady, J, Lake, J, Howdle, P. (eds). London: Harcourt Publishers Limited; 2000. p1

[83] Flejou JF, Barge J, Menu Y, Degott C, Bismuth H, Potet F, et al. Liver adenomatosis. An entity distinct from liver adenoma? Gastroenterology 1985;89(5): 1132-8.

[84] Labrune P, Trioche P, Duvaltier I, Chevalier P, Odievre M. Hepatocellular adenomas in glycogen storage disease type I and III: a series of 43 patients and review of the literature. J Pediatr Gastroenterol Nutr 1997;24(3): 276-9.

[85] Espat J, Chamberlain RS, Sklar C, Blumgart LH. Hepatic adenoma associated with recombinant human growth hormone therapy in a patient with Turner's syndrome. Dig Surg 2000;17(6): 640-3.

[86] Carrasco D, Prieto J, Pallardó L, Moll JL, Cruz JM, Munoz C, et al. Multiple hepatic adenomas after long term therapy with testosterone enanthate. Review of the literature. J Hepatol 1985;1(6): 573-8.

[87] Colli A, Fraquelli M, Massironi S, Colucci A, Paggi S, Conte D. Elective surgery for benign liver tumours. Cochrane Database sys Rev 2007;24(1): CD005164.

[88] Molina E, Schiff E. Benign solid lesions of the liver. In: Schiff E, Sorrell M, Maddrey W. (eds). Schiff's Disease of the liver8th ed. Philadelphia: Lippincott-Rave; 1999. p1245.

[89] Leese T, Farges O, Bismuth H. Liver cell adenomas. A 12-year surgical experience from a specialist hepato-biliary unit. Ann Surg 1988; 208(5): 558-64.

[90] Nagorney DM. Benign hepatic tumors: focal nodular hyperplasia and hepatocellular adenoma. World J Surg 1995;19(1): 13-8.

[91] Rubin RA, Mitchell DG. Evaluation of the solid hepatic mass. Med Clin North Am 1996;80(5): 907-28.

[92] Gyorffy EJ, Bredfeldt JE, Black WC. Transformation of hepatic cell adenoma to hepatocellular carcinoma due to oral contraceptive use. Ann Intern Med 1989;110(6): 489-90.

[93] Tesluk H, Lawrie J. Hepatocellular adenoma. Arch Pathol Lab Med 1981;105(6): 296-9.

[94] Mathieu D, Bruneton JN, Drouillard J, Pointreau CC, Vasile N. Hepatic adenomas and focal nodular hyperplasia: dynamic CT study. Radiology 1986;160(1): 53-8.

[95] Golli M, Van Nhieu JT, Mathieu D, Zafrani ES, Cherqui D, Dhumeaux D, et al. Hepatocellular adenoma: color Doppler US and pathologic correlations. Radiology 1994;190(3): 741-4.

[96] Grazoli L, Federle MP, Brancatelli G, Ichikawa T, Olivetti L, Blachar A. Hepatic adenomas: imaging and pathologic findings. Radiographics 2001;21(4): 877-92.

[97] Chung KY, Mayo-Smith WW, Saini S, Rahmouni A, Golli M, Mathieu D. Hepatocellular adenoma: MR imaging features with pathologic correlation. ARJ Am J Roentgenol 1994;163(2): 303-8.

[98] Paulson EK, McClellan JS, Washington K, Spritzer CE, Meyers WC, Baker ME. Hepatic adenoma: MR characteristics and correlation with pathological findings. AJR 1994;163(1): 113-6.

[99] Rubin RA, Lichenstein GR. Hepatic scintigraphy in the evaluation of solitary solid liver masses. J Nucl Med 1993;34(4): 697-705.

[100] Koffron A, Geller D, Gamblin TC, Abecassis M. Laparoscopic liver surgery; Shifting the management of liver tumors. J Hepatology 2006;44(6):1694-700.

[101] Bis KA, Waxman B. Rupture cf the liver associated with pregnancy: a review of the literature and report of 2 cases. Obstet Gynecol Surv 1976;31(11); 763-73.

[102] Moran CA, Ishak KG, Goodman ZD. Solitary fibrous tumor of the liver: a clinicopathologic and immunohistochemical study of nine cases. Ann Diagn Pathol 1998;2(1): 19-24.

[103] Pounder DJ. Hepatic angiomyolipoma. Am J Surg Pathol 1982;6(7): 677-81.

[104] Grases PJ, Matos-Villaobos M, Arcia-Romero F, Lecuna-Torres V. Mesenchymal hamartoma of the liver. Gastroenterology 1979;76(6): 1466-9.

[105] Stocker JT, Ishak KG. Mesenchymal hamartoma of the liver: report of 30 cases and review of the literature. Pediatr Pathol 1983;1(3): 245-67.

[106] Klaassen Z, Paragi PR, Chamberlain RS. Adult Mesenchymal hamartoma of the liver: Case report. Case Rep Gastroenterol 2010;4(1):84-92.

[107] Foster JH, Berman M. Solid Liver Tumors. Philadelphia, PA: WB Saunders; 1977.

[108] Yoon GS, Kang GH, Kim OJ. Primary myxoid leiomyoma of the liver. Arch Pathol Lab Med 1998;122(12): 1112-5.

[109] Ukiyama E, Endo M, Yoshida F. Hepatoduodenal ligament teratoma with hepatic artery running inside. Pediatr Surg Int. 2008;24(11): 1239-42.

[110] Prasad SR, Wang H, Rosas H, Menias CO, Narra VR, Middleton WD, et al. Fat-containing lesions of the liver: radiologic-pathologic correlation. Radiographics 2005;25(2): 321-31.

Chapter 13

Liver Resection for Hepatocellular Carcinoma

Mazen Hassanain, Faisal Alsaif,
Abdulsalam Alsharaabi and Ahmad Madkhali

Additional information is available at the end of the chapter

1. Introduction

Hepatocellular carcinoma (HCC), an epithelial tumor derived from hepatocytes, accounts for 80% of all primary liver cancers and ranks globally as the fourth leading cause of cancer-related deaths. Annual mortality rates of HCC remain comparable to its yearly incidence, making it one of the most lethal varieties of solid-organ cancer. Well-established risk factors for the development of HCC include hepatitis B carrier state, chronic hepatitis C infection, hereditary hemochromatosis, and cirrhosis of any etiology, as well as certain environmental toxins. HCC treatment is a multidisciplinary and a multimodal task with surgery in the form of liver resection and liver transplantation representing the only potentially curative modalities. Here we going to discuss the liver resection as treatment modality for HCC in detail.

1.1. Pathology of HCC

Three gross morphologic types of HCC have been identified: nodular, massive and diffuse. Nodular HCC is often associated with cirrhosis and is characterized by well-circumscribed nodules. The massive type of HCC, usually associated with a non-cirrhotic liver, occupies a large area with or without satellite nodules in the surrounding liver. The less common diffuse type is characterized by diffuse involvement of many small indistinct tumor nodules through-out the liver. *Histologically*, six growth forms of HCC can be differentiated. The most common form is the trabecular type, usually comprising highly differentiated carcinomas with polygonal tumor cells similar to hepatocytes; they grow in multilayered trabeculae and enclose blood spaces lined with endothelium (usually without Kupffer cells). The pseudoglandular type is generally found in combination with the trabecular form. It is characterized by the formation of gland-like structures containing detritus and bile or liquid material. The scirrhous type shows excessive deposits of sclerosed connective tissue, which is relatively low in cells. The

Surgical:
Liver resection
Liver transplantation
Locoregional:
- Ablation:
Radiofrequency ablation (RFA)
Percutaneous ethanol injection (PEI)
Cryotherapy
- Embolization:
Bland embolization
Transarterial chemoembolization (TACE)
Radioembolization
Systemic treatment:
Sorafenib
External beam radiation therapy and stereotactic radiotherapy

Table 1. Different treatment modality of HCC

moderately differentiated tumor cells lie between the septa, which resemble connective tissue. This type is mostly found after chemotherapy or radiation therapy. The solid type is an undifferentiated HCC, with the tumor cells displaying considerable cellular polymorphism; the trabecular tissue pattern has disappeared. The tumor is compact due to compression of the sinusoids. Differentiation is, however, only possible in rare cases, since there is often considerable heterogenicity within the tumor, i.e. different tissue types may be found in the same HCC. Fibrolamellar HCC is rare; it consists of solid cell trabeculae with connective-tissue septation and a capsule. Spindle cell-like differentiated HCC is likewise a very rare histological form with a fascicular-sarcomatous growth pattern. Prognosis is significantly poorer than with other forms of HCC.

1.2. Preoperative evaluation

The preoperative evaluation for resection of HCC should focus on the likelihood of disease being confined to the liver, and whether the anatomical location of the tumor and the underlying liver function will permit resection (table 2).

Preoperative checklist for HCC patient	
Chick points	**Remark**
History:	
Age	
Comorbidities	
Liver disease	Symptom of liver cirrhosis
Alcohol /smoking	

	Medication (warfarin, aspirin)	
	Previous surgery	
	Physical status	
Examination:		
	Stigmata of liver disease	
	Chest /CVS/abdomen	
	Nutritional assessment	
Laboratory:		
	CBC	Platelet level is an indirect reflection of portal hypertension in cirrhosis.
	LFT	Bilirubin level for synthetic liver function and for Child Pugh score
	Coagulation	PT/INR for synthetic liver function and for Child Pugh score
	Chemistry	For electrolyte imbalance in cirrhotic patient and assessment of renal function
	Albumin	Liver synthetic function, nutritional status and Child Pugh score
	AFP	Tumor marker for HCC
	Hepatitis screen	HBV and HCV
Radiology:		
	US	Screening image for high risk patient It evaluate liver lesion and liver parenchyma
	CT and /or MRI	- Number and size of lesion and it's relation to major vessel - If it is classic HCC appearance by one image no need to be supported by another imaging. - Assessment of future liver remnant - Role out metastasis
	Portal Portal vein pressure	If patient cirrhotic and he is candidate for surgery with questionable portal hypertension
	Portal vein embolization	If future liver remnant small
Biopsy:		
	Core biopsy (tumor)	If CT and /or MRI are not classical for HCC
	Liver biopsy	If cirrhosis is not clear and sometime to know the cause of cirrhosis
Others:		
	Child Pugh score	If patient cirrhotic
	ICG	For assessment of adequate liver reserve before major resection (in some centers)
	Pre anesthesia evaluation	

Table 2. Preoperative checklist for HCC patient

1.3. Determining the extent of tumor involvement

Anatomic delineation of tumor extent is best achieved with dynamic multiphase CT or MRI scanning. Arterial phase imaging detects 30 -40 % more tumor nodules than conventional CT and may be the only phase to demonstrate the tumor in 7 -10 % of cases. Typical picture of HCC on CT will be an enhanced lesion on arterial phase (figure 1(a)) and early washout of contrast on venous phase (figure 1(b)).

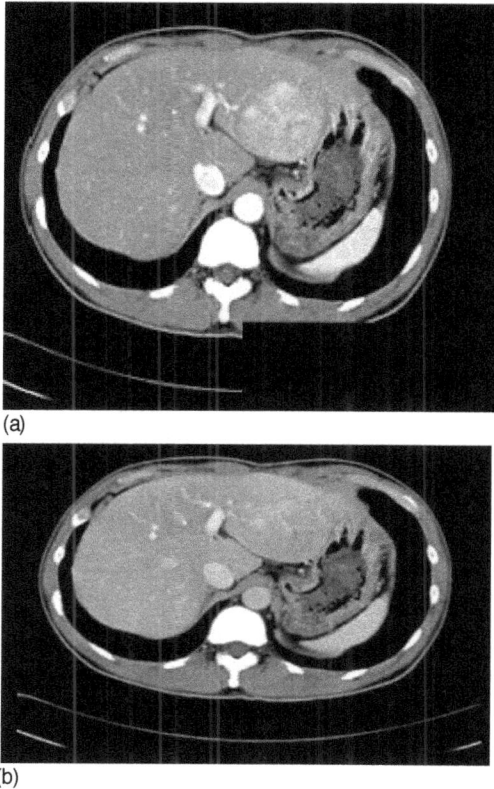

(a)

(b)

Figure 1. (a) Enhanced lesion on arterial phase in HCC (b) Early washout of contrast on venous phase in HCC

There is no general rule regarding tumor size for selection of patients for resection. Certainly, patients with smaller tumors are less likely to harbor occult vascular invasion and have a better outcome after therapy. Size alone is not a contraindication for resection of multinodular HCC.

Lymph node metastases are uncommon overall (between 1 - 8 %), but their presence portends a worse outcome. Preoperative detection of nodal metastases is limited by the frequent presence of benign nodal enlargement, most often involving the porta hepatis and portacaval space, in patients with cirrhosis. Highly suspicious nodes based on enhancement similar to the intrahepatic HCC lesions indicate the need for biopsy in a patient being considered for resection. However, involved nodes are not a contraindication to surgery for fibrolamellar HCC; these patients should have a formal lymph node dissection.

A chest CT is recommended to complete the staging evaluation and bone scan if suspicious bone pain or hypercalcemia. HCC has lower FDG accumulation in well-differentiated and low-grade tumors than in high-grade tumors. In a study by *Khan et al*, the sensitivity of PET in diagnosis of HCC was 55% compared with 90% for CT scanning, although some tumors (15 %) were detected by PET only (including distant metastases). So, PET imaging may help assess tumor differentiation and may be useful in the diagnosis, staging and prognostication of HCC as an adjunct to CT. However, the utility of PET scanning for detection primary and occult distant metastatic disease is uncertain, need to be explored further and not recommended in guidelines from the National Comprehensive Cancer Network (NCCN).

1.4. Assessment of hepatic reserve

Operative mortality is related to the severity of the underlying liver disease; it is 7- 25% in cirrhotic and less than 3% in non-cirrhotic patients. In patients with cirrhosis, surgical resection is most safely performed in those with Child-Pugh class A (table 3) disease who has a normal bilirubin and well preserved liver function. However, even Child-Pugh class A patients may develop rapid hepatic decompensation following surgery due to limited functional hepatic reserve.

CHILD – PUGH SCORE			
Clinical and laboratory parameter	**Scores**		
	1	**2**	**3**
Encephalopathy (grade)	None	1-2	3-4
Ascites	None	Slight	Moderate
Albumin (g/dL)	> 3.5	2.8-3.5	< 2.8
Prothrombin time prolonged (sec)	1-4	4-6	6
Bilirubin (mg/dL)	< 2	2-3	> 3
For primary biliary cirrhosis	< 4	4-10	> 10

Class A = 5–6 points; Class B = 7–9 points; Class C = 10–15 points.

Class A: Good operative risk

Class B: Moderate operative risk

Class C: Poor operative risk

Table 3. Child Pugh score

Although helpful, the Child-Pugh classification and other tools for assessing underlying liver disease, such as the Model for End-stage Liver Disease (MELD) score, are not adequate to select patients with sufficient hepatic reserve for major resection.

Multiple studies have demonstrated that a normal serum bilirubin level and the absence of clinically significant portal hypertension (i.e., hepatic venous pressure gradient <10 mm Hg) are the best available indicators of acceptably low risk of postoperative liver failure after liver resection. In the absence of an elevated serum bilirubin and portal hypertension, survival after PH can exceed 70% at 5 years. Survival after liver resection in patients with significant portal hypertension alone decreases to < 50% at 5 years. However, in patients with both an elevated serum bilirubin and significant portal hypertension, survival drops to < 30% at 5 years, regardless of Child-Pugh score. Direct measurement of portal pressure is not necessary in patients with clinical signs of severe portal hypertension, including esophageal varices, ascites, or splenomegaly associated with a platelet count less than 100,000/mL.

In many centers the Child-Pugh score may be supplemented by specialized investigations such as the indocyaninegreen (ICG) retention test, especially in marginal cases (e.g. Child-Pugh B, possible mild portal hypertension). ICG retention of 14% at 15 min is widely accepted (in Asia Pacific area) as a reflection of adequate functional reserves for major resection (defined as resection of > 2 Couinaud segments) (figure 2).

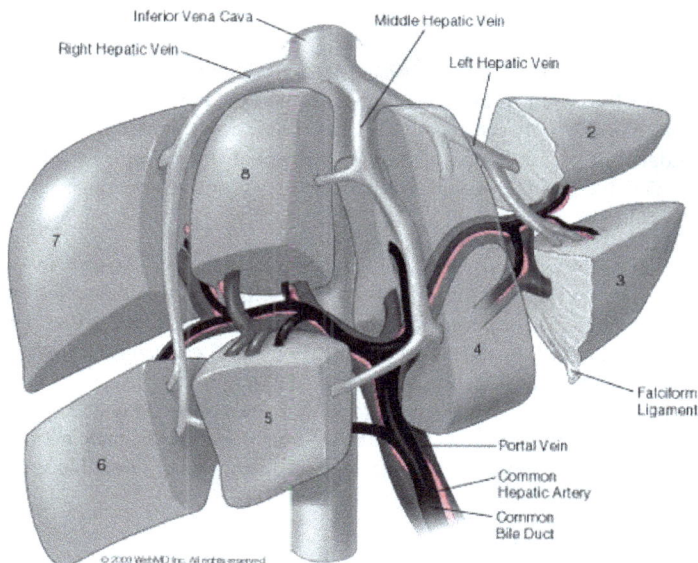

Source: ACS Surgery © 2003 WebMD Inc.

Figure 2. Couinaud liver segments

Assessment of the volume and function of residual liver should also be addressed by CT volumetry, particularly since portal vein embolization can be a valuable tool to increase the liver remnant volume and function prior to major hepatic resection, particularly for right sided tumors.

1.5. Portal vein embolization

Preoperative portal vein embolization (PVE) is a valuable adjunct to major liver resection. PVE can initiate hypertrophy of the anticipated future liver remnant to enable an extended resection in a patient with normal liver or major resection in a well compensated cirrhotic patient that would otherwise leave a remnant liver insufficient to support life following partial hepatectomy.

There are potential benefits to use of PVE:

• Reduce post-operative morbidity and mortality,

• Convert unresectable tumor due insufficient future liver remnant to resectable for potential cure.

• Subclinical disease or rapid progression may be detected prior to definitive surgery on post-embolization imaging studies, thus sparing the patient an unnecessary operation.

• The absence of compensatory hypertrophy identify patient with impaired liver regeneration,for that decrease post liver resection failure by preclude them from major liver resection.

The success of PVE was addressed by *Abulkhir A et al,* in a meta-analysis of data from 37 published series of PVE prior to liver resection. Four weeks after PVE, there was an overall increase in liver volume of between 10 and 12 % that was independent of technique, and 85 % of patients underwent planned laparotomy for attempted major hepatectomy. Following resection, only 23 patients had transient liver failure (2.5 %), and seven patients died of acute liver failure (0.8 %).

Liver regeneration usually peaks within the first 2 weeks after PVE. Studies in swine have shown that regeneration peaks within 7 days of PVE, with 14% of hepatocytes undergoing replication. Regeneration rates reported for humans are comparable to those found in animals. Non-cirrhotic livers demonstrate the fastest regeneration: 12–21 cm^3/day at 2 weeks after PVE, approximately 11 cm^3/day at 4 weeks, and 6 cm^3/day at 32 days. Livers in patients with cirrhosis or diabetes regenerate more slowly (approximately 9 cm^3/day at 2 weeks). Biliary obstruction, diabetes, chronic ethanol consumption, nutritional status, male gender, old age, and hepatitis all limit regeneration. Controlling these factors where possible is essential to maximize liver hypertrophy.

Two techniques can be utilized for PVE:

• The TIPE procedure is performed via a minilaparotomy and requires general anesthesia.

• The percutaneous approach (PTPE), which is more commonly used, can be performed in the radiology suite with local anesthesia and conscious sedation.

Volumetric assessment of the liver volume with CT imaging should be done before PVE and again before surgery. A standardized technique for measuring the future liver remnant, to select patients for PVE prior to a planned extended hepatectomy (trisegmentectomy) or hemihepatectomy in the setting of underlying liver disease, is strongly recommended. In considering the need for PVE, the ratio of future liver remnant and total estimated liver volume should be calculated. If the future liver remnant is < 20% in a patient with a normal liver or 40% in a patient with a cirrhotic liver, PVE should be considered.

Some reports have shown accelerated tumor growth in the liver after PVE. Problems with tumor growth are not seen when all of the tumor-bearing areas of the liver are embolized.

Transarterial chemoembolization (TACE) has been proposed as a complementary procedure prior to PVE in patients with HCC. TACE not only eliminates the arterial blood supply to the tumor, but it also embolizes potential arterioportal shunts in cirrhotic livers that attenuate the effects of PVE. Most reserve the "double embolization" procedure for patients with HCC in patients with liver disease who require right hepatectomy.

2. Surgery

Liver resection is a potentially curative therapy for patients with early-stage HCC (solitary tumor ≤5 cm in size, or ≤3 tumors each ≤ cm in size and no evidence of gross vascular invasion) in a Child-Pugh class A score and no evidence of portal hypertension (although a minor resection could be considered in some patients with portal hypertension) $_{Table 4}$. However, in highly selected cases, patients with a Child-Pugh class B score may be considered for limited liver resection, particularly if liver function tests are normal and no portal hypertension.

Optimal tumor characteristics for liver resection are solitary tumors without major vascular invasion. Although no limitation on the size of the tumor is specified for liver resection, the risk of vascular invasion and dissemination increases with size. However, in one study by *Pawlik TM*, no evidence of vascular invasion was seen in approximately one-third of patients with single HCC tumors of 10 cm or larger. Nevertheless, the presence of macro- or microscopic vascular invasion is considered to be a strong predictor of HCC recurrence.

Liver resection is controversial in patients with limited and multifocal disease as well as those with major vascular invasion. Multifocality is associated with lower survival, but does not exclude a good outcome in selected patients. In several studies, resection of multifocal HCC is associated with five-year survival rates of approximately 24 %. Patients with multinodular HCC who appear to benefit from resection are those with sufficient liver reserve to tolerate resection, without extrahepatic disease and without major vascular invasion. Liver resection in patients with major vascular invasion should only be performed in highly selected situations by experienced teams.

Despite even aggressive surgical approaches, most patients have HCC or liver disease too advance to permit treatment with "curative" intent. In high-incidence regions of the world, only 10 to 15 % of newly diagnosed patients are candidates for standard resection, whereas in low incidence areas, between 15 and 30 % of patients are potentially resectable. Furthermore, only one-half of patients referred for surgery actually have resectable tumors. Among the reasons for unresectability are the extent of intrahepatic disease, extrahepatic extension, inadequate functional hepatic reserve, and involvement of the confluence of the portal or hepatic veins.

Indication for liver resection:
Indicated:
- Solitary tumor ≤5 cm in size or ≤3 tumors each ≤3 cm in size and no evidence of gross vascular invasion.
- Solitary tumors (any size) without major vascular invasion.
Patient should be in a Child-Pugh class A and no evidence of portal hypertension
Controversial:
- Multifocal disease
- Major vascular invasion
- Child-Pugh class B score and no portal hypertension

Table 4. Indications for liver resection in hepatocellular carcinoma

2.1. Intraoperative staging

Laparoscopy and intraoperative ultrasound (IOUS) may improve the selection of patients for potentially curative resection. IOUS can accurately determine the size of the primary tumor and detect portal or hepatic vein involvement, which precludes curative resection. Another benefit of IOUS is the identification of major intrahepatic vascular structures, which can be used to guide segmental or non-anatomic resections.

2.2. Technique

In non-cirrhotic liver, an anatomical resection should be performed. Up to two-thirds of the functional parenchyma can be removed safely depending upon the age of the patient and his liver's regenerative capacity. However, for cirrhotic patients, because the capacity for liver regeneration is impaired in these patients, resection is generally limited to less than 25% of functional parenchyma. to maintain postoperative liver function. However, some patients maintain adequate functional hepatic reserve even after a formal hemi-hepatectomy, particularly if preoperative portal vein embolization (PVE) is used to induce compensatory hypertrophy in the future liver remnant. Both anatomic and wedge resection are acceptable, though some studies suggest portal-oriented resections enable longer overall and disease-free survival when feasible which might be because of the pattern of intrahepatic spread of liver cancer cells along segmental portal vein pedicle, so segmental resection may improve the chance of tumor clearance compared with a non-anatomical wedge resection.

Surgical outcomes in cirrhotic patients have improved over the past decade as a result of advances in surgical techniques, in particular the techniques that help to reduce bleeding during liver parenchyma transection and perioperative support. One of the most important advances is the thorough understanding of the segmental anatomy of the liver, which can be delineated using intraoperative ultrasound during operation. The delineation of a proper transaction plane is important not only for adequate tumor-free margin in resection of liver tumors but also to avoid inadvertent injuries to major intrahepatic vessels or bile duct pedicles. Use of the Pringle maneuver for vascular inflow occlusion as an alternative to total vascular occlusion has decrease deleterious effect on liver. Intermittent Pringle occlusion is well tolerated by cirrhotic patients for up to 60 minutes and is better tolerated than continuous clamping. The use of low CVP (less than 5mm Hg) anesthesia and newer instruments such as the ultrasonic dissector, hydrojet and vascular stapling devices has also significantly improved visualization, limited blood loss and decreased operative times.

2.3. Anterior technique

Some surgeons have advocated an anterior or "no touch" technique to resection of these tumors. This approach utilizes initial transection of the liver parenchyma to the inferior vena cava (IVC), and ligation of the inflow and outflow vessels before mobilization of the right liver lobe. The advocates of this technique hypothesize that separation of the right liver and the tumor from the IVC before mobilization avoids prolonged rotation and displacement of the hepatic lobes, therefore reducing the risk of vascular rupture. In addition, division of the vessels before tumor manipulation theoretically minimizes the potential for tumor cell dissemination caused by tumor compression.

2.4. Centrally located tumors

Surgical management of centrally located tumors (i.e., those in segments IV, V, and VIII) is especially problematic. Extended right or left hemi-hepatectomy is the treatment of choice if potentially curative surgery can be undertaken safely. An alternative segment-oriented approach, meso-hepatectomy (also called central hepatectomy), has been proposed in which the central liver segments IV and/or V, and VIII (with or without segment I) are removed and the lateral sectors remain intact.

While randomized trials have not been conducted, the available data suggest that meso-hepatectomy is a reasonable alternative to extended resection for centrally located tumors, providing acceptable oncologic outcomes with less hepatic parenchymal loss. However, in some centers, meso-hepatectomy is seldom used, partly because it is a complex and technically demanding procedure that requires two hepatic resection planes and bilateral biliary reconstruction. This results in a higher risk of bile leak and bleeding as well as long-term biliary stricture and biliary dysfunction. In addition, some data suggest that portal vein embolization followed by major hepatectomy might be safer.

2.5. Minimally invasive surgery

The success of minimally invasive resection of benign hepatic tumors has led to interest in laparoscopic approaches to surgery for HCC. The available literature is limited by the lack of prospective trials and the paucity of information on long-term oncologic outcomes.

Looking to the available literature, laparoscopic resection is feasible and safe in experienced hands. It is also highly technically demanding and should be undertaken only in high volume centers.

2.6. Tumor rupture

Approximately 10% of HCC spontaneously rupture. The clinical picture is that of acute abdominal pain and distension with drop in the hematocrit and hypotension. Initially, these patients' hemodynamic should be stabilized followed by trans-arterial embolization for control of bleeding. If unsuccessful, emergency surgery may be required.

Although the presence of a tumor rupture suggests a high likelihood of peritoneal seeding and usually a poor outcome from resection, this is not inevitable. If bleeding can be controlled (arterial embolization is recommended), a formal staging evaluation should be undertaken, followed by laparoscopic exploration and a subsequent attempt at resection, if feasible. Several retrospective series suggest a low, but defined long-term survival rate following resection in such situations.

In the largest series from Hong Kong by *Liu CL et al*, 154 of 1716 patients who were newly diagnosed with HCC between 1989 and 1998 presented with spontaneous rupture. The 30-day mortality rate following tumor rupture was 38 %. After initial stabilization and clinical evaluation, 33 underwent hepatic resection. Although the median survival after hepatectomy was worse in ruptured as compared to non-ruptured cases (26 *versus* 49 months) and the rate of extra-hepatic recurrence was higher (46 *versus* 26 %), 8 patients (24 %) remained alive without recurrent disease after a median follow-up period of 45 months.

2.7. Postoperative management

Postoperative management is primarily supportive. Those patients should be monitored in ICU with great attention to hydration status, not over or under hydrated with CVP monitor. The extent of postoperative morbidity is related to the extent of operative resection. Major postoperative complications include bile leak in 8% and pleural effusion in 7%, which are usually treated conservatively.

2.8. Perioperative mortality

The 30-day operative mortality rate in modern series of HCC resection ranges widely from 1 -24 %. Fewer than 10 % of perioperative deaths are due to uncontrolled intraoperative hemorrhage; most are due to postoperative liver failure. The presence of cirrhosis is the most important predictor of post-resection liver failure and death. The 30-day postoperative mortality for cirrhotic patients ranges from 14 -24 %, compared to 0.8 -7 % for non-cirrhotics.

Two additional factors influence the development of postoperative liver failure in cirrhotic patients are intraoperative blood loss of >1500 ml and postoperative infection of any type. Mortality can also be reduced by appropriate selection of patients and meticulous surgical technique, with the inclusion of preoperative volumetry and portal vein embolization when appropriate.

Consensus is growing that 30-day operative mortality is an inadequate indicator of risk, particularly of postoperative hepatic insufficiency and failure. Using an approach similar to liver transplantation reporting, 90-day mortality reporting appears to be a more valuable indicator of outcome of liver resection, especially in the cases of extended resection and resection in patients with diseased livers. This relates to the late development of slowly progressive jaundice, ascites, and eventual death, which typically occurs outside the hospital and well after 30 postoperative days in patients with marginal or inadequate liver remnants (post resection liver failure will be discussed in detail at end of this chapter).

2.9. Fast track surgery

Surgical pathway and 'fast-track' (FT) programs are structured interdisciplinary strategies that have been introduced to optimize perioperative care and accelerate post-operative recovery. The main aim of the FT protocol is to reduce the metabolic and inflammatory response to surgical stress and preserve vital functions. A review done by *Lidewij et al* showed primary hospital stay was significantly reduced after FT care in two out of the three studies. In one study, median hospital stay was 6 days in the FT group compared with 8 days in the control group ($P < 0.001$). In the other study, primary hospital stay was reduced from 11 days to 7 days ($P < 0.01$). There were no significant differences in rates of readmission, morbidity and mortality between FT and control groups. One trial found a significantly shorter time to successful resumption of a normal diet in the FT group (1 post-operative day for FT patients vs. 3 days for the control group).

3. Long-term outcomes

Results of large retrospective studies have shown 5-year survival rates of over 50% for patients undergoing liver resection for HCC, and some studies suggest that in carefully selected patients having no vascular invasion by tumor, solitary lesions without intrahepatic metastasis, tumor diameter ≤5 cm, and a negative surgical margin of >1 cm, five-year survival rates up to 78 %. However, HCC tumor recurrence rates at 5 years following liver resection have been reported up to 70%.

Palavecino M et al reported series of 54 patients with advanced HCC and significant tumor burden who were treated with PVE plus major hepatectomy, the five-year overall survival was 72 % and the five-year disease-free survival was 56 %.

3.1. Tumor-related prognostic factors

The most important tumor-related prognostic factors are presence and degree of vascular invasion, tumor number and size, and surgical margin status. Other poor prognostic indicators are absence of a tumor capsule, preoperative alpha fetoprotein (AFP) levels >10,000 ng/ml, and poor histologic grade of differentiation.

Both intrahepatic and extra-hepatic spread of HCC is more common with tumors >5 cm, particularly when associated with venous invasion. In a report by Zhou XD compared 1000 patients with tumors ≤5 cm and 1366 patients having tumors >5 cm, all of whom underwent hepatectomy over the same period, five-year survival rates were significantly better for patients with smaller tumors (63 %versus 37 %, respectively). Nevertheless, several series indicate five year survival rates ranging from 25 to 35 percent in selected patients undergoing resection for single HCC ≥10 cm. However, although increasing tumor size is associated with increased risk for vascular invasion, large, solitary tumors without vascular invasion have the same prognosis as small solitary tumors without vascular invasion.

The importance of wide resection margins is debated. In study by *Ozawa K* of 225 patients with HCC who underwent resection, three-year survival was significantly better when a >1 cm tumor-free margin was achieved (77 % *versus* 21%, respectively). However, larger series by *Poon RT* suggest that a negative margin of <1 cm is acceptable.

Gross or microscopic invasion of branches of the portal or hepatic veins is associated with a lower probability of survival following resection.

3.2. Underlying liver dysfunction

Preoperative liver dysfunction and cirrhosis are important negative prognostic factors. *Yamanaka N* reported a series of 295 patients undergoing resection of HCC, the four-year survival was more than twofold higher for non-cirrhotic compared to cirrhotic patients (81% versus 35 %). This difference in outcome may be related in part to the higher frequency of multicentric HCC in cirrhotic patients.

In patients with cirrhosis related to HBV infection, active hepatitis is also a poor prognostic factor. As a general rule, the severity of cirrhosis, rather than the presence of a small, early stage HCC, limits long-term survival in cirrhotic patients with HCC. Chronic liver disease provides a field that contributes to the development of second primary HCCs and a persisting risk of HCC-related death beyond five years.

3.3. Recurrences

Treatment of recurrence is a poorly investigated area. Solitary recurrence might benefit from repeat resection, but in most patients recurrence will be multifocal.It has been suggested, retrospective analyses, that patients with recurrence might be candidates for salvage trans-plantation.. Most of the recurrences and specially those that appear early during follow-up are due to tumor dissemination and have a more aggressive biological pattern as compared to primary tumors. Hence, only those patients in whom recurrence is due to de novo oncogenesis can be expected to benefit from salvage transplantation or repeated resection. While the most accurate predictors of recurrence due to dissemination (vascular invasion, satellites) may be identified on pathology, and since the results of transplantation in these patients is good, some authors have proposed that this category of patients should be listed immediately after resection. This might be more effective than waiting for recurrence to develop with excessive tumor burden possibly excluding liver transplantation. Organ allocation policies might have to be modified to take these findings into account. Other treatment modalities can provide disease control (i.e., trans-arterial arterial embolization of chemoembolization, radiofrequency ablations, sorafenib).

Fewer than 20 % of disease recurrences have an extra-hepatic with overall poor prognosis, and the benefit of systemic therapy is modest, at best.

3.4. Surveillance

Although data on the role of surveillance in patients with resected HCC are very limited, recommendations are based on the consensus that earlier identification of disease may facilitate patient eligibility for investigational studies or other forms of treatment. The NCCN panel recommends high-quality cross-sectional imaging every 3-6 months for 2 years, then every 6-12 months. AFP levels, if initially elevated, should be measured every 3 months for 2 years, then every 6-12 months. Re-evaluation according to the initial work-up should be considered in the event of disease recurrence.

3.5. Survival

Liver resection is a potentially curative therapy for patients with early-stage HCC (solitary tumor ≤5 cm in size, or ≤3 tumors each ≤ cm in size and no evidence of gross vascular invasion). 5-year survival rates of over 50% for patients undergoing liver resection for HCC, and some studies suggest that for selected patients with preserved liver function and early stage HCC, liver resection can achieve a 5-year survival rate of about 70%. However, HCC tumor recurrence rates at 5 years following liver resection have been reported to exceed 70%.

3.6. Post-Resection Liver Failure (PRLF)

PRLF is a devastating complication that is resource intensive and carries with it considerable morbidity and mortality. The reported incidence of PRLF ranges between 0.7 - 9.1%. An inadequate quantity and/or quality of residual liver mass are key events in its pathogenesis. Major risk factors are the presence of comorbid conditions, pre-existent liver disease and small Remnant Liver Volume (RLV). It is essential to identify these risk factors during the pre-operative assessment that includes evaluation of liver volume, anatomy and function.

There is no uniformity concerning the definition of PRLF. In general, PRLF is characterized as failure of one or more of the hepatic synthetic and excretory functions that include hyper-bilirubinemia, hypo-albuminemia, prolonged prothrombin time, elevated serum lactate and/or different grades of hepatic encephalopathy (HE). PRLF is defined by the so-called 50–50 criteria, which describe PRLF as prothrombin index less than 50% (equal to an international standardized ratio more than1.7) and serum bilirubin more than 50 mmol/L (2.9 mg/dL) on post-operative day 5. When these 50–50 criteria were fulfilled, patients had a 59% risk of mortality compared with 1.2% when they were not met (sensitivity 69.6% and specificity 98.5%). This rarely occurs in isolation and is often coupled with failure of multiple organs and/or features of sepsis.

3.7. Pathophysiology of PRLF

After resection of various amounts of functional liver mass, both death and regeneration of the remaining hepatocytes occur. Physiologically, regeneration outweighs hepatocyte death and both liver mass and function are restored rapidly. For example, during the first 10 days after right hepatectomy for living donor liver transplantation, restoration of liver mass up to 74% of the initial volume has been reported. This regeneration is triggered by an increased metabolic demand placed upon remnant hepatocytes. The ability of the liver remnant to surmount the effect of surgical resection depends on its capacity to limit hepatocyte death, to resist metabolic stress, to preserve or recover an adequate synthetic function and to enhance its regenerative power. These factors rely on both the quality and the quantity of remaining liver parenchyma. A variety of intraoperative as well as post-operative hits identified that may attribute to the development of PRLF. These include hepatic parenchymal congestion, ischemia–reperfusion injury (IRI) and reduced phagocytosis capacity. Liver failure could be defined as either "cholestatic" (characterized by regeneration of hepatocytes and fibrosis) or "non-regenerative" (characterized by pronounced apoptosis of hepatocytes).

3.8. Hepatic parenchymal congestion

Partial liver resection leads to a relatively augmented sinusoidal perfusion, leading to shear–stress and congestion of hepatic parenchyma and resulting in vascular and parenchymal damage similar to small-for-size syndrome after liver transplantation, although less severe. Moreover, inadequate venous drainage of the liver remnant induces hepatic venous congestion and functional hepatic volume loss. Hepatic parenchymal congestion may be less severe in patients with cirrhosis of the liver with preexisting portacaval collaterals.

3.9. Hepatic ischemia–reperfusion injury

Hepatic ischemia–reperfusion injury follows massive bleeding or hepatic in- or outflow occlusion during liver surgery. Although the resistance of the liver to warm ischemia is relatively high, hepatic ischemia and reperfusion activate a complex cascade that triggers the innate immune response by recruitment and activation of Kupffer cells, endothelial cells and the complement system. These express pro-inflammatory proteins [nuclear factor kB, tumour necrosis factor-a, interleukin-6], reactive oxygen species, chemokines, complement factors and vascular cell adhesion molecules. Subsequently, polymorphonuclear neutrophils are activated, which aggravate hepatic injury. Although these processes are primarily intended to maintain homoeostasis, unrestrained activation may become destructive.

3.10. Reduced phagocytosis capacity

Infection complicates the course of PLF either as a precipitant or during later stages. Partial hepatectomy reduced the phagocytosis capacity of the hepatic reticuloendothelial system. Nevertheless, the liver remnant has to clear bacteria and their products following bacterial translocation or intra-abdominal infection. Diminished hepatic clearance of bacteria might enhance the susceptibility for the development infections and PRLF.

3.11. Risk factors of PRLF

The extent of resection correlates most closely with rates of PRLF and death; and the incidence increases with the number of segments resected. The incidence of PLF is < 1 % in patients with no underlying parenchymal disease when 1-2 segments are resected, around 10 % when 4 segments are resected, and 30 % when5 segments or more are resected. However, the exact amount of residual liver mass required to preserve sufficient liver function is unknown. In general, an RLV ≥ 25–30% in otherwise healthy livers is consistent with a good post-resectional outcome. RLV below 25% in normal livers predicted PRLF with a positive predictive value of 90% (95% CI 68–99%) and a specificity of 98% (95% CI 92–100%). When liver function is restricted, RLV should be as high as 40% to guarantee adequate remnant liver function.

The use of vascular occlusive techniques and significant intraoperative blood loss can exacerbate the level of dysfunction. Vascular occlusive techniques induce ischemia in the liver remnant. These effects are greatest following total vascular exclusion (inflow + outflow occlusion), but also occur after prolonged intermittent inflow occlusion.

Intraoperative blood loss (> 1–1.2 liters) and the need for blood transfusion increase the risk of PLF and sepsis. This may relate to the immunosuppressive effects of blood transfusion or the initiation of the inflammatory response that accompanies significant hemorrhage.

Vascular reconstruction following *in situ en bloc* liver and inferior vena cava resection or *ex vivo* liver resection is associated with increased rates of PRLF. *Ex vivo* resection and reimplantation is associated with an unacceptably high mortality rate. Biliary reconstruction is associated with increased morbidity and mortality after liver resection but does not independently predict PRLF.

Underlying parenchymal disease reduces the functional and regenerative capacity of the liver remnant. In patients with cirrhosis but no functional impairment or portal hypertension, resection of up to 50 % is safe. In patients with Child–Pugh grade B or C disease, even small resections can result in PRLF. The high risk of developing PLF in patients with cirrhosis can be explained by the wide range of comorbid conditions like portal hypertension, diabetes mellitus, jaundice, malnutrition, hypersplenism and coagulopathy as well as frequent impaired preoperative liver function and hepatic functional reserve. Furthermore, patients with cirrhosis have an impaired hepatic regenerative capacity. NAFLD (non alcoholic fatty liver disease) represents a spectrum of disease ranging from steatosis to steatohepatitis (non-alcoholic steatohepatitis, NASH), fibrosis and cirrhosis. The grade of steatosis, correlates with rates of PRLF and death following major Resection. The presence of steatosis is hypothesized to be associated with impaired hepatic microcirculation, decreased resistance to ischemia–reperfusion injury, increased intrahepatic oxidative stress and dysfunction in mitochondrial adenosine triphosphate synthesis. Chemotherapy-induced liver injury is increasingly prevalent as more patients receive chemotherapy for colorectal liver metastases before liver resection. Cholestasis reduces both hepatic metabolic and regenerative capacities, and increases rates of liver dysfunction after major resection.

Other patient-based factors that predict PRLF are age, malnutrition, diabetes mellitus and male sex. Elderly patients (≥65) suffer frequently from comorbid conditions and have reduced regenerative capacity of hepatocytes. Approximately 65–90% of patients with advanced liver disease suffer from protein–calorie malnutrition. Malnutrition is associated with an altered immune response, reduced hepatic protein synthesis and a reduction in hepatocyte regenerative capacity. Diabetes mellitus is associated with increased morbidity and mortality after liver resection. This may be due to immune dysfunction or because insulin absence or resistance reduces regenerative capacity. PRLF is more common in males as testosterone may have immune-inhibitory effects, predisposing to septic complications.

3.12. Prevention of PRLF

Diabetes mellitus should be screened for and treated before surgery. Nutrition should be evaluated and consideration given to preoperative oral carbohydrate loading in order to reduce postoperative insulin resistance. There is no evidence to support delaying liver resection for a period of nutritional optimization, unless the patient is severely malnourished. It has been hypothesized that the nutritional status of depleted patients should be corrected via oral, enteral or parenteral methods before surgery. A meta-analysis on the effect of total parenteral nutrition compared with enteral nutrition on morbidity and mortality after liver resection revealed no superiority of either form of nutrition. However, a beneficial effect of additional parenteral nutrition has been demonstrated in a subgroup of patients who had cirrhosis and underwent major hepatectomy.

The risk of PRLF may be reduced by strategies to increase parenchymal volume and protect against parenchymal damage. Strategies available for volume manipulation for HCC patients include portal vein embolization alone or in combined with locoregional treatment (RFA or TAE). Portal vein embolization induces apoptosis in the ipsilateral lobe, and proliferation and growth of the contralateral lobe. This increases the functional capacity of the liver remnant, limits the effects of hepatic hyperperfusion that may occur in a small-for-size remnant, and predicts the regenerative response in the future remnant. Failure to proliferate after portal vein embolization can be used to select patients with impaired regenerative capacity in which major resection would not be tolerated. The primary concern over portal vein embolization is that it may increase tumor growth owing to an ipsilateral surge in hepatic arterial flow. Locoregional treatment can be used in combination with Portal vein embolization to control tumor load before resection.

In order to limit parenchymal damage and optimize regenerative capacity, a series of hepato-protective measures may be employed (intermittent portal clamping, ischemic preconditioning and hypothermic liver preservation). Total vascular occlusion should be avoided unless resection cannot be undertaken without it (for example a tumor at the cavohepatic intersection). If resection without vascular occlusion is not possible, inflow occlusion is preferable to total vascular exclusion. Intermittent portal clamping with intervals allowed for reperfusion is preferred to continuous clamping, usually applying a 15-min clamp–5-min release regimen. Ischemic preconditioning increases tolerance to prolonged hepatic ischemia and adenosine 5 -triphosphate depletion by exposing the parenchyma to short intervals of ischemia and reperfusion intraoperatively before resection. This downregulates ischemia–reperfusion injury and results in less hepatic injury. Ischemic preconditioning reduces the histological effects of ischemia–reperfusion injury, however, without improving clinical outcome. Hypothermic liver preservation in conjunction with total vascular exclusion attenuates ischemia–reperfusion injury. The future remnant is infused with a preservative fluid and surrounded by crushed ice to maintain the liver at 4 °C.

Data from living liver donors suffering from biopsy proven moderate steatosis revealed that a body weight reduction of 5% or intervention with a low-fat, high protein diet and exercise significantly improved hepatic steatosis. However, weight reduction before surgery may not be feasible because of time deficit and the often pre-existent malnutrition.

Patients with cirrhosis of the liver are more susceptible to the development of PRLF in case of resection of comparable tumor volumes. However, cirrhosis of the liver cannot be prevented and, therefore, prevention of PRLF in these patients can only be achieved by careful patient selection, adequate nutritional support and the use of an appropriate surgical technique.

3.13. Manifestation

PRLF reflects deregulation of the synthetic, excretory and detoxifying capacities of the liver remnant. In addition, the majority of patients suffering from PRLF will also meet the criteria of the systemic inflammatory response syndrome and experience multiple organ failure. Unfortunately, a substantial number of patients suffering from PRLF deteriorate, leading to a fatal outcome in approximately 80%. However, PRLF is a potentially reversible disorder because of the regenerative capacity of the liver remnant.

3.13.1. Liver

The clinical consequences of PRLF are jaundice, coagulopathy, ascites, edema and/or HE.

Ascites occurs as a result of surgery (portal hypertension, dissection, gross fluid overload), and may be difficult to assess it in the immediate postoperative period.

Data from Suc et al. and Balzan et al. concerning liver function on different days after uncomplicated hepatic resection showed an initial increase of serum bilirubin and a decrease of prothrombin time before normalization of these values on the seventh post-operative day.

3.13.2. Circulation

Circulatory failure occurring during PRLF resembles the circulatory failure of patients with sepsis. The pathophysiological changes usually observed are enhanced vascular permeability, diffuse intravascular coagulation and peripheral vasodilatation that are clinically represented by reduced peripheral resistance and hemodynamic instability.

3.13.3. Kidneys

Post-hepatic resection renal dysfunction can either result from perioperative disturbances in renal circulation inducing acute tubular necrosis or accompany PRLF. It is characterized by azotemia or oliguria and may cause ascites formation; pleural effusion and fluid overload requiring diuretics or hemofiltration. There is a distinct chance of reversibility of renal failure when there is recovery of PRLF. Furthermore, it can be hypothesized that the pivotal role of the kidney in ammonia excretion is impaired, leading to hyperammonemia and HE in patients suffering from PRLF.

3.13.4. Lung

Although moderate pulmonary edema seems to be a normal finding after partial hepatic resection owing to general hemodynamic alterations, this usually does not impair oxygen exchange. Severe remote lung injury, pulmonary edema and acute respiratory distress syndrome can develop as part of the multiple organ dysfunction syndromes that accompanies PRLF.

3.13.5. Hepatic encephalopathy

Hepatic encephalopathy is a potentially reversible neuropsychiatric disorder, characterized by varying degrees of confusion and disorientation. Hyperammonemia plays a central role in its development and has a direct toxic effect on neurotransmission and astrocyte function. Although hepatic encephalopathy are important markers for liver failure, altered mental state may occur in response to drugs such as opiates and may be difficult to assess it in the immediate postoperative period.

3.13.6. Treatment of PRLF

Large, randomized trials concerning the treatment of PRLF are lacking, and therefore, recommendations for treatment modalities are difficult to make. Management principles resemble those applied to patients with acute liver failure, acute-on-chronic liver failure or sepsis and focus on support of liver and end-organ function. Goal-directed therapy should be provided for circulatory disturbances, renal and ventilatory dysfunction, coagulopathy, malnutrition and HE (table 5). Patients should undergo clinical and laboratory assessment after liver resection, with the frequency of monitoring and level of care stratified according to risk.

It is normal for serum bilirubin levels and INR to rise in the first 48–72 h postresection. However, bilirubin concentration above 50 μmol/l (3 mg/dl) or INR greater than 1 7 beyond 5 days is unusual and usually reflects liver dysfunction. Serum bilirubin remains the most sensitive predictor of outcome in PLF. PT and INR are also valuable, but interpretation may be compromised if the patient has received clotting factors. Serum albumin, although an indicator of hepatic synthetic function, will vary in response to inflammation and administration of intravenous fluids. Increased levels of liver enzymes are common after liver resection and do not predict outcome. C-reactive protein levels are dampened after major liver resection, and day 1 levels inversely correlate with PRLF indices. Serum lactate has a prognostic value in severe sepsis and ALF, with a serum lactate level above 3 0 mmol/l after fluid resuscitation predicting death in ALF.

The systemic inflammatory response syndrome (SIRS) is present in more than 50 % of patients with ALF and predicts a negative outcome.The incidence of SIRS in patients with PLF has not been evaluated formally, but as in ALF it is likely to be implicated in sepsis, encephalopathy and end-organ dysfunction. Several studies have examined the role of postoperative functional assessment of the liver. The ICG15 predicts PRLF, but its value diminishes once liver failure is established because changes in hepatic blood flow also influence ICG15. Although PRLF is a potentially reversible condition, mortality rates remain high and currently there is little scope for therapeutic intervention.

Management of PRLF must be undertaken in conjunction with critical care, hepatology, infectious disease and radiology services. The pattern of organ dysfunction that occurs as a result of PRLF is similar to that in sepsis. Cardiovascular failure is characterized by reduced systemic vascular resistance and capillary leak. Acute lung injury, pulmonary oedema and acute respiratory distress syndrome may ensue. Acute kidney injury can progress rapidly in PRLF. Fluid balance should be managed judiciously with avoidance of salt and water overload. Identifying and treating underlying sepsis is a key in managing patients with PRLF. Sepsis may exacerbate PRLF, and bacterial infection is present in 80 % of patients with PRLF and in 90 % of those with ALF. Any acute deterioration should be attributed to sepsis until proven otherwise. Management of sepsis should be in accordance with the surviving sepsis guidelines. A trial of prophylactic antibiotics after liver resection failed to show a reduction in liver dysfunction or infective complications. However, the administration of antibiotics in patients suffering from acute liver failure is associated with a significant decrease in infectious complications and this may also be advantageous in patients suffering from PRLF. In critically ill patients with PRLF, chest radiography and cultures of blood, urine, sputum and drain site/ascetic fluid should be performed. Current guidelines for ALF propose that broad spectrum antibiotics should be administered empirically to patients with progression to grade 3 or 4 hepatic encephalopathy, renal failure and/or worsening SIRS parameters.

Coagulopathy may occur transiently after major resections and is found in all patients with PRLF. As in ALF, coagulation parameters can be used to chart the progress of PRLF, provided blood products have not been given. In a multinational review of fresh frozen plasma given for transient coagulopathy after resection, there was no consensus for its use. In the absence of bleeding it is not necessary to correct clotting abnormalities, except for invasive procedures or when coagulopathy is profound. The level at which a coagulopathy should be corrected before an interventional procedure in ALF has yet to be defined (the commonly used threshold for correction is an INR above 1 5). Vitamin K may be given, but this is not supported by clinical trials. Thrombocytopenia may complicate liver failure. Indications for platelet transfusion in ALF include bleeding, profound thrombocytopenia (< 20×10^{6} /L), or when an invasive procedure is planned. A platelet count above 70×10^{6} / L is deemed safe for interventional Procedures. Recombinant factor VIIa (rFVIIa) has been used to treat coagulopathy in patients with ALF.In a large controlled trial of rFVIIa following major liver resection, no reduction in bleeding events was observed. Its role in PRLF is yet to be defined.

Gastrointestinal hemorrhage is a recognized complication of liver failure. In ALF, H_2-receptor blockers and proton pump inhibitors (PPIs) reduce gastrointestinal hemorrhage in mechanically ventilated patients. In the non-ventilated patient an oral or sublingual PPI or oral H_2-receptor blocker is likely to protect against gastrointestinal hemorrhage. High risk patients or patients with established PRLF should therefore receive prophylaxis. Large-volume ascites may also complicate PRLF. As in ALF, when this causes severe abdominal discomfort and/or respiratory compromise, consideration should be given to therapeutic paracentesis with simultaneous volume replacement with a plasma expander (ideally 20 % salt-poor albumin solution). The ratio for replacement is 6-8 gram 20% albumin per liter ascites drained. Nutrition is important and supplementation should be established early in patients with liver failure. Enteral nutrition is the preferred route as it improves gut function and restores normal intestinal flora. Parenteral nutrition can be used when enteral feeding is not tolerated, but should be introduced with caution owing to the risk of infection. In critically ill patients ensuring euglycemia improves survival and reduces morbidity.

The role of imaging in PRLF is to assess hepatic blood flow, identify reversible causes of liver failure and locate sites of infection. Hepatic blood flow can be evaluated using non-invasive imaging. Doppler ultrasonography may identify portal vein, hepatic artery and hepatic vein thrombosis. Contrast CT or MRI can be used to establish hepatic blood flow, provide more details of vascular abnormalities and identify sites of infection. If patency of hepatic vessels is still in doubt on cross-sectional imaging, angiography is the 'gold standard'. Vascular disorders may complicate liver resection and induce PRLF, but are rare. Longitudinal exposure of hepatic veins and the use of ultrasonic dissection may lead to hepatic vein thrombosis. Portal vein thrombosis has also been implicated in the development of PRLF. In these rare cases of inflow and outflow thrombosis with PRLF, a decision must be made regarding the benefit of surgical or radiological thrombectomy or dissolution *versus* anticoagulation.

Cerebral edema and intracranial hypertension may occur as a result of PRLF. Cerebral edema is unlikely in patients with grade 1 or 2 hepatic encephalopathy. With progression to grade 3, a head CT should be performed to exclude intracranial hemorrhage or other causes of declining mental status. In patients with established ALF and encephalopathy, enteral lactulose might prevent or treat cerebral edema, although the benefits remain unproven. Progression to grade 3/4 encephalopathy warrants ventilation and may require intracranial pressure monitoring.

Stress ulcer	Proton pump inhibitor
Nutrition	Enteral energy supply of 2000 kcal/day
	Enteral preferred over total parenteral nutrition
	Maintain euglycemia
Sepsis	Serial chest X-ray, sputum, urine and blood culture
	Ascetic fluid from drain site
	Consider CT abdomen
	Broad spectrum antibiotic if progression of encephalopathy, renal failure or worsening SIRS parameters
Circulatory disturbances	CVP 8–12 mmHg
	MAP 70 mmHg
	Hematocrit >30%
	Pulmonary capillary wedge pressure ≤ 12–15 mmHg
Ventilatory dysfunction	Arterial oxygen saturation > 93%
	Central venous oxygen saturation > 70%
Renal dysfunction	Urine output > 0 5 mL/kg/hour
Coagulopathy	Correct if bleeding or interventional procedure (INR<1.5)
Thrombocytopenia	Correct if bleeding, profound thrombocytopenia (<20 × 10^6/L) or interventional procedure planned (<70 × 10^6/L)
Vascular inflow/outflow (thrombosis)	Doppler ultrasound
	CT/MR angiography
	If evidence of inflow/outflow occlusion consider anticoagulation/revascularization
Ascites	Paracentesis if severe pain/respiratory impairment
Encephalopathy	Lactulose
	If progression to grade 3–4 encephalopathy, CT head, ventilate and consider ICP monitoring

** ICP: IntraCranial Pressure, INR: International Normalized Ratio, MAP: Mean Arterial Pressure MR: Magnetic Resonance CT: Computed Tomography, CVP: Central Venous Pressure, SIRS: Systemic Inflammatory Response Syndrome

Table 5. Management of post resection liver failure.

The concept of hepatocyte transplantation has been investigated as a strategy to boost residual liver function. Intrahepatic hepatocyte transplantation has been used successfully to treat patients with metabolic disorders of the liver. The efficacy of orthotopic liver transplantation for PRLF has only recently been reported. However, no criteria are available for the selection of patients who will benefit from emergency liver transplantation for PRLF. Patient who have favorable tumor characteristics (i.e. R0 resection, low T and negative N status, HCC within Milan criteria and absence of extra-hepatic disease), without comorbid conditions and without a limited life expectancy because of other medical conditions considered to be good candidate for emergency transplantation.

Extracorporeal liver support (ELS) devices fall into two categories: artificial and bioartificial systems. Artificial devices use combinations of hemodialysis and adsorption over charcoal or albumin to detoxify plasma. Bioartificial devices use human or xenogenic hepatocytes maintained within a bioreactor to detoxify and provide synthetic function. These systems have not been evaluated extensively in patients with PRLF. A recent meta-analysis and systematic review showed that ELS might improve survival in patients with ALF, but not acute-on-chronic liver failure, in comparison with standard medical therapy.

Abbreviation

ALF Acute liver faliure

ICG Indocyaninegreen

HCC Hepatocellular carcinoma

FT Fast Track

RFA Radiofrequency ablation

PEI Percutaneous ethanol injection

SIRS Systemic Inflammatory Response Syndrome

PLRF Post-resection liver failure

TACE Transarterial chemoembolization

Author details

Mazen Hassanain*, Faisal Alsaif, Abdulsalam Alsharaabi and Ahmad Madkhali

*Address all correspondence to: mhassanain@ksu.edu.sa

Department of surgery, College of Medicine, Liver Disease Research Centre, King Saud University, Riyadh, Saudi Arabia

References

[1] Jordi Bruix, and Morris Sherman. Management of Hepatocellular Carcinoma: An Update. HEPATOLOGY, (2011)., 53(3)

[2] Peter Abrams, J. Wallis Marsh. Current Approach to Hepatocellular Carcinoma. Surg Clin N Am (2010)., 90(2010), 803-816.

[3] Erwin Kuntz. Malignant liver tumor. HEPATOLOGY TEXTBOOK AND ATLAS (2008). Part DOI:, 4, 795-835.

[4] Khan, M. A, Combs, C. S, Brunt, E. M, et al. Positron emission tomography scanning in the evaluation of hepatocellar carcinoma. J Hepatol (2000)., 32, 792-7.

[5] Hepatobiliary Cancers. National comprehensive cancer network 2.(2012). www.nccn.org

[6] Steven Curley, Carlton Barnett, Eddie Abdalla. Surgical resection for hepatocellular carcinoma.uptodate Feb.2. (2012). www.uptodate.com

[7] Pierce Kah-Hoe ChowResection for hepatocellular carcinoma: Is it justifiable to restrict this to the American Association for the Study of the Liver/Barcelona Clinic for Liver Cancer criteria? REVIEW. Journal of Gastroenterology and Hepatology (2012)., 27(2012), 452-457.

[8] Pawlik, T. M, Delman, K. A, Vauthey, J. N, et al. Tumor size predicts vascular invasion and histologic grade: Implications for selection of surgical treatment for hepatocellular carcinoma. Liver Transpl. (2005). Sep;, 11(9), 1086-92.

[9] Yokoyama, Y, Nagino, M, & Nimura, Y. Mechanisms of Hepatic Regeneration Following Portal Vein Embolization and Partial Hepatectomy: A Review.World J Surg ((2007).

[10] Abbdalla, E, Hicks, M, & Vauthey, J. Portal vien embolization: rational,technique and future prospect.British Journal of Surgery (2011)., 2011(88), 165-175.

[11] Eddie, K. Abdalla.Portal Vein Embolization Prior to Major Hepatectomy: The Evidence.Venous Embolization of the Liver.(2011). part, 6, 293-305.

[12] David, C. Transhepatic Portal Vein Embolization: Anatomy, Indications, and Technical Considerations. RadioGraphics (2002)., 22, 1063-1076.

[13] Abulkhir, A, Limongelli, P, Healey, A. J, et al. Preoperative portal vein embolization for major liver resection: a meta-analysis. Ann Surg. (2008).

[14] Liu, C. L, Fan, S. T, Lo, C. M, et al. Management of spontaneous rupture of hepatocellular carcinoma: single-center experience. J Clin Oncol. (2001).

[15] Lidewij Spelt et alFast-track programmes for hepatopancreatic resections: where do we stand?.Review. HPB (2011)., 2011(13), 833-838.

[16] De-Xin Lin *et al*Implementation of a Fast-Track Clinical Pathway Decreases Postoperative Length of Stay and Hospital Charges for Liver Resection. Cell Biochem Biophys ((2011).

[17] Zhou, X. D, Tang, Z. Y, Yang, B. H, et al. Experience of 1000 patients who underwent hepatectomy for small hepatocellular carcinoma. Cancer. (2001).

[18] Ozawa, K, Takayasu, T, Kumada, K, et al. Experience with 225 hepatic resections for hepatocellular carcinoma over a year period. Am J Surg. (1991). , 4.

[19] Poon, R. T, Fan, S. T, Ng, I. O, & Wong, J. Significance of resection margin in hepatectomy for hepatocellular carcinoma: A critical reappraisal. Ann Surg. (2000).

[20] Poon, R. T, Fan, S. T, Ng, I. O, & Wong, J. Significance of resection margin in hepatectomy for hepatocellular carcinoma: A critical reappraisal. Ann Surg. (2000).

[21] Giuseppe GarceaG. J. Maddern. Liver failure after major hepatic resection. J Hepatobiliary Pancreat Surg ((2009).

[22] Hammond, J. S. *et al*. Prediction, prevention and management of postresection liver failure. *British Journal of Surgery* (2011). , 98, 1188-1200.

[23] Maartje, A. J. van den Broek. Liver failure after partial hepatic resection: definition, pathophysiology, risk factors and treatment. Liver International ((2008).

[24] Suc, B, Panis, Y, Belghiti, J, & Fekete, F. Natural history' of hepatectomy. Br J Surg (1992). , 79, 39-42.

[25] Balzan, S, Belghiti, J, Farges, O, et al. The "50-50 criteria" on postoperative day 5: an accurate predictor of liver failure and death after hepatectomy. Ann Surg (2005). , 242, 824-8.

Chapter 14

Surgical Management of Primary Hepatocellular Carcinoma

Kun-Ming Chan and Ashok Thorat

Additional information is available at the end of the chapter

1. Introduction

Hepatocellular carcinoma (HCC) is among the most common malignancy and cause of cancer related death worldwide, with a high prevalence in Asia and south Africa as well as an increasing incidence in the western country. Patients with liver cirrhosis are at highest risk of developing this malignant disease, and the majority of HCC patients will develop the disease on the background of preexisting hepatitis virus infection. It is estimated 50–70% associated with hepatitis C virus in North America and Europe and 70% associated with hepatitis B virus in Asia and Africa [1], and the incidence of HCC is significantly higher in men than in women. However, surveillance programs for HCC in patients with cirrhosis and chronic hepatitis, and the advancement of diagnostic tools are likely to further increase the incidence of HCC and the detection of small lesions in the liver that prompted the proportion of patients diagnosed at a potentially curative stage of disease.

Several staging systems have so far been proposed for aiding assessment of treatment planning for HCC patients, but an overall consensus remains not exist for any of these staging systems.The Tumor-Node-Metastasis staging (TNM) system of the American Joint Committee on Cancer/Committee of the International Union Against Cancer (AJCC/UICC) has been widely used for numerous cancer staging in order to stratify patients into prognostic groups [2], but it is not perfectly applicable for HCC in terms of treatment assessment as the TNM staging does not consider the underlying liver functional reserve and seems only applicable to patients undergoing liver resection or liver transplantation. The Cancer of the Liver Italian Program (CLIP) classifications and the Okuda staging system were introduced not only considering tumor features but also liver functional reserve.The CLIP scoring system considers cirrhotic status in terms of Child-Turcotte-Pugh (CTP) class and several factors related to tumor features including tumor morphology, Alphafeto protein (AFP) level, and portal vein thrombosis [3]. Although the CLIP scoring system is probably helpful to identify patients with a poor prognosis, it might be inadequate to identify patients at early stages of disease. The Okuda system has also been found unsuitable for prognostic stratification of patients at an early stage of disease [4]. Therefore, the Japan Integrated Staging score that combines the CTP class with the Liver Cancer Study Group of Japan TNM stage was formulated to provide better stratification of patients with early HCC than that achieved by the CLIP score and Okuda system [5]. Additionally, the Barcelona Clinic Liver Cancer(BCLC) staging system was suggested as a modification of the Okuda system, and has been validated superior for prognostic stratification of patients with HCC than other staging systems [6-8]. The BCLC staging system involves factors related to underlying liver function, tumor characteristics, and patients' performance status, and was proposed as a means of predicting prognosis and as a guide to selecting appropriate therapy for HCC patients.

Generally, these staging systems was developed aiming to stratify patients into groups with similar prognoses and to serve as a guiding choice of therapy. Current popular treatments for HCC include liver resection, percutaneous ethanol injection (PEI) or radiofrequency ablation (RFA), transcatheter arterial chemoembolization (TACE), liver transplantation, and targeted therapy with novel biologic agent such as sorafenib. The selection of treatment modality for HCC patients should be based on the patient's prognosis, which is complex to assess, as it depends on three factors, namely, the tumor characteristics, the underlying liver functional reserve and the patient's physical condition. At present, only liver resection and liver transplantation are considered the best potential curative therapies. Nonetheless, because of underlying liver dysfunction, lack of liver donor availability, and/or late detection at advanced cancerous stage, only a small proportion of patients are eligible for these curative treatments. This chapter reviews the importance and clinical impact of surgical management in terms of liver resection and transplantation for patients with primary HCC and highlights their relative strengths and weakness.

2. Liver Resection

Liver resection remains the mainstay curative treatment for patients with HCC. However, the majority of HCC patients are often associated with liver cirrhosis due to hepatitis B or C viral infection, which might prohibit from liver resection because of impaired liver function. Moreover, many HCC patients present with advanced tumor stage and only approximately 20–30% of patients are candidates for liver resection on presentation [9-11]. In spite of this situation, the advancement in anesthetic and surgical techniques, as well as a thorough understanding of the liver anatomy, and better perioperative care, have contributed dramatically to the safety and effectiveness of liver resection for HCC.

Since the proposal of the finger fracture technique for hepatic lobectomy in 1953, transection of the hepatic parenchyma has evolved during the last 50 years. By finger fracture technique, the liver tissue is fractured and crushed by the thumb and index finger followed by isolating and ligating the resistant intrahepatic vascular and ductal structures [12]. However, there is some troublesome bleeding from the resection line which makes the surgeons fear for the safety of the finger fracture technique. To overcome this short coming of the finger fracture technique, many special instruments were invented to increase the successful rate and safety of liver resection ever since (Figure 1). Currently, Kelly clamp crushing technique is still one of the most widely used techniques for liver resection. However, in many centers, including the author's center, ultrasonic dissection using the Cavitron Ultrasonic Surgical Aspirator (CUSA) has become the standard technique of liver resection. Today, laparoscopic liver resection has become feasible in experienced centers due to improvement in instruments [13-15]. Additionally, modern concepts including the use of vascular inflow occlusion, anatomic resection, and low central venous pressure anesthesia, and surgical approaches such as the anterior approach and liver hanging maneuver have been developed along with using more effective instruments for transection of hepatic parenchyma [16-18]. As a result, liver resections are increasingly being performed and accepted as a safety procedure.

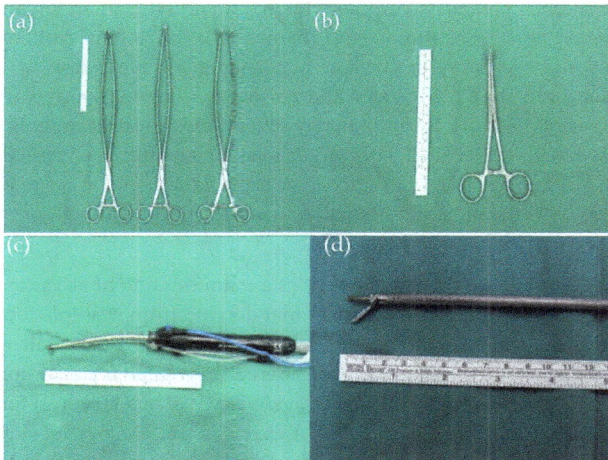

Figure 1. Liver resection instruments. (a). Lin's clamp designed by T.Y. Lin. (b). Kelly clamp. (c). Cavitron Ultrasonic Surgical Aspirator (CUSA). (d). Harmonic scalpel for laparoscopic liver resection.

2.1. Preoperative assessment

The major concern of liver resection in HCC patients is postoperative liver failure, which is particularly worrisome in patients requiring major resections and/or diseased background of cirrhotic liver. Therefore, a thorough evaluating of patients in terms of tumor features of radiologic examination, underlying liver function, and the patient's physical status is very important. Theoretically, a successful liver resection for HCC patients should be weighed against the balance of these three factors.

2.1.1. Evaluation of tumor status

The assessment of tumor status is the essential step for determining resectability and the appropriate type of liver resection. The routine radiologic imaging examination prior to liver resection should include a dynamic liver computed tomography (CT) scan, hepatic angiography, and/or magnetic resonance imaging (MRI) to confirm the diagnosis of HCC as well as tumor status in terms of size, number and location. Additionally, a chest X-ray or contrast CT scan of chest and abdomen could be performed to exclude lung or other extrahepatic metastasis. The CT scan provides important information not only on the tumor size, number, location, and any vascular invasion but also on the relationship between the tumor and major vasculature. Generally, pre-operative biopsy is not necessary as may risk needle track related tumor seeding.

Large HCC—Solitary HCC with diameter of less than 5 cm is the best candidate for liver resection because of favorable patients' outcome in terms of HCC recurrent-free survival [19, 20]. However, numerous patients continue to be diagnosed HCC at an advanced stage that sometimes presented with large tumor with diameter exceed 10 cm. Although liver resection for patients with large tumor can be a great challenge for liver surgeons, liver resection for large HCCs has been shown to be safe and reasonable long-term survival results can be achieved that appear to be much better than any other nonsurgical treatments [21-23]. The 5-year survival rate in patients with tumors larger than 10 cm after liver resection is approximately 21–27.5% [23-25]. Additionally, since liver transplantation and local ablation are not indicated for these patients, surgical resection remains the only treatment of choice that provides potential cure of patients with large HCC.

Multiple HCCs—Multiple HCCs may represent as a manifestation of advanced disease with intrahepatic metastasis or independent tumors that derived from multiple foci of hepatocarcinogenesis, which could be an event associated with a poor prognosis. Patients with multiple HCCs more than 3 nodules have been considered unsuitable for resection. However, it had been shown that liver resection still can provide survival benefits even for patients with multiple tumors in a background of CTP class A cirrhosis, and the overall survival rates can up to 58% at 5 years [26]. Additionally, combined resection and radiofrequency ablation is considered a new strategy to increase the chance of curative treatment for patients with bilobar multiple HCCs. For example, resection of the large tumor in one lobe and ablation of smaller tumors in the other lobe can be performed, or resection of peripheral lesions and ablation of central lesions for patients with multifocal tumors associated with cirrhosis and borderline liver function can be performed [27, 28]. The results showed patients who underwent surgical resection for multiple HCCs had better survival outcomes as compared with those who received nonsurgical therapy. Hence, when clearance of all tumor nodules is feasible and liver function permits, surgical resection or plus effective local ablative therapy should be considered for patients with bilobar or multiple HCCs.

HCC involving major portal and hepatic veins—HCCs with major portal or hepatic veins involvement represent an aggressive tumor behavior and frequently associated with multifocal tumors. Although HCCs with vascular invasion are not considered as favorable surgical candidates, studies from experienced liver surgical groups have shown that surgical resection for such tumors seems justified as it still results in better survival rates as compared with that of nonsurgical treatment [29, 30]. The overall survival rates at 5 years were ranged from 23% to 42% in selected patient who has no liver cirrhosis or impaired liver function.

2.1.2. Evaluation of liver function

Preoperative proper assessment of liver function is fundamental to the safe of liver resection for HCC patients, but there is no individual test accurately predicting liver function.The CTP classification is the most common measure to assess liver function, and it combines different parameters and provides a rough evaluation of the gross synthetic and excretory capacity of the liver. Generally, patients with CTP class A are considered good candidates for liver resection. Patients with CTP class B may be only suitable for minor liver resection such as wedge resection or single segmentectomy [31], whereas patients with CTP class C are contraindicated for resection. The risk of death after liver resection increases with each CTP class. However, this classification is a crude measure and has proven insufficient to stratify the surgical risk of patient with liver cirrhosis.

Portal hypertension is usually defined by that the portal venous pressure is greater than 10 mmHg, in which the normal value ranges from 5 to 8 mmHg. Patients with portal hypertension undergoing liver resection may lead to severe complications, such as variceal bleeding, endotoxemia, and even hepatic failure in the postoperative period [32]. However, measurement of portal venous pressure prior to liver resection is difficult, and portal hypertension could only be roughly assessed by clinical and radiologic signs including splenomegaly, abdominal collaterals, thrombocytopenia with platelet count less than $100,000/mm^3$, or esophagogastric varices. Although portal hypertension is considered a relatively contraindication of liver resection, study had shown that liver resection is also capable of providing survival benefits to patients with a background of portal hypertension [26]. Additionally, patients with abnormal elevation of liver function tests in terms of serum aspartate and alanine aminotransferase levels might have a higher risk of postoperative complication and mortality rates, and are considered to be poor candidates for major liver resection [33, 34]. Therefore, patients with abnormal liver function tests should be carefully assessed and selected prior to liver resection.

Additionally, several hepatobiliary centers have employed more sophisticated quantitative liver function tests, such as the lidocaine monoethylglycinexylidide test, aminopyrine breath test, galactose elimination capacity, and indocyanine green (ICG) clearance test to evaluate the hepatic metabolic function and to predict the risk of postoperative liver failure [35-37]. However, these specific tests reflect the function of the whole liver, whereas the risk of postoperative liver failure relies on the liver function reserve of the remnant liver. Among the various methods, the ICG test is the most widely used to assess liver function prior to liver resection. The ICG is an organic dye that is taken up by the hepatocytes and excreted via the bile in an adenosine triphosphate (ATP) dependent manner without been metabolized and undergoing enterohepatic circulation. Thus, the clearance of ICG from systemic circulation is merely a measure of hepatic blood flow and function. This test evaluates the retention ratio of ICG from the peripheral blood at definitive time point after injection of 0.5 mg ICG/kg (usually 15 minutes, ICG-15), and Makuuchi et al. have incorporated the ICG-15 and two clinical features in terms of serum bilirubin level and the presence of ascites into an algorithm of liver resection (Figure 2) [38]. In patients with bilirubin levels less than 1.0mg/dL and the absence of ascites, ICG-15 is used to predict the extent of liver segments that can be safely removed. In general, an ICG-15 of 10–20% is usually considered a safety upper limit for major liver resection. Accordingly, the algorithm has been validated toward zero surgical mortality after liver resection by several hepatobiliary centers [39, 40].

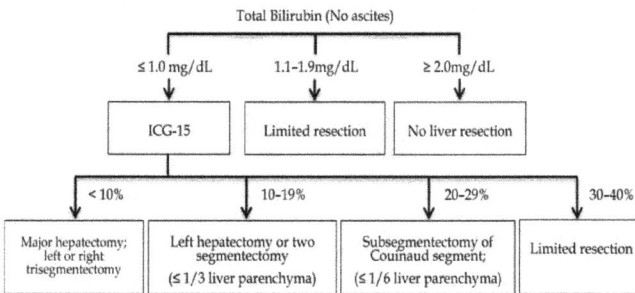

Figure 2. Makuuchi's algorithm for liver resection in patients with HCC [38]. Limited resection means enucleation of the tumor (usually ≤ 5cm) and less than 1 cm of liver tissue surrounding the tumor was removed.

ICG-15 (%)	Safe resection ratio of liver volume
0	< 63.3%
~ 5	< 53.4%
~ 10	<43.5%
~ 15	<33.6%
~ 20	<23.7%
~ 25	<13.8%
~ 30	<3.9%
≥ 32	0%

Table 1. The safe resection ratio of liver volume based on the ICG test [43].

Although the assessment of hepatic function and liver volume to be resected is crucial for a safe liver resection, volumetric analysis of the future liver remnant (FLR) has also been suggested. The FLR can be measured directly by computer-assisted models of contrast-enhanced spiral CT. However, it remains controversial regarding which index of the FLR volume should be used. Some surgeons use the actual total liver volume minus liver volume to be removed on CT images as the FLR volume, while others use the estimated ideal liver volume that is calculated by a formula based on body surface area as a standard for calculation of the FLR. Nonetheless, the exactly number of the adequate FLR volume in cirrhotic patients is also no consensus, and at least an FLR of 40% is recommended in patients with chronic liver disease [41, 42]. In the authors' center, we have established an equation to reveal the relationship between the ratio of FLR volume and ICG-15 values as well as references for determining a safe resection ratio of the liver volume (Table 1)[43].

2.2. Preoperative therapy

Since not all patients with HCC are amenable to surgical resection, several strategies such as preoperative TACE that might be used to downsize large HCC or portal vein embolization (PVE) to increase the FLR have been suggested. However, the efficacy of these preoperative approaches in terms of HCC oncologic viewpoint remains the subject of debate.

2.2.1. Portal vein embolization

The concept of PVE was introduced on the basis of the idea that an increase in the FLR will reduce the risk of liver failure after major liver resection for hilar bile duct carcinoma in 1982 [44]. By occluding portal venous branch of the tumor-bearing liver, PVE induces atrophy of the resection part and hypertrophy of the FLR. Although the ability of liver regeneration in cirrhotic liver is impaired, PVE may induce clinically sufficient hypertrophy in these patients as well.Currently, PVE could be considered for patients with liver cirrhosis when the FLR is expected less than 40% of the total liver volume [45, 46]. PVE may also be used as a dynamic liver function test, in which inadequate hypertrophy of the FLR or intolerance of the patient after PVE indicate that major liver resection is contraindicated. In general, PVE is a relatively safe procedure, and it may increase the resectability of initial unresectable HCC and reduce the risk of post hepatectomy liver failure. Additionally, it seems no adverse effect on the oncologic outcome of HCC patients undergoing major liver resection [47, 48]. However, the potential for progression of the primary tumor after PVE remains a major concern, whereas a combination of TACE as a complementary procedure to PVE could be considered in order to improve the outcome of HCC patients [49].

2.2.2. Preoperative transcatheter arterial chemoembolization (TACE)

The use of TACE as a neoadjuvant treatment for HCC was proposed in a variety of settings such as palliative treatment for unresectable HCC, to improve the resectability of intial unresectable HCC, to downstage the primary tumor for liver transplantation or for delay surgery. The major goal of TACE is aimed at inducing tumor necrosis and shrinkage as well as preventing the dissemination of the primary tumor (Figure 3). Theroretically, the use of neoadjuvant TACE in the setting of resectable HCC might be capable of improving survival by reducing tummor recurrences. Nonethelesss, the fact is that most studies show conflict out-comes of TACE as a neoadjuvant therapy and do not support routine use of preoperative TACE before liver resection [50-53]. Moreover, preoperative TACE for resectable large HCC is not recommended because it does not provide complete necrosis of the large tumor and-may actually result in progression of the primary tumor owing to delay surgery and compli-cate the operation during the process of liver mobilization due to the presence of perihepatic adhesions after TACE.

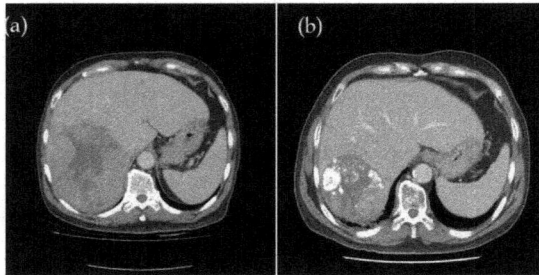

Figure 3. TACE induces remarkable shrinkage of tumor mass. (a) A huge liver tumor around 15 cm in size located at right lobe liver. (b) The tumor was decreased by half in size after three courses of TACE. (4 months after HCC diag-nosed)

Clinically, spontaneous tumor rupture accompanied by hemorrhaging has been seen in small portion of patients with HCC at initial presentation, which might lead to a life-threat-ening condition depending on the severity of hemorrhage. Transcatheter arterial emboliza-tion (TAE) should be performed for ruptured HCC to control tumor bleeding as well as stabilizing clinical condition of patients. Liver resection then can be evaluated after the pa-tient has recovery from shock status and post-TAE damage of the liver according to the cri-teria of liver resection. Generally, TAE followed by staged liver resection of tumor seems to be a rational treatment strategy for patients with ruptured HCC and hemorrhage if the le-sion is resectable, and long-term survival could be expected [54, 55].

2.3. Outcome of liver resection

The operative mortality of liver resection has been reduced to less than 5% with some cen-ters approaching to zero mortality in recent years [39, 40, 56]. The improvement is primarily resulting from advances in surgical techniques, perioperative management, and more cau-tious patient selection. However, the postoperative morbidity rate remains high that ranges from 25 to 50% even in experienced centers [11, 57, 58]. Ascites and pulmonary complica-tions are the most common complications, but serious complications such as liver failure, postoperative hemorrhage, bile leakage, and intra-abdominal sepsis are less frequent nowa-days. Apart from that, the long-term survival after resection of HCC have much improved-lately, but HCC recurrence remains a major concern for patients undergoing liver resection.

2.3.1. HCC recurrence

The high incidence of postoperative recurrence, estimated excess of 70% at 5 years, is the greatest frustration in treating patient with HCC. Recurrent HCCs are mostly intrahepatic that accounts for approximately 80–90% of cases after liver resection. There are two peaks of HCC recurrence after liver resection:The first peak occurs at approximately 1 year posthepatectomy and about 40% of recurrence within the period, in which metastatic dissemination of the primary tumor is mainly responsible for this early peak. The second peak is observed at the 4th postoperative year with a 3E% of recurrent rate per year, and the majority of the second peak is more likely attributable to new tumors development related to the carcinogenic effect of underlying chronic liver disease [59].

Currently, there is no well-established adjuvant therapy to reduce the risk of recurrence after curative liver resection. Although numerous studies have demonstrated the efficacy of some new modalities including acyclic retinoid, polyprenoic acid [60], intra-arterial iodine-131-labelled lipiodol [61], and adaptive immunotherapy [62] as adjuvant therapy in the prevention of HCC recurrence after liver resection, the sample size of these individual studies was rather small and further validated by randomized trials with large sample size is required. Additionally, interferon has been proposed as adjuvant therapy in patients with HCC and viral hepatitis after liver resection and shown beneficial for reducing recurrence and prolonging survival [63, 64]. Nonetheless, a more recent cohort study based on a phase III randomized trial of adjuvant interferon alfa-2b in HCC after curative resection does not support the benefit of interferon in reducing postoperative recurrence of viral hepatitis-related HCC [65]. Apart from that, another potential approach is to use molecular targeted therapy such as sorafenib that is applied for advanced HCC and may inhibit HCC cell proliferation and angiogenesis. However, further trial is indicated to test the efficacy of these targeting drugs as adjuvant therapy after resection of HCC.

Compared with the development of postoperative adjuvant therapy, the risk factors for HCC recurrence after liver resection have been extensively explored and established. The risk factors for tumor recurrence can be categorized into three core groups, related to host factors, tumor factors, and surgical factors [66]. The tumor factors including vascular invasion, satellite nodules, large tumor, elevation of AFP, poor differentiated histologic grade, tumor rupture, and advanced tumor stage are frequently reported risk factor for HCC recurrence after liver resection. The host factors are the patient's characteristics and underlying liver diseases such as cirrhosis and viral hepatitis. Both tumor and host factors are determined before operation, and the surgeon can only control surgical factors including negative resection margin, anatomic resection, meticulous liver mobilization, and less blood transfusion.

The treatment strategy for HCC recurrence after liver resection should be the same as that for primary HCC. Although repeat hepatectomy could be a difficulty owing to perihepatic adhesion related to first operation, surgical resection remains a preferred treatment whenever the tumor is considered to be resectable [67-69].

2.3.2. Survival of patients

Despite the high incidence of postoperative HCC recurrence, current strategy of aggressive multimodality treatments for recurrent tumors using TACE, RFA, or liver transplantation has largely improved the overall outcomes of patients even after the development of recurrent HCC. Moreover, surgical resection of recurrent HCC presenting as extrahepatic metastases could be considered in selected patients who are with isolated extrahepatic metastases and has otherwise good performance status, good hepatic functional reserve, and well-treated intrahepatic HCC, and a survival benefit can be expected from this aggressive approach [70]. Generally, the overall 5-year survival after resection of HCC reported in the literature from large series is mostly near 50% or even better in recent years (Table 2).

Authors (Years)	Study period	Subgroups	No. of patients	5-year RFS/OS
Hanazaki et al. (2000) [71]	1983–1997		386	23.3%/34.4%
Zhou et al. (2001) [58]	1967–1998	size≤5cm	1000	—/62.7%
		size>5cm	1366	—/37.1%
Wang et al. (2010) [72]	1991–2004		438	—/43.3%
Fan et al. (2011) [73]	1989–2008	1989–1998	390	24%/42.1%
		1999–2008	808	34.8%/54.8%
Sakamoto et al. (2011) [74]	1988–2010	Caudate lobe	46	44%/76%
		Other sites	737	40%/64%
Nara et al. (2012) [75]	1990–2007	SM>1mm	374	40.0%/72.2%
		SM≤1mm	165	28.1%/63.5%
		SM-postive	31	7.4%/36%
Chan et al. (2012) [76]	2001–2005		651	33.9%/51.7%
Giuliante et al. (2012) [77]	1992–2008	Tumor ≤3cm	588	32.4%/52.8%
Shrager et al. (2012) [78]	1992–2008	Non-cirrhosis	206	39%/46.3%
		cirrhosis	462	—/—
Altekruse et al. (2012) [79]	1998–2008	SEER–13	1348	—/47%

Table 2. Long-term survival of patients undergoing liver resection for HCC reported from large series in recent years. RFS, recurrence-free survival; OS, overall survival; SM, surgical margin; SEER-13, Surveillance Epidemiology and End Results of USA.

3. Liver transplantation

Recurrence remains a major problem after liver resection for HCC even after margin-negative resection. Most of the patients with HCC have underlying cirrhosis that provides potential field for development of hepatocellular carcinoma. Since majority hepatic malignancies are HCC and almost 80% of them have underlying cirrhosis, resection is option in only small number of patients and in such patients, recurrence rate is high after resection. Liver transplantation practically offers greater chance of cure by removing underlying liver cirrhosis and HCC. Also, HCC is multifocal especially with hepatitis C, and total hepatectomy removes the source of potential possibility of later-developing tumors whereas partial hepatic resection does not.

However, liver transplantation for HCC did not yield satisfactory results initially. Recurrence rates were up to 80% and long term survival rates were unacceptably below that of patients who underwent liver transplantation for non-malignant causes. These recurrences usually appeared within 2 years of transplant, most common site being liver allograft that led to a decline in enthusiasm and a serious concern about using precious donor livers for treatment [80].It was Bismuth who initially reported good outcomes with liver transplantation for small HCC [81] and subsequently, Mazzaferro et al introduced the Milan criteria reporting liver transplantation for HCC with equivalent outcomes to non-HCC patients[82].

Liver transplantation has become now potential curative treatment and it is presently the treatment of choice for patients with CTP class B or C cirrhosis and early hepatocellular carcinoma. Compared with surgical resection, liver transplantation is associated with better overall and recurrence-free survival in well selected patients [83-85]. The improved overall-results after liver transplantation are thought to be due to better patient selection and the emergence of various locoregional therapies for HCC that prevent tumor progression while patient is waitlisted for liver transplantation, thus preventing drop out.

3.1. Patient selection criteria

A major goal of liver transplant team is to select the patients with HCC and cirrhosis at earlier stage of their disease in order to achieve survival duration comparable with that of other patients with benign liver disease receiving transplants, so as to justify or prioritize the allocation of a liver graft. Liver transplant candidates with HCC must meet the Milan criteria to qualify for exceptional HCC waiting list consideration. Also, several other extended criteria such as UCSF (University of California at San Francisco) criteria are used for patient selection in highly specialized transplant centres.

3.1.1. Milan's criteria

In 1996, a prospective cohort study defined restrictive selection criteria that led to superior survival for transplant patients in comparison with any other previous experience with transplantation or other options for HCC. Since then, these selection criteria have become universally known as the Milan criteria in recognition of their origin (Table 3) [82]. These criteria have been widely applied in the selection of patients with HCC for liver transplantation.In North America as well as in many other world regions, patients within Milan criteria HCC are given priority to liver transplantation. Generally, a 4-year overall and recurrence-free survival rates of 85% and 92%, respectively, can be achieved using this selection criteria.

Criteria of liver transplantation for patients with HCC
Single lesion ≤5 cm.
Up to three separate lesions, none larger than 3 cm.
No evidence of gross vascular invasion.
No regional nodal or distant metastases.

Table 3. Milan's criteria for liver transplantation.

3.1.2. Extended Criteria

Considerable interest has arisen in expansion of usual transplant criteria in highly specialized centres to offer liver transplantation to broader group of patients with HCC as investigators argued that Milan's criteria are too restrictive and limit liver transplantation at the
time when incidence of HCC is on the rise. Using explant pathologic data, Yao and co-work-
ers at the University of California, San Francisco (UCSF) reported 5-year post-transplantation survival of 75% in patients with tumors as large as 6.5 cm and cumulative tumor
burden ≤8 cm (Table 4)[86].

Extended criteria of liver transplantation for patients with HCC
Solitary tumor up to 6.5 cm.
A maximum of 3 tumor nodules each up to 4.5 cm.
A total tumor diameter not exceeding 8 cm.
No regional nodal or distant metastases.

Table 4. UCSF criteria for liver transplantation.

The UCSF criteria have been shown to be associated with long -term survival similar to Milan criteria when based on explant pathology [87, 88]. However, because of the small sample
size and use of retrospective explant tumor pathology, the results of these studies were chal-
lenged and also several groups advised caution in expanding the criteria.

Additionally, a recent multicentre study led by the Milan's group had retrospectively reviewed patients who underwent transplantation for HCC in order to explore the survival of
patients with tumors that exceed the Milan criteria. Accordingly, a prognostic model of
overall survival based on tumor characteristics in terms of size and number was derived,
and an expanded criterion termed "up-to-seven criteria" was introduced [89]. Patients who
fell within the criteria that the sum of the largest tumor size and the number of tumors does
not exceed seven could achieve a 5-year overall survival of 71.2% after liver transplantation
enabling more patients to qualify as transplant candidates.

3.2. Prognostic Indicators

Several studies have identified patient and tumor-related variables associated with prognosis following liver transplantation for HCC. The majority of prognostic factors are similar to that of liver resection for patients with HCC.

3.2.1. Tumor related factors

Important prognostic factors in most of scientific studies include tumor number, size, and location (especially bilobar distribution). The most consistent association is with tumor size. Other factors are histologic grade of differentiation, stage of disease according to the American Liver Tumor Study Group (ALTSG) modification of the TNM staging criteria, the presence of macrovascular and microvascular invasion, absolute level of serum AFP, and extrahepatic spread. Tumor size predicts both the likelihood of vascular invasion and tumor grade, but the relationship is nonlinear and a significant proportion of small tumors have unfavourable histology, whereas some larger ones do not [90, 91].

3.2.2. Patient related factors

Patients with HCV infection tend to have severe underlying liver disease and more advanced HCC at presentation as compared to HBV infection and underlying alcoholic cirrhosis. Hence, the recurrence of HCC is more common among the HCV recipients and thus reduced survival [92]. The immunosuppressive treatment after liver transplantation is associated with increased risk of tumor recurrence. Thus, immunosuppressant should be reduced to minimum effective levels. Several studies have shown lower recurrence with sirolimus which is attributed to its anti-proliferative effects on HCC [93-95]. But there is need for large randomized controlled trials to conclude sirolimus as most appropriate immunosuppressant for patients undergoing liver transplantation for HCC.

3.3. Deceased Donor Liver Transplantation (DDLT)

3.3.1. Graft Allocation

The shortage of donor livers has necessitated the development of allocation system, whereby priority for donor organs is given to the most severely ill patients. The prolonged waiting period frequently results in tumor progression to an extent beyond the transplantable criteria, leading to a patient's removal or dropout from the waiting list [96]. Allocation of deceased donor livers for both adults and children is based upon the "model for end stage liver disease" or MELD score, a statistical model based upon predicted survival in patients with cirrhosis. As a result of the high dropout rate for patients with HCC, the Organ Procurement and Transplantation Network (OPTN) of the U.S. has reconsidered the priority of liver graft allocation. While waiting list priority was determined primarily by liver disease severity based on the Model for End-Stage Liver Disease (MELD) score, patients with HCC that fulfilled the Milan criteria were registered with an adjusted score and were subsequently assigned additional scores at regular intervals to reflect their risk for dropout as a result of tumor progression.

3.3.2. Listing Criteria of transplantation candidates

In an attempt to ensure that preoperative assessment is as accurate as possible, UNOS provides a set of specific requirements for listing patients with HCC for orthotopic liver transplantation.

I.   The diagnosis must be confirmed by thorough assessment by imaging modalities such as ultrasound, dynamic CT and /or MRI. Tumor numbers, size, presence or absence of extrahepatic disease and major vascular disease must be documented.

II.   Patient must have one of the following:

   1.   An Alfa fetoprotein level > 200 ng/mL.

   2.   Celiac angiography showing tumor blush corresponding to the site shown by CT/MRI/ultrasonography.

   3.   A biopsy confirming HCC

   4.   History of RFA, TACE or other locoregional therapy.

III.   Must be within Milan's criteria.

IV.   Continued documentation of the tumor is required every three months by CT or MRI to ensure continued eligibility for liver transplantation.

Patients will be given priority MELD score depending upon the state of underlying disease. Prioritization scores for patients with HCC are based upon tumor size and number. With this new organ allocation policy, waiting time for the patients with HCC to receive a deceased-donor liver has decreased significantly.

3.4. Living Donor Liver Transplantation (LDLT)

The shortage of organs from deceased donors has curtailed the adoption of living donor liver transplantation. Living donors can potentially provide an essentially unlimited source of liver grafts for a planned transplant operation as soon as the diagnosis of HCC is made, thus decreasing the uncertainty of long waiting periods and reducing possibility of tumor progression [97]. The living donor can be from adult-to-adult or adult-to-child. In children mostly left lateral segment of the liver harvested and donors are usually ABO-compatible parents. While in adult-to-adult, right or left liver can be harvested that depends upon pre-transplant evaluation of donor and CT volumetry of liver. The GRWR (graft to recipient weight ratio) must be more than 0.8%. Donor not meeting these criteria is rejected for the fear of small-for-size syndrome and subsequent graft failure [98, 99].

Because a live donor graft is a dedicated gift that is directed exclusively to a particular recipient, there is no need for an objective allocation system based on a prioritization scheme. Presently LDLT comprises almost >90% of liver transplants in Asia as compared to <5% in

US. Unlike in the U.S., where recipients with malignancies receive extra prioritization in the deceased donor organ allocation scheme, HCC patients in Asia do not. HCC patients in Asia have a dismal chance of receiving a deceased donor graft and LDLT is often the only option.

3.5. Pretransplant locoregional therapies

Pretransplant locoregional therapy has been adopted by the liver transplant community worldwide. This concept, known as "bridging therapy" is meant to limit tumor progression and dropout rate while patients are on the transplant wait list. The most popular techniques include TACE, transarterial drug-eluting beads, transarterial radio-embolizationand RFA.In the transplant setting, TACE is currently the most popular neo-adjuvant treatment. It is indicated in Child–Pugh A or B cirrhotic patients to downstage tumors into the Milan criteria or to prevent tumor progression. For patients with small HCC confined to the liver, recent data also indicate that transplantation when used with multimodal therapy using locoregional procedures and neoadjuvant systemic chemotherapy, results in improved recurrence-free survival [100, 101]. Apart from that, it is also important to know the wait list dropout rate and bridging therapy-associated complication rate, because the benefit of preventing wait list dropout should outweigh the risk of bridging therapy.Patient-individualized treatment strategy should be based on the performance status, hepatic reserve, tumor burden, and tumor vascularity pattern.

3.6. Outcome of liver transplantation

To date, orthotopic liver transplantation is no doubt the best therapeutic option for early, unresectable HCC, although it is limited by graft shortage and the need for appropriate patient selection. Since the introduction of the milan's criteria, the liver transplantation for primary HCC is on rise with promising recurrence-free survival and overall survival. Excellent 5-year post-transplant patient survival of at least 70% has been reported from many centers [102]. Furthermore, better definition of the prognostic factors and more rigorous patient selection have resulted in significant improvement in 5-year survival for patients receiving transplants for HCC in the past decade.

However, a tendency for higher HCC recurrence has been reported for patients who underwent LDLT than patients who underwent DDLT [103, 104]. The reasons for this difference are not completely answered by current studies. Possible explanations can be related to the selection bias for clinical characteristics associated with aggressive tumor behavior, elimination of natural selection during the waiting period, and enhancement of tumor growth and invasiveness by small-for-size graft injury and regeneration [105, 106]. Additionally, more clinical studies with long-term follow-up are needed to evaluate the role of LDLT for early HCC. At present, if a suitable and willing donor is identified, LDLT is a reasonable alternative to waiting 6 to 12 months for a deceased donor graft in patients with HCC who are otherwise eligible for liver transplantation.

Although liver transplantation is the only option for the cure in majority of the patients with HCC complicated by underlying cirrhosis precluding resection, identification of prognostic factors and refinement of selection criteria will improve the outcomes of liver transplantation for this otherwise fatal disease. Nonetheless, liver transplantation may also pose a risk of post transplant lymphoproliferative disorders and other de novo malignacy associated with long term immunosuppression.

4. Conclusion

The management of patients with HCC remains complex and challenging. Although liver resection and liver transplantation are the curative treatments for HCC at present, there is considerable controversy as to whether patients with HCC are better served with liver transplantation versus liver resection. Liver transplantation removes HCC with underlying cirrhosis and thus sounds best option; however, technical challenges associated with transplantation and/or immunosuppression should be taken into consideration for selecting transplant cadidates. Currently, most studies suggest that liver resection should be a priority in patients who are candidates for either liver resection or transplantation [102, 107]. Despite a better cancer cure rate for liver transplatation, liver resection remains superior for patients in terms of limited organ availability and transplantation-associated morbidity and mortality. Therefore, the optimal treatment for patients with preservd liver function should always be resection whenever the tumor is resectable, and liver transplantation could be reserved as a salvage therapy for patients who encounter HCC recurrence after primary liver resection. Theoretically, this strategy will not only improve patient survival but relieve the growing demand of available donor livers.

Author details

Kun-Ming Chan* and Ashok Thorat*

*Address all correspondence to: chankunming@adm.cgmh.org.tw

Division of Liver and Organ Transplantation Surgery, Department of General Surgery, Chang Gung Memorial Hospital at Linkou, Chang Gung University College of Medicine, Taiwan, Republic of China

References

[1] Forner, A., Llovet, J. M., & Bruix, J. (2012). Hepatocellular Carcinoma. *Lancet*, 379(9822), 1245-1255.

[2] Sobin, L. H., Gospodarowicz, M. K., & wittekind, C. (2009). Tnm Classification of Malignant Tumours. *John Wiley & Sons*.

[3] Anonymous. (1998). New A.PrognosticSystem.forHepatocellular.CarcinomaA Retrospective Study of 435 Patients: The Cancer of the Liver Italian Program (Clip) Investigators. *Hepatology*, 28(3), 751-755.

[4] Okuda, K., Ohtsuki, T., Obata, H., Tomimatsu, M., Okazaki, N., Hasegawa, H., Nakajima, Y., & Ohnishi, K. (1985) Natural History of Hepatocellular Carcinoma and Prognosis in Relation to Treatment. Study of 850 Patients. *Cancer*, 56(4), 918-928.

[5] Kudo, M., Chung, H., & Osaki, Y. (2003). Prognostic Staging System for Hepatocellular Carcinoma (Clip Score): Its Value and Limitations, and a Proposal for a New Staging System, the Japan Integrated Staging Score (Jis Score). *J Gastroenterol*, 38(3), 207-215.

[6] Llovet, J. M., Bru, C., & Bruix, J. (1999). Prognosis of Hepatocellular Carcinoma: The Bclc Staging Classification. *Semin Liver Dis*, 19(3), 329-338.

[7] Marrero, J. A., Fontana, R. J., Barrat, A., Askari, F., Conjeevaram, H. S., Su, G. L., & Lok, A. S. (2005). Prognosis of Hepatocellular Carcinoma: Comparison of 7 Staging Systems in an American Cohort. *Hepatology*, 41(4), 707-16.

[8] Bruix, J., & Llovet, J. M. (2009). Major Achievements in Hepatocellular Carcinoma. *Lancet*, 373(9664), 614-616.

[9] Cance, W. G., Stewart, A. K., & Menck, H. R. (2000). The National Cancer Data Base Report on Treatment Patterns for Hepatocellular Carcinomas: Improved Survival of Surgically Resected Patients, 1935-1996. *Cancer*, 88(4), 912-920.

[10] Liu, C. L., & Fan, S. T. (1997). Nonresectional Therapies for Hepatocellular Carcinoma. *Am J Surg*, 173(4), 358-365.

[11] Fong, Y., Sun, R. L., Jarnagin, W., & Blumgart, L. H. (1999). An Analysis of 412 Cases of Hepatocellular Carcinoma at a Western Center. *Ann Surg*, 229(6), 790-799, discussion 9-800.

[12] Lin, T. Y. (1974). A Simplified Technique for Hepatic Resection: The Crush Method. *Ann Surg*, 180(3), 285-290.

[13] Buell, J. F., Thomas, M. T., Rudich, S., Marvin, M., Nagubandi, R., Ravindra, K. V., Brock, G., & Mc Masters, K. M. (2008). Experience with More Than 500 Minimally Invasive Hepatic Procedures. *Ann Surg*, 248(3), 475-486.

[14] Cherqui, D., Husson, E., Hammoud, R., Malassagne, B., Stephan, F., Bensaid, S., Rotman, N., & Fagniez, P. L. (2000). Laparoscopic Liver Resections: A Feasibility Study in 30 Patients. *Ann Surg*, 232(6), 753-762.

[15] Vibert, E., Perniceni, T., Levard, H., Denet, C., Shahri, N. K., & Gayet, B. (2006). Laparoscopic Liver Resection. *Br J Surg*, 93(1), 67-72.

[16] Lai, E. C., Fan, S. T., Lo, C. M., Chu, K. M., & Liu, C. L. (1996). Anterior Approach for Difficult Major Right Hepatectomy. World J Surg discussion 8., 20(3), 314-317.

[17] Wang, W. D., Liang, L. J., Huang, X. Q., & Yin, X. Y. (2006). Low Central Venous Pressure Reduces Blood Loss in Hepatectomy. *World J Gastroenterol*, 12(6), 935-939.

[18] Wu, T. J., Wang, F., Lin, Y. S., Chan, K. M., Yu, M. C., & Lee, W. C. (2010). Right Hepatectomy by the Anterior Method with Liver Hanging Versus Conventional Approach for Large Hepatocellular Carcinomas. *Br J Surg*, 97(7), 1070-1078.

[19] Dahiya, D., Wu, T. J., Lee, C. F., Chan, K. M., Lee, W. C., & Chen, M. F. (2010). Minor Versus Major Hepatic Resection for Small Hepatocellular Carcinoma (Hcc) in Cirrhotic Patients: A 20-Year Experience. *Surgery*, 147(5), 676-685.

[20] Wayne, J. D., Lauwers, G. Y., Ikai, I., Doherty, D. A., Belghiti, J., Yamaoka, Y., Regimbeau, J. M., Nagorney, D. M., Do, K. A., Ellis, L. M., Curley, S. A., Pollock, R. E., & Vauthey, J. N. (2002). Preoperative Predictors of Survival after Resection of Small Hepatocellular Carcinomas. *Ann Surg*, 235(5), 722-730, discussion 30-1.

[21] Ng, K. K., Vauthey, J. N., Pawlik, T. M., Lauwers, G. Y., Regimbeau, J. M., Belghiti, J., Ikai, I., Yamaoka, Y., Curley, S. A., Nagorney, D. M., Ng, I. O., Fan, S. T., & Poon, R. T. (2005). Is Hepatic Resection for Large or Multinodular Hepatocellular Carcinoma Justified? Results from a Multi-Institutional Database. *Ann Surg Oncol*, 12(5), 364-373.

[22] Regimbeau, J. M., Farges, O., Shen, B. Y., Sauvanet, A., & Belghiti, J. (1999). Is Surgery for Large Hepatocellular Carcinoma Justified? *J Hepatol*, 31(6), 1062-1068.

[23] Yeh, C. N., Lee, W. C., & Chen, M. F. (2003). Hepatic Resection and Prognosis for Patients with Hepatocellular Carcinoma Larger Than 10 Cm: Two Decades of Experience at Chang Gung Memorial Hospital. *Ann Surg Oncol*, 10(9), 1070-1076.

[24] Huang, J. F., Wu, S. M., Wu, T. H., Lee, C. F., Wu, T. J., Yu, M. C., Chan, K. M., & Lee, W. C. (2012). Liver Resection for Complicated Hepatocellular Carcinoma: Challenges but Opportunity for Long-Term Survivals. J Surg Oncol. (In press)

[25] Poon, R. T., Fan, S. T., & Wong, J. (2002). Selection Criteria for Hepatic Resection in Patients with Large Hepatocellular Carcinoma Larger Than 10 Cm in Diameter. *J Am Coll Surg*, 194(5), 592-602.

[26] Ishizawa, T, Hasegawa, K, Aoki, T, Takahashi, M, Inoue, Y, Sano, K, Imamura, H, Sugawara, Y, Kokudo, N, & Makuuchi, M. (2008). Neither Multiple Tumors nor Portal Hypertension Are Surgical Contraindications for Hepatocellular Carcinoma. *Gastroenterology*, 134(7), 1908-1916.

[27] Choi, D., Lim, H. K., Joh, J. W., Kim, S. J., Kim, M. J., Rhim, H., Kim, Y. S., Yoo, B. C., Paik, S. W., & Park, C. K. (2007). Combined Hepatectomy and Radiofrequency Ablation for Multifocal Hepatocellular Carcinomas: Long-Term Follow-up Results and Prognostic Factors. *Ann Surg Oncol*, 14(12), 3510-3518.

[28] Liu, C. L., Fan, S. T., Lo, C. M., Ng, I. O., Poon, R. T., & Wong, J. (2003). Hepatic Resection for Bilobar Hepatocellular Carcinoma: Is It Justified? *Arch Surg*, 138(1), 100-104.

[29] Pawlik, T. M., Poon, R. T., Abdalla, E. K., Ikai, I., Nagorney, D. M., Belghiti, J., Kian-manesh, R., Ng, I. O., Curley, S. A., Yamaoka, Y., Lauwers, G. Y., & Vauthey, J. N. (2005). Hepatectomy for Hepatocellular Carcinoma with Major Portal or Hepatic Vein Invasion: Results of a Multicenter Study. *Surgery*, 137(4), 403-410.

[30] Minagawa, M., Makuuchi, M., Takayama, T., & Ohtomo, K. (2001). Selection Criteria for Hepatectomy in Patients with Hepatocellular Carcinoma and Portal Vein Tumor Thrombus. *Ann Surg*, 233(3), 379-384.

[31] Kuroda, S., Tashiro, H., Kobayashi, T., Oshita, A., Amano, H., & Ohdan, H. (2011). Selection Criteria for Hepatectomy in Patients with Hepatocellular Carcinoma Classi-fied as Child-Pugh Class B. *World J Surg*, 35(4), 834-841.

[32] Bruix, J., Castells, A., Bosch, J., Feu, F., Fuster, J., Garcia-Pagan, J. C., Visa, J., Bru, C., & Rodes, J. (1996). Surgical Resection of Hepatocellular Carcinoma in Cirrhotic Pa-tients: Prognostic Value of Preoperative Portal Pressure. *Gastroenterology*, 111(4), 1018-1022.

[33] Noun, R., Jagot, P., Farges, O., Sauvanet, A., & Belghiti, J. (1997). High Preoperative Serum Alanine Transferase Levels: Effect on the Risk of Liver Resection in Child Grade a Cirrhotic Patients. *World J Surg* discussion 5., 21(4), 390-394.

[34] Poon, R. T., Fan, S. T., Lo, C. M., Liu, C. L., Ng, I. O., & Wong, J. (2000). Long-Term Prognosis after Resection of Hepatocellular Carcinoma Associated with Hepatitis B-Related Cirrhosis. *J Clin Oncol*, 18(5), 1094-1101.

[35] Ercolani, G., Grazi, G. L., Calliva, R., Pierangeli, F., Cescon, M., Cavallari, A., & Maz-ziotti, A. (2000). The Lidocaine (Megx) Test as an Index of Hepatic Function: Its Clini-cal Usefulness in Liver Surgery. *Surgery*, 127(4), 464-471.

[36] Merkel, C., Gatta, A., Zoli, M., Bolognesi, M., Angeli, P., Iervese, T., Marchesini, G., & Ruol, A. (1991). Prognostic Value of Galactose Elimination Capacity, Aminopyrine Breath Test, and Icg Clearance in Patients with Cirrhosis. Comparison with the Pugh Score. *Dig Dis Sci*, 36(9), 1197-21103.

[37] Redaelli, C. A., Dufour, J. F., Wagner, M., Schilling, M., Husler, J., Krahenbuhl, L., Buchler, M. W., & Reichen, J. (2002). Preoperative Galactose Elimination Capacity Predicts Complications and Survival after Hepatic Resection. *Ann Surg*, 235(1), 77-85.

[38] Makuuchi, M., Kosuge, T., Takayama, T., Yamazaki, S., Kakazu, T., Miyagawa, S., & Kawasaki, S. (1993). Surgery for Small Liver Cancers. *Semin Surg Oncol*, 9(4), 298-304.

[39] Imamura, H., Seyama, Y., Kokudo, N., Maema, A., Sugawara, Y., Sano, K., Takaya-ma, T., & Makuuchi, M. (2003). One Thousand Fifty-Six Hepatectomies without Mor-tality in 8 Years. *Arch Surg*, 138(11), 1198-206, discussion 206.

[40] Torzilli, G., Makuuchi, M., Inoue, K., Takayama, T., Sakamoto, Y., Sugawara, Y., Ku-bota, K., & Zucchi, A. (1999). No-Mortality Liver Resection for Hepatocellular Carci-noma in Cirrhotic and Noncirrhotic Patients: Is There a Way? A Prospective Analysis of Our Approach. *Arch Surg*, 134(9), 984-992.

[41] Kubota, K., Makuuchi, M., Kusaka, K., Kobayashi, T., Miki, K., Hasegawa, K., Hari-hara, Y., & Takayama, T. (1997). Measurement of Liver Volume and Hepatic Func-tional Reserve as a Guide to Decision-Making in Resectional Surgery for Hepatic Tumors. *Hepatology*, 26(5), 1176-1181.

[42] Shirabe, K., Shimada, M., Gion, T., Hasegawa, H., Takenaka, K., Utsunomiya, T., & Sugimachi, K. (1999). Postoperative Liver Failure after Major Hepatic Resection for Hepatocellular Carcinoma in the Modern Era with Special Reference to Remnant Liv-er Volume. *J Am Coll Surg*, 188(3), 304-309.

[43] Lee, C. F., Yu, M. C., Kuo, L. M., Chan, K. M., Jan, Y. Y., Chen, M. F., & Lee, W. C. (2007). Using Indocyanine Green Test to Avoid Post-Hepatectomy Liver Dysfunc-tion. *Chang Gung Med J*, 30(4), 333-338.

[44] Makuuchi, M., Thai, B. L., Takayasu, K., Takayama, T., Kosuge, T., Gunven, P., Ya-mazaki, S., Hasegawa, H., & Ozaki, H. (1990). Preoperative Portal Embolization to In-crease Safety of Major Hepatectomy for Hilar Bile Duct Carcinoma: A Preliminary Report. *Surgery*, 107(5), 521-527.

[45] Farges, O., Belghiti, J., Kianmanesh, R., Regimbeau, J. M., Santoro, R., Vilgrain, V., Denys, A., & Sauvanet, A. (2003). Portal Vein Embolization before Right Hepatecto-my: Prospective Clinical Trial. *Ann Surg*, 237(2), 208-217.

[46] Kokudo, N., & Makuuchi, M. (2004). Current Role of Portal Vein Embolization/ Hepatic Artery Chemoembolization. *Surg Clin North Am*, 84(2), 643-657.

[47] Seo, D. D., Lee, H. C., Jang, M. K., Min, H. J., Kim, K. M., Lim, Y. S., Chung, Y. H., Lee, Y. S., Suh, D. J., Ko, G. Y., Lee, Y. J., & Lee, S. G. (2007). Preoperative Portal Vein Embolization and Surgical Resection in Patients with Hepatocellular Carcinoma and Small Future Liver Remnant Volume: Comparison with Transarterial Chemoemboli-zation. *Ann Surg Oncol*, 14(12), 3501-3509.

[48] Siriwardana, R. C., Lo, C. M., Chan, S. C., & Fan, S. T. (2012). Role of Portal Vein Em-bolization in Hepatocellular Carcinoma Management and Its Effect on Recurrence: A Case-Control Study. *World J Surg*, 36(7), 1640-6.

[49] Yoo, H., Kim, J. H., Ko, G. Y., Kim, K. W., Gwon, D. I., Lee, S. G., & Hwang, S. (2011). Sequential Transcatheter Arterial Chemoembolization and Portal Vein Embolization Versus Portal Vein Embolization Only before Major Hepatectomy for Patients with Hepatocellular Carcinoma. *Ann Surg Oncol*, 18(5), 1251-1257.

[50] Wu, C. C., Ho, Y. Z., Ho, W. L., Wu, T. C., Liu, T. J., & P'Eng, F. K. (1995). Preopera-tive Transcatheter Arterial Chemoembolization for Resectable Large Hepatocellular Carcinoma: A Reappraisal. *Br J Surg*, 82(1), 122-126.

[51] Yamasaki, S., Hasegawa, H., Kinoshita, H., Furukawa, M., Imaoka, S., Takasaki, K., Kakumoto, Y., Saitsu, H., Yamada, R., Oosaki, Y., Arii, S., Okamoto, E., Monden, M., Ryu, M., Kusano, S., Kanematsu, T., Ikeda, K., Yamamoto, M., Saoshiro, T., & Tsuzu-ki, T. (1996). A Prospective Randomized Trial of the Preventive Effect of Pre-Opera-

tive Transcatheter Arterial Embolization against Recurrence of Hepatocellular Carcinoma. *Jpn J Cancer Res*, 87(2), 206-211.

[52] Zhang, Z., Liu, Q., He, J., Yang, J., Yang, G., & Wu, M. (2000). The Effect of Preoperative Transcatheter Hepatic Arterial Chemoembolization on Disease-Free Survival after Hepatectomy for Hepatocellular Carcinoma. *Cancer*, 89(12), 2606-2612.

[53] Zhou, W. P., Lai, E. C., Li, A. J., Fu, S. Y., Zhou, J. P., Pan, Z. Y., Lau, W. Y., & Wu, M. C. (2009). A Prospective, Randomized, Controlled Trial of Preoperative Transarterial Chemoembolization for Resectable Large Hepatocellular Carcinoma. *Ann Surg*, 249(2), 195-202.

[54] Hwang, T. L., Chen, M. F., Lee, T. Y., Chen, T. J., Lin, D. Y., & Liaw, Y. F. (1987). Resection of Hepatocellular Carcinoma after Transcatheter Arterial Embolization. Reevaluation of the Advantages and Disadvantages of Preoperative Embolization. *Arch Surg*, 122(7), 756-759.

[55] Liu, C. L., Fan, S. T., Lo, C. M., Tso, W. K., Poon, R. T., Lam, C. M., & Wong, J. (2001). Management of Spontaneous Rupture of Hepatocellular Carcinoma: Single-Center Experience. *J Clin Oncol*, 19(17), 3725-3232.

[56] Grazi, G. L., Ercolani, G., Pierangeli, F., Del Gaudio, M., Cescon, M., Cavallari, A., & Mazziotti, A. (2001). Improved Results of Liver Resection for Hepatocellular Carcinoma on Cirrhosis Give the Procedure Added Value. *Ann Surg*, 234(1), 71-78.

[57] Wei, A. C., Tung-Ping, Poon. R. Fan, S. T., & Wong, J. (2003). Risk Factors for Perioperative Morbidity and Mortality after Extended Hepatectomy for Hepatocellular Carcinoma. *Br J Surg*, 90(1), 33-41.

[58] Zhou, X. D., Tang, Z. Y., Yang, B. H., Lin, Z. Y., Ma, Z. C., Ye, S. L., Wu, Z. Q., Fan, J., Qin, L. X., & Zheng, B. H. (2001). Experience of 1000 Patients Who Underwent Hepatectomy for Small Hepatocellular Carcinoma. *Cancer*, 91(8), 1479-1486.

[59] Imamura, H., Matsuyama, Y., Tanaka, E., Ohkubo, T., Hasegawa, K., Miyagawa, S., Sugawara, Y., Minagawa, M., Takayama, T., Kawasaki, S., & Makuuchi, M. (2003). Risk Factors Contributing to Early and Late Phase Intrahepatic Recurrence of Hepatocellular Carcinoma after Hepatectomy. *J Hepatol*, 38(2), 200-207.

[60] Muto, Y., Moriwaki, H., Ninomiya, M., Adachi, S., Saito, A., Takasaki, K. T., Tanaka, T., Tsurumi, K., Okuno, M., Tomita, E., Nakamura, T., & Kojima, T. (1996). Prevention of Second Primary Tumors by an Acyclic Retinoid, Polyprenoic Acid, in Patients with Hepato cellular Carcinoma. Hepatoma Prevention Study Group. *N Engl J Med*, 334(24), 1561-1567.

[61] Lau, W. Y., Leung, T. W., Ho, S. K., Chan, M., Machin, D., Lau, J., Chan, A. T., Yeo, W., Mok, T. S., Yu, S. C., Leung, N. W., & Johnson, P. J. (1999). Adjuvant Intra-Arterial Iodine-131-Labelled Lipiodol for Resectable Hepatocellular Carcinoma: A Prospective Randomised Trial. *Lancet*, 353(9155), 797-801.

[62] Takayama, T., Sekine, T., Makuuchi, M., Yamasaki, S., Kosuge, T., Yamamoto, J., Shimada, K., Sakamoto, M., Hirohashi, S., Ohashi, Y., & Kakizoe, T. (2000). Adoptive Immunotherapy to Lower Postsurgical Recurrence Rates of Hepatocellular Carcinoma: A Randomised Trial. *Lancet*, 356(9232), 802-807.

[63] Huang, J. F., Yu, M. L., Huang, C. F., Chiu, C. F., Dai, C. Y., Huang, C. I., Yeh, M. L., Yang, J. F., Hsieh, M. Y., Hou, N. J., & LinChenWangChuang, Z. Y.S. C.L. Y.W. L. (2011). The Efficacy and Safety of Pegylated Interferon Plus Ribavirin Combination Therapy in Chronic Hepatitis C Patients with Hepatocellular Carcinoma Post Curative Therapies- a Multicenter Prospective Trial. *J Hepatol*, 54(2), 219-26.

[64] Shen, Y. C., Hsu, C., Chen, L. T., Cheng, C. C., Hu, F. C., & Cheng, A. L. (2010). Adjuvant Interferon Therapy after Curative Therapy for Hepatocellular Carcinoma (Hcc): A Meta-Regression Approach. *J Hepatol*, 52(6), 889-894.

[65] Chen, L. T., Chen, M. F., Li, L. A., Lee, P. H., Jeng, L. B., Lin, D. Y., Wu, C. C., Mok, K. T., Chen, C. L., Lee, W. C., Chau, G. Y., Chen, Y. S., Lui, W. Y., Hsiao, C. F., Whang-Peng, J., & Chen, P. J. (2012). Long-Term Results of a Randomized, Observation-Controlled, Phase Iii Trial of Adjuvant Interferon Alfa-2b in Hepatocellular Carcinoma after Curative Resection. *Ann Surg*, 255(1), 8-17.

[66] Tung-Ping, Poon. R., Fan, S. T., & Wong, J. (2000). Risk Factors, Prevention, and Management of Postoperative Recurrence after Resection of Hepatocellular Carcinoma. *Ann Surg*, 232(1), 10-24.

[67] Itamoto, T., Nakahara, H., Amano, H., Kohashi, T., Ohdan, H., Tashiro, H., & Asahara, T. (2007). Repeat Hepatectomy for Recurrent Hepatocellular Carcinoma. *Surgery*, 141(5), 589-597.

[68] Lee, P. H., Lin, W. J., Tsang, Y. M., Hu, R. H., Sheu, J. C., Lai, M. Y., Hsu, H. C., May, W., & Lee, C. S. (1995). Clinical Management of Recurrent Hepatocellular Carcinoma. *Ann Surg*, 222(5), 670-676.

[69] Wu, C. C., Cheng, S. B., Yeh, D. C., Wang, J., & P'Eng, F. K. (2009). Second and Third Hepatectomies for Recurrent Hepatocellular Carcinoma Are Justified. *Br J Surg*, 96(9), 1049-1057.

[70] Chan, K. M., Yu, M. C., Wu, T. J., Lee, C. F., Chen, T. C., Lee, W. C., & Chen, M. F. (2009). Efficacy of Surgical Resection in Management of Isolated Extrahepatic Metastases of Hepatocellular Carcinoma. *World J Gastroenterol*, 15(43), 5481-5488.

[71] Hanazaki, K., Kajikawa, S., Shimozawa, N., Mihara, M., Shimada, K., Hiraguri, M., Koide, N., Adachi, W., & Amano, J. (2000). Survival and Recurrence after Hepatic Resection of 386 Consecutive Patients with Hepatocellular Carcinoma. *J Am Coll Surg*, 191(4), 381-388.

[72] Wang, J., Xu, L. B., Liu, C., Pang, H. W., Chen, Y. J., & Ou, Q. J. (2010). Prognostic Factors and Outcome of 438 Chinese Patients with Hepatocellular Carcinoma Underwent Partial Hepatectomy in a Single Center. *World J Surg*, 34(10), 2434-2441.

[73] Fan, S. T., Mau, Lo. C., Poon, R. T., Yeung, C., Leung, Liu. C., Yuen, W. K., Ming, Lam. C., Ng, K. K., & Ching, Chan. S. (2011). Continuous Improvement of Survival Outcomes of Resection of Hepatocellular Carcinoma: A 20-Year Experience. *Ann Surg*, 253(4), 745-758.

[74] Sakamoto, Y., Nara, S., Hata, S., Yamamoto, Y., Esaki, M., Shimada, K., & Kosuge, T. (2011). Prognosis of Patients Undergoing Hepatectomy for Solitary Hepatocellular Carcinoma Originating in the Caudate Lobe. *Surgery*, 150(5), 959-967.

[75] Nara, S., Shimada, K., Sakamoto, Y., Esaki, M., Kishi, Y., Kosuge, T., & Ojima, H. (2012). Prognostic Impact of Marginal Resection for Patients with Solitary Hepatocellular Carcinoma: Evidence from 570 Hepatectomies. *Surgery*, 151(4), 526-536.

[76] Chan, K. M., Lee, C. F., Wu, T. J., Chou, H. S., Yu, M. C., Lee, W. C., & Chen, M. F. (2012). Adverse Outcomes in Patients with Postoperative Ascites after Liver Resection for Hepatocellular Carcinoma. *World J Surg*, 36(2), 392-400.

[77] Giuliante, F., Ardito, F., Pinna, A. D., Sarno, G., Giulini, S. M., Ercolani, G., Portolani, N., Torzilli, G., Donadon, M., Aldrighetti, L., Pulitano, C., Guglielmi, A., Ruzzenente, A., Capussotti, L., Ferrero, A., Calise, F., Scuderi, V., Federico, B., & Nuzzo, G. (2012). Liver Resection for Hepatocellular Carcinoma </=3 Cm: Results of an Italian Multicenter Study on 588 Patients. *J Am Coll Surg*.

[78] Shrager, B., Jibara, G., Schwartz, M., & Roayaie, S. (2012). Resection of Hepatocellular Carcinoma without Cirrhosis. *Ann Surg*, 255(6), 1135-1143.

[79] Altekruse, S. F., McGlynn, K. A., Dickie, L. A., & Kleiner, D. E. (2012). Hepatocellular Carcinoma Confirmation, Treatment, and Survival in Surveillance, Epidemiology, and End Results Registries, 1992-2008. *Hepatology*, 55(2), 476-482.

[80] Pichlmayr, R. (1988). Is There a Place for Liver Grafting for Malignancy? *Transplant Proc*, 20(1, 1), 478-482.

[81] Bismuth, H., Chiche, L., Adam, R., Castaing, D., Diamond, T., & Dennison, A. (1993). Liver Resection Versus Transplantation for Hepatocellular Carcinoma in Cirrhotic Patients. *Ann Surg*, 218(2), 145-151.

[82] Mazzaferro, V., Regalia, E., Doci, R., Andreola, S., Pulvirenti, A., Bozzetti, F., Montalto, F., Ammatuna, M., Morabito, A., & Gennari, L. (1996). Liver Transplantation for the Treatment of Small Hepatocellular Carcinomas in Patients with Cirrhosis. *N Engl J Med*, 334(11), 693-9.

[83] Onaca, N., Davis, G. L., Goldstein, R. M., Jennings, L. W., & Klintmalm, G. B. (2007). Expanded Criteria for Liver Transplantation in Patients with Hepatocellular Carcinoma: A Report from the International Registry of Hepatic Tumors in Liver Transplantation. *Liver Transpl*, 13(3), 391-399.

[84] Takada, Y., Ito, T., Ueda, M., Sakamoto, S., Haga, H., Maetani, Y., Ogawa, K., Ogura, Y., Oike, F., Egawa, H., & Uemoto, S. (2007). Living Donor Liver Transplantation for

Patients with Hcc Exceeding the Milan Criteria: A Proposal of Expanded Criteria. *Dig Dis*, 25(4), 299-302.

[85] Toso, C., Asthana, S., Bigam, D. L., Shapiro, A. M., & Kneteman, N. M. (2009). Reassessing Selection Criteria Prior to Liver Transplantation for Hepatocellular Carcinoma Utilizing the Scientific Registry of Transplant Recipients Database. *Hepatology*, 49(3), 832-838.

[86] Yao, F. Y., Ferrell, L., Bass, N. M., Watson, J. J., Bacchetti, P., Venook, A., Ascher, N. L., & Roberts, J. P. (2001). Liver Transplantation for Hepatocellular Carcinoma: Expansion of the Tumor Size Limits Does Not Adversely Impact Survival. *Hepatology*, 33(6), 1394-1403.

[87] Sotiropoulos, G. C., Molmenti, E. P., Omar, O. S., Bockhorn, M., Brokalaki, E. I., Lang, H., Frilling, A., Broelsch, C. E., & Malago, M. (2006). Liver Transplantation for Hepatocellular Carcinoma in Patients Beyond the Milan but within the Ucsf Criteria. *Eur J Med Res*, 11(11), 467-470.

[88] Yao, F. Y., Ferrell, L., Bass, N. M., Bacchetti, P., Ascher, N. L., & Roberts, J. P. (2002). Liver Transplantation for Hepatocellular Carcinoma: Comparison of the Proposed Ucsf Criteria with the Milan Criteria and the Pittsburgh Modified Tnm Criteria. *Liver Transpl*, 8(9), 765-774.

[89] Mazzaferro, V., Llovet, J. M., Miceli, R., Bhoori, S., Schiavo, M., Mariani, L., Camerini, T., Roayaie, S., Schwartz, M. E., Grazi, G. L., Adam, R., Neuhaus, P., Salizzoni, M., Bruix, J., Forner, A., De Carlis, L., Cillo, U., Burroughs, A. K., Troisi, R., Rossi, M., Gerunda, G. E., Lerut, J., Belghiti, J., Boin, I., Gugenheim, J., Rochling, F., Van Hoek, B., & Majno, P. (2009). Predicting Survival after Liver Transplantation in Patients with Hepatocellular Carcinoma Beyond the Milan Criteria: A Retrospective, Exploratory Analysis. *Lancet Oncol*, 10(1), 35-43.

[90] Klintmalm, G. B. (1998). Liver Transplantation for Hepatocellular Carcinoma: A Registry Report of the Impact of Tumor Characteristics on Outcome. *Ann Surg*, 228(4), 479-490.

[91] Pawlik, T. M., Delman, K. A., Vauthey, J. N., Nagorney, D. M., Ng, I. O., Ikai, I., Yamaoka, Y., Belghiti, J., Lauwers, G. Y., Poon, R. T., & Abdalla, E. K. (2005). Tumor Size Predicts Vascular Invasion and Histologic Grade: Implications for Selection of Surgical Treatment for Hepatocellular Carcinoma. *Liver Transpl*, 11(9), 1086-1092.

[92] Roayaie, S., Haim, M. B., Emre, S., Fishbein, T. M., Sheiner, P. A., Miller, C. M., & Schwartz, M. E. (2000). Comparison of Surgical Outcomes for Hepatocellular Carcinoma in Patients with Hepatitis B Versus Hepatitis C: A Western Experience. *Ann Surg Oncol*, 7(10), 764-770.

[93] Kneteman, N. M., Oberholzer, J., Al Saghier, M., Meeberg, G. A., Blitz, M., Ma, M. M., Wong, W. W., Gutfreund, K., Mason, A. L., Jewell, L. D., Shapiro, A. M., Bain, V. G., & Bigam, D. L. (2004). Sirolimus-Based Immunosuppression for Liver Transplan-

tation in the Presence of Extended Criteria for Hepatocellular Carcinoma. *Liver Transpl*, 10(10), 1301-1311.

[94] Toso, C., Meeberg, G. A., Bigam, D. L., Oberholzer, J., Shapiro, A. M., Gutfreund, K., Mason, A. L., Wong, W. W., Bain, V. G., & Kneteman, N. M. (2007). De Novo Sirolimus-Based Immunosuppression after Liver Transplantation for Hepatocellular Carcinoma: Long-Term Outcomes and Side Effects. *Transplantation*, 83(9), 1162-1168.

[95] Zimmerman, M. A., Trotter, J. F., Wachs, M., Bak, T., Campsen, J., Skibba, A., & Kam, I. (2008). Sirolimus-Based Immunosuppression Following Liver Transplantation for Hepatocellular Carcinoma. *Liver Transpl*, 14(5), 633-638.

[96] Yao, F. Y., Bass, N. M., Nikolai, B., Davern, T. J., Kerlan, R., Wu, V., Ascher, N. L., & Roberts, J. P. (2002). Liver Transplantation for Hepatocellular Carcinoma: Analysis of Survival According to the Intention-to-Treat Principle and Dropout from the Waiting List. *Liver Transpl*, 8(10), 873-883.

[97] Lo, C. M., & Fan, S. T. (2004). Liver Transplantation for Hepatocellular Carcinoma. *Br J Surg*, 91(2), 131-3.

[98] Dahm, F., Georgiev, P., & Clavien, P. A. (2005). Small-for-Size Syndrome after Partial Liver Transplantation: Definition, Mechanisms of Disease and Clinical Implications. *Am J Transplant*, 5(11), 2605-2610.

[99] Emond, J. C., Renz, J. F., Ferrel, L. D., Rosenthal, P., Lim, R. C., Roberts, J. P., Lake, J. R., & Ascher, N. L. (1996). Functional Analysis of Grafts from Living Donors. Implications for the Treatment of Older Recipients. *Ann Surg*, 224(4), 544-552, discussion 52-4.

[100] Yao, F. Y., Kerlan, R. K., Jr, Hirose, R., Davern, T. J., 3rd, Bass, N. M., Feng, S., Peters, M., Terrault, N., Freise, C. E., Ascher, N. L., & Roberts, J. P. (2008). Excellent Outcome Following Down-Staging of Hepatocellular Carcinoma Prior to Liver Transplantation: An Intention-to-Treat Analysis. *Hepatology*, 48(3), 819-827.

[101] Chan, K. M., Yu, M. C., Chou, H. S., Wu, T. J., Lee, C. F., & Lee, W. C. (2011). Significance of Tumor Necrosis for Outcome of Patients with Hepatocellular Carcinoma Receiving Locoregional Therapy Prior to Liver Transplantation. *Ann Surg Oncol*, 18(9), 2638-2646.

[102] Rahbari, N. N., Mehrabi, A., Mollberg, N. M., Muller, S. A., Koch, M., Buchler, M. W., & Weitz, J. (2011). Hepatocellular Carcinoma: Current Management and Perspectives for the Future. *Ann Surg*, 253(3), 453-469.

[103] Fisher, R. A., Kulik, L. M., Freise, C. E., Lok, A. S., Shearon, T. H., Brown, R. S., Jr., Ghobrial, R. M., Fair, J. H., Olthoff, K. M., Kam, I., & Berg, C. L. (2007). Hepatocellular Carcinoma Recurrence and Death Following Living and Deceased Donor Liver Transplantation. *Am J Transplant*, 7(6), 1601-1608.

[104] Poon, R. T., Fan, S. T., Lo, C. M., Liu, C. L., & Wong, J. (2007). Difference in Tumor Invasiveness in Cirrhotic Patients with Hepatocellular Carcinoma Fulfilling the Mi-

lan Criteria Treated by Resection and Transplantation: Impact on Long-Term Survival. *Ann Surg*, 245(1), 51-58.

[105] Jonas, S., Bechstein, W. O., Steinmuller, T., Herrmann, M., Radke, C., Berg, T., Settmacher, U., & Neuhaus, P. (2001). Vascular Invasion and Histopathologic Grading Determine Outcome after Liver Transplantation for Hepatocellular Carcinoma in Cirrhosis. *Hepatology*, 33(5), 1080-1086.

[106] Yang, Z. F., Poon, R. T., Luo, Y., Cheung, C. K., Ho, D. W., Lo, C. M., & Fan, S. T. (2004). Up-Regulation of Vascular Endothelial Growth Factor (Vegf) in Small-for-Size Liver Grafts Enhances Macrophage Activities through Vegf Receptor 2-Dependent Pathway. *J Immunol*, 173(4), 2507-2515.

[107] Koniaris, L. G., Levi, D. M., Pedroso, F. E., Franceschi, D., Tzakis, A. G., Santamaria-Barria, J. A., Tang, J., Anderson, M., Misra, S., Solomon, N. L., Jin, X., Di Pasco, P. J., Byrne, M. M., & Zimmers, T. A. (2011). Is Surgical Resection Superior to Transplantation in the Treatment of Hepatocellular Carcinoma? *Ann Surg*, 254(3), 527-537, discussion 37-8.

Permissions

The contributors of this book come from diverse backgrounds, making this book a truly international effort. This book will bring forth new frontiers with its revolutionizing research information and detailed analysis of the nascent developments around the world.

We would like to thank Hesham Abdeldayem, MD, for lending his expertise to make the book truly unique. He has played a crucial role in the development of this book. Without his invaluable contribution this book wouldn't have been possible. He has made vital efforts to compile up to date information on the varied aspects of this subject to make this book a valuable addition to the collection of many professionals and students.

This book was conceptualized with the vision of imparting up-to-date information and advanced data in this field. To ensure the same, a matchless editorial board was set up. Every individual on the board went through rigorous rounds of assessment to prove their worth. After which they invested a large part of their time researching and compiling the most relevant data for our readers. Conferences and sessions were held from time to time between the editorial board and the contributing authors to present the data in the most comprehensible form. The editorial team has worked tirelessly to provide valuable and valid information to help people across the globe.

Every chapter published in this book has been scrutinized by our experts. Their significance has been extensively debated. The topics covered herein carry significant findings which will fuel the growth of the discipline. They may even be implemented as practical applications or may be referred to as a beginning point for another development. Chapters in this book were first published by InTech; hereby published with permission under the Creative Commons Attribution License or equivalent.

The editorial board has been involved in producing this book since its inception. They have spent rigorous hours researching and exploring the diverse topics which have resulted in the successful publishing of this book. They have passed on their knowledge of decades through this book. To expedite this challenging task, the publisher supported the team at every step. A small team of assistant editors was also appointed to further simplify the editing procedure and attain best results for the readers.

Our editorial team has been hand-picked from every corner of the world. Their multi-ethnicity adds dynamic inputs to the discussions which result in innovative outcomes. These outcomes are then further discussed with the researchers and contributors who give their valuable feedback and opinion regarding the same. The feedback is then collaborated with the researches and they are edited in a comprehensive manner to aid the understanding of the subject.

Apart from the editorial board, the designing team has also invested a significant amount of their time in understanding the subject and creating the most relevant covers. They scrutinized every image to scout for the most suitable representation of the subject and create an appropriate cover for the book.

The publishing team has been involved in this book since its early stages. They were actively engaged in every process, be it collecting the data, connecting with the contributors or procuring relevant information. The team has been an ardent support to the editorial, designing and production team. Their endless efforts to recruit the best for this project, has resulted in the accomplishment of this book. They are a veteran in the field of academics and their pool of knowledge is as vast as their experience in printing. Their expertise and guidance has proved useful at every step. Their uncompromising quality standards have made this book an exceptional effort. Their encouragement from time to time has been an inspiration for everyone.

The publisher and the editorial board hope that this book will prove to be a valuable piece of knowledge for researchers, students, practitioners and scholars across the globe.

List of Contributors

J.H.M.B. Stoot
Department of Surgery, Maastricht University Medical Centre, Maastricht, The Netherlands
Department of Surgery, Orbis Medical Centre, Sittard, The Netherlands
Department of Surgery, Atrium Medical Centre, Heerlen, The Netherlands

R.J.S. Coelen
Department of Surgery, Maastricht University Medical Centre, Maastricht, The Netherlands
Department of Surgery, Orbis Medical Centre, Sittard, The Netherlands

J.L.A. van Vugt
Department of Surgery, Orbis Medical Centre, Sittard, The Netherlands

C.H.C. Dejong
Department of Surgery, Maastricht University Medical Centre, Maastricht, The Netherlands
NUTRIM School for Nutrition, Metabolism and Toxicology, Maastricht University Medical Centre, Maastricht, The Netherlands

Aparna Dalal and John D. Jr. Lang
The University of Washington School of Medicine, Department of Anesthesiology & Pain Medicine, NE Pacific, Seattle, WA, USA

Ashok Thorat and Wei-Chen Lee
Division of Liver and Transplantation Surgery, Department of General Surgery, Chang-Gung Memorial Hospital at Linkou, Chang-Gung University College of Medicine, Taiwan

Ronald S. Chamberlain
Department of Surgery, Saint Barnabas Medical Center, Livingston, NJ, USA
Department of Surgery, University of Medicine and Dentistry of New Jersey, Newark, NJ, USA
Saint George's University School of Medicine, Grenada, West Indies

M.B. Jiménez-Castro, M. Elias-Miró and A. Casillas-Ramírez
August Pi i Sunyer Biomedical Research Institute, Barcelona, Spain

C. Peralta
August Pi i Sunyer Biomedical Research Institute, Barcelona, Spain
Networked Biomedical Research Center of Hepatic and Digestive Diseases, Barcelona, Spain

Hiroshi Sadamori, Takahito Yagi and Toshiyoshi Fujiwara
Department of Gastroenterological Surgery, Okayama University Graduate School of Medicine, Dentistry and Pharmaceutical Sciences, Okayama, Japan

Fabrizio Romano, Mattia Garancini, Fabio Uggeri, Luca Gianotti, Luca Nespoli, Angelo Nespoli and Franco Uggeri
Department of Surgery, University of Milan Bicocca, San Gerardo Hospital Monza, Milan, Italy

Vittorio Giardini
Department of General Surgery, Ospedale San Gerardo, Monza, Italy

Guido Torzilli
Third Department of Surgery, University of Milan School of Medicine, IRCCS Istituto Clinico Humanitas, Rozzano, Milan, Italy

Hideaki Uchiyama, Shinji Itoh and Kenji Takenaka
Department of Surgery, Fukuoka City Hospital, Japan

O. Al-Jiffry Bilal
Surgery, Taif University, Taif, Saudi Arabia
Surgery, AlHada Military Hospital, Taif, Saudi Arabia

Khayat H. Samah
Surgery, AlHada Military Hospital, Taif, Saudi Arabia

Hiromichi Ishii, Shimpei Ogino, Koki Ikemoto, Kenichi Takemoto, Atsushi Toma, Kenji Nakamura and Tsuyoshi Itoh
Division of Surgery, Kyoto Prefectural Yosanoumi Hospital, Japan

Kim Oelhafen
Saint George's University School of Medicine, Grenada, West Indies, Grenada

Mazen Hassanain, Faisal Alsaif, Abdulsalam Alsharaabi and Ahmad Madkhali
Department of surgery, College of Medicine, Liver Disease Research Centre, King Saud University, Riyadh, Saudi Arabia

Kun-Ming Chan
Division of Liver and Organ Transplantation Surgery, Department of General Surgery, Chang Gung Memorial Hospital at Linkou, Chang Gung University College of Medicine, Taiwan, Republic of China